AN INTRODUCTION TO PROBABILITY AND STATISTICS

WILEY SERIES IN PROBABILITY AND STATISTICS

Established by WALTER A. SHEWHART and SAMUEL S. WILKS

A complete list of the titles in this series appears at the end of this volume.

AN INTRODUCTION TO PROBABILITY AND STATISTICS

Third Edition

VIJAY K. ROHATGI

A. K. Md. EHSANES SALEH

WILEY

Published by John Wiley & Sons, Inc., Hoboken, New Jersey
Published simultaneously in Canada

Library of Congress Cataloging-in-Publication Data:

Rohatgi, V. K., 1939-
An introduction to probability theory and mathematical statistics / Vijay K. Rohatgi and A. K. Md. Ehsanes Saleh. – 3rd edition.
pages cm
Includes index.
ISBN 978-1-118-79964-2 (cloth)
1. Probabilities. 2. Mathematical statistics. I. Saleh, A. K. Md. Ehsanes. II. Title.
QA273.R56 2015
519.5–dc23

2015004848

Set in 10/12pts Times Lt Std by SPi Global, Pondicherry, India

10 9 8 7 6 5 4 3 2 1

3 2015

To Bina and Shahidara.

CONTENTS

PREFACE TO THE THIRD EDITION xiii

PREFACE TO THE SECOND EDITION xv

PREFACE TO THE FIRST EDITION xvii

ACKNOWLEDGMENTS xix

ENUMERATION OF THEOREMS AND REFERENCES xxi

1 Probability 1

1.1 Introduction, 1
1.2 Sample Space, 2
1.3 Probability Axioms, 7
1.4 Combinatorics: Probability on Finite Sample Spaces, 20
1.5 Conditional Probability and Bayes Theorem, 26
1.6 Independence of Events, 31

2 Random Variables and Their Probability Distributions 39

2.1 Introduction, 39
2.2 Random Variables, 39
2.3 Probability Distribution of a Random Variable, 42
2.4 Discrete and Continuous Random Variables, 47
2.5 Functions of a Random Variable, 55

3 Moments and Generating Functions **67**

3.1 Introduction, 67
3.2 Moments of a Distribution Function, 67
3.3 Generating Functions, 83
3.4 Some Moment Inequalities, 93

4 Multiple Random Variables **99**

4.1 Introduction, 99
4.2 Multiple Random Variables, 99
4.3 Independent Random Variables, 114
4.4 Functions of Several Random Variables, 123
4.5 Covariance, Correlation and Moments, 143
4.6 Conditional Expectation, 157
4.7 Order Statistics and Their Distributions, 164

5 Some Special Distributions **173**

5.1 Introduction, 173
5.2 Some Discrete Distributions, 173
5.2.1 Degenerate Distribution, 173
5.2.2 Two-Point Distribution, 174
5.2.3 Uniform Distribution on n Points, 175
5.2.4 Binomial Distribution, 176
5.2.5 Negative Binomial Distribution (Pascal or Waiting Time Distribution), 178
5.2.6 Hypergeometric Distribution, 183
5.2.7 Negative Hypergeometric Distribution, 185
5.2.8 Poisson Distribution, 186
5.2.9 Multinomial Distribution, 189
5.2.10 Multivariate Hypergeometric Distribution, 192
5.2.11 Multivariate Negative Binomial Distribution, 192
5.3 Some Continuous Distributions, 196
5.3.1 Uniform Distribution (Rectangular Distribution), 199
5.3.2 Gamma Distribution, 202
5.3.3 Beta Distribution, 210
5.3.4 Cauchy Distribution, 213
5.3.5 Normal Distribution (the Gaussian Law), 216
5.3.6 Some Other Continuous Distributions, 222
5.4 Bivariate and Multivariate Normal Distributions, 228
5.5 Exponential Family of Distributions, 240

6 Sample Statistics and Their Distributions **245**

6.1 Introduction, 245
6.2 Random Sampling, 246
6.3 Sample Characteristics and Their Distributions, 249

6.4 Chi-Square, t-, and F-Distributions: Exact Sampling Distributions, 262
6.5 Distribution of $(\overline{X}, S^2)$ in Sampling from a Normal Population, 271
6.6 Sampling from a Bivariate Normal Distribution, 276

7 Basic Asymptotics: Large Sample Theory **285**

7.1 Introduction, 285
7.2 Modes of Convergence, 285
7.3 Weak Law of Large Numbers, 302
7.4 Strong Law of Large Numbers, 308
7.5 Limiting Moment Generating Functions, 316
7.6 Central Limit Theorem, 321
7.7 Large Sample Theory, 331

8 Parametric Point Estimation **337**

8.1 Introduction, 337
8.2 Problem of Point Estimation, 338
8.3 Sufficiency, Completeness and Ancillarity, 342
8.4 Unbiased Estimation, 359
8.5 Unbiased Estimation (*Continued*): A Lower Bound for the Variance of An Estimator, 372
8.6 Substitution Principle (Method of Moments), 386
8.7 Maximum Likelihood Estimators, 388
8.8 Bayes and Minimax Estimation, 401
8.9 Principle of Equivariance, 418

9 Neyman–Pearson Theory of Testing of Hypotheses **429**

9.1 Introduction, 429
9.2 Some Fundamental Notions of Hypotheses Testing, 429
9.3 Neyman–Pearson Lemma, 438
9.4 Families with Monotone Likelihood Ratio, 446
9.5 Unbiased and Invariant Tests, 453
9.6 Locally Most Powerful Tests, 459

10 Some Further Results on Hypotheses Testing **463**

10.1 Introduction, 463
10.2 Generalized Likelihood Ratio Tests, 463
10.3 Chi-Square Tests, 472
10.4 t-Tests, 484
10.5 F-Tests, 489
10.6 Bayes and Minimax Procedures, 491

11 Confidence Estimation **499**

11.1 Introduction, 499
11.2 Some Fundamental Notions of Confidence Estimation, 499
11.3 Methods of Finding Confidence Intervals, 504
11.4 Shortest-Length Confidence Intervals, 517
11.5 Unbiased and Equivariant Confidence Intervals, 523
11.6 Resampling: Bootstrap Method, 530

12 General Linear Hypothesis **535**

12.1 Introduction, 535
12.2 General Linear Hypothesis, 535
12.3 Regression Analysis, 543
12.3.1 Multiple Linear Regression, 543
12.3.2 Logistic and Poisson Regression, 551
12.4 One-Way Analysis of Variance, 554
12.5 Two-Way Analysis of Variance with One Observation Per Cell, 560
12.6 Two-Way Analysis of Variance with Interaction, 566

13 Nonparametric Statistical Inference **575**

13.1 Introduction, 575
13.2 *U*-Statistics, 576
13.3 Some Single-Sample Problems, 584
13.3.1 Goodness-of-Fit Problem, 584
13.3.2 Problem of Location, 590
13.4 Some Two-Sample Problems, 599
13.4.1 Median Test, 601
13.4.2 Kolmogorov–Smirnov Test, 602
13.4.3 The Mann–Whitney–Wilcoxon Test, 604
13.5 Tests of Independence, 608
13.5.1 Chi-square Test of Independence—Contingency Tables, 608
13.5.2 Kendall's Tau, 611
13.5.3 Spearman's Rank Correlation Coefficient, 614
13.6 Some Applications of Order Statistics, 619
13.7 Robustness, 625
13.7.1 Effect of Deviations from Model Assumptions on Some Parametric Procedures, 625
13.7.2 Some Robust Procedures, 631

FREQUENTLY USED SYMBOLS AND ABBREVIATIONS **637**

REFERENCES **641**

STATISTICAL TABLES 647

ANSWERS TO SELECTED PROBLEMS 667

AUTHOR INDEX 677

SUBJECT INDEX 679

PREFACE TO THE THIRD EDITION

The *Third Edition* contains some new material. More specifically, the chapter on large sample theory has been reorganized, repositioned, and re-titled in recognition of the growing role of asymptotic statistics. In Chapter 12 on General Linear Hypothesis, the section on regression analysis has been greatly expanded to include multiple regression and logistic and Poisson regression.

Some more problems and remarks have been added to illustrate the material covered. The basic character of the book, however, remains the same as enunciated in the Preface to the first edition. It remains a solid introduction to first-year graduate students or advanced seniors in mathematics and statistics as well as a reference to students and researchers in other sciences.

We are grateful to the readers for their comments on this book over the past 40 years and would welcome any questions, comments, and suggestions. You can communicate with Vijay K. Rohatgi at vrohatg@bgsu.edu and with A. K. Md. Ehsanes Saleh at esaleh@math.carleton.ca.

Solana Beach, CA
Ottawa, Canada

Vijay K. Rohatgi
A. K. Md. Ehsanes Saleh

PREFACE TO THE SECOND EDITION

There is a lot that is different about this second edition. First, there is a co-author without whose help this revision would not have been possible. Second, we have benefited from countless letters from readers and colleagues who have pointed out errors and omissions and have made valuable suggestions over the past 25 years. These communications make this revision worth the effort. Third, we have tried to update the content of the book while striving to preserve the character and spirit of the first edition.

Here are some of the numerous changes that have been made.

1. The Introduction section has been removed. We have also removed Chapter 14 on sequential statistical inference.
2. Many parts of the book have gone substantial rewriting. For example, Chapter 4 has many changes, such as inclusion of exchangeability. In Chapter 3, an introduction to characteristic functions has been added. In Chapter 5 some new distributions have been added while in Chapter 6 there have been many changes in proofs.
3. The statistical inference part of the book (Chapters 8 to 13) has been updated. Thus in Chapter 8 we have expanded the coverage of invariance and have included discussions of ancillary statistics and conjugate prior distributions.
4. Similar changes have been made in Chapter 9. A new section on locally most powerful tests has been added.
5. Chapter 11 has been greatly revised and a discussion of invariant confidence intervals has been added.
6. Chapter 13 has been completely rewritten in the light of increased emphasis on nonparametric inference. We have expanded the discussion of U-statistics. Later sections show the connection between commonly used tests and U-statistics.
7. In Chapter 12, the notation has been changed to confirm to the current convention.

8. Many problems and examples have been added.
9. More figures have been added to illustrate examples and proofs.
10. Answers to selected problems have been provided.

We are truly grateful to the readers of the first edition for countless comments and suggestions and hope we will continue to hear from them about this edition.

Special thanks are due Ms. Gillian Murray for her superb word processing of the manuscript, and Dr. Indar Bhatia for figures that appear in the text. Dr. Bhatia spent countless hours preparing the diagrams for publication. We also acknowledge the assistance of Dr. K. Selvavel.

Vijay K. Rohatgi
A. K. Md. Ehsanes Saleh

PREFACE TO THE FIRST EDITION

This book on probability theory and mathematical statistics is designed for a three-quarter course meeting 4 hours per week or a two-semester course meeting 3 hours per week. It is designed primarily for advanced seniors and beginning graduate students in mathematics, but it can also be used by students in physics and engineering with strong mathematical backgrounds. Let me emphasize that this is a mathematics text and not a "cookbook." It should not be used as a text for service courses.

The mathematics prerequisites for this book are modest. It is assumed that the reader has had basic courses in set theory and linear algebra and a solid course in advanced calculus. No prior knowledge of probability and/or statistics is assumed.

My aim is to provide a solid and well-balanced introduction to probability theory and mathematical statistics. It is assumed that students who wish to do graduate work in probability theory and mathematical statistics will be taking, concurrently with this course, a measure-theoretic course in analysis if they have not already had one. These students can go on to take advanced-level courses in probability theory or mathematical statistics after completing this course.

This book consists of essentially three parts, although no such formal divisions are designated in the text. The first part consists of Chapters 1 through 6, which form the core of the probability portion of the course. The second part, Chapters 7 through 11, covers the foundations of statistical inference. The third part consists of the remaining three chapters on special topics. For course sequences that separate probability and mathematical statistics, the first part of the book can be used for a course in probability theory, followed by a course in mathematical statistics based on the second part and, possibly, one or more chapters on special topics.

The reader will find here a wealth of material. Although the topics covered are fairly conventional, the discussions and special topics included are not. Many presentations give far more depth than is usually the case in a book at this level. Some special features of the book are the following:

1. A well-referenced chapter on the preliminaries.
2. About 550 problems, over 350 worked-out examples, about 200 remarks, and about 150 references.
3. An advance warning to reader wherever the details become too involved. They can skip the later portion of the section in question on first reading without destroying the continuity in any way.
4. Many results on characterizations of distributions (Chapter 5).
5. Proof of the central limit theorem by the method of operators and proof of the strong law of large numbers (Chapter 6).
6. A section on minimal sufficient statistics (Chapter 8).
7. A chapter on special tests (Chapter 10).
8. A careful presentation of the theory of confidence intervals, including Bayesian intervals and shortest-length confidence intervals (Chapter 11).
9. A chapter on the general linear hypothesis, which carries linear models through to their use in basic analysis of variance (Chapter 12).
10. Sections on nonparametric estimation and robustness (Chapter 13).
11. Two sections on sequential estimation (Chapter 14).

The contents of this book were used in a 1-year (two-semester) course that I taught three times at the Catholic University of America and once in a three-quarter course at Bowling Green State University. In the fall of 1973 my colleague, Professor Eugene Lukacs, taught the first quarter of this same course on the basis of my notes, which eventually became this book. I have always been able to cover this book (with few omissions) in a 1-year course, lecturing 3 hours a week. An hour-long problem session every week is conducted by a senior graduate student.

In a book of this size there are bound to be some misprints, errors, and ambiguities of presentation. I shall be grateful to any reader who brings these to my attention.

Bowling Green, Ohio V. K. ROHATGI
February 1975

ACKNOWLEDGMENTS

We take this opportunity to thank many correspondents whose comments and criticisms led to improvements in the *Third Edition*. The list below is far from complete since it does not include the names of countless students whose reactions to the book as a text helped the authors in this revised edition. We apologize to those whose names may have been inadvertently omitted from the list because we were not diligent enough to keep a complete record of all the correspondence. For the third edition we wish to thank Professors Yue-Cune Chang, Anirban Das Gupta, A. G. Pathak, Arno Weiershauser, and many other readers who sent their questions and comments. We also wish to acknowledge the assistance of Dr. Pooplasingam Sivakumar in preparation of the manuscript. For the second edition: Barry Arnold, Lennart Bondesson, Harry Cohn, Frank Connonito, Emad El-Neweihi, Ulrich Faigle, Pier Alda Ferrari, Martin Feuerrnan, Xavier Fernando, Z. Govindarajulu, Arjun Gupta, Hassein Hamedani, Thomas Hem, Jin-Sheng Huang, Bill Hudson, Barthel Huff, V. S. Huzurbazar, B. K. Kale, Sam Kotz, Bansi Lal, Sri Gopal Mohanty, M. V. Moorthy, True Nguyen, Tom O'Connor, A. G. Pathak, Edsel Pena, S. Perng, Madan Puri, Prem Puri, J. S. Rao, Bill Raser, Andrew Rukhin, K. Selvavel, Rajinder Singh, R. J. Tomkins; for the first edition, Ralph Baty, Ralph Bradley, Eugene Lukacs, Kae Lea Main, Tom and Carol O'Connor, M. S. Scott Jr., J. Sethuraman, Beatrice Shube, Jeff Spielman, and Robert Tortora.

We thank the publishers of the *American Mathematical Monthly*, the *SIAM Review*, and the *American Statistician* for permission to include many examples and problems that appeared in these journals. Thanks are also due to the following for permission to include tables: Professors E. S. Pearson and L. R. Verdooren (Table ST11), Harvard University Press (Table ST1), Hafner Press (Table ST3), Iowa State University Press (Table ST5), Rand Corporation (Table ST6), the American Statistical Association (Tables ST7 and ST10), the Institute of Mathematical Statistics (Tables ST8 and ST9), Charles Griffin & Co., Ltd. (Tables ST12 and ST13), and John Wiley & Sons (Tables ST1, ST2, ST4, ST10, and ST11).

ENUMERATION OF THEOREMS AND REFERENCES

This book is divided into 13 chapters, numbered 1 through 13. Each chapter is divided into several sections. Lemmas, theorems, equations, definitions, remarks, figures, and so on, are numbered consecutively within each section. Thus Theorem $i.j.k$ refers to the kth theorem in Section j of Chapter i, Section $i.j$ refers to the jth section of Chapter i, and so on. Theorem j refers to the jth theorem of the section in which it appears. A similar convention is used for equations except that equation numbers are enclosed in parentheses. Each section is followed by a set of problems for which the same numbering system is used.

References are given at the end of the book and are denoted in the text by numbers enclosed in square brackets, [ ]. If a citation is to a book, the notation ([i, $p.j$]) refers to the jth page of the reference numbered [i].

A word about the proofs of results stated without proof in this book. If a reference appears immediately following or preceding the statement of a result, it generally means that the proof is beyond the scope of this text. If no reference is given, it indicates that the proof is left to the reader. Sometimes the reader is asked to supply the proof as a problem.

1

PROBABILITY

1.1 INTRODUCTION

The theory of probability had its origin in gambling and games of chance. It owes much to the curiosity of gamblers who pestered their friends in the mathematical world with all sorts of questions. Unfortunately this association with gambling contributed to a very slow and sporadic growth of probability theory as a mathematical discipline. The mathematicians of the day took little or no interest in the development of any theory but looked only at the combinatorial reasoning involved in each problem.

The first attempt at some mathematical rigor is credited to Laplace. In his monumental work, *Theorie analytique des probabilités* (1812), Laplace gave the classical definition of the probability of an event that can occur only in a finite number of ways as the proportion of the number of favorable outcomes to the total number of all possible outcomes, provided that all the outcomes are *equally likely*. According to this definition, the computation of the probability of events was reduced to combinatorial counting problems. Even in those days, this definition was found inadequate. In addition to being circular and restrictive, it did not answer the question of what probability is, it only gave a practical method of computing the probabilities of some simple events.

An extension of the classical definition of Laplace was used to evaluate the probabilities of sets of events with infinite outcomes. The notion of *equal likelihood* of certain events played a key role in this development. According to this extension, if Ω is some region with a well-defined measure (length, area, volume, etc.), the probability that a point chosen *at random* lies in a subregion A of Ω is the ratio $\text{measure}(A)/\text{measure}(\Omega)$. Many problems of geometric probability were solved using this extension. The trouble is that one can

An Introduction to Probability and Statistics, Third Edition. Vijay K. Rohatgi and A.K. Md. Ehsanes Saleh.

define "at random" in any way one pleases, and different definitions therefore lead to different answers. Joseph Bertrand, for example, in his book *Calcul des probabilités* (Paris, 1889) cited a number of problems in geometric probability where the result depended on the method of solution. In Example 9 we will discuss the famous Bertrand paradox and show that in reality there is nothing paradoxical about Bertrand's paradoxes; once we define "probability spaces" carefully, the paradox is resolved. Nevertheless difficulties encountered in the field of geometric probability have been largely responsible for the slow growth of probability theory and its tardy acceptance by mathematicians as a mathematical discipline.

The mathematical theory of probability, as we know it today, is of comparatively recent origin. It was A. N. Kolmogorov who axiomatized probability in his fundamental work, *Foundations of the Theory of Probability* (Berlin), in 1933. According to this development, random events are represented by sets and probability is just a *normed measure* defined on these sets. This measure-theoretic development not only provided a logically consistent foundation for probability theory but also, at the same time, joined it to the mainstream of modern mathematics.

In this book we follow Kolmogorov's axiomatic development. In Section 1.2 we introduce the notion of a sample space. In Section 1.3 we state Kolmogorov's axioms of probability and study some simple consequences of these axioms. Section 1.4 is devoted to the computation of probability on finite sample spaces. Section 1.5 deals with conditional probability and Bayes's rule while Section 1.6 examines the independence of events.

1.2 SAMPLE SPACE

In most branches of knowledge, experiments are a way of life. In probability and statistics, too, we concern ourselves with special types of experiments. Consider the following examples.

Example 1. A coin is tossed. Assuming that the coin does not land on the side, there are two possible outcomes of the experiment: heads and tails. On any performance of this experiment one does not know what the outcome will be. The coin can be tossed as many times as desired.

Example 2. A roulette wheel is a circular disk divided into 38 equal sectors numbered from 0 to 36 and 00. A ball is rolled on the edge of the wheel, and the wheel is rolled in the opposite direction. One bets on any of the 38 numbers or some combinations of them. One can also bet on a color, red or black. If the ball lands in the sector numbered 32, say, anybody who bet on 32 or combinations including 32 wins, and so on. In this experiment, all possible outcomes are known in advance, namely 00, 0, 1, 2, ..., 36, but on any performance of the experiment there is uncertainty as to what the outcome will be, provided, of course, that the wheel is not rigged in any manner. Clearly, the wheel can be rolled any number of times.

Example 3. A manufacturer produces footrules. The experiment consists in measuring the length of a footrule produced by the manufacturer as accurately as possible. Because of errors in the production process one does not know what the true length of the footrule selected will be. It is clear, however, that the length will be, say, between 11 and 13 in., or, if one wants to be safe, between 6 and 18 in.

Example 4. The length of life of a light bulb produced by a certain manufacturer is recorded. In this case one does not know what the length of life will be for the light bulb selected, but clearly one is aware in advance that it will be some number between 0 and ∞ hours.

The experiments described above have certain common features. For each experiment, we know in advance all possible outcomes, that is, there are no surprises in store after the performance of any experiment. On any performance of the experiment, however, we do not know what the specific outcome will be, that is, there is uncertainty about the outcome on any performance of the experiment. Moreover, the experiment can be repeated under identical conditions. These features describe a *random* (or a *statistical*) *experiment.*

Definition 1. A random (or a statistical) experiment is an experiment in which

(a) all outcomes of the experiment are known in advance,
(b) any performance of the experiment results in an outcome that is not known in advance, and
(c) the experiment can be repeated under identical conditions.

In probability theory we study this uncertainty of a random experiment. It is convenient to associate with each such experiment a set Ω, the set of all possible outcomes of the experiment. To engage in any meaningful discussion about the experiment, we associate with Ω a σ-field $\mathcal{S}$, of subsets of Ω. We recall that a σ-field is a nonempty class of subsets of Ω that is closed under the formation of countable unions and complements and contains the null set Φ.

Definition 2. The sample space of a statistical experiment is a pair $(\Omega, \mathcal{S})$, where

(a) Ω is the set of all possible outcomes of the experiment and
(b) $\mathcal{S}$ is a σ-field of subsets of Ω.

The elements of Ω are called *sample points*. Any set $A \in \mathcal{S}$ is known as an *event*. Clearly A is a collection of sample points. We say that an event A happens if the outcome of the experiment corresponds to a point in A. Each one-point set is known as a *simple* or an *elementary event*. If the set Ω contains only a finite number of points, we say that $(\Omega, \mathcal{S})$ is a *finite sample space*. If Ω contains at most a countable number of points, we call $(\Omega, \mathcal{S})$ a *discrete sample space*. If, however, Ω contains uncountably many points, we say that $(\Omega, \mathcal{S})$ is an *uncountable sample space*. In particular, if $\Omega = \mathcal{R}_k$ or some rectangle in $\mathcal{R}_k$, we call it a *continuous sample space*.

Remark 1. The choice of $\mathcal{S}$ is an important one, and some remarks are in order. If Ω contains at most a countable number of points, we can always take $\mathcal{S}$ to be the class of all subsets of Ω. This is certainly a σ-field. Each one point set is a member of $\mathcal{S}$ and is the fundamental object of interest. Every subset of Ω is an event. If Ω has uncountably many points, the class of all subsets of Ω is still a σ-field, but it is much too large a class of sets to be of interest. It may not be possible to choose the class of all subsets of Ω as $\mathcal{S}$. One of the most important examples of an uncountable sample space is the case in which $\Omega = \mathcal{R}$ or Ω is an interval in $\mathcal{R}$. In this case we would like all one-point subsets of Ω and all intervals (closed, open, or semiclosed) to be events. We use our knowledge of analysis to specify $\mathcal{S}$. We will not go into details here except to recall that the class of all semiclosed intervals $(a,b]$ generates a class $\mathfrak{B}_1$ which is a σ-field on $\mathcal{R}$. This class contains all one-point sets and all intervals (finite or infinite). We take $\mathcal{S} = \mathfrak{B}_1$. Since we will be dealing mostly with the one-dimensional case, we will write $\mathfrak{B}$ instead of $\mathfrak{B}_1$. There are many subsets of $\mathcal{R}$ that are not in $\mathfrak{B}_1$, but we will not demonstrate this fact here. We refer the reader to Halmos [42], Royden [96], or Kolmogorov and Fomin [54] for further details.

Example 5. Let us toss a coin. The set Ω is the set of symbols H and T, where H denotes head and T represents tail. Also, $\mathcal{S}$ is the class of all subsets of Ω, namely, $\{\{\mathrm{H}\},\{\mathrm{T}\},\{\mathrm{H},\mathrm{T}\},\Phi\}$. If the coin is tossed two times, then

$$\Omega = \{(\mathrm{H},\mathrm{H}),(\mathrm{H},\mathrm{T}),(\mathrm{T},\mathrm{H}),(\mathrm{T},\mathrm{T})\}, \quad \mathcal{S} = \{\emptyset,\{(\mathrm{H},\mathrm{H})\},$$
$$\{(\mathrm{H},\mathrm{T})\},\{(\mathrm{T},\mathrm{H})\},\{(\mathrm{T},\mathrm{T})\},\{(\mathrm{H},\mathrm{H}),(\mathrm{H},\mathrm{T})\},\{(\mathrm{H},\mathrm{H}),(\mathrm{T},\mathrm{H})\},$$
$$\{(\mathrm{H},\mathrm{H}),(\mathrm{T},\mathrm{T})\},\{(\mathrm{H},\mathrm{T}),(\mathrm{T},\mathrm{H})\},\{(\mathrm{T},\mathrm{T}),(\mathrm{T},\mathrm{H})\},\{(\mathrm{T},\mathrm{T}),$$
$$(\mathrm{H},\mathrm{T})\},\{(\mathrm{H},\mathrm{H}),(\mathrm{H},\mathrm{T}),(\mathrm{T},\mathrm{H})\},\{(\mathrm{H},\mathrm{H}),(\mathrm{H},\mathrm{T}),(\mathrm{T},\mathrm{T})\},$$
$$\{(\mathrm{H},\mathrm{H}),(\mathrm{T},\mathrm{H}),(\mathrm{T},\mathrm{T})\},\{(\mathrm{H},\mathrm{T}),(\mathrm{T},\mathrm{H}),(\mathrm{T},\mathrm{T})\},\Omega\},$$

where the first element of a pair denotes the outcome of the first toss and the second element, the outcome of the second toss. The event *at least one head* consists of sample points (H,H), (H,T), (T,H). The event *at most one head* is the collection of sample points (H,T), (T,H), (T,T).

Example 6. A die is rolled n times. The sample space is the pair $(\Omega,\mathcal{S})$, where Ω is the set of all n-tuples $(x_1,x_2,\ldots,x_n)$, $x_i \in \{1,2,3,4,5,6\}$, $i = 1,2,\ldots,n$, and $\mathcal{S}$ is the class of all subsets of Ω. Ω contains 6^n elementary events. The event A that 1 shows at least once is the set

$$\begin{aligned} A &= \{(x_1,x_2,\ldots,x_n) : \text{at least one of } x_i\text{'s is } 1\} \\ &= \Omega - \{(x_1,x_2,\ldots,x_n) : \text{none of the } x_i\text{'s is } 1\} \\ &= \Omega - \{(x_1,x_2,\ldots,x_n) : x_i \in \{2,3,4,5,6\},\ i = 1,2,\ldots,n\}. \end{aligned}$$

Example 7. A coin is tossed until the first head appears. Then

$$\Omega = \{\mathrm{H},(\mathrm{T},\mathrm{H}),(\mathrm{T},\mathrm{T},\mathrm{H}),(\mathrm{T},\mathrm{T},\mathrm{T},\mathrm{H}),\ldots\},$$

and $\mathcal{S}$ is the class of all subsets of Ω. An equivalent way of writing Ω would be to look at the number of tosses required for the first head. Clearly, this number can take values

$1, 2, 3, \ldots$, so that Ω is the set of all positive integers. The $\mathcal{S}$ is the class of all subsets of positive integers.

Example 8. Consider a pointer that is free to spin about the center of a circle. If the pointer is spun by an impulse, it will finally come to rest at some point. On the assumption that the mechanism is not rigged in any manner, each point on the circumference is a possible outcome of the experiment. The set Ω consists of all points $0 \leq x < 2\pi r$, where r is the radius of the circle. Every one-point set $\{x\}$ is a simple event, namely, that the pointer will come to rest at x. The events of interest are those in which the pointer stops at a point belonging to a specified arc. Here $\mathcal{S}$ is taken to be the Borel σ-field of subsets of $[0, 2\pi r)$.

Example 9. A rod of length l is thrown onto a flat table, which is ruled with parallel lines at distance $2l$. The experiment consists in noting whether the rod intersects one of the ruled lines.

Let r denote the distance from the center of the rod to the nearest ruled line, and let θ be the angle that the axis of the rod makes with this line (Fig. 1). Every outcome of this experiment corresponds to a point (r, θ) in the plane. As Ω we take the set of all points (r, θ) in $\{(r, \theta) \colon 0 \leq r \leq l, 0 \leq \theta < \pi\}$. For $\mathcal{S}$ we take the Borel σ-field, $\mathfrak{B}_2$, of subsets of Ω, that is, the smallest σ-field generated by rectangles of the form

$$\{(x, y) \colon a < x \leq b, \quad c < y \leq d, \quad 0 \leq a < b \leq l, \quad 0 \leq c < d < \pi\}.$$

Clearly the rod will intersect a ruled line if and only if the center of the rod lies in the area enclosed by the locus of the center of the rod (while one end touches the nearest line) and the nearest line (shaded area in Fig. 2).

Remark 2. From the discussion above it should be clear that in the discrete case there is really no problem. Every one-point set is also an event, and $\mathcal{S}$ is the class of all subsets of Ω.

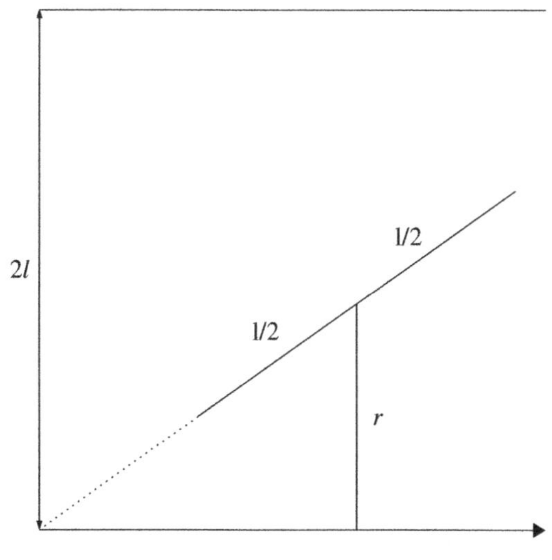

Fig. 1

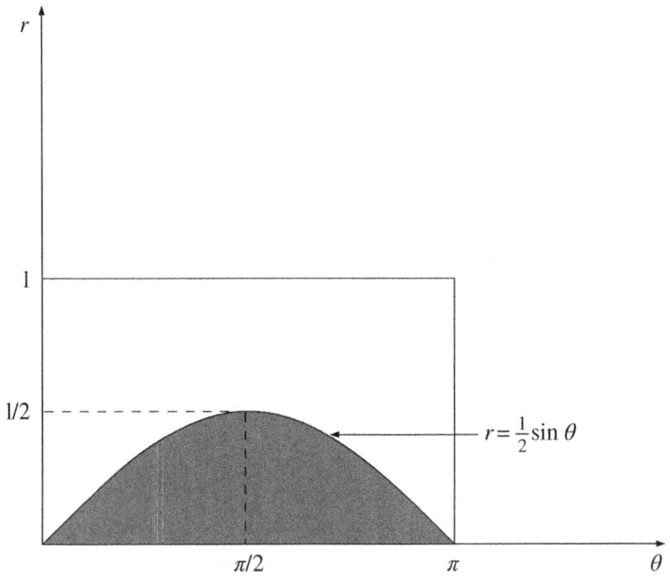

Fig. 2

The problem, if there is any, arises only in regard to uncountable sample spaces. The reader has to remember only that in this case not all subsets of Ω are events. The case of most interest is the one in which $\Omega = \mathcal{R}_k$. In this case, roughly all sets that have a well-defined volume (or area or length) are events. Not every set has the property in question, but sets that lack it are not easy to find and one does not encounter them in practice.

PROBLEMS 1.2

1. A club has five members A, B, C, D, and E. It is required to select a chairman and a secretary. Assuming that one member cannot occupy both positions, write the sample space associated with these selections. What is the event that member A is an office holder?

2. In each of the following experiments, what is the sample space?

(a) In a survey of families with three children, the sexes of the children are recorded in increasing order of age.

(b) The experiment consists of selecting four items from a manufacturer's output and observing whether or not each item is defective.

(c) A given book is opened to any page, and the number of misprints is counted.

(d) Two cards are drawn (i) with replacement and (ii) without replacement from an ordinary deck of cards.

3. Let A, B, C be three arbitrary events on a sample space $(\Omega, \mathcal{S})$. What is the event that only A occurs? What is the event that at least two of A, B, C occur? What is the event that both A and C, but not B, occur? What is the event that at most one of A, B, C occurs?

1.3 PROBABILITY AXIOMS

Let $(\Omega, \mathcal{S})$ be the sample space associated with a statistical experiment. In this section we define a probability set function and study some of its properties.

Definition 1. Let $(\Omega, \mathcal{S})$ be a sample space. A set function P defined on $\mathcal{S}$ is called a probability measure (or simply probability) if it satisfies the following conditions:

(i) $P(A) \geq 0$ for all $A \in \mathcal{S}$.

(ii) $P(\Omega) = 1$.

(iii) Let $\{A_j\}$, $A_j \in \mathcal{S}$, $j = 1, 2, \ldots$, be a disjoint sequence of sets, that is, $A_j \cap A_k = \Phi$ for $j \neq k$ where Φ is the null set. Then

$$P\left(\sum_{j=1}^{\infty} A_j\right) = \sum_{j=1}^{\infty} P(A_j), \tag{1}$$

where we have used the notation $\sum_{j=1}^{\infty} A_j$ to denote union of disjoint sets A_j.

We call $P(A)$ the *probability of event A*. If there is no confusion, we will write PA instead of $P(A)$. Property (iii) is called *countable additivity*. That $P\Phi = 0$ and P is also finitely additive follows from it.

Remark 1. If Ω is discrete and contains at most n ($< \infty$) points, each single-point set $\{\omega_j\}$, $j = 1, 2, \ldots, n$, is an elementary event, and it is sufficient to assign probability to each $\{\omega_j\}$. Then, if $A \in \mathcal{S}$, where $\mathcal{S}$ is the class of all subsets of Ω, $PA = \sum_{\omega \in A} P\{\omega\}$. One such assignment is the *equally likely* assignment or the assignment of *uniform* probabilities. According to this assignment, $P\{\omega_j\} = 1/n, j = 1, 2, \ldots, n$. Thus $PA = m/n$ if A contains m elementary events, $1 \leq m \leq n$.

Remark 2. If Ω is discrete and contains a countable number of points, one cannot make an equally likely assignment of probabilities. It suffices to make the assignment for each elementary event. If $A \in \mathcal{S}$, where $\mathcal{S}$ is the class of all subsets of Ω, define $PA = \sum_{\omega \in A} P\{\omega\}$.

Remark 3. If Ω contains uncountably many points, each one-point set is an elementary event, and again one cannot make an equally likely assignment of probabilities. Indeed, one cannot assign positive probability to each elementary event without violating the axiom $P\Omega = 1$. In this case one assigns probabilities to compound events consisting of intervals. For example, if $\Omega = [0, 1]$ and $\mathcal{S}$ is the Borel σ-field of all subsets of Ω, the assignment $P[I] =$ length of I, where I is a subinterval of Ω, defines a probability.

Definition 2. The triple $(\Omega, \mathcal{S}, P)$ is called a probability space.

Definition 3. Let $A \in \mathcal{S}$. We say that the odds for A are a to b if $PA = a/(a+b)$, and then the odds against A are b to a.

In many games of chance, probability is often stated in terms of odds against an event. Thus in horse racing a two dollar bet on a horse to win with odds of 2 to 1 (against) pays approximately six dollars if the horse wins the race. In this case the probability of winning is 1/3.

Example 1. Let us toss a coin. The sample space is $(\Omega, \mathcal{S})$, where $\Omega = \{\mathrm{H}, \mathrm{T}\}$, and $\mathcal{S}$ is the σ-field of all subsets of Ω. Let us define P on $\mathcal{S}$ as follows.

$$P\{\mathrm{H}\} = 1/2, \quad P\{\mathrm{T}\} = 1/2.$$

Then P clearly defines a probability. Similarly, $P\{\mathrm{H}\} = 2/3$, $P\{\mathrm{T}\} = 1/3$, and $P\{\mathrm{H}\} = 1$, $P\{\mathrm{T}\} = 0$ are probabilities defined on $\mathcal{S}$. Indeed,

$$P\{\mathrm{H}\} = p \quad \text{and} \quad P\{\mathrm{T}\} = 1 - p \quad (0 \le p \le 1)$$

defines a probability on $(\Omega, \mathcal{S})$.

Example 2. Let $\Omega = \{1, 2, 3, \ldots\}$ be the set of positive integers, and let $\mathcal{S}$ be the class of all subsets of Ω. Define P on $\mathcal{S}$ as follows:

$$P\{i\} = \frac{1}{2^i}, \quad i = 1, 2, \ldots.$$

Then $\sum_{i=1}^{\infty} P\{i\} = 1$, and P defines a probability.

Example 3. Let $\Omega = (0, \infty)$ and $\mathcal{S} = \mathfrak{B}$, the Borel σ-Field on Ω. Define P as follows: for each interval $I \subseteq \Omega$,

$$PI = \int_I e^{-x}\, dx.$$

Clearly $PI \ge 0$, $P\Omega = 1$, and P is countably additive by properties of integrals.

Theorem 1. P is monotone and subtractive; that is, if $A, B \in \mathcal{S}$ and $A \subseteq B$, then $PA \le PB$ and $P(B - A) = PB - PA$, where $B - A = B \cap A^c$, A^c being the complement of the event A.

Proof. If $A \subseteq B$, then

$$B = (A \cap B) + (B - A) = A + (B - A).$$

and it follows that $PB = PA + P(B - A)$.

Corollary. For all $A \in \mathcal{S}$, $0 \le PA \le 1$.

Remark 4. We wish to emphasize that, if $PA = 0$ for some $A \in \mathcal{S}$, we call A an event with *zero probability* or a *null* event. However, it does not follow that $A = \Phi$. Similarly, if $PB = 1$ for some $B \in \mathcal{S}$, we call B a *certain event* but it does not follow that $B = \Omega$.

Theorem 2 (The Addition Rule). If $A, B \in \mathcal{S}$, then

$$P(A \cup B) = PA + PB - P(A \cap B). \tag{2}$$

Proof. Clearly

$$A \cup B = (A - B) + (B - A) + (A \cap B)$$

and

$$A = (A \cap B) + (A - B), B = (A \cap B) + (B - A).$$

The result follows by countable additivity of P.

Corollary 1. P is subadditive, that is, if $A, B \in \mathcal{S}$, then

$$P(A \cup B) \leq PA + PB. \tag{3}$$

Corollary 1 can be extended to an arbitrary number of events A_j,

$$P\left(\bigcup_j A_j\right) \leq \sum_j PA_j. \tag{4}$$

Corollary 2. If $B = A^c$, then A and B are disjoint and

$$PA = 1 - PA^c. \tag{5}$$

The following generalization of (2) is left as an exercise.

Theorem 3 (The Principle of Inclusion–Exclusion). Let $A_1, A_2, \ldots, A_n \in \mathcal{S}$. Then

$$\begin{aligned} P\left(\bigcup_{k=1}^n A_k\right) &= \sum_{k=1}^n PA_k - \sum_{k_1 < k_2}^n P(A_{k_1} \cap A_{k_2}) \\ &\quad + \sum_{k_1 < k_2 < k_3}^n P(A_{k_1} \cap A_{k_2} \cap A_{k_3}) \\ &\quad + \cdots + (-1)^{n+1} P\left(\bigcap_{k=1}^n A_k\right). \end{aligned} \tag{6}$$

Example 4. A die is rolled twice. Let all the elementary events in $\Omega = \{(i,j): i,j = 1,2,\ldots,6\}$ be assigned the same probability. Let A be the event that the first throw shows a number ≤ 2, and B, the event that the second throw shows at least 5. Then

$$\begin{aligned} A &= \{(i,j): 1 \leq i \leq 2,\ j = 1,2,\ldots,6\}, \\ B &= \{(i,j): 5 \leq j \leq 6,\ i = 1,2,\ldots,6\}, \\ A \cap B &= \{(1,5),(1,6),(2,5),(2,6)\}; \end{aligned}$$

$$\begin{aligned} P(A \cup B) &= PA + PB - P(A \cap B) \\ &= \tfrac{1}{3} + \tfrac{1}{3} - \tfrac{4}{36} = \tfrac{5}{9}. \end{aligned}$$

Example 5. A coin is tossed three times. Let us assign equal probability to each of the 2^3 elementary events in Ω. Let A be the event that at least one head shows up in three throws. Then

$$\begin{aligned} P(A) &= 1 - P(A^c) \\ &= 1 - P(\text{no heads}) \\ &= 1 - P(\text{TTT}) = \tfrac{7}{8}. \end{aligned}$$

We next derive two useful inequalities.

Theorem 4 (Bonferroni's Inequality). Given n (> 1) events $A_1, A_2, \ldots, A_n$,

$$\sum_{i=1}^{n} PA_i - \sum_{i<j} P(A_i \cap A_j) \leq P\left(\bigcup_{i=1}^{n} A_i\right) \leq \sum_{i=1}^{n} PA_i. \tag{7}$$

Proof. In view of (4) it suffices to prove the left side of (7). The proof is by induction. The inequality on the left is true for $n = 2$ since

$$PA_1 + PA_2 - P(A_1 \cap A_2) = P(A_1 \cup A_2).$$

For $n = 3$,

$$P\left(\bigcup_{i=1}^{3} A_i\right) = \sum_{i=1}^{3} PA_i - \sum_{i<j} P(A_i \cap A_j) + P(A_1 \cap A_2 \cap A_3),$$

and the result holds. Assuming that (7) holds for $3 < m \leq n-1$, we show that it holds also for $m+1$:

$$\begin{aligned} P\left(\bigcup_{i=1}^{m+1} A_i\right) &= P\left(\left(\bigcup_{i=1}^{m} A_i\right) \cup A_{m+1}\right) \\ &= P\left(\bigcup_{i=1}^{m} A_i\right) + PA_{m+1} - P\left(A_{m+1} \cap \left(\bigcup_{1}^{m} A_i\right)\right) \end{aligned}$$

$$\geq \sum_{i=1}^{m+1} PA_i - \sum_{i<j}^{m} P(A_i \cap A_j) - P\left(\bigcup_{i=1}^{m}(A_i \cap A_{m+1})\right)$$

$$\geq \sum_{i=1}^{m+1} PA_i - \sum_{i<j}^{m} P(A_i \cap A_j) - \sum_{i=1}^{m} P(A_i \cap A_{m+1})$$

$$= \sum_{i=1}^{m+1} PA_i - \sum_{i<j}^{m+1} P(A_i \cap A_j).$$

Theorem 5 (Boole's Inequality). For any two events, A and B,

$$P(A \cap B) \geq 1 - PA^c - PB^c. \tag{8}$$

Corollary 1. Let $\{A_j\}, j = 1, 2, \ldots,$ be a countable sequence of events; then

$$P(\cap A_j) \geq 1 - \sum P(A_j^c). \tag{9}$$

Proof. Take

$$B = \bigcap_{j=2}^{\infty} A_j \quad \text{and} \quad A = A_1$$

in (8).

Corollary 2 (The Implication Rule). If $A, B, C \in \mathcal{S}$ and A and B imply C, then

$$PC^c \leq PA^c + PB^c. \tag{10}$$

Let $\{A_n\}$ be a sequence of sets. The set of all points $\omega \in \Omega$ that belong to A_n for infinitely many values of n is known as the *limit superior* of the sequence and is denoted by

$$\limsup_{n\to\infty} A_n \quad \text{or} \quad \overline{\lim_{n\to\infty}} A_n.$$

The set of all points that belong to A_n for all but a finite number of values of n is known as the *limit inferior* of the sequence $\{A_n\}$ and is denoted by

$$\liminf_{n\to\infty} A_n \quad \text{or} \quad \underline{\lim}_{n\to\infty} A_n.$$

If

$$\underline{\lim}_{n\to\infty} A_n = \overline{\lim_{n\to\infty}} A_n,$$

we say that the limit exists and write $\lim_{n\to\infty} A_n$ for the common set and call it the *limit set*.

We have

$$\varliminf_{n\to\infty} A_n = \bigcup_{n=1}^{\infty}\bigcap_{k=n}^{\infty} A_k \subseteq \bigcap_{n=1}^{\infty}\bigcup_{k=n}^{\infty} A_k = \varlimsup_{n\to\infty} A_n.$$

If the sequence $\{A_n\}$ is such that $A_n \subseteq A_{n+1}$, for $n = 1,2,\ldots$, it is called *nondecreasing*; if $A_n \supseteq A_{n+1}$, $n = 1,2,\ldots$, it is called *nonincreasing*. If the sequence A_n is nondecreasing, we write $A_n \downarrow$; if A_n is *nonincreasing*, we write $A_n \uparrow$. Clearly, if $A_n \uparrow$ or $A_n \downarrow$, the limit exists and we have

$$\lim_n A_n = \bigcup_{n=1}^{\infty} A_n \quad \text{if } A_n \downarrow$$

and

$$\lim_n A_n = \bigcap_{n=1}^{\infty} A_n \quad \text{if } A_n \uparrow .$$

Theorem 6. Let $\{A_n\}$ be a nondecreasing sequence of events in $\mathcal{S}$, that is, $A_n \in \mathcal{S}$, $n = 1,2,\ldots$, and

$$A_n \supseteq A_{n-1}, \quad n = 2,3,\ldots.$$

Then

$$\lim_{n\to\infty} PA_n = P\left(\lim_{n\to\infty} A_n\right) = P\left(\bigcup_{n=1}^{\infty} A_n\right). \tag{11}$$

Proof. Let

$$A = \bigcup_{j=1}^{\infty} A_j.$$

Then

$$A = A_n + \sum_{j=n}^{\infty}(A_{j+1} - A_j).$$

By countable additivity we have

$$PA = PA_n + \sum_{j=n}^{\infty} P(A_{j+1} - A_j).$$

and letting $n \to \infty$, we see that

$$PA = \lim_{n\to\infty} PA_n + \lim_{n\to\infty}\sum_{j=n}^{\infty} P(A_{j+1} - A_j).$$

The second term on the right tends to 0 as $n \to \infty$ since the sum $\sum_{j=1}^{\infty} P(A_{j+1} - A_j) \leq 1$ and each summand is nonnegative. The result follows.

Corollary. Let $\{A_n\}$ be a nonincreasing sequence of events in $\mathcal{S}$. Then

$$\lim_{n\to\infty} PA_n = P\left(\lim_{n\to\infty} A_n\right) = P\left(\bigcap_{n=1}^{\infty} A_n\right). \tag{12}$$

Proof. Consider the nondecreasing sequence of events $\{A_n^c\}$. Then

$$\lim_{n\to\infty} A_n^c = \bigcup_{j=1}^{\infty} A_j^c = A^c.$$

It follows from Theorem 6 that

$$\lim_{n\to\infty} PA_n^c = P\left(\lim_{n\to\infty} A_n^c\right) = P\left(\bigcup_{j=1}^{\infty} A_n^c\right) = P(A^c).$$

In other words,

$$\lim_{n\to\infty} (1 - PA_n) = 1 - PA,$$

as asserted.

Remark 5. Theorem 6 and its corollary will be used quite frequently in subsequent chapters. Property (11) is called the *continuity of P from below*, and (12) is known as the *continuity of P from above*. Thus Theorem 6 and its corollary assure us that the set function P is continuous from above and below.

We conclude this section with some remarks concerning the use of the word "random" in this book. In probability theory "random" has essentially three meanings. First, in sampling from a finite population a sample is said to be a *random sample* if at each draw all members available for selection have the same probability of being included. We will discuss sampling from a finite population in Section 1.4. Second, we speak of a *random sample from a probability distribution*. This notion is formalized in Section 6.2. The third meaning arises in the context of geometric probability, where statements such as "a point is randomly chosen from the interval (a, b)" and "a point is picked randomly from a unit square" are frequently encountered. Once we have studied random variables and their distributions, problems involving geometric probabilities may be formulated in terms of problems involving independent uniformly distributed random variables, and these statements can be given appropriate interpretations.

Roughly speaking, these statements involve a certain assignment of probability. The word "random" expresses our desire to assign equal probability to sets of equal lengths, areas, or volumes. Let $\Omega \subseteq \mathcal{R}_n$ be a given set, and A be a subset of Ω. We are interested in the probability that a "randomly chosen point" in Ω falls in A. Here "randomly chosen" means that the point may be any point of Ω and that the probability of its falling in some subset A of Ω is proportional to the measure of A (independently of the location and shape of A). Assuming that both A and Ω have well-defined finite measures (length, area, volume, etc.), we define

$$PA = \frac{\text{measure}(A)}{\text{measure}(\Omega)}.$$

(In the language of measure theory we are assuming that Ω is a measurable subset of $\mathcal{R}_n$ that has a finite, positive Lebesque measure. If A is any measurable set, $PA = \mu(A)/\mu(\Omega)$, where μ is the n-dimensional Lebesque measure.) Thus, if a point is chosen at random from the interval (a,b), the probability that it lies in the interval (c,d), $a \leq c < d \leq b$, is $(d-c)/(b-a)$. Moreover, the probability that the randomly selected point lies in any interval of length $(d-c)$ is the same.

We present some examples.

Example 6. A point is picked "at random" from a unit square. Let $\Omega = \{(x,y): 0 \leq x \leq 1, 0 \leq y \leq 1\}$. It is clear that all rectangles and their unions must be in $\mathcal{S}$. So too should all circles in the unit square, since the area of a circle is also well defined. Indeed, every set that has a well-defined area has to be in $\mathcal{S}$. We choose $\mathcal{S} = \mathfrak{B}_2$, the Borel σ-field generated by rectangles in Ω. As for the probability assignment, if $A \in \mathcal{S}$, we assign PA to A, where PA is the area of the set A. If $A = \{(x,y): 0 \leq x \leq 1/2, 1/2 \leq y \leq 1\}$, then $PA = 1/4$. If B is a circle with center $(1/2, 1/2)$ and radius $1/2$, then $PB = \pi(1/2)^2 = \pi/4$. If C is the set of all points which are at most a unit distance from the origin, then $PC = \pi/4$ (see Figs. 1–3).

***Example 7* (Buffon's Needle Problem).** We return to Example 1.2.9. A needle (rod) of length l is tossed at random on a plane that is ruled with a series of parallel lines at distance

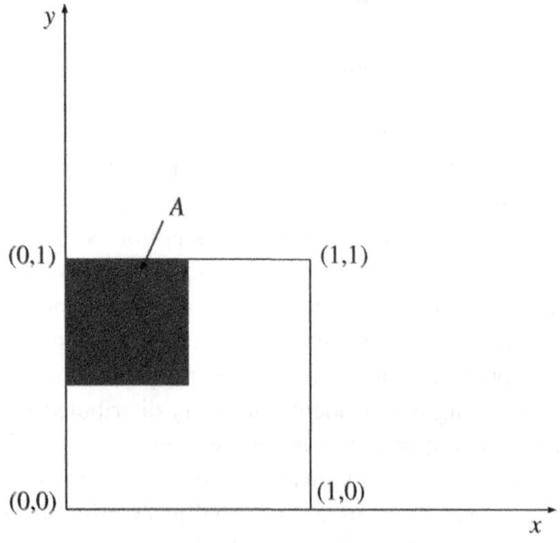

Fig. 1 $A = \{(x,y): 0 \leq x \leq 1/2, 1/2 \leq y \leq 1\}$.

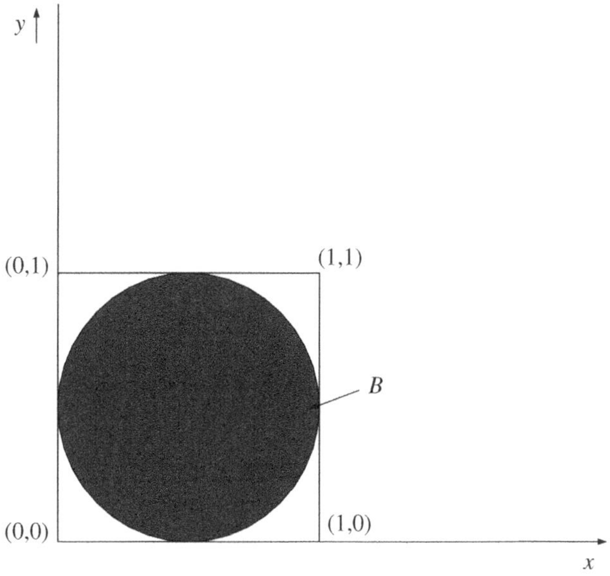

Fig. 2 $B = \{(x,y) : (x-1/2)^2 + (y-1/2)^2 = 1\}$.

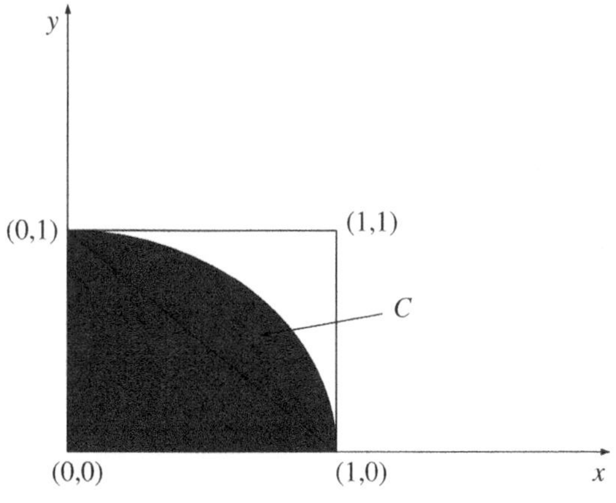

Fig. 3 $C = \{(x,y) : (x^2 + y^2 \leq 1\}$.

$2l$ apart. We wish to find the probability that the needle will intersect one of the lines. Denoting by r the distance from the center of the needle to the closest line and by θ the angle that the needle forms with this line, we see that a necessary and sufficient condition for the needle to intersect the line is that $r \leq (l/2)\sin\theta$. The needle will intersect the nearest line if and only if its center falls in the shaded region in Fig. 1.2.2. We assign probability to an event A as follows:

$$PA = \frac{\text{area of set } A}{l\pi}.$$

Thus the required probability is

$$\frac{1}{l\pi}\int_0^{\pi}\frac{l}{2}\sin\theta\,d\theta=\frac{1}{\pi}.$$

Here we have interpreted "at random" to mean that the position of the needle is characterized by a point (r,θ) which lies in the rectangle $0\le r\le l$, $0\le\theta\le\pi$. We have assumed that the probability that the point (r,θ) lies in any arbitrary subset of this rectangle is proportional to the area of this set. Roughly, this means that "all positions of the midpoint of the needle are assigned the same weight and all directions of the needle are assigned the same weight."

Example 8. An interval of length 1, say (0, 1), is divided into three intervals by choosing two points at random. What is the probability that the three line segments form a triangle?

It is clear that a necessary and sufficient condition for the three segments to form a triangle is that the length of any one of the segments be less than the sum of the other two. Let x,y be the abscissas of the two points chosen at random. Then we must have either

$$0<x<\frac{1}{2}<y<1 \quad\text{and}\quad y-x<\frac{1}{2}$$

or

$$0<y<\frac{1}{2}<x<1 \quad\text{and}\quad x-y<\frac{1}{2}.$$

This is precisely the shaded area in Fig. 4. It follows that the required probability is 1/4.

If it is specified in advance that the point x is chosen at random from $(0,1/2)$, and the point y at random from $(1/2,1)$, we must have

$$0<x<\frac{1}{2},\quad \frac{1}{2}<y<1,$$

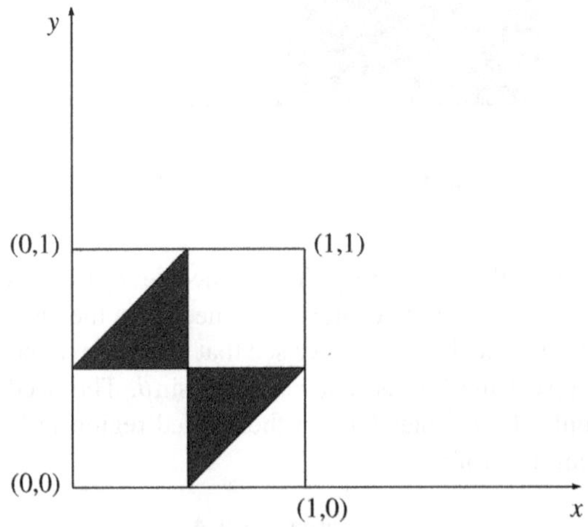

Fig. 4 $\{(x,y): 0<x<1/2<y<1, \text{and } (y-x)<1/2 \text{ or } 0<y<1/2<x<1, \text{and } (x-y)<1/2\}$.

and

$$y - x < x + 1 - y \quad \text{or} \quad 2(y - x) < 1.$$

In this case the area bounded by these lines is the shaded area in Fig. 5, and it follows that the required probability is 1/2.

Note the difference in sample spaces in the two computations made above.

***Example 9* (Bertrand's Paradox).** A chord is drawn at random in the unit circle. What is the probability that the chord is longer than the side of the equilateral triangle inscribed in the circle?

We present here three solutions to this problem, depending on how we interpret the phrase "at random." The paradox is resolved once we define the probability spaces carefully.

SOLUTION 1. Since the length of a chord is uniquely determined by the position of its midpoint, choose a point C at random in the circle and draw a line through C and O, the center of the circle (Fig. 6). Draw the chord through C perpendicular to the line OC. If l_1 is the length of the chord with C as midpoint, $l_1 > \sqrt{3}$ if and only if C lies inside the circle with center O and radius $1/2$. Thus $PA = \pi(1/2)^2/\pi = 1/4$.

In this case Ω is the circle with center O and radius 1, and the event A is the concentric circle with center O and radius $\frac{1}{2}$. $\mathcal{S}$ is the usual Borel σ-field of subsets of Ω.

SOLUTION 2. Because of symmetry, we may fix one end point of the chord at some point P and then choose the other end point P_1 at random. Let the probability that P_1 lies on an arbitrary arc of the circle be proportional to the length of this arc. Now the inscribed equilateral triangle having P as one of its vertices divides the circumference into three

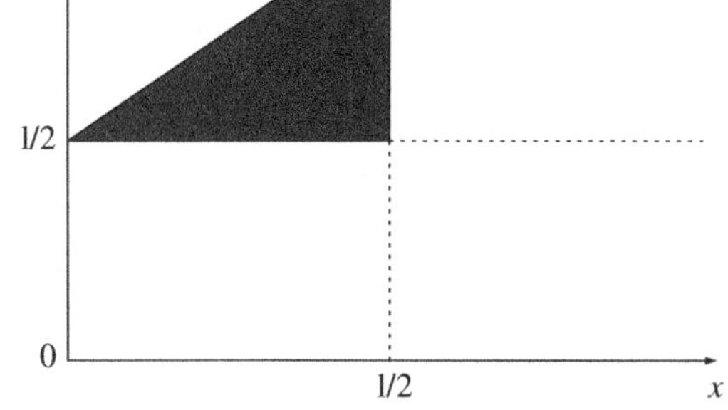

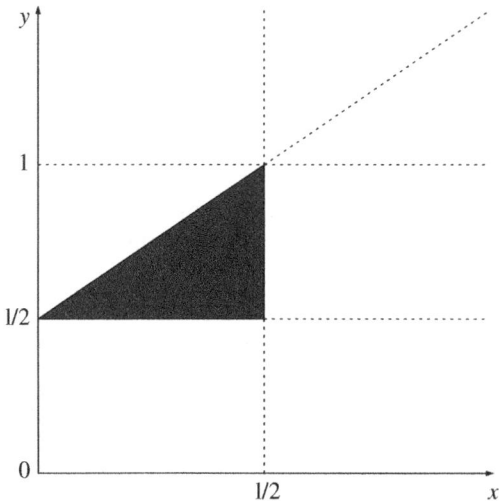

Fig. 5 $\{(x,y) : 0 < x < 1/2,\ 1/2 < y < 1 \text{ and } 2(y-x) < 1\}$.

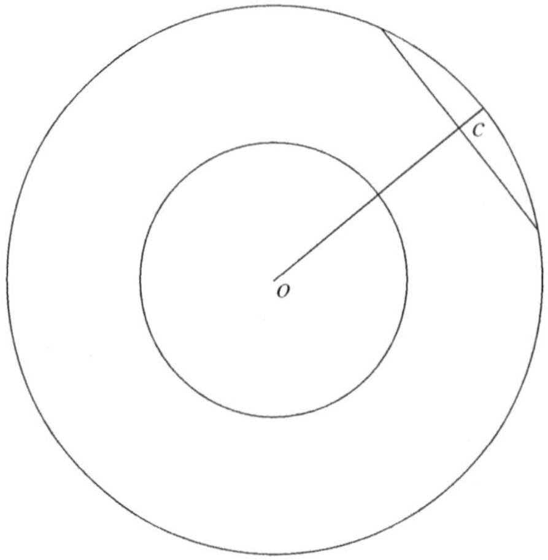

Fig. 6

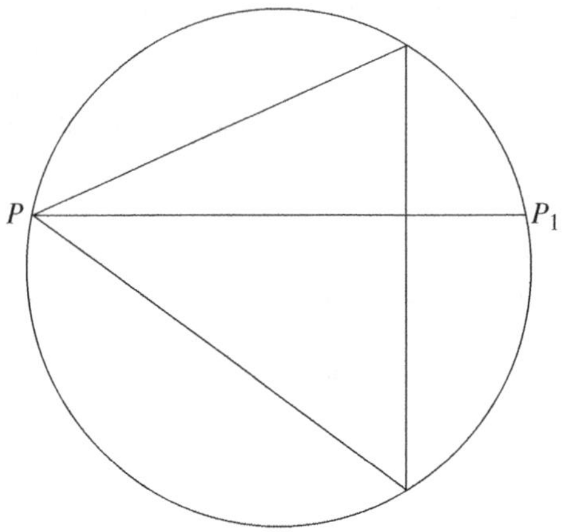

Fig. 7

equal parts. A chord drawn through P will be longer than the side of the triangle if and only if the other end point P_1 (Fig. 7) of the chord lies on that one third of the circumference that is opposite to P. It follows that the required probability is $1/3$. In this case $\Omega = [0, 2\pi]$, $\mathcal{S} = \mathfrak{B}_1 \cap \Omega$ and $A = [2\pi/3, 4\pi/3]$.

Solution 3. Note that the length of a chord is uniquely determined by the distance of its midpoint from the center of the circle. Due to the symmetry of the circle, we assume that the midpoint of the chord lies on a fixed radius, OM, of the circle (Fig. 8). The probability that the midpoint M lies in a given segment of the radius through M is then proportional to the length of this segment. Clearly, the length of the chord will be longer than the side of the inscribed equilateral triangle if the length of OM is less than $radius/2$. It follows that the required probability is $1/2$.

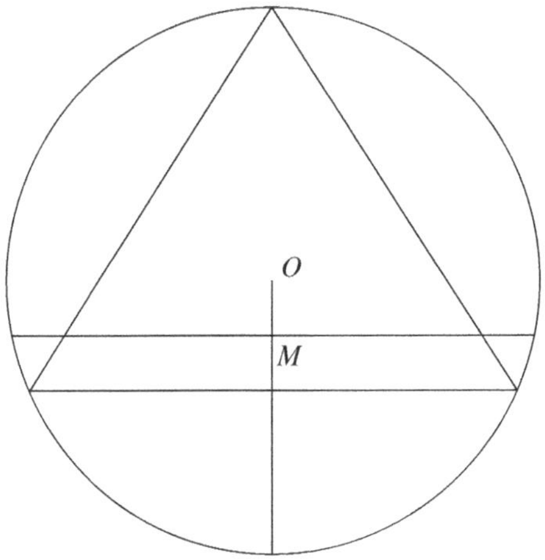

Fig. 8

PROBLEMS 1.3

1. Let Ω be the set of all nonnegative integers and $\mathcal{S}$ the class of all subsets of Ω. In each of the following cases does P define a probability on $(\Omega, \mathcal{S})$?

(a) For $A \in \mathcal{S}$, let

$$PA = \sum_{x \in A} \frac{e^{-\lambda}\lambda^x}{x!}, \quad \lambda > 0.$$

(b) For $A \in \mathcal{S}$, let

$$PA = \sum_{x \in A} p(1-p)^x, \quad 0 < p < 1.$$

(c) For $A \in \mathcal{S}$, let $PA = 1$ if A has a finite number of elements, and $PA = 0$ otherwise.

2. Let $\Omega = \mathcal{R}$ and $\mathcal{S} = \mathfrak{B}$. In each of the following cases does P define a probability on $(\Omega, \mathcal{S})$?

(a) For each interval I, let

$$PI = \int_I \frac{1}{\pi} \cdot \frac{1}{1+x^2}\, dx.$$

(b) For each interval I, let $PI = 1$ if I is an interval of finite length and $PI = 0$ if I is an infinite interval.

(c) For each interval I, let $PI = 0$ if $I \subseteq (-\infty, 1)$ and $PI = \int_I (1/2)\, dx$ if $I \subseteq [1, \infty]$. (If $I = I_1 + I_2$, where $I_1 \subseteq (-\infty, 1)$ and $I_2 \subseteq [1, \infty)$, then $PI = PI_2$.)

3. Let A and B be two events such that $B \supseteq A$. What is $P(A \cup B)$? What is $P(A \cap B)$? What is $P(A - B)$?

4. In Problems 1(a) and (b), let $A = \{\text{all integers} > 2\}$, $B = \{\text{all nonnegative integers} < 3\}$, and $C = \{\text{all integers } x,\ 3 < x < 6\}$. Find PA, PB, PC, $P(A \cap B)$, $P(A \cup B)$, $P(B \cup C)$, $P(A \cap C)$, and $P(B \cap C)$.
5. In Problem 2(a) let A be the event $A = \{x\colon x \geq 0\}$. Find PA. Also find $P\{x\colon x > 0\}$.
6. A box contains 1000 light bulbs. The probability that there is at least 1 defective bulb in the box is 0.1, and the probability that there are at least 2 defective bulbs is 0.05. Find the probability in each of the following cases:
   (a) The box contains no defective bulbs.
   (b) The box contains exactly 1 defective bulb.
   (c) The box contains at most 1 defective bulb.
7. Two points are chosen at random on a line of unit length. Find the probability that each of the three line segments so formed will have a length $>1/4$.
8. Find the probability that the sum of two randomly chosen positive numbers (both ≤ 1) will not exceed 1 and that their product will be $\leq 2/9$.
9. Prove Theorem 3.
10. Let $\{A_n\}$ be a sequence of events such that $A_n \to A$ as $n \to \infty$. Show that $PA_n \to PA$ as $n \to \infty$.
11. The base and the altitude of a right triangle are obtained by picking points randomly from $[0,a]$ and $[0,b]$, respectively. Show that the probability that the area of the triangle so formed will be less than $ab/4$ is $(1 + \ell\text{n } 2)/2$.
12. A point X is chosen at random on a line segment AB. (i) Show that the probability that the ratio of lengths AX/BX is smaller than a $(a > 0)$ is $a/(1+a)$. (ii) Show that the probability that the ratio of the length of the shorter segment to that of the larger segment is less than 1/3 is 1/2.

1.4 COMBINATORICS: PROBABILITY ON FINITE SAMPLE SPACES

In this section we restrict attention to sample spaces that have at most a finite number of points. Let $\Omega = \{\omega_1, \omega_2, \ldots, \omega_n\}$ and $\mathcal{S}$ be the σ-field of all subsets of Ω. For any $A \in \mathcal{S}$,

$$PA = \sum_{\omega_j \in A} P\{\omega_j\}.$$

Definition 1. An assignment of probability is said to be equally likely (or uniform) if each elementary event in Ω is assigned the same probability. Thus, if Ω contains n points ω_j, $P\{\omega_j\} = 1/n, j = 1, 2, \ldots, n$.

With this assignment

$$PA = \frac{\text{number of elementary events in } A}{\text{total number of elementary events in } \Omega}. \tag{1}$$

Example 1. A coin is tossed twice. The sample space consists of four points. Under the uniform assignment, each of four elementary events is assigned probability $1/4$.

Example 2. Three dice are rolled. The sample space consists of 6^3 points. Each one-point set is assigned probability $1/6^3$.

In games of chance we usually deal with finite sample spaces where uniform probability is assigned to all simple events. The same is the case in sampling schemes. In such instances the computation of the probability of an event A reduces to a combinatorial counting problem. We therefore consider some rules of counting.

Rule 1. Given a collection of n_1 elements $a_{11}, a_{12}, \ldots, a_{1n_1}$, n_2 elements $a_{21}, a_{22}, \ldots, a_{2n_2}$, and so on, up to n_k elements $a_{k1}, a_{k2}, \ldots, a_{kn_k}$, it is possible to form $n_1 \cdot n_2 \cdot \cdots \cdot n_k$ ordered k-tuples $(a_{1j_1}, a_{2j_2}, \ldots, a_{kj_k})$ containing one element of each kind, $1 \le j_i \le n_i$, $i = 1, 2, \ldots, k$.

Example 3. Here r distinguishable balls are to be placed in n cells. This amounts to choosing one cell for each ball. The sample space consists of n^r r-tuples $(i_1, i_2, \ldots, i_r)$, where i_j is the cell number of the jth ball, $j = 1, 2, \ldots, r$, $(1 \le i_j \le n)$.

Consider r tossings with a coin. There are 2^r possible outcomes. The probability that no heads will show up in r throws is $(1/2)^r$. Similarly, the probability that no 6 will turn up in r throws of a die is $(5/6)^r$.

Rule 2 is concerned with *ordered samples*. Consider a set of n elements $a_1, a_2, \ldots, a_n$. Any ordered arrangement $(a_{i_1}, a_{i_2}, \ldots, a_{i_r})$ of r of these n symbols is called an *ordered sample* of size r. If elements are selected one by one, there are two possibilities:

1. *Sampling with replacement* In this case repetitions are permitted, and we can draw samples of an arbitrary size. Clearly there are n^r samples of size r.
2. *Sampling without replacement* In this case an element once chosen is not replaced, so that there can be no repetitions. Clearly the sample size cannot exceed n, the size of the population. There are $n(n-1)\cdots(n-r+1) = {}_nP_r$, say, possible samples of size r. Clearly ${}_nP_r = 0$ for integers $r > n$. If $r = n$, then ${}_nP_r = n!$.

Rule 2. If ordered samples of size r are drawn from a population of n elements, there are n^r different samples with replacement and ${}_nP_r$ samples without replacement.

Corollary. The number of permutations of n objects is $n!$.

Remark 1. We will frequently use the term "random sample" in this book to describe the equal assignment of probability to all possible samples in sampling from a finite population. Thus, when we speak of a random sample of size r from a population of n elements, it means that each of n^r samples, in sampling with replacement, has the same probability $1/n^r$ or that each of ${}_nP_r$ samples, in sampling without replacement, is assigned probability $1/{}_nP_r$.

Example 4. Consider a set of n elements. A sample of size r is drawn at random with replacement. Then the probability that no element appears more than once is clearly ${}_nP_r/n^r$.

Thus, if n balls are to be randomly placed in n cells, the probability that each cell will be occupied is $n!/n^n$.

Example 5. Consider a class of r students. The birthdays of these r students form a sample of size r from the 365 days in the year. Then the probability that all r birthdays are different is ${}_{365}P_r/(365)^r$. One can show that this probability is $<1/2$ if $r = 23$.

The following table gives the values of $q_r = {}_{365}P_r/(365)^r$ for some selected values of r.

r	20	23	25	30	35	60
q_r	0.589	0.493	0.431	0.294	0.186	0.006

Next suppose that each of the r students is asked for his birth date in order, with the instruction that as soon as a student hears his birth date he is to raise his hand. Let us compute the probability that a hand is first raised when the kth ($k = 1,2,\ldots,r$) student is asked his birth date. Let p_k be the probability that the procedure terminates at the kth student. Then

$$p_1 = 1 - \left(\frac{364}{365}\right)^{r-1}$$

and

$$p_k = \frac{{}_{365}P_{k-1}}{(365)^{k-1}}\left(1 - \frac{k-1}{365}\right)^{r-k+1}\left[1 - \left(\frac{365-k}{365-k+1}\right)^{r-k}\right], \qquad k = 2,3,\ldots,r.$$

Example 6. Let Ω be the set of all permutations of n objects. Let A_i be the set of all permutations that leave the ith object unchanged. Then the set $\cup_{i=1}^{n} A_i$ is the set of permutations with at least one fixed point. Clearly

$$PA_i = \frac{(n-1)!}{n!}, \qquad i = 1,2,\ldots,n,$$

$$P(A_i \cap A_j) = \frac{(n-2)!}{n!}, \qquad i < j; i,j = 1,2,\ldots,n, \text{ etc.}$$

By Theorem 1.3.3 we have

$$P\left(\bigcup_{i=1}^{n} A_i\right) = \left(1 - \frac{1}{2!} + \frac{1}{3!} - \cdots \pm \frac{1}{n!}\right).$$

As an application consider an absent-minded secretary who places n letters in n envelopes at random. Then the probability that she will misplace every letter is

$$1 - \left(1 - \frac{1}{2!} + \frac{1}{3!} - \cdots \pm \frac{1}{n!}\right).$$

It is easy to see that this last probability $\longrightarrow e^{-1} = 0.3679$ as $n \to \infty$.

Rule 3. There are $\binom{n}{r}$ different subpopulations of size $r \le n$ from a population of n elements, where

$$\binom{n}{r} = \frac{n!}{r!(n-r)!}. \tag{2}$$

Example 7. Consider the random distribution of r balls in n cells. Let A_k be the event that a specified cell has exactly k balls, $k = 0, 1, 2, \ldots, r$; k balls can be chosen in $\binom{r}{k}$ ways. We place k balls in the specified cell and distribute the remaining $r-k$ balls in the $n-1$ cells in $(n-1)^{r-k}$ ways. Thus

$$PA_k = \binom{r}{k}\frac{(n-1)^{r-k}}{n^r} = \binom{r}{k}\left(\frac{1}{n}\right)^k\left(1-\frac{1}{n}\right)^{r-k}.$$

Example 8. There are $\binom{52}{13} = 635{,}013{,}559{,}600$ different hands at bridge, and $\binom{52}{5} = 2{,}598{,}960$ hands at poker.

The probability that all 13 cards in a bridge hand have different face values is $4^{13}/\binom{52}{13}$.

The probability that a hand at poker contains five different face values is $\binom{13}{5}4^5/\binom{52}{5}$.

Rule 4. Consider a population of n elements. The number of ways in which the population can be partitioned into k subpopulations of sizes $r_1, r_2, \ldots, r_k$, respectively, $r_1 + r_2 + \cdots + r_k = n$, $0 \le r_i \le n$, is given by

$$\binom{n}{r_1, r_2, \ldots, r_k} = \frac{n!}{r_1!\,r_2!\cdots r_k!}. \tag{3}$$

The numbers defined in (3) are known as *multinomial coefficients*.

Proof. For the proof of Rule 4 one uses Rule 3 repeatedly. Note that

$$\binom{n}{r_1, r_2, \ldots, r_k} = \binom{n}{r_1}\binom{n-r_1}{r_2}\cdots\binom{n-r_1\cdots-r_{k-2}}{r_{k-1}}. \tag{4}$$

Example 9. In a game of bridge the probability that a hand of 13 cards contains 2 spades, 7 hearts, 3 diamonds, and 1 club is

$$\frac{\binom{13}{2}\binom{13}{7}\binom{13}{3}\binom{13}{1}}{\binom{52}{13}}.$$

Example 10. An urn contains 5 red, 3 green, 2 blue, and 4 white balls. A sample of size 8 is selected at random without replacement. The probability that the sample contains 2 red, 2 green, 1 blue, and 3 white balls is

$$\frac{\binom{5}{2}\binom{3}{2}\binom{2}{1}\binom{4}{3}}{\binom{14}{8}}.$$

PROBLEMS 1.4

1. How many different words can be formed by permuting letters of the word "Mississippi"? How many of these start with the letters "Mi"?
2. An urn contains R red and W white marbles. Marbles are drawn from the urn one after another without replacement. Let A_k be the event that a red marble is drawn for the first time on the kth draw. Show that

$$PA_k = \left(\frac{R}{R+W-k+1}\right)\prod_{j=1}^{k-1}\left(1-\frac{R}{R+W-j+1}\right).$$

   Let p be the proportion of red marbles in the urn before the first draw. Show that $PA_k \to p(1-p)^{k-1}$ as $R+W \to \infty$. Is this to be expected?
3. In a population of N elements, R are red and $W = N-R$ are white. A group of n elements is selected at random. Find the probability that the group so chosen will contain exactly r red elements.
4. Each permutation of the digits 1, 2, 3, 4, 5, 6 determines a six-digit number. If the numbers corresponding to all possible permutations are listed in increasing order of magnitude, find the 319th number on this list.
5. The numbers $1,2,\ldots,n$ are arranged in random order. Find the probability that the digits $1,2,\ldots,k$ $(k<n)$ appear as neighbors in that order.
6. A pin table has seven holes through which a ball can drop. Five balls are played. Assuming that at each play a ball is equally likely to go down any one of the seven holes, find the probability that more than one ball goes down at least one of the holes.
7. If $2n$ boys are divided into two equal subgroups find the probability that the two tallest boys will be (a) in different subgroups and (b) in the same subgroup.
8. In a movie theater that can accommodate $n+k$ people, n people are seated. What is the probability that $r \le n$ given seats are occupied?
9. Waiting in line for a Saturday morning movie show are $2n$ children. Tickets are priced at a quarter each. Find the probability that nobody will have to wait for change if, before a ticket is sold to the first customer, the cashier has $2k$ $(k<n)$ quarters. Assume that it is equally likely that each ticket is paid for with a quarter or a half-dollar coin.
10. Each box of a certain brand of breakfast cereal contains a small charm, with k distinct charms forming a set. Assuming that the chance of drawing any particular charm is equal to that of drawing any other charm, show that the probability of finding at least one complete set of charms in a random purchase of $N \ge k$ boxes equals

$$1-\binom{k}{1}\left(\frac{k-1}{k}\right)^N+\binom{k}{2}\left(\frac{k-2}{k}\right)^N-\binom{k}{3}\left(\frac{k-3}{k}\right)^N$$
$$+\cdots+(-1)^{k-1}\binom{k}{k-1}\left(\frac{1}{k}\right)^N.$$

[*Hint*: Use (1.3.6).]

11. Prove Rules 1–4.

12. In a five-card poker game, find the probability that a hand will have:
- (a) A royal flush (ace, king, queen, jack, and 10 of the same suit).
- (b) A straight flush (five cards in a sequence, all of the same suit; ace is high but A, 2, 3, 4, 5 is also a sequence) excluding a royal flush.
- (c) Four of a kind (four cards of the same face value).
- (d) A full house (three cards of the same face value x and two cards of the same face value y).
- (e) A flush (five cards of the same suit excluding cards in a sequence).
- (f) A straight (five cards in a sequence).
- (g) Three of a kind (three cards of the same face value and two cards of different face values).
- (h) Two pairs.
- (i) A single pair.

13. (a) A married couple and four of their friends enter a row of seats in a concert hall. What is the probability that the wife will sit next to her husband if all possible seating arrangements are equally likely?
- (b) In part (a), suppose the six people go to a restaurant after the concert and sit at a round table. What is the probability that the wife will sit next to her husband?

14. Consider a town with N people. A person sends two letters to two separate people, each of whom is asked to repeat the procedure. Thus for each letter received, two letters are sent out to separate persons chosen at random (irrespective of what happened in the past). What is the probability that in the first n stages the person who started the chain letter game will not receive a letter?

15. Consider a town with N people. A person tells a rumor to a second person, who in turn repeats it to a third person, and so on. Suppose that at each stage the recipient of the rumor is chosen at random from the remaining $N-1$ people. What is the probability that the rumor will be repeated n times
- (a) Without being repeated to any person.
- (b) Without being repeated to the originator.

16. There were four accidents in a town during a seven-day period. Would you be surprised if all four occurred on the same day? Each of the four occurred on a different day?

17. While Rules 1 and 2 of counting deal with ordered samples with or without replacement, Rule 3 concerns unordered sampling without replacement. The most difficult rule of counting deals with unordered with replacement sampling. Show that there are $\binom{n+r-1}{r}$ possible unordered samples of size r from a population of n elements when sampled with replacement.

1.5 CONDITIONAL PROBABILITY AND BAYES THEOREM

So far, we have computed probabilities of events on the assumption that no information was available about the experiment other than the sample space. Sometimes, however, it is known that an event H has happened. How do we use this information in making a statement concerning the outcome of another event A? Consider the following examples.

Example 1. Let urn 1 contain one white and two black balls, and urn 2, one black and two white balls. A fair coin is tossed. If a head turns up, a ball is drawn at random from urn 1 otherwise, from urn 2. Let E be the event that the ball drawn is black. The sample space is $\Omega = \{\mathrm{H}b_{11}, \mathrm{H}b_{12}, \mathrm{H}w_{11}, \mathrm{T}b_{21}, \mathrm{T}w_{21}, \mathrm{T}w_{22}\}$, where H denotes head, T denotes tail, b_{ij} denotes jth black ball in ith urn, $i = 1,2$, and so on. Then

$$PE = P\{\mathrm{H}b_{11}, \mathrm{H}b_{12}, \mathrm{T}b_{21}\} = \tfrac{3}{6} = \tfrac{1}{2}.$$

If, however, it is known that the coin showed a head, the ball could not have been drawn from urn 2. Thus, the probability of E, conditional on information H, is $\frac{2}{3}$. Note that this probability equals the ratio $P\{\text{Head and ball drawn black}\}/P\{\text{Head}\}$.

Example 2. Let us toss two fair coins. Then the sample space of the experiment is $\Omega = \{\mathrm{HH}, \mathrm{HT}, \mathrm{TH}, \mathrm{TT}\}$. Let event $A = \{\text{both coins show same face}\}$ and $B = \{\text{at least one coin shows H}\}$. Then $PA = 2/4$. If B is known to have happened, this information assures that TT cannot happen, and $P\{A \text{ conditional on the information that } B \text{ has happened}\} = \frac{1}{3} = \frac{1}{4}/\frac{3}{4} = P(A \cap B)/PB$.

Definition 1. Let $(\Omega, \mathcal{S}, P)$ be a probability space, and let $H \in \mathcal{S}$ with $PH > 0$. For an arbitrary $A \in \mathcal{S}$ we shall write

$$P\{A \mid H\} = \frac{P(A \cap H)}{PH} \tag{1}$$

and call the quantity so defined the conditional probability of A, given H. Conditional probability remains undefined when $PH = 0$.

Theorem 1. Let $(\Omega, \mathcal{S}, P)$ be a probability space, and let $H \in \mathcal{S}$ with $PH > 0$. Then $(\Omega, \mathcal{S}, P_H)$, where $P_H(A) = P\{A \mid H\}$ for all $A \in \mathcal{S}$, is a probability space.

Proof. Clearly $P_H(A) = P\{A \mid H\} \geq 0$ for all $A \in \mathcal{S}$. Also, $P_H(\Omega) = P(\Omega \cap H)/PH = 1$. If $A_1, A_2, \ldots$ is a disjoint sequence of sets in $\mathcal{S}$, then

$$P_H\left(\sum_{i=1}^{\infty} A_i\right) = P\left\{\sum_{i=1}^{\infty} A_i \mid H\right\} = \frac{P\{(\sum_1^\infty A_i)\cap H\}}{PH}$$
$$= \frac{\sum_{i=1}^\infty P(A_i\cap H)}{PH}$$
$$= \sum_{i=1}^{\infty} P_H(A_i).$$

Remark 1. What we have done is to consider a new sample space consisting of the basic set H and the σ-field $\mathcal{S}_H = \mathcal{S}\cap H$, of subsets $A\cap H$, $A\in\mathcal{S}$, of H. On this space we have defined a set function P_H by multiplying the probability of each event by $(PH)^{-1}$. Indeed, $(H,\mathcal{S}_H,P_H)$ is a probability space.

Let A and B be two events with $PA > 0$, $PB > 0$. Then it follows from (1) that

$$\begin{cases} P(A\cap B) = PA\cdot P\{B \mid A\}, \\ P(A\cap B) = PB\cdot P\{A \mid B\}. \end{cases} \tag{2}$$

Equation (2) may be generalized to any number of events. Let $A_1,A_2,\ldots,A_n \in \mathcal{S}$, $n\geq 2$, and assume that $P(\bigcap_{j=1}^{n-1} A_i) > 0$. Since

$$A_1 \supset (A_1\cap A_2) \supset (A_1\cap A_2\cap A_3) \supset\cdots\supset \left(\bigcap_{j=1}^{n-2} A_j\right) \supset \left(\bigcap_{j=1}^{n-1} A_j\right),$$

we see that

$$PA_1 > 0,\quad P(A_1\cap A_2) > 0,\ldots,\quad P\left(\bigcap_{j=1}^{n-2} A_j\right) > 0.$$

It follows that $P\{A_k \mid \cap_{j=1}^{k-1} A_j\}$ are well defined for $k=2,3,\ldots,n$.

Theorem 2 (The Multiplication Rule). Let $(\Omega,\mathcal{S},P)$ be a probability space and $A_1,A_2,\ldots,$ $A_n\in\mathcal{S}$, with $P(\cap_{j=1}^{n-1} A_j) > 0$. Then

$$P\left\{\bigcap_{j=1}^{n} A_j\right\} = P(A_1)P\{A_2 \mid A_1\}P\{A_3 \mid A_1\cap A_2\}\cdots P\left\{A_n \mid \bigcap_{j=1}^{n-1} A_j\right\}. \tag{3}$$

Proof. The proof is simple.

Let us suppose that $\{H_j\}$ is a countable collection of events in $\mathcal{S}$ such that $H_j\cap H_k = \Phi$, $j\neq k$, and $\sum_{j=1}^\infty H_j = \Omega$. Suppose that $PH_j > 0$ for all j. Then

$$PB = \sum_{j=1}^{\infty} P(H_j)P\{B \mid H_j\} \quad \text{for all } B\in\mathcal{S}. \tag{4}$$

For the proof we note that

$$B = \sum_{j=1}^{\infty} (B \cap H_j),$$

and the result follows. Equation (4) is called the *total probability rule.*

Example 3. Consider a hand of five cards in a game of poker. If the cards are dealt at random, there are $\binom{52}{5}$ possible hands of five cards each. Let $A = \{\text{at least 3 cards of spades}\}$ and $B = \{\text{all 5 cards of spades}\}$. Then

$$\begin{aligned} P(A \cap B) &= P\{\text{all 5 cards of spades}\} \\ &= \frac{\binom{13}{5}}{\binom{52}{5}} \end{aligned}$$

and

$$\begin{aligned} P\{B \mid A\} &= \frac{P(A \cap B)}{PA} \\ &= \frac{\binom{13}{5} \Big/ \binom{52}{5}}{\left[\binom{13}{3}\binom{39}{2} + \binom{13}{4}\binom{39}{1} + \binom{13}{5}\right] \Big/ \binom{52}{5}}. \end{aligned}$$

Example 4. Urn 1 contains one white and two black marbles, urn 2 contains one black and two white marbles, and urn 3 contains three black and three white marbles. A die is rolled. If a 1, 2, or 3 shows up, urn 1 is selected; if a 4 shows up, urn 2 is selected; and if a 5 or 6 shows up, urn 3 is selected. A marble is then drawn at random from the selected urn. Let A be the event that the marble drawn is white. If U, V, W, respectively, denote the events that the urn selected is 1, 2, 3, then

$$\begin{aligned} A &= (A \cap U) + (A \cap V) + (A \cap W), \\ P(A \cap U) &= P(U) \cdot P\{A \mid U\} = \tfrac{3}{6} \cdot \tfrac{1}{3}, \\ P(A \cap V) &= P(V) \cdot P\{A \mid V\} = \tfrac{1}{6} \cdot \tfrac{2}{3}, \\ P(A \cap W) &= P(W) \cdot P\{A \mid W\} = \tfrac{2}{6} \cdot \tfrac{3}{6}. \end{aligned}$$

It follows that

$$PA = \tfrac{1}{6} + \tfrac{1}{9} + \tfrac{1}{6} = \tfrac{4}{9}.$$

A simple consequence of the total probability rule is the Bayes rule, which we now prove.

Theorem 3 (Bayes Rule). Let $\{H_n\}$ be a disjoint sequence of events such that $PH_n > 0$, $n = 1,2,\ldots$, and $\sum_{n=1}^{\infty} H_n = \Omega$. Let $B \in \mathcal{S}$ with $PB > 0$. Then

$$P\{H_j \mid B\} = \frac{P(H_j)P\{B \mid H_j\}}{\sum_{i=1}^{\infty} P(H_i)P\{B \mid H_i\}}, \quad j = 1,2,\ldots. \tag{5}$$

Proof. From (2)

$$P\{B \cap H_j\} = P(B)P\{H_j \mid B\} = PH_j P\{B \mid H_j\},$$

and it follows that

$$P\{H_j \mid B\} = \frac{PH_j P\{B \mid H_j\}}{PB}.$$

The result now follows on using (4).

Remark 2. Suppose that $H_1, H_2, \ldots$ are all the "causes" that lead to the outcome of a random experiment. Let H_j be the set of outcomes corresponding to the jth cause. Assume that the probabilities $PH_j, j = 1,2,\ldots$, called the *prior* probabilities, can be assigned. Now suppose that the experiment results in an event B of positive probability. This information leads to a reassessment of the prior probabilities. The conditional probabilities $P\{H_j \mid B\}$ are called the *posterior* probabilities. Formula (5) can be interpreted as a rule giving the probability that observed event B was due to cause or hypothesis H_j.

Example 5. In Example 4 let us compute the conditional probability $P\{V \mid A\}$. We have

$$\begin{aligned} P\{V \mid A\} &= \frac{PV\,P\{A \mid V\}}{PU\,P\{A \mid U\} + PV\,P\{A \mid V\} + PW\,P\{A \mid W\}} \\ &= \frac{\frac{1}{6}\cdot\frac{2}{3}}{\frac{3}{6}\cdot\frac{1}{3} + \frac{1}{6}\cdot\frac{2}{3} + \frac{2}{6}\cdot\frac{3}{6}} = \frac{\frac{1}{9}}{\frac{4}{9}} = \frac{1}{4}. \end{aligned}$$

PROBLEMS 1.5

1. Let A and B be two events such that $PA = p_1 > 0$, $PB = p_2 > 0$, and $p_1 + p_2 > 1$. Show that $P\{B \mid A\} \geq 1 - [(1 - p_2)/p_1]$.
2. Two digits are chosen at random without replacement from the set of integers $\{1,2,3,4,5,6,7,8\}$.
   (a) Find the probability that both digits are greater than 5.
   (b) Show that the probability that the sum of the digits will be equal to 5 is the same as the probability that their sum will exceed 13.
3. The probability of a family chosen at random having exactly k children is αp^k, $0 < p < 1$. Suppose that the probability that any child has blue eyes is b, $0 < b < 1$,

independently of others. What is the probability that a family chosen at random has exactly r ($r \geq 0$) children with blue eyes?

4. In Problem 3 let us write

$$p_k = \text{probability of a randomly chosen family having exactly } k \text{ children} = \alpha p^k,$$
$$k = 1, 2, \ldots,$$
$$p_0 = 1 - \frac{\alpha p}{(1-p)}.$$

Suppose that all sex distributions of k children are equally likely. Find the probability that a family has exactly r boys, $r \geq 1$. Find the conditional probability that a family has at least two boys, given that it has at least one boy.

5. Each of $(N+1)$ identical urns marked $0, 1, 2, \ldots, N$ contains N balls. The kth urn contains k black and $N-k$ white balls, $k = 0, 1, 2, \ldots, N$. An urn is chosen at random, and n random drawings are made from it, the ball drawn being always replaced. If all the n draws result in black balls, find the probability that the $(n+1)$th draw will also produce a black ball. How does this probability behave as $N \to \infty$?

6. Each of n urns contains four white and six black balls, while another urn contains five white and five black balls. An urn is chosen at random from the $(n+1)$ urns, and two balls are drawn from it, both being black. The probability that five white and three black balls remain in the chosen urn is $1/7$. Find n.

7. In answering a question on a multiple choice test, a candidate either knows the answer with probability p ($0 \leq p < 1$) or does not know the answer with probability $1-p$. If he knows the answer, he puts down the correct answer with probability 0.99, whereas if he guesses, the probability of his putting down the correct result is $1/k$ (k choices to the answer). Find the conditional probability that the candidate knew the answer to a question, given that he has made the correct answer. Show that this probability tends to 1 as $k \to \infty$.

8. An urn contains five white and four black balls. Four balls are transferred to a second urn. A ball is then drawn from this urn, and it happens to be black. Find the probability of drawing a white ball from among the remaining three.

9. Prove Theorem 2.

10. An urn contains r red and g green marbles. A marble is drawn at random and its color noted. Then the marble drawn, together with $c > 0$ marbles of the same color, are returned to the urn. Suppose n such draws are made from the urn? Find the probability of selecting a red marble at any draw.

11. Consider a bicyclist who leaves a point P (see Fig. 1), choosing one of the roads PR_1, PR_2, PR_3 at random. At each subsequent crossroad he again chooses a road at random.

(a) What is the probability that he will arrive at point A?

(b) What is the conditional probability that he will arrive at A via road PR_3?

12. Five percent of patients suffering from a certain disease are selected to undergo a new treatment that is believed to increase the recovery rate from 30 percent to 50 percent. A person is randomly selected from these patients after the completion of

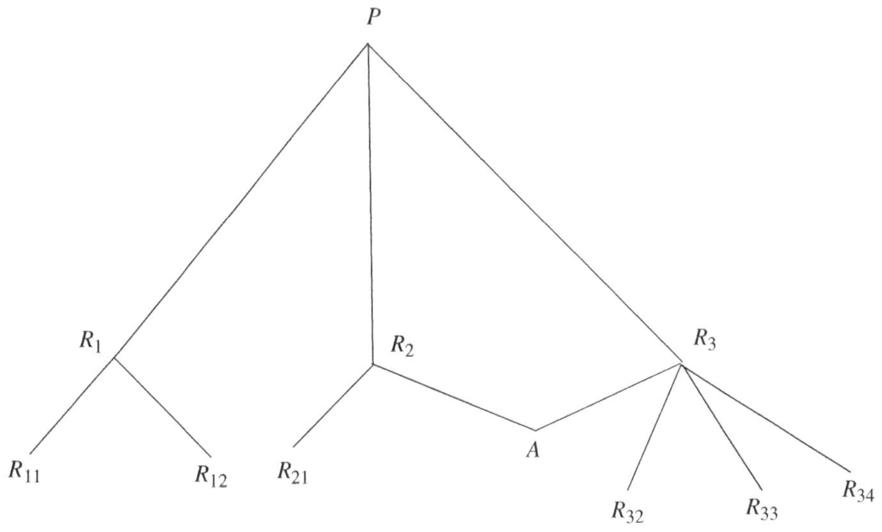

Fig. 1 Map for Problem 11.

the treatment and is found to have recovered. What is the probability that the patient received the new treatment?

13. Four roads lead away from the county jail. A prisoner has escaped from the jail and selects a road at random. If road I is selected, the probability of escaping is 1/8; if road II is selected, the probability of success is 1/6; if road III is selected, the probability of escaping is 1/4; and if road IV is selected, the probability of success is 9/10.

(a) What is the probability that the prisoner will succeed in escaping?

(b) If the prisoner succeeds, what is the probability that the prisoner escaped by using road IV? Road I?

14. A diagnostic test for a certain disease is 95 percent accurate; in that if a person has the disease, it will detect it with a probability of 0.95, and if a person does not have the disease, it will give a negative result with a probability of 0.95. Suppose only 0.5 percent of the population has the disease in question. A person is chosen at random from this population. The test indicates that this person has the disease. What is the (conditional) probability that he or she does have the disease?

1.6 INDEPENDENCE OF EVENTS

Let $(\Omega, \mathcal{S}, P)$ be a probability space, and let $A, B \in \mathcal{S}$, with $PB > 0$. By the multiplication rule we have

$$P(A \cap B) = P(B)P\{A \mid B\}.$$

In many experiments the information provided by B does not affect the probability of event A, that is, $P\{A \mid B\} = P\{A\}$.

Example 1. Let two fair coins be tossed, and let $A = \{\text{head on the second throw}\}$, $B = \{\text{head on the first throw}\}$. Then

$$P(A) = P\{\text{HH}, \text{TH}\} = \tfrac{1}{2}, \quad P(B) = \{\text{HH}, \text{HT}\} = \tfrac{1}{2},$$

and

$$P\{A \mid B\} = \frac{P(A \cap B)}{P(B)} = \frac{\frac{1}{4}}{\frac{1}{2}} = \tfrac{1}{2} = P(A)$$

Thus

$$P(A \cap B) = P(A)\,P(B).$$

In the following, we will write $A \cap B = AB$.

Definition 1. Two events, A and B, are said to be independent if and only if

$$P(AB) = P(A)\,P(B). \tag{1}$$

Note that we have not placed any restriction on $P(A)$ or $P(B)$. Thus conditional probability is not defined when $P(A)$ or $P(B) = 0$ but independence is. Clearly, if $P(A) = 0$, then A is independent of every $E \in \mathcal{S}$. Also, any event $A \in \mathcal{S}$ is independent of Φ and Ω.

Theorem 1. If A and B are independent events, then

$$P\{A \mid B\} = P(A) \quad \text{if } P(B) > 0$$

and

$$P\{B \mid A\} = P(B) \quad \text{if } P(A) > 0.$$

Theorem 2. If A and B are independent, so are A and B^c, A^c and B, and A^c and B^c.

Proof.

$$\begin{aligned} P(A^cB) &= P(B - (A \cap B)) \\ &= P(B) - P(A \cap B) \quad \text{since } B \supseteq (A \cap B) \\ &= P(B)\{1 - P(A)\} \\ &= P(A^c)\,P(B). \end{aligned}$$

Similarly, one proves that (i) A^c and B^c and (ii) A and B^c are independent.

We wish to emphasize that independence of events is not to be confused with disjoint or mutually exclusive events. If two events, each with nonzero probability, are mutually exclusive, they are obviously dependent since the occurrence of one will automatically preclude the occurrence of the other. Similarly, if A and B are independent and $PA > 0$, $PB > 0$, then A and B cannot be mutually exclusive.

Example 2. A card is chosen at random from a deck of 52 cards. Let A be the event that the card is an ace and B, the event that it is a club. Then

$$P(A) = \tfrac{4}{52} = \tfrac{1}{13}, \quad P(B) = \tfrac{13}{52} = \tfrac{1}{4},$$
$$P(AB) = P\{\text{ace of clubs}\} = \tfrac{1}{52},$$

so that A and B are independent.

Example 3. Consider families with two children, and assume that all four possible distributions of sex—BB, BG, GB, GG, where B stands for boy and G for girl—are equally likely. Let E be the event that a randomly chosen family has at most one girl and F, the event that the family has children of both sexes. Then

$$P(E) = \tfrac{3}{4}, \quad P(F) = \tfrac{1}{2}, \quad \text{and} \quad P(EF) = \tfrac{1}{2},$$

so that E and F are not independent.

Now consider families with three children. Assuming that each of the eight possible sex distributions is equally likely, we have

$$P(E) = \tfrac{4}{8}, \quad P(F) = \tfrac{6}{8}, \quad P(EF) = \tfrac{3}{8},$$

so that E and F are independent.

An obvious extension of the concept of independence between two events A and B to a given collection $\mathfrak{U}$ of events is to require that any two distinct events in $\mathfrak{U}$ be independent.

Definition 2. Let $\mathfrak{U}$ be a family of events from $\mathcal{S}$. We say that the events $\mathfrak{U}$ are pairwise independent if and only if, for every pair of distinct events $A, B \in \mathfrak{U}$,

$$P(AB) = PA\,PB.$$

A much stronger and more useful concept is *mutual* or *complete independence*.

Definition 3. A family of events $\mathfrak{U}$ is said to be a mutually or completely independent family if and only if, for every finite sub collection $\{A_{i_1}, A_{i_2}, \ldots, A_{i_k}\}$ of $\mathfrak{U}$, the following relation holds:

$$P(A_{i_1} \cap A_{i_2} \cap \cdots \cap A_{i_k}) = \prod_{j=1}^{k} PA_{i_j}. \tag{2}$$

In what follows we will omit the adjective "mutual" or "complete" and speak of independent events. It is clear from Definition 3 that in order to check the independence of n events $A_1, A_2, \ldots, A_n \in \mathcal{S}$ we must check the following $2^n - n - 1$ relations.

$$
\begin{aligned}
P(A_iA_j) &= PA_i\,PA_j, & i \neq j;\ i,j = 1,2,\ldots,n,\\
P(A_iA_jA_k) &= PA_i\,PA_j\,PA_k, & i \neq j \neq k;\ i,j,k = 1,2,\ldots,n,\\
&\ \ \vdots & \\
P(A_1A_2\cdots A_n) &= PA_1\,PA_2\cdots PA_n. &
\end{aligned}
$$

The first of these requirements is pairwise independence. Independence therefore implies pairwise independence, but not conversely.

Example 4 (Wong [120]). Take four identical marbles. On the first, write symbols $A_1A_2A_3$. On each of the other three, write A_1, A_2, A_3, respectively. Put the four marbles in an urn and draw one at random. Let E_i denote the event that the symbol A_i appears on the drawn marble. Then

$$
\begin{aligned}
P(E_1) = P(E_2) = P(E_3) &= \tfrac{1}{2},\\
P(E_1E_2) = P(E_2E_3) = P(E_1E_3) &= \tfrac{1}{4},
\end{aligned}
$$

and

$$P(E_1E_2E_3) = \tfrac{1}{4}. \tag{3}$$

It follows that although events E_1, E_2, E_3 are not independent, they are pairwise independent.

Example 5 (Kac [48], pp. 22–23). In this example $P(E_1E_2E_3) = P(E_1)P(E_2)P(E_3)$, but E_1, E_2, E_3 are not pairwise independent and hence not independent. Let $\Omega = \{1,2,3,4\}$, and let p_i be the probability assigned to $\{i\}$, $i = 1,2,3,4$. Let $p_1 = \frac{\sqrt{2}}{2} - \frac{1}{4}$, $p_2 = \frac{1}{4}$, $p_3 = \frac{3}{4} - \frac{\sqrt{2}}{2}$, $p_4 = \frac{1}{4}$. Let $E_1 = \{1,3\}$, $E_2 = \{2,3\}$, $E_3 = \{3,4\}$. Then

$$
\begin{aligned}
P(E_1E_2E_3) = P\{3\} = \frac{3}{4} - \frac{\sqrt{2}}{2} &= \frac{1}{2}\left(1 - \frac{\sqrt{2}}{2}\right)\left(1 - \frac{\sqrt{2}}{2}\right)\\
&= (p_1 + p_3)(p_2 + p_3)(p_3 + p_4)\\
&= P(E_1)P(E_2)P(E_3).
\end{aligned}
$$

But $P(E_1E_2) = \frac{3}{4} - \frac{\sqrt{2}}{2} \neq PE_1PE_2$, and it follows that E_1, E_2, E_3 are not independent.

Example 6. A die is rolled repeatedly until a 6 turns up. We will show that event A, that "a 6 will eventually show up," is certain to occur. Let A_k be the event that a 6 will show up for the first time on the kth throw. Let $A = \sum_{k=1}^{\infty} A_k$. Then

$$PA_k = \frac{1}{6}\left(\frac{5}{6}\right)^{k-1}, \quad k = 1,2,\ldots,$$

and

$$PA = \frac{1}{6}\sum_{k=1}^{\infty}\left(\frac{5}{6}\right)^{k-1} = \frac{1}{6}\frac{1}{1-\frac{5}{6}} = 1.$$

Alternatively, we can use the corollary to Theorem 1.3.6. Let B_n be the event that a 6 does not show up on the first n trials. Clearly $B_{n+1} \subseteq B_n$, and we have $A^c = \cap_{n=1}^{\infty} B_n$. Thus

$$1 - PA = PA^c = P\left(\bigcap_{n=1}^{\infty} B_n\right) = \lim_{n\to\infty} P(B_n) = \lim_{n\to\infty}\left(\frac{5}{6}\right)^n = 0.$$

Example 7. A slip of paper is given to person A, who marks it with either a plus or a minus sign; the probability of her writing a plus sign is $1/3$. A passes the slip to B, who may either leave it alone or change the sign before passing it to C. Next, C passes the slip to D after perhaps changing the sign; finally, D passes it to a referee after perhaps changing the sign. The referee sees a plus sign on the slip. It is known that B, C, and D each change the sign with probability $2/3$. We shall compute the probability that A originally wrote a plus.

Let N be the event that A wrote a plus sign, and M, the event that she wrote a minus sign. Let E be the event that the referee saw a plus sign on the slip. We have

$$P\{N \mid E\} = \frac{P(N)P\{E \mid N\}}{P(M)P\{E \mid M\} + P(N)P\{E \mid N\}}.$$

Now

$$\begin{aligned} P\{E \mid N\} &= P\{\text{the plus sign was either not changed or changed exactly twice}\} \\ &= \left(\frac{1}{3}\right)^3 + 3\left(\frac{2}{3}\right)^2\left(\frac{1}{3}\right) \end{aligned}$$

and

$$\begin{aligned} P\{E \mid M\} &= P\{\text{the minus sign was changed either once or three times}\} \\ &= 3\left(\frac{2}{3}\right)\left(\frac{1}{3}\right)^2 + \left(\frac{2}{3}\right)^3. \end{aligned}$$

It follows that

$$\begin{aligned} P\{N \mid E\} &= \frac{(\frac{1}{3})[(\frac{1}{3})^3 + 3(\frac{2}{3})^2(\frac{1}{3})]}{(\frac{1}{3})[(\frac{1}{3})^3 + 3(\frac{2}{3})^2(\frac{1}{3})] + (\frac{2}{3})[3(\frac{2}{3})(\frac{1}{3})^2 + (\frac{2}{3})^3]} \\ &= \frac{\frac{13}{18}}{\frac{41}{81}} = \frac{13}{41}. \end{aligned}$$

PROBLEMS 1.6

1. A biased coin is tossed until a head appears for the first time. Let p be the probability of a head, $0 < p < 1$. What is the probability that the number of tosses required is odd? Even?
2. Let A and B be two independent events defined on some probability space, and let $PA = 1/3$, $PB = 3/4$. Find (a) $P(A \cup B)$, (b) $P\{A \mid A \cup B\}$, and (c) $P\{B \mid A \cup B\}$.
3. Let A_1, A_2, and A_3 be three independent events. Show that A_1^c, A_2^c, and A_3^c are independent.
4. A biased coin with probability p, $0 < p < 1$, of success (heads) is tossed until for the first time the same result occurs three times in succession (i.e., three heads or three tails in succession). Find the probability that the game will end at the seventh throw.
5. A box contains 20 black and 30 green balls. One ball at a time is drawn at random, its color is noted, and the ball is then replaced in the box for the next draw.
   (a) Find the probability that the first green ball is drawn on the fourth draw.
   (b) Find the probability that the third and fourth green balls are drawn on the sixth and ninth draws, respectively.
   (c) Let N be the trial at which the fifth green ball is drawn. Find the probability that the fifth green ball is drawn on the nth draw. (Note that N take values $5, 6, 7, \ldots$.)
6. An urn contains four red and four black balls. A sample of two balls is drawn at random. If both balls drawn are of the same color, these balls are set aside and a new sample is drawn. If the two balls drawn are of different colors, they are returned to the urn and another sample is drawn. Assume that the draws are independent and that the same sampling plan is pursued at each stage until all balls are drawn.
   (a) Find the probability that at least n samples are drawn before two balls of the same color appear.
   (b) Find the probability that after the first two samples are drawn four balls are left, two black and two red.
7. Let A, B, and C be three boxes with three, four, and five cells, respectively. There are three yellow balls numbered 1 to 3, four green balls numbered 1 to 4, and five red balls numbered 1 to 5. The yellow balls are placed at random in box A, the green in B, and the red in C, with no cell receiving more than one ball. Find the probability that only one of the boxes will show no matches.
8. A pond contains red and golden fish. There are 3000 red and 7000 golden fish, of which 200 and 500, respectively, are tagged. Find the probability that a random sample of 100 red and 200 golden fish will show 15 and 20 tagged fish, respectively.
9. Let $(\Omega, \mathcal{S}, P)$ be a probability space. Let A, B, $C \in \mathcal{S}$ with PB and $PC > 0$. If B and C are independent show that

$$P\{A \mid B\} = P\{A \mid B \cap C\}PC + P\{A \mid B \cap C^c\}PC^c.$$

   Conversely, if this relation holds, $P\{A \mid BC\} \neq P\{A \mid B\}$, and $PA > 0$, then B and C are independent (Strait [111]).

10. Show that the converse of Theorem 2 also holds. Thus A and B are independent if, and only if, A and B^c are independent, and so on.

11. A lot of five identical batteries is life tested. The probability assignment is assumed to be

$$P(A) = \int_A (1/\lambda)e^{-x/\lambda}dx$$

for any event $A \subseteq [0, \infty)$, where $\lambda > 0$ is a known constant. Thus the probability that a battery fails after time t is given by

$$P(t, \infty) = \int_t^\infty (1/\lambda)e^{-x/\lambda}dx, \; t \geq 0.$$

If the times to failure of the batteries are independent, what is the probability that at least one battery will be operating after t_0 hours?

12. On $\Omega = (a, b)$, $-\infty < a < b < \infty$, each subinterval is assigned a probability proportional to the length of the interval. Find a necessary and sufficient condition for two events to be independent.

13. A game of craps is played with a pair of fair dice as follows. A player rolls the dice. If a sum of 7 or 11 shows up, the player wins; if a sum of 2, 3, or 12 shows up, the player loses. Otherwise the player continues to roll the pair of dice until the sum is either 7 or the first number rolled. In the former case the player loses and in the latter the player wins.

(a) Find the probability that the player wins on the nth roll.

(b) Find the probability that the player wins the game.

(c) What is the probability that the game ends on: (i) the first roll, (ii) second roll, and (iii) third roll?

2

RANDOM VARIABLES AND THEIR PROBABILITY DISTRIBUTIONS

2.1 INTRODUCTION

In Chapter 1 we dealt essentially with random experiments which can be described by finite sample spaces. We studied the assignment and computation of probabilities of events. In practice, one observes a function defined on the space of outcomes. Thus, if a coin is tossed n times, one is not interested in knowing which of the 2^n n-tuples in the sample space has occurred. Rather, one would like to know the number of heads in n tosses. In games of chance one is interested in the net gain or loss of a certain player. Actually, in Chapter 1 we were concerned with such functions without defining the term *random variable*. Here we study the notion of a random variable and examine some of its properties.

In Section 2.2 we define a random variable, while in Section 2.3 we study the notion of probability distribution of a random variable. Section 2.4 deals with some special types of random variables, and Section 2.5 considers functions of a random variable and their induced distributions.

The fundamental difference between a random variable and a real-valued function of a real variable is the associated notion of a probability distribution. Nevertheless our knowledge of advanced calculus or real analysis is the basic tool in the study of random variables and their probability distributions.

2.2 RANDOM VARIABLES

In Chapter 1 we studied properties of a set function P defined on a sample space $(\Omega, \mathcal{S})$. Since P is a set function, it is not very easy to handle; we cannot perform arithmetic or

---

An Introduction to Probability and Statistics, Third Edition. Vijay K. Rohatgi and A.K. Md. Ehsanes Saleh.

algebraic operations on sets. Moreover, in practice one frequently observes some function of elementary events. When a coin is tossed repeatedly, which replication resulted in heads is not of much interest. Rather one is interested in the number of heads, and consequently the number of tails, that appear in, say, n tossings of the coin. It is therefore desirable to introduce a point function on the sample space. We can then use our knowledge of calculus or real analysis to study properties of P.

Definition 1. Let $(\Omega, \mathcal{S})$ be a sample space. A finite, single-valued function which maps Ω into $\mathcal{R}$ is called a random variable (RV) if the inverse images under X of all Borel sets in $\mathcal{R}$ are events, that is, if

$$X^{-1}(B) = \{\omega : X(\omega) \in B\} \in \mathcal{S} \qquad \text{for all } B \in \mathfrak{B}. \tag{1}$$

In order to verify whether a real-valued function on $(\Omega, \mathcal{S})$ is an RV, it is not necessary to check that (1) holds for all Borel sets $B \in \mathfrak{B}$. It suffices to verify (1) for any class $\mathfrak{A}$ of subsets of $\mathcal{R}$ which generates $\mathfrak{B}$. By taking $\mathfrak{A}$ to be the class of semiclosed intervals $(-\infty, x]$, $x \in \mathcal{R}$ we get the following result.

Theorem 1. X is an RV if and only if for each $x \in \mathcal{R}$

$$\{\omega : X(\omega) \leq x\} = \{X \leq x\} \in \mathcal{S}. \tag{2}$$

Remark 1. Note that the notion of probability does not enter into the definition of an RV.

Remark 2. If X is an RV, the sets $\{X = x\}$, $\{a < X \leq b\}$, $\{X < x\}$, $\{a \leq X < b\}$, $\{a < X < b\}$, $\{a \leq X \leq b\}$ are all events. Moreover, we could have used any of these intervals to define an RV. For example, we could have used the following equivalent definition: X is an RV if and only if

$$\{\omega : X(\omega) < x\} \in \mathcal{S} \qquad \text{for all } x \in \mathcal{R}. \tag{3}$$

We have

$$\{X < x\} = \bigcup_{n=1}^{\infty} \left\{X \leq x - \frac{1}{n}\right\} \tag{4}$$

and

$$\{X \leq x\} = \bigcap_{n=1}^{\infty} \left\{X < x + \frac{1}{n}\right\}. \tag{5}$$

Remark 3. In practice (1) or (2) is a technical condition in the definition of an RV which the reader may ignore and think of RVs simply as real-valued functions defined on Ω. It should be emphasized though that there do exist subsets of $\mathcal{R}$ which do not belong to $\mathfrak{B}$ and hence there exist real-valued functions defined on Ω which are not RVs but the reader will not encounter them in practical applications.

Example 1. For any set $A \subseteq \Omega$, define

$$I_A(\omega) = \begin{cases} 0, & \omega \notin A, \\ 1, & \omega \in A. \end{cases}$$

$I_A(\omega)$ is called the *indicator function* of set A. I_A is an RV if and only if $A \in \mathcal{S}$.

Example 2. Let $\Omega = \{\mathrm{H}, \mathrm{T}\}$ and $\mathcal{S}$ be the class of all subsets of Ω. Define X by $X(\mathrm{H}) = 1$, $X(\mathrm{T}) = 0$. Then

$$X^{-1}(-\infty, x] = \begin{cases} \phi & \text{if } x < 0, \\ \{\mathrm{T}\} & \text{if } 0 \leq x < 1, \\ \{\mathrm{H}, \mathrm{T}\} & \text{if } 1 \leq x, \end{cases}$$

and we see that X is an RV.

Example 3. Let $\Omega = \{\mathrm{HH}, \mathrm{TT}, \mathrm{HT}, \mathrm{TH}\}$ and $\mathcal{S}$ be the class of all subsets of Ω. Define X by

$$X(\omega) = \text{number of H's in } \omega.$$

Then $X(\mathrm{HH}) = 2$, $X(\mathrm{HT}) = X(\mathrm{TH}) = 1$, and $X(\mathrm{TT}) = 0$.

$$X^{-1}(-\infty, x] = \begin{cases} \phi, & x < 0, \\ \{\mathrm{TT}\}, & 0 \leq x < 1, \\ \{\mathrm{TT}, \mathrm{HT}, \mathrm{TH}\}, & 1 \leq x < 2, \\ \Omega, & 2 \leq x. \end{cases}$$

Thus X is an RV.

Remark 4. Let $(\Omega, \mathcal{S})$ be a discrete sample space; that is, let Ω be a countable set of points and $\mathcal{S}$ be the class of all subsets of Ω. Then every numerical valued function defined on $(\Omega, \mathcal{S})$ is an RV.

Example 4. Let $\Omega = [0, 1]$ and $\mathcal{S} = \mathfrak{B} \cap [0, 1]$ be the σ-field of Borel sets on $[0, 1]$. Define X on Ω by

$$X(\omega) = \omega, \qquad \omega \in [0, 1].$$

Clearly X is an RV. Any Borel subset of Ω is an event.

Remark 5. Let X be an RV defined on $(\Omega, \mathcal{S})$ and a, b be constants. Then $aX + b$ is also an RV on $(\Omega, \mathcal{S})$. Moreover, X^2 is an RV and so also is $1/X$, provided that $\{X = 0\} = \phi$. For a general result see Theorem 2.5.1.

PROBLEMS 2.2

1. Let X be the number of heads in three tosses of a coin. What is Ω? What are the values that X assigns to points of Ω? What are the events $\{X \leq 2.75\}$, $\{0.5 \leq X \leq 1.72\}$?
2. A die is tossed two times. Let X be the sum of face values on the two tosses and Y be the absolute value of the difference in face values. What is Ω? What values do X and Y assign to points of Ω? Check to see whether X and Y are random variables.
3. Let X be an RV. Is $|X|$ also an RV? If X is an RV that takes only nonnegative values, is $\sqrt{X}$ also an RV?
4. A die is rolled five times. Let X be the sum of face values. Write the events $\{X = 4\}$, $\{X = 6\}$, $\{X = 30\}$, $\{X \geq 29\}$.
5. Let $\Omega = [0, 1]$ and $\mathcal{S}$ be the Borel σ-field of subsets of Ω. Define X on Ω as follows: $X(\omega) = \omega$ if $0 \leq \omega \leq 1/2$, and $X(\omega) = \omega - 1/2$ if $1/2 < \omega \leq 1$. Is X an RV? If so, what is the event $\{\omega: X(\omega) \in (1/4, 1/2)\}$?
6. Let $\mathfrak{A}$ be a class of subsets of $\mathcal{R}$ which generates $\mathfrak{B}$. Show that X is an RV on Ω if and only if $X^{-1}(A) \in \mathcal{R}$ for all $A \in \mathfrak{A}$.

2.3 PROBABILITY DISTRIBUTION OF A RANDOM VARIABLE

In Section 2.2 we introduced the concept of an RV and noted that the concept of probability on the sample space was not used in this definition. In practice, however, random variables are of interest only when they are defined on a probability space. Let $(\Omega, \mathcal{S}, P)$ be a probability space, and let X be an RV defined on it.

Theorem 1. The RV X defined on the probability space $(\Omega, \mathcal{S}, P)$ induces a probability space $(\mathcal{R}, \mathfrak{B}, Q)$ by means of the correspondence

$$Q(B) = P\{X^{-1}(B)\} = P\{\omega: X(\omega) \in B\} \qquad \text{for all } B \in \mathfrak{B}. \tag{1}$$

We write $Q = PX^{-1}$ and call Q or PX^{-1} the (probability) *distribution* of X.

Proof. Clearly $Q(B) \geq 0$ for all $B \in \mathfrak{B}$, and also $Q(\mathcal{R}) = P\{X \in \mathcal{R}\} = P(\Omega) = 1$.

Let $B_i \in \mathcal{B}$, $i = 1, 2, \ldots$ with $B_i \cap B_j = \phi$, $i \neq j$. Since the inverse image of a disjoint union of Borel sets is the disjoint union of their inverse images, we have

$$\begin{aligned} Q\left(\sum_{i=1}^{\infty} B_i\right) &= P\left\{X^{-1}\left(\sum_{i=1}^{\infty} B_i\right)\right\} \\ &= P\left\{\sum_{i=1}^{\infty} X^{-1}(B_i)\right\} \\ &= \sum_{i=1}^{\infty} PX^{-1}(B_i) = \sum_{i=1}^{\infty} Q(B_i). \end{aligned}$$

It follows that $(\mathcal{R}, \mathfrak{B}, Q)$ is a probability space, and the proof is complete.

We note that Q is a set function, and set functions are not easy to handle. It is therefore more practical to use (2.2.2) since then $Q(-\infty,x]$ is a point function. Let us first introduce and study some properties of a special point function on $\mathcal{R}$.

Definition 1. A real-valued function F defined on $(-\infty,\infty)$ that is nondecreasing, right continuous, and satisfies

$$F(-\infty)=0 \qquad \text{and} \qquad F(+\infty)=1$$

is called a distribution function (DF).

Remark 1. Recall that if F is a nondecreasing function on $\mathcal{R}$, then $F(x-)=\lim_{t\uparrow x}F(t)$, $F(x+)=\lim_{t\downarrow x}F(t)$ exist and are finite. Also, $F(+\infty)$ and $F(-\infty)$ exist as $\lim_{t\uparrow+\infty}F(t)$ and $\lim_{t\downarrow-\infty}F(t)$, respectively. In general,

$$F(x-)\le F(x)\le F(x+),$$

and x is a jump point of F if and only if $F(x+)$ and $F(x-)$ exist but are unequal. Thus a nondecreasing function F has only jump discontinuities. If we define

$$F^*(x)=F(x+) \qquad \text{for all } x,$$

we see that F^* is nondecreasing and right continuous on $\mathcal{R}$. Thus in Definition 1 the nondecreasing part is very important. Some authors demand left continuity in the definition of a DF instead of right continuity.

Theorem 2. The set of discontinuity points of a DF F is at most countable.

Proof. Let $(a,b]$ be a finite interval with at least n discontinuity points:

$$a<x_1<x_2<\cdots<x_n\le b.$$

Then

$$F(a)\le F(x_1-)<F(x_1)\le\cdots\le F(x_n-)<F(x_n)\le F(b).$$

Let $p_k=F(x_k)-F(x_k-)$, $k=1,2,\ldots,n$. Clearly,

$$\sum_{k=1}^{n}p_k\le F(b)-F(a),$$

and it follows that the number of points x in $(a,b]$ with jump $p(x)>\varepsilon>0$ is at most $\varepsilon^{-1}\{F(b)-F(a)\}$. Thus, for every integer N, the number of discontinuity points with jump greater than $1/N$ is finite. It follows that there are no more than a countable number of discontinuity points in every finite interval $(a,b]$. Since $\mathcal{R}$ is a countable union of such intervals, the proof is complete.

Definition 2. Let X be an RV defined on $(\Omega, \mathcal{S}, P)$. Define a point function $F(.)$ on $\mathcal{R}$ by using (1), namely,

$$F(x) = Q(-\infty, x] = P\{\omega\colon X(\omega) \leq x\} \qquad \text{for all } x \in \mathcal{R}. \tag{2}$$

The function F is called the distribution function of RV X.

If there is no confusion, we will write

$$F(x) = P\{X \leq x\}.$$

The following result justifies our calling F as defined by (2) a DF.

Theorem 3. The function F defined in (2) is indeed a DF.

Proof. Let $x_1 < x_2$. Then $(-\infty, x_1] \subset (-\infty, x_2]$, and we have

$$F(x_1) = P\{X \leq x_1\} \leq P\{X \leq x_2\} = F(x_2).$$

Since F is nondecreasing, it is sufficient to show that for any sequence of numbers $x_n \downarrow x$, $x_1 > x_2 > \cdots > x_n > \cdots > x$, $F(x_n) \to F(x)$. Let $A_k = \{\omega\colon X(\omega) \in (x, x_k]\}$. Then $A_k \in \mathcal{S}$ and $A_k \downarrow \emptyset$. Also,

$$\lim_{k\to\infty} A_k = \bigcap_{k=1}^{\infty} A_k = \phi,$$

since none of the intervals $(x, x_k]$ contains x. It follows that $\lim_{k\to\infty} P(A_k) = 0$. But,

$$\begin{aligned} P(A_k) &= P\{X \leq x_k\} - P\{X \leq x\} \\ &= F(x_k) - F(x), \end{aligned}$$

so that

$$\lim_{k\to\infty} F(x_k) = F(x)$$

and F is right continuous.

Finally, let $\{x_n\}$ be a sequence of numbers decreasing to $-\infty$. Then,

$$\{X \leq x_n\} \supseteq \{X \leq x_{n+1}\} \qquad \text{for each } n$$

and

$$\lim_{n\to\infty} \{X \leq x_n\} = \bigcap_{n=1}^{\infty} \{X \leq x_n\} = \phi.$$

Therefore,

$$F(-\infty) = \lim_{n\to\infty} P\{X \leq x_n\} = P\left\{\lim_{n\to\infty} \{X \leq x_n\}\right\} = 0.$$

Similarly,

$$F(+\infty) = \lim_{x_n \to \infty} P\{X \le x_n\} = 1,$$

and the proof is complete.

The next result, stated without proof, establishes a correspondence between the induced probability Q on $(\mathcal{R}, \mathfrak{B})$ and a point function F defined on $\mathcal{R}$.

Theorem 4. Given a probability Q on $(\mathcal{R}, \mathfrak{B})$, there exists a distribution function F satisfying

$$Q(-\infty, x] = F(x) \qquad \text{for all } x \in \mathcal{R}, \tag{3}$$

and, conversely, given a DF F, there exists a unique probability Q defined on $(\mathcal{R}, \mathfrak{B})$ that satisfies (3).

For proof see Chung [15, pp. 23–24].

Theorem 5. Every DF is the DF of an RV on some probability space.

Proof. Let F be a DF. From Theorem 4 it follows that there exists a unique probability Q defined on $\mathcal{R}$ that satisfies

$$Q(-\infty, x] = F(x) \qquad \text{for all } x \in \mathcal{R}.$$

Let $(\mathcal{R}, \mathfrak{B}, Q)$ be the probability space on which we define

$$X(\omega) = \omega, \qquad \omega \in \mathcal{R}.$$

Then

$$Q\{\omega \colon X(\omega) \le x\} = Q(-\infty, x] = F(x),$$

and F is the DF of RV X.

Remark 2. If X is an RV on $(\Omega, \mathcal{S}, P)$, we have seen (Theorem 3) that $F(x) = P\{X \le x\}$ is a DF associated with X. Theorem 5 assures us that to every DF F we can associate some RV. Thus, given an RV, there exists a DF, and conversely. In this book, when we speak of an RV we will assume that it is defined on some probability space.

Example 1. Let X be defined on $(\Omega, \mathcal{S}, P)$ by

$$X(\omega) = c \qquad \text{for all } \omega \in \Omega.$$

Then

$$P\{X=c\}=1,$$
$$F(x)=Q(-\infty,x]=P\{X^{-1}(-\infty,x]\}=0 \qquad \text{if } x<c$$

and

$$F(x)=1 \qquad \text{if } x\geq c.$$

Example 2. Let $\Omega=\{\text{H},\text{T}\}$ and X be defined by

$$X(\text{H})=1, \qquad X(\text{T})=0.$$

If P assigns equal mass to $\{\text{H}\}$ and $\{\text{T}\}$, then

$$P\{X=0\}=\frac{1}{2}=P\{X=1\}$$

and

$$F(x)=Q(-\infty,x]=\begin{cases}0, & x<0,\\ \frac{1}{2}, & 0\leq x<1,\\ 1, & 1\leq x.\end{cases}$$

Example 3. Let $\Omega=\{(i,j)\colon i,j\in\{1,2,3,4,5,6\}\}$ and $\mathcal{S}$ be the set of all subsets of Ω. Let $P\{(i,j)\}=1/6^2$ for all 6^2 pairs (i,j) in Ω. Define

$$X(i,j)=i+j, \qquad 1\leq i,j\leq 6.$$

Then,

$$F(x)=Q(-\infty,x]=P\{X\leq x\}=\begin{cases}0, & x<2,\\ \frac{1}{36}, & 2\leq x<3,\\ \frac{3}{36}, & 3\leq x<4,\\ \frac{6}{36}, & 4\leq x<5,\\ \vdots & \\ \frac{35}{36}, & 11\leq x<12,\\ 1, & 12\leq x.\end{cases}$$

Example 4. We return to Example 2.2.4. For every subinterval I of $[0,1]$ let $P(I)$ be the length of the interval. Then $(\Omega,\mathcal{S},P)$ is a probability space, and the DF of RV $X(\omega)=\omega$, $\omega\in\Omega$ is given by $F(x)=0$ if $x<0$, $F(x)=P\{\omega\colon X(\omega)\leq x\}=P([0,x])=x$ if $x\in[0,1]$, and $F(x)=1$ if $x\geq 1$.

PROBLEMS 2.3

1. Write the DF of RV X defined in Problem 2.2.1, assuming that the coin is fair.
2. What is the DF of RV Y defined in Problem 2.2.2, assuming that the die is not loaded?
3. Do the following functions define DFs?
   (a) $F(x) = 0$ if $x < 0$, $= x$ if $0 \le x < 1/2$, and $= 1$ if $x \ge \frac{1}{2}$.
   (b) $F(x) = (1/\pi)\tan^{-1} x$, $-\infty < x < \infty$.
   (c) $F(x) = 0$ if $x \le 1$, and $= 1 - (1/x)$ if $1 < x$.
   (d) $F(x) = 1 - e^{-x}$ if $x \ge 0$, and $= 0$ if $x < 0$.
4. Let X be an RV with DF F.
   (a) If F is the DF defined in Problem 3(a), find $P\{X > \frac{1}{4}\}$, $P\{\frac{1}{3} < X \le \frac{3}{8}\}$.
   (b) If F is the DF defined in Problem 3(d), find $P\{-\infty < X < 2\}$.

2.4 DISCRETE AND CONTINUOUS RANDOM VARIABLES

Let X be an RV defined on some fixed, but otherwise arbitrary, probability space $(\Omega, \mathcal{S}, P)$, and let F be the DF of X. In this book, we shall restrict ourselves mainly to two cases, namely, the case in which the RV assumes at most a countable number of values and hence its DF is a step function and that in which the DF F is (absolutely) continuous.

Definition 1. An RV X defined on $(\Omega, \mathcal{S}, P)$ is said to be of the discrete type, or simply discrete, if there exists a countable set $E \subseteq \mathcal{R}$ such that $P\{X \in E\} = 1$. The points of E which have positive mass are called jump points or points of increase of the DF of X, and their probabilities are called jumps of the DF.

Note that $E \in \mathfrak{B}$ since every one-point is in $\mathfrak{B}$. Indeed, if $x \in \mathcal{R}$, then

$$\{x\} = \bigcap_{n=1}^{\infty} \left\{ \left(x - \frac{1}{n} < x \le x + \frac{1}{n} \right) \right\}. \tag{1}$$

Thus $\{X \in E\}$ is an event. Let X take on the value x_i with probability p_i $(i = 1, 2, \ldots)$. We have

$$P\{\omega \colon X(\omega) = x_i\} = p_i, \qquad i = 1, 2, \ldots, \quad p_i \ge 0 \text{ for all } i.$$

Then $\sum_{i=1}^{\infty} p_i = 1$.

Definition 2. The collection of numbers $\{p_i\}$ satisfying $P\{X = x_i\} = p_i \ge 0$, for all i and $\sum_{i=1}^{\infty} p_i = 1$, is called the probability mass function (pmf) of RV X.

The DF F of X is given by

$$F(x) = P\{X \le x\} = \sum_{x_i \le x} p_i. \tag{2}$$

If I_A denotes the indicator function of the set A, we may write

$$X(\omega) = \sum_{i=1}^{\infty} x_i I_{[X=x_i]}(\omega). \tag{3}$$

Let us define a function $\varepsilon(x)$ as follows:

$$\varepsilon(x) = \begin{cases} 1, & x \geq 0, \\ 0, & x < 0. \end{cases}$$

Then we have

$$F(x) = \sum_{i=1}^{\infty} p_i \varepsilon(x - x_i). \tag{4}$$

Example 1. The simplest example is that of an RV X *degenerate* at c, $P\{X = c\} = 1$:

$$F(x) = \varepsilon(x - c) = \begin{cases} 0, & x < c, \\ 1, & x \geq c. \end{cases}$$

Example 2. A box contains good and defective items. If an item drawn is good, we assign the number 1 to the drawing; otherwise, the number 0. Let p be the probability of drawing at random a good item. Then

$$P\left\{X = \begin{matrix} 0 \\ 1 \end{matrix}\right\} = \begin{cases} 1 - p \\ p, \end{cases}$$

and

$$F(x) = P\{X \leq x\} = \begin{cases} 0, & x < 0, \\ 1 - p, & 0 \leq x < 1, \\ 1, & 1 \leq x. \end{cases}$$

Example 3. Let X be an RV with PMF

$$P\{X = k\} = \frac{6}{\pi^2} \cdot \frac{1}{k^2}, \qquad k = 1, 2, \ldots.$$

Then,

$$F(x) = \frac{6}{\pi^2} \sum_{k=1}^{\infty} \frac{1}{k^2} \varepsilon(x - k).$$

Theorem 1. Let $\{p_k\}$ be a collection of nonnegative real numbers such that $\sum_{k=1}^{\infty} p_k = 1$. Then $\{p_k\}$ is the PMF of some RV X.

We next consider RVs associated with DFs that have no jump points. The DF of such an RV is continuous. We shall restrict our attention to a special subclass of such RVs.

Definition 3. Let X be an RV defined on $(\Omega, \mathcal{S}, P)$ with DF F. Then X is said to be of the continuous type (or, simply, continuous) if F is absolutely continuous, that is, if there exists a nonnegative function $f(x)$ such that for every real number x we have

$$F(x) = \int_{-\infty}^{x} f(t)\, dt. \tag{5}$$

The function f is called the probability density function (PDF) of the RV X.

Note that $f \geq 0$ and satisfies $\lim_{x\to+\infty} F(x) = F(+\infty) = \int_{-\infty}^{\infty} f(t)\, dt = 1$. Let a and b be any two real numbers with $a < b$. Then

$$\begin{aligned} P\{a < X \leq b\} &= F(b) - F(a) \\ &= \int_{a}^{b} f(t)\, dt. \end{aligned}$$

In view of remarks following Definition 2.2.1, the following result holds.

Theorem 2. Let X be an RV of the continuous type with PDF f. Then for every Borel set $B \in \mathfrak{B}$

$$P(B) = \int_{B} f(t)\, dt. \tag{6}$$

If F is absolutely continuous and f is continuous at x, we have

$$F'(x) = \frac{dF(x)}{dx} = f(x). \tag{7}$$

Theorem 3. Every nonnegative real function f that is integrable over $\mathcal{R}$ and satisfies

$$\int_{-\infty}^{\infty} f(x)\, dx = 1$$

is the PDF of some continuous type RV X.

Proof. In view of Theorem 2.3.5 it suffices to show that there corresponds a DF F to f. Define

$$F(x) = \int_{-\infty}^{x} f(t)\, dt, \qquad x \in \mathcal{R}.$$

Then $F(-\infty) = 0$, $F(+\infty) = 1$, and, if $x_2 > x_1$,

$$F(x_2) = \left(\int_{-\infty}^{x_1} + \int_{x_1}^{x_2} \right) f(t)\, dt \geq \int_{-\infty}^{x_1} f(t)\, dt = F(x_1).$$

Finally, F is (absolutely) continuous and hence continuous from the right.

Remark 1. In the discrete case, $P\{X = a\}$ is the probability that X takes the value a. In the continuous case, $f(a)$ is not the probability that X takes the value a. Indeed, if X is of the continuous type, it assumes every value with probability 0.

Theorem 4. Let X be any RV. Then

$$P\{X = a\} = \lim_{\substack{t \to a \\ t < a}} P\{t < X \leq a\}. \tag{8}$$

Proof. Let $t_1 < t_2 < \cdots < a, \quad t_n \to a$, and write

$$A_n = \{t_n < X \leq a\}.$$

Then A_n is a nonincreasing sequence of events which converges to $\bigcap_{n=1}^{\infty} A_n = \{X = a\}$. It follows that $\lim_{n \to \infty} PA_n = P\{X = a\}$.

Remark 2. Since $P\{t < X \leq a\} = F(a) - F(t)$, it follows that

$$\lim_{\substack{t \to a \\ t < a}} P\{t < X \leq a\} = P\{X = a\} = F(a) - \lim_{\substack{t \to a \\ t < a}} F(t)$$
$$= F(a) - F(a-).$$

Thus F has a jump discontinuity at a if and only if $P\{X = a\} > 0$, that is, F is continuous at a if and only if $P\{X = a\} = 0$. If X is an RV of the continuous type, $P\{X = a\} = 0$ for all $a \in \mathcal{R}$. Moreover,

$$P\{X \in \mathcal{R} - \{a\}\} = 1.$$

This justifies Remark 1.3.4.

Remark 3. The set of real numbers x for which a DF F increases is called the *support* of F. Let X be the RV with DF F, and let S be the support of F. Then $P(X \in S) = 1$ and $P(X \in S^c) = 0$. The set of positive integers is the support of the DF in Example 3, and the open interval $(0, 1)$ is the support of F in Example 4 below.

Example 4. Let X be an RV with DF F given by (Fig. 1)

$$F(x) = \begin{cases} 0, & x \leq 0, \\ x, & 0 < x \leq 1, \\ 1, & 1 < x. \end{cases}$$

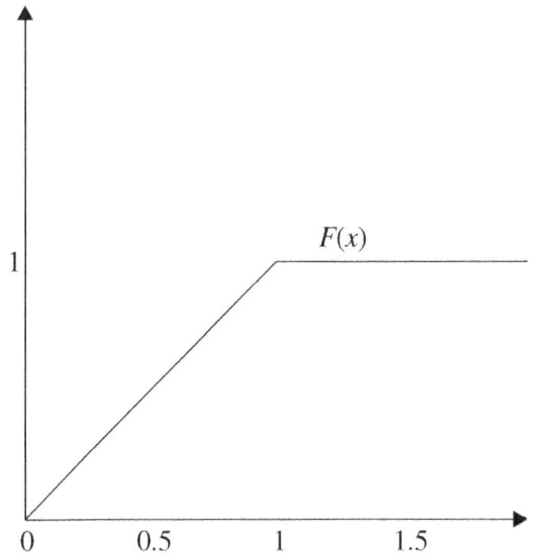

Fig. 1

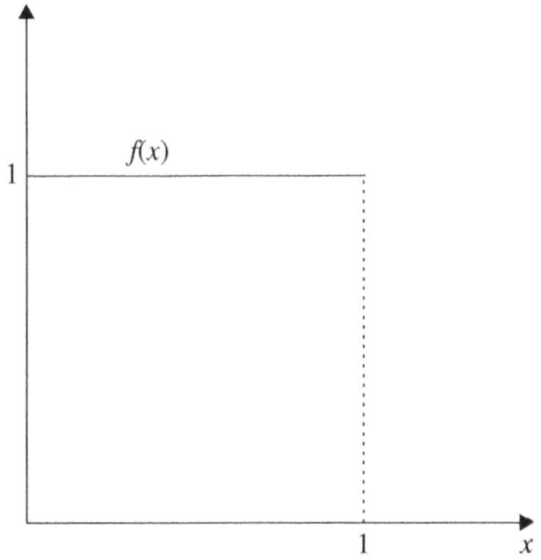

Fig. 2

Differentiating F with respect to x at continuity points of f, we get

$$f(x) = F'(x) = \begin{cases} 0, & x < 0 \text{ or } x > 1, \\ 1, & 0 < x < 1. \end{cases}$$

The function f is not continuous at $x = 0$ or at $x = 1$ (Fig. 2). We may define $f(0)$ and $f(1)$ in any manner. Choosing $f(0) = f(1) = 0$, we have

$$f(x) = \begin{cases} 1, & 0 < x < 1, \\ 0, & \text{otherwise}. \end{cases}$$

Then

$$P\{0.4 < X \le 0.6\} = F(0.6) - F(0.4) = 0.2.$$

Example 5. Let X have the *triangular* PDF (Fig. 3)

$$f(x) = \begin{cases} x, & 0 < x \le 1, \\ 2 - x, & 1 \le x \le 2, \\ 0, & \text{otherwise.} \end{cases}$$

It is easy to check that f is a PDF. For the DF F of X we have (Fig. 4)

$$F(x) = 0 \qquad \text{if } x \le 0,$$

$$F(x) = \int_0^x t\,dt = \frac{x^2}{2} \qquad \text{if } 0 < x \le 1,$$

$$F(x) = \int_0^1 t\,dt + \int_1^x (2-t)\,dt = 2x - \frac{x^2}{2} - 1 \qquad \text{if } 1 < x \le 2,$$

and

$$F(x) = 1 \qquad \text{if } x \ge 2.$$

Then

$$\begin{aligned} P\{0.3 < X \le 1.5\} &= P\{X \le 1.5\} - P\{X \le 0.3\} \\ &= 0.83. \end{aligned}$$

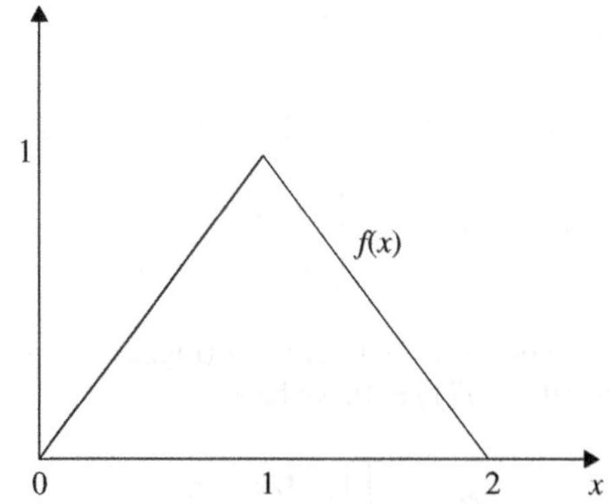

Fig. 3 Graph of f.

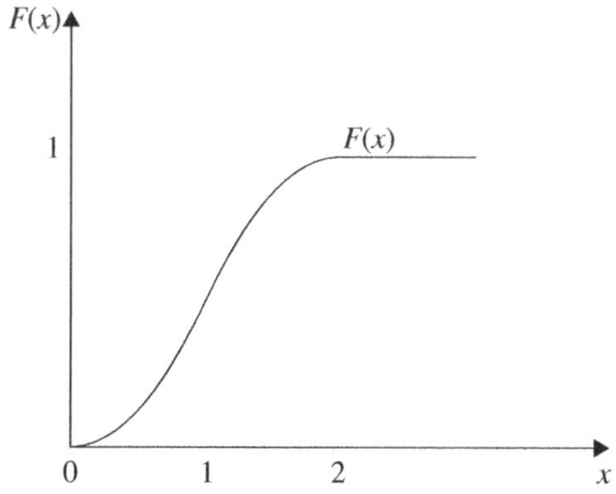

Fig. 4 Graph of F.

Example 6. Let $k > 0$ be a constant, and

$$f(x) = \begin{cases} kx(1-x), & 0 < x < 1, \\ 0, & \text{otherwise.} \end{cases}$$

Then $\int_0^1 f(x)\,dx = k/6$. It follows that $f(x)$ defines a PDF if $k = 6$. We have

$$P\{X > 0.3\} = 1 - 6\int_0^{.3} x(1-x)\,dx = 0.784.$$

We conclude this discussion by emphasizing that the two types of RVs considered above form only a part of the class of all RVs. These two classes, however, contain practically all the random variables that arise in practice. We note without proof (see Chung [15, p. 9]) that every DF F can be decomposed into two parts according to

$$F(x) = aF_d(x) + (1-a)F_c(x). \tag{9}$$

Here F_d and F_c are both DFs; F_d is the DF of a discrete RV, while F_c is a continuous (not necessarily absolutely continuous) DF. In fact, F_c can be further decomposed, but we will not go into that (see Chung [15, p.11]).

Example 7. Let X be an RV with DF

$$F(x) = \begin{cases} 0, & x < 0, \\ \frac{1}{2}, & x = 0, \\ \frac{1}{2} + \frac{x}{2}, & 0 < x < 1, \\ 1, & 1 \le x. \end{cases}$$

Note that the DF F has a jump at $x = 0$ and F is continuous (in fact, absolutely continuous) in the interval $(0,1)$. F is the DF of an RV X that is neither discrete nor continuous. We can write

$$F(x) = \frac{1}{2}F_d(x) + \frac{1}{2}F_c(x),$$

where

$$F_d(x) = \begin{cases} 0, & x < 0, \\ 1, & x \geq 0 \end{cases}$$

and

$$F_c(x) = \begin{cases} 0, & x \leq 0, \\ x, & 0 < x < 1, \\ 1, & 1 \leq x. \end{cases}$$

Here $F_d(x)$ is the DF of the RV degenerate at $x = 0$, and $F_c(x)$ is the DF with PDF

$$f_c(x) = \begin{cases} 1, & 0 < x < 1, \\ 0, & \text{otherwise.} \end{cases}$$

PROBLEMS 2.4

1. Let

$$p_k = p(1-p)^k, \qquad k = 0,1,2,\ldots, \quad 0 < p < 1.$$

   Does $\{p_k\}$ define the PMF of some RV? What is the DF of this RV? If X is an RV with PMF $\{p_k\}$, what is $P\{n \leq X \leq N\}$, where n, N $(N > n)$ are positive integers?
2. In Problem 2.3.3, find the PDF associated with the DFs of parts (b), (c), and (d).
3. Does the function $f_\theta(x) = \theta^2 x e^{-\theta x}$ if $x > 0$, and $= 0$ if $x \leq 0$, where $\theta > 0$, define a PDF? Find the DF associated with $f_\theta(x)$; if X is an RV with PDF $f_\theta(x)$, find $P\{X \geq 1\}$.
4. Does the function $f_\theta(x) = \{(x+1)/[\theta(\theta+1)]\}e^{-x/\theta}$ if $x > 0$, and $= 0$ otherwise, where $\theta > 0$ define a PDF? Find the corresponding df.
5. For what values of K do the following functions define the PMF of some RV?
   (a) $f(x) = K(\lambda^x/x!)$, $x = 0,1,2,\ldots$, $\lambda > 0$.
   (b) $f(x) = K/N$, $x = 1,2,\ldots,N$.
6. Show that the function

$$f(x) = \frac{1}{2}e^{-|x|}, \qquad -\infty < x < \infty,$$

   is a PDF. Find its DF.
7. For the PDF $f(x) = x$ if $0 \leq x < 1$, and $= 2 - x$ if $1 \leq x < 2$, find $P\{1/6 < X \leq 7/4\}$.

8. Which of the following functions are density functions:

(a) $f(x) = x(2-x)$, $0 < x < 2$, and 0 elsewhere.

(b) $f(x) = x(2x-1)$, $0 < x < 2$, and 0 elsewhere.

(c) $f(x) = \frac{1}{\lambda}\exp\{-(x-\theta)/\lambda\}$, $x > \theta$, and 0 elsewhere, $\lambda > 0$.

(d) $f(x) = \sin x$, $0 < x < \pi/2$, and 0 elsewhere.

(e) $f(x) = 0$ for $x < 0$, $= (x+1)/9$ for $0 \le x < 1$, $= 2(2x-1)/9$ for $1 \le x < 3/2$, $= 2(5-2x)/9$ for $3/2 \le x < 1$, $= 4/27$ for $2 \le x < 5$, and 0 elsewhere.

(f) $f(x) = 1/[\pi(1+x^2)]$, $x \in \mathcal{R}$.

9. Are the following functions distribution functions? If so, find the corresponding density or probability functions.

(a) $F(x) = 0$ for $x \le 0$, $= x/2$ for $0 \le x < 1$, $= 1/2$ for $1 \le x < 2$, $= x/4$ for $2 \le x < 4$ and $= 1$ for $x \ge 4$.

(b) $F(x) = 0$ if $x < -\theta$, $= \frac{1}{2}\left(\frac{x}{\theta}+1\right)$ if $|x| \le \theta$, and 1 for $x > \theta$ where $\theta > 0$.

(c) $F(x) = 0$ if $x < 0$, and $= 1-(1+x)\exp(-x)$ if $x \ge 0$.

(d) $F(x) = 0$ if $x < 1$, $= (x-1)^2/8$ if $1 \le x < 3$, and 1 for $x \ge 3$.

(e) $F(x) = 0$ if $x < 0$, and $= 1 - e^{-x^2}$ if $x \ge 0$.

10. Suppose $P(X \ge x)$ is given for a random variable X (of the continuous type) for all x. How will you find the corresponding density function? In particular find the density function in each of the following cases:

(a) $P(X \ge x) = 1$ if $x \le 0$, and $P(X \ge x) = e^{-\lambda x}$ for $x > 0$, $\lambda > 0$ is a constant.

(b) $P(X \ge x) = 1$ if $x < 0$, and $= (1+x/\lambda)^{-\lambda}$, for $x \ge 0$, $\lambda > 0$ is a constant.

(c) $P(X \ge x) = 1$ if $x \le 0$, and $= 3/(1+x)^2 - 2/(1+x)^3$ if $x > 0$.

(d) $P(X > x) = 1$ if $x \le x_0$, and $= (x_0/x)^\alpha$ if $x > x_0$; $x_0 > 0$ and $\alpha > 0$ are constants.

2.5 FUNCTIONS OF A RANDOM VARIABLE

Let X be an RV with a known distribution, and let g be a function defined on the real line. We seek the distribution of $Y = g(X)$, provided that Y is also an RV. We first prove the following result.

Theorem 1. Let X be an RV defined on $(\Omega, \mathcal{S}, P)$. Also, let g be a Borel-measurable function on $\mathcal{R}$. Then $g(X)$ is also an RV.

Proof. For $y \in \mathcal{R}$, we have

$$\{g(X) \le y\} = \{X \in g^{-1}(-\infty, y]\},$$

and since g is Borel-measurable, $g^{-1}(-\infty, y]$ is a Borel set. It follows that $\{g(X) \le y\} \in \mathcal{S}$, and the proof is complete.

Theorem 2. Given an RV X with a known DF, the distribution of the RV $Y = g(X)$, where g is a Borel-measurable function, is determined.

Proof. Indeed, for all $y \in \mathcal{R}$

$$P\{Y \leq y\} = P\{X \in g^{-1}(-\infty, y]\}. \tag{1}$$

In what follows, we will always assume that the functions under consideration are Borel-measurable.

Example 1. Let X be an RV with DF F. Then $|X|$, $aX+b$ (where $a \neq 0$ and b are constants), X^k (where $k \geq 0$ is an integer), and $|X|^\alpha$ ($\alpha > 0$) are all RVs. Define

$$X^+ = \begin{cases} X, & X \geq 0, \\ 0, & X < 0, \end{cases}$$

and

$$X^- = \begin{cases} X, & X \leq 0, \\ 0, & X > 0. \end{cases}$$

Then X^+, X^- are also RVs. We have

$$\begin{aligned} P\{|X| \leq y\} &= P\{-y \leq X \leq y\} = P\{X \leq y\} - P\{X < -y\} \\ &= F(y) - F(-y) + P\{X = -y\}, \qquad y > 0; \\ P\{aX+b \leq y\} &= P\{aX \leq y-b\} \\ &= \begin{cases} P\left\{X \leq \dfrac{y-b}{a}\right\} & \text{if } a > 0, \\ P\left\{X \geq \dfrac{y-b}{a}\right\} & \text{if } a < 0; \end{cases} \\ P\{X^+ \leq y\} &= \begin{cases} 0 & \text{if } y < 0, \\ P\{X \leq 0\} & \text{if } y = 0, \\ P\{X < 0\} + P\{0 \leq X \leq y\} & \text{if } y > 0. \end{cases} \end{aligned}$$

Similarly,

$$P\{X^- \leq y\} = \begin{cases} 1 & \text{if } y \geq 0, \\ P\{X \leq y\} & \text{if } y < 0. \end{cases}$$

Let X be an RV of the discrete type, and A be the countable set such that $P\{X \in A\} = 1$ and $P\{X = x\} > 0$ for $x \in A$. Let $Y = g(X)$ be a one-to-one mapping from A onto some set B. Then the inverse map, g^{-1}, is a single-valued function of y. To find $P\{Y = y\}$, we note that

$$\begin{gathered} P\{Y = y\} = P\{g(X) = y\} = P\{X = g^{-1}(y)\}, \quad y \in B, \\ \text{and} \qquad P\{Y = y\} = 0, \quad y \in B^c. \end{gathered}$$

Example 2. Let X be a *Poisson* RV with PMF

$$P\{X=k\}=\begin{cases} e^{-\lambda}\dfrac{\lambda^k}{k!}, & k=0,1,2,\ldots;\ \lambda>0,\\ 0, & \text{otherwise.}\end{cases}$$

Let $Y=X^2+3$. Then $y=x^2+3$ maps $A=\{0,1,2,\ldots\}$ onto $B=\{3,4,7,12,19,28,\ldots\}$. The inverse map is $x=\sqrt{(y-3)}$, and since there are no negative values in A we take the positive square root of $y-3$. We have

$$P\{Y=y\}=P\{X=\sqrt{y-3}\}=\frac{e^{-\lambda}\lambda^{\sqrt{y-3}}}{\sqrt{(y-3)}!}, \qquad y\in B,$$

and $P\{Y=y\}=0$ elsewhere.

Actually the restriction to a single-valued inverse on g is not necessary. If g has a finite (or even a countable) number of inverses for each y, from countable additivity of P we have

$$\begin{aligned}P\{Y=y\}=P\{g(X)=y\}&=P\left\{\bigcup_a [X=a,g(a)=y]\right\}\\ &=\sum_a P\{X=a,g(a)=y\}.\end{aligned}$$

Example 3. Let X be an RV with PMF

$$P\{X=-2\}=\frac{1}{5},\qquad P\{X=-1\}=\frac{1}{6},\qquad P\{X=0\}=\frac{1}{5},$$
$$P\{X=1\}=\frac{1}{15},\qquad \text{and}\qquad P\{X=2\}=\frac{11}{30}.$$

Let $Y=X^2$. Then

$$A=\{-2,-1,0,1,2\}\qquad \text{and}\qquad B=\{0,1,4\}.$$

We have

$$P\{Y=y\}=\begin{cases}\frac{1}{5} & y=0,\\ \frac{1}{6}+\frac{1}{15}=\frac{7}{30}, & y=1,\\ \frac{1}{5}+\frac{11}{30}=\frac{17}{30}, & y=4.\end{cases}$$

The case in which X is an RV of the continuous type is not as simple. First we note that if X is a continuous type RV and g is some Borel-measurable function, $Y=g(X)$ may not be an RV of the continuous type.

Example 4. Let X be an RV with *uniform* distribution on $[-1,1]$, that is, the PDF of X is $f(x) = 1/2$, $-1 \le x \le 1$, and $= 0$ elsewhere. Let $Y = X^+$. Then, from Example 1,

$$P\{Y \le y\} = \begin{cases} 0, & y < 0, \\ \frac{1}{2}, & y = 0, \\ \frac{1}{2} + \frac{1}{2}y, & 1 \ge y > 0, \\ 1, & y > 1. \end{cases}$$

We see that the DF of Y has a jump at $y = 0$ and that Y is neither discrete nor continuous. Note that all we require is that $P\{X < 0\} > 0$ for X^+ to be of the mixed type.

Example 4 shows that we need some conditions on g to ensure that $g(X)$ is also an RV of the continuous type whenever X is continuous. This is the case when g is a continuous monotonic function. A sufficient condition is given in the following theorem.

Theorem 3. Let X be an RV of the continuous type with PDF f. Let $y = g(x)$ be differentiable for all x and either $g'(x) > 0$ for all x or $g'(x) < 0$ for all x. Then $Y = g(X)$ is also an RV of the continuous type with PDF given by

$$h(y) = \begin{cases} f[g^{-1}(y)] \left| \frac{d}{dy} g^{-1}(y) \right|, & \alpha < y < \beta, \\ 0, & \text{otherwise,} \end{cases} \tag{2}$$

where $\alpha = \min\{g(-\infty), g(+\infty)\}$ and $\beta = \max\{g(-\infty), g(+\infty)\}$.

Proof. If g is differentiable for all x and $g'(x) > 0$ for all x, then g is continuous and strictly increasing, the limits α, β exist (may be infinite), and the inverse function $x = g^{-1}(y)$ exists, is strictly increasing, and is differentiable. The DF of Y for $\alpha < y < \beta$ is given by

$$P\{Y \le y\} = P\{X \le g^{-1}(y)\}.$$

The PDF of g is obtained on differentiation. We have

$$\begin{aligned} h(y) &= \frac{d}{dy} P\{Y \le y\} \\ &= f[g^{-1}(y)] \frac{d}{dy} g^{-1}(y). \end{aligned}$$

Similarly, if $g' < 0$, then g is strictly decreasing and we have

$$\begin{aligned} P\{Y \le y\} &= P\{X \ge g^{-1}(y)\} \\ &= 1 - P\{X \le g^{-1}(y)\} \qquad (X \text{ is a continuous type RV}) \end{aligned}$$

so that

$$h(y) = -f[g^{-1}(y)] \cdot \frac{d}{dy} g^{-1}(y).$$

Since g and g^{-1} are both strictly decreasing, $(d/dy)\,g^{-1}(y)$ is negative and (2) follows.

Note that

$$\frac{d}{dy}g^{-1}(y) = \left.\frac{1}{dg(x)/dx}\right|_{x=g^{-1}(y)},$$

so that (2) may be rewritten as

$$h(y) = \left.\frac{f(x)}{|dg(x)/dx|}\right|_{x=g^{-1}(y)}, \qquad \alpha < y < \beta. \tag{3}$$

Remark 1. The key to computation of the induced distribution of $Y = g(X)$ from the distribution of X is (1). If the conditions of Theorem 3 are satisfied, we are able to identify the set $\{X \in g^{-1}(-\infty, y]\}$ as $\{X \le g^{-1}(y)\}$ or $\{X \ge g^{-1}(y)\}$, according to whether g is increasing or decreasing. In practice Theorem 3 is quite useful, but whenever the conditions are violated one should return to (1) to compute the induced distribution. This is the case, for example, in Examples 7 and 8 and Theorem 4 below.

Remark 2. If the PDF f of X vanishes outside an interval $[a,b]$ of finite length, we need only to assume that g is differentiable in (a,b) and either $g'(x) > 0$ or $g'(x) < 0$ throughout the interval. Then we take

$$\alpha = \min\{g(a), g(b)\} \qquad \text{and} \qquad \beta = \max\{g(a), g(b)\}$$

in Theorem 3.

Example 5. Let X have the density $f(x) = 1$, $0 < x < 1$, and $= 0$ otherwise. Let $Y = e^X$. Then $X = \log Y$, and we have

$$h(y) = \left|\frac{1}{y}\right| \cdot 1, \qquad 0 < \log y < 1,$$

that is,

$$h(y) = \begin{cases} \dfrac{1}{y}, & 1 < y < e, \\ 0, & \text{otherwise.} \end{cases}$$

If $y = -2\log x$, then $x = e^{-y/2}$ and

$$\begin{aligned} h(y) &= \left|-\frac{1}{2}e^{-y/2}\right| \cdot 1, \qquad 0 < e^{-y/2} < 1, \\ &= \begin{cases} \frac{1}{2}e^{-y/2}, & 0 < y < \infty, \\ 0, & \text{otherwise.} \end{cases} \end{aligned}$$

Example 6. Let X be a nonnegative RV of the continuous type with PDF f, and let $\alpha > 0$. Let $Y = X^\alpha$. Then

$$P\{X^\alpha \le y\} = \begin{cases} P\{X \le y^{1/\alpha}\} & \text{if } y \ge 0, \\ 0 & \text{if } y < 0. \end{cases}$$

The PDF of Y is given by

$$\begin{aligned} h(y) &= f(y^{1/\alpha}) \left| \frac{d}{dy} y^{1/\alpha} \right| \\ &= \begin{cases} \frac{1}{\alpha} y^{1/\alpha - 1} f(y^{1/\alpha}), & y > 0, \\ 0, & y \le 0. \end{cases} \end{aligned}$$

Example 7. Let X be an RV with PDF

$$f(x) = \frac{1}{\sqrt{2\pi}} e^{-x^2/2}, \qquad -\infty < x < \infty.$$

Let $Y = X^2$. In this case, $g'(x) = 2x$ which is > 0 for $x > 0$, and < 0 for $x < 0$, so that the conditions of Theorem 3 are not satisfied. But for $y > 0$

$$\begin{aligned} P\{Y \le y\} &= P\{-\sqrt{y} \le X \le \sqrt{y}\} \\ &= F(\sqrt{y}) - F(-\sqrt{y}), \end{aligned}$$

where F is the DF of X. Thus the PDF of Y is given by

$$h(y) = \begin{cases} \frac{1}{2\sqrt{y}} \{f(\sqrt{y}) + f(-\sqrt{y})\}, & y > 0, \\ 0, & y \le 0. \end{cases}$$

Thus

$$h(y) = \begin{cases} \frac{1}{\sqrt{2\pi y}} e^{-y/2}, & 0 < y, \\ 0, & y \le 0. \end{cases}$$

Example 8. Let X be an RV with PDF

$$f(x) = \begin{cases} \frac{2x}{\pi^2}, & 0 < x < \pi, \\ 0, & \text{otherwise.} \end{cases}$$

Let $Y = \sin X$. In this case $g'(x) = \cos x > 0$ for x in $(0, \pi/2)$ and < 0 for x in $(\pi/2, \pi)$, so that the conditions of Theorem 3 are not satisfied. To compute the PDF of Y we return to (1) and see that (Fig. 1) the DF of Y is given by

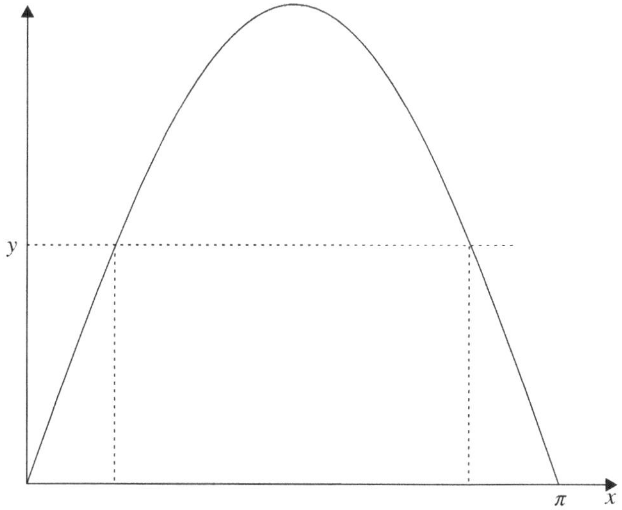

Fig. 1 $y = \sin x,\ 0 \le x \le \pi.$

$$\begin{aligned} P\{Y \le y\} &= P\{\sin X \le y\}, \qquad 0 < y < 1, \\ &= P\{[0 \le X \le x_1] \cup [x_2 \le X \le \pi]\}, \end{aligned}$$

where $x_1 = \sin^{-1} y$ and $x_2 = \pi - \sin^{-1} y$. Thus

$$\begin{aligned} P\{Y \le y\} &= \int_0^{x_1} f(x)\,dx + \int_{x_2}^{\pi} f(x)\,dx \\ &= \left(\frac{x_1}{\pi}\right)^2 + 1 - \left(\frac{x_2}{\pi}\right)^2, \end{aligned}$$

and the PDF of Y is given by

$$\begin{aligned} h(y) &= \frac{d}{dy}\left(\frac{\sin^{-1} y}{\pi}\right)^2 + \frac{d}{dy}\left[1 - \left(\frac{\pi - \sin^{-1} y}{\pi}\right)^2\right] \\ &= \begin{cases} \dfrac{2}{\pi\sqrt{1-y^2}}, & 0 < y < 1, \\ 0, & \text{otherwise.} \end{cases} \end{aligned}$$

In Examples 7 and 8 the function $y = g(x)$ can be written as the sum of two monotone functions. We applied Theorem 3 to each of these monotonic summands. These two examples are special cases of the following result.

Theorem 4. Let X be an RV of the continuous type with PDF f. Let $y = g(x)$ be differentiable for all x, and assume that $g'(x)$ is continuous and nonzero at all but a finite number of values of x. Then, for every real number y,

(a) there exist a positive integer $n = n(y)$ and real numbers (inverses) $x_1(y), x_2(y), \ldots, x_n(y)$ such that

$$g[x_k(y)] = y, \qquad g'[x_k(y)] \neq 0, \qquad k = 1, 2, \ldots, n(y),$$

or

(b) there does not exist any x such that $g(x) = y$, $g'(x) \neq 0$, in which case we write $n(y) = 0$.

Then Y is a continuous RV with PDF given by

$$h(y) = \begin{cases} \sum_{k=1}^{n} f[x_k(y)]\,|g'[x_k(y)]|^{-1} & \text{if } n > 0, \\ 0 & \text{if } n = 0. \end{cases}$$

Example 9. Let X be an RV with PDF f, and let $Y = |X|$. Here $n(y) = 2$, $x_1(y) = y$, $x_2(y) = -y$ for $y > 0$, and

$$h(y) = \begin{cases} f(y) + f(-y), & y > 0, \\ 0, & y \leq 0. \end{cases}$$

Thus, if $f(x) = 1/2$, $-1 \leq x \leq 1$, and $= 0$ otherwise, then

$$h(y) = \begin{cases} 1, & 0 \leq y \leq 1, \\ 0, & \text{otherwise}. \end{cases}$$

If $f(x) = (1/\sqrt{2\pi})e^{-(x^2/2)}$, $-\infty < x < \infty$, then

$$h(y) = \begin{cases} \dfrac{2}{\sqrt{2\pi}} e^{-(y^2/2)}, & y > 0, \\ 0, & \text{otherwise}. \end{cases}$$

Example 10. Let X be an RV of the continuous type with PDF f, and let $Y = X^{2m}$, where m is a positive integer. In this case $g(x) = x^{2m}$, $g'(x) = 2mx^{2m-1} > 0$ for $x > 0$ and $g'(x) < 0$ for $x < 0$. Writing $n = 2m$, we see that, for any $y > 0$, $n(y) = 2$, $x_1(y) = -y^{1/n}$, $x_2(y) = y^{1/n}$. It follows that

$$\begin{aligned} h(y) &= f[x_1(y)] \cdot \frac{1}{ny^{1-1/n}} + f[x_2(y)] \frac{1}{ny^{1-1/n}} \\ &= \begin{cases} \dfrac{1}{ny^{1-1/n}} \{f(y^{1/n}) + f(-y^{1/n})\} & \text{if } y > 0, \\ 0 & \text{if } y \leq 0. \end{cases} \end{aligned}$$

In particular, if f is the PDF given in Example 7, then

$$h(y) = \begin{cases} \dfrac{2}{\sqrt{2\pi}\, n y^{1-1/n}} \exp\left\{-\dfrac{y^{2/n}}{2}\right\} & \text{if } y > 0, \\ 0 & \text{if } y \le 0. \end{cases}$$

Remark 3. The basic formula (1) and the countable additivity of probability allow us to compute the distribution of $Y = g(X)$ in some instances even if g has a countable number of inverses. Let $A \subseteq \mathcal{R}$ and g map A into $B \subseteq \mathcal{R}$. Suppose that A can be represented as a countable union of disjoint sets A_k, $k = 1, 2, \ldots$. Then the DF of Y is given by

$$\begin{aligned} P\{Y \le y\} &= P\{X \in g^{-1}(-\infty, y]\} \\ &= P\left\{X \in \sum_{k=1}^{\infty} [\{g^{-1}(-\infty, y]\} \cap A_k]\right\} \\ &= \sum_{k=1}^{\infty} P\{X \in A_k \cap \{g^{-1}(-\infty, y]\}\}. \end{aligned}$$

If the conditions of Theorem 3 are satisfied by the restriction of g to each A_k, we may obtain the PDF of Y on differentiating the DF of Y. We remind the reader that term-by-term differentiation is permissible if the differentiated series is uniformly convergent.

Example 11. Let X be an RV with PDF

$$f(x) = \begin{cases} \theta e^{-\theta x}, & x > 0, \\ 0, & x \le 0, \end{cases} \qquad \theta > 0.$$

Let $Y = \sin X$, and let $\sin^{-1} y$ be the principal value. Then (Fig. 2), for $0 < y < 1$,

$$\begin{aligned} &P\{\sin X \le y\} \\ &\quad = P\{0 < X \le \sin^{-1} y \text{ or } (2n-1)\pi - \sin^{-1} y \le X \le 2n\pi + \sin^{-1} y \\ &\qquad\qquad \text{for all integers } n \ge 1\} \\ &\quad = P\{0 < X \le \sin^{-1} y\} + \sum_{n=1}^{\infty} P\{(2n-1)\pi - \sin^{-1} y \le X \le 2n\pi + \sin^{-1} y\} \\ &\quad = 1 - e^{-\theta \sin^{-1} y} + \sum_{n=1}^{\infty} [e^{-\theta[(2n-1)\pi - \sin^{-1} y]} - e^{-\theta(2n\pi + \sin^{-1} y)}] \\ &\quad = 1 - e^{-\theta \sin^{-1} y} + (e^{\theta\pi + \theta \sin^{-1} y} - e^{-\theta \sin^{-1} y}) \sum_{n=1}^{\infty} e^{-(2\theta\pi)n} \\ &\quad = 1 - e^{-\theta \sin^{-1} y} + (e^{\theta\pi + \theta \sin^{-1} y} - e^{-\theta \sin^{-1} y}) \left(\frac{e^{-2\theta\pi}}{1 - e^{-2\theta\pi}}\right) \\ &\quad = 1 + \frac{e^{-\theta\pi + \theta \sin^{-1} y} - e^{-\theta \sin^{-1} y}}{1 - e^{-2\pi\theta}}. \end{aligned}$$

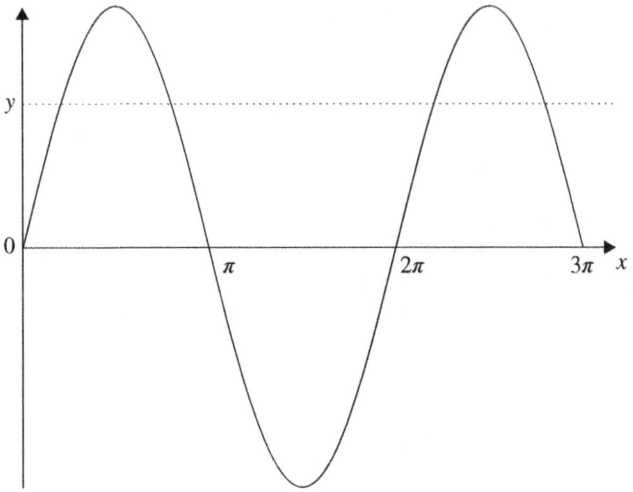

Fig. 2 $y = \sin x,\ x \geq 0.$

A similar computation can be made for $y < 0$. It follows that the PDF of Y is given by

$$h(y) = \begin{cases} \theta e^{-\theta\pi}(1-e^{-2\theta\pi})^{-1}(1-y^2)^{-1/2}[e^{\theta \sin^{-1} y} + e^{-\theta\pi - \theta \sin^{-1} y}] & \text{if } -1 < y < 0, \\ \theta(1-e^{-2\theta\pi})^{-1}(1-y^2)^{-1/2}[e^{-\theta \sin^{-1} y} + e^{-\theta\pi + \theta \sin^{-1} y}] & \text{if } 0 < y < 1, \\ 0 & \text{otherwise.} \end{cases}$$

PROBLEMS 2.5

1. Let X be a random variable with probability mass function

$$P\{X = r\} = \binom{n}{r} p^r (1-p)^{n-r}, \qquad r = 0, 1, 2, \ldots, n, \quad 0 \leq p \leq 1.$$

Find the PMFs of the RVs (a) $Y = aX + b$, (b) $Y = X^2$, and (c) $Y = \sqrt{X}$.

2. Let X be an RV with PDF

$$f(x) = \begin{cases} 0 & \text{if } x \leq 0, \\ \dfrac{1}{2} & \text{if } 0 < x \leq 1, \\ \dfrac{1}{2x^2} & \text{if } 1 < x < \infty. \end{cases}$$

Find the PDF of the RV $1/X$.

3. Let X be a positive RV of the continuous type with PDF $f(\cdot)$. Find the PDF of the RV $U = X/(1+X)$. If, in particular, X has the PDF

$$f(x) = \begin{cases} 1, & 0 \le x \le 1, \\ 0, & \text{otherwise}, \end{cases}$$

what is the PDF of U?

4. Let X be an RV with PDF f defined by Example 11. Let $Y = \cos X$ and $Z = \tan X$. Find the DFs and PDFs of Y and Z.

5. Let X be an RV with PDF

$$f_\theta(x) = \begin{cases} \theta e^{-\theta x} & \text{if } x \ge 0, \\ 0 & \text{otherwise}, \end{cases}$$

where $\theta > 0$. Let $Y = [X - 1/\theta]^2$. Find the PDF of Y.

6. A point is chosen at random on the circumference of a circle of radius r with center at the origin, that is, the polar angle θ of the point chosen has the PDF

$$f(\theta) = \frac{1}{2\pi}, \qquad \theta \in (-\pi, \pi).$$

Find the PDF of the abscissa of the point selected.

7. For the RV X of Example 7 find the PDF of the following RVs: (a) $Y_1 = e^X$, (b) $Y_2 = 2X^2 + 1$, and (c) $Y_3 = g(X)$, where $g(x) = 1$ if $x > 0$, $= 1/2$ if $x = 0$, and $= -1$ if $x < 0$.

8. Suppose that a projectile is fired at an angle θ above the earth with a velocity V. Assuming that θ is an RV with PDF

$$f(\theta) = \begin{cases} \dfrac{12}{\pi} & \text{if } \dfrac{\pi}{6} < \theta < \dfrac{\pi}{4}, \\ 0 & \text{otherwise}, \end{cases}$$

find the PDF of the range R of the projectile, where $R = V^2 \sin 2\theta / g$, g being the gravitational constant.

9. Let X be an RV with PDF $f(x) = 1/(2\pi)$ if $0 < x < 2\pi$, and $= 0$ otherwise. Let $Y = \sin X$. Find the DF and PDF of Y.

10. Let X be an RV with PDF $f(x) = 1/3$ if $-1 < x < 2$, and $= 0$ otherwise. Let $Y = |X|$. Find the PDF of Y.

11. Let X be an RV with PDF $f(x) = 1/(2\theta)$ if $-\theta \le x \le \theta$, and $= 0$ otherwise. Let $Y = 1/X^2$. Find the PDF of Y.

12. Let X be an RV of the continuous type, and let $Y = g(X)$ be defined as follows:

(a) $g(x) = 1$ if $x > 0$, and $= -1$ if $x \le 0$.

(b) $g(x) = b$ if $x \ge b$, $= x$ if $|x| < b$, and $= -b$ if $x \le -b$.

(c) $g(x) = x$ if $|x| \ge b$, and $= 0$ if $|x| < b$.

Find the distribution of Y in each case.

3

MOMENTS AND GENERATING FUNCTIONS

3.1 INTRODUCTION

The study of probability distributions of a random variable is essentially the study of some numerical characteristics associated with them. These so-called *parameters of the distribution* play a key role in mathematical statistics. In Section 3.2 we introduce some of these parameters, namely, moments and order parameters, and investigate their properties. In Section 3.3 the idea of generating functions is introduced. In particular, we study probability generating functions, moment generating functions, and characteristic functions. Section 3.4 deals with some moment inequalities.

3.2 MOMENTS OF A DISTRIBUTION FUNCTION

In this section we investigate some numerical characteristics, called *parameters*, associated with the distribution of an RV X. These parameters are (a) *moments* and their functions and (b) *order parameters*. We will concentrate mainly on moments and their properties.

Let X be a random variable of the discrete type with probability mass function $p_k = P\{X = x_k\}$, $k = 1, 2, \ldots$. If

$$\sum_{k=1}^{\infty} |x_k| p_k < \infty, \tag{1}$$

An Introduction to Probability and Statistics, Third Edition. Vijay K. Rohatgi and A.K. Md. Ehsanes Saleh.

we say that the *expected value* (or the *mean* or the *mathematical expectation*) of X exists and write

$$\mu = EX = \sum_{k=1}^{\infty} x_k p_k. \tag{2}$$

Note that the series $\sum_{k=1}^{\infty} x_k p_k$ may converge but the series $\sum_{k=1}^{\infty} |x_k| p_k$ may not. In that case we say that EX does not exist.

Example 1. Let X have the PMF given by

$$p_j = P\left\{X = (-1)^{j+1}\frac{3^j}{j}\right\} = \frac{2}{3^j}, \qquad j = 1, 2, \ldots.$$

Then

$$\sum_{j=1}^{\infty} |x_j| p_j = \sum_{j=1}^{\infty} \frac{2}{j} = \infty,$$

and EX does not exist, although the series

$$\sum_{j=1}^{\infty} x_j p_j = \sum_{j=1}^{\infty} (-1)^{j+1} \frac{2}{j}$$

is convergent.

If X is of the continuous type and has PDF f, we say that EX exists and equals $\int x f(x)\,dx$, provided that

$$\int |x| f(x)\,dx < \infty.$$

A similar definition is given for the mean of any Borel-measurable function $h(X)$ of X. Thus, if X is of the continuous type and has PDF f, we say that $Eh(X)$ exists and equals $\int h(x) f(x)\,dx$, provided that

$$\int |h(x)| f(x)\,dx < \infty.$$

We emphasize that the condition $\int |x| f(x)\,dx < \infty$ must be checked before it can be concluded that EX exists and equals $\int x f(x)\,dx$. Moreover, it is worthwhile to recall at this point that the integral $\int_{-\infty}^{\infty} \varphi(x)\,dx$ exists, provided that the limit $\lim_{b \to \infty}^{a \to \infty} \int_{-b}^{a} \varphi(x)\,dx$ exists. It is quite possible for the limit $\lim_{a \to \infty} \int_{-a}^{a} \varphi(x)\,dx$ to exist without the existence of $\int_{-\infty}^{\infty} \varphi(x)\,dx$. As an example, consider the Cauchy PDF:

$$f(x) = \frac{1}{\pi}\frac{1}{1+x^2}, \qquad -\infty < x < \infty.$$

Clearly

$$\lim_{a\to\infty}\int_{-a}^{a}\frac{x}{\pi}\frac{1}{1+x^2}\,dx=0.$$

However, EX does not exist since the integral $(1/\pi)\int_{-\infty}^{\infty}|x|/(1+x^2)\,dx$ diverges.

Remark 1. Let $X(\omega)=I_A(\omega)$ for some $A\in\mathcal{S}$. Then $EX=P(A)$.

Remark 2. If we write $h(X)=|X|$, we see that EX exists if and only if $E|X|$ does.

Remark 3. We say that an RV X is *symmetric* about a point α if

$$P\{X\geq\alpha+x\}=P\{X\leq\alpha-x\}\qquad\text{for all } x.$$

In terms of DF F of X, this means that, if

$$F(\alpha-x)=1-F(\alpha+x)+P\{X=\alpha+x\}$$

holds for all $x\in\mathcal{R}$, we say that the DF F (or the RV X) is symmetric with α as the *center* of *symmetry*. If $\alpha=0$, then for every x

$$F(-x)=1-F(x)+P\{X=x\}.$$

In particular, if X is an RV of the continuous type, X is symmetric with center α if and only if the PDF f of X satisfies

$$f(\alpha-x)=f(\alpha+x)\qquad\text{for all } x.$$

If $\alpha=0$, we will say simply that X is symmetric (or that F is symmetric).

As an immediate consequence of this definition we see that, if X is symmetric with α as the center of symmetry and $E|X|<\infty$, then $EX=\alpha$. A simple example of a symmetric distribution is the Cauchy PDF considered above (before Remark 1). We will encounter many such distributions later.

Remark 4. If a and b are constants and X is an RV with $E|X|<\infty$, then $E|aX+b|<\infty$ and $E\{aX+b\}=aEX+b$. In particular, $E\{X-\mu\}=0$, a fact that should not come as a surprise.

Remark 5. If X is bounded, that is, if $P\{|X|<M\}=1$, $0<M<\infty$, then EX exists.

Remark 6. If $\{X\geq 0\}=1$, and EX exists, then $EX\geq 0$.

Theorem 1. Let X be an RV, and g be a Borel-measurable function on $\mathcal{R}$. Let $Y=g(X)$. If X is of discrete type then

$$EY=\sum_{j=1}^{\infty}g(x_j)P\{X=x_j\}\tag{3}$$

in the sense that, if either side of (3) exists, so does the other, and then the two are equal. If X is of continuous type with PDF f then $EY = \int g(x)f(x)dx$ in the sense that, if either of the two integrals converges absolutely, so does the other, and the two are equal.

Remark 7. Let X be a discrete RV. Then Theorem 1 says that

$$\sum_{j=1}^{\infty} g(x_j)P\{X = x_j\} = \sum_{k=1}^{\infty} y_k P\{Y = y_k\}$$

in the sense that, if either of the two series converges absolutely, so does the other, and the two sums are equal. If X is of the continuous type with PDF f, let $h(y)$ be the PDF of $Y = g(X)$. Then, according to Theorem 1,

$$\int g(x)f(x)\,dx = \int y h(y)\,dy,$$

provided that $E|g(X)| < \infty$.

Proof of Theorem 1. In the discrete case, suppose that $P\{X \in A\} = 1$. If $y = g(x)$ is a one-to-one mapping of A onto some set B, then

$$P\{Y = y\} = P\{X = g^{-1}(y)\}, \qquad y \in B.$$

We have

$$\sum_{x \in A} g(x)P\{X = x\} = \sum_{y \in B} yP\{Y = y\}.$$

In the continuous case, suppose g satisfies the conditions of Theorem 2.5.3. Then

$$\int g(x)f(x)\,dx = \int_{\alpha}^{\beta} yf[g^{-1}(y)]|\frac{d}{dy}g^{-1}(y)|dy$$

by changing the variable to $y = g(x)$. Thus

$$\int g(x)f(x)\,dx = \int_{\alpha}^{\beta} y h(y)\,dy.$$

The functions $h(x) = x^n$, where n is a positive integer, and $h(x) = |x|^{\alpha}$, where α is a positive real number, are of special importance. If EX^n exists for some positive integer n, we call EX^n the *nth moment* of (the distribution function of) X *about the origin*. If $E|X|^{\alpha} < \infty$ for some positive real number α, we call $E|X|^{\alpha}$ the αth absolute moment of X. We shall use the following notation:

$$m_n = EX^n \qquad \beta_{\alpha} = E|X|^{\alpha}, \tag{4}$$

whenever the expectations exist.

Example 2. Let X have the *uniform* distribution on the first N natural numbers, that is, let

$$P\{X=k\}=\frac{1}{N}, \qquad k=1,2,\ldots,N.$$

Clearly, moments of all order exist:

$$EX=\sum_{k=1}^{N}k\cdot\frac{1}{N}=\frac{N+1}{2},$$
$$EX^2=\sum_{k=1}^{N}k^2\cdot\frac{1}{N}=\frac{(N+1)(2N+1)}{6}.$$

Example 3. Let X be an RV with PDF

$$f(x)=\begin{cases}\dfrac{2}{x^3}, & x\geq 1,\\ 0, & x<1.\end{cases}$$

Then

$$EX=\int_1^\infty \frac{2}{x^2}\,dx=2.$$

But

$$EX^2=\int_1^\infty \frac{2}{x}\,dx$$

does not exist. Indeed, it is easily possible to construct examples of random variables for which all moments of a specified order exist by no higher-order moments do.

Example 4. Two players, A and B, play a coin-tossing game. A gives B one dollar if a head turns up; otherwise B pays A one dollar. If the probability that the coin shows a head is p, find the expected gain of A.

Let X denote the gain of A. Then

$$P\{X=1\}=P\{\text{Tails}\}=1-p, \qquad P\{X=-1\}=p$$

and

$$EX=1-p-p=1-2p\begin{cases}>0 & \text{if and only if } p<\frac{1}{2},\\ =0 & \text{if and only if } p=\frac{1}{2},\end{cases}$$

Thus $EX=0$ if and only if the coin is *fair*.

Theorem 2. If the moment of order t exists for an RV X, moments of order $0 < s < t$ exist.

Proof. Let X be of the continuous type with PDF f. We have

$$\begin{aligned} E|X|^s &= \int_{|x|^s \le 1} |x|^s f(x)\,dx + \int_{|x|^s > 1} |x|^s f(x)\,dx \\ &\le P\{|X|^s \le 1\} + E|X|^t < \infty. \end{aligned}$$

A similar proof can be given when X is a discrete RV.

Theorem 3. Let X be an RV on a probability space $(\Omega, \mathcal{S}, P)$. Let $E|X|^k < \infty$ for some $k > 0$. Then

$$n^k P\{|X| > n\} \to 0 \qquad \text{as } n \to \infty.$$

Proof. We provide the proof for the case in which X is of the continuous type with density f. We have

$$\infty > \int |x|^k f(x)\,dx = \lim_{n\to\infty} \int_{|x|\le n} |x|^k f(x)\,dx.$$

It follows that

$$\lim_{n\to\infty} \int_{|x|>n} |x|^k f(x)\,dx \to 0 \qquad \text{as } n \to \infty.$$

But

$$\int_{|x|>n} |x|^k f(x)\,dx \ge n^k P\{|X| > n\},$$

completing the proof.

Remark 8. Probabilities of the type $P\{|X| > n\}$ or either of its components, $P\{X > n\}$ or $P\{X < -n\}$, are called *tail probabilities*. The result of Theorem 3, therefore, gives the rate at which $P\{|X| > n\}$ converges to 0 as $n \to \infty$.

Remark 9. The converse of Theorem 3 does not hold in general, that is,

$$n^k P\{|X| > n\} \to 0 \qquad \text{as } n \to \infty \text{ for some } k$$

does not necessarily imply that $E|X|^k < \infty$, for the RV

$$P\{X = n\} = \frac{c}{n^2 \log n}, \qquad n = 2, 3, \ldots,$$

where c is a constant determined from

$$\sum_{n=2}^{\infty} \frac{c}{n^2 \log n} = 1.$$

We have

$$P\{X > n\} \approx c \int_n^{\infty} \frac{1}{x^2 \log x}\, dx \approx cn^{-1}(\log n)^{-1}$$

and $nP\{X > n\} \to 0$ as $n \to \infty$. (Here and subsequently $\approx$ means that the ratio of two sides $\to 1$ as $n \to \infty$.) But

$$EX = \sum \frac{c}{n \log n} = \infty.$$

In fact, we need

$$n^{k+\delta} P\{|X| > n\} \to 0 \qquad \text{as } n \to 0$$

for some $\delta > 0$ to ensure that $E|X|^k < \infty$. A condition such as this is called a *moment condition*.

For the proof we need the following lemma.

Lemma 1. Let X be a nonnegative RV with distribution function F. Then

$$EX = \int_0^{\infty} [1 - F(x)]\, dx, \tag{5}$$

in the sense that, if either side exists, so does the other and the two are equal.

Proof. If X is of the continuous type with density f and $EX < \infty$, then

$$EX = \int_0^{\infty} xf(x)\, dx = \lim_{n \to \infty} \int_0^n xf(x)\, dx.$$

On integration by parts we obtain

$$\begin{aligned}\int_0^n xf(x)\, dx &= nF(n) - \int_0^n F(x)\, dx \\ &= -n[1 - F(n)] + \int_0^n [1 - F(x)]\, dx.\end{aligned}$$

But

$$\begin{aligned} n[1-F(n)] &= n\int_n^{\infty} f(x)\,dx \\ &< \int_n^{\infty} xf(x)\,dx, \end{aligned}$$

and, since $E|X| < \infty$, it follows that

$$n[1-F(n)] \to 0 \qquad \text{as } n \to \infty.$$

We have

$$\begin{aligned} EX &= \lim_{n\to\infty}\int_0^n xf(x)\,dx = \lim_{n\to\infty}\int_0^n [1-F(x)]\,dx \\ &= \int_0^{\infty} [1-F(x)]\,dx. \end{aligned}$$

If $\int_0^{\infty}[1-F(x)]\,dx < \infty$, then

$$\int_0^n xf(x)\,dx \le \int_0^n [1-F(x)]\,dx \le \int_0^{\infty} [1-F(x)]\,dx,$$

and it follows that $E|X| < \infty$.

We leave the reader to complete the proof in the discrete case.

Corollary. For any RV X, $E|X| < \infty$ if and only if the integrals $\int_{-\infty}^{0} P\{X \le x\}\,dx$ and $\int_0^{\infty} P\{X > x\}\,dx$ both converge, and in that case

$$EX = \int_0^{\infty} P\{X > x\}\,dx - \int_{-\infty}^{0} P\{X \le x\}\,dx.$$

Actually we can get a little more out of Lemma 1 than the above corollary. In fact,

$$E|X|^{\alpha} = \int_0^{\infty} P\{|X|^{a} > x\}\,dx = \alpha\int_0^{\infty} x^{\alpha-1}P\{|X| > x\}\,dx,$$

and we see that *an RV X possesses* an *absolute moment of order* $\alpha > 0$ *if and only if* $|x|^{\alpha-1}P\{|X| > x\}$ is integrable over $(0,\infty)$.

A simple application of the integral test leads to the following *moments lemma*.

Lemma 2.

$$E|X|^{\alpha} < \infty \Leftrightarrow \sum_{n=1}^{\infty} P\{|X| > n^{1/\alpha}\} < \infty. \tag{6}$$

Note that an immediate consequence of Lemma 2 is Theorem 3. We are now ready to prove the following result.

Theorem 4. Let X be an RV with a distribution satisfying $n^{\alpha}P\{|X| > n\} \to 0$ as $n \to \infty$ for some $\alpha > 0$. Then $E|X|^{\beta} < \infty$ for $0 < \beta < \alpha$.

Proof. Given $\varepsilon > 0$, we can choose an $N = N(\varepsilon)$ such that

$$P\{|X| > n\} < \frac{\varepsilon}{n^{\alpha}} \qquad \text{for all } n \geq N.$$

It follows that for $0 < \beta < \alpha$

$$\begin{aligned} E|X|^{\beta} &= \beta \int_0^N x^{\beta-1} P\{|X| > x\}\, dx + \beta \int_N^{\infty} x^{\beta-1} P\{|X| > x\}\, dx \\ &\leq N^{\beta} + \beta\varepsilon \int_N^{\infty} x^{\beta-\alpha-1}\, dx \\ &< \infty. \end{aligned}$$

Remark 10. Using Theorems 3 and 4, we demonstrate the existence of random variables for which moments of any order do not exist, that is, for which $E|X|^{\alpha} = \infty$ for every $\alpha > 0$. For such an RV $n^{\alpha}P\{|X| > n\} \nrightarrow 0$ as $n \to \infty$ for any $\alpha > 0$. Consider, for example, the RV X with PDF

$$f(x) = \begin{cases} \dfrac{1}{2|x|(\log|x|)^2} & \text{for } |x| > e \\ 0 & \text{otherwise.} \end{cases}$$

The DF of X is given by

$$F(x) = \begin{cases} \dfrac{1}{2\log|x|} & \text{if } x \leq -e \\ \dfrac{1}{2} & \text{if } -e < x < e, \\ 1 - \dfrac{1}{2\log x} & \text{if } x \geq e. \end{cases}$$

Then for $x > e$

$$\begin{aligned} P\{|X| > x\} &= 1 - F(x) + F(-x) \\ &= \frac{1}{2\log x}, \end{aligned}$$

and $x^{\alpha}P\{|X| > x\} \to \infty$ as $x \to \infty$ for any $\alpha > 0$. It follows that $E|X|^{\alpha} = \infty$ for every $\alpha > 0$. In this example we see that $P\{|X| > cx\}/P\{|X| > x\} \to 1$ as $x \to \infty$ for every $c > 0$. A positive function $L(\cdot)$ defined on $(0, \infty)$ is said to be a function of *slow variation* if and only if $L(cx)/L(x) \to 1$ as $x \to \infty$ for every $c > 0$. For such a function $x^{\alpha}L(x) \to \infty$ for every $\alpha > 0$ (see Feller [26, pp. 275–279]). It follows that, if $P\{|X| > x\}$ is slowly varying, $E|X|^{\alpha} = \infty$ for every $\alpha > 0$. Functions of slow variation play an important role in the theory of probability.

Random variables for which $P\{|X| > x\}$ is slowly varying are clearly excluded from the domain of the following result.

Theorem 5. Let X be an RV satisfying

$$\frac{P\{|X| > cx\}}{P\{|X| > x\}} \to 0 \qquad \text{as } x \to \infty \quad \text{for all } c > 1; \tag{7}$$

then X possesses moments of all orders. (Note that, if $c = 1$, the limit in (7) is 1, whereas if $c < 1$, the limit will not go to 0 since $P\{|X| > cx\} \geq P\{|X| > x\}$.)

Proof. Let $\varepsilon > 0$ (we will choose ε later), choose x_0 so large that

$$\frac{P\{|X| > cx\}}{P\{|X| > x\}} < \varepsilon \qquad \text{for all } x \geq x_0, \tag{8}$$

and choose x_1 so large that

$$P\{|X| > x\} < \varepsilon \qquad \text{for all } x \geq x_1. \tag{9}$$

Let $N = \max(x_0, x_1)$. We have, for a fixed positive integer r,

$$\frac{P\{|X| > c^r x\}}{P\{|X| > x\}} = \prod_{p=1}^{r} \frac{P\{|X| > c^p x\}}{P\{|X| > c^{p-1}x\}} \leq \varepsilon^r \tag{10}$$

for $x \geq N$. Thus for $x \geq N$ we have, in view of (9),

$$P\{|X| > c^r x\} \leq \varepsilon^{r+1}. \tag{11}$$

Next note that, for any fixed positive integer n,

$$\begin{aligned} E|X|^n &= n\int_0^{\infty} x^{n-1}P\{|X| > x\}\,dx \\ &= n\int_0^{N} x^{n-1}P\{|X| > x\}\,dx + n\int_N^{\infty} x^{n-1}P\{|X| > x\}\,dx. \end{aligned} \tag{12}$$

Since the first integral in (12) is finite, we need only show that the second integral is also finite. We have

$$\begin{aligned}\int_N^\infty x^{n-1}P\{|X|>x\}\,dx &= \sum_{r=1}^{\infty}\int_{c^{r-1}N}^{c^rN} x^{n-1}P\{|X|>x\}\,dx\\ &\le \sum_{r=1}^{\infty}(c^rN)^{n-1}\varepsilon^r\cdot 2c^rN\\ &= 2N^n\sum_{r=1}^{\infty}(\varepsilon c^n)^r\\ &= 2N^n\frac{\varepsilon c^n}{1-\varepsilon c^n}<\infty,\end{aligned}$$

provided that we choose ε such that $\varepsilon c^n < 1$. It follows that $E|X|^n < \infty$ for $n = 1,2,\ldots$. Actually we have shown that (7) implies $E|X|^\delta < \infty$ for all $\delta > 0$.

Theorem 6. If $h_1, h_2, \ldots, h_n$ are Borel-measurable functions of an RV X and $Eh_i(X)$ exists for $i = 1,2,\ldots,n$, then $E\left\{\sum_{i=1}^n h_i(X)\right\}$ exists and equals $\sum_{i=1}^n Eh_i(X)$.

Definition 1. Let k be a positive integer and c be a constant. If $E(X-c)^k$ exists, we call it the *moment of order* k about the point c. If we take $c = EX = \mu$, which exists since $E|X| < \infty$, we call $E(X-\mu)^k$ the *central moment of order* k or the moment of order k about the mean. We shall write

$$\mu_k = E\{X-\mu\}^k.$$

If we know $m_1, m_2, \ldots, m_k$, we can compute $\mu_1, \mu_2, \ldots, \mu_k$, and conversely. We have

$$\mu_k = E\{X-\mu\}^k = m_k - \binom{k}{1}\mu m_{k-1} + \binom{k}{2}\mu^2 m_{k-2} - \cdots + (-1)^k\mu^k \tag{13}$$

and

$$m_k = E\{X-\mu+\mu\}^k = \mu_k + \binom{k}{1}\mu\mu_{k-1} + \binom{k}{2}\mu^2\mu_{k-2} + \cdots + \mu^k. \tag{14}$$

The case $k = 2$ is of special importance.

Definition 2. If EX^2 exists, we call $E\{X-\mu\}^2$ the *variance* of X, and we write $\sigma^2 = \text{var}(X) = E\{X-\mu\}^2$. The quantity σ is called the *standard deviation* (SD) of X.

From Theorem 6 we see that

$$\sigma^2 = \mu_2 = EX^2 - (EX)^2. \tag{15}$$

Variance has some important properties.

Theorem 7. $\text{Var}(X) = 0$ if and only if X is degenerate.

Theorem 8. $\text{Var}(X) < E(X-c)^2$ for any $c \neq EX$.

Proof. We have

$$\text{var}(X) = E\{X-\mu\}^2 = E\{X-c\}^2 + (c-\mu)^2.$$

Note that

$$\text{var}(aX+b) = a^2\,\text{var}(X).$$

Let $E|X|^2 < \infty$. Then we define

$$Z = \frac{X - EX}{\sqrt{\text{var}(X)}} = \frac{X-\mu}{\sigma} \tag{16}$$

and see that $EZ = 0$ and $\text{var}(Z) = 1$. We call Z a *standardized* RV.

Example 5. Let X be an RV with *binomial* PMF

$$P\{X=k\} = \binom{n}{k} p^k (1-p)^{n-k}, \qquad k = 0,1,2,\ldots,n; \quad 0<p<1.$$

Then

$$\begin{aligned}
EX &= \sum_{k=0}^{\infty} k \binom{n}{k} p^k (1-p)^{n-k} \\
&= np \sum \binom{n-1}{k-1} p^{k-1}(1-p)^{n-k} \\
&= np; \\
EX^2 &= E\{X(X-1)+X\} \\
&= \sum k(k-1)\binom{n}{k} p^k (1-p)^{n-k} + np \\
&= n(n-1)p^2 + np; \\
\text{var}(X) &= n(n-1)p^2 + np - n^2p^2 \\
&= np(1-p); \\
EX^3 &= E\{X(X-1)(X-2) + 3X(X-1) + X\} \\
&= n(n-1)(n-2)p^3 + 3n(n-1)p^2 + np;
\end{aligned}$$

$$\begin{aligned}
\mu_3 &= m_3 - 3\mu m_2 + 2\mu^3 \\
&= n(n-1)(n-2)p^3 + 3n(n-1)p^2 + np - 3np[n(n-1)p^2 + np] + 2n^3p^3 \\
&= np(1-p)(1-2p).
\end{aligned}$$

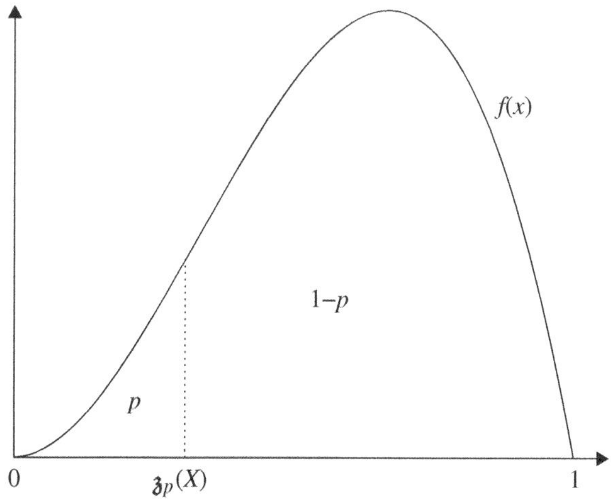

Fig. 1 Quantile of order p.

In the above example we computed *factorial moments* $EX(X-1)(X-2)\cdots(X-k+1)$ for various values of k. For some discrete integer-valued RVs whose PMF contains factorials or binomial coefficients it may be more convenient to compute factorial moments.

We have seen that for some distributions even the mean does not exist. We next consider some parameters, called *order parameters*, which always exist.

Definition 3. A number x (Fig. 1) satisfying

$$P\{X \le x\} \ge p, \qquad P\{X \ge x\} \ge 1-p, \quad 0<p<1, \tag{17}$$

is called a quantile of order p [or $(100p)$th percentile] for the RV X (or, for the DF F of X). We write $\mathfrak{z}_p(X)$ for a quantile of order p for the RV X.

If x is a quantile of order p for an RV X with DF F, then

$$p \le F(x) \le p + P\{X=x\}. \tag{18}$$

If $P\{X=x\}=0$, as is the case—in particular, if X is of the continuous type—a quantile of order p is a solution of the equation

$$F(x) = p. \tag{19}$$

If F is strictly increasing, (19) has a unique solution. Otherwise (Fig. 2) there may be many (even uncountably many) solutions of (19), each of which is then called a quantile of order p. Quantiles are of great deal of interest in testing of hypotheses.

Definition 4. Let X be an RV with DF F. A number x satisfying

$$\frac{1}{2} \le F(x) \le \frac{1}{2} + P\{X=x\} \tag{20}$$

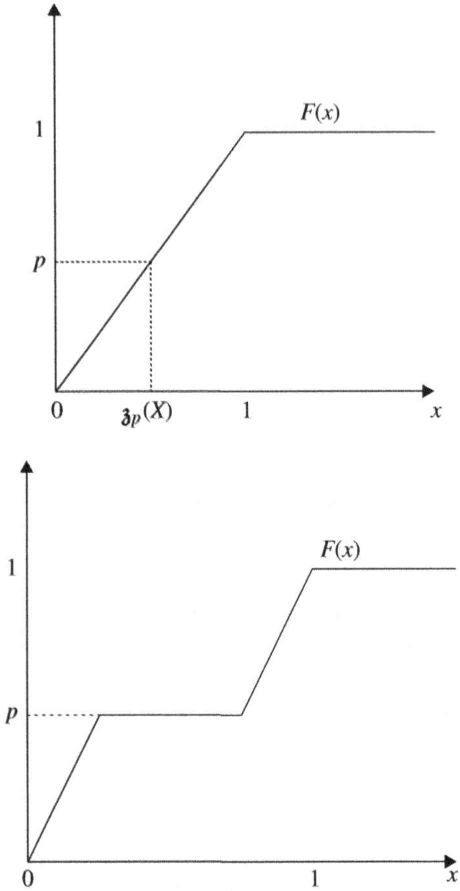

Fig. 2 (a) Unique quantile and (b) infinitely many solutions of $F(x) = p$.

or, equivalently,

$$P\{X \le x\} \ge \frac{1}{2} \quad \text{and} \quad P\{X \ge x\} \ge \frac{1}{2} \tag{21}$$

is called a *median* of X (or F).

Again we note that there may be many values that satisfy (20) or (21). Thus a median is not necessarily unique.

If F is a symmetric DF, the center of symmetry is clearly the median of the DF F. The median is an important centering constant especially in cases where the mean of the distribution does not exist.

Example 6. Let X be an RV with *Cauchy* PDF

$$f(x) = \frac{1}{\pi}\frac{1}{1+x^2}, \qquad -\infty < x < \infty.$$

Then $E|X|$ is not finite but $E|X|^{\delta} < \infty$ for $0 < \delta < 1$. The median of the RV X is clearly $x = 0$.

Example 7. Let X be an RV with PMF

$$P\{X = -2\} = P\{X = 0\} = \frac{1}{4}, \qquad P\{X = 1\} = \frac{1}{3}, \qquad P\{X = 2\} = \frac{1}{6}.$$

Then

$$P\{X \le 0\} = \frac{1}{2} \qquad \text{and} \qquad P\{X \ge 0\} = \frac{3}{4} > \frac{1}{2}.$$

In fact, if x is any number such that $0 < x < 1$, then

$$P\{X \le x\} = P\{X = -2\} + P\{X = 0\} = \frac{1}{2}$$

and

$$P\{X \ge x\} = P\{X = 1\} + P\{X = 2\} = \frac{1}{2},$$

and it follows that every x, $0 \le x < 1$, is a median of the RV X.

If $p = 0.2$, the quantile of order p is $x = -2$, since

$$P\{X \le -2\} = \frac{1}{4} > p \qquad \text{and} \qquad P\{X \ge -2\} = 1 > 1 - p.$$

PROBLEMS 3.2

1. Find the expected number of throws of a fair die until a 6 is obtained.
2. From a box containing N identical tickets numbered 1 through N, n tickets are drawn with replacement. Let X be the largest number drawn. Find EX.
3. Let X be an RV with PDF

$$f(x) = \frac{c}{(1 + x^2)^m}, \qquad -\infty < x < \infty, \quad m \ge 1,$$

   where $c = \Gamma(m)/[\Gamma(1/2)\Gamma(m - 1/2)]$. Show that EX^{2r} exists if and only if $2r < 2m - 1$. What is EX^{2r} if $2r < 2m - 1$?
4. Let X be an RV with PDF

$$f(x) = \begin{cases} \dfrac{ka^k}{(x + a)^{k+1}} & \text{if } x \ge 0, \\ 0 & \text{otherwise } (a > 0). \end{cases}$$

   Show that $E|X|^{\alpha} < \infty$ for $\alpha < k$. Find the quantile of order p for the RV X.

5. Let X be an RV such that $E|X| < \infty$. Show that $E|X-c|$ is minimized if we choose c equal to the median of the distribution of X.

6. *Pareto's distribution* with parameters α and β (both α and β positive) is defined by the PDF

$$f(x) = \begin{cases} \dfrac{\beta\alpha^\beta}{x^{\beta+1}} & \text{if } x \geq \alpha, \\ 0 & \text{if } x < \alpha. \end{cases}$$

Show that the moment of order n exists if and only if $n < \beta$. Let $\beta > 2$. Find the mean and the variance of the distribution.

7. For an RV X with PDF

$$f(x) = \begin{cases} \frac{1}{2}x & \text{if } 0 \leq x < 1, \\ \frac{1}{2} & \text{if } 1 < x \leq 2, \\ \frac{1}{2}(3-x) & \text{if } 2 < x \leq 3, \end{cases}$$

show that moments of all order exist. Find the mean and the variance of X.

8. For the PMF of Example 5 show that

$$EX^4 = np + 7n(n-1)p^2 + 6n(n-1)(n-2)p^3 + n(n-1)(n-2)(n-3)p^4$$

and

$$\mu_4 = 3(npq)^2 + npq(1-6pq),$$

where $0 \leq p \leq 1$, $q = 1-p$.

9. For the *Poisson* RV X with PMF

$$P\{X = x\} = e^{-\lambda}\frac{\lambda^x}{x!}, \qquad x = 0, 1, 2, \ldots,$$

show that $EX = \lambda$, $EX^2 = \lambda + \lambda^2$, $EX^3 = \lambda + 3\lambda^2 + \lambda^3$, $EX^4 = \lambda + 7\lambda^2 + 6\lambda^3 + \lambda^4$, and $\mu_2 = \mu_3 = \lambda$, $\mu_4 = \lambda + 3\lambda^2$.

10. For any RV X with $E|X|^4 < \infty$ define

$$\alpha_3 = \frac{\mu_3}{(\mu_2)^{3/2}}, \qquad \alpha_4 = \frac{\mu_4}{\mu_2^2}.$$

Here α_3 is known as the *coefficient of skewness* and is sometimes used as a measure of asymmetry, and α_4 is known as *kurtosis* and is used to measure the peakedness ("flatness of the top") of a distribution.

Compute α_3 and α_4 for the PMFs of Problems 8 and 9.

11. For a positive RV X define the negative moment of order n by EX^{-n}, where $n > 0$ is an integer. Find $E\{1/(X+1)\}$ for the PMFs of Example 5 and Problem 9.

12. Prove Theorem 6.

13. Prove Theorem 7.

14. In each of the following cases, compute EX, $\text{var}(X)$, and EX^n (for $n \geq 0$, an integer) whenever they exist.

(a) $f(x) = 1$, $-1/2 \leq x \leq 1/2$, and 0 elsewhere.

(b) $f(x) = e^{-x}$, $x \geq 0$, and 0 elsewhere.

(c) $f(x) = (k-1)/x^k$, $x \geq 1$, and 0 elsewhere; $k > 1$ is a constant.

(d) $f(x) = 1/[\pi(1+x^2)]$, $-\infty < x < \infty$.

(e) $f(x) = 6x(1-x)$, $0 < x < 1$, and 0 elsewhere.

(f) $f(x) = xe^{-x}$, $x \geq 0$, and 0 elsewhere.

(g) $P(X = x) = p(1-p)^{x-1}$, $x = 1, 2, \ldots$, and 0 elsewhere: $0 < p < 1$.

15. Find the quantile of order $p(0 < p < 1)$ for the following distributions.

(a) $f(x) = 1/x^2$, $x \geq 1$, and 0 elsewhere.

(b) $f(x) = 2x\exp(-x^2)$, $x \geq 0$, and 0 otherwise.

(c) $f(x) = 1/\theta$, $0 \leq x \leq \theta$, and 0 elsewhere.

(d) $P(X = x) = \theta(1-\theta)^{x-1}$, $x = 1, 2, \ldots$, and 0 otherwise; $0 < \theta < 1$.

(e) $f(x) = (1/\beta^2)x\exp(-x/\beta)$, $x > 0$, and 0 otherwise; $\beta > 0$.

(f) $f(x) = (3/b^3)(b-x)^2$, $0 < x < b$, and 0 elsewhere.

3.3 GENERATING FUNCTIONS

In this section we consider some functions that generate probabilities or moments of an RV. The simplest type of generating function in probability theory is the one associated with integer-valued RVs. Let X be an RV, and let

$$p_k = P\{X = k\}, \qquad k = 0, 1, 2, \ldots$$

with $\sum_{k=0}^{\infty} p_k = 1$.

Definition 1. The function defined by

$$P(s) = \sum_{k=0}^{\infty} p_k s^k, \tag{1}$$

which surely converges for $|s| \leq 1$, is called the probability generating function (PGF) of X.

Example 1. Consider the Poisson RV with PMF

$$P\{X = k\} = e^{-\lambda}\frac{\lambda^k}{k!}, \qquad k = 0, 1, 2, \ldots.$$

We have

$$P(s) = \sum_{k=0}^{\infty} (s\lambda)^k \frac{e^{-\lambda}}{k!} = e^{-\lambda} e^{s\lambda} = e^{-\lambda(1-s)}, \qquad \text{for all } s.$$

Example 2. Let X be an RV with *geometric* distribution, that is, let

$$P\{X = k\} = pq^k, \qquad k = 0, 1, 2, \ldots; \quad 0 < p < 1, \quad q = 1 - p.$$

Then

$$P(s) = \sum_{k=0}^{\infty} s^k pq^k = p \frac{1}{1 - sq}, \qquad |s| \le 1.$$

Remark 1. Since $P(1) = 1$, series (1) is uniformly and absolutely convergent in $|s| \le 1$ and the PGF P is a continuous function of s. It determines the PGF uniquely, since $P(s)$ can be represented in a unique manner as a power series.

Remark 2. Since a power series with radius of convergence r can be differentiated termwise any number of times in $(-r, r)$, it follows that

$$P^{(k)}(s) = \sum_{n=k}^{\infty} n(n-1)\cdots(n-k+1)P(X = n)s^{n-k},$$

where $P^{(k)}$ is the kth derivative of P. The series converges at least for $-1 < s < 1$. For $s = 1$ the right side reduces formally to $E\{X(X-1)\cdots(X-k+1)\}$ which is the kth factorial moment of X whenever it exists. In particular, if $EX < \infty$ then $P'(1) = EX$, and if $EX^2 < \infty$ then $P''(1) = EX(X-1)$ and $\text{Var}(X) = EX^2 - (EX)^2 = P''(1) - [P'(1)]^2 + P'(1)$.

Example 3. In Example 1 we found that $P(s) = e^{-\lambda(1-s)}$, $|s| \le 1$, for a Poisson RV. Thus

$$P'(s) = \lambda e^{-\lambda(1-s)},$$
$$P''(s) = \lambda^2 e^{-\lambda(1-s)}.$$

Also, $EX = \lambda$, $E\{X^2 - X\} = \lambda^2$, so that $\text{var}(X) = EX^2 - (EX)^2 = \lambda^2 + \lambda - \lambda^2 = \lambda$.

In Example 2 we computed $P(s) = p/(1 - sq)$, so that

$$P'(s) = \frac{pq}{(1-sq)^2} \qquad \text{and} \qquad P''(s) = \frac{2pq^2}{(1-sq)^3}.$$

Thus

$$EX = \frac{q}{p}, \qquad EX^2 = \frac{q}{p} + \frac{2pq^2}{p^3}, \qquad \text{var}(X) = \frac{q^2}{p^2} + \frac{q}{p} = \frac{q}{p^2}.$$

Example 4. Consider the PGF

$$P(s) = [(1+s)/2]^n, \quad -\infty < s < \infty.$$

Expanding the right side into a power series we get

$$P(s) = \sum_{k=0}^{n} \frac{1}{2^n} \binom{n}{k} s^{n-k} = \sum_{k=0}^{n} p_k s^k,$$

and it follows that

$$P(X = k) = p_k = \frac{\binom{n}{k}}{2^n}, \quad k = 0, 1, \ldots, n.$$

We note that the PGF, being defined only for discrete integer-valued RVS, has limited utility. We next consider a generating function which is quite useful in probability and statistics.

Definition 2. Let X be an RV defined on $(\Omega, \mathcal{S}, P)$. The function

$$M(s) = Ee^{sX} \tag{2}$$

is known as the moment generating function (MGF) of the RV X if the expectation on the right side of (2) exists in some neighborhood of the origin.

Example 5. Let X have the PMF

$$f(x) = \begin{cases} \dfrac{6}{\pi^2} \cdot \dfrac{1}{k^2}, & k = 1, 2, \ldots, \\ 0, & \text{otherwise.} \end{cases}$$

Then $(1/\pi^2) \sum_{k=1}^{\infty} e^{sk}/k^2$, is infinite for every $s > 0$. We see that the MGF of X does not exist. In fact, $EX = \infty$.

Example 6. Let X have the PDF

$$f(x) = \begin{cases} \frac{1}{2} e^{-x/2}, & x > 0, \\ 0, & \text{otherwise.} \end{cases}$$

Then

$$\begin{aligned} M(s) &= \frac{1}{2} \int_0^{\infty} e^{(s-1/2)x}\, dx \\ &= \frac{1}{1-2s}, \qquad s < \frac{1}{2}. \end{aligned}$$

Example 7. Let X have the PMF

$$P\{X = k\} = \begin{cases} e^{-\lambda} \dfrac{\lambda^k}{k!}, & k = 0, 1, 2, \ldots, \\ 0, & \text{otherwise.} \end{cases}$$

Then

$$
\begin{aligned}
M(s) = Ee^{sX} &= e^{-\lambda}\sum_{k=0}^{\infty} e^{sk}\frac{\lambda^k}{k!}\\
&= e^{-\lambda(1-e^s)} \qquad \text{for all } s.
\end{aligned}
$$

The following result will be quite useful subsequently.

Theorem 1. The MGF uniquely determines a DF and, conversely, if the MGF exists, it is unique.

For the proof we refer the reader to Widder [117, p. 460], or Curtiss [19]. Theorem 2 explains why we call $M(s)$ an MGF.

Theorem 2. If the MGF $M(s)$ of an RV X exists for s in $(-s_0, s_0)$ say, $s_0 > 0$, the derivatives of all order exist at $s = 0$ and can be evaluated under the integral sign, that is,

$$M^{(k)}(s)\big|_{s=0} = EX^k \text{ for positive integral } k. \tag{3}$$

For the proof of Theorem 2 we refer to Widder [117, pp. 446–447]. See also Problem 9.

Remark 3. Alternatively, if the MGF $M(s)$ exists for s in $(-s_0, s_0)$ say $s_0 > 0$, one can express $M(s)$ (uniquely) in a Maclaurin series expansion:

$$M(s) = M(0) + \frac{M'(0)}{1!}s + \frac{M''(0)}{2!}s^2 + \cdots, \tag{4}$$

so that EX^k is the coefficient of $s^k/k!$ in expansion (4).

Example 8. Let X be an RV with PDF $f(x) = (1/2)e^{-x/2}, x > 0$. From Example 6, $M(s) = 1/(1-2s)$ for $s < 1/2$. Thus

$$M'(s) = \frac{2}{(1-2s)^2} \qquad \text{and} \qquad M''(s) = \frac{4\cdot 2}{(1-2s)^3}, \qquad s < \frac{1}{2}.$$

It follows that

$$EX = 2, \qquad EX^2 = 8, \qquad \text{and} \qquad \operatorname{var}(X) = 4.$$

Example 9. Let X be an RV with PDF $f(x) = 1, 0 \le x \le 1$, and $= 0$ otherwise. Then

$$
\begin{aligned}
M(s) &= \int_0^1 e^{sx}\,dx = \frac{e^s - 1}{s}, \qquad \text{all } s,\\
M'(s) &= \frac{e^s\cdot s - (e^s - 1)\cdot 1}{s^2},\\
EX = M'(0) &= \lim_{s\to 0}\frac{se^s - e^s + 1}{s^2} = \frac{1}{2}.
\end{aligned}
$$

We emphasize that the expectation Ee^{sX} does not exist unless s is carefully restricted. In fact, the requirement that $M(s)$ exists in a neighborhood of zero is a very strong requirement that is not satisfied by some common distributions. We next consider a generating function which exists for all distributions.

Definition 3. Let X be an RV. The complex-valued function ϕ defined on $\mathcal{R}$ by

$$\phi(t) = E(e^{itX}) = E(\cos tX) + iE(\sin tX), \quad t \in \mathcal{R}$$

where $i = \sqrt{(-1)}$ is the imaginary unit, is called the *characteristic function* (CF) of RV X.

Clearly

$$\phi(t) = \sum_k (\cos tk + i \sin tk) P(X = k)$$

in the discrete case, and

$$\phi(t) = \int_{-\infty}^{\infty} \cos tx f(x) dx + i \int_{-\infty}^{\infty} \sin tx\, f(x)\, dx$$

in the continuous case.

Example 10. Let X be a normal RV with PDF

$$f(x) = \left(\frac{1}{\sqrt{2\pi}}\right) \exp\left(\frac{-x^2}{2}\right), \quad x \in \mathcal{R}.$$

Then

$$\phi(t) = \left(\frac{1}{\sqrt{2\pi}}\right) \int_{-\infty}^{\infty} \cos tx\, e^{-x^2/2} dx + \left(\frac{i}{\sqrt{2\pi}}\right) \int_{-\infty}^{\infty} \sin tx\, e^{-x^2/2} dx.$$

Note that $\sin tx$ is an odd function and so also is $\sin tx\, e^{-x^2/2}$. Thus the second integral on the right-side vanishes and we have

$$\begin{aligned} \phi(t) &= \left(\frac{1}{\sqrt{2\pi}}\right) \int_{-\infty}^{\infty} \cos tx\, e^{-x^2/2} dx \\ &= \left(\frac{2}{\sqrt{2\pi}}\right) \int_{0}^{\infty} \cos tx\, e^{-x^2/2} dx = e^{-t^2/2}, \quad t \in \mathcal{R}. \end{aligned}$$

Remark 4. Unlike an MGF which may not exist for some distributions, a CF always exists which makes it a much more convenient tool. In fact, it is easy to see that ϕ is continuous on $\mathcal{R}$, $|\phi(t)| \leq 1$ for all t, and $\phi(-t) = \overline{\phi}(t)$ where $\overline{\phi}$ is the complex-conjugate of ϕ. Thus $\overline{\phi}$ is the CF of $-X$. Moreover, ϕ uniquely determines the DF of RV X. For these and many other properties of characteristic functions we need a comprehensive knowledge of complex variable theory, well beyond the scope of this book. We refer the reader to Lukacs [69].

Finally, we consider the problem of characterizing a distribution from its moments. Given a set of constants $\{\mu_0 = 1, \mu_1, \mu_2, \ldots\}$ the problem of moments asks if they can be moments of a distribution function F. At this point it will be worthwhile to take note of some facts.

First, we have seen that if the $M(s) = Ee^{sX}$ exists for some X for s in some neighborhood of zero, then $E|X|^n < \infty$ for all $n \geq 1$. Suppose, however, that $E|X|^n < \infty$ for all $n \geq 1$. It does not follow that the MGF of X exists.

Example 11. Let X be an RV with PDF

$$f(x) = ce^{-|x|^\alpha}, \qquad 0 < \alpha < 1, \quad -\infty < x < \infty,$$

where c is a constant determined from

$$c\int_{-\infty}^{\infty} e^{-|x|^\alpha}\, dx = 1.$$

Let $s > 0$. Then

$$\int_0^\infty e^{sx}e^{-x^\alpha}\, dx = \int_0^\infty e^{x(s-x^{\alpha-1})}\, dx$$

and since $\alpha - 1 < 0$, $\int_0^\infty s^{sx}e^{-x^\alpha}\, dx$ is not finite for any $s > 0$. Hence the MGF does not exist. But

$$E|X|^n = c\int_{-\infty}^{\infty} |x|^n e^{-|x|^\alpha}\, dx = 2c\int_0^\infty x^n e^{-x^\alpha}\, dx < \infty \qquad \text{for each } n,$$

as is easily checked by substituting $y = x^\alpha$.

Second, two (or more) RVs may have the same set of moments.

Example 12. Let X have *lognormal* PDF

$$f(x) = (x\sqrt{2\pi})^{-1}e^{-(\log x)^2/2}, \quad x > 0,$$

and $f(x) = 0$ for $x \leq 0$. Let X_ε, $|\varepsilon| \leq 1$, have PDF

$$f_\varepsilon(x) = f(x)[1 + \varepsilon \sin(2\pi \log x)], \quad x \in \mathcal{R}.$$

(Note that $f_\varepsilon \geq 0$ for all ε, $|\varepsilon| \leq 1$, and $\int_\infty^\infty f_\varepsilon(x)dx = 1$, so f_ε is a PDF.) Since, however,

$$\begin{aligned}\int_0^\infty x^k f(x)\sin(2\pi \log x)dx &= \left(\frac{1}{\sqrt{2\pi}}\right)\int_{-\infty}^\infty e^{-(t^2/2)+kt}\sin(2\pi t)dt \\ &= \left(\frac{1}{\sqrt{2\pi}}\right)e^{k^2/2}\int_{-\infty}^\infty e^{-y^2/2}\sin(2\pi y)dy \\ &= 0,\end{aligned}$$

we see that

$$\int_0^\infty x^k f(x)dx = \int_0^\infty x^k f_\varepsilon(x)dx$$

for all ε, $|\varepsilon| \leq 1$, and $k = 0, 1, 2, \ldots$. But $f(x) \neq f_\varepsilon(x)$.

Third, moments of any RV X necessarily satisfy certain conditions. For example, if $\beta_\nu = E|X|^\nu$, we will see (Theorem 3.4.3) that $(\beta_\nu)^{1/\nu}$ is an increasing function of ν. Similarly, the quadratic form

$$E\left(\sum_{i=1}^n X^{\alpha_i} t_i\right)^2 \geq 0$$

yields a relation between moments of various orders of X.

The following result, which we will not prove here, gives a sufficient condition for unique determination of F from its moments.

Theorem 3. Let $\{m_k\}$ be the moment sequence of an RV X. If the series

$$\sum_{k=1}^\infty \frac{m_k}{k!}s^k \tag{5}$$

converges absolutely for some $s > 0$, then $\{m_k\}$ uniquely determines the DF F of X.

Example 13. Suppose X has PDF

$$f(x) = e^{-x}, \quad \text{for } x \geq 0, \text{ and } = 0 \text{ for } x < 0.$$

Then $EX^k = \int_0^\infty x^k e^{-x}dx = k!$ and from Theorem 3

$$\sum_{k=1}^\infty \frac{m_k}{k!}s^k = \sum_{k=1}^\infty s^k = s/(1-s)$$

for $0 < s < 1$ so that $\{m_k\}$ determines F uniquely. In fact, from Remark 3

$$M(s) = \sum_{k=0}^{\infty} \frac{m_k s^k}{k!} = \sum_{k=0}^{\infty} s^k = \frac{1}{(1-s)},$$

$0 < s < 1$, which is the MGF of X.

In particular if for some constant c

$$|m_k| \le c^k, \qquad k = 1, 2, \ldots,$$

then

$$\sum_{k=1}^{\infty} \frac{|m_k|}{k!} s^k \le \sum_{1}^{\infty} \frac{(cs)^k}{k!} < e^{cs} \qquad \text{for } s > 0,$$

and the DF of X is uniquely determined. Thus if $P\{|X| \le c\} = 1$ for some $c > 0$, then all moments of X exist, satisfying $|m_k| \le c^k$, $k \ge 1$, and the DF of X is uniquely determined from its moments.

Finally, we mention some sufficient conditions for a moment sequence to determine a unique DF:

(i) The range of the RV is finite.
(ii) (Carleman) $\sum_{k=1}^{\infty} (m_{2k})^{-1/2k} = \infty$ when the range of the RV is $(-\infty, \infty)$. If the range is $(0, \infty)$, a sufficient condition is $\sum_{k=1}^{\infty} (m_k)^{-1/2k} = \infty$.
(iii) $\lim_{n\to\infty} \{(m_{2n})^{1/2n}/2n\}$ is finite.

PROBLEMS 3.3

1. Find the PGF of the RVs with the following PMFs:
   (a) $P\{X = k\} = \binom{n}{k} p^k (1-p)^{n-k}$, $k = 0, 1, 2, \ldots, 0 \le p \le 1$.
   (b) $P\{X = k\} = [e^{-\lambda}/(1-e^{-\lambda})](\lambda^k/k!)$, $k = 1, 2, \ldots; \lambda > 0$.
   (c) $P\{X = k\} = pq^k (1-q^{N+1})^{-1}$, $k = 0, 1, 2, \ldots, N; 0 < p < 1, q = 1 - p$.
2. Let X be an integer-valued RV with PGF $P(s)$. Let α and β be nonnegative integers, and write $Y = \alpha X + \beta$. Find the PGF of Y.
3. Let X be an integer-valued RV with PGF $P(s)$, and suppose that the mgf $M(s)$ exists for $s \in (-s_0, s_0)$, $s_0 > 0$. How are $M(s)$ and $P(s)$ related? Using $M^{(k)}(s)|_{s=0} = EX^k$ for positive integral k, find EX^k in terms of the derivatives of $P(s)$ for values of $k = 1, 2, 3, 4$.
4. For the Cauchy PDF

$$f(x) = \frac{1}{\pi} \frac{1}{1+x^2}, \qquad -\infty < x < \infty,$$

does the MGF exist?

5. Let X be an RV with PMF

$$P\{X=j\}=p_j, \qquad j=0,1,2,\ldots.$$

Set $P\{X>j\}=q_j, j=0,1,2,\ldots$. Clearly $q_j=p_{j+1}+p_{j+2}+\cdots, j\geq 0$. Write $Q(s)=\sum_{j=0}^{\infty} q_j s^j$. Then the series for $Q(s)$ converges in $|s|<1$. Show that

$$Q(s)=\frac{1-P(s)}{1-s} \qquad \text{for } |s|<1,$$

where $P(s)$ is the PGF of X. Find the mean and the variance of X (when they exist) in terms of Q and its derivatives.

6. For the PMF

$$P\{X=j\}=\frac{a_j\theta^j}{f(\theta)}, \qquad j=0,1,2,\ldots, \quad \theta>0,$$

where $a_j\geq 0$ and $f(\theta)=\sum_{j=0}^{\infty} a_j\theta^j$, find the PGF and the MGF in terms of f.

7. For the *Laplace* PDF

$$f(x)=\frac{1}{2\lambda}e^{-|x-\mu|/\lambda}, \qquad -\infty<x<\infty; \quad \lambda>0, \quad -\infty<\mu<\infty,$$

show that the MGF exists and equals

$$M(t)=(1-\lambda^2t^2)^{-1}e^{\mu t}, \qquad |t|<\frac{1}{\lambda}.$$

8. For any integer-valued RV X, show that

$$\sum_{n=0}^{\infty} s^n P\{X\leq n\}=(1-s)^{-1}P(s),$$

where P is the PGF of X.

9. Let X be an RV with MGF $M(t)$, which exists for $t\in(-t_0,t_0)$, $t_0>0$. Show that

$$E|X|^n<n!s^{-n}[M(s)+M(-s)]$$

for any fixed s, $0<s<t_0$, and for each integer $n\geq 1$. Expanding e^{tx} in a power series, show that, for $t\in(-s,s)$, $0<s<t_0$,

$$M(t)=\sum_{n=0}^{\infty} t^n\frac{EX^n}{n!}.$$

(Since a power series can be differentiated term by term within the interval of convergence, it follows that for $|t| < s$,

$$M^{(k)}(t)|_{t=0} = EX^k$$

for each integer $k \geq 1$.) (Roy, LePage, and Moore [95])

10. Let X be an integer-valued random variable with

$$\mathcal{E}\{X(X-1)\cdots(X-k+1)\} = \begin{cases} k!\binom{n}{k} & \text{if } k = 0,1,2,\ldots,n \\ 0 & \text{if } k > n. \end{cases}$$

Show that X must be degenerate at n.
[*Hint:* Prove and use the fact that if $EX^k < \infty$ for all k, then

$$P(s) = \sum_{k=0}^{\infty} \frac{(s-1)^k}{k!} E\{X(X-1)\cdots(X-k+1)\}.$$

Write $P(s)$ as

$$P(s) = \sum_{k=0}^{\infty} P(X=k)s^k = \sum_{k=0}^{\infty} P(X=k)\sum_{i=0}^{k}(s-1)^i$$
$$= \sum_{i=0}^{\infty}(s-1)^i \sum_{k=i}^{\infty}\binom{k}{i}P(X=k).\ ]$$

11. Let $p(n,k) = f(n,k)/n!$ where $f(n,k)$ is given by

$$f(n+1,k) = f(n,k) + f(n,k-1) + \cdots + f(n,k-n),$$

for $k = 0,1,\ldots,\binom{n}{2}$ and

$$f(n,k) = 0 \quad \text{for } k < 0, f(1,0) = 1, f(1,k) = 0 \text{ otherwise.}$$

Let

$$P_n(s) = \frac{1}{n!}\sum_{k=0}^{\infty} s^k f(n,k)$$

be the probability generating function of $p(n,k)$. Show that

$$P_n(s) = (n!)^{-1}\prod_{k=2}^{n}\frac{1-s^k}{1-s} \quad |s| < 1.$$

(P_n is the generating function of Kendall's τ-statistic.)

12. For $k = 0, 1, \ldots, \binom{n}{2}$ let $u_n(k)$ be defined recursively by

$$u_n(k) = u_{n-1}(k-n) + u_{n-1}(k)$$

with $u_0(0) = 1$, $u_0(k) = 0$ otherwise, and $u_n(k) = 0$ for $k < 0$. Let $P_n(s) = \sum_{k=0}^{\infty} s^k u_n(k)$ be the generating function of $\{u_n\}$. Show that

$$P_n(s) = \prod_{j=1}^{n}(1+s^j) \quad \text{for } |s| < 1.$$

If $p_n(k) = u_n(k)/2^n$, find $\{p_n(k)\}$ for $n = 2, 3, 4$. (P_n is the generating function of one-sample Wilcoxon test statistic.)

3.4 SOME MOMENT INEQUALITIES

In this section we derive some inequalities for moments of an RV. The main result of this section is Theorem 1 (and its corollary), which gives a bound for tail probability in terms of some moment of the random variable.

Theorem 1. Let $h(X)$ be a nonnegative Borel-measurable function of an RV X. If $Eh(X)$ exists, then, for every $\varepsilon > 0$,

$$P\{h(X) \geq \varepsilon\} \leq \frac{Eh(X)}{\varepsilon}. \tag{1}$$

Proof. We prove the result when X is discrete. Let $P\{X = x_k\} = p_k$, $k = 1, 2, \ldots$. Then

$$\begin{aligned} Eh(X) &= \sum_k h(x_k)p_k \\ &= \left(\sum_A + \sum_{A^c}\right) h(x_k)p_k, \end{aligned}$$

where

$$A = \{k : h(x_k) \geq \varepsilon\}.$$

Then

$$\begin{aligned} Eh(X) &\geq \sum_A h(x_k)p_k \geq \varepsilon \sum_A p_k \\ &= \varepsilon P\{h(X) \geq \varepsilon\}. \end{aligned}$$

Corollary. Let $h(X) = |X|^r$ and $\varepsilon = K^r$, where $r > 0$ and $K > 0$. Then

$$P\{|X| \geq K\} \leq \frac{E|X|^r}{K^r}, \tag{2}$$

which is *Markov's inequality*. In particular, if we take $h(X) = (X-\mu)^2$, $\varepsilon = K^2\sigma^2$, we get *Chebychev–Bienayme inequality*:

$$P\{|X-\mu| \geq K\sigma\} \leq \frac{1}{K^2}, \tag{3}$$

where $EX = \mu$, $\text{var}(X) = \sigma^2$.

Remark 1. The inequality (3) is generally attributed to Chebychev although recent research has shown that credit should also go to I.J. Bienayme.

Remark 2. If we wish to be consistent with our definition of a DF as $F_X(x) = P(X \leq x)$, then we may want to reformulate (1) in the following form.

$$P\{h(X) > \varepsilon\} < Eh(X)/\varepsilon.$$

For RVs with finite second-order moments one cannot do better than the inequality in (3).

Example 1.

$$\begin{aligned} &P\{X=0\} = 1 - \frac{1}{K^2} \\ & \qquad\qquad\qquad\qquad K > 1, \text{ constant}, \\ &P\{X = \mp 1\} = \frac{1}{2K^2} \\ &EX = 0, \qquad EX^2 = \frac{1}{K^2}, \qquad \sigma = \frac{1}{K}, \\ &P\{|X| \geq K\sigma\} = P\{|X| \geq 1\} = \frac{1}{K^2}, \end{aligned}$$

so that equality is achieved.

Example 2. Let X be distributed with PDF $f(x) = 1$ if $0 < x < 1$, and $= 0$ otherwise. Then

$$EX = \frac{1}{2}, \qquad EX^2 = \frac{1}{3}, \ \text{var}(X) = \frac{1}{3} - \frac{1}{4} = \frac{1}{12},$$

$$P\left\{|X - \frac{1}{2}| < 2\sqrt{\frac{1}{12}}\right\} = P\left\{\frac{1}{2} - \frac{1}{\sqrt{3}} < X < \frac{1}{2} + \frac{1}{\sqrt{3}}\right\} = 1.$$

From Chebychev's inequality

$$P\left\{|X - \frac{1}{2}| < 2\sqrt{\frac{1}{12}}\right\} \geq 1 - \frac{1}{4} = 0.75.$$

In Fig. 1 we compare the upper bound for $P\{|X - 1/2| \geq k/\sqrt{12}\}$ with the exact probability.

It is possible to improve upon Chebychev's inequality, at least in some cases, if we assume the existence of higher order moments. We need the following lemma.

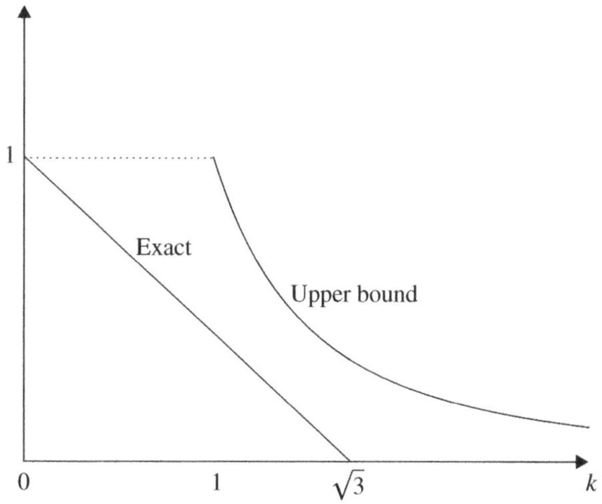

Fig. 1 Chebychev upper bound versus exact probability.

Lemma 1. Let X be an RV with $EX = 0$ and $\text{var}(X) = \sigma^2$. Then

$$P\{X \geq x\} \leq \frac{\sigma^2}{\sigma^2 + x^2} \qquad \text{if } x > 0, \tag{4}$$

$$P\{X \geq x\} \geq \frac{x^2}{\sigma^2 + x^2} \qquad \text{if } x < 0, \tag{5}$$

Proof. Let $h(t) = (t + c)^2$, $c > 0$. Then $h(t) \geq 0$ for all t and

$$h(t) \geq (x + c)^2 \qquad \text{for } t \geq x > 0.$$

It follows that

$$\begin{aligned} P\{X \geq x\} &\leq P\{h(X) \geq (x + c)^2\} \\ &\leq \frac{E(X + c)^2}{(x + c)^2} \qquad \text{for all } c > 0, \quad x > 0. \end{aligned} \tag{6}$$

Since $EX = 0$, $EX^2 = \sigma^2$, and the right side of (6) is minimum when $c = \sigma^2/x$, we have

$$P\{X \geq x\} \leq \frac{\sigma^2}{\sigma^2 + x^2}, \qquad x > 0.$$

Similar proof holds for (5).

Remark 3. Inequalities (4) and (5) cannot be improved (Problem 3).

Theorem 2. Let $E|X|^4 < \infty$, and let $EX = 0$, $EX^2 = \sigma^2$. Then

$$P\{|X| \geq K\sigma\} \leq \frac{\mu_4 - \sigma^4}{\mu_4 + \sigma^4 K^4 - 2K^2\sigma^4} \qquad \text{for } K > 1, \tag{7}$$

where $\mu_4 = EX^4$.

Proof. For the proof let us substitute $(X^2-\sigma^2)/(K^2\sigma^2-\sigma^2)$ for X and take $x=1$ in (4). Then

$$\begin{aligned} P\{X^2-\sigma^2 \geq K^2\sigma^2-\sigma^2\} &\leq \frac{\text{var}\{(X^2-\sigma^2)/(K^2\sigma^2-\sigma^2)\}}{1+\text{var}\{(X^2-\sigma^2)/(K^2\sigma^2-\sigma^2)\}} \\ &= \frac{\mu_4-\sigma^4}{\sigma^4(K^2-1)^2+\mu_4-\sigma^4} \\ &= \frac{\mu_4-\sigma^4}{\mu_4+\sigma^4K^4-2K^2\sigma^4}, \qquad K>1, \end{aligned}$$

as asserted.

Remark 4. Bound (7) is better than bound (3) if $K^2 \geq \mu_4/\sigma^4$ and worse if $1 \leq K^2 < \mu_4/\sigma^4$ (Problem 5).

Example 3. Let X have the uniform density

$$f(x) = \begin{cases} 1 & \text{if } 0<x<1, \\ 0 & \text{otherwise.} \end{cases}$$

Then

$$EX = \frac{1}{2}, \qquad \text{var}(X) = \frac{1}{12}, \qquad \mu_4 = E\left\{X-\frac{1}{2}\right\}^4 = \frac{1}{80},$$

and

$$P\left\{\left|X-\frac{1}{2}\right| \geq 2\sqrt{\frac{1}{12}}\right\} \leq \frac{\frac{1}{80}-\frac{1}{144}}{\frac{1}{80}+\frac{1}{144}\cdot 16-8\frac{1}{144}} = \frac{4}{49},$$

that is,

$$P\left\{\left|X-\frac{1}{2}\right| < 2\sqrt{\frac{1}{12}}\right\} \geq \frac{45}{49} \approx 0.92,$$

which is much better than the bound given by Chebychev's inequality (Example 2).

Theorem 3 (Lyapunov Inequality). Let $\beta_n = E|X|^n < \infty$. Then for arbitrary k, $2 \leq k \leq n$, we have

$$\beta_{k-1}^{1/(k-1)} \leq \beta_k^{1/k}. \tag{8}$$

Proof. Consider the quadratic form

$$Q(u,v) = \int_{-\infty}^{\infty} (u|x|^{(k-1)/2}+v|x|^{(k+1)/2})^2 f(x)\,dx,$$

where we have assumed that X is continuous with PDF f. We have

$$Q(u,v) = u^2\beta_{k-1} + 2uv\beta_k + \beta_{k+1}v^2.$$

Clearly $Q \geq 0$ for all u, v real. It follows that

$$\begin{vmatrix} \beta_{k-1} & \beta_k \\ \beta_k & \beta_{k+1} \end{vmatrix} \geq 0,$$

implying that

$$\beta_k^{2k} \leq \beta_{k-1}^k \beta_{k+1}^k.$$

Thus

$$\beta_1^2 \leq \beta_0^1\beta_2^1, \qquad \beta_2^4 \leq \beta_1^2\beta_3^2, \ldots, \beta_{n-1}^{2(n-1)} \leq \beta_{n-2}^{n-1}\beta_n^{n-1},$$

where $\beta_0 = 1$. Multiplying successive $k-1$ of these, we have

$$\beta_{k-1}^k \leq \beta_k^{k-1} \qquad \text{or} \qquad \beta_{k-1}^{1/(k-1)} \leq \beta_k^{1/k}.$$

It follows that

$$\beta_1 \leq \beta_2^{1/2} \leq \beta_3^{1/3} \leq \cdots \leq \beta_n^{1/n}.$$

The equality holds if and only if

$$\beta_k^{1/k} = \beta_{k+1}^{1/(k+1)} \qquad \text{for } k = 1, 2, \ldots,$$

that is, $\{\beta_k^{1/k}\}$ is a constant sequence of numbers, which happens if and only if $|X|$ is degenerate, that is, for some c, $P\{|X| = c\} = 1$.

PROBLEMS 3.4

1. For the RV with PDF

$$f(x;\lambda) = \frac{e^{-x}x^{\lambda}}{\lambda!}, \qquad x > 0,$$

where $\lambda \geq 0$ is an integer, show that

$$P\{0 < X < 2(\lambda+1)\} > \frac{\lambda}{\lambda+1}.$$

2. Let X be any RV, and suppose that the MGF of X, $M(t) = Ee^{tx}$, exists for every $t > 0$. Then for any $t > 0$

$$P\{tX > s^2 + \log M(t)\} < e^{-s^2}.$$

3. Construct an example to show that inequalities (4) and (5) cannot be improved.

4. Let $g(.)$ be a function satisfying $g(x) > 0$ for $x > 0$, $g(x)$ increasing for $x > 0$, and $E|g(X)| < \infty$. Show that

$$P\{|X| > \varepsilon\} < \frac{Eg(|X|)}{g(\varepsilon)} \qquad \text{for every } \varepsilon > 0.$$

5. Let X be an RV with $EX = 0$, $\text{var}(X) = \sigma^2$, and $EX^4 = \mu_4$. Let K be any positive real number. Show that

$$P\{|X| \geq K\sigma\} \leq \begin{cases} 1 & \text{if } K^2 < 1, \\ \frac{1}{K^2} & \text{if } 1 \leq K^2 < \frac{\mu_4}{\sigma^4}, \\ \dfrac{\mu_4 - \sigma^4}{\mu_4 + \sigma^4 K^4 - 2K^2\sigma^4} & \text{if } K^2 \geq \frac{\mu_4}{\sigma^4}. \end{cases}$$

In other words, show that bound (7) is better than bound (3) if $K^2 \geq \mu_4/\sigma^4$ and worse if $1 \leq K^2 < \mu_4/\sigma^4$. Construct an example to show that the last inequalities cannot be improved.

6. Use Chebychev's inequality to show that for any $k > 1$, $e^{k+1} \geq k^2$.

7. For any RV X, show that

$$P\{X \geq 0\} \leq \inf\{\varphi(t) : t \geq 0\} \leq 1,$$

where $\varphi(t) = Ee^{tX}$, $0 < \varphi(t) \leq \infty$.

8. Let X be an RV such that $P(a \leq X \leq b) = 1$ where $-\infty < a < b < \infty$. Show that $\text{var}(X) \leq (b-a)^2/4$.

4

MULTIPLE RANDOM VARIABLES

4.1 INTRODUCTION

In many experiments an observation is expressible, not as a single numerical quantity, but as a family of several separate numerical quantities. Thus, for example, if a pair of distinguishable dice is tossed, the outcome is a pair (x, y), where x denotes the face value on the first die, and y, the face value on the second die. Similarly, to record the height and weight of every person in a certain community we need a pair (x, y), where the components represent, respectively, the height and weight of a particular individual. To be able to describe such experiments mathematically we must study the *multidimensional random variables*.

In Section 4.2 we introduce the basic notations involved and study joint, marginal, and conditional distributions. In Section 4.3 we examine independent random variables and investigate some consequences of independence. Section 4.4 deals with functions of several random variables and their induced distributions. Section 4.5 considers moments, covariance, and correlation, and in Section 4.6 we study conditional expectation. The last section deals with ordered observations.

4.2 MULTIPLE RANDOM VARIABLES

In this section we study multidimensional RVs. Let $(\Omega, \mathcal{S}, P)$ be a fixed but otherwise arbitrary probability space.

An Introduction to Probability and Statistics, Third Edition. Vijay K. Rohatgi and A.K. Md. Ehsanes Saleh.

Definition 1. The collection $\mathbf{X} = (X_1, X_2, \ldots, X_n)$ defined on $(\Omega, \mathcal{S}, P)$ into $\mathcal{R}_n$ by

$$\mathbf{X}(\omega) = (X_1(\omega), X_2(\omega), \ldots, X_n(\omega)), \qquad \omega \in \Omega,$$

is called an n-dimensional RV if the inverse image of every n-dimensional interval

$$I = \{(x_1, x_2, \ldots, x_n) \colon -\infty < x_i \le a_i, a_i \in \mathcal{R}, i = 1, 2, \ldots, n\}$$

is also in $\mathcal{S}$, that is, if

$$\mathbf{X}^{-1}(I) = \{\omega \colon X_1(\omega) \le a_1, \ldots, X_n(\omega) \le a_n\} \in \mathcal{S} \qquad \text{for } a_i \in \mathcal{R}.$$

Theorem 1. Let $X_1, X_2, \ldots, X_n$ be n RVs on $(\Omega, \mathcal{S}, P)$. Then $\mathbf{X} = (X_1, X_2, \ldots, X_n)$ is an n-dimensional RV on $(\Omega, \mathcal{S}, P)$.

Proof. Let $I = \{(x_1, x_2, \ldots, x_n) \colon -\infty < x_i \le a_i,\ i = 1, 2, \ldots, n\}$. Then

$$\begin{aligned}\{(X_1, X_2, \ldots, X_n) \in I\} &= \{\omega \colon X_1(\omega) \le a_1, X_2(\omega) \le a_2, \ldots, X_n(\omega) \le a_n\}\\ &= \bigcap_{k=1}^{n} \{\omega \colon X_k(\omega) \le a_k\} \in \mathcal{S}\end{aligned}$$

as asserted.

From now on we will restrict attention mainly to two-dimensional random variables. The discussion for the n-dimensional $(n > 2)$ case is similar except when indicated. The development follows closely the one-dimensional case.

Definition 2. The function $F(\cdot, \cdot)$, defined by

$$F(x, y) = P\{X \le x, Y \le y\}, \qquad \text{all } (x, y) \in \mathcal{R}_2, \tag{1}$$

is known as the DF of the RV (X, Y).

Following the discussion in Section 2.3, it is easily shown that

(i) $F(x, y)$ is nondecreasing and continuous from the right with respect to each coordinate and

(ii)
$$\begin{aligned}&\lim_{\substack{x \to +\infty \\ y \to +\infty}} F(x, y) = F(+\infty, +\infty) = 1,\\ &\lim_{y \to -\infty} F(x, y) = F(x, -\infty) = 0 \qquad \text{for all } x,\\ &\lim_{x \to -\infty} F(x, y) = F(-\infty, y) = 0 \qquad \text{for all } y.\end{aligned}$$

But (i) and (ii) are not sufficient conditions to make any function $F(\cdot, \cdot)$ a DF.

Example 1. Let F be a function (Fig. 1) of two variables defined by

$$F(x, y) = \begin{cases} 0, & x < 0 \text{ or } x + y < 1 \text{ or } y < 0,\\ 1, & \text{otherwise.}\end{cases}$$

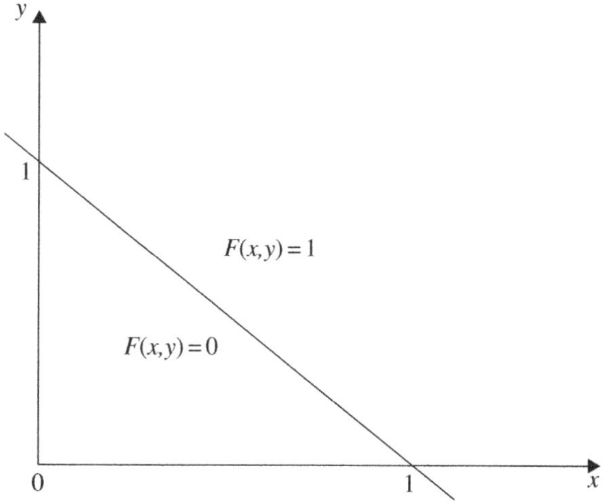

Fig. 1

Then F satisfies both (i) and (ii) above. However, F is not a DF since

$$\begin{aligned} P\{\tfrac{1}{3} < X \le 1, \tfrac{1}{3} < Y \le 1\} &= F(1,1) + F(\tfrac{1}{3},\tfrac{1}{3}) - F(1,\tfrac{1}{3}) - F(\tfrac{1}{3},1) \\ &= 1 + 0 - 1 - 1 = -1 \ngeq 0. \end{aligned}$$

Let $x_1 < x_2$ and $y_1 < y_2$. We have

$$\begin{aligned} P\{x_1 < X \le x_2, y_1 < Y \le y_2\} &= P\{X \le x_2, Y \le y_2\} + P\{X \le x_1, Y \le y_1\} \\ &\quad - P\{X \le x_1, Y \le y_2\} - P\{X \le x_2, Y \le y_1\} \\ &= F(x_2,y_2) + F(x_1,y_1) - F(x_1,y_2) - F(x_2,y_1) \\ &\ge 0 \end{aligned}$$

for all pairs (x_1,y_1), (x_2,y_2) with $x_1 < x_2$, $y_1 < y_2$ (see Fig. 2).

Theorem 2. A function F of two variables is a DF of some two-dimensional RV if and only if it satisfies the following conditions:

(i) F is nondecreasing and right continuous with respect to both arguments;
(ii) $F(-\infty,y) = F(x,-\infty) = 0$ and $F(+\infty,+\infty) = 1$; and
(iii) for every $(x_1,y_1),(x_2,y_2)$ with $x_1 < x_2$ and $y_1 < y_2$ the inequality

$$F(x_2,y_2) - F(x_2,y_1) + F(x_1,y_1) - F(x_1,y_2) \ge 0 \tag{2}$$

holds.

The "if" part of the theorem has already been established. The "only if" part will not be proved here (see Tucker [114, p. 26]).

Theorem 2 can be generalized to the n-dimensional case in the following manner.

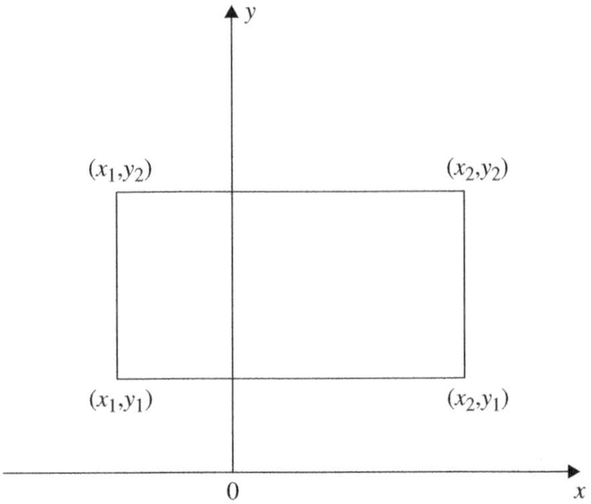

Fig. 2 $\{x_1 < x < x_2, y_1 < y \le y_2\}$.

Theorem 3. A function $F(x_1, x_2, \ldots, x_n)$ is the joint DF of some n-dimensional RV if and only if F is nondecreasing and continuous from the right with respect to all the arguments $x_1, x_2, \ldots, x_n$ and satisfies the following conditions:

(i)
$$\begin{aligned} F(-\infty, x_2, \ldots, x_n) &= F(x_1, -\infty, x_3, \ldots, x_n) \cdots \\ &= F(x_1, \ldots, x_{n-1}, -\infty) = 0, \\ F(+\infty, +\infty, \ldots, +\infty) &= 1. \end{aligned}$$

(ii) For every $(x_1, x_2, \ldots, x_n) \in \mathcal{R}_n$ and all $\varepsilon_i > 0 (i = 1, 2, \ldots, n)$ the inequality

$$\begin{aligned} &F(x_1+\varepsilon_1, x_2+\varepsilon_2, \ldots, x_n+\varepsilon_n) \\ &\quad - \sum_{i=1}^{n} F(x_1+\varepsilon_1, \ldots, x_{i-1}+\varepsilon_{i-1}, x_i, x_{i+1}+\varepsilon_{i+1}, \ldots, x_n+\varepsilon_n) \\ &\quad + \sum_{\substack{i,j=1\\ i<j}}^{n} F(x_1+\varepsilon_1, \ldots, x_{i-1}+\varepsilon_{i-1}, x_i, x_{i+1}+\varepsilon_{i+1}, \ldots, x_{j-1}+\varepsilon_{j-1}, \\ &\quad\quad x_j, x_{j+1}+\varepsilon_{j+1}, \ldots, x_n+\varepsilon_n) \\ &\quad + \cdots \\ &\quad + (-1)^n F(x_1, x_2, \ldots, x_n) \ge 0 \end{aligned} \tag{3}$$

holds.

We restrict ourselves here to two-dimensional RVs of the discrete or the continuous type, which we now define.

Definition 3. A two-dimensional (or bivariate) RV (X,Y) is said to be of the *discrete* type if it takes on pairs of values belonging to a countable set of pairs A with probability 1. We call every pair (x_i,y_j) that is assumed with positive probability p_{ij} a *jump point* of the DF of (X,Y) and call p_{ij} the *jump* at (x_i,y_j). Here A is the support of the distribution of (X,Y).

Clearly $\sum_{ij} p_{ij} = 1$. As for the DF of (X,Y), we have

$$F(x,y) = \sum_{B} p_{ij},$$

where $B = \{(i,j) : x_i \le x, y_j \le y\}$.

Definition 4. Let (X,Y) be an RV of the discrete type that takes on pairs of values $(x_i,y_j), i = 1,2,\ldots,$ and $j = 1,2,\ldots$. We call

$$p_{ij} = P\{X = x_i, Y = y_j\}, \qquad i = 1,2,\ldots, j = 1,2,\ldots,$$

the *joint probability mass function* (PMF) of (X,Y).

Example 2. A fair die is rolled, and a fair coin is tossed independently. Let X be the face value on the die, and let $Y = 0$ if a tail turns up and $Y = 1$ if a head turns up. Then

$$A = \{(1,0),(2,0),\ldots,(6,0),(1,1),(2,1),\ldots,(6,1)\},$$
$$p_{ij} = \frac{1}{12} \qquad \text{for } i = 1,2,\ldots,6;\ j = 0,1.$$

The DF of (X,Y) is given by

$$F(x,y) = \begin{cases} 0, & x < 1, -\infty < y < \infty; -\infty < x < \infty, y < 0, \\ \frac{1}{12}, & 1 \le x < 2, 0 \le y < 1, \\ \frac{1}{6}, & 2 \le x < 3, 0 \le y < 1; 1 \le x < 2, 1 \le y, \\ \frac{1}{4}, & 3 \le x < 4, 0 \le y < 1, \\ \frac{1}{3}, & 4 \le x < 5, 0 \le y < 1; 2 \le x < 3, 1 \le y, \\ \frac{5}{12}, & 5 \le x < 6, 0 \le y < 1, \\ \frac{1}{2}, & 6 \le x, 0 \le y < 1; 3 \le x < 4, 1 \le y, \\ \frac{2}{3}, & 4 \le x < 5, 1 \le y, \\ \frac{5}{6}, & 5 \le x < 6, 1 \le y, \\ 1, & 6 \le x, 1 \le y. \end{cases}$$

Theorem 4. A collection of nonnegative numbers $\{p_{ij}: i = 1,2,\ldots; j = 1,2,\ldots\}$ satisfying $\sum_{i,j=1}^{\infty} p_{ij} = 1$ is the PMF of some RV.

Proof. The proof of Theorem 4 is easy to construct with the help of Theorem 2.

Definition 5. A two-dimensional RV (X,Y) is said to be of the *continuous* type if there exists a nonnegative function $f(\cdot,\cdot)$ such that for every pair $(x,y) \in \mathcal{R}_2$ we have

$$F(x,y) = \int_{-\infty}^{x} \left[\int_{-\infty}^{y} f(u,v)\,dv \right] du, \tag{4}$$

where F is the DF of (X,Y). The function f is called the (joint) PDF of (X,Y).

Clearly,

$$\begin{aligned} F(+\infty,+\infty) &= \lim_{\substack{x\to+\infty \\ y\to+\infty}} \int_{-\infty}^{x} \int_{-\infty}^{y} f(u,v)\,dv\,du \\ &= \int_{-\infty}^{\infty} \int_{-\infty}^{\infty} f(u,v)\,dv\,du = 1. \end{aligned}$$

If f is continuous at (x,y), then

$$\frac{\partial^2 F(x,y)}{\partial x \partial y} = f(x,y). \tag{5}$$

Example 3. Let (X,Y) be an RV with joint PDF (Fig. 3) given by

$$f(x,y) = \begin{cases} e^{-(x+y)}, & 0 < x < \infty, \quad 0 < y < \infty, \\ 0, & \text{otherwise.} \end{cases}$$

Then

$$F(x,y) = \begin{cases} (1-e^{-x})(1-e^{-y}), & 0 < x < \infty, \quad 0 < y < \infty, \\ 0, & \text{otherwise.} \end{cases}$$

Theorem 5. If f is a nonnegative function satisfying $\int_{-\infty}^{\infty} \int_{-\infty}^{\infty} f(x,y)\,dx\,dy = 1$, then f is the joint density function of some RV.

Proof. For the proof define

$$F(x,y) = \int_{-\infty}^{x} \left[\int_{-\infty}^{y} f(u,v)\,dv \right] du$$

and use Theorem 2.

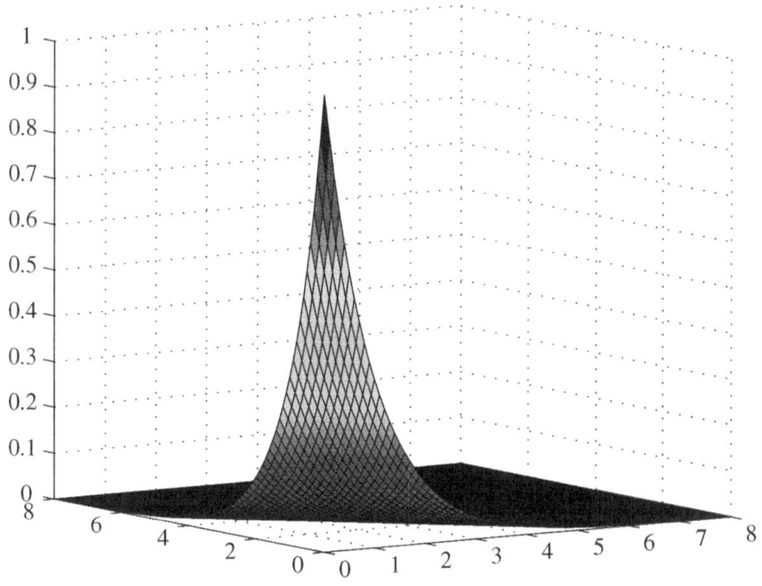

Fig. 3 $f(x,y) = \exp\{-(x+y)\}, x > 0, y > 0.$

Let (X, Y) be a two-dimensional RV with PMF

$$p_{ij} = P\{X = x_i, Y = y_j\}.$$

Then

$$\sum_{i=1}^{\infty} p_{ij} = \sum_{i=1}^{\infty} P\{X = x_i, Y = y_j\} = P\{Y = y_j\} \tag{6}$$

and

$$\sum_{j=1}^{\infty} p_{ij} = \sum_{j=1}^{\infty} P\{X = x_i, Y = y_j\} = P\{X = x_i\}. \tag{7}$$

Let us write

$$p_{i\cdot} = \sum_{j=1}^{\infty} p_{ij} \quad \text{and} \quad p_{\cdot j} = \sum_{i=1}^{\infty} p_{ij}. \tag{8}$$

Then $p_{i\cdot} \geq 0$ and $\sum_{i=1}^{\infty} p_{i\cdot} = 1, p_{\cdot j} \geq 0$ and $\sum_{j=1}^{\infty} p_{\cdot j} = 1$, and $\{p_{i\cdot}\}, \{p_{\cdot j}\}$ represent PMFs.

Definition 6. The collection of numbers $\{p_{i\cdot}\}$ is called the *marginal* PMF of X, and the collection $\{p_{\cdot j}\}$, the marginal PMF of Y.

Example 4. A fair coin is tossed three times. Let $X =$ number of heads in three tossings, and $Y =$ difference, in absolute value, between number of heads and number of tails. The joint PMF of (X,Y) is given in the following table:

$Y \backslash X$	0	1	2	3	$P\{Y=y\}$
1	0	$\frac{3}{8}$	$\frac{3}{8}$	0	$\frac{6}{8}$
3	$\frac{1}{8}$	0	0	$\frac{1}{8}$	$\frac{2}{8}$
$P\{X=x\}$	$\frac{1}{8}$	$\frac{3}{8}$	$\frac{3}{8}$	$\frac{1}{8}$	1

The marginal PMF of Y is shown in the column representing row totals, and the marginal PMF of X, in the row representing column totals.

If (X,Y) is an RV of the continuous type with PDF f, then

$$f_1(x) = \int_{-\infty}^{\infty} f(x,y)\,dy \tag{9}$$

and

$$f_2(y) = \int_{-\infty}^{\infty} f(x,y)\,dx \tag{10}$$

satisfy $f_1(x) \geq 0, f_2(y) \geq 0$, and $\int_{-\infty}^{\infty} f_1(x)\,dx = 1, \int_{-\infty}^{\infty} f_2(y)\,dy = 1$. It follows that $f_1(x)$ and $f_2(y)$ are PDFs.

Definition 7. The functions $f_1(x)$ and $f_2(y)$, defined in (9) and (10), are called the *marginal* PDF of X and the marginal PDF of Y, respectively.

Example 5. Let (X,Y) be jointly distributed with PDF $f(x,y) = 2, 0 < x < y < 1$, and, $= 0$ otherwise (Fig. 4). Then

$$f_1(x) = \int_x^1 2\,dy = \begin{cases} 2-2x, & 0<x<1 \\ 0, & \text{otherwise} \end{cases}$$

and

$$f_2(y) = \int_0^y 2\,dx = \begin{cases} 2y, & 0<y<1 \\ 0, & \text{otherwise} \end{cases}$$

are the two marginal density functions.

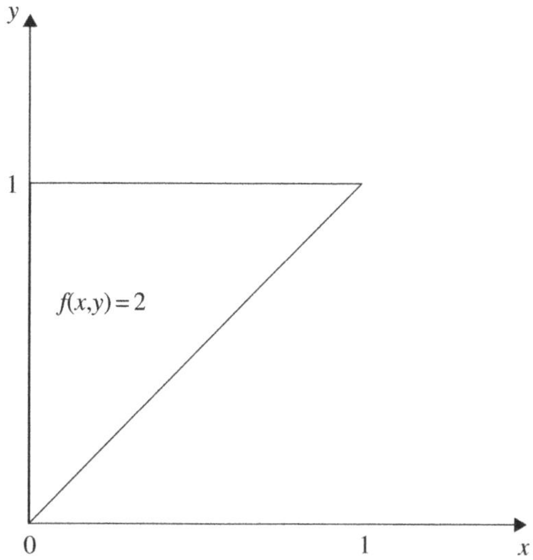

Fig. 4 $f(x,y) = 2, 0 < x < y < 1$.

Definition 8. Let (X,Y) be an RV with DF F. Then the *marginal* DF of X is defined by

$$\begin{aligned} F_1(x) = F(x,\infty) &= \lim_{y\to\infty} F(x,y) \\ &= \begin{cases} \sum_{x_i \le x} p_{i\cdot} & \text{if } (X,Y) \text{ is discrete,} \\ \int_{-\infty}^{x} f_1(t)\,dt & \text{if } (X,Y) \text{ is continuous.} \end{cases} \end{aligned} \tag{11}$$

A similar definition is given for the marginal DF of Y.

In general, given a DF $F(x_1,x_2,\ldots,x_n)$ of an n-dimensional RV $(X_1,X_2,\ldots,X_n)$, one can obtain any k-dimensional $(1 \le k \le n-1)$ marginal DF from it. Thus the marginal DF of $(X_{i_1},X_{i_2},\ldots X_{i_k})$, where $1 \le i_1 < i_2 < \cdots < i_k \le n$, is given by

$$\begin{aligned} &\lim_{\substack{x_i\to\infty \\ i\neq i_1,i_2,\ldots,i_k}} F(x_1,x_2,\ldots,x_n) \\ &= F(+\infty,\ldots,+\infty,x_{i_1},+\infty,\ldots,+\infty,\ldots,x_{i_k},+\infty,\ldots,+\infty). \end{aligned}$$

We now consider the concept of *conditional distributions*. Let (X,Y) be an RV of the discrete type with PMF $p_{ij} = P\{X = x_i, Y = y_j\}$. The marginal PMFs are $p_{i\cdot} = \sum_{j=1}^{\infty}$ and $p_{\cdot j} = \sum_{i=1}^{\infty} p_{ij}$. Recall that, if $A,B \in \mathcal{S}$ and $PB > 0$, the conditional probability of A, given B, is defined by

$$P\{A \mid B\} = \frac{P(AB)}{P(B)}.$$

Take $A=\{X=x_i\}=\{(x_i,y):\ -\infty<y<\infty\}$ and $B=\{Y=y_j\}=\{(x,y_j);-\infty<x<\infty\}$, and assume that $PB=P\{Y=y_j\}=p_{\cdot j}>0$. Then $A\cap B=\{X=x_i,Y=y_j\}$ and

$$P\{A\mid B\}=P\{X=x_i\mid Y=y_j\}=\frac{p_{ij}}{p_{\cdot j}}.$$

For fixed j, the function $P\{X=x_i\mid Y=y_j\}\geq 0$ and $\sum_{i=1}^{\infty}P\{X=x_i\mid Y=y_j\}=1$. Thus $P\{X=x_i\mid Y=y_j\}$, for fixed j, defines a PMF.

Definition 9. Let (X,Y) be an RV of the discrete type. If $P\{Y=y_j\}>0$, the function

$$P\{X=x_i\mid Y=y_j\}=\frac{P\{X=x_i,Y=y_j\}}{P\{Y=y_j\}} \tag{12}$$

for fixed j is known as the *conditional* PMF of X, given $Y=y_j$. A similar definition is given for $P\{Y=y_j\mid X=x_i\}$, the conditional PMF of Y, given $X=x_i$, provided that $P\{X=x_i\}>0$.

Example 6. For the joint PMF of Example 4, we have for $Y=1$

$$P\{X=i\mid Y=1\}=\begin{cases}0, & i=0,3,\\ \frac{1}{2}, & i=1,2.\end{cases}$$

Similarly

$$P\{X=i\mid Y=3\}=\begin{cases}\frac{1}{2}, & \text{if } i=0,3,\\ 0, & \text{if } i=1,2,\end{cases}$$

$$P\{Y=j\mid X=0\}=\begin{cases}0, & \text{if } j=1,\\ 1, & \text{if } j=3,\end{cases}$$

and so on.

Next suppose that (X,Y) is an RV of the continuous type with joint PDF f. Since $P\{X=x\}=0$, $P\{Y=y\}=0$ for any x,y, the probability $P\{X\leq x\mid Y=y\}$ or $P\{Y\leq y\mid X=x\}$ is not defined. Let $\varepsilon>0$, and suppose that $P\{y-\varepsilon<Y\leq y+\varepsilon\}>0$. For every x and every interval $(y-\varepsilon,y+\varepsilon]$, consider the conditional probability of the event $\{X\leq x\}$, given that $Y\in(y-\varepsilon,y+\varepsilon]$. We have

$$P\{X\leq x\mid y-\varepsilon<Y\leq y+\varepsilon\}=\frac{P\{X\leq x,y-\varepsilon<Y\leq y+\varepsilon\}}{P\{Y\in(y-\varepsilon,y+\varepsilon]\}}.$$

For any fixed interval $(y-\varepsilon,y+\varepsilon]$, the above expression defines the conditional DF of X, given that $Y\in(y-\varepsilon,y+\varepsilon]$, provided that $P\{Y\in(y-\varepsilon,y+\varepsilon]\}>0$. We shall be interested in the case where the limit

$$\lim_{\varepsilon\to 0+}P\{X\leq x\mid Y\in(y-\varepsilon,y+\varepsilon]\}$$

exists.

Definition 10. The *conditional* DF of an RV X, given $Y = y$, is defined as the limit

$$\lim_{\varepsilon\to 0+} P\{X \le x \mid Y \in (y-\varepsilon, y+\varepsilon]\}, \tag{13}$$

provided that the limit exists. If the limit exists, we denote it by $F_{X|Y}(x \mid y)$ and define the conditional density function of X, given $Y = y, f_{X|Y}(x \mid y)$, as a nonnegative function satisfying

$$F_{X|Y}(x \mid y) = \int_{-\infty}^{x} f_{X|Y}(t \mid y)\, dt \qquad \text{for all } x \in \mathcal{R}. \tag{14}$$

For fixed y, we see that $f_{X|Y}(x \mid y) \ge 0$ and $\int_{-\infty}^{\infty} f_{X|Y}(x \mid y)\, dx = 1$. Thus $f_{X|Y}(x \mid y)$ is a PDF for fixed y.

Suppose that (X, Y) is an RV of the continuous type with PDF f. At every point (x, y) where f is continuous and the marginal PDF $f_2(y) > 0$ and is continuous, we have

$$\begin{aligned} F_{X|Y}(x \mid y) &= \lim_{\varepsilon\to 0+} \frac{P\{X \le x, Y \in (y-\varepsilon, y+\varepsilon]\}}{P\{Y \in (y-\varepsilon, y+\varepsilon]\}} \\ &= \lim_{\varepsilon\to 0+} \frac{\int_{-\infty}^{x} \left\{\int_{y-\varepsilon}^{y+\varepsilon} f(u,v)\, dv\right\} du}{\int_{y-\varepsilon}^{y+\varepsilon} f_2(v)\, dv}. \end{aligned}$$

Dividing numerator and denominator by 2ε and passing to the limit as $\varepsilon \to 0+$, we have

$$\begin{aligned} F_{X|Y}(x \mid y) &= \frac{\int_{-\infty}^{x} f(u,y)\, du}{f_2(y)} \\ &= \int_{-\infty}^{x} \left\{\frac{f(u,y)}{f_2(y)}\right\} du. \end{aligned}$$

It follows that there exists a conditional PDF of X, given $Y = y$, that is expressed by

$$f_{X|Y}(x \mid y) = \frac{f(x,y)}{f_2(y)}, \qquad f_2(y) > 0.$$

We have thus proved the following theorem.

Theorem 6. Let f be the PDF of an RV (X, Y) of the continuous type, and let f_2 be the marginal PDF of Y. At every point (x, y) at which f is continuous and $f_2(y) > 0$ and is continuous, the conditional PDF of X, given $Y = y$, exists and is expressed by

$$f_{X|Y}(x \mid y) = \frac{f(x,y)}{f_2(y)}. \tag{15}$$

Note that

$$\int_{-\infty}^{x} f(u,y)\, du = f_2(y) F_{X|Y}(x \mid y),$$

so that

$$F_1(x) = \int_{-\infty}^{\infty}\left\{\int_{-\infty}^{x} f(u,y)\,du\right\}dy = \int_{-\infty}^{\infty} f_2(y)F_{X|Y}(x \mid y)\,dy, \tag{16}$$

where F_1 is the marginal DF of X.

It is clear that similar definitions may be made for the conditional DF and conditional PDF of the RV Y, given $X = x$, and an analog of Theorem 6 holds.

In the general case, let $(X_1, X_2, \ldots, X_n)$ be an n-dimensional RV of the continuous type with PDF $f_{X_1,X_2,\ldots,X_n}(x_1, x_2, \ldots, x_n)$. Also, let $\{i_1 < i_2 < \cdots < i_k, j_1 < j_2 < \cdots < j_l\}$ be a subset of $\{1, 2, \ldots, n\}$. Then

$$F(x_{i_1}, x_{i_2}, \ldots, x_{i_k} \mid x_{j_1}, x_{j_2}, \ldots, x_{j_l}) \tag{17}$$
$$= \frac{\int_{-\infty}^{x_{i_1}} \cdots \int_{-\infty}^{x_{ik}} f_{X_{i1}\ldots,X_{ik},X_{j1},\ldots,X_{jl}}(u_{i_1}, \ldots, u_{i_k}, x_{j_1}, \ldots, x_{j_l}) \prod_{p=1}^{k} du_{i_p}}{\int_{-\infty}^{\infty} \cdots \int_{-\infty}^{\infty} f_{X_{i1},\ldots,X_{1k},X_{j1}\ldots,X_{jl}}(u_{i_1}, \ldots, u_{i_k}, x_{j_1}, \ldots, x_{j_l}) \prod_{p=1}^{k} du_{i_p}}$$

provided that the denominator exceeds 0. Here $f_{X_{i_1},\ldots,X_{i_k},X_{j_1},\ldots,X_{j_l}}$ is the joint marginal PDF of $(X_{i_1}, X_{i_2}, \ldots, X_{i_k}, X_{j_1}, X_{j_2}, \ldots, X_{j_l})$. The conditional densities are obtained in a similar manner.

The case in which $(X_1, X_2, \ldots, X_n)$ is of the discrete type is similarly treated.

Example 7. For the joint PDF of Example 5 we have

$$f_{Y|X}(y \mid x) = \frac{f(x,y)}{f_1(x)} = \frac{1}{1-x}, \qquad x < y < 1,$$

so that the conditional PDF $f_{Y|X}$ is uniform on $(x, 1)$. Also,

$$f_{X|Y}(x \mid y) = \frac{1}{y}, \qquad 0 < x < y,$$

which is uniform on $(0, y)$. Thus

$$P\left\{Y \geq \frac{1}{2} \mid x = \frac{1}{2}\right\} = \int_{1/2}^{1} \frac{1}{1-\frac{1}{2}}\,dy = 1,$$
$$P\left\{X \geq \frac{1}{3} \middle| y = \frac{2}{3}\right\} = \int_{1/3}^{2/3} \frac{1}{\frac{2}{3}}\,dx = \frac{1}{2}.$$

We conclude this section with a discussion of a technique called *truncation*. We consider two types of truncation each with a different objective. In probabilistic modeling we use *truncated distributions* when sampling from an incomplete population.

Definition 11. Let X be an RV on $(\Omega, \mathcal{S}, P)$, and $T \in \mathfrak{B}$ such that $0 < P\{X \in T\} < 1$. Then the conditional distribution $P\{X \leq x \mid X \in T\}$, defined for any real x, is called the *truncated distribution* of X.

If X is a discrete RV with PMF $p_i = P\{X = x_i\},\ i = 1,2,\ldots$, the truncated distribution of X is given by

$$P\{X = x_i \mid X \in T\} = \frac{P\{X = x_i, X \in T\}}{P\{X \in T\}} = \begin{cases} \dfrac{p_i}{\sum_{x_j \in T} p_j} & \text{if } x_i \in T, \\ 0 & \text{otherwise.} \end{cases} \tag{18}$$

If X is of the continuous type with PDF f, then

$$P\{X \le x \mid X \in T\} = \frac{P\{X \le x, X \in T\}}{P\{X \in T\}} = \frac{\int_{(-\infty,x]\cap T} f(y)\,dy}{\int_T f(y)\,dy}. \tag{19}$$

The PDF of the truncated distribution is given by

$$h(x) = \begin{cases} \dfrac{f(x)}{\int_T f(y)\,dy}, & x \in T, \\ 0, & x \notin T. \end{cases} \tag{20}$$

Here T is not necessarily a bounded set of real numbers. If we write Y for the RV with distribution function $P\{X \le x \mid X \in T\}$, then Y has support T.

Example 8. Let X be an RV with *standard normal* PDF

$$f(x) = \frac{1}{\sqrt{2\pi}} e^{-x^2/2}.$$

Let $T = (-\infty, 0]$. Then $P\{X \in T\} = 1/2$, since X is symmetric and continuous. For the truncated PDF, we have

$$h(x) = \begin{cases} 2f(x), & -\infty < x \le 0, \\ 0, & x > 0. \end{cases}$$

Some other examples are the truncated Poisson distribution

$$P\{X = k\} = \frac{e^{-\lambda}}{1 - e^{-\lambda}} \frac{\lambda^k}{k!}, \quad k = 1,2,\ldots,$$

where $T = \{X \ge 1\}$, and the truncated uniform distribution

$$f(x) = 1/\theta, \quad 0 < x < \theta, \text{ and } = 0 \text{ otherwise,}$$

where $T = \{X < \theta\}$, $\theta > 0$.

The second type of truncation is very useful in probability limit theory specially when the DF F in question does not have a finite mean. Let $a < b$ be finite real numbers. Define RV X^* by

$$X^* = \begin{cases} X, & \text{if } a \le X \le b \\ 0, & \text{if } X < a, \text{ or } X > b. \end{cases}$$

This method produces an RV for which $P(a \le X^* \le b) = 1$ so that X^* has moments of all orders. The special case when $b = c > 0$ and $a = -c$ is quite useful in probability limit theory when we wish to approximate X through bounded rvs. We say that X^c is *X truncated at c* if $X^c = X$ for $|X| \le c$, and $= 0$ for $|X| > c$. Then $E|X^c|^k \le c^k$. Moreover,

$$P\{X \neq X^c\} = P\{|X| > c\}$$

so that c can be selected sufficiently large to make $P\{|X| > c\}$ arbitrarily small. For example, if $E|X|^2 < \infty$ then

$$P\{|X| > c\} \le E|X|^2/c^2$$

and given $\varepsilon > 0$, we can choose c such that $E|X|^2/c^2 < \varepsilon$.

The distribution of X^c is no longer the truncated distribution $P\{X \le x \mid |X| \le c\}$. In fact,

$$F^c(y) = \begin{cases} 0, & y \le -c \\ F(y) - F(-c), & -c < y < 0 \\ 1 - F(c) + F(y), & 0 \le y < c \\ 1 & y > c, \end{cases}$$

where F is the DF of X and F^c, that of X^c.

A third type of truncation, sometimes called Winsorization, sets

$$X^* = X, \text{ if } a < X < b, \ = a \text{ if } X \le a, \text{ and } = b \text{ if } X \ge b.$$

This method also produces an RV for which $P(a \le X^* \le b) = 1$, moments of all orders for X^* exist but its DF is given by

$$F^*(y) = 0 \text{ for } y < a, \ = F(y) \text{ for } a \le y < b, \ = 1 \text{ for } y \ge b.$$

PROBLEMS 4.2

1. Let $F(x,y) = 1$ if $x + 2y \ge 1$, and $= 0$ if $x + 2y < 1$. Does F define a DF in the plane?
2. Let T be a closed triangle in the plane with vertices $(0,0)$, $(0,\sqrt{2})$, and $(\sqrt{2},\sqrt{2})$. Let $F(x,y)$ denote the elementary area of the intersection of T with $\{(x_1,x_2): x_1 \le x, x_2 \le y\}$. Show that F defines a DF in the plane, and find its marginal DFs.
3. Let (X,Y) have the joint PDF f defined by $f(x,y) = 1/2$ inside the square with corners at the points $(1,0)$, $(0,1)$, $(-1,0)$, and $(0,-1)$ in the (x,y)-plane, and $= 0$ otherwise. Find the marginal PDFs of X and Y and the two conditional PDFs.
4. Let $f(x,y,z) = e^{-x-y-z}$, $x > 0$, $y > 0$, $z > 0$, and $= 0$ otherwise, be the joint PDF of (X,Y,Z). Compute $P\{X < Y < Z\}$ and $P\{X = Y < Z\}$.
5. Let (X,Y) have the joint PDF $f(x,y) = \frac{4}{3}[xy + (x^2/2)]$ if $0 < x < 1$, $0 < y < 2$, and $= 0$ otherwise. Find $P\{Y < 1 \mid X < 1/2\}$.

6. For DFs $F, F_1, F_2, \ldots, F_n$ show that

$$1 - \sum_{i=1}^{n} \{1 - F_i(x_i)\} \le F(x_1, x_2, \ldots, x_n) \le \min_{1 \le i \le n} F_i(x_i)$$

for all real numbers $x_1, x_2, \ldots, x_n$ if and only if F_i's are marginal DFs of F.

7. For the *bivariate negative binomial* distribution

$$P\{X = x, Y = y\} = \frac{(x+y+k-1)!}{x!\,y!\,(k-1)!} p_1^x p_2^y (1 - p_1 - p_2)^k,$$

where $x, y = 0, 1, 2, \ldots, k \ge 1$ is an integer, $0 < p_1 < 1, 0 < p_2 < 1$, and $p_1 + p_2 < 1$, find the marginal PMFs of X and Y and the conditional distributions.

In Problems 8–10 the bivariate distributions considered are not unique generalizations of the corresponding univariate distributions.

8. For the *bivariate Cauchy* RV (X, Y) with PDF

$$f(x,y) = \frac{c}{2\pi}(c^2 + x^2 + y^2)^{-3/2}, -\infty < x < \infty, -\infty < y < \infty, c > 0,$$

find the marginal PDFs of X and Y. Find the conditional PDF of Y given $X = x$.

9. For the *bivariate beta* RV (X, Y) with PDF

$$f(x,y) = \frac{\Gamma(p_1 + p_2 + p_3)}{\Gamma(p_1)\Gamma(p_2)\Gamma(p_3)} x^{p_1 - 1} y^{p_2 - 1} (1 - x - y)^{p_3 - 1},$$
$$x \ge 0, y \ge 0, x + y \le 1,$$

where p_1, p_2, p_3 are positive real numbers, find the marginal PDFs of X and Y and the conditional PDFs. Find also the conditional PDF of $Y/(1-X)$, given $X = x$.

10. For the *bivariate gamma* RV (X, Y) with PDF

$$f(x,y) = \frac{\beta^{\alpha+\gamma}}{\Gamma(\alpha)\Gamma(\gamma)} x^{\alpha-1}(y-x)^{\gamma-1} e^{-\beta y}, \qquad 0 < x < y;\ \alpha, \beta, \gamma > 0,$$

find the marginal PDFs of X and Y and the conditional PDFs. Also, find the conditional PDF of $Y - X$, given $X = x$, and the conditional distribution of X/Y, given $Y = y$.

11. For the *bivariate hypergeometric* RV (X, Y) with PMF

$$P\{X = x, Y = y\} = \binom{N}{n}^{-1} \binom{Np_1}{x} \binom{Np_2}{y} \binom{N - Np_1 - Np_2}{n - x - y},$$
$$x, y = 0, 1, 2, \ldots, n,$$

where $x \le Np_1$, $y \le Np_2$, $n - x - y \le N(1 - p_1 - p_2)$, N, n integers with $n \le N$, and $0 < p_1 < 1, 0 < p_2 < 1$ so that $p_1 + p_2 \le 1$, find the marginal PMFs of X and Y and the conditional PMFs.

12. Let X be an RV with PDF $f(x) = 1$ if $0 \le x \le 1$, and $= 0$ otherwise. Let $T = \{x\colon 1/3 < x \le 1/2\}$. Find the PDF of the truncated distribution of X, its means, and its variance.

13. Let X be an RV with PMF

$$P\{X = x\} = e^{-\lambda}\frac{\lambda^x}{x!}, \quad x = 0, 1, 2, \ldots, \lambda > 0.$$

Suppose that the value $x = 0$ cannot be observed. Find the PMF of the truncated RV, its mean, and its variance.

14. Is the function

$$f(x,y,z,u) = \begin{cases} \exp(-u), & 0 < x < y < z < u < \infty \\ 0 & \text{elsewhere} \end{cases}$$

a joint density function? If so, find $P(X \le 7)$, where (X,Y,Z,U) is a random variable with density f.

15. Show that the function defined by

$$f(x,y,z,u) = \frac{24}{(1+x+y+z+u)^5}, \quad x > 0, y > 0, z > 0, u > 0$$

and 0 elsewhere is a joint density function.

(a) Find $P(X > Y < Z > U)$.

(b) Find $P(X + Y + Z + U \ge 1)$.

16. Let (X,Y) have joint density function f and joint distribution function F. Suppose that

$$f(x_1,y_1)f(x_2,y_2) \le f(x_1,y_2)f(x_2,y_1)$$

holds for $x_1 \le a \le x_2$ and $y_1 \le b \le y_2$. Show that

$$F(a,b) \le F_1(a)F_2(b).$$

17. Suppose (X,Y,Z) are jointly distributed with density

$$f(x,y,z) = \begin{cases} g(x)g(y)g(z), & x > 0, y > 0, z > 0 \\ 0 & \text{elsewhere.} \end{cases}$$

Find $P(X > Y > Z)$. Hence find the probability that $(x,y,z) \notin \{X > Y > Z\}$ or $\{X < Y < Z\}$. (Here g is density function on $\mathcal{R}$.)

4.3 INDEPENDENT RANDOM VARIABLES

We recall that the joint distribution of a multiple RV uniquely determines the marginal distributions of the component random variables, but, in general, knowledge of marginal distributions is not enough to determine the joint distribution. Indeed, it is quite possible to have an infinite collection of joint densities f_α with given marginal densities.

Example 1. (Gumbel [39]). Let f_1,f_2,f_3 be three PDFs with corresponding DFs F_1,F_2,F_3, and let α be a constant, $|\alpha|\le 1$. Define

$$\begin{aligned} f_\alpha(x_1,x_2,x_3) = {} & f_1(x_1)f_2(x_2)f_3(x_3) \\ & \cdot\{1+\alpha[2F_1(x_1)-1][2F_2(x_2)-1][2F_3(x_3)-1]\}. \end{aligned}$$

We show that F_α is a PDF for each α in $[-1,1]$ and that the collection of densities $\{f_\alpha; -1\le\alpha\le 1\}$ has the same marginal densities f_1,f_2,f_3. First note that

$$|[2F_1(x_1)-1][2F_2(x_2)-1][2F_3(x_3)-1]|\le 1,$$

so that

$$1+\alpha[2F_1(x_1)-1][2F_2(x_2)-1][2F_3(x_3)-1]\ge 0.$$

Also,

$$\begin{aligned} & \iiint f_\alpha(x_1,x_2,x_3)\,dx_1\,dx_2\,dx_3 \\ & = 1+\alpha\left(\int [2F_1(x_1)-1]f_1(x_1)\,dx_1\right)\left(\int [2F_2(x_2)-1]f_2(x_2)\,dx_2\right) \\ & \quad \cdot\left(\int [2F_3(x_3)-1]f_3(x_3)\,dx_3\right) \\ & = 1+\alpha\{[F_1^2(x_1)|_{-\infty}^{\infty}-1][F_2^2(x_2)|_{-\infty}^{\infty}-1][F_3^2(x_3)|_{-\infty}^{\infty}-1]\} \\ & = 1. \end{aligned}$$

It follows that f_α is a density function. That f_1,f_2,f_3 are the marginal densities of f_α follows similarly.

In this section we deal with a very special class of distributions in which the marginal distributions uniquely determine the joint distribution of a multiple RV. First we consider the bivariate case.

Let $F(x,y)$ and $F_1(x),F_2(y)$, respectively, be the joint DF of (X,Y) and the marginal DFs of X and Y.

Definition 1. We say that X and Y are *independent* if and only if

$$F(x,y)=F_1(x)F_2(y) \quad \text{for all } (x,y)\in\mathcal{R}_2. \tag{1}$$

Lemma 1. If X and Y are independent and $a<c, b<d$ are real numbers, then

$$P\{a<X\le c, b<Y\le d\}=P\{a<X\le c\}P\{b<Y\le d\}. \tag{2}$$

Theorem 1. (a) A necessary and sufficient condition for RVs X,Y of the discrete type to be independent is that

$$P\{X=x_i,Y=y_j\}=P\{X=x_i\}P\{Y=y_j\} \tag{3}$$

for all pairs (x_i, y_j). (b) Two RVs X and Y of the continuous type are independent if and only if

$$f(x,y) = f_1(x)f_2(y) \quad \text{for all } (x,y) \in \mathcal{R}_2, \tag{4}$$

where f, f_1, f_2, respectively, are the joint and marginal densities of X and Y, and f is everywhere continuous.

Proof. (a) Let X, Y be independent. Then from Lemma 1, letting $a \to c$ and $b \to d$, we get

$$P\{X = c, Y = d\} = P\{X = c\}P\{Y = d\}.$$

Conversely,

$$F(x,y) = \sum_B P\{X = x_i, Y = y_j\},$$

where

$$B = \{(i,j) \colon x_i \leq x, y_j \leq y\}.$$

Then

$$\begin{aligned} F(x,y) &= \sum_B P\{X = x_i\}P\{Y = y_j\} \\ &= \sum_{x_i \leq x} [\sum_{y_j \leq y} P\{Y = y_j\}] P\{X = x_i\} = F(x)F(y). \end{aligned}$$

The proof of part (b) is left as an exercise.

Corollary. Let X and Y be independent RVs. Then $F_{Y|X}(y \mid x) = F_Y(y)$ for all y, and $F_{X|Y}(x \mid y) = F_X(x)$ for all x.

Theorem 2. The RVs X and Y are independent if and only if

$$P\{X \in A_1, Y \in A_2\} = P\{X \in A_1\}P\{Y \in A_2\} \tag{5}$$

for all Borel sets A_1 on the x-axis and A_2 on the y-axis.

Theorem 3. Let X and Y be independent RVs and f and g be Borel-measurable functions. Then $f(X)$ and $g(Y)$ are also independent.

Proof. We have

$$\begin{aligned} P\{f(X) \le x, g(Y) \le y\} &= P\{X \in f^{-1}(-\infty, x], Y \in g^{-1}(-\infty, y]\} \\ &= P\{X \in f^{-1}(-\infty, x]\} P\{Y \in g^{-1}(-\infty, y]\} \\ &= P\{f(X) \le x\} P\{g(Y) \le y\}. \end{aligned}$$

Note that a degenerate RV is independent of any RV.

Example 2. Let X and Y be jointly distributed with PDF

$$f(x,y) = \begin{cases} \dfrac{1+xy}{4}, & |x| < 1, |y| < 1, \\ 0, & \text{otherwise.} \end{cases}$$

Then X and Y are not independent since $f_1(x) = 1/2, |x| < 1$, and $f_2(y) = 1/2, |y| < 1$ are the marginal densities of X and Y, respectively. However, the RVs X^2 and Y^2 are independent. Indeed,

$$\begin{aligned} P\{X^2 \le u, Y^2 \le v\} &= \int_{-v^{1/2}}^{v^{1/2}} \int_{-u^{1/2}}^{u^{1/2}} f(x,y)\,dx\,dy \\ &= \frac{1}{4} \int_{-v^{1/2}}^{v^{1/2}} \left[\int_{-u^{1/2}}^{u^{1/2}} (1+xy)\,dx \right] dy \\ &= u^{1/2} v^{1/2} \\ &= P\{X^2 \le u\} P\{Y^2 \le v\}. \end{aligned}$$

Note that $\phi(X^2)$ and $\psi(Y^2)$ are independent where ϕ and ψ are Borel–measurable functions. But X is not a Borel-measurable function of X^2.

Example 3. We return to Buffon's needle problem, discussed in Examples 1.2.9 and 1.3.7. Suppose that the RV R, which represents the distance from the center of the needle to the nearest line, is uniformly distributed on $(0, l]$. Suppose further that Θ, the angle that the needle forms with this line, is uniformly distributed on $[0, \pi)$. If R and Θ are assumed to be independent, the joint PDF is given by

$$f_{R,\Theta}(r,\theta) = f_R(r) f_\Theta(\theta) = \begin{cases} \dfrac{1}{l} \cdot \dfrac{1}{\pi} & \text{if } 0 < r \le l, 0 \le \pi, \\ 0 & \text{otherwise.} \end{cases}$$

The needle will intersect the nearest line if and only if

$$\frac{l}{2} \sin\Theta \ge R.$$

Therefore, the required probability is given by

$$P\left\{\sin\Theta \geq \frac{2R}{l}\right\} = \int_0^{\pi}\int_0^{(\frac{l}{2})\sin\theta} f_{R,\Theta}(r,\theta)\,dr\,d\theta$$
$$= \frac{1}{l\pi}\int_0^{\pi}\frac{l}{2}\sin\theta\,d\theta = \frac{1}{\pi}.$$

Definition 2. A collection of jointly distributed RVs $X_1, X_2, \ldots, X_n$ is said to be *mutually or completely independent* if and only if

$$F(x_1, x_2, \ldots, x_n) = \prod_{i=1}^{n} F_i(x_i), \quad \text{for all } (x_1, x_2, \ldots, x_n) \in \mathcal{R}_n, \tag{6}$$

where F is the joint DF of $(X_1, X_2, \ldots, X_n)$, and $F_i (i = 1, 2, \ldots, n)$ is the marginal DF of X_i. $X_1, \ldots, X_n$, which are said to be *pairwise independent* if and only if every pair of them are independent.

It is clear that an analog of Theorem 1 holds, but we leave the reader to construct it.

Example 4. In Example 1 we cannot write

$$f_\alpha(x_1, x_2, x_3) = f_1(x_1) f_2(x_2) f_3(x_3)$$

except when $\alpha = 0$. It follows that X_1, X_2, and X_3 are not independent except when $\alpha = 0$.

The following result is easy to prove.

Theorem 4. If $X_1, X_2, \ldots, X_n$ are independent, every subcollection $X_{i_1}, X_{i_2}, \ldots, X_{i_k}$ of $X_1, X_2, \ldots, X_n$ is also independent.

Remark 1. It is quite possible for RVs $X_1, X_2, \ldots X_n$ to be *pairwise independent* without being mutually independent. Let (X, Y, Z) have the joint PMF defined by

$$P\{X = x, Y = y, Z = z\} = \begin{cases} \dfrac{3}{16} & \text{if } (x,y,z) \in \{(0,0,0),(0,1,1), \\ & \quad (1,0,1),(1,1,0)\}, \\ \dfrac{1}{16} & \text{if } (x,y,z) \in \{(0,0,1),(0,1,0), \\ & \quad (1,0,0),(1,1,1)\}. \end{cases}$$

Clearly, X, Y, Z are not independent (why?). We have

$$P\{X=x,Y=y\}=\frac{1}{4}, \quad (x,y)\in\{(0,0),(0,1),(1,0),(1,1)\},$$

$$P\{Y=y,Z=z\}=\frac{1}{4}, \quad (y,z)\in\{(0,0),(0,1),(1,0),(1,1)\},$$

$$P\{X=x,Z=z\}=\frac{1}{4}, \quad (x,z)\in\{(0,0),(0,1),(1,0),(1,1)\},$$

$$P\{X=x\}=\frac{1}{2}, \qquad x=0, x=1,$$

$$P\{Y=y\}=\frac{1}{2}, \qquad y=0, y=1,$$

$$P\{Z=z\}=\frac{1}{2}, \qquad z=0, z=1.$$

It follows that X and Y, Y and Z, and X and Z are pairwise independent.

Definition 3. A sequence $\{X_n\}$ of RVs is said to be independent if for every $n=2,3,4,\ldots$ the RVs $X_1,X_2,\ldots,X_n$ are independent.

Similarly, one can speak of an independent family of RVs.

Definition 4. We say that RVs X and Y are *identically distributed* if X and Y have the same DF, that is,

$$F_X(x)=F_Y(x) \qquad \text{for all } x\in\mathcal{R}$$

where F_X and F_Y are the DF's of X and Y, respectively.

Definition 5. We say that $\{X_n\}$ is a sequence of *independent, identically distributed* (iid) RVs with common law $\mathcal{L}(X)$ if $\{X_n\}$ is an independent sequence of RVs and the distribution of $X_n (n=1,2\ldots)$ is the same as that of X.

According to Definition 4, X and Y are identically distributed if and only if they have the same distribution. It does not follow that $X=Y$ with probability 1 (see Problem 7). If $P\{X=Y\}=1$, we say that X and Y are *equivalent* RVs. All Definition 4 says is that X and Y are identically distributed if and only if

$$P\{X\in A\}=P\{Y\in A\} \qquad \text{for all } A\in\mathfrak{B}.$$

Nothing is said about the equality of events $\{X\in A\}$ and $\{Y\in A\}$.

Definition 6. Two multiple RVs $(X_1,X_2,\ldots,X_m)$ and $(Y_1,Y_2,\ldots,Y_n)$ are said to be independent if

$$F(x_1,x_2,\ldots,x_m,y_1,y_2,\ldots,y_n)=F_1(x_1,x_2,\ldots,x_m)F_2(y_1,y_2,\ldots,y_n) \tag{7}$$

for all $(x_1,x_2,\ldots,x_m,y_1,y_2,\ldots,y_n) \in \mathcal{R}_{m+n}$, where F, F_1, F_2 are the joint distribution functions of $(X_1,X_2,\ldots,X_m,Y_1,Y_2,\ldots,Y_n)$, $(X_1,X_2,\ldots,X_m)$, and $(Y_1,Y_2,\ldots,Y_n)$, respectively.

Of course, the independence of $(X_1,X_2,\ldots,X_m)$ and $(Y_1,Y_2,\ldots,Y_n)$ does not imply the independence of components $X_1,X_2,\ldots,X_m$ of $\mathbf{X}$ or components $Y_1,Y_2,\ldots,Y_n$ of $\mathbf{Y}$.

Theorem 5. Let $\mathbf{X} = (X_1,X_2,\ldots,X_m)$ and $\mathbf{Y} = (Y_1,Y_2,\ldots,Y_n)$ be independent RVs. Then the component X_j of $\mathbf{X}(j = 1,2,\ldots,m)$ and the component Y_k of $\mathbf{Y}(k = 1,2,\ldots,n)$ are independent RVs. If h and g are Borel-measurable functions, $h(X_1,X_2,\ldots,X_m)$ and $g(Y_1,Y_2,\ldots,Y_n)$ are independent.

Remark 2. It is possible that an RV X may be independent of Y and also of Z, but X may not be independent of the random vector (Y,Z). See the example in Remark 1.

Let $X_1,X_2,\ldots,X_n$ be independent and identically distributed RVs with common DF F. Then the joint DF G of $(X_1,X_2,\ldots,X_n)$ is given by

$$G(x_1,x_2,\ldots,x_n) = \prod_{j=1}^{n} F(x_j).$$

We note that for any of the $n!$ permutations $(x_{i_1},x_{i_2},\ldots,x_{i_n})$ of $(x_1,x_2,\ldots,x_n)$

$$G(x_1,x_2,\ldots,x_n) = \prod_{j=1}^{n} F(x_{i_j}) = G(x_{i_1},x_{i_2},\ldots,x_{i_n})$$

so that G is a symmetric function of $x_1,x_2,\ldots,x_n$. Thus $(X_1,X_2,\ldots,X_n) \stackrel{d}{=} (X_{i_1},X_{i_2},\ldots,X_{i_n})$, where $\mathbf{X} \stackrel{d}{=} \mathbf{Y}$ means that $\mathbf{X}$ and $\mathbf{Y}$ are identically distributed RVs.

Definition 7. The RVs $X_1,X_2,\ldots,X_n$ are said to be *exchangeable* if

$$(X_1,X_2,\ldots,X_n) \stackrel{d}{=} (X_{i_1},X_{i_2},\ldots,X_{i_n})$$

for all $n!$ permutations $(i_1,i_2,\ldots,i_n)$ of $(1,2,\ldots,n)$. The RVs in the sequence $\{X_n\}$ are said to be exchangeable if $X_1,X_2,\ldots,X_n$ are exchangeable for each n.

Clearly if $X_1,X_2,\ldots,X_n$ are exchangeable, then X_i are identically distributed but not necessarily independent.

Example 5. Suppose $X,\ Y,\ Z$ have joint PDF

$$f(x,y,z) = \begin{cases} \frac{2}{3}(x+y+z), & 0<x<1,\ 0<y<1, 0<z<1 \\ 0, & \text{otherwise.} \end{cases}$$

Then $X,\ Y,\ Z$ are exchangeable but not independent.

Example 6. Let $X_1, X_2, \ldots, X_n$ be iid RVs. Let $S_n = \sum_{j=1}^{n} X_j$, $n = 1, 2, \ldots$ and $Y_k = X_k - S_n/n$, $k = 1, 2, \ldots, n-1$. Then $Y_1, Y_2, \ldots, Y_{n-1}$ are exchangeable.

Theorem 6. Let X, Y be exchangeable RVs. Then $X - Y$ has a symmetric distribution.

Definition 8. Let X be an RV, and let X' be an RV that is independent of X and $X' \stackrel{d}{=} X$. We call the RV

$$X^s = X - X'$$

the *symmetrized* X.

In view of Theorem 6, X^s is symmetric about 0 so that

$$P\{X^s \geq 0\} \geq \frac{1}{2} \quad \text{and} \quad P\{X^s \leq 0\} \geq \frac{1}{2}.$$

If $E|X| < \infty$, then $E|X^s| \leq 2E|X| < \infty$, and $EX^s = 0$.

The technique of symmetrization is an important tool in the study of probability limit theorems. We will need the following result later. The proof is left to the reader.

Theorem 7. For $\varepsilon > 0$,

(a) $P\{|X^s| > \varepsilon\} \leq 2P\{|X| > \varepsilon/2\}$.
(b) If $a \geq 0$ such that $P\{X \geq a\} \leq 1 - p$ and $P\{X \leq -a\} \leq 1 - p$, then

$$P\{|X^s| \geq \varepsilon\} \geq P\{|X| > a + \varepsilon\},$$

for $\varepsilon > 0$.

PROBLEMS 4.3

1. Let A be a set of k numbers, and Ω be the set of all ordered samples of size n from A with replacement. Also, let $\mathcal{S}$ be the set of all subsets of Ω, and P be a probability defined on $\mathcal{S}$. Let $X_1, X_2, \ldots, X_n$ be RVs defined on $(\Omega, \mathcal{S}, P)$ by setting

$$X_i(a_1, a_2, \ldots, a_n) = a_i, \qquad (i = 1, 2, \ldots, n).$$

Show that $X_1, X_2, \ldots, X_n$ are independent if and only if each sample point is equally likely.

2. Let X_1, X_2 be iid RVs with common PMF

$$P\{X = \pm 1\} = \frac{1}{2}.$$

Write $X_3 = X_1 X_2$. Show that X_1, X_2, X_3 are pairwise independent but not independent.

3. Let (X_1,X_2,X_3) be an RV with joint PMF

$$\begin{aligned} f(x_1,x_2,x_3) &= \frac{1}{4} \qquad \text{if } (x_1,x_2,x_3) \in A, \\ &= 0 \qquad \text{otherwise,} \end{aligned}$$

where

$$A = \{(1,0,0),(0,1,0),(0,0,1),(1,1,1)\}.$$

Are X_1,X_2,X_3 independent? Are X_1,X_2,X_3 pairwise independent? Are X_1+X_2 and X_3 independent?

4. Let X and Y be independent RVs such that XY is degenerate at $c \neq 0$. That is, $P(XY=c)=1$. Show that X and Y are also degenerate.

5. Let $(\Omega,\mathcal{S},P)$ be a probability space and $A,B \in \mathcal{S}$. Define X and Y so that

$$X(\omega) = I_A(\omega), \qquad Y(\omega) = I_B(\omega) \text{ for all } \omega \in \Omega.$$

Show that X and Y are independent if and only if A and B are independent.

6. Let $X_1,X_2,\ldots,X_n$ be a set of exchangeable RVs. Then

$$E\left\{\frac{X_1+X_2+\cdots+X_k}{X_1+X_2+\cdots+X_n}\right\} = \frac{k}{n}, \qquad 1 \leq k \leq n.$$

7. Let X and Y be identically distributed. Construct an example to show that X and Y need not be equal, that is, $P\{X=Y\}$ need not equal 1.

8. Prove Lemma 1.

9. Let $X_1,X_2,\ldots,X_n$ be RVs with joint PDF f, and let f_j be the marginal PDF of $X_j (j = 1,2,\ldots,n)$. Show that $X_1,X_2,\ldots,X_n$ are independent if and only if

$$f(x_1,x_2,\ldots,x_n) = \prod_{j=1}^{n} f_j(x_j) \qquad \text{for all } (x_1,x_2,\ldots x_n) \in \mathcal{R}_n.$$

10. Suppose two buses, A and B, operate on a route. A person arrives at a certain bus stop on this route at time 0. Let X and Y be the arrival times of buses A and B, respectively, at this bus stop. Suppose X and Y are independent and have density functions given, respectively, by

$$f_1(x) = \frac{1}{a}, \quad 0 \leq x \leq a, \text{ and 0 elsewhere,}$$

$$f_2(y) = \frac{1}{b}, \quad 0 \leq y \leq b, \text{ and 0 otherwise.}$$

What is the probability that bus A will arrive before bus B?

11. Consider two batteries, one of Brand A and the other of Brand B. Brand A batteries have a length of life with density function

$$f(x) = 3\lambda x^2 \exp(-\lambda x^3), \quad x > 0, \text{ and 0 elsewhere,}$$

whereas Brand B batteries have a length of life with density function given by

$$g(x) = 3\mu y^2 \exp(-\mu y^3), \quad y > 0, \text{ and } 0 \text{ elsewhere}$$

Brand A and Brand B batteries operate independently and are put to a test. What is the probability that Brand B battery will outlast Brand A? In particular, what is the probability if $\lambda = \mu$?

12. (a) Let (X, Y) have joint density f. Show that X and Y are independent if and only if for some constant $k > 0$ and nonnegative functions f_1 and f_2

$$f(x,y) = kf_1(x)f_2(y)$$

for all $x, y \in \mathcal{R}$.

(b) Let $A = \{f_X(x) > 0\}$, $B = \{f_Y(y) > 0\}$, and f_X, f_Y are marginal densities of X and Y, respectively. Show that if X and Y are independent then $\{f > 0\} = A \times B$.

13. If ϕ is the CF of X, show that the CF of X^s is real and even.

14. Let X, Y be jointly distributed with PDF $f(x,y) = (1 - x^3 y)/4$ for $|x| < 1$, $|y| < 1$, and $= 0$ otherwise. Show that $X \stackrel{d}{=} Y$ and $X - Y$ has a symmetric distribution.

4.4 FUNCTIONS OF SEVERAL RANDOM VARIABLES

Let $X_1, X_2, \ldots, X_n$ be RVs defined on a probability space $(\Omega, \mathcal{S}, P)$. In practice we deal with functions of $X_1, X_2, \ldots, X_n$ such as $X_1 + X_2$, $X_1 - X_2$, X_1X_2, $\min(X_1, \ldots, X_n)$, and so on. Are these also RVs? If so, how do we compute their distribution given the joint distribution of $X_1, X_2, \ldots, X_n$?

What functions of $(X_1, X_2, \ldots, X_n)$ are RVs?

Theorem 1. Let $g : \mathcal{R}_n \to \mathcal{R}_m$ be a Borel-measurable function, that is, if $B \in \mathfrak{B}_m$, then $g^{-1}(B) \in \mathfrak{B}_n$. If $\mathbf{X} = (X_1, X_2, \ldots, X_n)$ is an n-dimensional RV ($n \geq 1$), then $g(\mathbf{X})$ is an m-dimensional RV.

Proof. For $B \in \mathfrak{B}_m$

$$\{g(X_1, X_2, \ldots, X_n) \in B\} = \{(X_1, X_2, \ldots, X_n) \in g^{-1}(B)\},$$

and, since $g^{-1}(B) \in \mathfrak{B}_n$, it follows that $\{(X_1, X_2, \ldots, X_n) \in g^{-1}(B)\} \in \mathcal{S}$, which concludes the proof.

In particular, if $g : R_n \to R_m$ is a continuous function, then $g(X_1, X_2, \ldots, X_n)$ is an RV.

How do we compute the distribution of $g(X_1, X_2, \ldots, X_n)$? There are several ways to go about it. We first consider the method of distribution functions. Suppose that $Y = g(X_1, \ldots, X_n)$ is real-valued, and let $y \in \mathcal{R}$. Then

$$P\{Y \le y\} = P(g(X_1,\ldots,X_n) \le y)$$
$$= \begin{cases} \displaystyle\sum_{\{(x_1,\ldots,x_n):g(x_1,\ldots,x_n)\le y\}} P(X_1 = x_1,\ldots,x_n = X_n) & \text{in the discrete case} \\ \displaystyle\int_{\{(x_1,\ldots,x_n):g(x_1,\ldots,x_n)\le y\}} f(x_1,\ldots,x_n)dx_1\ldots dx_n & \text{in the continuous case,} \end{cases}$$

where in the continuous case f is the joint PDF of $(X_1,\ldots,X_n)$.

In the continuous case we can obtain the PDF of $Y = g(X_1,\ldots,X_n)$ by differentiating the DF $P\{Y \le y\}$ with respect to y provided that Y is also of the continuous type. In the discrete case it is easier to compute $P\{g(X_1,\ldots,X_n) = y\}$.

We take a few examples,

Example 1. Consider the bivariate negative binomial distribution with PMF

$$P\{X = x, Y = y\} = \frac{(x+y+k-1)!}{x!\,y!\,(k-1)!} p_1^x p_2^y (1-p_1-p_2)^k,$$

where $x,y = 0,1,2,\ldots; k \ge 1$ is an integer; $p_1,p_2 \in (0,1)$; and $p_1 + p_2 < 1$. Let us find the PMF of $U = X+Y$. We introduce an RV $V = Y$ (see Remark 1 below) so that $u = x+y, v = y$ represents a one-to-one mapping of $A = \{(x,y) : x,y = 0,1,2,\ldots\}$ onto the set $B = \{(u,v) : v = 0,1,2,\ldots,u;\ u = 0,1,2,\ldots\}$ with inverse mapping $x = u-v, y = v$. It follows that the joint PMF of (U,V) is given by

$$P\{U = u, V = v\} = \begin{cases} \dfrac{(u+k-1)!}{(u-v)!\,v!\,(k-1)!} p_1^{u-v} p_2^v (1-p_1-p_2)^k & \text{for } (u,v) \in B, \\ 0 & \text{otherwise.} \end{cases}$$

The marginal PMF of U is given by

$$\begin{aligned} P\{U = u\} &= \frac{(u+k-1)!\,(1-p_1-p_2)^k}{(k-1)!\,u!} \sum_{v=0}^{u} \binom{u}{v} p_1^{u-v} p_2^v \\ &= \frac{(u+k-1)!\,(1-p_1-p_2)^k}{(k-1)!\,u!} (p_1+p_2)^u \\ &= \binom{u+k-1}{u} (p_1+p_2)^u (1-p_1-p_2)^k \qquad (u = 0,1,2,\ldots). \end{aligned}$$

Example 2. Let (X_1,X_2) have uniform distribution on the triangle $\{0 \le x_1 \le x_2 \le 1\}$, that is, (X_1,X_2) has joint density function

$$f(x_1,x_2) = \begin{cases} 2, & 0 \le x_1 \le x_2 \le 1 \\ 0, & \text{elsewhere.} \end{cases}$$

Let $Y = X_1 + X_2$. Then for $y < 0$, $P(Y \le y) = 0$, and for $y > 2$, $P(Y \le y) = 1$. For $0 \le y \le 2$, we have

$$P(Y \le y) = P(X_1 + X_2 \le y) = \iint_{\substack{0 \le x_1 \le x_2 \le 1 \\ x_1 + x_2 \le y}} f(x_1,x_2)dx_1dx_2.$$

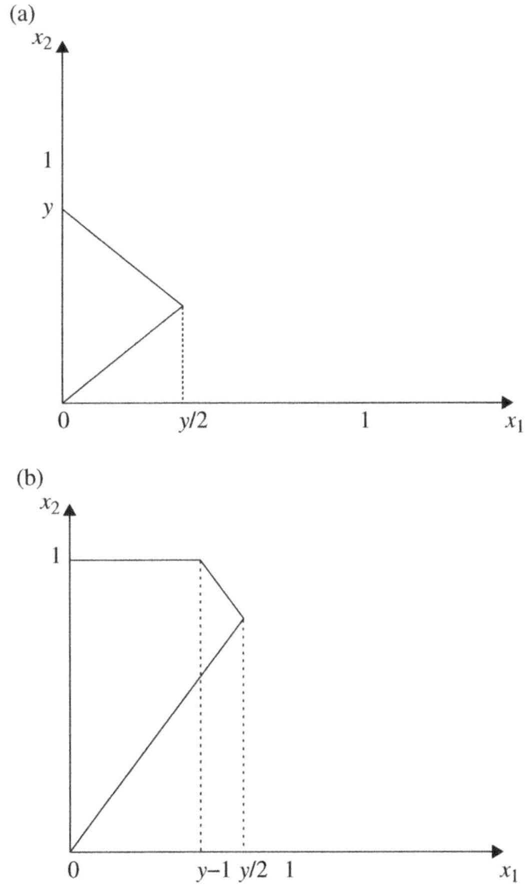

Fig. 1 (*a*) $\{x_1+x_2 \leq y, 0<x_1 \leq x_2 \leq 1, 0<y \leq 1\}$ and (*b*) $\{x_1+x_2 \leq y, 0 \leq x_1 \leq x_2 \leq 1 \leq y \leq 2\}$.

There are two cases to consider according to whether $0 \leq y \leq 1$ or $1 \leq y \leq 2$ (Fig. 1a and 1b). In the former case,

$$P(Y \leq y) = \int_{x_1=0}^{y/2} \left(\int_{x_2=x_1}^{y-x_1} 2dx_2 \right) dx_1 = 2\int_0^{y/2} (y-2x_1)dx_1 = y^2/2$$

and in the latter case,

$$\begin{aligned} P(Y \leq y) &= 1 - P(Y > y) + 1 - \int_{x_2=y/2}^{1} \left(\int_{x_1=y-x_2}^{x_2} 2dx_1 \right) dx_2 \\ &= 1 - 2\int_{y/2}^{1} (2x_2 - y)dx_1 = 1 - \frac{(y-2)^2}{2}. \end{aligned}$$

Hence the density function of Y is given by

$$f_Y(y) = \begin{cases} y, & 0 \le y \le 1 \\ 2-y, & 1 \le y \le 2 \\ 0, & \text{elsewhere.} \end{cases}$$

The method of distribution functions can also be used in the case when g takes values in $\mathcal{R}_m$, $1 \le m \le n$, but the integration becomes more involved.

Example 3. Let X_1 be the time that a customer takes from getting in line at a service desk in a bank to completion of service, and let X_2 be the time she waits in line before she reaches the service desk. Then $X_1 \ge X_2$ and $X_1 - X_2$ is the service time of the customer. Suppose the joint density of (X_1, X_2) is given by

$$f(x_1,x_2) = \begin{cases} e^{-x_1}, & 0 \le x_2 \le x_1 < \infty \\ 0, & \text{elsewhere.} \end{cases}$$

Let $Y_1 = X_1 + X_2$ and $Y_2 = X_1 - X_2$. Then the joint distribution of (Y_1, Y_2) is given by

$$P(Y_1 \le y_1,\ Y_2 \le y_2) = \int\int_A f(x_1,x_2)dx_1dx_2,$$

where $A = \{(x_1,x_2) : x_1 + x_2 \le y_1,\ x_1 - x_2 \le y_2,\ 0 \le x_2 \le x_1 < \infty\}$. Clearly, $x_1 + x_2 \ge x_1 - x_2$ so that the set A is as shown in Fig. 2. It follows that

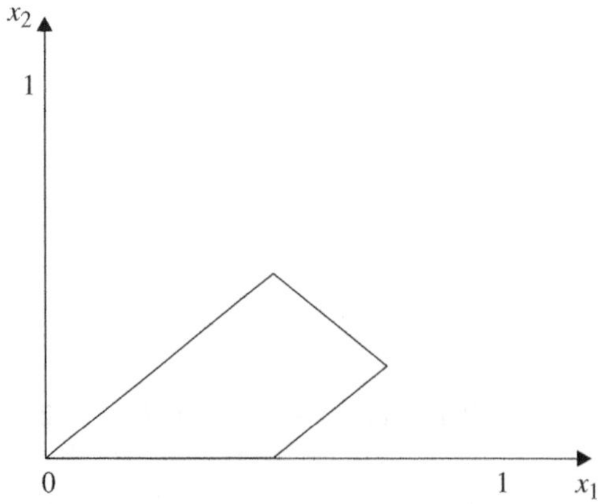

Fig. 2 $\{x_1 + x_2 \le y_1, x_1 - x_2 \le y_2, 0 \le x_2 \le x_1 < \infty\}$.

$$\begin{aligned}
P(Y_1 \le y_1,\ y_2 \le y_2) &= \int_{x_2=0}^{(y_1-y_2)/2} \left(\int_{x_1=x_2}^{x_2+y_2} e^{-x_1} dx_1 \right) dx_2 \\
&\quad + \int_{x_2=(y_1-y_2)/2}^{y_1/2} \left(\int_{x_1=x_2}^{y_1-x_2} e^{-x_1} dx_1 \right) dx_2 \\
&= \int_0^{(y_1-y_2)/2} e^{-x_2}(1-e^{-y_2}) dx_2 \\
&\quad + \int_{(y_1-y_2)/2}^{y_1/2} (e^{-x_2} - e^{-y_1+x_2}) dx_2 \\
&= (1-e^{-y_2})(1-e^{-(y_1-y_2)/2}) \\
&\quad + (e^{-(y_1-y_2)/2} - e^{-y_1/2}) - e^{-y_1}(e^{y_1/2} - e^{(y_1-y_2)/2}) \\
&= 1 - e^{-y_2} - 2e^{-y_1/2} + 2e^{-(y_1+y_2)/2}.
\end{aligned}$$

Hence the joint density of Y_1, Y_2 is given by

$$f_{Y_1,Y_2}(y_1,y_2) = \begin{cases} \frac{1}{2}e^{-(y_1+y_2)/2}, & 0 \le y_2 \le y_1 < \infty \\ 0, & \text{elsewhere.} \end{cases}$$

The marginal densities of Y_1, Y_2 are easily obtained as

$$\begin{aligned}
f_{y_1}(y_1) &= e^{-y_1} \quad \text{for } y_1 \ge 0, \text{ and 0 elsewhere;} \\
f_{y_2}(y_2) &= e^{-y_2/2}(1-e^{-y_2/2}), \quad \text{for } y_2 \ge 0, \text{ and 0 elsewhere.}
\end{aligned}$$

We next consider the method of transformations. Let $(X_1,\ldots,X_n)$ be jointly distributed with continuous PDF $f(x_1,x_2,\ldots,x_n)$, and let $\mathbf{y} = \mathbf{g}(x_1,x_2,\ldots,x_n) = (y_1,y_2,\ldots,y_n)$, where

$$y_i = g_i(x_1,x_2,\ldots,x_n), \quad i = 1,2,\ldots,n$$

be a mapping of $\mathcal{R}_n$ to R_n. Then

$$\begin{aligned}
P\{(Y_1,Y_2,\ldots,Y_n) \in B\} &= P\{(X_1,X_2,\ldots,X_n) \in \mathbf{g}^{-1}(B)\} \\
&= \int_{\mathbf{g}^{-1}(B)} f(x_1,x_2,\ldots,x_n) \prod_{i=1}^{n} dx_i,
\end{aligned}$$

where $\mathbf{g}^{-1}(B) = \{\mathbf{x} = (x_1,x_2,\ldots,x_n) \in \mathcal{R}_n\colon\ \mathbf{g}(\mathbf{x}) \in B\}$. Let us choose B to be the n-dimensional interval

$$B = B_{\mathbf{y}} = \{(y_1',y_2',\ldots,y_n')\colon\ -\infty < y_i' \le y_i, i = 1,2,\ldots,n\}.$$

Then the joint DF of $\mathbf{Y}$ is given by

$$P\{\mathbf{Y} \in B_{\mathbf{y}}\} = G_{\mathbf{Y}}(\mathbf{y}) = P\{g_1(\mathbf{X}) \le y_1, g_2(\mathbf{X}) \le y_2, \ldots, g_n(\mathbf{X}) \le y_n\}$$
$$= \int \underset{\mathbf{g}^{-1}(B_{\mathbf{y}})}{\cdots} \int f(x_1, x_2, \ldots, x_n) \prod_{i=1}^{n} dx_i,$$

and (if $G_{\mathbf{Y}}$ is absolutely continuous) the PDF of $\mathbf{Y}$ is given by

$$w(\mathbf{y}) = \frac{\partial^n G_{\mathbf{Y}}(\mathbf{y})}{\partial y_1 \partial y_2 \cdots \partial y_n}$$

at every continuity point $\mathbf{y}$ of w. Under certain conditions it is possible to write w in terms of f by making a change of variable in the multiple integral.

Theorem 2. Let $(X_1, X_2, \ldots, X_n)$ be an n-dimensional RV of the continuous type with PDF $f(x_1, x_2, \ldots, x_n)$.

(a) Let

$$\begin{aligned} y_1 &= g_1(x_1, x_2, \ldots, x_n), \\ y_2 &= g_2(x_1, x_2, \ldots, x_n), \\ &\vdots \\ y_n &= g_n(x_1, x_2, \ldots, x_n), \end{aligned}$$

be a one-to-one mapping of $\mathcal{R}_n$ into itself, that is, there exists the inverse transformation

$$x_1 = h_1(y_1, y_2, \ldots, y_n), \qquad x_2 = h_2(y_1, y_2, \ldots, y_n), \ldots, \qquad x_n = h_n(y_1, y_2, \ldots, y_n)$$

defined over the range of the transformation.

(b) Assume that both the mapping and its inverse are continuous.

(c) Assume that the partial derivatives

$$\frac{\partial x_i}{\partial y_j}, \qquad 1 \le i \le n, 1 \le j \le n,$$

exist and are continuous.

(d) Assume that the Jacobian J of the inverse transformation

$$J = \frac{\partial(x_1, \ldots, x_n)}{\partial(y_1, \ldots, y_n)} = \begin{vmatrix} \frac{\partial x_1}{\partial y_1} & \frac{\partial x_1}{\partial y_2} & \cdots & \frac{\partial x_1}{\partial y_n} \\ \frac{\partial x_2}{\partial y_1} & \frac{\partial x_2}{\partial y_2} & \cdots & \frac{\partial x_2}{\partial y_n} \\ \vdots & \vdots & & \vdots \\ \frac{\partial x_n}{\partial y_1} & \frac{\partial x_n}{\partial y_2} & \cdots & \frac{\partial x_n}{\partial y_n} \end{vmatrix}$$

is different from 0 for $(y_1, y_2, \ldots, y_n)$ in the range of the transformation.

Then $(Y_1, Y_2, \ldots, Y_n)$ has a joint absolutely continuous DF with PDF given by

$$w(y_1,y_2,\ldots,y_n) = |J|f(h_1(y_1,\ldots y_n),\ldots,h_n(y_1,\ldots,y_n)). \tag{1}$$

Proof. For $(y_1,y_2,\ldots,y_n) \in \mathcal{R}_n$, let

$$B = \{(y_1',y_2',\ldots,y_n') \in \mathcal{R}_n\colon\ -\infty < y_i' \le y_i, \quad i = 1,2,\ldots,n\}.$$

Then

$$\mathbf{g}^{-1}(B) = \{\mathbf{x} \in \mathcal{R}_n\colon \mathbf{g}(\mathbf{x}) \in B\} = \{(x_1,x_2,\ldots,x_n)\colon g_i(\mathbf{x}) \le y_i, \quad i = 1,2,\ldots,n\}$$

and

$$\begin{aligned} G_{\mathbf{Y}}(\mathbf{y}) &= P\{\mathbf{Y} \in B\} = P\{\mathbf{X} \in \mathbf{g}^{-1}(B)\} \\ &= \int \underset{\mathbf{g}^{-1}(B)}{\cdots} \int f(x_1,x_2,\ldots,x_n)\,dx_1\,dx_2\cdots dx_n \\ &= \int_{-\infty}^{y_1} \cdots \int_{-\infty}^{y_n} f(h_1(\mathbf{y}),\ldots,h_n(\mathbf{y})) \left| \frac{\partial(x_1,x_2,\ldots,x_n)}{\partial(y_1,y_2,\ldots,y_n)} \right| dy_1\cdots dy_n. \end{aligned}$$

Result (1) now follows on differentiation of DF $G_{\mathbf{Y}}$.

Remark 1. In actual applications we will not know the mapping from $x_1,x_2,\ldots,x_n$ to $y_1,y_2,\ldots,y_n$ completely, but one or more of the functions g_i will be known. If only $k, 1 \le k < n$, of the g_i's are known, we introduce arbitrarily $n-k$ functions such that the conditions of the theorem are satisfied. To find the joint marginal density of these k variables we simply integrate the w function over all the $n-k$ variables that were arbitrarily introduced.

Remark 2. An analog of Theorem 2.5.4 holds, which we state without proof.

Let $\mathbf{X} = (X_1,X_2,\ldots,X_n)$ be an RV of the continuous type with joint PDF f, and let $y_i = g_i(x_1,x_2,\ldots,x_n)$, $i = 1,2,\ldots,n$, be a mapping of $\mathcal{R}_n$ into itself. Suppose that for each $\mathbf{y}$ the transformation g has a finite number $k = k(\mathbf{y})$ of inverses. Suppose further that $\mathcal{R}_n$ can be partitioned into k disjoint sets $A_1,A_2,\ldots,A_k$, such that the transformation g from $A_i (i = 1,2,\ldots,n)$ into $\mathcal{R}_n$ is one-to-one with inverse transformation

$$x_1 = h_{1_i}(y_1,y_2,\ldots,y_n),\ldots, \quad x_n = h_{n_i}(y_1,y_2,\ldots,y_n), \quad i = 1,2,\ldots,k.$$

Suppose that the first partial derivatives are continuous and that each Jacobian

$$J_i = \begin{vmatrix} \frac{\partial h_{1i}}{\partial y_1} & \frac{\partial h_{1i}}{\partial y_2} & \cdots & \frac{\partial h_{1i}}{\partial y_n} \\ \frac{\partial h_{2i}}{\partial y_1} & \frac{d h_{2i}}{\partial y_2} & \cdots & \frac{\partial h_{2i}}{\partial y_n} \\ \vdots & \vdots & & \vdots \\ \frac{d h_{ni}}{d y_1} & \frac{\partial h_{ni}}{\partial y_2} & \cdots & \frac{\partial h_{ni}}{\partial y_n} \end{vmatrix}$$

is different from 0 in the range of the transformation. Then the joint PDF of $\mathbf{Y}$ is given by

$$w(y_1,y_2,\ldots,y_n)=\sum_{i=1}^{k}|J_i|f(h_{1i}(y_1,y_2,\ldots,y_n),\ldots,h_{ni}(y_1,y_2,\ldots,y_n)).$$

Example 4. Let X_1,X_2,X_3 be iid RVs with common *exponential* density function

$$f(x)=\begin{cases} e^{-x} & \text{if } x>0,\\ 0 & \text{otherwise.}\end{cases}$$

Also, let

$$Y_1=X_1+X_2+X_3,\quad Y_2=\frac{X_1+X_2}{X_1+X_2+X_3},\quad Y_3=\frac{X_1}{X_1+X_2}.$$

Then

$$\begin{aligned}&x_1=y_1y_2y_3, x_2=y_1y_2-x_1=y_1y_2(1-y_3)\quad\text{and}\\ &x_3=y_1-y_1y_2=y_1(1-y_2).\end{aligned}$$

The Jacobian of transformation is given by

$$J=\begin{vmatrix} y_2y_3 & y_1y_3 & y_1y_2\\ y_2(1-y_3) & y_1(1-y_3) & -y_1y_2\\ 1-y_2 & -y_1 & 0\end{vmatrix}=-y_1^2y_2.$$

Note that $0<y_1<\infty$, $0<y_2<1$, and $0<y_3<1$. Thus the joint PDF of Y_1,Y_2,Y_3 is given by

$$\begin{aligned}w(y_1,y_2,y_3)&=y_1^2y_2e^{-y_1}\\ &=(2y_2)\left(\frac{1}{2}y_1^2e^{-y_1}\right),\qquad 0<y_1<\infty, 0<y_2,y_3<1.\end{aligned}$$

It follows that Y_1, Y_2, and Y_3 are independent.

Example 5. Let X_1,X_2 be independent RVs with common density given by

$$f(x)=\begin{cases} 1 & \text{if } 0<x<1,\\ 0 & \text{otherwise.}\end{cases}$$

Let $Y_1=X_1+X_2, Y_2=X_1-X_2$. Then the Jacobian of the transformation is given by

$$J=\begin{vmatrix}\frac{1}{2} & \frac{1}{2}\\ \frac{1}{2} & -\frac{1}{2}\end{vmatrix}=-\frac{1}{2},$$

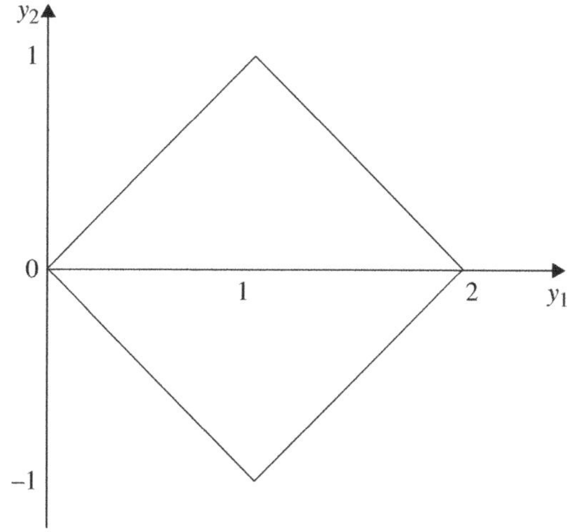

Fig. 3 $\{0 < y_1 + y_2 < 2, 0 < y_1 - y_2 < 2\}$.

and the joint density of Y_1, Y_2 (Fig. 3) is given by

$$
\begin{aligned}
f_{Y_1,Y_2}(y_1,y_2) &= \frac{1}{2} f\left(\frac{y_1+y_2}{2}\right) f\left(\frac{y_1-y_2}{2}\right) \\
&\qquad \text{if } 0 < \frac{y_1+y_2}{2} < 1, 0 < \frac{y_1-y_2}{2} < 1, \\
&= \frac{1}{2} \qquad \text{if } (y_1,y_2) \in \{0 < y_1 + y_2 < 2, 0 < y_1 - y_2 < 2\}.
\end{aligned}
$$

The marginal PDFs of Y_1 and Y_2 are given by

$$
f_{Y_1}(y_1) = \begin{cases} \int_{-y_1}^{y_1} \frac{1}{2} dy_2 = y_1, & 0 < y_1 \le 1, \\ \int_{y_1-2}^{2-y_1} \frac{1}{2} dy_2 = 2 - y_1, & 1 < y_1 < 2, \\ 0, & \text{otherwise;} \end{cases}
$$

$$
f_{Y_2}(y_2) = \begin{cases} \int_{-y_2}^{y_2+2} \frac{1}{2} dy_1 = y_2 + 1, & -1 < y_2 \le 0, \\ \int_{y_2}^{2-y_2} \frac{1}{2} dy_1 = 1 - y_2, & 0 < y_2 < 1, \\ 0, & \text{otherwise.} \end{cases}
$$

Example 6. Let X_1, X_2, X_3 be iid RVs with common PDF

$$
f(x) = \frac{1}{\sqrt{2\pi}} e^{-x^2/2}, \qquad -\infty < x < \infty.
$$

Let $Y_1 = (X_1 - X_2)/\sqrt{2}, Y_2 = (X_1 + X_2 - 2X_3)/\sqrt{6}$, and $Y_3 = (X_1 + X_2 + X_3)/\sqrt{3}$. Then

$$x_1 = \frac{y_1}{\sqrt{2}} + \frac{y_2}{\sqrt{6}} + \frac{y_3}{\sqrt{3}},$$
$$x_2 = -\frac{y_1}{\sqrt{2}} + \frac{y_2}{\sqrt{6}} + \frac{y_3}{\sqrt{3}}$$
$$x_3 = -\frac{\sqrt{2}y_2}{\sqrt{3}} + \frac{y_3}{\sqrt{3}}.$$

The Jacobian of transformation is given by

$$J = \begin{vmatrix} \frac{1}{\sqrt{2}} & \frac{1}{\sqrt{6}} & \frac{1}{\sqrt{3}} \\ \frac{-1}{\sqrt{2}} & \frac{1}{\sqrt{6}} & \frac{1}{\sqrt{3}} \\ 0 & \frac{-\sqrt{2}}{\sqrt{3}} & \frac{1}{\sqrt{3}} \end{vmatrix} = 1.$$

The joint PDF of X_1, X_2, X_3 is given by

$$g(x_1, x_2, x_3) = \frac{1}{(\sqrt{2\pi})^3} \exp\left\{-\frac{x_1^2 + x_2^2 + x_3^2}{2}\right\}, \qquad x_1, x_2, x_3 \in \mathcal{R}.$$

It is easily checked that

$$x_1^2 + x_2^2 + x_3^2 = y_1^2 + y_2^2 + y_3^2,$$

so that the joint PDF of Y_1, Y_2, Y_3 is given by

$$w(y_1, y_2, y_3) = \frac{1}{(\sqrt{2\pi})^3} \exp\left\{-\frac{y_1^2 + y_2^2 + y_3^2}{2}\right\}.$$

It follows that Y_1, Y_2, Y_3 are also iid RVs with common PDF f.

In Example 6 the transformation used is orthogonal and is known as *Helmert's transformation.* In fact, we will show in Section 6.5 that under orthogonal transformations iid RVs with PDF f defined above are transformed into iid RVs with the same PDF.

It is easily verified that

$$y_1^2 + y_2^2 = \sum_{j=1}^{3} \left(x_j - \frac{x_1 + x_2 + x_3}{3}\right)^2.$$

We have therefore proved that $(X_1 + X_2 + X_3)$ is independent of $\sum_{j=1}^{3}\{X_j - [(X_1 + X_2 + X_3)/3]\}^2$. This is a very important result in mathematical statistics, and we will return to it in Section 7.7.

Example 7. Let (X,Y) be a *bivariate normal* RV with joint PDF

$$f(x,y)=\frac{1}{2\pi\sigma_1\sigma_2(1-\rho^2)^{1/2}}$$
$$\cdot\exp\left\{-\frac{1}{2(1-\rho^2)}\left[\frac{(x-\mu_1)^2}{\sigma_1^2}-\frac{2\rho(x-\mu_1)(y-\mu_2)}{\sigma_1\sigma_2}+\frac{(y-\mu_2)^2}{\sigma_2^2}\right]\right\},$$
$$-\infty<x<\infty,\quad -\infty<y<\infty;\ \mu_1\in\mathcal{R},\mu_2\in\mathcal{R};$$
$$\text{and } \sigma_1>0,\sigma_2>0,|\rho|<1.$$

Let

$$U_1=\sqrt{X^2+Y^2},\qquad U_2=\frac{X}{Y}.$$

For $u_1>0$,

$$\sqrt{x^2+y^2}=u_1\qquad\text{and}\qquad\frac{x}{y}=u_2$$

have two solutions:

$$x_1=\frac{u_1u_2}{\sqrt{1+u_2^2}},\quad y_1=\frac{u_1}{\sqrt{1+u_2^2}},\qquad\text{and}\qquad x_2=-x_1,\quad y_2=-y_1$$

for any $u_2\in\mathcal{R}$. The Jacobians are given by

$$J_1=J_2=\begin{vmatrix}\dfrac{u_2}{\sqrt{1+u_2^2}} & \dfrac{u_1}{(1+u_2^2)^{3/2}}\\ \dfrac{1}{\sqrt{1+u_2^2}} & -\dfrac{u_1u_2}{(1+u_2^2)^{3/2}}\end{vmatrix}=-\frac{u_1}{1+u_2^2}.$$

It follows from the result in Remark 2 that the joint PDF of (U_1,U_2) is given by

$$w(u_1,u_2)=\begin{cases}\dfrac{u_1}{1+u_2^2}\left[f\left(\dfrac{u_1u_2}{\sqrt{1+u_2^2}},\dfrac{u_1}{\sqrt{1+u_2^2}}\right)\right.\\ \qquad\left.+f\left(\dfrac{-u_1u_2}{\sqrt{1+u_2^2}},\dfrac{-u_1}{\sqrt{1+u_2^2}}\right)\right] & \text{if } u_1>0,u_2\in\mathcal{R},\\ 0 & \text{otherwise.}\end{cases}$$

In the special case where $\mu_1=\mu_2=0$, $\rho=0$, and $\sigma_1=\sigma_2=\sigma$, we have

$$f(x,y)=\frac{1}{2\pi\sigma^2}e^{-[(x^2+y^2)/2\sigma^2]}$$

so that X and Y are independent. Moreover,

$$f(x,y)=f(-x,-y),$$

and it follows that when X and Y are independent

$$w(u_1,u_2)=\begin{cases}\dfrac{1}{2\pi\sigma^2}\dfrac{2u_1}{1+u_2^2}e^{-u_1^2/2\sigma^2}, & u_1>0,\ -\infty<u_2<\infty,\\ 0, & \text{otherwise.}\end{cases}$$

Since

$$w(u_1,u_2)=\frac{1}{\pi(1+u_2^2)}\frac{u_1}{\sigma^2}e^{-u_1^2/2\sigma^2},$$

it follows that U_1 and U_2 are independent with marginal PDFs given by

$$w_1(u_1)=\begin{cases}\dfrac{u_1}{\sigma^2}e^{-u_1^2/2\sigma^2}, & u_1>0,\\ 0, & u_1\le 0\end{cases}$$

and

$$w_2(u_2)=\frac{1}{\pi(1+u_2^2)}, \qquad -\infty<u_2<\infty,$$

respectively.

An important application of the result in Remark 2 will appear in Theorem 4.7.2.

Theorem 3. Let (X,Y) be an RV of the continuous type with PDF f. Let

$$Z=X+Y,\ U=X-Y,\ V=XY;$$

and let $W=X/Y$. Then the PDFs of Z, U, V, and W are, respectively, given by

$$f_Z(z)=\int_{-\infty}^{\infty}f(x,z-x)\,dx, \tag{2}$$

$$f_U(u)=\int_{-\infty}^{\infty}f(u+y,y)\,dy, \tag{3}$$

$$f_V(v)=\int_{-\infty}^{\infty}f\left(x,\frac{v}{x}\right)\frac{1}{|x|}\,dx, \tag{4}$$

$$f_W(w)=\int_{-\infty}^{\infty}f(xw,x)|x|\,dx. \tag{5}$$

Proof. The proof is left as an exercise.

Corollary. If X and Y are independent with PDFs f_1 and f_2, respectively, then

$$f_Z(z) = \int_{-\infty}^{\infty} f_1(x) f_2(z-x)\, dx, \tag{6}$$

$$f_U(u) = \int_{-\infty}^{\infty} f_1(u+y) f_2(y)\, dy, \tag{7}$$

$$f_V(v) = \int_{-\infty}^{\infty} f_1(x) f_2\left(\frac{v}{x}\right) \frac{1}{|x|}\, dx, \tag{8}$$

$$f_W(w) = \int_{-\infty}^{\infty} f_1(xw) f_2(x) |x|\, dx. \tag{9}$$

Remark 3. Let F and G be two absolutely continuous DFs; then

$$H(x) = \int_{-\infty}^{\infty} F(x-y) G'(y)\, dy = \int_{-\infty}^{\infty} G(x-y) F'(y)\, dy$$

is also an absolutely continuous DF with PDF

$$H'(x) = \int_{-\infty}^{\infty} F'(x-y) G'(y)\, dy = \int_{-\infty}^{\infty} G'(x-y) F'(y)\, dy.$$

If

$$F(x) = \sum_k p_k \varepsilon(x - x_k) \qquad \text{and} \qquad G(x) = \sum_j q_j \varepsilon(x - y_j)$$

are two DFs, then

$$H(x) = \sum_k \sum_j p_k q_j \varepsilon(x - x_k - y_j)$$

is also a DF of an RV of the discrete type. The DF H is called the *convolution* of F and G, and we write $H = F * G$. Clearly, the operation is commutative and associative; that is, if F_1, F_2, F_3 are DFs, $F_1 * F_2 = F_2 * F_1$ and $(F_1 * F_2) * F_3 = F_1 * (F_2 * F_3)$. In this terminology, if X and Y are independent RVs with DFs F and G, respectively, $X + Y$ has the convolution DF $H = F * G$. Extension to an arbitrary number of independent RVs is obvious.

Finally, we consider a technique based on MGF or CF which can be used in certain situations to determine the distribution of a function $g(X_1, X_2, \ldots, X_n)$ of $X_1, X_2, \ldots, X_n$.

Let $(X_1, X_2, \ldots, X_n)$ be an n-variate RV, and g be a Borel-measurable function from $\mathcal{R}_n$ to $\mathcal{R}_1$.

Definition 1. If $(X_1,X_2,\ldots,X_n)$ is discrete type and $\sum_{x_1,\ldots,x_n}|g(x_1,x_2,\ldots,x_n)|P\{X_1 = x_1,X_2 = x_2,\ldots,X_n = x_n\} < \infty$, then the series

$$Eg(X_1,X_2,\ldots,X_n) = \sum_{x_1,\ldots,x_n} g(x_1,x_2,\ldots,x_n)P\{X_1 = x_1,X_2 = x_2,\ldots,X_n = x_n\}$$

is called the *expected value* of $g(X_1,X_2,\ldots,X_n)$. If $(X_1,X_2,\ldots,X_n)$ is a continuous type RV with joint PDF f, and if

$$\int_{-\infty}^{\infty}\int_{-\infty}^{\infty}\cdots\int_{-\infty}^{\infty}|g(x_1,x_2,\ldots,x_n)|f(x_1,x_2,\ldots,x_n)\prod_{i=1}^{n}dx_i < \infty,$$

then

$$Eg(X_1,X_2,\ldots,X_n) = \int_{-\infty}^{\infty}\int_{-\infty}^{\infty}\cdots\int_{-\infty}^{\infty}g(x_1,x_2,\ldots,x_n)f(x_1,x_2,\ldots,x_n)\prod_{i=1}^{n}dx_i$$

is called the expected value of $g(X_1,X_2,\ldots,X_n)$.

Let $Y = g(X_1,X_2,\ldots,X_n)$, and let $h(y)$ be its PDF. If $E|Y| < \infty$ then

$$EY = \int_{-\infty}^{\infty} yh(y)dy.$$

An analog of Theorem 3.2.1 holds. That is,

$$\int_{-\infty}^{\infty} yh(y)dy = \int_{-\infty}^{\infty}\int_{-\infty}^{\infty}\cdots\int_{-\infty}^{\infty}g(x_1,x_2,\ldots,x_n)f(x_1,x_2,\ldots,x_n)\prod_{i=1}^{n}dx_i$$

in the sense that if either integral exists so does the other and the two are equal. The result also holds in the discrete case.

Some special functions of interest are $\sum_{j=1}^{n}x_j$, $\prod_{j=1}^{n}x_j^{k_j}$, where $k_1,k_2,\ldots,k_n$ are non-negative integers, $e^{\sum_{j=1}^{n}t_jx_j}$, where $t_1,t_2,\ldots,t_n$ are real numbers, and $e^{i\sum_{j=1}^{n}t_jx_j}$, where $i=\sqrt{-1}$.

Definition 2. Let $X_1,X_2,\ldots,X_n$ be *jointly distributed.* If $Ee^{\sum_{j=1}^{n}t_jX_j}$ exists for $|t_j| \le h_j$, $j = 1,2,\ldots,n$, for some $h_j > 0, j = 1,2,\ldots,n$, we write

$$M(t_1,t_2,\ldots,t_n) = E\left(e^{t_1X_1+t_2X_2+\cdots+t_nX_n}\right) \tag{10}$$

and call it the MGF of the joint distribution of $(X_1,X_2,\ldots,X_n)$ or, simply, the MGF of $(X_1,X_2,\ldots,X_n)$.

Definition 3. Let $t_1,t_2,\ldots,t_n$ be real numbers and $i = \sqrt{-1}$. Then the CF of $(X_1,X_2,\ldots,X_n)$ is defined by

$$\begin{aligned}\phi(t_1,t_2,\ldots,t_n) &= E\left\{\exp\left(i\sum_{j=1}^{n}t_jX_j\right)\right\}\\ &= E\left\{\cos\left(\sum_{j=1}^{n}t_jX_j\right)\right\}+iE\left\{\sin\left(\sum_{j=1}^{n}t_jX_j\right)\right\}\end{aligned} \tag{11}$$

As in the univariate case $\phi(t_1,t_2,\ldots,t_n)$ always exists.

We will mostly deal with MGF even though the condition that it exist for $|t_j| \le h_j$, $j=1,2,\ldots,n$ restricts its application considerably. The multivariate MGF (CF) has properties similar to the univariate MGF discussed earlier. We state some of these without proof. For notational convenience we restrict ourselves to the bivariate case.

Theorem 4. The MGF $M(t_1,t_2)$ uniquely determines the joint distribution of (X,Y), and conversely, if the MGF exists it is unique.

Corollary. The MGF $M(t_1,t_2)$ completely determines the marginal distributions of X and Y. Indeed,

$$M(t_1,0) = E(e^{t_1X}) = M_X(t_1), \tag{12}$$

$$M(0,t_2) = E(e^{t_2Y}) = M_Y(t_2). \tag{13}$$

Theorem 5. If $M(t_1,t_2)$ exists, the moments of all orders of (X,Y) exist and may be obtained from

$$\left.\frac{\partial^{m+n}M(t_1,t_2)}{\partial t_1^m \partial t_2^n}\right|_{t_1=t_2=0} = E(X^mY^n). \tag{14}$$

Thus,

$$\frac{\partial M(0,0)}{\partial t_1} = EX, \qquad \frac{\partial M(0,0)}{\partial t_2} = EY,$$

$$\frac{\partial^2 M(0,0)}{\partial t_1^2} = EX^2, \qquad \frac{\partial^2 M(0,0)}{\partial t_2^2} = EY^2,$$

$$\frac{\partial^2 M(0,0)}{\partial t_1 \partial t_2} = E(XY),$$

and so on.

A formal definition of moments in the multivariate case will be given in Section 4.5.

Theorem 6. X and Y are independent RVs if and only if

$$M(t_1,t_2) = M(t_1,0)M(0,t_2) \qquad \text{for all } t_1,t_2 \in \mathfrak{R}. \tag{15}$$

Proof. Let X and Y be independent. Then,

$$M(t_1,t_2) = E\{e^{t_1X+t_2Y}\} = E(e^{t_1X})E(e^{t_2Y}) = M(t_1,0)M(0,t_2).$$

Conversely, if

$$M(t_1,t_2)=M(t_1,0)M(0,t_2),$$

then, in the continuous case,

$$\iint e^{t_1x+t_2y}f(x,y)\,dx\,dy=\left[\int e^{t_1x}f_1(x)\,dx\right]\left[\int e^{t_2y}f_2(y)\,dy\right],$$

that is,

$$\iint e^{t_1x+t_2y}f(x,y)\,dx\,dy=\iint e^{t_1x+t_2y}f_1(x)f_2(y)\,dx\,dy.$$

By the uniqueness of the MGF (Theorem 4) we must have

$$f(x,y)=f_1(x)f_2(y) \qquad \text{for all } (x,y)\in\mathcal{R}_2.$$

It follows that X and Y are independent. A similar proof is given in the case where (X,Y) is of the discrete type.

The MGF technique uses the uniqueness property of Theorem 4. In order to find the distribution (DF, PDF, or PMF) of $Y=g(X_1,X_2,\ldots,X_n)$ we compute the MGF of Y using definition. If this MGF is one of the known kind then Y must have this kind of distribution. Although the technique applies to the case when Y is an m-dimensional RV, $1\le k\le n$, we will mostly use it for the $m=1$ case.

Example 8. Let us first consider a simple case when X is normal PDF

$$f(x)=\frac{1}{\sqrt{2\pi}}e^{-x^2/2}, \quad -\infty<x<-\infty.$$

Let $Y=X^2$. Then

$$\begin{aligned}M_Y(s)=Ee^{sX^2}&=\frac{1}{\sqrt{2\pi}}\int_{-\infty}^{\infty}e^{\frac{1}{2}(1-2s)x^2}\,dx\\&=\frac{1}{\sqrt{1-2s}}, \quad \text{for } x<1/2.\end{aligned}$$

It follows (see Section 5.3 and also Example 2.5.7) that Y has a *chi-square* PDF

$$w(y)=\frac{(e^{-y/2})}{\sqrt{y\pi}}, \quad y>0.$$

Example 9. Suppose X_1 and X_2 are independent with common PDF f of Example 8. Let $Y_1=X_1-X_2$. There are three equivalent ways to use MGF technique here. Let $Y_2=X_2$. Then rather than compute

$$M(s_1,s_2)=Ee^{s_1Y_1+s_2Y_2},$$

it is simpler to recognize that Y_1 is univariate so

$$\begin{aligned} M_{Y_1}(s) &= Ee^{s(X_1-X_2)} \\ &= (Ee^{sX_1})(Ee^{-sX_2}) \\ &= e^{s^2/2}e^{s^2/2} = e^{s^2}. \end{aligned}$$

It follows that Y_1 has PDF

$$f(x) = \frac{1}{\sqrt{4\pi}} e^{-s^2/4}, \qquad -\infty < x < \infty.$$

Note that $M_{Y_1}(s) = M(s,0)$.

Let $Y_3 = X_1 + X_2$. Let us find the joint distribution of $Y_1$and Y_3. Indeed

$$\begin{aligned} E\left(e^{s_1 Y_1 + s_2 Y_3}\right) &= E\left(e^{(s_1+s_2)X_1} \cdot e^{(s_1-s_2)X_2}\right) \\ &= E\left(e^{(s_1+s_2)X_1}\right) E\left(e^{(s_1-s_2)X_2}\right) \\ &= e^{(s_1+s_2)^2/2} \cdot e^{(s_1-s_2)^2/2} = e^{s_1^2} \cdot e^{s_2^2} \end{aligned}$$

and it follows that Y_1 and Y_3 are independent RVs with common PDF f defined above.

The following result has many applications as we will see. Example 9 is a special case.

Theorem 7. Let $X_1, X_2, \ldots, X_n$ be independent RVs with respective MGFs $M_i(s)$, $i = 1, 2, \ldots, n$. Then the MGF of $Y = \sum_{i=1}^n a_i X_i$ for real numbers $a_1, a_2, \ldots, a_n$ is given by

$$M_Y(s) = \prod_{i=1}^n M_i(a_i s).$$

Proof. If M_i exists for $|s| \le h_i$, $h_i > 0$, then M_Y exists for $|s| \le \min(h_1, \ldots, h_n)$ and

$$M_Y(s) = Ee^{s\sum_{i=1}^n a_i X_i} = \prod_{i=1}^n Ee^{sa_i X_i} = \prod_{i=1}^n M_i(a_i s).$$

Corollary. If X_i's are iid, then the MGF of $Y = \sum_1^n X_i$ is given by $M_Y(s) = [M(s)]^n$.

Remark 4. The converse of Theorem 7 does not hold. We leave the reader to construct an example illustrating this fact.

Example 10. Let $X_1, X_2, \ldots, X_m$ be iid RVs with common PMF

$$P\{X = k\} = \binom{n}{k} p^k (1-p)^{n-k}, \qquad k = 0, 1, 2, \ldots, n; \; 0 < p < 1.$$

Then the MGF of X_i is given by

$$M(t) = (1-p+pe^t)^n.$$

It follows that the MGF of $S_m = X_1 + X_2 + \cdots + X_m$ is

$$\begin{aligned} M_{S_m}(t) &= \prod_1^m (1-p+pe^t)^n \\ &= (1-p+pe^t)^{nm}, \end{aligned}$$

and we see that S_m has the PMF

$$P\{S_m = s\} = \binom{mn}{s} p^s (1-p)^{mn-s}, \qquad s = 0,1,2,\ldots,mn.$$

From these examples it is clear that to use this technique effectively one must be able to recognize the MGF of the function under consideration. In Chapter 5 we will study a number of commonly occurring probability distributions and derive their MGFs (whenever they exist). We will have occasion to use Theorem 7 quite frequently.

For integer-valued RVs one can sometimes use PGFs to compute the distribution of certain functions of a multiple RV.

We emphasize the fact that a CF always exists and analogs of Theorems 4–7 can be stated in terms of CFs.

PROBLEMS 4.4

1. Let F be a DF and ε be a positive real number. Show that

$$\Psi_1(x) = \frac{1}{\varepsilon} \int_x^{x+\varepsilon} F(x)\,dx$$

and

$$\Psi_2(x) = \frac{1}{2\varepsilon} \int_{x-\varepsilon}^{x+\varepsilon} F(x)\,dx$$

are also distribution functions.

2. Let X, Y be iid RVs with common PDF

$$f(x) = \begin{cases} e^{-x} & \text{if } x > 0, \\ 0 & \text{if } x \le 0. \end{cases}$$

(a) Find the PDF of RVs $X+Y$, $X-Y$, XY, X/Y, $\min\{X,Y\}$, $\max\{X,Y\}$, $\min\{X,Y\}/\max\{X,Y\}$, and $X/(X+Y)$.

(b) Let $U = X + Y$ and $V = X - Y$. Find the conditional PDF of V, given $U = u$, for some fixed $u > 0$.

(c) Show that U and $Z = X/(X+Y)$ are independent.

3. Let X and Y be independent RVs defined on the space $(\Omega, \mathcal{S}, P)$. Let X be uniformly distributed on $(-a, a), a > 0$, and Y be an RV of the continuous type with density f, where f is continuous and positive on $\mathcal{R}$. Let F be the DF of Y. If $u_0 \in (-a, a)$ is a fixed number, show that

$$f_{Y|X+Y}(y \mid u_0) = \begin{cases} \dfrac{f(y)}{F(u_0+a) - F(u_0-a)} & \text{if } u_0 - a < y < u_0 + a, \\ 0 & \text{otherwise,} \end{cases}$$

where $f_{Y|X+Y}(y \mid u_0)$ is the conditional density function of Y, given $X + Y = u_0$.

4. Let X and Y be iid RVs with common PDF

$$f(x) = \begin{cases} 1 & \text{if } 0 \le x \le 1, \\ 0 & \text{otherwise.} \end{cases}$$

Find the PDFs of RVs XY, X/Y, $\min\{X,Y\}$, $\max\{X,Y\}$, $\min\{X,Y\}/\max\{X,Y\}$.

5. Let X_1, X_2, X_3 be iid RVs with common density function

$$f(x) = \begin{cases} 1 & \text{if } 0 \le x \le 1; \\ 0 & \text{otherwise.} \end{cases}$$

Show that the PDF of $U = X_1 + X_2 + X_3$ is given by

$$g(u) = \begin{cases} \dfrac{u^2}{2}, & 0 \le u < 1, \\ 3u - u^2 - \dfrac{3}{2}, & 1 \le u < 2, \\ \dfrac{(u-3)^2}{2}, & 2 \le u \le 3, \\ 0, & \text{elsewhere.} \end{cases}$$

An extension to the n-variate case holds.

6. Let X and Y be independent RVs with common geometric PMF

$$P\{X = k\} = \pi(1-\pi)^k, \qquad k = 0, 1, 2, \ldots; 0 < \pi < 1.$$

Also, let $M = \max\{X,Y\}$. Find the joint distribution of M and X, the marginal distribution of M, and the conditional distribution of X, given M.

7. Let X be a nonnegative RV of the continuous type. The integral part, Y, of X is distributed with PMF $P\{Y = k\} = \lambda^k e^{-\lambda}/k!$, $k = 0, 1, 2, \ldots, \lambda > 0$; and the fractional part, Z, of X has PDF $f_z(z) = 1$ if $0 \le z \le 1$, and $= 0$ otherwise. Find the PDF of X, assuming that Y and Z are independent.

8. Let X and Y be independent RVs. If at least one of X and Y is of the continuous type, show that $X+Y$ is also continuous. What if X and Y are not independent?

9. Let X and Y be independent integral RVs. Show that

$$P(t) = P_X(t)P_Y(t),$$

where P, P_X, and P_Y, respectively, are the PGFs of $X+Y$, X, and Y.

10. Let X and Y be independent nonnegative RVs of the continuous type with PDFs f and g, respectively. Let $f(x) = e^{-x}$ if $x > 0$, and $= 0$ if $x \leq 0$, and let g be arbitrary. Show that the MGF $M(t)$ of Y, which is assumed to exist, has the property that the DF of X/Y is $1 - M(-t)$.

11. Let X, Y, Z have the joint PDF

$$f(x,y,z) = \begin{cases} 6(1+x+y+z)^{-4} & \text{if } 0 < x, 0 < y, 0 < z, \\ 0 & \text{otherwise.} \end{cases}$$

Find the PDF of $U = X + Y + Z$.

12. Let X and Y be iid RVs with common PDF

$$f(x) = \begin{cases} (x\sqrt{2\pi})^{-1}e^{-(1/2)(\log x)^2}, & x > 0, \\ 0, & x \leq 0. \end{cases}$$

Find the PDF of $Z = XY$.

13. Let X and Y be iid RVs with common PDF f defined in Example 8. Find the joint PDF of U and V in the following cases:
(a) $U = \sqrt{X^2+Y^2}$, $V = \tan^{-1}(X/Y)$, $-(\pi/2) < V \leq (\pi/2)$.
(b) $U = (X+Y)/2$, $V = (X-Y)^2/2$.

14. Construct an example to show that even when the MGF of $X+Y$ can be written as a product of the MGF of X and the MGF of Y, X and Y need not be independent.

15. Let $X_1, X_2, \ldots, X_n$ be iid with common PDF

$$f(x) = \frac{1}{(b-a)}, \quad a < x < b, \quad = 0 \quad \text{otherwise.}$$

Using the distribution function technique show that
(a) The joint PDF of $X_{(n)} = \max(X_1, X_2, \ldots, X_n)$, and $X_{(1)} = \min(X_1, X_2, \ldots, X_n)$ is given by

$$u(x,y) = \frac{n(n-1)(x-y)^{n-2}}{(b-a)^n}, \quad a < y < x < b,$$

and $= 0$ otherwise.
(b) The PDF of $X_{(n)}$ is given by

$$g(z) = \frac{n(z-a)^n}{(b-a)^n}, \quad a < z < b, \quad = 0 \quad \text{otherwise}$$

and that of $X_{(1)}$ by

$$h(z) = \frac{n(b-z)^{n-1}}{(b-a)^n}, \quad a < z < b, \quad = 0 \quad \text{otherwise.}$$

16. Let X_1, X_2 be iid with common Poisson PMF

$$P(X_i = x) = e^{-\lambda}\frac{\lambda^x}{x!}, \; x = 0, 1, 2, \ldots, i = 1, 2,$$

where $\lambda > 0$ is a constant. Let $X_{(2)} = \max(X_1, X_2)$ and $X_{(1)} = \min(X_1, X_2)$. Find the PMF of $X_{(2)}$.

17. Let X have the binomial PMF

$$P(X = k) = \binom{n}{k} p^k(1-p)^{n-k}, \quad k = 0, 1, \ldots, n; \; 0 < p < 1.$$

Let Y be independent of X and $Y \stackrel{d}{=} X$. Find PMF of $U = X + Y$ and $W = X - Y$.

4.5 COVARIANCE, CORRELATION AND MOMENTS

Let X and Y be jointly distributed on $(\Omega, \mathcal{S}, P)$. In Section 4.4 we defined $Eg(X,Y)$ for Borel functions g on $\mathcal{R}_2$. Functions of the form $g(x,y) = x^j y^k$ where j and k are nonnegative integers are of interest in probability and statistics.

Definition 1. If $E|X^j Y^k| < \infty$ for nonnegative integers j and k, we call $E(X^j Y^k)$ a *moment of order* $(j+k)$ of (X,Y) and write

$$m_{jk} = E(X^j Y^k). \tag{1}$$

Clearly,

$$\left.\begin{array}{lll} m_{10} = EX, & m_{01} = EY & \\ m_{20} = EX^2, & m_{11} = EXY, & m_{02} = EY^2. \end{array}\right\} \tag{2}$$

Definition 2. If $E\left|(X-EX)^j(Y-EY)^k\right| < \infty$ for nonnegative integers j and k, we call $E\left\{(X-EX)^j(Y-EY)^k\right\}$ a *central moment of order* $(j+k)$ and write

$$\mu_{jk} = E\left\{(X-EX)^j(Y-EY)^k\right\}. \tag{3}$$

Clearly,

$$\left.\begin{array}{l} \mu_{10} = \mu_{01} = 0, \quad \mu_{20} = \text{var}(X), \quad \mu_{02} = \text{var}(Y), \\ \mu_{11} = E\{(X-m_{10})(Y-m_{01})\}. \end{array}\right\} \tag{4}$$

We see easily that

$$\mu_{11} = E(XY) - EX\,EY. \tag{5}$$

Note that if X and Y increase (or decrease) together then $(X-EX)(Y-EY)$ should be positive, whereas if X decreases while Y increases (and conversely) then the product should be negative. Hence the average value of $(X-EX)(Y-EY)$, namely μ_{11}, provides a measure of association or joint variation between X and Y.

Definition 3. If $E\{(X-EX)(Y-EY)\}$ exists, we call it the *covariance* between X and Y and write

$$\operatorname{cov}(X,Y)=E\{(X-EX)(Y-EY)\}=E(XY)-EXEY. \tag{6}$$

Recall (Theorem 3.2.8) that $E\{Y-a\}^2$ is minimized when we choose $a=EY$ so that EY may be interpreted as the best constant predictor of Y. If instead, we choose to predict Y by a linear function of X, say $aX+b$, and measure the error in this prediction by $E\{Y-aX-b\}^2$, then we should choose a and b to minimize this so-called *mean square error*. Clearly, $E(Y-aX-b)^2$ is minimized, for any a, by choosing $b=E(Y-aX)=EY-aEX$. With this choice of b, we find a such that

$$\begin{aligned}E(Y-aX-b)^2&=E\{(Y-EY)-a(X-EX)\}^2\\&=\sigma_Y^2-2a\mu_{11}+a^2\sigma_X^2\end{aligned}$$

is minimum. An easy computation shows that the minimum occurs if we choose

$$a=\frac{\mu_{11}}{\sigma_X^2}, \tag{7}$$

provided $\sigma_X^2>0$. Moreover,

$$\begin{aligned}\min_{a,b}E(Y-aX-b)^2&=\min_a\left\{\sigma_Y^2-2a\mu_{11}+a^2\sigma_X^2\right\}\\&=\frac{\sigma_Y^2-\mu_{11}^2}{\sigma_X^2}\\&=\sigma_Y^2\left\{1-\left[\frac{\mu_{11}}{(\sigma_X\sigma_Y)}\right]^2\right\}.\end{aligned} \tag{8}$$

Let us write

$$\rho=\frac{\mu_{11}}{\sigma_X\sigma_Y}. \tag{9}$$

Then (8) shows that predicting Y by a linear function of X reduces the prediction error from σ_Y^2 to $\sigma_Y^2(1-\rho^2)$. We may therefore think of ρ as a measure of the *linear dependence* between RVs X and Y.

Definition 4. If EX^2, EY^2 exist, we define the *correlation coefficient* between X and Y as

$$\rho=\frac{\operatorname{cov}(X,Y)}{\mathrm{SD}(X)SD(Y)}=\frac{EXY-EXEY}{\sqrt{EX^2-(EX)^2}\sqrt{EY^2-(EY)^2}}, \tag{10}$$

where $\mathrm{SD}(X)$ denotes the standard deviation of RV X.

We note that for any two real numbers a and b

$$|ab| \leq \frac{a^2+b^2}{2},$$

so that $E|XY| < \infty$ if $EX^2 < \infty$ and $EY^2 < \infty$.

Definition 5. We say that RVs X and Y are *uncorrelated* if $\rho = 0$, or equivalently, $\text{cov}(X,Y) = 0$.

If X and Y are independent, then from (5) $\text{cov}(X,Y) = 0$, and X and Y are uncorrelated. If, however, $\rho = 0$ then X and Y may not necessarily be independent.

Example 1. Let U and V be two RVs with common mean and common variance. Let $X = U+V$ and $Y = U-V$. Then

$$\text{cov}(X,Y) = E(U^2 - V^2) - E(U+V)E(U-V) = 0$$

so that X and Y are uncorrelated but not necessarily independent. See Example 4.4.9.

Let us now study some properties of the correlation coefficient. From the definition we see that ρ (and also $\text{cov}(X,Y)$) is symmetric in X and Y.

Theorem 1.

(a) The correlation coefficient ρ between two RVs X and Y satisfies

$$|\rho| \leq 1. \tag{11}$$

(b) The equality $|\rho| = 1$ holds if and only if there exist constants $a \neq 0$ and b such that $P\{aX + b = 1\} = 1$.

Proof. From (8) since $E(Y - aX - b)^2 \geq 0$, we must have $1 - \rho^2 \geq 0$, or equivalently, (11) holds.

Equality in (11) holds if and only if $\rho^2 = 1$, or equivalently, $E(Y - aX - b)^2 = 0$ holds. This implies and is implied by $P(Y = aX + b) = 1$. Here $a \neq 0$.

Remark 1. From (7) and (9) we note that the signs of a and ρ are the same so if $\rho = 1$ then $P(Y = aX + b)$ where $a > 0$, and if $\rho = -1$ then $a < 0$.

Theorem 2. Let $EX^2 < \infty$, $EY^2 < \infty$, and let $U = aX + b$, $V = cY + d$. Then,

$$\rho_{X,Y} = \pm\rho_{U,V},$$

where $\rho_{X,Y}$ and $\rho_{U,V}$, respectively, are the correlation coefficients between X and Y and U and V.

Proof. The proof is simple and is left as an exercise.

Example 2. Let X, Y be identically distributed with common PMF

$$P\{X=k\}=\frac{1}{N}, \qquad k=1,2,\ldots,N(N>1).$$

Then

$$EX=EY=\frac{N+1}{2}, \qquad EX^2=EY^2=\frac{(N+1)(2N+1)}{6},$$

so that

$$\text{var}(X)=\text{var}(Y)=\frac{N^2-1}{12}.$$

Also,

$$\begin{aligned}E(XY)&=\frac{1}{2}\{EX^2+EY^2-E(X-Y)^2\}\\&=\frac{(N+1)(2N+1)}{6}-\frac{E(X-Y)^2}{2}.\end{aligned}$$

Thus,

$$\begin{aligned}\text{cov}(X,Y)&=\frac{(N+1)(2N+1)}{6}-\frac{E(X-Y)^2}{2}-\frac{(N+1)^2}{4}\\&=\frac{(N+1)(N-1)}{12}-\frac{1}{2}E(X-Y)^2\end{aligned}$$

and

$$\begin{aligned}\rho_{X,Y}&=\frac{(N^2-1)/12-E(X-Y)^2/2}{(N^2-1)/12}\\&=1-\frac{6E(X-Y)^2}{N^2-1}.\end{aligned}$$

If $P\{X=Y\}=1$, then $\rho=1$, and conversely. If $P\{Y=N+1-X\}=1$, then

$$\begin{aligned}E(X-Y)^2&=E(2X-N-1)^2\\&=4\frac{(N+1)(2N+1)}{6}-4\frac{(N+1)^2}{2}+(N+1)^2,\end{aligned}$$

and it follows that $\rho_{XY}=-1$. Conversely, if $\rho_{X,Y}=-1$, from Remark 1 it follows that $Y=-aX+b$ with probability 1 for some $a>0$ and some real number b. To find a and b, we note that $EY=-aEX+b$ so that $b=[(N+1)/2](1+a)$. Also $EY^2=E(b-aX)^2$, which yields

$$(1-a^2)EX^2+2abEX-b^2=0.$$

Substituting for b in terms of a and the values of EX^2 and EX, we see that $a^2=1$, so that $a=1$. Hence, $b=N+1$, and it follows that $Y=N+1-X$ with probability 1.

Example 3. Let (X,Y) be jointly distributed with density function

$$f(x,y) = \begin{cases} x+y, & 0<x<1, 0<y<1. \\ 0, & \text{otherwise}. \end{cases}$$

Then

$$\begin{aligned} EX^lY^m &= \int_0^1 \int_0^1 x^l y^m (x+y)\, dx\, dy \\ &= \int_0^1 \int_0^1 x^{l+1} y^m\, dx\, dy + \int_0^1 \int_0^1 x^l y^{m+1}\, dx\, dy \\ &= \frac{1}{(l+2)(m+1)} + \frac{1}{(l+1)(m+2)}, \end{aligned}$$

where l and m are positive integers. Thus

$$\begin{aligned} EX = EY &= \frac{7}{12}, \\ EX^2 = EY^2 &= \frac{5}{12}, \\ \text{var}(X) = \text{var}(Y) &= \frac{5}{12} - \frac{49}{144} = \frac{11}{144}, \\ \text{cov}(X,Y) &= \frac{1}{3} - \frac{49}{144} = -\frac{1}{144}, \quad \rho = -1/11. \end{aligned}$$

Theorem 3. Let $X_1, X_2, \ldots, X_n$ be RVs such that $E|X_i| < \infty, i = 1,2,\ldots,n$. Let $a_1, a_2, \ldots, a_n$ be real numbers, and write

$$S = a_1X_1 + a_2X_2 + \cdots + a_nX_n.$$

Then ES exists, and we have

$$ES = \sum_{j=1}^{n} a_j EX_j. \tag{12}$$

Proof. If $(X_1, X_2, \ldots, X_n)$ is of the discrete type, then

$$\begin{aligned} ES &= \sum_{i_1, i_2, \ldots, i_n} (a_1 x_{i_1} + a_2 x_{i_2} + \cdots + a_n x_{i_n}) P\{X_1 = x_{i_1}, X_2 = x_{i_2}, \cdots, X_n = x_{i_n}\} \\ &= a_1 \sum_{i_1} x_{i_1} \sum_{i_2, \ldots, i_n} P\{X_1 = x_{i_1}, \ldots, X_n = x_{i_n}\} \\ &\quad + \cdots + a_n \sum_{i_n} x_{i_n} \sum_{i_1, \ldots, i_{n-1}} P\{X_1 = x_{i_1}, \ldots, X_n = x_{i_n}\} \\ &= a_1 \sum_{i_1} x_{i_1} P\{X_1 = x_{i_1}\} + \cdots + a_n \sum_{i_n} P\{X_n = x_{i_n}\} \\ &= a_1 EX_1 + \cdots + a_n EX_n. \end{aligned}$$

The existence of ES follows easily by replacing each a_j by $|a_j|$ and each x_{ij} by $|x_{ij}|$ and remembering that $E|X_j| < \infty, j = 1,2,\ldots,n$. The case of continuous type $(X_1,X_2,\ldots,X_n)$ is similarly treated.

Corollary. Take $a_1 = a_2 = \cdots = a_n = 1/n$. Then

$$E\left(\frac{X_1+X_2+\cdots+X_n}{n}\right) = \frac{1}{n}\sum_{i=1}^{n} EX_i,$$

and if $EX_1 = EX_2 = \cdots = EX_n = \mu$, then

$$E\left(\frac{X_1+X_2+\cdots X_n}{n}\right) = \mu.$$

Theorem 4. Let $X_1,X_2,\ldots,X_n$ be independent RVs such that $E|X_i| < \infty, i = 1,2,\ldots,n$. Then $E(\prod_{i=1}^{n} X_i)$ exists and

$$E\left(\prod_{i=1}^{n} X_i\right) = \prod_{i=1}^{n} EX_i. \tag{13}$$

Let X and Y be independent, and $g_1(\cdot)$ and $g_2(\cdot)$ be Borel-measurable functions. Then we know (Theorem 4.3.3) that $g_1(X)$ and $g_2(Y)$ are independent. If $E\{g_1(X)\}$, $E\{g_2(Y)\}$, and $E\{g_1(X)g_2(Y)\}$ exist, it follows from Theorem 4 that

$$E\{g_1(X)g_2(Y)\} = E\{g_1(X)\}E\{g_2(Y)\}. \tag{14}$$

Conversely, if for any Borel sets A_1 and A_2 we take $g_1(X) = 1$ if $X \in A_1$, and $= 0$ otherwise, and $g_2(Y) = 1$ if $Y \in A_2$, and $= 0$ otherwise, then

$$E\{g_1(X)g_2(Y)\} = P\{X \in A_1, Y \in A_2\}$$

and $E\{g_1(X)\} = P\{X \in A_1\}$, $E\{g_2(Y)\} = P\{Y \in A_2\}$. Relation (14) implies that for any Borel sets A_1 and A_2 of real numbers

$$P\{X \in A_1, Y \in A_2\} = P\{X \in A_1\}P\{Y \in A_2\}.$$

It follows that X and Y are independent if (14) holds. We have thus proved the following theorem.

Theorem 5. Two RVs X and Y are independent if and only if for every pair of Borel-measurable functions g_1 and g_2 the relation

$$E\{g_1(X)g_2(Y)\} = E\{g_1(X)\}E\{g_2(Y)\} \tag{15}$$

holds, provided that the expectations on both sides of (15) exist.

Theorem 6. Let $X_1, X_2, \ldots, X_n$ be RVs with $E|X_i|^2 < \infty$ for $i = 1, 2, \ldots, n$. Let $a_1, a_2, \ldots, a_n$ be real numbers and write $S = \sum_{i=1}^n a_i X_i$. Then the variance of S exists and is given by

$$\operatorname{var}(S) = \sum_{i=1}^n a_i^2 \operatorname{var}(X_i) + \sum_{i=1}^n \sum_{\substack{j=1 \\ i \neq j}}^n a_i a_j \operatorname{cov}(X_i, X_j). \tag{16}$$

If, in particular, $X_1, X_2, \ldots, X_n$ are such that $\operatorname{cov}(X_i, X_j) = 0$ for $i, j = 1, 2, \ldots, n, i \neq j$, then

$$\operatorname{var}(S) = \sum_{i=1}^n a_i^2 \operatorname{var}(X_i). \tag{17}$$

Proof. We have

$$\begin{aligned}
\operatorname{var}(S) &= E\left\{\sum_{i=1}^n a_i X_i - \sum_{i=1}^n a_i E X_i\right\}^2 \\
&= E\left\{\sum_{i=1}^n a_i^2 (X_i - EX_i)^2 + \sum_{i \neq j} a_i a_j (X_i - EX_i)(X_j - EX_j)\right\} \\
&= \sum_{i=1}^n a_i^2 E(X_i - EX_i)^2 + \sum_{i \neq j} a_i a_j E\{(X_i - EX_i)(X_j - EX_j)\}.
\end{aligned}$$

If the X_i's satisfy

$$\operatorname{cov}(X_i, X_j) = 0 \qquad \text{for } i, j = 1, 2, \ldots, n;\ i \neq j,$$

the second term on the right side of (16) vanishes, and we have (17).

Corollary 1. Let $X_1, X_2, \ldots, X_n$ be exchangeable RVs with $\operatorname{var}(X_i) = \sigma^2$, $i = 1, 2, \ldots, n$. Then

$$\operatorname{var}\left(\sum_{i=1}^n a_i X_i\right) = \sigma^2 \sum_{i=1}^n a_i^2 + \rho\sigma^2 \sum_{i \neq j}^n a_i a_j,$$

where ρ is the correlation coefficient between X_i and X_j, $i \neq j$. In particular,

$$\operatorname{var}\left(\sum_{i=1}^n \frac{X_i}{n}\right) = \frac{\sigma^2}{n} + \frac{n-1}{n}\rho\sigma^2.$$

Corollary 2. If $X_1, X_2, \ldots, X_n$ are exchangeable and uncorrelated then

$$\operatorname{var}\left(\sum_{i=1}^n a_i X_i\right) = \sigma^2 \sum_{i=1}^n a_i^2,$$

and

$$\operatorname{var}\left(\sum_{i=1}^{n}\frac{X_i}{n}\right)=\frac{\sigma^2}{n}.$$

Theorem 7. Let $X_1,X_2,\ldots,X_n$ be iid RVs with common variance σ^2. Also, let $a_1,a_2,\ldots,a_n$ be real numbers such that $\sum_1^n a_i=1$, and let $S=\sum_{i=1}^n a_iX_i$. Then the variance of S is least if we choose $a_i=1/n, i=1,2,\ldots,n$.

Proof. We have

$$\operatorname{var}(S)=\sigma^2\sum_{i=1}^{n}a_i^2,$$

which is least if and only if we choose the a_i's so that $\sum_{i=1}^n a_i^2$ is smallest, subject to the condition $\sum_{i=1}^n a_i=1$. We have

$$\begin{aligned}\sum_{i=1}^{n}a_i^2&=\sum_{i=1}^{n}\left(a_i-\frac{1}{n}+\frac{1}{n}\right)^2\\&=\sum_{i=1}^{n}\left(a_i-\frac{1}{n}\right)^2+\frac{2}{n}\sum_{i=1}^{n}\left(a_i-\frac{1}{n}\right)+\frac{1}{n}\\&=\sum_{i=1}^{n}\left(a_i-\frac{1}{n}\right)^2+\frac{1}{n},\end{aligned}$$

which is minimized for the choice $a_i=1/n$, $i=1,2,\ldots,n$.

Note that the result holds if we replace independence by the condition that X_i's are exchangeable and uncorrelated.

Example 4. Suppose that r balls are drawn one at a time without replacement from a bag containing n white and m black balls. Let S_r be the number of black balls drawn.

Let us define RVs X_k as follows:

$$\begin{aligned}X_k&=1\quad\text{if the }k\text{th ball drawn is black}\\&=0\quad\text{if the }k\text{th ball drawn is white}\end{aligned}\qquad k=1,2,\ldots,r.$$

Then

$$S_r=X_1+X_2+\cdots+X_r.$$

Also

$$P\{X_k=1\}=\frac{m}{m+n},\quad P\{X_k=0\}=\frac{n}{m+n}.\tag{18}$$

Thus $EX_k = m/(m+n)$ and

$$\operatorname{var}(X_k) = \frac{m}{m+n} - \frac{m^2}{(m+n)^2} = \frac{mn}{(m+n)^2}.$$

To compute $\operatorname{cov}(X_j, X_k), j \neq k$, note that the RV $X_jX_k = 1$ if the jth and the kth balls drawn are black, and $= 0$ otherwise. Thus

$$E(X_jX_k) = P\{X_j = 1, X_k = 1\} = \frac{m}{m+n}\frac{m-1}{m+n-1} \tag{19}$$

and

$$\operatorname{cov}(X_j, X_k) = -\frac{mn}{(m+n)^2(m+n-1)}.$$

Thus

$$ES_r = \sum_{k=1}^{r} EX_k = \frac{mr}{m+n}$$

and

$$\begin{aligned}\operatorname{var}(S_r) &= r\frac{mn}{(m+n)^2} - r(r-1)\frac{mn}{(m+n)^2(m+n-1)}\\ &= \frac{mnr}{(m+n)^2(m+n+1)}(m+n-r).\end{aligned}$$

The reader is asked to satisfy himself that (18) and (19) hold.

Example 5. Let $X_1, X_2, \ldots, X_n$ be independent, and $a_1, a_2, \ldots, a_n$ be real numbers such that $\sum a_i = 1$. Assume that $E|X_i^2| < \infty, i = 1, 2, \ldots, n$, and let $\operatorname{var}(X_i) = \sigma_i^2, i = 1, 2, \ldots, n$. Write $S = \sum_{i=1}^{n} a_i X_i$. Then $\operatorname{var}(S) = \sum_{i=1}^{n} a_i^2 \sigma_i^2 = \sigma$, say. To find weights a_i such that σ is minimum, we write

$$\sigma = a_1^2\sigma_1^2 + a_2^2\sigma_2^2 + \cdots + (1 - a_1 - a_2 - \cdots - a_{n-1})^2\sigma_n^2$$

and differentiate partially with respect to $a_1, a_2, \ldots, a_{n-1}$, respectively. We get

$$\begin{aligned}\frac{\partial\sigma}{\partial a_1} &= 2a_1\sigma_1^2 - 2(1 - a_1 - a_2 - \cdots - a_{n-1})\sigma_n^2 = 0,\\ &\vdots\\ \frac{\partial\sigma}{\partial a_{n-1}} &= 2a_{n-1}\sigma_{n-1}^2 - 2(1 - a_1 - a_2 - \cdots - a_{n-1})\sigma_n^2 = 0.\end{aligned}$$

It follows that

$$a_j\sigma_j^2 = a_n\sigma_n^2, \qquad j = 1, 2, \ldots, n-1,$$

that is, the weights $a_j, j = 1, 2, \ldots, n$, should be chosen proportional to $1/\sigma_j^2$. The minimum value of σ is then

$$\sigma_{\min} = \sum_{i=1}^{n} \frac{k^2}{\sigma_i^4} \sigma_i^2 = k^2 \sum_{i=1}^{n} \frac{1}{\sigma_i^2},$$

where k is given by $\sum_{j=1}^{n} (k/\sigma_j^2) = 1$. Thus

$$\sigma_{\min} = \frac{1}{\sum_{j=1}^{n} (1/\sigma_j^2)} = \frac{H}{n},$$

where H is the harmonic mean of the σ_j^2.

We conclude this section with some important moment inequalities. We begin with the simple inequality

$$|a+b|^r \le c_r(|a|^r + |b|^r), \tag{20}$$

where $c_r = 1$ for $0 \le r \le 1$ and $= 2^{r-1}$ for $r > 1$. For $r = 0$ and $r = 1$, (20) is trivially true.

First note that it is sufficient to prove (20) when $0 < a \le b$. Let $0 < a \le b$, and write $x = a/b$. Then

$$\frac{(a+b)^r}{a^r + b^r} = \frac{(1+x)^r}{1+x^r}.$$

Writing $f(x) = (1+x)^r/(1+x^r)$, we see that

$$f'(x) = \frac{r(1+x)^{r-1}}{(1+x^r)^2}(1 - x^{r-1}),$$

where $0 < x \le 1$. It follows that $f'(x) > 0$ if $r > 1$, $= 0$ if $r = 1$, and < 0 if $r < 1$. Thus

$$\max_{0 \le x \le 1} f(x) = f(0) = 1 \qquad \text{if } r \le 1,$$

while

$$\max_{0 \le x \le 1} f(x) = f(1) = 2^{r-1} \qquad \text{if } r \ge 1.$$

Note that $|a+b|^r \le 2^r(|a|^r + |b|^r)$ is trivially true since

$$|a+b| \le \max(2|a|, 2|b|).$$

An immediate application of (20) is the following result.

Theorem 8. Let X and Y be RVs and $r > 0$ be a fixed number. If $E|X|^r$, $E|Y|^r$ are both finite, so also is $E|X+Y|^r$.

Proof. Let $a = X$ and $b = Y$ in (20). Taking the expectation on both sides, we see that

$$E|X+Y|^r \le c_r(E|X|^r + E|Y|^r),$$

where $c_r = 1$ if $0 < r \le 1$ and $= 2^{r-1}$ if $r > 1$.

Next we establish Hölder's inequality,

$$|xy| \le \frac{|x|^p}{p} + \frac{|y|^q}{q}, \tag{21}$$

where p and q are positive real numbers such that $p > 1$ and $1/p + 1/q = 1$. Note that for $x > 0$ the function $w = \log x$ is concave. It follows that for $x_1, x_2 > 0$

$$\log[tx_1 + (1-t)x_2] \ge t \log x_1 + (1-t) \log x_2.$$

Taking antilogarithms, we get

$$x_1^t x_2^{1-t} \ge tx_1 + (1-t)x_2.$$

Now we choose $x_1 = |x|^p, x_2 = |y|^q, t = 1/p, 1-t = 1/q$, where $p > 1$ and $1/p + 1/q = 1$, to get (21).

Theorem 9. Let $p > 1, q > 1$ so that $1/p + 1/q = 1$. Then

$$E|XY| \le (E|X|^p)^{1/p}(E|Y|^q)^{1/q}. \tag{22}$$

Proof. By Hölder's inequality, letting $x = X\{E|X|^p\}^{-1/p}, y = Y\{E|Y|^q\}^{-1/q}$, we get

$$\begin{aligned}|XY| \le\; & p^{-1}|X|^p\{E|X|^p\}^{1/p-1}\{E|Y|^q\}^{1/q} \\ & + q^{-1}|Y|^q\{E|Y|^q\}^{1/q-1}\{E|X|^p\}^{1/p}.\end{aligned}$$

Taking the expectation on both sides leads to (22).

Corollary. Taking $p = q = 2$, we obtain the *Cauchy–Schwarz inequality*,

$$E|XY| \le E^{1/2}|X|^2 E^{1/2}|Y|^2.$$

The final result of this section is an inequality due to Minkowski.

Theorem 10. For $p \ge 1$,

$$\{E|X+Y|^p\}^{1/p} \le \{E|X|^p\}^{1/p} + \{E|Y|^p\}^{1/p}. \tag{23}$$

Proof. We have, for $p > 1$,

$$|X+Y|^p \le |X|\,|X+Y|^{p-1} + |Y|\,|X+Y|^{p-1}.$$

Taking expectations and using Hölder's inequality with Y replaced by $|X+Y|^{p-1} (p>1)$, we have

$$\begin{aligned} E|X+Y|^p &\le \{E|X|^p\}^{1/p}\{E|X+Y|^{(p-1)q}\}^{1/q} \\ &\quad + \{E|Y|^p\}^{1/p}\{E|X+Y|^{(p-1)q}\}^{1/q} \\ &= [\{E|X|^p\}^{1/p} + \{E|Y|^p\}^{1/p}]\cdot\{E|X+Y|^{(p-1)q}\}^{1/q}. \end{aligned}$$

Excluding the trivial case in which $E|X+Y|^p = 0$, and noting that $(p-1)q = p$, we have, after dividing both sides of the last inequality by $\{E|X+Y|^p\}^{1/q}$,

$$\{E|X+Y|^p\}^{1/p} \le \{E|X|^p\}^{1/p} + \{E|Y|^p\}^{1/p}, \qquad p > 1.$$

The case $p = 1$ being trivial, this establishes (23).

PROBLEMS 4.5

1. Suppose that the RV (X,Y) is uniformly distributed over the region $R = \{(x,y): 0 < x < y < 1\}$. Find the covariance between X and Y.
2. Let (X,Y) have the joint PDF given by

$$f(x,y) = \begin{cases} x^2 + \dfrac{xy}{3} & \text{if } 0 < x < 1, 0 < y < 2, \\ 0 & \text{otherwise.} \end{cases}$$

   Find all moments of order 2.
3. Let (X,Y) be distributed with joint density

$$f(x,y) = \begin{cases} \dfrac{1}{4}[1 + xy(x^2 - y^2)] & \text{if } |x| \le 1, |y| \le 1, \\ 0 & \text{otherwise.} \end{cases}$$

   Find the MGF of (X,Y). Are X, Y independent? If not, find the covariance between X and Y.
4. For a positive RV X with finite first moment show that (1) $E\sqrt{X} \le \sqrt{EX}$ and (2) $E\{1/X\} \ge 1/EX$.
5. If X is a nondegenerate RV with finite expectation and such that $X \ge a > 0$, then

$$E\{\sqrt{X^2 - a^2}\} < \sqrt{(EX)^2 - a^2}.$$

   (Kruskal [56])
6. Show that for $x > 0$

$$\left(\int_x^\infty t e^{-t^2/2}\,dt\right)^2 \le \int_x^\infty e^{-t^2/2}\,dt \int_x^\infty t^2 e^{-t^2/2}\,dt,$$

and hence that

$$\int_x^\infty e^{-t^2/2}\,dt \geq \frac{1}{2}[(4+x^2)^{1/2}-x]e^{-x^2/2}.$$

7. Given a PDF f that is nondecreasing in the interval $a \leq x \leq b$, show that for any $s > 0$

$$\int_a^b x^{2s} f(x)\,dx \geq \frac{b^{2s+1}-a^{2s+1}}{(2s+1)(b-a)} \int_a^b f(x)\,dx,$$

with the inequality reversed if f is nonincreasing.

8. Derive the Lyapunov inequality (Theorem 3.4.3)

$$\{E|X|^r\}^{1/r} \leq \{E|X|^s\}^{1/s}, \qquad 1 < r < s < \infty,$$

from Hölder's inequality (22).

9. Let X be an RV with $E|X|^r < \infty$ for $r > 0$. Show that the function $\log E|X|^r$ is a convex function of r.

10. Show with the help of an example that Theorem 9 is not true for $p < 1$.

11. Show that the converse of Theorem 8 also holds for independent RVs, that is, if $E|X+Y|^r < \infty$ for some $r > 0$ and X and Y are independent, then $E|X|^r < \infty$, $E|Y|^r < \infty$.
[*Hint:* Without loss of generality assume that the median of both X and Y is 0. Show that, for any $t > 0$, $P\{|X+Y| > t\} > (1/2)P\{|X| > t\}$. Now use the remarks preceding Lemma 3.2.2 to conclude that $E|X|^r < \infty$.]

12. Let $(\Omega, \mathcal{S}, P)$ be a probability space, and $A_1, A_2, \ldots, A_n$ be events in $\mathcal{S}$ such that $P(\cup_{k=1}^n A_k) > 0$. Show that

$$2 \sum_{1 \leq j < k < n} P(A_j A_k) \geq \frac{(\sum_{k=1}^n PA_k)^2 - \sum_{k=1}^n PA_k}{P(\cup_{k=1}^n A_k)}.$$

(Chung and Erdös [14])

[*Hint:* Let X_k be the indicator function of A_k, $k = 1, 2, \ldots, n$. Use the Cauchy–Schwarz inequality.]

13. Let $(\Omega, \mathcal{S}, P)$ be a probability space, and $A, B, \in \mathcal{S}$ with $0 < PA < 1, 0 < PB < 1$. Define $\rho(A,B)$ by $\rho(A,B) =$ correlation coefficient between RVs I_A, and I_B, where I_A, I_B, are the indicator functions of A and B, respectively. Express $\rho(A,B)$ in terms of PA, PB, and $P(AB)$ and conclude that $\rho(A,B) = 0$ if and only if A and B are independent. What happens if $A = B$ or if $A = B^c$?

(a) Show that

$$\rho(A,B) > 0 \Leftrightarrow P\{A \mid B\} > P(A) \Leftrightarrow P\{B \mid A\} > P(B)$$

and

$$\rho(A,B) < 0 \Leftrightarrow P\{A \mid B\} < PA \Leftrightarrow P\{B \mid A\} < PB.$$

(b) Show that

$$\rho(A,B) = \frac{P(AB)\,P(A^cB^c) - P(AB^c)\,P(A^cB)}{(PA\,PA^c \cdot PB\,PB^c)^{1/2}}.$$

14. Let $X_1, X_2, \ldots, X_n$ be iid RVs and define

$$\bar{X} = \frac{\sum_{i=1}^{n} X_i}{n}, \qquad S^2 = \frac{\sum_{i=1}^{n}(X_i - \bar{X})^2}{(n-1)}.$$

Suppose that the common distribution is symmetric. Assuming the existence of moments of appropriate order, show that $\text{cov}(\bar{X}, S^2) = 0$.

15. Let X, Y be iid RVs with common standard normal density

$$f(x) = \frac{1}{\sqrt{2\pi}} e^{-x^2/2}, \qquad -\infty < x < \infty.$$

Let $U = X + Y$ and $V = X^2 + Y^2$. Find the MGF of the random variable (U, V). Also, find the correlation coefficient between U and V. Are U and V independent?

16. Let X and Y be two discrete RVs:

$$P\{X = x_1\} = p_1, \qquad P\{X = x_2\} = 1 - p_1;$$

and

$$P\{Y = y_1\} = p_2, \qquad P\{Y = y_2\} = 1 - p_2.$$

Show that X and Y are independent if and only if the correlation coefficient between X and Y is 0.

17. Let X and Y be dependent RVs with common means 0, variances 1, and correlation coefficient ρ. Show that

$$E\{\max(X^2, Y^2)\} \le 1 + \sqrt{1 - \rho^2}.$$

18. Let X_1, X_2 be independent normal RVs with density functions

$$f_i(x) = \frac{1}{\sigma_i\sqrt{2\pi}} \exp\left\{-\frac{1}{2}\left(\frac{x - \mu_i}{\sigma_i}\right)^2\right\}, \qquad -\infty < x < \infty;\ i = 1, 2.$$

Also let

$$Z = X_1 \cos\theta + X_2 \sin\theta \quad \text{and} \quad W = X_2 \cos\theta - X_1 \sin\theta.$$

Find the correlation coefficient between Z and W and show that

$$0 \le \rho^2 \le \left(\frac{\sigma_1^2 - \sigma_2^2}{\sigma_1^2 + \sigma_2^2}\right)^2,$$

where ρ denotes the correlation coefficient between Z and W.

19. Let $(X_1, X_2, \ldots, X_n)$ be an RV such that the correlation coefficient between each pair $X_i, X_j, i \ne j$, is ρ. Show that $-(n-1)^{-1} \le \rho \le 1$.

20. Let $X_1, X_2, \ldots, X_{m+n}$ be iid RVs with finite second moment. Let $S_k = \sum_{j=1}^{k} X_j$, $k = 1, 2, \ldots, m+n$. Find the correlation coefficient between S_n and $S_{m+n} - S_m$, where $n > m$.

21. Let f be the PDF of a positive RV, and write

$$g(x,y) = \begin{cases} \dfrac{f(x+y)}{x+y} & \text{if } x > 0, y > 0, \\ 0 & \text{otherwise.} \end{cases}$$

Show that g is a density function in the plane. If the mth moment of f exists for some positive integer m, find EX^m. Compute the means and variances of X and Y and the correlation coefficient between X and Y in terms of moments of f. (Adapted from Feller [26, p. 100].)

22. A die is thrown $n+2$ times. After each throw a $+$ sign is recorded for 4, 5, or 6, and a $-$ sign for 1, 2, or 3, the signs forming an ordered sequence. Each sign, except the first and the last, is attached to a characteristic RV that assumes the value 1 if both the neighboring signs differ from the one between them and 0 otherwise. Let $X_1, X_2, \ldots, X_n$ be these characteristic RVs, where X_i corresponds to the $(i+1)$st sign $(i = 1, 2, \ldots, n)$ in the sequence. Show that

$$E\left\{\sum_1^n X_i\right\} = \frac{n}{4} \quad \text{and} \quad \operatorname{var}\left\{\sum_1^n X_i\right\} = \frac{5n-2}{16}.$$

23. Let (X, Y) be jointly distributed with PDF f defined by $f(x,y) = \frac{1}{2}$ inside the square with corners at the points $(0,1), (1,0), (-1,0), (0,-1)$ in the (x,y)-plane, and $f(x,y) = 0$ otherwise. Are X, Y independent? Are they uncorrelated?

4.6 CONDITIONAL EXPECTATION

In Section 4.2 we defined the conditional distribution of an RV X, given Y. We showed that, if (X, Y) is of the discrete type, the conditional PMF of X, given $Y = y_j$, where $P\{Y = y_j\} > 0$, is a PMF when considered as a function of the x_i's (for fixed y_j). Similarly, if (X, Y) is an RV of the continuous type with PDF $f(x,y)$ and marginal densities f_1 and f_2, respectively, then, at every point (x, y) at which f is continuous and at which $f_2(y) > 0$ and is continuous, a conditional density function of X, given Y, exists and may be defined by

$$f_{X|Y}(x \mid y) = \frac{f(x,y)}{f_2(y)}.$$

We also showed that $f_{X|Y}(x \mid y)$, for fixed y, when considered as a function of x is a PDF in its own right. Therefore, we can (and do) consider the moments of this conditional distribution.

Definition 1. Let X and Y be RVs defined on a probability space $(\Omega, \mathcal{S}, P)$, and let h be a Borel-measurable function. Then the *conditional expectation* of $h(X)$, given Y, written as $E\{h(X) \mid Y\}$, is an RV that takes the value $E\{h(X) \mid y\}$, defined by

$$E\{h(X) \mid y\} = \begin{cases} \sum_x h(x)P\{X = x \mid Y = y\} & \text{if } (X,Y) \text{ is of the discrete} \\ & \text{type and } P\{Y = y\} > 0, \\ \int_{-\infty}^{\infty} h(x) f_{X|Y}(x \mid y)\,dx & \text{if } (X,Y) \text{ is of the contain-} \\ & \text{nous type and } f_2(y) > 0. \end{cases} \tag{1}$$

when the RV Y assumes the value y.

Needless to say, a similar definition may be given for the conditional expectation $E\{h(Y) \mid X\}$.

It is immediate that $E\{h(X) \mid Y\}$ satisfies the usual properties of an expectation provided we remember that $E\{h(X) \mid Y\}$ is not a constant but an RV. The following results are easy to prove. We assume existence of indicated expectations.

$$E\{c \mid Y\} = c, \quad \text{for any constant } c \tag{2}$$
$$E\{[a_1 g_1(X) + a_2 g_2(X)] \mid Y\} = a_1 E\{g_1(X) \mid Y\} + a_2 E\{g_2(X) \mid Y\}, \tag{3}$$

for any Borel functions g_1, g_2.

$$P(X \geq 0) = 1 \Longrightarrow E\{X \mid Y\} \geq 0 \tag{4}$$
$$P(X_1 \geq X_2) = 1 \Longrightarrow E\{X_1 \mid Y\} \geq E\{X_2 \mid Y\}. \tag{5}$$

The statements in (3), (4), and (5) should be understood to hold with probability 1.

$$E\{X \mid Y\} = E(X), \quad E\{Y \mid X\} = E(Y) \tag{6}$$

for independent RVs X and Y.

If $\phi(X,Y)$ is a function of X and Y, then

$$E\{\phi(X,Y) \mid y\} = E\{\phi(X,y) \mid y\} \tag{7}$$
$$E\{\psi(X)\phi(X,Y) \mid X\} = \psi(X)E\{\phi(X,Y) \mid X\} \tag{8}$$

for any Borel functions ψ and ϕ.

Again (8) should be understood as holding with probability 1. Relation (7) is useful as a computational device. See Example 3 below.

The moments of a conditional distribution are defined in the usual manner. Thus, for $r \geq 0$, $E\{X^r \mid Y\}$ defines the rth moment of the conditional distribution. We can define the central moments of the conditional distribution and, in particular, the variance. There is no difficulty in generalizing these concepts for n-dimensional distributions when $n > 2$. We leave the reader to furnish the details.

Example 1. An urn contains three red and two green balls. A random sample of two balls is drawn (a) with replacement and (b) without replacement. Let $X = 0$ if the first ball drawn is green, $= 1$ if the first ball drawn is red, and let $Y = 0$ if the second ball drawn is green, $= 1$ if the second ball drawn is red.

The joint PMF of (X, Y) is given in the following tables:

(a) With Replacement

$Y \backslash X$	0	1	
0	$\frac{4}{25}$	$\frac{6}{25}$	$\frac{2}{5}$
1	$\frac{6}{25}$	$\frac{9}{25}$	$\frac{3}{5}$
	$\frac{2}{5}$	$\frac{3}{5}$	1

(b) Without Replacement

$Y \backslash X$	0	1	
0	$\frac{2}{20}$	$\frac{6}{20}$	$\frac{2}{5}$
1	$\frac{6}{20}$	$\frac{6}{20}$	$\frac{3}{5}$
	$\frac{2}{5}$	$\frac{3}{5}$	1

The conditional PMFs and the conditional expectations are as follows:

(a)
$$P\{X = x \mid 0\} = \begin{cases} \frac{2}{5}, & x = 0, \\ \frac{3}{5}, & x = 1, \end{cases} \qquad P\{Y = y \mid 0\} = \begin{cases} \frac{2}{5}, & y = 0, \\ \frac{3}{5}, & y = 1, \end{cases}$$

$$P\{X = x \mid 1\} = \begin{cases} \frac{2}{5}, & x = 0, \\ \frac{3}{5}, & x = 1, \end{cases} \qquad P\{Y = y \mid 1\} = \begin{cases} \frac{2}{5}, & y = 1, \\ \frac{3}{5}, & y = 1, \end{cases}$$

$$E\{X \mid Y\} = \begin{cases} \frac{3}{5}, & y = 0 \\ \frac{3}{5}, & y = 1, \end{cases} \qquad E\{Y \mid X\} = \begin{cases} \frac{3}{5}, & x = 0, \\ \frac{3}{5}, & x = 1; \end{cases}$$

(b)
$$P\{X = x \mid 0\} = \begin{cases} \frac{1}{4}, & x = 0, \\ \frac{3}{4}, & x = 1, \end{cases} \qquad P\{Y = y \mid 0\} = \begin{cases} \frac{1}{4}, & y = 0, \\ \frac{3}{4}, & y = 1, \end{cases}$$

$$P\{X = x \mid 1\} = \begin{cases} \frac{1}{2}, & x = 0, \\ \frac{1}{2}, & x = 1, \end{cases} \qquad P\{Y = y \mid 1\} = \begin{cases} \frac{1}{2}, & y = 0, \\ \frac{1}{2}, & y = 1, \end{cases}$$

$$E\{X \mid Y\} = \begin{cases} \frac{3}{4}, & y = 0, \\ \frac{1}{2}, & y = 1, \end{cases} \qquad E\{Y \mid X\} = \begin{cases} \frac{3}{4}, & x = 0, \\ \frac{1}{2}, & x = 1. \end{cases}$$

Example 2. For the RV (X, Y) considered in Examples 4.2.5 and 4.2.7

$$E\{Y \mid x\} = \int_x^1 y f_{Y|X}(y \mid x)\, dy = \frac{1}{2} \frac{1 - x^2}{1 - x} = \frac{1 + x}{2} \qquad 0 < x < 1$$

and

$$E\{X \mid y\} = \int_0^y x f_{X|Y}(x \mid y)\, dx = \frac{y}{2}, \qquad 0 < y < 1.$$

Also,

$$E\{X^2 \mid y\} = \int_0^y x^2 \frac{1}{y}\, dx = \frac{y^2}{3}, \qquad 0 < y < 1$$

and

$$\begin{aligned} \operatorname{var}\{X \mid y\} &= E\{X^2 \mid y\} - [E\{X \mid y\}]^2 \\ &= \frac{y^2}{3} - \frac{y^2}{4} = \frac{y^2}{12}, \qquad 0 < y < 1. \end{aligned}$$

Theorem 1. Let $Eh(X)$ exist. Then,

$$Eh(X) = E\{E\{h(X) \mid Y\}\}, \tag{9}$$

Proof. Let (X, Y) be of the discrete type. Then,

$$\begin{aligned} E\{E\{h(X) \mid Y\}\} &= \sum_y \left\{ \sum_x h(x) P\{X = x \mid Y = y\} \right\} P\{Y = y\} \\ &= \sum_y \left\{ \sum_x h(x) P\{X = x, Y = y\} \right\} \\ &= \sum_x h(x) \sum_y P\{X = x, Y = y\} \\ &= Eh(X). \end{aligned}$$

The proof in the continuous case is similar.

Theorem 1 is quite useful in computation of $Eh(X)$ in many applications.

Example 3. Let X and Y be independent continuous type RVs with respective PDF f and g, and DF's F and G. Then $P\{X < Y\}$ is of interest in many statistical applications. In view of Theorem 1

$$P(X < Y) = EI_{\{X<Y\}} = E\left\{E\left\{I_{\{X<Y\}} | Y\right\}\right\},$$

where I_A is the indicator function of event A. Now

$$\begin{aligned} E\left\{I_{\{X<Y\}} | Y = y\right\} &= E\left\{I_{[X<y]} \mid y\right\} \\ &= E\left(I_{[X<y]}\right) = F(y) \end{aligned}$$

and it follows that

$$P\{X < Y\} = E\{F(Y)\} = \int_{-\infty}^{\infty} F(y) g(y) dy.$$

If, in particular, $X \stackrel{d}{=} Y$, then

$$P\{X < Y\} = \int_{-\infty}^{\infty} F(y)f(y)dy = \frac{1}{2}.$$

More generally,

$$\begin{aligned}P\{X - Y \leq z\} &= E\{E\{I_{\{X-Y\leq z\}} \mid Y\}\} = E\{F(Y+z)\} \\ &= \int_{-\infty}^{\infty} F(y+z)g(y)dy\end{aligned}$$

gives the DF of $Z = X - Y$ as computed in corollary to Theorem 4.4.3.

Example 4. Consider the joint PDF

$$f(x,y) = xe^{-x(1+y)}, \; x \geq 0, \; y \geq 0, \text{ and 0 otherwise}$$

of (X,Y). Then

$$\begin{aligned}f_X(x) &= e^{-x}, \; x \geq 0, \quad \text{and 0 otherwise} \\ f_Y(y) &= \frac{1}{(1+y)^2}, \; y \geq 0, \quad \text{and 0 otherwise.}\end{aligned}$$

Clearly, EY does not exist but

$$E\{Y \mid x\} = \int_0^{\infty} yxe^{-xy}dy = \frac{1}{x}.$$

Theorem 2. If $EX^2 < \infty$, then

$$\operatorname{var}(X) = \operatorname{var}(E\{X \mid Y\}) + E(\operatorname{var}\{X \mid Y\}). \tag{10}$$

Proof. The right-hand side of (10) equals, by definition,

$$\begin{aligned}&\{E(E\{X \mid Y\})^2 - [E(E\{X \mid Y\})]^2\} + E(E\{X^2 \mid Y\} - (E\{X \mid Y\})^2) \\ &= \{E(E\{X \mid Y\})^2 - (EX)^2\} + EX^2 - E(E\{X \mid Y\})^2 \\ &= \operatorname{var}(X).\end{aligned}$$

Corollary. If $EX^2 < \infty$, then

$$\operatorname{var}(X) \geq \operatorname{var}(E\{X \mid Y\}) \tag{11}$$

with equality if and only if X is a function of Y.

Equation (11) follows immediately from (10). The equality in (11) holds if and only if

$$E(\text{var}\{X \mid Y\}) = E(X - E\{X \mid Y\})^2 = 0,$$

which holds if and only if with probability 1

$$X = E\{X \mid Y\}. \tag{12}$$

Example 5. Let $X_1, X_2, \ldots$ be iid RVs and let N be a positive integer-valued RV. Let $S_N = \sum_{k=1}^{N} X_k$ and suppose that the X's and N are independent. Then,

$$E(S_N) = E\{E\{S_N \mid N\}\}.$$

Now,

$$E\{S_N \mid N = n\} = E\{S_n \mid N = n\} = nEX_1$$

so that

$$E(S_N) = E\{NEX_1\} = (EN)(EX_1).$$

Again, we have assumed above and below that all indicated expectations exist. Also,

$$\text{var}(S_N) = \text{var}(E\{S_N \mid N\}) + E(\text{var}\{S_N \mid N\}).$$

First,

$$\text{var}(E\{S_N \mid N\}) = \text{var}(NEX_1) = (EX_1)^2\,\text{var}(N).$$

Second,

$$\text{var}\{S_N \mid N = n\} = n\,\text{var}(X_1)$$

so

$$E(\text{var}\{S_N \mid N\}) = (EN)\,\text{var}(X_1).$$

It follows that

$$\text{Var}(S_N) = (EX_1)^2\,\text{var}(N) + (EN)\,\text{var}(X_1).$$

PROBLEMS 4.6

1. Let X be an RV with PDF given by

$$f(x) = \frac{1}{\sigma\sqrt{2\pi}} \exp\left\{-\frac{1}{2}\frac{(x-\mu)^2}{\sigma^2}\right\}, \quad -\infty < x < \infty, -\infty < \mu < \infty, \sigma > 0.$$

Find $E\{X \mid a < X < b\}$, where a and b are constants.

2. (a) Let (X, Y) be jointly distributed with density

$$f(x,y) = \begin{cases} y(1+x)^{-4}e^{-y(1+x)^{-1}}, & x,y \geq 0, \\ 0, & \text{otherwise.} \end{cases}$$

Find $E\{Y \mid X\}$.

(b) Do the same for the joint density

$$\begin{aligned} f(x,y) &= \frac{4}{5}(x+3y)e^{-x-2y}, & x,y \geq 0, \\ &= 0, & \text{otherwise.} \end{aligned}$$

3. Let (X, Y) be jointly distributed with bivariate normal density

$$f(x,y) = \frac{1}{2\pi\sigma_1\sigma_2\sqrt{1-\rho^2}} \cdot \exp\left\{\frac{1}{2(1-\rho^2)}\left[\left(\frac{x-\mu_1}{\sigma_1}\right)^2 - 2\rho\left(\frac{x-\mu_1}{\sigma_1}\right)\left(\frac{y-\mu_2}{\sigma_2}\right) + \left(\frac{y-\mu_2}{\sigma_2}\right)^2\right]\right\}.$$

Find $E\{X \mid y\}$ and $E\{Y \mid x\}$. (Here, $\mu_1, \mu_2 \in \mathcal{R}, \sigma_1, \sigma_2 > 0$, and $|\rho| < 1$.)

4. Find $E\{Y - E\{Y \mid X\}\}^2$.

5. Show that $E(Y - \phi(X))^2$ is minimized by choosing $\phi(X) = E\{Y \mid X\}$.

6. Let X have PMF

$$P_\lambda(X = x) = \lambda^x e^{-\lambda}/x!, \quad x = 0, 1, 2, \ldots$$

and suppose that λ is a realization of a RV Λ with PDF

$$f(\lambda) = e^{-\lambda},\ \lambda > 0.$$

Find $E\{e^{-\Lambda} \mid X = 1\}$.

7. Find $E(XY)$ by conditioning on X or Y for the following cases:

(a) $f(x,y) = xe^{-x(1+y)}$, $x > 0$, $y > 0$, and 0 otherwise.

(b) $f(x,y) = 2$, $0 \leq y \leq x \leq 1$, and zero otherwise.

8. Suppose X has uniform PDF $f(x) = 1, 0 \leq x \leq 1$, and 0 otherwise. Let Y be chosen from interval $(0, X]$ according to PDF

$$g(y \mid x) = \frac{1}{x}, \quad 0 < y \leq x, \text{ and 0 otherwise}$$

Find $E\{Y^k \mid X\}$ and EY^k for any fixed constant $k > 0$.

4.7 ORDER STATISTICS AND THEIR DISTRIBUTIONS

Let $(X_1,X_2,\ldots,X_n)$ be an n-dimensional random variable and $(x_1,x_2,\ldots,x_n)$ be an n-tuple assumed by $(X_1,X_2,\ldots,X_n)$. Arrange $(x_1,x_2,\ldots,x_n)$ in increasing order of magnitude so that

$$x_{(1)} \le x_{(2)} \le \cdots \le x_{(n)},$$

where $x_{(1)} = \min(x_1,x_2,\ldots,x_n)$, $x_{(2)}$ is the second smallest value in $x_1,x_2,\ldots,x_n$, and so on, $x_{(n)} = \max(x_1,x_2,\ldots,x_n)$. If any two x_i,x_j are equal, their order does not matter.

Definition 1. The function $X_{(k)}$ of $(X_1,X_2,\ldots,X_n)$ that takes on the value $x_{(k)}$ in each possible sequence $(x_1,x_2,\ldots,x_n)$ of values assumed by $(X_1,X_2,\ldots,X_n)$ is known as the *kth order statistic or statistic of order k*. $\{X_{(1)},X_{(2)},\ldots,X_{(n)}\}$ is called the set of order statistics for $(X_1,X_2,\ldots,X_n)$.

Example 1. Let X_1,X_2,X_3 be three RVs of the discrete type. Also, let X_1,X_3 take on values 0, 1, and X_2 take on values 1, 2, 3. Then the RV (X_1,X_2,X_3) assumes these triplets of values: $(0,1,0)$, $(0,2,0)$, $(0,3,0)$, $(0,1,1)$, $(0,2,1)$, $(0,3,1)$, $(1,1,0)$, $(1,2,0)$, $(1,3,0)$, $(1,1,1)$, $(1,2,1)$, $(1,3,1)$; $X_{(1)}$ takes on values 0, 1; $X_{(2)}$ takes on values 0, 1; and $X(3)$ takes on values 1, 2, 3.

Theorem 1. Let $(X_1,X_2,\ldots,X_n)$ be an n-dimensional RV. Let $X_{(k)}, 1 \le k \le n$, be the order statistic of order k. Then $X_{(k)}$ is also an RV.

Statistical considerations such as sufficiency, completeness, invariance, and ancillarity (Chapter 8) lead to the consideration of order statistics in problems of statistical inference. Order statistics are particularly useful in nonparametric statistics (Chapter 13) where, for example, many test procedures are based on ranks of observations. Many of these methods require the distribution of the ordered observations which we now study.

In the following we assume that $X_1,X_2,\ldots,X_n$ are iid RVs. In the discrete case there is no magic formula to compute the distribution of any $X_{(j)}$ or any of the joint distributions. A direct computation is the best course of action.

Example 2. Suppose X_n's are iid with geometric PMF

$$p_k = P(X=k) = pq^{k-1},\ k=1,2,\ldots,0<p<1,\ q=1-p.$$

Then for any integers $x \ge 1$ and $r \ge 1$

$$P\{X_{(r)} = x\} = P\{X_{(r)} \le x\} - P\{X_{(r)} \le x-1\}.$$

Now

$$P\{X_{(r)} \le x\} = P\{\text{At least } r \text{ of } X\text{'s are } \le x\}$$

$$= \sum_{i=1}^{r} \binom{n}{i} [P(X_1 \le x)]^i [P(X_1 > x)]^{n-i}$$

and

$$P\{X_1 \ge x) = \sum_{k=x}^{\infty} pq^{k-1} = (1-p)^{x-1}.$$

It follows that

$$P\{X_{(r)} = x\} = \sum_{i=r}^{n} \binom{n}{i} q^{(x-1)(n-i)} \{q^{n-i}[1-q^x]^i - [1-q^{x-1}]^i\},$$

$x = 1, 2, \ldots$. In particular, let $n = r = 2$. Then,

$$P\{X_{(2)} = x\} = pq^{x-1}\{pq^{x-1} + 2 - 2q^{x-1}\}, \quad x \ge 1.$$

Also for integers $x, y \ge 1$ we have

$$\begin{aligned} &P\left\{X_{(1)} = x, X_{(2)} - X_{(1)} = y\right\} = P\{X_{(1)} = x, X_{(2)} = x + y\} \\ &= P\{X_1 = x, X_2 = x + y\} + P\{X_1 = x + y, X_2 = x\} \\ &= 2pq^{x-1} \cdot pq^{x+y-1} \\ &= 2pq^{2x-2} \cdot pq^y \end{aligned}$$

and

$$P\{X_{(1)} = 1, X_{(2)} - X_{(1)} = 0\} = P\{X_{(1)} = X_{(2)} = 1\} = p^2.$$

It follows that $X_{(1)}$ and $X_{(2)} - X_{(1)}$ are independent RVs.

In the following we assume that $X_1, X_2, \ldots, X_n$ are iid RVs of the continuous type with PDF f. Let $\{X_{(1)}, X_{(2)}, \ldots, X_{(n)}\}$ be the set of order statistics for $X_1, X_2, \ldots, X_n$. Since the X_i are all continuous type RVs, it follows with probability 1 that

$$X_{(1)} < X_{(2)} < \cdots < X_{(n)}.$$

Theorem 2. The joint PDF of $(X_{(1)}, X_{(2)}, \ldots, X_{(n)})$ is given by

$$g(x_{(1)}, x_{(2)}, \ldots, x_{(n)}) = \begin{cases} n! \prod_{i=1}^{n} f(x_{(i)}), & x_{(1)} < x_{(2)} < \cdots < x_{(n)}, \\ 0, & \text{otherwise}. \end{cases} \tag{1}$$

Proof. The transformation from $(X_1, X_2, \ldots, X_n)$ to $(X_{(1)}, X_{(2)}, \ldots, X_{(n)})$ is not one-to-one. In fact, there are $n!$ possible arrangements of $x_1, x_2, \ldots, x_n$ in increasing order of magnitude. Thus there are $n!$ inverses to the transformation. For example, one of the $n!$ permutations might be

$$x_4 < x_1 < x_{n-1} < x_3 < \cdots < x_n < x_2,$$

then the corresponding inverse is

$$x_4 = x_{(1)}, \quad x_1 = x_{(2)}, \quad x_{n-1} = x_{(3)}, \quad x_3 = x_{(4)}, \ldots, x_n = x_{(n-1)},$$
$$x_2 = x_{(n)}.$$

The Jacobian of this transformation is the determinant of an $n \times n$ identity matrix with rows rearranged, since each $x_{(i)}$ equals one and only one of $x_1, x_2, \ldots, x_n$. Therefore $J = \pm 1$ and

$$g(x_{(2)}, x_{(n)}, x_{(4)}, x_{(1)}, \ldots, x_{(3)}, x_{(n-1)}) = |J| \prod_{i=1}^{n} f(x_{(i)}),$$
$$x_{(1)} < x_{(2)} < \cdots < x_{(n)}.$$

The same expression holds for each of the $n!$ arrangements.

It follows (see Remark 2) that

$$g(x_{(1)}, x_{(2)}, \ldots, x_{(n)}) = \sum_{\substack{\text{all } n! \\ \text{inverses}}} \prod_{i=1}^{n} f(x_{(i)})$$
$$= \begin{cases} n! f(x_{(1)}) f(x_{(2)}) \cdots f(x_{(n)}) & \text{if } x_{(1)} < x_{(2)} \cdots < x_{(n)}. \\ 0 & \text{otherwise.} \end{cases}$$

Example 3. Let X_1, X_2, X_3, X_4 be iid RVs with PDF f. The joint PDF of $X_{(1)}, X_{(2)}, X_{(3)}, X_{(4)}$ is

$$g(y_1, y_2, y_3, y_4) = \begin{cases} 4! f(y_1) f(y_2) f(y_3) f(y_4), & y_1 < y_2 < y_3 < y_4, \\ 0, & \text{otherwise.} \end{cases}$$

Let us compute the marginal PDF of $X_{(2)}$. We have

$$\begin{aligned} g_2(y_2) &= 4! \iiint f(y_1) f(y_2) f(y_3) f(y_4)\, dy_1\, dy_3\, dy_4 \\ &= 4! f(y_2) \int_{-\infty}^{y_2} \int_{y_2}^{\infty} \left[\int_{y_3}^{\infty} f(y_4)\, dy_4 \right] f(y_3) f(y_1)\, dy_3\, dy_1 \\ &= 4! f(y_2) \int_{-\infty}^{y_2} \left\{ \int_{y_2}^{\infty} [1 - F(y_3)] f(y_3)\, dy_3 \right\} f(y_1)\, dy_1 \\ &= 4! f(y_2) \int_{-\infty}^{y_2} \frac{[1 - F(y_2)]^2}{2} f(y_1)\, dy_1 \\ &= 4! f(y_2) \frac{[1 - F(y_2)]^2}{2!} F(y_2), \qquad y_2 \in \mathcal{R}. \end{aligned}$$

The procedure for computing the marginal PDF of $X_{(r)}$, the rth-order statistic of $X_1, X_2, \ldots, X_n$ is similar. The following theorem summarizes the result.

Theorem 3. The marginal PDF of $X_{(r)}$ is given by

$$g_r(y_r) = \frac{n!}{(r-1)!\,(n-r)!}[F(y_r)]^{r-1}[1-F(y_r)]^{n-r}f(y_r), \tag{2}$$

where F is the common DF of $X_1, X_2, \ldots, X_n$.

Proof.

$$\begin{aligned}
g_r(y_r) &= n!f(y_r)\int_{-\infty}^{y_r}\int_{-\infty}^{y_{r-1}}\cdots\int_{-\infty}^{y_2}\int_{y_r}^{\infty}\int_{y_{r+1}}^{\infty}\cdots\int_{y_{n-1}}^{\infty}\prod_{i\neq r}^{n} f(y_i)\,dy_n\cdots dy_{r+1}dy_1\cdots dy_{r-1}\\
&= n!f(y_r)\frac{[1-F(y_r)]^{n-r}}{(n-r)!}\int_{-\infty}^{y_2}\cdots\int_{-\infty}^{y_r}\prod_{i=1}^{r-1}[f(y_i)\,dy_i]\\
&= n!f(y_r)\frac{[1-F(y_r)]^{n-r}}{(n-r)!}\;\frac{[F(y_r)]^{r-1}}{(r-1)!}
\end{aligned}$$

as asserted.

We now compute the joint PDF of $X_{(j)}$ and $X_{(k)}$, $1 \le j < k \le n$.

Theorem 4. The joint PDF of $X_{(j)}$ and $X_{(k)}$ is given by

$$g_{jk}(y_j, y_k) = \begin{cases} \dfrac{n!}{(j-1)!\,(k-j-1)!\,(n-k)!}F^{j-1}(y_j)[F(y_k)- \\ F(y_j)]^{k-j-1}[1-F(y_k)]^{n-k}f(y_j)f(y_k) & \text{if } y_j < y_k, \\ 0 & \text{otherwise.} \end{cases} \tag{3}$$

Proof.

$$\begin{aligned}
g_{jk}(y_j, y_k) &= \int_{-\infty}^{y_j}\cdots\int_{-\infty}^{y_2}\int_{y_j}^{y_k}\cdots\int_{y_{k-2}}^{y_k}\int_{y_k}^{\infty}\cdots\int_{y_{n-1}}^{\infty} n!f(y_1)\cdots f(y_n)\\
&\quad \cdot dy_n\cdots dy_{k+1}\,dy_{k-1}\cdots dy_{j+1}\,dy_1\cdots dy_{j-1}\\
&= n!\int_{-\infty}^{y_j}\cdots\int_{-\infty}^{y_2}\int_{y_j}^{y_k}\cdots\int_{y_{k-2}}^{y_k}\frac{[1-F(y_k)]^{n-k}}{(n-k)!}f(y_1)f(y_2)\cdots f(y_k)\\
&\quad \cdot dy_{k-1}\cdots dy_{j+1}dy_1\cdots dy_{j-1}\\
&= n!\,\frac{[1-F(y_k)]^{n-k}}{(n-k)!}f(y_k)\int_{-\infty}^{y_j}\cdots\int_{-\infty}^{y_2}\frac{[F(y_k)-F(y_j)]^{k-j-1}}{(k-j-1)!}\\
&\quad \cdot f(y_1)f(y_2)\cdots f(y_j)\,dy_1\cdots dy_{j-1}
\end{aligned}$$

$$= \frac{n!}{(n-k)!\,(k-j-1)!}[1-F(y_k)]^{n-k}[F(y_k)-F(y_j)]^{k-j-1}$$
$$\cdot f(y_k)f(y_j)\frac{[F(y_j)]^{j-1}}{(j-1)!}, \qquad y_j < y_k,$$

as asserted.

In a similar manner we can show that the joint PDF of $X_{(j_1)},\ldots,X_{(j_k)}, 1 \le j_1 < j_2 < \cdots < j_k \le n, 1 \le k \le n$, is given by

$$g_{j_1,j_2,\ldots,j_k}(y_1,y_2\ldots,y_k) = \frac{n!}{(j_1-1)!\,(j_2-j_1-1)!\cdots(n-j_k)!}$$
$$\cdot F^{j_1-1}(y_1)f(y_1)[F(y_2)-F(y_1)]^{j_2-j_1-1}$$
$$f(y_2)\cdots[1-F(y_k)]^{n-j_k}f(y_k)$$

for $y_1 < y_2 < \cdots < y_k$, and $= 0$ otherwise.

Example 4. Let $X_1,X_2,\ldots,X_n$ be iid RVs with common PDF

$$f(x) = \begin{cases} 1 & \text{if } 0 < x < 1, \\ 0 & \text{otherwise.} \end{cases}$$

Then,

$$g_r(y_r) = \begin{cases} \dfrac{n!}{(r-1)!\,(n-r)!}y_r^{r-1}(1-y_r)^{n-r}, & 0 < y_r < 1 \\ & (1 \le r \le n), \\ 0, & \text{otherwise.} \end{cases}$$

The joint distribution of $X_{(j)}$ and $X_{(k)}$ is given by

$$g_{jk}(y_j,y_k) = \begin{cases} \dfrac{n!}{(j-1)!\,(k-j-1)!\,(n-k)!}y_j^{j-1}(y_k-y_j)^{k-j-1}(1-y_k)^{n-k}, & \\ & 0 < y_j < y_k < 1, \\ 0, & \text{otherwise,} \end{cases}$$

where $1 \le j < k \le n$.

The joint PDF of $X_{(1)}$ and $X_{(n)}$ is given by

$$g_{1n}(y_1,y_n) = n(n-1)(y_n-y_1)^{n-2} \quad 0 < y_1 < y_n < 1$$

and that of the range $R_n = X_{(n)} - X_{(1)}$ by

$$g_{R_n}(w) = \begin{cases} n(n-1)w^{n-2}(1-w), & 0 < w < 1, \\ 0, & \text{otherwise.} \end{cases}$$

Example 5. Let $X_{(1)}, X_{(2)}, X_{(3)}$ be the order statistics of iid RVs X_1, X_2, X_3 with common PDF

$$f(x) = \begin{cases} \beta e^{-x\beta}, & x > 0, \\ 0, & \text{otherwise}, \end{cases} \qquad (\beta > 0).$$

Let $Y_1 = X_{(3)} - X_{(2)}$ and $Y_2 = X_{(2)}$. We show that Y_1 and Y_2 are independent. The joint PDF of $X_{(2)}$ and $X_{(3)}$ is given by

$$g_{23}(x,y) = \begin{cases} \dfrac{3!}{1!0!0!}(1 - e^{-\beta x})\beta e^{-\beta x}\beta e^{-\beta y}, & x < y, \\ 0, & \text{otherwise}. \end{cases}$$

The PDF of (Y_1, Y_2) is

$$\begin{aligned} f(y_1, y_2) &= 3!\,\beta^2(1 - e^{-\beta y_2})e^{-\beta y_2}e^{-(y_1+y_2)\beta} \\ &= \begin{cases} \{3!\,\beta e^{-2\beta y_2}(1 - e^{-\beta y_2})\}\{\beta e^{-\beta y_1}\}, & 0 < y_1 < \infty, 0 < y_2 < \infty, \\ 0, & \text{otherwise}. \end{cases} \end{aligned}$$

It follows that Y_1 and Y_2 are independent.

Finally, we consider the moments, namely, the means, variances, and covariances of order statistics. Suppose $X_1, X_2, \ldots, X_n$ are iid RVs with common DF F. Let g be a Borel function on $\mathcal{R}$ such that $E|g(X)| < \infty$, where X has DF F. Then for $1 \le r \le n$

$$\begin{aligned} &\left| \int_{-\infty}^{\infty} g(x) \frac{n!}{(n-r)!(r-1)!} [F(x)]^{r-1}[1 - F(x)]^{n-r} f(x)dx \right| \\ &\le n\binom{n-1}{r-1} \int_{-\infty}^{\infty} |g(x)| f(x)dx \quad (0 \le F \le 1) \\ &< \infty \end{aligned}$$

and we write

$$Eg(X_{(r)}) = \int_{-\infty}^{\infty} g(y) g_r(y) dy,$$

for $r = 1, 2, \ldots, n$. The converse also holds. Suppose $E|g(X_{(r)})| < \infty$ for $r = 1, 2, \ldots, n$. Then,

$$n\binom{n-1}{r-1} \int_{-\infty}^{\infty} |g(x)| F^{r-1}(x)[1 - F(x)]^{n-r} f(x)dx < \infty,$$

for $r = 1,2,\ldots,n$ and hence

$$n\int_{-\infty}^{\infty}\left\{\sum_{r=1}^{n}\binom{n-1}{r-1}F^{r-1}(x)[1-F(x)]^{n-r}\right\}|g(x)|f(x)dx$$
$$= n\int_{-\infty}^{\infty}|g(x)|f(x)dx < \infty.$$

Moreover, it also follows that

$$\sum_{r=1}^{n} Eg(X_{(r)}) = nEg(X).$$

As a consequence of the above remarks we note that if $E|g(X_{(r)})| = \infty$ for some r, $1 \leq r \leq n$, then $E|g(X)| = \infty$ and conversely, if $E|g(X)| = \infty$ then $E|g|X_{(r)})| = \infty$ for some r, $1 \leq r \leq n$.

Example 6. Let $X_1, X_2, \ldots, X_n$ be iid with *Pareto* PDF $f(x) = 1/x^2$, if $x \geq 1$, and $= 0$ otherwise. Then $EX = \infty$. Now for $1 \leq r \leq n$

$$EX_{(r)} = n\binom{n-1}{r-1}\int_{1}^{\infty} x\left(1-\frac{1}{x}\right)^{r-1}\frac{1}{x^{n-r}}\frac{dx}{x^2}$$
$$= n\binom{n-1}{r-1}\int_{0}^{1} y^{r-1}(1-y)^{n-r-1}dy.$$

Since the integral on the right side converges for $1 \leq r \leq n-1$ and diverges for $r > n-k$, we see that $EX_{(r)} = \infty$ for $r = n-k+1,\ldots,n$.

PROBLEMS 4.7

1. Let $X_{(1)}, X_{(2)}, \ldots X_{(n)}$ be the set of order statistics of independent RVs $X_1, X_2, \ldots, X_n$ with common PDF

$$f(x) = \begin{cases} \beta e^{-x\beta} & \text{if } x \geq 0, \\ 0 & \text{otherwise.} \end{cases}$$

(a) Show that $X_{(r)}$ and $X_{(s)} - X_{(r)}$ are independent for any $s > r$.

(b) Find the PDF of $X_{(r+1)} - X_{(r)}$.

(c) Let $Z_1 = nX_{(1)}, Z_2 = (n-1)(X_{(2)} - X_{(1)}), Z_3 = (n-2)(X_{(3)} - X_{(2)}), \ldots, Z_n = (X_{(n)} - X_{(n-1)}$. Show that $(Z_1, Z_2, \ldots, Z_n)$ and $(X_1, X_2, \ldots, X_n)$ are identically distributed.

2. Let $X_1, X_2, \ldots, X_n$ be iid from PMF

$$p_k = 1/N, \quad k = 1,2,\ldots,N.$$

Find the marginal distributions of $X_{(1)}$, $X_{(n)}$, and their joint PMF.

3. Let $X_1, X_2, \ldots, X_n$ be iid with a DF

$$f(y) = \begin{cases} y^\alpha & \text{if } 0 < y < 1, \\ 0 & \text{otherwise,} \end{cases} \qquad \alpha > 0.$$

Show that $X_{(i)}/X_{(n)}, i = 1, 2, \ldots, n-1$, and $X_{(n)}$ are independent.

4. Let $X_1, X_2, \ldots, X_n$ be iid RVs with common Pareto DF $f(x) = \alpha\sigma^\alpha/x^{\alpha+1}$, $x > \sigma$ where $\alpha > 0$, $\sigma > 0$. Show that

(a) $X_{(1)}$ and $(X_{(2)}/X_{(1)}, \ldots, X_{(n)}/X_{(1)})$ are independent,

(b) $X_{(1)}$ has Pareto $(\sigma, n\alpha)$ distribution, and

(c) $\sum_{j=1}^{n} \ell n(X_{(j)}/X_{(1)})$ has PDF

$$f(x) = \frac{x^{n-2}e^{-\alpha x}}{(n-2)!}, \qquad x > 0.$$

5. Let $X_1, X_2, \ldots, X_n$ be iid nonnegative RVs of the continuous type. If $E|X| < \infty$, show that $E|X_{(r)}| < \infty$. Write $M_n = X_{(n)} = \max(X_1, X_2, \ldots, X_n)$. Show that

$$EM_n = EM_{n-1} + \int_0^\infty F^{n-1}(x)[1 - F(x)]\,dx, \qquad n = 2, 3, \ldots.$$

Find EM_n in each of the following cases:

(a) X_i have the common DF

$$F(x) = 1 - e^{-\beta x}, \qquad x \geq 0.$$

(b) X_i have the common DF

$$F(x) = x, 0 < x < 1.$$

6. Let $X_{(1)}, X_{(2)}, \ldots, X_{(n)}$ be the order statistics of n independent RVs $X_1, X_2, \ldots, X_n$ with common PDF $f(x) = 1$ if $0 < x < 1$, and $= 0$ otherwise. Show that $Y_1 = X_{(1)}/X_{(2)}$, $Y_2 = X_{(2)}/X_{(3)}, \ldots, Y_{(n-1)} = X_{(n-1)}/X_{(n)}$, and $Y_n = X_{(n)}$ are independent. Find the PDFs of $Y_1, Y_2, \ldots, Y_n$.

7. For the PDF in Problem 4 find $EX_{(r)}$.

8. An urn contains N identical marbles numbered 1 through N. From the urn n marbles are drawn and let $X_{(n)}$ be the largest number drawn. Show that $P(X_{(n)} = k) = \binom{k-1}{n-1}/\binom{N}{n}$, $k = n, n+1, \ldots, N$, and $EX_{(n)} = n(N+1)/(n+1)$.

5

SOME SPECIAL DISTRIBUTIONS

5.1 INTRODUCTION

In preceding chapters we studied probability distributions in general. In this chapter we will study some commonly occurring probability distributions and investigate their basic properties. The results of this chapter will be of considerable use in theoretical as well as practical applications. We begin with some discrete distributions in Section 5.2 and follow with some continuous models in Section 5.3. Section 5.4 deals with bivariate and multivariate normal distributions and in Section 5.5 we discuss the exponential family of distributions.

5.2 SOME DISCRETE DISTRIBUTIONS

In this section we study some well-known univariate and multivariate discrete distributions and describe their important properties.

5.2.1 Degenerate Distribution

The simplest distribution is that of an RV X *degenerate* at point k, that is, $P\{X = k\} = 1$ and $= 0$ elsewhere. If we define

$$\varepsilon(x) = \begin{cases} 0 & \text{if } x < 0, \\ 1 & \text{if } x \geq 0, \end{cases} \tag{1}$$

An Introduction to Probability and Statistics, Third Edition. Vijay K. Rohatgi and A.K. Md. Ehsanes Saleh.

the DF of the RV X is $\varepsilon(x-k)$. Clearly, $EX^l = k^l, l = 1,2,\ldots$, and $M(t) = e^{tk}$. In particular, $\text{var}(X) = 0$. This property characterizes a degenerate RV. As we shall see, the degenerate RV plays an important role in the study of limit theorems.

5.2.2 Two-Point Distribution

We say that an RV X has a *two-point distribution* if it takes two values, x_1 and x_2, with probabilities

$$P\{X = x_1\} = p, \qquad P\{X = x_2\} = 1-p, \qquad 0 < p < 1.$$

We may write

$$X = x_1\, I_{[X=x_1]} + x_2 I_{[X=x_2]}, \tag{2}$$

where I_A is the indicator function of A. The DF of X is given by

$$F(x) = p\varepsilon(x-x_1) + (1-p)\varepsilon(x-x_2). \tag{3}$$

Also

$$EX^k = px_1^k + (1-p)x_2^k, \qquad k = 1,2,\ldots, \tag{4}$$

$$M(t) = pe^{tx_i} + (1-p)e^{tx_2} \qquad \text{for all } t. \tag{5}$$

In particular,

$$EX = px_1 + (1-p)x_2 \tag{6}$$

and

$$\text{var}(X) = p(1-p)(x_1-x_2)^2. \tag{7}$$

If $x_1 = 1$, $x_2 = 0$, we get the important *Bernoulli* RV:

$$P\{X = 1\} = p, \qquad P\{X = 0\} = 1-p, \qquad 0 < p < 1. \tag{8}$$

For a Bernoulli RV X with parameter p, we write $X \sim b(1,p)$ and have

$$EX = p, \qquad \text{var}(X) = p(1-p), \qquad M(t) = 1 + p(e^t - 1), \quad \text{all } t. \tag{9}$$

Bernoulli RVs occur in practice, for example, in coin-tossing experiments. Suppose that $P\{\text{H}\} = p$, $0 < p < 1$, and $P\{\text{T}\} = 1-p$. Define RV X so that $X(\text{H}) = 1$ and $X(\text{T}) = 0$. Then $P\{X = 1\} = p$ and $P\{X = 0\} = 1-p$. Each repetition of the experiment will be called a *trial*. More generally, any nontrivial experiment can be dichotomized to yield a Bernoulli model. Let $(\Omega, \mathcal{S}, P)$ be the sample space of an experiment, and let $A \in \mathcal{S}$ with $P(A) = p > 0$. Then $P(A^c) = 1-p$. Each performance of the experiment is a Bernoulli trial. It will be convenient to call the occurrence of event A a *success* and the occurrence of A^c, a *failure*.

Example 1 (Sabharwal [97]). In a sequence of n Bernoulli trials with constant probability p of success (S), and $1-p$ of failure (F), let Y_n denote the number of times the combination SF occurs. To find EY_n and $\text{var}(Y_n)$, let X_i represent the event that occurs on the ith trial, and define RVs

$$f(X_i, X_{i+1}) = \begin{cases} 1 & \text{if } X_i = S,\ X_{i+1} = F, \\ 0 & \text{otherwise.} \end{cases} \qquad (i = 1, 2, \ldots, n-1).$$

Then

$$Y_n = \sum_{i=1}^{n-1} f(X_i, X_{i+1})$$

and

$$EY_n = (n-1)p(1-p).$$

Also,

$$\begin{aligned} EY_n^2 &= E\left\{\sum_{i=1}^{n-1} f^2(X_i, X_{i+1})\right\} + E\left\{\sum\sum_{i \neq j} f(X_i, X_{i+1}) f(X_j X_{j+1})\right\} \\ &= (n-1)p(1-p) + (n-2)(n-3)p^2(1-p)^2, \end{aligned}$$

so that

$$\text{var}(Y_n) = (n-1)p(1-p)(1-p+p^2).$$

If $p = 1/2$, then

$$EY_n = \frac{n-1}{4} \qquad \text{and} \qquad \text{var}(Y_n) = \frac{3(n-1)}{16}.$$

5.2.3 Uniform Distribution on n Points

X is said to have a *uniform distribution* on n points $\{x_1, x_2, \ldots, x_n\}$ if its PMF is of the form

$$P\{X = x_i\} = \frac{1}{n}, \qquad i = 1, 2, \ldots, n. \tag{10}$$

Thus we may write

$$X = \sum_{i=1}^{n} x_i I_{[X = x_i]} \qquad \text{and} \qquad F(x) = \frac{1}{n} \sum_{i=1}^{n} \varepsilon(x - x_i),$$

$$EX = \frac{1}{n} \sum_{i=1}^{n} x_i, \tag{11}$$

$$EX^l = \frac{1}{n} \sum_{i=1}^{n} x_i^l, \qquad l = 1, 2, \ldots, \tag{12}$$

and

$$\operatorname{var}(X) = \frac{1}{n}\sum_{i=1}^{n} x_i^2 - \left(\frac{1}{n}\sum x_i\right)^2 = \frac{1}{n}\sum_{i=1}^{n}(x_i - \bar{x})^2 \tag{13}$$

if we write $\bar{x} = \sum_{i=1}^{n} x_i/n$. Also,

$$M(t) = \frac{1}{n}\sum_{i=1}^{n} e^{tx_i} \qquad \text{for all } t. \tag{14}$$

If, in particular, $x_i = i$, $i = 1, 2, \ldots, n$,

$$EX = \frac{n+1}{2}, \qquad EX^2 = \frac{(n+1)(2n+1)}{6}, \tag{15}$$

$$\operatorname{var}(X) = \frac{n^2 - 1}{12}. \tag{16}$$

Example 2. A box contains tickets numbered 1 to N. Let X be the largest number drawn in n random drawings with replacement.

Then $P\{X \le k\} = (k/N)^n$, so that

$$\begin{aligned} P\{X = k\} &= P\{X \le k\} - P\{X \le k-1\} \\ &= \left(\frac{k}{N}\right)^n - \left(\frac{k-1}{N}\right)^n. \end{aligned}$$

Also,

$$\begin{aligned} EX &= N^{-n}\sum_{1}^{N}[k^{n+1} - (k-1)^{n+1} - (k-1)^n] \\ &= N^{-n}\left[N^{n+1} - \sum_{1}^{N}(k-1)^n\right]. \end{aligned}$$

5.2.4 Binomial Distribution

We say that X has a *binomial distribution* with parameter p if its PMF is given by

$$p_k = P\{X = k\} = \binom{n}{k} p^k (1-p)^{n-k}, \qquad k = 0, 1, 2, \ldots, n; \quad 0 \le p \le 1. \tag{17}$$

Since $\sum_{k=0}^{n} p_k = [p + (1-p)]^n = 1$, the p_k's indeed define a PMF. If X has PMF (17), we will write $X \sim b(n,p)$. This is consistent with the notation for a Bernoulli RV. We have

$$F(x) = \sum_{k=0}^{n} \binom{n}{k} p^k (1-p)^{n-k} \varepsilon(x-k).$$

In Example 3.2.5 we showed that

$$EX = np, \tag{18}$$
$$EX^2 = n(n-1)p^2 + np, \tag{19}$$

and

$$\text{var}(X) = np(1-p) = npq, \tag{20}$$

where $q = 1-p$. Also,

$$\begin{aligned} M(t) &= \sum_{k=0}^{n} e^{tk}\binom{n}{k}p^k(1-p)^{n-k} \\ &= (q+pe^t)^n \qquad \text{for all } t. \end{aligned} \tag{21}$$

The PGF of $X \sim b(n,p)$ is given by $P(s) = \{1-p(1-s)\}^n$, $|s| \le 1$.

Binomial distribution can also be considered as the distribution of the sum of n independent, identically distributed $b(1,p)$ random variables. If we toss a coin, with constant probability p of heads and $1-p$ of tails, n times, the distribution of the number of heads is given by (17). Alternatively, if we write

$$X_k = \begin{cases} 1 & \text{if } k\text{th toss results in a head,} \\ 0 & \text{otherwise,} \end{cases}$$

the number of heads in n trials is the sum $S_n = X_1 + X_2 + \cdots + X_n$. Also

$$P\{X_k = 1\} = p, \qquad P\{X_k = 0\} = 1-p, \qquad k = 1,2,\ldots,n.$$

Thus

$$ES_n = \sum_{1}^{n} EX_i = np,$$
$$\text{var}(S_n) = \sum_{1}^{n} \text{var}(X_i) = np(1-p),$$

and

$$\begin{aligned} M(t) &= \prod_{i=1}^{n} Ee^{tX_i} \\ &= (q+pe^t)^n. \end{aligned}$$

Theorem 1. Let $X_i (i = 1,2,\ldots,k)$ be independent RVs with $X_i \sim b(n_i,p)$. Then $S_k = \sum_{i=1}^{k} X_i$ has a $b(n_1 + n_2 + \cdots + n_k,\ p)$ distribution.

Corollary. If $X_i (i = 1,2,\ldots,k)$ are iid RVs with common PMF $b(n,p)$, then S_k has a $b(nk,p)$ distribution.

Actually, the additive property described in Theorem 1 characterizes the binomial distribution in the following sense. Let X and Y be two independent, nonnegative, finite integer-valued RVs and let $Z = X + Y$. Then Z is a binomial RV with parameter p if and only if X and Y are binomial RVs with the same parameter p. The "only if" part is due to Shanbhag and Basawa [103] and will not be proved here.

Example 3. A fair die is rolled n times. The probability of obtaining exactly one 6 is $n(\frac{1}{6})(\frac{5}{6})^{n-1}$, the probability of obtaining no 6 is $(\frac{5}{6})^n$, and the probability of obtaining at least one 6 is $1-(\frac{5}{6})^n$.

The number of trials needed for the probability of at least one 6 to be $\geq 1/2$ is given by the smallest integer n such that

$$1-\left(\frac{5}{6}\right)^n \geq \frac{1}{2}$$

so that

$$n \geq \frac{\log 2}{\log 1.2} \approx 3.8.$$

Example 4. Here r balls are distributed in n cells so that each of n^r possible arrangements has probability n^{-r}. We are interested in the probability p_k that a specified cell has exactly k balls ($k = 0, 1, 2, \ldots, r$). Then the distribution of each ball may be considered as a trial. A success results if the ball goes to the specified cell (with probability $1/n$); otherwise the trial results in a failure (with probability $1-1/n$). Let X denote the number of successes in r trials. Then

$$p_k = P\{X=k\} = \binom{r}{k}\left(\frac{1}{n}\right)^k\left(1-\frac{1}{n}\right)^{r-k}, \qquad k = 0, 1, 2, \ldots, n.$$

5.2.5 Negative Binomial Distribution (Pascal or Waiting Time Distribution)

Let $(\Omega, \mathcal{S}, P)$ be a probability space of a given statistical experiment, and let $A \in \mathcal{S}$ with $P(A) = p$. On any performance of the experiment, if A happens we call it a success, otherwise a failure. Consider a succession of trials of this experiment, and let us compute the probability of observing exactly r successes, where $r \geq 1$ is a fixed integer. If X denotes the number of failures that precede the rth success, $X + r$ is the total number of replications needed to produce r successes. This will happen if and only if the last trial results in a success and among the previous $(r + X - 1)$ trials there are exactly X failures. It follows by independence that

$$P\{X=x\} = \binom{x+r-1}{x} p^r (1-p)^x, \qquad x = 0, 1, 2, \ldots. \tag{22}$$

Rewriting (22) in the form

$$P\{X=x\} = \binom{-r}{x} p^r (-q)^x, \qquad x = 0, 1, 2, \ldots; \quad q = 1-p, \tag{23}$$

we see that

$$\sum_{x=0}^{\infty}\binom{-r}{x}(-q)^x = (1-q)^{-r} = p^{-r}. \tag{24}$$

It follows that

$$\sum_{x=0}^{\infty} P\{X=x\} = 1.$$

Definition 1. For a fixed positive integer $r \geq 1$ and $0 < p < 1$, an RV with PMF given by (22) is said to have a *negative binomial distribution*. We will use the notation $X \sim NB(r;p)$ to denote that X has a negative binomial distribution.

We may write

$$X = \sum_{x=0}^{\infty} x I_{[X=x]} \qquad \text{and} \qquad F(x) = \sum_{k=0}^{\infty}\binom{k+r-1}{k} p^r(1-p)^k \varepsilon(x-k).$$

For the MGF of X we have

$$\begin{aligned} M(t) &= \sum_{x=0}^{\infty}\binom{x+r-1}{x} p^r(1-p)^x e^{tx} \\ &= p^r \sum_{x=0}^{\infty}(qe^t)^x \binom{x+r-1}{x} \qquad (q = 1-p) \\ &= p^r(1-qe^t)^{-r} \qquad \text{for } qe^t < 1. \end{aligned} \tag{25}$$

The PGF is given by $P(s) = p^r(1-sq)^{-r}$, $|s| \leq 1$. Also,

$$\begin{aligned} EX &= \sum_{x=0}^{\infty} x\binom{x+r-1}{x} p^r q^x \\ &= rp^r \sum_{x=0}^{\infty}\binom{x+r}{x} q^{x+1} \\ &= rp^r q(1-q)^{-r-1} = \frac{rq}{p}. \end{aligned} \tag{26}$$

Similarly, we can show that

$$\operatorname{var}(X) = \frac{rq}{p^2}. \tag{27}$$

If, however, we are interested in the distribution of the number of trials required to get r successes, we have, writing $Y = X + r$,

$$P\{Y=y\} = \binom{y-1}{r-1} p^r(1-p)^{y-r}, \qquad y = r, r+1, \ldots, \tag{28}$$

$$\begin{cases} EY = EX + r = \dfrac{r}{p}, \\ \operatorname{var}(Y) = \operatorname{var}(X) = \dfrac{rq}{p^2}, \end{cases} \tag{29}$$

and

$$M_Y(t) = (pe^t)^r(1-qe^t)^{-r} \qquad \text{for } qe^t < 1. \tag{30}$$

Let X be a $b(n,p)$ RV, and let Y be the RV defined in (28). If there are r or more successes in the first n trials, at most n trials were required to obtain the first r of these successes. We have

$$P\{X \geq r\} = P\{Y \leq n\} \tag{31}$$

and also

$$P\{X < r\} = P\{Y > n\}. \tag{32}$$

In the special case when $r = 1$, the distribution of X is given by

$$P\{X = x\} = pq^x, \qquad x = 0,1,2,\ldots. \tag{33}$$

An RV X with PMF (33) is said to have a *geometric distribution*. Clearly, for the geometric distribution, we have

$$\begin{cases} M(t) = p(1-qe^t)^{-1}, \\ EX = \dfrac{q}{p}, \\ \operatorname{var}(X) = \dfrac{q}{p^2}. \end{cases} \tag{34}$$

Example 5 **(Banach's matchbox problem).** A mathematician carries one matchbox each in his right and left pockets. When he wants a match, he selects the left pocket with probability p and the right pocket with probability $1-p$. Suppose that initially each box contains N matches. Consider the moment when the mathematician discovers that a box is empty. At that time the other box may contain $0,1,2\ldots,N$ matches. Let us identify success with the choice of the left pocket. The left-pocket box will be empty at the moment when the right-pocket box contains exactly r matches if and only if exactly $N-r$ failures precede the $(N+1)$st success. A similar argument applies to the right pocket, and we have

$$\begin{aligned} p_r &= \text{probability that the mathematician discovers a box empty while} \\ &\quad\ \text{the other contains } r \text{ matches} \\ &= \binom{2N-r}{N-r} p^{N+1} q^{N-r} + \binom{2N-r}{N-r} q^{N+1} p^{N-r}. \end{aligned}$$

Example 6. A fair die is rolled repeatedly. Let us compute the probability of event A that a 2 will show up before a 5. Let A_j be the event that a 2 shows up on the jth trial $(j = 1,2,\ldots)$

for the first time, and a 5 does not show up on the previous $j-1$ trials. Then $PA = \sum_{j=1}^{\infty} PA_j$, where $PA_j = \frac{1}{6}(\frac{4}{6})^{j-1}$. It follows that

$$P(A) = \sum_{j=1}^{\infty} \left(\frac{1}{6}\right)\left(\frac{4}{6}\right)^{j-1} = \frac{1}{2}.$$

Similarly the probability that a 2 will show up before a 5 or a 6 is $1/3$, and so on.

Theorem 2. Let $X_1, X_2, \ldots, X_k$ be independent $NB(r_i; p)$ RV's, $i = 1, 2, \ldots, k$, respectively. Then $S_k = \sum_{i=1}^{k} X_i$ is distributed as $NB(r_1 + r_2 + \cdots + r_k; p)$.

Corollary. If $X_1, X_2, \ldots, X_k$ are iid geometric RVs, then S_k is an $NB(k; p)$ RV.

Theorem 3. Let X and Y be independent RVs with PMFs $NB(r_1; p)$ and $NB(r_2; p)$, respectively. Then the conditional PMF of X, given $X + Y = t$, is expressed by

$$P\{X = x | X + Y = t\} = \frac{\binom{x+r_1-1}{x}\binom{t+r_2-x-1}{t-x}}{\binom{t+r_1+r_2-1}{t}}.$$

If, in particular, $r_1 = r_2 = 1$, the conditional distribution is uniform on $t+1$ points.

Proof. By Theorem 2, $X + Y$ is an $NB(r_1 + r_2; p)$ RV. Thus

$$\begin{aligned} P\{X = x | X + Y = t\} &= \frac{P\{X = x,\ Y = t - x\}}{P\{X + Y = t\}} \\ &= \frac{\binom{x+r_1-1}{x} p^{r_1}(1-p)^x \binom{t-x+r_2-1}{t-x} p^{r_2}(1-p)^{t-x}}{\binom{t+r_1+r_2-1}{t} p^{r_1+r_2}(1-p)^t} \\ &= \frac{\binom{x+r_1-1}{x}\binom{t+r_2-x-1}{t-x}}{\binom{t+r_1+r_2-1}{t}}, \qquad t = 0, 1, 2, \ldots. \end{aligned}$$

If $r_1 = r_2 = 1$, that is, if X and Y are independent geometric RVs, then

$$P\{X = x | X + Y = t\} = \frac{1}{t+1}, \qquad x = 0, 1, 2, \ldots, t; \quad t = 0, 1, 2, \ldots. \tag{35}$$

Theorem 4 (Chatterji [13]). Let X and Y be iid RVs, and let

$$P\{X = k\} = p_k > 0, \qquad k = 0, 1, 2, \ldots.$$

If

$$P\{X = t | X + Y = t\} = P\{X = t - 1 | X + Y = t\} = \frac{1}{t+1}, \qquad t \geq 0, \tag{36}$$

then X and Y are geometric RVs.

Proof. We have

$$P\{X=t|X+Y=t\}=\frac{p_t p_0}{\sum_{k=0}^{t} p_k p_{t-k}}=\frac{1}{t+1} \tag{37}$$

and

$$P\{X=t-1|X+Y=t\}=\frac{p_{t-1} p_1}{\sum_{k=0}^{t} p_k p_{t-k}}=\frac{1}{t+1}. \tag{38}$$

It follows that

$$\frac{p_t}{p_{t-1}}=\frac{p_1}{p_0}$$

and by iteration $p_t=(p_1/p_0)^t p_0$. Since $\sum_{t=0}^{\infty} p_t=1$, we must have $(p_1/p_0)<1$. Moreover,

$$1=p_0\frac{1}{1-(p_1/p_0)},$$

so that $(p_1/p_0)=1-p_0$, and the proof is complete.

Theorem 5. If X has a geometric distribution, then, for any two nonnegative integers m and n,

$$P\{X>m+n|X>m\}=P\{X\geq n\}. \tag{39}$$

Proof. The proof is left as an exercise.

Remark 1. Theorem 5 says that the geometric distribution has *no memory*, that is, the information of no successes in m trials is forgotten in subsequent calculations.

The converse of Theorem 5 is also true.

Theorem 6. Let X be a nonnegative integer-valued RV satisfying

$$P\{X>m+1|X>m\}=P\{X\geq 1\}$$

for any nonnegative integer m. Then X must have a geometric distribution.

Proof. Let the PMF of X be written as

$$P\{X=k\}=p_k, \qquad k=0,1,2,\ldots.$$

Then

$$P\{X\geq n\}=\sum_{k=n}^{\infty} p_k$$

and

$$P\{X > m\} = \sum_{m+1}^{\infty} p_k = q_m, \quad \text{say},$$

$$P\{X > m+1 | X > m\} = \frac{P\{X > m+1\}}{P\{X > m\}} = \frac{q_{m+1}}{q_m}.$$

Thus

$$q_{m+1} = q_m q_0,$$

where $q_0 = P\{X > 0\} = p_1 + p_2 + \cdots = 1 - p_0$. It follows that $q_k = (1-p_0)^{k+1}$, and hence $p_k = q_{k-1} - q_k = (1-p_0)^k p_0$, as asserted.

Theorem 7. Let $X_1, X_2, \ldots, X_n$ be independent geometric RVs with parameters $p_1, p_2, \ldots, p_n$, respectively. Then $X_{(1)} = \min(X_1, X_2, \ldots, X_n)$ is also a geometric RV with parameter

$$p = 1 - \prod_{i=1}^{n} (1 - p_i).$$

Proof. The proof is left as an exercise.

Corollary. Iid RVs $X_1, X_2, \ldots, X_n$ are $NB(1; p)$ if and only if $X_{(1)}$ is a geometric RV with parameter $1-(1-p)^n$.

Proof. The necessity follows from Theorem 7. For the sufficiency part of the proof let

$$P\{X_{(1)} \leq k\} = 1 - P\{X_{(1)} > k\} = 1 - (1-p)^{n(k+1)}.$$

But

$$\begin{aligned} P\{X_{(1)} \leq k\} &= 1 - P\{X_1 > k, X_2 > k, \ldots, X_n > k\} \\ &= 1 - [1 - F(k)]^n, \end{aligned}$$

where F is the common DF of $X_1, X_2, \ldots, X_n$. It follows that

$$[1 - F(k)] = (1-p)^{k+1},$$

so that $P\{X_1 > k\} = (1-p)^{k+1}$, which completes the proof.

5.2.6 Hypergeometric Distribution

A box contains N marbles. Of these, M are drawn at random, marked, and returned to the box. The contents of the box are then thoroughly mixed. Next, n marbles are drawn at random from the box, and the marked marbles are counted. If X denotes the number of marked marbles, then

$$P\{X = x\} = \binom{N}{n}^{-1} \binom{M}{x} \binom{N-M}{n-x}. \tag{40}$$

Since x cannot exceed M or n, we must have

$$x \leq \min(M, n). \tag{41}$$

Also $x \geq 0$ and $N - M \geq n - x$, so that

$$x \geq \max(0, M + n - N). \tag{42}$$

Note that

$$\sum_{k=0}^{n} \binom{a}{k} \binom{b}{n-k} = \binom{a+b}{n}$$

for arbitrary numbers a, b and positive integer n. It follows that

$$\sum_{x} P\{X = x\} = \binom{N}{n}^{-1} \sum_{x} \binom{M}{x} \binom{N-M}{n-x} = 1.$$

Definition 2. An RV X with PMF given by (47) is called a *hypergeometric* RV.

It is easy to check that

$$EX = \frac{n}{N} M, \tag{43}$$

$$EX^2 = \frac{M(M-1)}{N(N-1)} n(n-1) + \frac{nM}{N}, \tag{44}$$

and

$$\text{var}(X) = \frac{nM}{N^2(N-1)} (N - M)(N - n). \tag{45}$$

Example 7. A lot consisting of 50 bulbs is inspected by taking at random 10 bulbs and testing them. If the number of defective bulbs is at most 1, the lot is accepted; otherwise, it is rejected. If there are, in fact, 10 defective bulbs in the lot, the probability of accepting the lot is

$$\frac{\binom{10}{1}\binom{40}{9}}{\binom{50}{10}} + \frac{\binom{40}{10}}{\binom{50}{10}} = .3487.$$

Example 8. Suppose that an urn contains b white and c black balls, $b + c = N$. A ball is drawn at random, and before drawing the next ball, $s + 1$ balls of the same color are added to the urn. The procedure is repeated n times. Let X be the number of white balls drawn in n draws, $X = 0, 1, 2, \ldots, n$. We shall find the PMF of X.

First note that the probability of drawing k white balls in successive draws is

$$\left(\frac{b}{N}\right)\left(\frac{b+s}{N+s}\right)\left(\frac{b+2s}{N+2s}\right)\cdots\left[\frac{b+(k-1)s}{N+(k-1)s}\right],$$

and the probability of drawing k white balls in the first k draws and then $n-k$ black balls in the next $n-k$ draws is

$$p_k = \left(\frac{b}{N}\right)\left(\frac{b+s}{N+s}\right)\cdots\left[\frac{b+(k-1)s}{N+(k-1)s}\right]\left(\frac{c}{N+ks}\right)\left[\frac{c+s}{N+(k+1)s}\right] \cdots\left[\frac{c+(n-k-1)s}{N+(n-1)s}\right]. \tag{46}$$

Here p_k also gives the probability of drawing k white and $n-k$ black balls in any given order. It follows that

$$P\{X=k\} = \binom{n}{k} p_k. \tag{47}$$

An RV X with PMF given by (47) is said to have a *Polya distribution*. Let us write

$$Np = b, \qquad N(1-p) = c, \qquad \text{and} \qquad N\alpha = s.$$

Then with $q = 1-p$, we have

$$P\{X=k\} = \binom{n}{k} \frac{p(p+\alpha)\cdots[p+(k-1)\alpha]q(q+\alpha)\cdots[q+(n-k-1)\alpha]}{1(1+\alpha)\cdots[1+(n-1)\alpha]}.$$

Let us take $s=-1$. This means that the ball drawn at each draw is not replaced in the urn before drawing the next ball. In this case $\alpha = -1/N$, and we have

$$\begin{aligned} P\{X=k\} &= \binom{n}{k}\frac{Np(Np-1)\cdots[Np-(k-1)]c(c-1)\cdots[c-(n-k-1)]}{N(N-1)\cdots[N-(n-1)]} \\ &= \frac{\binom{Np}{k}\binom{Nq}{n-k}}{\binom{N}{n}}, \end{aligned} \tag{48}$$

which is a hypergeometric distribution. Here

$$\max(0, n-Nq) \le k \le \min(n, Np). \tag{49}$$

Theorem 8. Let X and Y be independent RVs with PMFs $b(m,p)$ and $b(n,p)$, respectively. Then the conditional distribution of X, given $X+Y$, is hypergeometric.

5.2.7 Negative Hypergeometric Distribution

Consider the model of Section 5.2.6. A box contains N marbles, M of these are marked (or say defective) and $N-M$ are unmarked. A sample of size n is taken and let X denote the number of defective marbles in the sample. If the sample is drawn without replacement we saw that X has a hypergeometric distribution with PMF (40). If, on the other hand, the sample is drawn with replacement then $X \sim b(n,p)$ where $p = M/N$.

Let Y denote the number of draws needed to draw the rth defective marble. If the draws are made with replacement then Y has the negative binomial distribution given in (22) with $p = M/N$. What if the draws are made without replacement? In that case in order that the kth draw $(k \geq r)$ be the rth defective marble drawn, the kth draw must produce a defective marble, whereas the previous $k-1$ draws must produce $r-1$ defectives. It follows that

$$P(Y = k) = \frac{\binom{M}{r-1}\binom{N-M}{k-r}}{\binom{N}{k-1}} \cdot \frac{M-r+1}{N-k+1}$$

for $k = r, r+1, \ldots, N$. Rewriting we see that

$$P(Y = k) = \binom{k-1}{r-1}\frac{\binom{N-k}{m-r}}{\binom{N}{M}}. \tag{50}$$

An RV Y with PMF (50) is said to have a *negative hypergeometric distribution*.

It is easy to see that

$$EY = r\frac{N+1}{M+1}, \ EY(Y+1) = \frac{r(r+1)(N+1)(N+2)}{(M+1)(M+2)},$$

and

$$\text{var}(Y) = \frac{r(N-M)(N+1)(M+1-r)}{(M+1)^2(M+2)}.$$

Also, if $r/N \to 0$, and $k/N \to 0$ as $N \to \infty$, then

$$\binom{k-1}{r-1}\binom{N-k}{M-r}\Big/\binom{N}{M} \longrightarrow \binom{k-1}{r-1}\left(\frac{M}{N}\right)^r\left(1-\frac{M}{N}\right)^{k-r}$$

which is (22).

5.2.8 Poisson Distribution

Definition 3. An RV X is said to be a *Poisson* RV with parameter $\lambda > 0$ if its PMF is given by

$$P\{X = k\} = \frac{e^{-\lambda}\lambda^k}{k!}, \qquad k = 0, 1, 2, \ldots. \tag{51}$$

We first check to see that (51) indeed defines a PMF. We have

$$\sum_{k=0}^{\infty} P\{X = k\} = e^{-\lambda}\sum_{k=0}^{\infty}\frac{\lambda^k}{k!} = e^{-\lambda}e^{\lambda} = 1.$$

If X has the PMF given by (51), we will write $X \sim P(\lambda)$. Clearly,

$$X = \sum_{k=0}^{\infty} kI_{[X=k]}$$

and

$$F(x) = \sum_{k=0}^{\infty} e^{-\lambda} \frac{\lambda^k}{k!} \varepsilon(x-k).$$

The mean and the variance are given by (see Problem 3.2.9)

$$EX = \lambda, \qquad EX^2 = \lambda + \lambda^2, \tag{52}$$

and

$$\text{var}(X) = \lambda. \tag{53}$$

The MGF of X is given by (see Example 3.3.7)

$$Ee^{tX} = \exp\{\lambda(e^t - 1)\}, \tag{54}$$

and the PGF by $P(s) = e^{-\lambda(1-s)}$, $|s| \leq 1$.

Theorem 9. Let $X_1, X_2, \ldots, X_n$ be independent Poisson RVs with $X_k \sim P(\lambda_k)$, $k = 1, 2, \ldots, n$. Then $S_n = X_1 + X_2 + \cdots + X_n$ is a $P(\lambda_1 + \lambda_2 + \cdots + \lambda_n)$ RV.

The converse of Theorem 9 is also true. Indeed, Raikov [84] showed that if $X_1, X_2, \ldots, X_n$ are independent and $S_n = \sum_{i=1}^{n} X_i$ has a Poisson distribution, each of the RVs $X_1, X_2, \ldots, X_n$ has a Poisson distribution.

Example 9. The number of female insects in a given region follows a Poisson distribution with mean λ. The number of eggs laid by each insect is a $P(\mu)$ RV. We are interested in the probability distribution of the number of eggs in the region.

Let F be the number of female insects in the given region. Then

$$P\{F = f\} = \frac{e^{-\lambda}\lambda^f}{f!}, \qquad f = 0, 1, 2, \ldots.$$

Let Y be the number of eggs laid by each insect. Then

$$\begin{aligned} P\{Y = y, F = f\} &= P\{F = f\} \cdot P\{Y = y | F = f\} \\ &= \frac{e^{-\lambda}\lambda^f}{f!} \frac{(f\mu)^y e^{-\mu f}}{y!}. \end{aligned}$$

Thus

$$P\{Y = y\} = \frac{e^{-\lambda}\mu^y}{y!} \sum_{f=0}^{\infty} \frac{(\lambda e^{-\mu})^f f^y}{f!}$$

The MGF of Y is given by

$$\begin{aligned}M(t) &= \sum_{f=0}^{\infty} \frac{\lambda^f e^{-\lambda}}{f!} \sum_{y=0}^{\infty} \frac{e^{yt}(f\mu)^y}{y!} e^{-\mu f} \\ &= \sum_{f=0}^{\infty} \frac{\lambda^f e^{-\lambda}}{f!} \exp\{f\mu(e^t - 1)\} \\ &= e^{-\lambda} \sum_{f=0}^{\infty} \frac{\{\lambda e^{\mu(e^t-1)}\}^f}{f!} \\ &= e^{-\lambda} \exp\{\lambda e^{\mu(e^t-1)}\}.\end{aligned}$$

Theorem 10. Let X and Y be independent RVs with PMFs $P(\lambda_1)$ and $P(\lambda_2)$, respectively. Then the conditional distribution of X, given $X+Y$, is binomial.

Proof. For nonnegative integers m and n, $m < n$, we have

$$\begin{aligned}P\{X=m|X+Y=n\} &= \frac{P\{X=m,\ Y=n-m\}}{P\{X+Y=n\}} \\ &= \frac{e^{-\lambda_1}(\lambda_1^m/m!)e^{-\lambda_2}(\lambda_2^{n-m}/(n-m)!)}{e^{-(\lambda_1+\lambda_2)}(\lambda_1+\lambda_2)^n/n!} \\ &= \binom{n}{m} \frac{\lambda_1^m \lambda_2^{n-m}}{(\lambda_1+\lambda_2)^n} \\ &= \binom{n}{m} \left(\frac{\lambda_1}{\lambda_1+\lambda_2}\right)^m \left(1 - \frac{\lambda_1}{\lambda_1+\lambda_2}\right)^{n-m}, \\ & \quad m = 0,1,2,\ldots,n,\end{aligned}$$

and the proof is complete.

Remark 2. The converse of this result is also true in the following sense. If X and Y are independent nonnegative integer-valued RVs such that $P\{X=k\} > 0$, $P\{Y=k\} > 0$, for $k = 0,1,2,\ldots$, and the conditional distribution of X, given $X+Y$, is binomial, both X and Y are Poisson. This result is due to Chatterji [13]. For the proof see Problem 13.

Theorem 11. If $X \sim P(\lambda)$ and the conditional distribution of Y, given $X = x$, is $b(x,p)$, then Y is a $P(\lambda p)$ RV.

Example 10. (Lamperti and Kruskal [60]). Let N be a nonnegative integer-valued RV. Independently of each other, N balls are placed either in urn A with probability p $(0 < p < 1)$ or in urn B with probability $1-p$, resulting in N_A balls in urn A and $N_B = N - N_A$ balls in urn B. We will show that the RVs N_A and N_B are independent if and only if N has a Poisson distribution. We have

$$P\{N_A = a \text{ and } N_B = b|N = a+b\} = \binom{a+b}{a} p^a(1-p)^b,$$

where a, b, are integers ≥ 0. Thus

$$P\{N_A = a,\ N_B = b\} = \binom{a+b}{a} p^a q^b P\{N = n\}, \qquad q = 1-p, \quad n = a+b.$$

If N has a Poisson (λ) distribution, then

$$\begin{aligned} P\{N_A = a,\ N_B = b\} &= \frac{(a+b)!}{a!b!} p^a q^b \frac{e^{-\lambda}\lambda^{a+b}}{(a+b)!} \\ &= \left(\frac{p^a \lambda^a e^{-\lambda/2}}{a!}\right)\left(\frac{q^b \lambda^b}{b!} e^{-\lambda/2}\right) \end{aligned}$$

so that N_A and N_B are independent.

Conversely, if N_A and N_B are independent, then

$$P\{N = n\}n! = f(a)g(b)$$

for some functions f and g. Clearly, $f(0) \neq 0$, $g(0) \neq 0$ because $P\{N_A = 0,\ N_B = 0\} > 0$. Thus there is a function h such that $h(a+b) = f(a)g(b)$ for all nonnegative integers a, b. It follows that

$$\begin{aligned} h(1) &= f(1)g(0) = f(0)g(1), \\ h(2) &= f(2)g(0) = f(1)g(1) = f(0)g(2), \end{aligned}$$

and so on. By induction,

$$f(a) = f(1)\left[\frac{g(1)}{g(0)}\right]^{a-1}, \qquad g(b) = g(1)\left[\frac{f(1)}{f(0)}\right]^{b-1}.$$

We may write, for some α_1, α_2, λ,

$$\begin{aligned} f(a) &= \alpha_1 e^{-a\lambda}, \qquad g(b) = \alpha_2 e^{-b\lambda}, \\ P\{N = n\} &= \alpha_1 \alpha_2 \frac{e^{-\lambda(a+b)}}{(a+b)!} \end{aligned}$$

so that N is a Poisson RV.

5.2.9 Multinomial Distribution

The binomial distribution is generalized in the following natural fashion. Suppose that an experiment is repeated n times. Each replication of the experiment terminates in one of k mutually exclusive and exhaustive events $A_1, A_2, \ldots, A_k$. Let p_j be the probability that the experiment terminates in A_j, $j = 1, 2, \ldots, k$, and suppose that p_j $(j = 1, 2, \ldots, k)$ remains constant for all n replications. We assume that the n replications are independent.

Let $x_1, x_2, \ldots, x_{k-1}$ be nonnegative integers such that $x_1 + x_2 + \cdots + x_{k-1} \leq n$. Then the probability that exactly x_i trials terminate in A_i, $i = 1, 2, \ldots, k-1$ and hence that $x_k = n - (x_1 + x_2 + \cdots + x_{k-1})$ trials terminate in A_k is clearly

$$\frac{n!}{x_1! x_2! \cdots x_k!} p_1^{x_1} p_2^{x_2} \cdots p_k^{x_k}.$$

If $(X_1, X_2, \ldots, X_k)$ is a random vector such that $X_j = x_j$ means that event A_j has occurred x_j times, $x_j = 0, 1, 2, \ldots, n$, the joint PMF of $(X_1, X_2, \ldots, X_k)$ is given by

$$P\{X_1 = x_1,\ X_2 = x_2, \ldots,\ X_k = x_k\} \tag{55}$$
$$= \begin{cases} \dfrac{n!}{x_1! x_2! \cdots x_k!} p_1^{x_1} p_2^{x_2} \ldots p_k^{x_k} & \text{if } n = \sum_1^k x_i, \\ 0 & \text{otherwise.} \end{cases}$$

Definition 4. An RV $(X_1, X_2, \ldots, X_{k-1})$ with joint PMF given by

$$P\{X_1 = x_1,\ X_2 = x_2, \ldots,\ X_{k-1} = x_{k-1}\} \tag{56}$$
$$= \begin{cases} \dfrac{n!}{x_1! x_2! \ldots (n - x_1 - \cdots - x_{k-1})!} p_1^{x_1} p_2^{x_2} \ldots p_k^{n - x_1 - \cdots - x_{k-1}} \\ \qquad \text{if } x_1 + x_2 + \cdots + x_{k-1} \leq n, \\ 0 \qquad \text{otherwise} \end{cases}$$

is said to have a *multinomial distribution*.

For the MGF of $(X_1, X_2, \ldots, X_{k-1})$ we have

$$\begin{aligned} M(t_1, t_2, \ldots, t_{k-1}) &= E e^{t_1 X_1 + t_2 X_2 + \cdots + t_{k-1} X_{k-1}} \\ &= \sum_{\substack{x_1, x_2, \ldots, x_{k-1} = 0 \\ x_1 + x_2 + \cdots x_{k-1} \leq n}}^{n} e^{t_1 x_1 + \cdots + t_{k-1} x_{k-1}} \frac{n! p_1^{x_1} p_2^{x_2} \ldots p_k^{x_k}}{x_1! x_2! \cdots x_k!} \\ &= \sum_{\substack{x_1, x_2, \ldots, x_{k-1} = 0 \\ x_1 + x_2 + \cdots x_{k-1} \leq n}}^{n} \frac{n!}{x_1! x_2! \ldots x_k!} (p_1 e^{t_1})^{x_1} (p_2 e^{t_2})^{x_2} \ldots \\ &\quad \cdot (p_{k-1} e^{t_{k-1}})^{x_{k-1}} p_k^{x_k} \\ &= (p_1 e^{t_1} + p_2 e^{t_2} + \cdots + p_{k-1} e^{t_{k-1}} + p_k)^n \\ &\quad \text{for all } t_1, t_2, \ldots, t_{k-1} \in \mathfrak{R}. \end{aligned} \tag{57}$$

Clearly,

$$M(t_1, 0, 0, \ldots, 0) = (p_1 e^{t_1} + p_2 + \cdots + p_k)^n = (1 - p_1 + p_1 e^{t_1})^n,$$

which is binomial. Indeed, the marginal PMF of each X_i, $i = 1, 2, \ldots, k-1$, is binomial. Similarly, the joint MGF of X_i, X_j, $i, j = 1, 2, \ldots, k-1$ $(i \neq j)$, is

$$M(0, 0, \ldots, 0, t_i, 0, \ldots, 0, t_j, 0, \ldots, 0) = [p_i e^{t_i} + p_j e^{t_j} + (1 - p_i - p_j)]^n,$$

which is the MGF of a *trinomial distribution* with PMF

$$f(x_i,x_j) = \frac{n!}{x_i!x_j!(n-x_i-x_j)!}p_i^{x_i}p_j^{x_j}p_k^{n-x_i-x_j}, \qquad p_k = 1-p_i-p_j. \tag{58}$$

Note that the RVs $X_1,X_2,\ldots,X_{k-1}$ are dependent.

From the MGF of $(X_1,X_2,\ldots,X_{k-1})$ or directly from the marginal PMFs we can compute the moments. Thus

$$EX_j = np_j \quad \text{and} \quad \text{var}(X_j) = np_j(1-p_j), \qquad j=1,2,\ldots,k-1, \tag{59}$$

and for $j=1,2,\ldots,k-1$, and $i \neq j$,

$$\text{cov}(X_i,X_j) = E\{(X_i-np_i)(X_j-np_j)\} = -np_ip_j, \tag{60}$$

It follows that the correlation coefficient between X_i and X_j is given by

$$\rho_{ij} = -\left[\frac{p_ip_j}{(1-p_i)(1-p_j)}\right]^{1/2}, \qquad i,j=1,2,\ldots,k-1 \quad (i\neq j). \tag{61}$$

Example 11. Consider the trinomial distribution with PMF

$$P\{X=x,Y=y\} = \frac{n!}{x!y!(n-x-y)!}p_1^xp_2^yp_3^{n-x-y},$$

where x, y are nonnegative integers such that $x+y\leq n$, and $p_1, p_2, p_3 > 0$ with $p_1+p_2+p_3=1$. The marginal PMF of X is given by

$$P\{X=x\} = \binom{n}{x}p_1^x(1-p_1)^{n-x}, \qquad x=0,1,2,\ldots,n.$$

It follows that

$$P\{Y=y|X=x\} = \begin{cases} \dfrac{(n-x)!}{y!(n-x-y)!}\left(\dfrac{p_2}{1-p_1}\right)\left(\dfrac{p_3}{1-p_1}\right)^{n-x-y} & \text{if } y=0,1,2,\ldots,n-x, \\ 0 & \text{otherwise,} \end{cases} \tag{62}$$

which is $b(n-x,p_2/(1-p_1))$. Thus

$$E\{Y|x\} = (n-x)\frac{p_2}{1-p_1}. \tag{63}$$

Similarly,

$$E\{X|y\} = (n-y)\frac{p_1}{1-p_2}. \tag{64}$$

Finally, we note that, if $\mathbf{X} = (X_1, X_2, \ldots, X_k)$ and $\mathbf{Y} = (Y_1, Y_2, \ldots, Y_k)$ are two independent multinomial RVs with common parameter $(p_1, p_2, \ldots, p_k)$, then $\mathbf{Z} = \mathbf{X} + \mathbf{Y}$ is also a multinomial RV with probabilities $(p_1, p_2, \ldots, p_k)$. This follows easily if one employs the MGF technique, using (57). Actually this property characterizes the multinomial distribution. If $\mathbf{X}$ and $\mathbf{Y}$ are k-dimensional, nonnegative, independent random vectors, and if $\mathbf{Z} = \mathbf{X} + \mathbf{Y}$ is a multinomial random vector with parameter $(p_1, p_2, \ldots, p_k)$, then $\mathbf{X}$ and $\mathbf{Y}$ also have multinomial distribution with the same parameter. This result is due to Shanbhag and Basawa [103] and will not be proved here.

5.2.10 Multivariate Hypergeometric Distribution

Consider an urn containing N items divided into k categories containing $n_1, n_2, \ldots, n_k$ items, where $\sum_{j=1}^{k} n_j = N$. A random sample, without replacement, of size n is taken from the urn. Let X_i = number of items in sample of type i. Then

$$P\{X_1 = x_1, X_2 = x_2, \ldots, X_k = x_k\} = \prod_{j=1}^{k} \binom{n_j}{x_j} \Big/ \binom{N}{n}, \tag{65}$$

where $x_j = 0, 1, \ldots, \min(n, n_j)$ and $\sum_{j=1}^{k} x_j = n$.

We say that $(X_1, X_2, \ldots, X_{k-1})$ has *multivariate hypergeometric distribution* if its joint PMF is given by (65). It is clear that each X_j has a marginal hypergeometric distribution. Moreover, the conditional distributions are also hypergeometric. Thus

$$P\{X_i = x_i | X_j = x_j\} = \frac{\binom{n_i}{n_i}\binom{N-n_i-x_j}{n-x_i-x_j}}{\binom{N-n_j}{n-x_j}},$$

and

$$P\{X_i = x_i | X_j = x_j, X_\ell = x_\ell\} = \frac{\binom{n_i}{x_i}\binom{N-n_i-n_j-n_\ell}{n-x_i-x_j-x_\ell}}{\binom{N-n_j-n_\ell}{n-x_j-x_\ell}},$$

and so on. It is therefore easy to write down the marginal and conditional means and variances. We leave the reader to show that

$$EX_j = n\,\frac{n_j}{N},$$
$$\operatorname{var}(X_j) = n\frac{n_j}{n}\left(\frac{N-n_j}{N}\right)\left(\frac{N-n}{N-1}\right),$$

and

$$\operatorname{cov}(X_i, X_j) = -\frac{N-n}{N-1}n\left(\frac{n_j}{N}\right)^2.$$

5.2.11 Multivariate Negative Binomial Distribution

Consider the setup of Section 5.2.9 where each replication of an experiment terminates in one of k mutually exclusive and exhaustive events $A_1, A_2, \ldots, A_k$. Let $p_j = P(A_j)$,

$j = 1,2,\ldots,k$. Suppose the experiment is repeated until event A_k is observed for the rth time, $r \geq 1$. Then

$$P(X_1 = x_1,\, X_2 = x_2, \ldots, X_k = r)$$
$$= \frac{(x_1 + x_2 + \cdots + x_{k-1} + r - 1)!}{\left(\prod_{j=1}^{k-1} x_j!\right)(r-1)!} p_k^r \prod_{j=1}^{k-1} p_j^{x_j}, \tag{66}$$

for $x_i = 0,1,2,\ldots$ $(i = 1,2,\ldots k-1)$, $1 \leq r < \infty$, $0 < p_i < 1$, $\sum_{i=1}^{k-1} p_i < 1$, and $p_k = 1 - \sum_{j=1}^{k-1} p_j$.

We say that $(X_1, X_2, \ldots, X_{k-1})$ has a *multivariate negative binomial (or negative multinomial) distribution* if its joint PMF is given by (66).

It is easy to see the marginal PMF of any subset of $\{X_1, X_2, \ldots, X_{k-1}\}$ is negative multinomial. In particular, each X_j has a negative binomial distribution.

We will leave the reader to show that

$$M(s_1, s_2, \ldots, s_{k-1}) = Ee^{\sum_{j=1}^{k-1} s_j X_j} = p_k^r \left(1 - \sum_{j=1}^{k-1} s_j p_j\right)^{-r}, \tag{67}$$

and

$$\operatorname{cov}(X_i, X_j) = \frac{r p_i p_j}{p_k^2}. \tag{68}$$

PROBLEMS 5.2

1. (a) Let us write

$$b(k; n, p) = \binom{n}{k} p^k (1-p)^{n-k}, \qquad k = 0,1,2,\ldots,n.$$

Show that, as k goes from 0 to n, $b(k;n,p)$ first increases monotonically and then decreases monotonically. The greatest value is assumed when $k = m$, where m is an integer such that

$$(n+1)p - 1 < m \leq (n+1)p$$

except that $b(m-1;n,p) = b(m;n,p)$ when $m = (n+1)p$.

(b) If $k \geq np$, then

$$P\{X \geq k\} \leq b(k;n,p) \frac{(k+1)(1-p)}{k+1-(n+1)p};$$

and if $k \leq np$, then

$$P\{X \leq k\} \leq b(k;n,p) \frac{(n-k+1)p}{(n+1)p-k}.$$

2. Generalize the result in Theorem 10 to n independent Poisson RVs, that is, if $X_1, X_2, \ldots, X_n$ are independent RVs with $X_i \sim P(\lambda_i)$, $i = 1,2,\ldots,n$, the conditional distribution of $X_1, X_2, \ldots, X_n$, given $\sum_{i=1}^n X_i = t$, is multinomial with parameters t, $\lambda_1/\sum_i^n \lambda_i, \ldots, \lambda_n/\sum_1^n \lambda_i$.
3. Let X_1, X_2 be independent RVs with $X_i \sim b(n_i, \frac{1}{2})$, $i = 1,2$. What is the PMF of $X_1 - X_2 + n_2$?
4. A box contains N identical balls numbered 1 through N. Of these balls, n are drawn at a time. Let $X_1, X_2, \ldots, X_n$ denote the numbers on the n balls drawn. Let $S_n = \sum_{i=1}^n X_i$. Find $\text{var}(S_n)$.
5. From a box containing N identical balls marked 1 through N, M balls are drawn one after another without replacement. Let X_i denote the number on the ith ball drawn, $i = 1,2,\ldots,M$, $1 \le M \le N$. Let $Y = \max(X_1, X_2, \ldots, X_M)$. Find the DF and the PMF of Y. Also find the conditional distribution of $X_1, X_2, \ldots, X_M$, given $Y = y$. Find EY and $\text{var}(Y)$.
6. Let $f(x; r, p)$, $x = 0,1,2,\ldots$, denote the PMF of an $NB(r;p)$ RV. Show that the terms $f(x;r,p)$ first increase monotonically and then decrease monotonically. When is the greatest value assumed?
7. Show that the terms

$$P_\lambda\{X = k\} = e^{-\lambda}\frac{\lambda^k}{k!}, \qquad k = 0,1,2,\ldots,$$

of the Poisson PMF reach their maxima when k is the largest integer $\le \lambda$ and at $(\lambda - 1)$ and λ if λ is an integer.
8. Show that

$$\binom{n}{k} p^k (1-p)^{n-k} \to e^{-\lambda}\frac{\lambda^k}{k!}$$

as $n \to \infty$ and $p \to 0$, so that $np = \lambda$ remains constant.
[*Hint:* Use Stirling's approximation, namely, $n! \approx \sqrt{2\pi}\, n^{n+1/2} e^{-n}$ as $n \to \infty$.]
9. A biased coin is tossed indefinitely. Let p $(0 < p < 1)$ be the probability of success (heads). Let Y_1 denote the length of the first run, and Y_2, the length of the second run. Find the PMFs of Y_1 and Y_2 and show that $EY_1 = q/p + p/q$, $EY_2 = 2$. If Y_n denotes the length of the nth run, $n \ge 1$, what is the PMF of Y_n? Find EY_n.
10. Show that

$$\binom{N}{n}^{-1}\binom{Np}{k}\binom{N(1-p)}{n-k} \to \binom{n}{k} p^k (1-p)^{n-k}$$

as $N \to \infty$.
11. Show that

$$\binom{r+k-1}{k} p^r (1-p)^k \to e^{-\lambda}\frac{\lambda^k}{k!}$$

as $p \to 1$ and $r \to \infty$ in such a way that $r(1-p) = \lambda$ remains fixed.

12. Let X and Y be independent geometric RVs. Show that min (X,Y) and $X-Y$ are independent.

13. Let X and Y be independent RVs with PMFs $P\{X=k\}=p_k$, $P\{Y=k\}=q_k$, $k=0,1,2,\ldots$, where $p_k,q_k>0$ and $\sum_{k=0}^{\infty}p_k=\sum_{k=0}^{\infty}q_k=1$. Let

$$P\{X=k|X+Y=t\}=\binom{t}{k}\alpha_t^k(1-\alpha_t)^{t-k}, \qquad 0\le k\le t.$$

Then $\alpha_t=\alpha$ for all t, and

$$p_k=\frac{e^{-\theta\beta}(\theta\beta)^k}{k!}, \qquad q_k=\frac{e^{-\theta}\theta^k}{k!},$$

where $\beta=\alpha/(1-\alpha)$, and $\theta>0$ is arbitrary. (Chatterji [13])

14. Generalize the result of Example 10 to the case of k urns, $k\ge 3$.

15. Let $(X_1,X_2,\ldots,X_{k-1})$ have a multinomial distribution with parameters $n,p_1,p_2,\ldots,p_{k-1}$. Write

$$Y=\sum_{i=1}^{k}\frac{(X_i-np_i)^2}{np_i},$$

where $p_k=1-p_1-\cdots-p_{k-1}$, and $X_k=n-X_1-\cdots-X_{k-1}$. Find EY and $\text{var}(Y)$.

16. Let X_1, X_2 be iid RVs with common DF F, having positive mass at $0,1,2,\ldots$. Also, let $U=\max(X_1,X_2)$ and $V=X_1-X_2$. Then

$$P\{U=j,\ V=0\}=P\{U=j\}P\{V=0\}$$

for all j if and only if F is a geometric distribution. (Srivastava [109])

17. Let X and Y be mutually independent RVs, taking nonnegative integer values. Then

$$P\{X\le n\}-P\{X+Y\le n\}=\alpha P\{X+Y=n\}$$

holds for $n=0,1,2,\ldots$ and some $\alpha>0$ if and only if

$$P\{Y=n\}=\frac{1}{1+\alpha}\left(\frac{\alpha}{1+\alpha}\right)^n, \qquad n=0,1,2,\ldots.$$

[*Hint:* Use Problem 3.3.8.] (Puri [83]]

18. Let $X_1,X_2,\ldots$ be a sequence of independent $b(1,p)$ RVs with $0<p<1$. Also, let $Z_N=\sum_{i=1}^{N}X_i$, where N is a $P(\lambda)$ RV which is independent of the X_i's. Show that Z_N and $N-Z_N$ are independent.

19. Prove Theorems 5, 7, 8, and 11.

20. In Example 2 show that

(a)

$$P\left(X_{(1)}=k\right)=pq^{2k-2}(1+q), \qquad k=1,2,\ldots.$$

(b)

$$P\left(X_{(2)} - X_{(1)} = k\right) = \frac{p}{(1+q)} \qquad \text{for } k = 0$$
$$= \frac{2pq^k}{(1+q)} \qquad \text{for } k = 1, 2, \ldots.$$

5.3 SOME CONTINUOUS DISTRIBUTIONS

In this section we study some most frequently used absolutely continuous distributions and describe their important properties. Before we introduce specific distributions it should be remarked that associated with each PDF f there is an *index* or a *parameter* θ (may be multidimensional) which takes values in an index set Θ. For any particular choice of $\theta \in \Theta$ we obtain a specific PDF f_θ from the family of PDFs $\{f_\theta, \theta \in \Theta\}$.

Let X be an RV with PDF $f_\theta(x)$, where θ is a real-valued parameter. We say that θ is a *location* parameter and $\{f_\theta\}$ is a *location family* if $X - \theta$ has PDF $f(x)$ which does not depend on θ. The parameter θ is said to be a *scale* parameter and $\{f_\theta\}$ is a *scale family* of PDFs if X/θ has PDF $f(x)$ which is free of θ. If $\theta = (\mu, \sigma)$ is two-dimensional, we say that θ is a *location-scale* parameter if the PDF of $(X - \mu)/\sigma$ is free of μ and σ. In that case $\{f_\theta\}$ is known as a *location-scale* family.

It is easily seen that θ is a location parameter if and only if $f_\theta(x) = f(x - \theta)$, a scale parameter if and only $f_\theta(x) = (1/\theta)f(x)$, and a location-scale parameter if $f_\theta(x) = (1/\sigma)f((x-\mu)/\sigma)$, $\sigma > 0$ for some PDF f. The density f is called the *standard* PDF for the family $\{f_\theta, \theta \in \Theta\}$.

A location parameter simply relocates or shifts the graph of PDF f without changing its shape. A scale parameter stretches (if $\theta > 1$) or contracts (if $\theta < 1$) the graph of f. A location-scale parameter, on the other hand, stretches or contracts the graph of f with the scale parameter and then shifts the graph to locate at μ. (see Fig. 1.)

Some PDFs also have a shape parameter. Changing its value alters the shape of the graph. For the Poisson distribution λ is a shape parameter.

For the following PDF

$$f(x; \mu, \beta, \alpha) = \frac{1}{\beta\Gamma(\alpha)}\left(\frac{x-\mu}{\beta}\right)^{\alpha-1} \exp\{-(x-\mu)/\beta\}, \quad x > \mu$$

and $= 0$ otherwise, μ is a location, β, a scale, and α, a shape parameter. The standard density for this location-scale family is

$$f(x) = \frac{1}{\Gamma(\alpha)}x^{\alpha-1}e^{-x}, \quad x > 0$$

and $= 0$ otherwise. For the standard PDF f, α is a shape parameter.

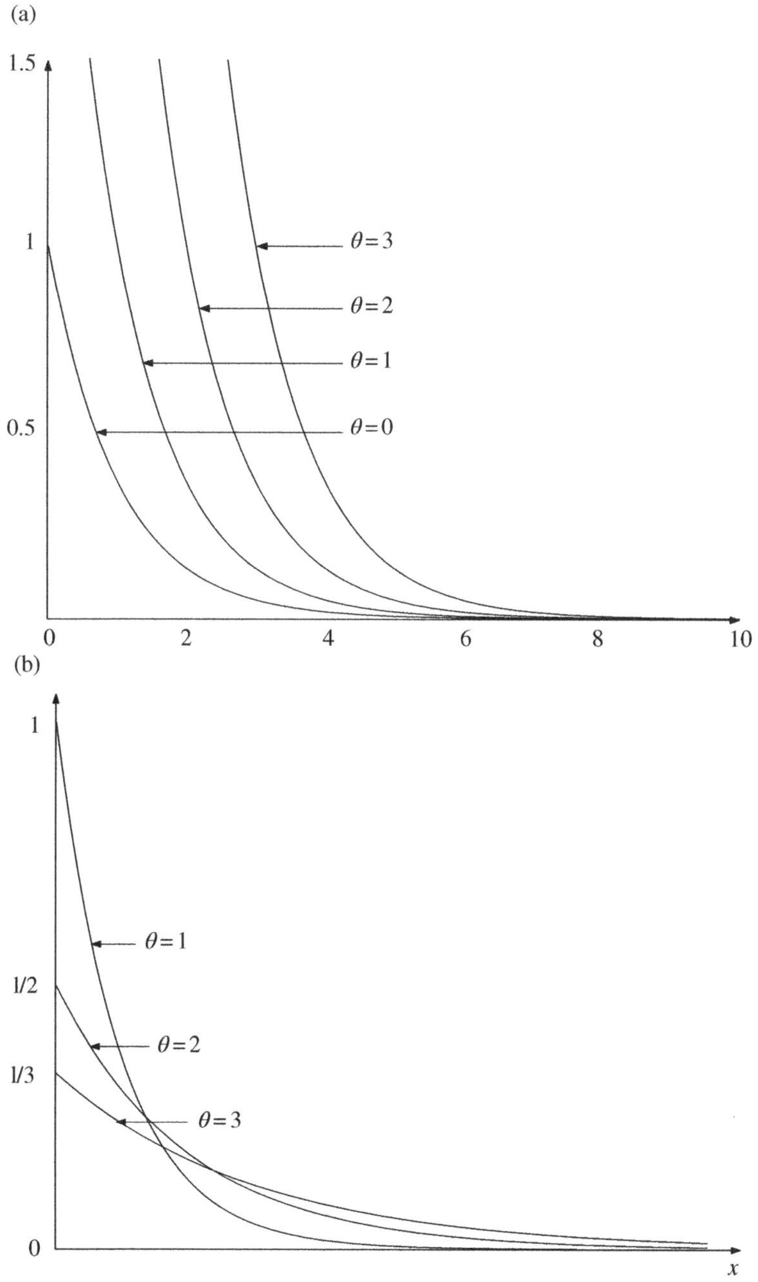

Fig. 1 (*a*) Exponential location family; (*b*) exponential scale family; (*c*) normal location-scale family; and (*d*) shape parameter family $f_\theta(x) = \theta x\theta - 1$.

(c)

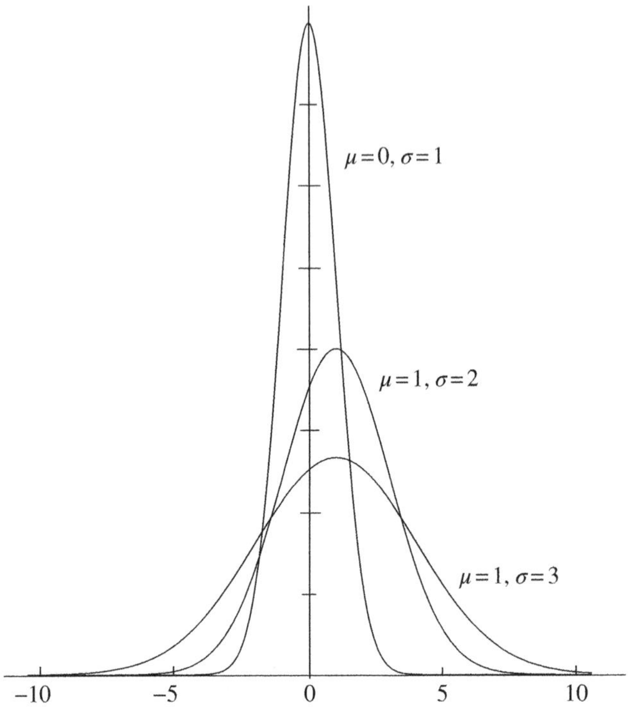

(d)

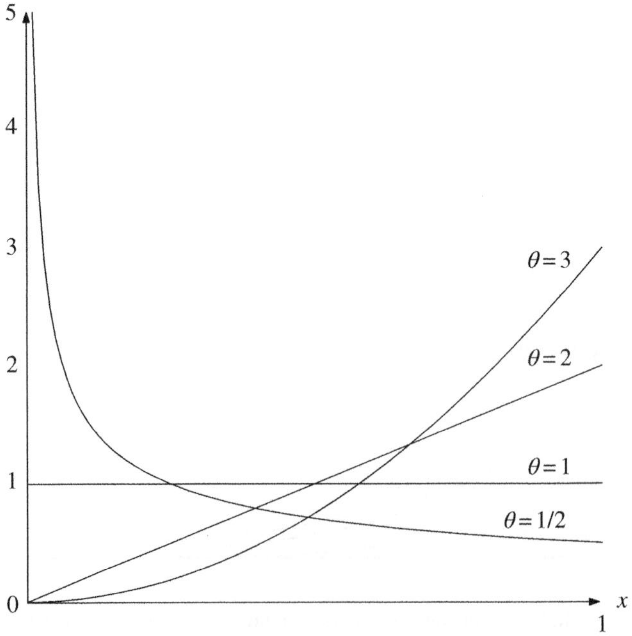

Fig. 1 (continued).

5.3.1 Uniform Distribution (Rectangular Distribution)

Definition 1. An RV X is said to have a *uniform distribution* on the interval $[a,b]$, $-\infty < a < b < \infty$ if its PDF is given by

$$f(x) = \begin{cases} \dfrac{1}{b-a}, & a \le x \le b, \\ 0, & \text{otherwise}. \end{cases} \tag{1}$$

We will write $X \sim U[a,b]$ if X has a uniform distribution on $[a,b]$.

The end point a or b or both may be excluded. Clearly,

$$\int_{-\infty}^{\infty} f(x)dx = 1,$$

so that (1) indeed defines a PDF. The DF of X is given by

$$F(x) = \begin{cases} 0, & x < a, \\ \dfrac{x-a}{b-a}, & a \le x < b, \\ 1, & b \le x; \end{cases} \tag{2}$$

$$EX = \frac{a+b}{2}, \qquad EX^k = \frac{b^{k+1} - a^{k+1}}{(k+1)(b-a)}, \qquad k > 0 \text{ is an integer;} \tag{3}$$

$$\operatorname{var}(X) = \frac{(b-a)^2}{12}; \tag{4}$$

$$M(t) = \frac{1}{t(b-a)}(e^{tb} - e^{ta}), \qquad t \ne 0. \tag{5}$$

Example 1. Let X have PDF given by

$$f(x) = \begin{cases} \lambda e^{-\lambda x}, & 0 < x < \infty, \quad \lambda > 0, \\ 0, & \text{otherwise}. \end{cases}$$

Then

$$F(x) = \begin{cases} 0 & x \le 0, \\ 1 - e^{-\lambda x}, & x > 0. \end{cases}$$

Let $Y = F(X) = 1 - e^{-\lambda X}$. The PDF of Y is given by

$$\begin{aligned} f_Y(y) &= \frac{1}{\lambda} \cdot \frac{1}{1-y} \lambda e^{-\lambda\left(-\frac{1}{\lambda}\right)\log(1-y)} \qquad 0 \le y < 1, \\ &= 1, \qquad 0 \le y < 1. \end{aligned}$$

Let us define $f_Y(y) = 1$ at $y = 1$. Then we see that Y has density function

$$f_Y(y) = \begin{cases} 1, & 0 \le y \le 1, \\ 0, & \text{otherwise}, \end{cases}$$

which is the $U[0,1]$ distribution. That this is not a mere coincidence is shown in the following theorem.

Theorem 1 (Probability Integral Transformation). Let X be an RV with a continuous DF F. Then $F(X)$ has the uniform distribution on $[0,1]$.

Proof. The proof is left as an exercise.

The reader is asked to consider what happens in the case where F is the DF of a discrete RV. In the converse direction the following result holds.

Theorem 2. Let F be any df, and let X be a $U[0,1]$ RV. Then there exists a function h such that $h(X)$ has DF F, that is,

$$P\{h(X) \le x\} = F(x) \qquad \text{for all } x \in (-\infty, \infty). \tag{6}$$

Proof. If F is the DF of a discrete RV Y, let

$$P\{Y = y_k\} = p_k, \qquad k = 1, 2, \ldots.$$

Define h as follows:

$$h(x) = \begin{cases} y_1 & \text{if } 0 \le x < p_1, \\ y_2 & \text{if } p_1 \le x < p_1 + p_2, \\ \vdots & \vdots \end{cases}$$

Then

$$P\{h(X) = y_1\} = P\{0 \le X < p_1\} = p_1,$$

$$P\{h(X) = y_2\} = P\{p_1 \le X < p_1 + p_2\} = p_2,$$

and, in general,

$$P\{h(X) = y_k\} = p_k, \qquad k = 1, 2, \ldots.$$

Thus $h(X)$ is a discrete RV with DF F.

If F is continuous and strictly increasing F^{-1} is well defined, and we take $h(X) = F^{-1}(X)$. We have

$$\begin{aligned} P\{h(X) \le x\} &= P\{F^{-1}(X) \le x\} \\ &= P\{X \le F(x)\} \\ &= F(x), \end{aligned}$$

as asserted.

In general, define

$$F^{-1}(y) = \inf\{x : F(x) \ge y\}, \tag{7}$$

and let $h(X) = F^{-1}(X)$. Then we have

$$\{F^{-1}(y) \le x\} = \{y \le F(x)\}. \tag{8}$$

Indeed, $F^{-1}(y) \le x$ implies, that, for every $\varepsilon > 0$, $y \le F(x+\varepsilon)$. Since $\varepsilon > 0$ is arbitrary and F is continuous on the right, we let $\varepsilon \to 0$ and conclude that $y \le F(x)$. Since $y \le F(x)$ implies $F^{-1}(y) \le x$ by definition (7), it follows that (8) holds generally. Thus

$$P\{F^{-1}(X) \le x\} = P\{X \le F(x)\} = F(x).$$

Theorem 2 is quite useful in generating samples with the help of the uniform distribution.

Example 2. Let F be the DF defined by

$$F(x) = \begin{cases} 0, & x \le 0 \\ 1 - e^{-x}, & x > 0. \end{cases}$$

Then the inverse to $y = 1 - e^{-x}$, $x > 0$, is $x = -\log(1-y)$, $0 < y < 1$. Thus

$$h(y) = -\log(1-y),$$

and $-\log(1-X)$ has the required distribution, where X is a $U[0,1]$ RV.

Theorem 3. Let X be an RV defined on $[0,1]$. If $P\{x < X \le y\}$ depends only on $y - x$ for all $0 \le x \le y \le 1$, then X is $U[0,1]$.

Proof. Let $P\{x < X \le y\} = f(y-x)$ then $f(x+y) = P\{0 < X \le x+y\} = P\{0 < X \le x\} + P\{x < X \le x+y\} = f(x) + f(y)$. Note that f is continuous from the right. We have

$$f(x) = f(x) + f(0),$$

so that

$$f(0) = 0.$$

We will show that $f(x) = cx$ for some constant c. It suffices to prove the result for positive x. Let m be an integer then

$$f(mx) = f(x) + \cdots + f(x) = mf(x).$$

Letting $x = n/m$, we get

$$f\left(m \cdot \left(\frac{n}{m}\right)\right) = mf\left(\frac{n}{m}\right),$$

so that

$$f\left(\frac{n}{m}\right) = \frac{1}{m}f(n) = \frac{n}{m}f(1),$$

for positive integers n and m. Letting $f(1) = c$, we have proved that

$$f(x) = cx$$

for rational numbers x.

To complete the proof we consider the case where x is a positive irrational number. Then we can find a decreasing sequence of positive rationals $x_1, x_2, \ldots$ such that $x_n \to x$. Since f is right continuous,

$$f(x) = \lim_{x_n \downarrow x} f(x_n) = \lim_{x_n \downarrow x} cx_n = cx.$$

Now, for $0 \le x \le 1$,

$$\begin{aligned} F(x) &= P\{X \le 0\} + P\{0 < X \le x\} \\ &= F(0) + P\{0 < X \le x\} \\ &= f(x) \\ &= cx, \qquad 0 \le x \le 1. \end{aligned}$$

Since $F(1) = 1$, we must have $c = 1$, so that

$$F(x) = x, \qquad 0 \le x \le 1.$$

This completes the proof.

5.3.2 Gamma Distribution

The integral

$$\Gamma(\alpha) = \int_{0+}^{\infty} x^{\alpha-1} e^{-x}\, dx \tag{9}$$

converges or diverges according as $\alpha > 0$ or ≤ 0. For $\alpha > 0$ the integral in (9) is called the *gamma function*. In particular, if $\alpha = 1$, $\Gamma(1) = 1$. If $\alpha > 1$, integration by parts yields

$$\Gamma(\alpha) = (\alpha - 1)\int_0^{\infty} x^{\alpha-2} e^{-x}\, dx = (\alpha - 1)\Gamma(\alpha - 1). \tag{10}$$

If $\alpha = n$ is a positive integer, then

$$\Gamma(n) = (n-1)!. \tag{11}$$

Also writing $x = y^2/2$ in $\Gamma\left(\frac{1}{2}\right)$ we see that

$$\Gamma\left(\frac{1}{2}\right) = \frac{1}{\sqrt{2}}\int_0^\infty e^{-y^2/2}dy.$$

Now consider the integral $I = \int_0^\infty e^{-y^2/2}dy$. We have

$$I^2 = \int_0^\infty \int_0^\infty \exp\{\frac{-(x^2+y^2)}{2}\}dxdy,$$

and changing to polar coordinates we get

$$I^2 = \int_0^{2\pi} \int_0^\infty r\exp(\frac{-r^2}{2})dr\, d\theta = 2\pi.$$

It follows that $\Gamma\left(\frac{1}{2}\right) = \sqrt{\pi}$.

Let us write $x = y/\beta$, $\beta > 0$, in the integral in (9). Then

$$\Gamma(\alpha) = \int_0^\infty \frac{y^{\alpha-1}}{\beta^\alpha} e^{-y/\beta}\, dy, \tag{12}$$

so that

$$\int_0^\infty \frac{1}{\Gamma(\alpha)\beta^\alpha} y^{\alpha-1} e^{-y/\beta}\, dy = 1. \tag{13}$$

Since the integrand in (13) is positive for $y > 0$, it follows that the function

$$f(y) = \begin{cases} \dfrac{1}{\Gamma(\alpha)\beta^\alpha} y^{\alpha-1} e^{-y/\beta} & 0 < y < \infty, \\ 0, & y \leq 0 \end{cases} \tag{14}$$

defines a PDF for $\alpha > 0$, $\beta > 0$.

Definition 2. An RV X with PDF defined by (14) is said to have a *gamma distribution* with parameters α and β. We will write $X \sim G(\alpha, \beta)$.

Figure 2 gives graphs of some gamma PDFs.

The DF of a $G(\alpha, \beta)$ RV is given by

$$F(x) = \begin{cases} 0, & x \leq 0, \\ \dfrac{1}{\Gamma(\alpha)\beta^\alpha} \displaystyle\int_0^x y^{\alpha-1} e^{-y/\beta}\, dy, & x > 0. \end{cases} \tag{15}$$

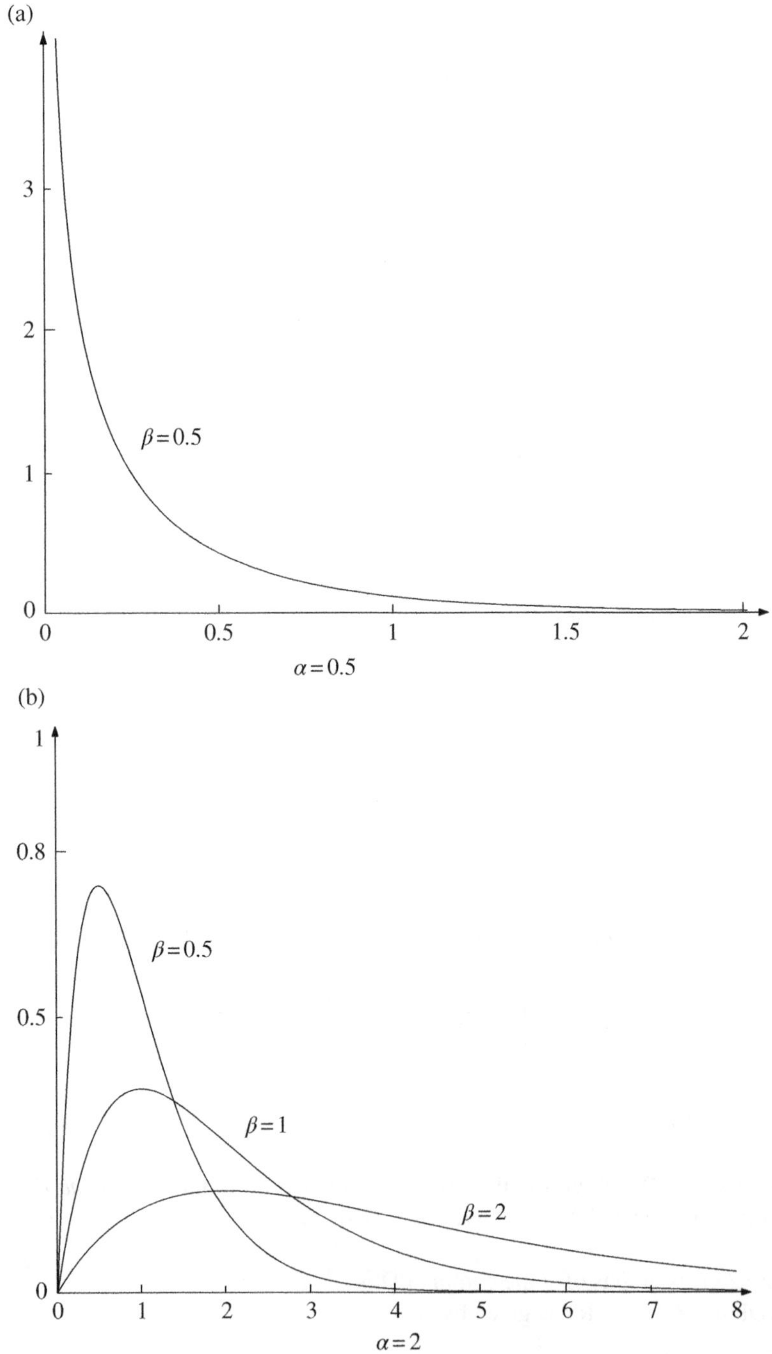

Fig. 2 Gamma density functions.

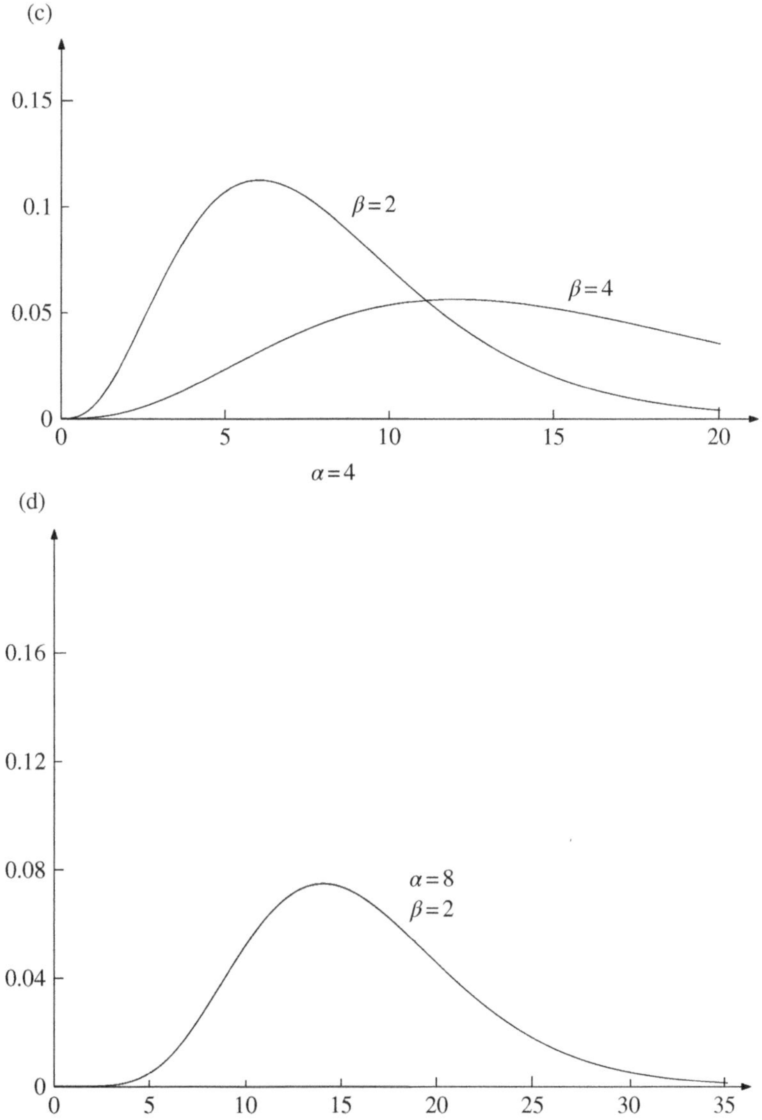

Fig. 2 (continued).

The MGF of X is easily computed. We have

$$
\begin{aligned}
M(t) &= \frac{1}{\Gamma(\alpha)\beta^{\alpha}} \int_0^{\infty} e^{x(t-1/\beta)} x^{\alpha-1}\, dx \\
&= \left(\frac{1}{1-\beta t}\right)^{\alpha} \int_0^{\infty} \frac{y^{\alpha-1} e^{-y}}{\Gamma(\alpha)}\, dy, \qquad t < \frac{1}{\beta} \\
&= (1-\beta t)^{-\alpha} \qquad t < \frac{1}{\beta}.
\end{aligned} \tag{16}
$$

It follows that

$$EX = M'(t)|_{t=0} = \alpha\beta, \tag{17}$$

$$EX^2 = M''(t)|_{t=0} = \alpha(\alpha+1)\beta^2, \tag{18}$$

so that

$$\text{var}(X) = \alpha\beta^2. \tag{19}$$

Indeed, we can compute the moment of order n such that $\alpha + n > 0$ directly from the density. We have

$$\begin{aligned} EX^n &= \frac{1}{\Gamma(\alpha)\beta^\alpha}\int_0^\infty e^{-x/\beta}x^{\alpha+n-1}\,dx \\ &= \beta^n\frac{\Gamma(\alpha+n)}{\Gamma(\alpha)} \\ &= \beta^n(\alpha+n-1)(\alpha+n-2)\cdots\alpha. \end{aligned} \tag{20}$$

The special case when $\alpha = 1$ leads to the *exponential distribution with parameter* β. The PDF of an exponentially distributed RV is therefore

$$f(x) = \begin{cases} \beta^{-1}e^{-x/\beta}, & x > 0, \\ 0, & \text{otherwise}. \end{cases} \tag{21}$$

Note that we can speak of the exponential distribution on $(-\infty, 0)$. The PDF of such an RV is

$$f(x) = \begin{cases} \beta^{-1}e^{x/\beta}, & x < 0, \\ 0, & x \geq 0. \end{cases} \tag{22}$$

Clearly, if $X \sim G(1, \beta)$, we have

$$EX^n = n!\beta^n \tag{23}$$

$$EX = \beta \quad \text{and} \quad \text{var}(X) = \beta^2, \tag{24}$$

$$M(t) = (1 - \beta t)^{-1} \quad \text{for } t < \beta^{-1}. \tag{25}$$

Another special case of importance is when $\alpha = n/2$, $n > 0$ (an integer), and $\beta = 2$.

Definition 3. An RV X is said to have a *chi-square distribution* (χ^2-distribution) with n degrees of freedom where n is a positive integer if its PDF is given by

$$f(x) = \begin{cases} \dfrac{1}{\Gamma(n/2)2^{n/2}}e^{-x/2}x^{n/2-1}, & 0 < x < \infty, \\ 0, & x \leq 0. \end{cases} \tag{26}$$

We will write $X \sim \chi^2(n)$ for a χ^2 RV with n degrees of freedom (d.f.). [Note the difference in the abbreviations of distribution function (DF) and *degrees of freedom* (d.f.).]

If $X \sim \chi^2(n)$, then

$$EX = n, \qquad \text{var}(X) = 2n, \tag{27}$$

$$EX^k = \frac{2^k\Gamma[(n/2)+k]}{\Gamma(n/2)}, \tag{28}$$

and

$$M(t) = (1-2t)^{-n/2} \qquad \text{for } t < \frac{1}{2}. \tag{29}$$

Theorem 4. Let $X_1, X_2, \ldots, X_n$ be independent RVs such that $X_j \sim G(\alpha_j, \beta), j = 1, 2, \ldots, n$. Then $S_n = \sum_{k=1}^n X_k$ is a $G(\sum_{j=1}^n \alpha_j, \beta)$ RV.

Corollary 1. Let $X_1, X_2, \ldots, X_n$ be iid RVs, each with an exponential distribution with parameter β. Then S_n is a $G(n, \beta)$ RV.

Corollary 2. If $X_1, X_2, \ldots, X_n$ are independent RVs such that $X_j \sim \chi^2(r_j), j = 1, 2, \ldots, n$, then S_n is a $\chi^2(\sum_{i=1}^n r_j)$ RV.

Theorem 5. Let $X \sim U(0,1)$. Then $Y = -2\log X$ is $\chi^2(2)$.

Corollary. Let $X_1, X_2, \ldots, X_n$ be iid RVs with common distribution $U(0,1)$. Then $-2\sum_{i=1}^n \log X_i = 2\log(1/\prod_{i=1}^n X_i)$ is $\chi^2(2n)$.

Theorem 6. Let $X \sim G(\alpha_1, \beta)$ and $Y \sim G(\alpha_2, \beta)$ be independent RVs. Then $X+Y$ and X/Y are independent.

Corollary. Let $X \sim G(\alpha_1, \beta)$ and $Y \sim G(\alpha_2, \beta)$ be independent RVs. Then $X+Y$ and $X/(X+Y)$ are independent.

The converse of Theorem 6 is also true. The result is due to Lukacs [68], and we state it without proof.

Theorem 7. Let X and Y be two nondegenerate RVs that take only positive values. Suppose that $U = X+Y$ and $V = X/Y$ are independent. Then X and Y have gamma distribution with the same parameter β.

Theorem 8. Let $X \sim G(1, \beta)$. Then the RV X has "no memory," that is,

$$P\{X > r+s | X > s\} = P\{X > r\}, \tag{30}$$

for any two positive real numbers r and s.

Proof. The proof is left as an exercise.

The converse of Theorem 8 is also true in the following sense.

Theorem 9. Let F be a DF such that $F(x) = 0$ if $x < 0$, $F(x) < 1$ if $x > 0$, and

$$\frac{1 - F(x+y)}{1 - F(y)} = 1 - F(x) \qquad \text{for all } x, y > 0. \tag{31}$$

Then there exists a constant $\beta > 0$ such that

$$1 - F(x) = e^{-x\beta}, \qquad x > 0. \tag{32}$$

Proof. Equation (31) is equivalent to

$$g(x+y) = g(x) + g(y)$$

if we write $g(x) = \log\{1 - F(x)\}$. From the proof of Theorem 3 it is clear that the only right continuous solution is $g(x) = cx$. Hence $F(x) = 1 - e^{cx}$, $x \geq 0$. Since $F(x) \to 1$ as $x \to \infty$, it follows that $c < 0$ and the proof is complete.

Theorem 10. Let $X_1, X_2, \ldots, X_n$ be iid RVs. Then $X_i \sim G(1, n\beta)$, $i = 1, 2, \ldots, n$, if and only if $X_{(1)}$ is $G(1, \beta)$.

Note that if $X_1, X_2, \ldots, X_n$ are independent with $X_i \sim G(1, \beta_i)$, $i = 1, 2, \ldots, n$, then $X_{(1)}$ is a $G\left(1, 1/\sum_{i=1}^{b} \beta_i^{-1}\right)$ RV.

The following result describes the relationship between exponential and Poisson RVs.

Theorem 11. Let $X_1, X_2, \ldots$ be a sequence of iid RVs having common exponential density with parameter $\beta > 0$. Let $S_n = \sum_{k=1}^{n} X_k$ be the nth partial sum, $n = 1, 2, \ldots$, and suppose that $t > 0$. If $Y =$ number of $S_n \in [0, t]$, then Y is a $P(t/\beta)$ RV.

Proof. We have

$$P\{Y = 0\} = P\{S_1 > t\} = \frac{1}{\beta} \int_t^{\infty} e^{-x/\beta}\, dx = e^{-t/\beta},$$

so that the assertion holds for $Y = 0$. Let n be a positive integer. Since the X_i's are nonnegative, S_n is nondecreasing, and

$$P\{Y = n\} = P\{S_n \leq t,\ S_{n+1} > t\}. \tag{33}$$

Now

$$P\{S_n \leq t\} = P\{S_n \leq t,\ S_{n+1} > t\} + P\{S_{n+1} \leq t\}. \tag{34}$$

It follows that

$$P\{Y=n\}=P\{S_n \le t\}-P\{S_{n+1} \le t\}, \tag{35}$$

and, since $S_n \sim G(n,\beta)$, we have

$$\begin{aligned} P\{Y=n\} &= \int_0^t \frac{1}{\Gamma(n)\beta^n} x^{n-1} e^{-x/\beta}\,dx - \int_0^t \frac{1}{\Gamma(n+1)\beta^{n+1}} x^n e^{-x/\beta}\,dx \\ &= \frac{t^n e^{-t/\beta}}{\beta^n n!}, \end{aligned}$$

as asserted.

Theorem 12. If X and Y are independent exponential RVs with parameter β, then $Z = X/(X+Y)$ has a $U(0,1)$ distribution.

Note that, in view of Theorem 7, Theorem 12 characterizes the exponential distribution in the following sense. Let X and Y be independent RVs that are nondegenerate and take only positive values. Suppose that $X+Y$ and X/Y are independent. If $X/(X+Y)$ is $U(0,1)$, X and Y both have the exponential distribution with parameter β. This follows since, by Theorem 7, X and Y must have the gamma distribution with parameter β. Thus $X/(X+Y)$ must have (see Theorem 14) the PDF

$$f(x) = \frac{\Gamma(\alpha_1+\alpha_2)}{\Gamma(\alpha_1)\Gamma(\alpha_2)} x^{\alpha_1-1}(1-x)^{\alpha_2-1}, \qquad 0<x<1,$$

and this is the uniform density on $(0,1)$ if and only if $\alpha_1 = \alpha_2 = 1$. Thus X and Y both have the $G(1,\beta)$ distribution.

Theorem 13. Let X be a $P(\lambda)$ RV. Then

$$P\{X \le K\} = \frac{1}{K!}\int_\lambda^\infty e^{-x}x^K\,dx \tag{36}$$

expresses the DF of X in terms of an incomplete gamma function.

Proof.

$$\begin{aligned} \frac{d}{d\lambda}P\{X \le K\} &= \sum_{j=0}^{K} \frac{1}{j!}\{je^{-\lambda}\lambda^{j-1} - \lambda^j e^{-\lambda}\} \\ &= \frac{-\lambda^K e^{-\lambda}}{K!}, \end{aligned}$$

and it follows that

$$P\{X \le K\} = \frac{1}{K!}\int_\lambda^\infty e^{-x}x^K\,dx,$$

as asserted.

An alternative way of writing (36) is the following:

$$P\{X \leq K\} = P\{Y \geq 2\lambda\},$$

where $X \sim P(\lambda)$ and $Y \sim \chi^2(2K+2)$.

5.3.3 Beta Distribution

The integral

$$B(\alpha,\beta) = \int_{0+}^{1-} x^{\alpha-1}(1-x)^{\beta-1}\,dx \tag{37}$$

converges for $\alpha > 0$ and $\beta > 0$ and is called a *beta function*. For $\alpha \leq 0$ or $\beta \leq 0$ the integral in (37) diverges. It is easy to see that for $\alpha > 0$ and $\beta > 0$

$$B(\alpha,\beta) = B(\beta,\alpha), \tag{38}$$

$$B(\alpha,\beta) = \int_{0+}^{\infty} x^{\alpha-1}(1+x)^{-\alpha-\beta}dx, \tag{39}$$

and

$$B(\alpha,\beta) = \frac{\Gamma(\alpha)\Gamma(\beta)}{\Gamma(\alpha+\beta)}. \tag{40}$$

It follows that

$$f(x) = \begin{cases} \dfrac{x^{\alpha-1}(1-x)^{\beta-1}}{B(\alpha,\beta)}, & 0 < x < 1, \\ 0, & \text{otherwise}, \end{cases} \tag{41}$$

defines a pdf.

Definition 4. An RV X with PDF given by (41) is said to have a *beta distribution* with parameters α and β, $\alpha > 0$ and $\beta > 0$. We will write $X \sim B(\alpha,\beta)$ for a beta variable with density (41).

Figure 3 gives graphs of some beta PDFs.

The DF of a $B(\alpha,\beta)$ RV is given by

$$F(x) = \begin{cases} 0, & x \leq 0, \\ [B(\alpha,\beta)]^{-1}\displaystyle\int_{0+}^{x} y^{\alpha-1}(1-y)^{\beta-1}\,dy, & 0 < x < 1, \\ 1, & x \geq 1. \end{cases} \tag{42}$$

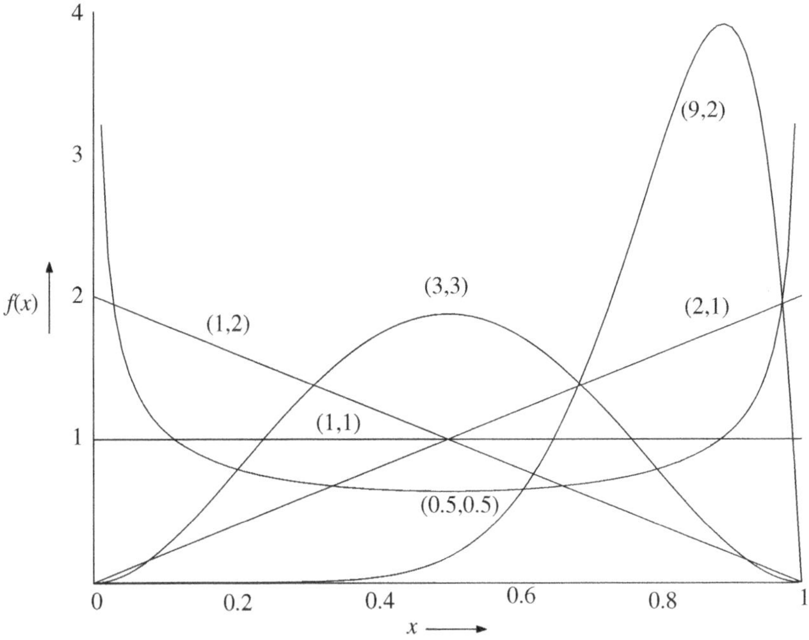

Fig. 3 Beta density functions.

If n is a positive number, then

$$EX^n = \frac{1}{B(\alpha,\beta)}\int_0^1 x^{n+\alpha-1}(1-x)^{\beta-1}\,dx$$
$$= \frac{B(n+\alpha,\beta)}{B(\alpha,\beta)} = \frac{\Gamma(n+\alpha)\Gamma(\alpha+\beta)}{\Gamma(\alpha)\Gamma(n+\alpha+\beta)}, \tag{43}$$

using (40). In particular,

$$EX = \frac{\alpha}{\alpha+\beta} \tag{44}$$

and

$$\text{var}(X) = \frac{\alpha\beta}{(\alpha+\beta)^2(\alpha+\beta+1)}. \tag{45}$$

For the MGF of $X \sim B(\alpha,\beta)$, we have

$$M(t) = \frac{1}{B(\alpha,\beta)}\int_0^1 e^{tx}x^{\alpha-1}(1-x)^{\beta-1}\,dx. \tag{46}$$

Since moments of all order exist, and $E|X|^j < 1$ for all j, we have

$$\begin{aligned} M(t) &= \sum_{j=0}^{\infty} \frac{t^j}{j!} EX^j \\ &= \sum_{j=0}^{\infty} \frac{t^j}{\Gamma(j+1)} \frac{\Gamma(\alpha+j)\Gamma(\alpha+\beta)}{\Gamma(\alpha+\beta+j)\Gamma(\alpha)}. \end{aligned} \tag{47}$$

Remark 1. Note that in the special case where $\alpha = \beta = 1$ we get the uniform distribution on $(0,1)$.

Remark 2. If X is a beta RV with parameters α and β, then $1-X$ is a beta variate with parameters β and α. In particular, X is $B(\alpha,\alpha)$ if and only if $1-X$ is $B(\alpha,\alpha)$. A special case is the uniform distribution on $(0,1)$. If X and $1-X$ have the same distribution, it does not follow that X has to be $B(\alpha,\alpha)$. All this entails is that the PDF satisfies

$$f(x) = f(1-x), \qquad 0 < x < 1.$$

Take

$$f(x) = \frac{1}{B(\alpha,\beta)+B(\beta,\alpha)}[x^{\alpha-1}(1-x)^{\beta-1} + (1-x)^{\alpha-1}x^{\beta-1}], \qquad 0 < x < 1.$$

Example 3. Let X be distributed with PDF

$$f(x) = \begin{cases} \frac{1}{12}x^2(1-x), & 0 < x < 1, \\ 0, & \text{otherwise.} \end{cases}$$

Then $X \sim B(3,2)$ and

$$\begin{aligned} EX^n &= \frac{\Gamma(n+3)\Gamma(5)}{\Gamma(3)\Gamma(n+5)} = \frac{4!}{2!} \cdot \frac{(n+2)!}{(n+4)!} = \frac{12}{(n+4)(n+3)}, \\ EX &= \frac{12}{20}, \qquad \text{var}(X) = \frac{6}{5^2 \cdot 6} = \frac{1}{25}, \\ M(t) &= \sum_{j=0}^{\infty} \frac{t^j}{j!} \cdot \frac{(j+2)!}{(j+4)!} \frac{4!}{2!}, \\ &= \sum_{j=0}^{\infty} \frac{12}{(j+4)(j+3)} \cdot \frac{t^j}{j!}, \end{aligned}$$

and

$$\begin{aligned} P\{0.2 < X < 0.5\} &= \frac{1}{12} \int_{0.2}^{0.5} (x^2 - x^3)\, dx \\ &= 0.023. \end{aligned}$$

Theorem 14. Let X and Y be independent $G(\alpha_1,\beta)$ and $G(\alpha_2,\beta)$, respectively, RVs. Then $X/(X+Y)$ is a $B(\alpha_1,\alpha_2)$ RV.

Let $X_1,X_2,\ldots,X_n$ be iid RVs with the uniform distribution on $[0,1]$. Let $X_{(k)}$ be the kth-order statistic.

Theorem 15. The RV $X_{(k)}$ has a beta distribution with parameters $\alpha=k$ and $\beta=n-k+1$.

Proof. Let X be the number of X_i's that lie in $[0,t]$. Then X is $b(n,t)$. We have

$$\begin{aligned} P\{X_{(k)} \le t\} &= P\{X \ge k\} \\ &= \sum_{j=k}^{n} \binom{n}{j} t^j (1-t)^{n-j}. \end{aligned}$$

Also

$$\begin{aligned} \frac{d}{dt}P\{X \ge k\} &= \sum_{j=k}^{n} \binom{n}{j} \{jt^{j-1}(1-t)^{n-j} - (n-j)t^j(1-t)^{n-j-1}\} \\ &= \sum_{j=k}^{n} \left\{ n\binom{n-1}{j-1} t^{j-1}(1-t)^{n-j} - n\binom{n-1}{j} t^j (1-t)^{n-j-1} \right\} \\ &= n\binom{n-1}{k-1} t^{k-1}(1-t)^{n-k}. \end{aligned}$$

On integration, we get

$$P\{X_{(k)} \le t\} = n\binom{n-1}{k-1} \int_0^t x^{k-1}(1-x)^{n-k}\,dx,$$

as asserted.

Remark 3. Note that we have shown that if X is $b(n,p)$, then

$$1 - P\{X < k\} = n\binom{n-1}{k-1} \int_0^p x^{k-1}(1-x)^{n-k}\,dx, \tag{48}$$

which expresses the DF of X in terms of the DF of a $B(k,n-k+1)$ RV.

Theorem 16. Let $X_1,X_2,\ldots,X_n$ be independent RVs. Then $X_1,X_2,\ldots,X_n$ are iid $B(\alpha,1)$ RVs if and only if $X_{(n)} \sim B(\alpha n,1)$.

5.3.4 Cauchy Distribution

Definition 5. An RV X is said to have a *Cauchy distribution* with parameters μ and θ if its PDF is given by

$$f(x) = \frac{\mu}{\pi}\frac{1}{\mu^2+(x-\theta)^2}, \qquad -\infty < x < \infty, \quad \mu > 0. \tag{49}$$

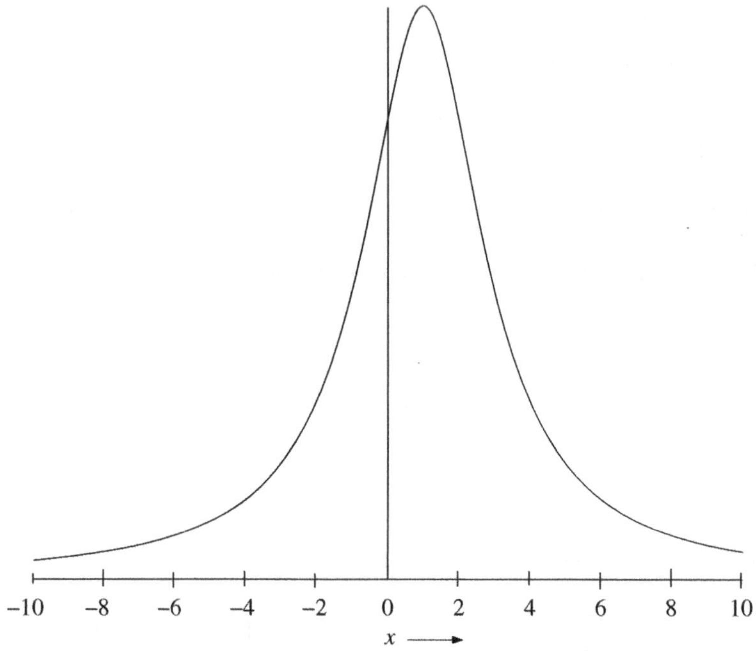

Fig. 4 Cauchy density function.

We will write $X \sim \mathcal{C}(\mu, \theta)$ for a Cauchy RV with density (49).

Figure 4 gives graph of a Cauchy PDF.

We first check that (49) in fact defines a PDF. Substituting $y = (x-\theta)/\mu$, we get

$$\int_{-\infty}^{\infty} f(x)\,dx = \frac{1}{\pi}\int_{-\infty}^{\infty} \frac{dy}{1+y^2} = \frac{2}{\pi}(\tan^{-1} y)_0^{\infty} = 1.$$

The DF of a $\mathcal{C}(1,0)$ RV is given by

$$F(x) = \frac{1}{2} + \frac{1}{\pi}\tan^{-1} x, \qquad -\infty < x < \infty. \tag{50}$$

Theorem 17. Let X be a Cauchy RV with parameters μ and θ. The moments of order < 1 exist, but the moments of order ≥ 1 do not exist for the RV X.

Proof. It suffices to consider the PDF

$$f(x) = \frac{1}{\pi}\frac{1}{1+x^2}, \qquad -\infty < x < \infty.$$

$$E|X|^{\alpha} = \frac{2}{\pi}\int_0^{\infty} x^{\alpha}\frac{1}{1+x^2}\,dx,$$

and letting $z = 1/(1+x^2)$ in the integral, we get

$$E|X|^\alpha = \frac{1}{\pi}\int_0^1 z^{(1-\alpha)/2-1}(1-z)^{[(\alpha+1)/2]-1}\,dz,$$

which converges for $\alpha < 1$ and diverges for $\alpha \geq 1$. This completes the proof of the theorem.

It follows from Theorem 17 that the MGF of a Cauchy RV does not exist. This creates some manipulative problems. We note, however, that the CF of $X \sim \mathcal{C}(\mu, 0)$ is given by

$$\phi(t) = e^{-\mu|t|}. \tag{51}$$

Theorem 18. Let $X \sim \mathcal{C}(\mu_1, \theta_1)$ and $Y \sim \mathcal{C}(\mu_2, \theta_2)$ be independent RVs. Then $X+Y$ is a $\mathcal{C}(\mu_1+\mu_2,\ \theta_1+\theta_2)$ RV.

Proof. For notational convenience we will prove the result in the special case where $\mu_1 = \mu_2 = 1$ and $\theta_1 = \theta_2 = 0$, that is, where X and Y have the common PDF

$$f(x) = \frac{1}{\pi}\cdot\frac{1}{1+x^2}, \qquad -\infty < x < \infty.$$

The proof in the general case follows along the same lines. If $Z = X+Y$, the PDF of Z is given by

$$f_Z(z) = \frac{1}{\pi^2}\int_{-\infty}^{\infty}\frac{1}{1+x^2}\cdot\frac{1}{1+(z-x)^2}\,dx.$$

Now

$$\begin{aligned}
&\frac{1}{(1+x^2)[1+(z-x)^2]}\\
&\quad = \frac{1}{z^2(z^2+4)}\left[\frac{2zx}{1+x^2}+\frac{z^2}{1+x^2}+\frac{2z^2-2zx}{1+(z-x)^2}+\frac{z^2}{1+(z-x)^2}\right],
\end{aligned}$$

so that

$$\begin{aligned}
f_Z(z) &= \frac{1}{\pi^2}\frac{1}{z^2(z^2+4)}\left[z\log\frac{1+x^2}{1+(z-x)^2}+z^2\tan^{-1}x+z^2\tan^{-1}(x-z)\right]_{-\infty}^{\infty}\\
&= \frac{1}{\pi}\frac{2}{z^2+2^2}, \qquad -\infty < z < \infty.
\end{aligned}$$

It follows that, if X and Y are iid $\mathcal{C}(1,0)$ RVs, then $X+Y$ is a $\mathcal{C}(2,0)$ RV. We note that the result follows effortlessly from (51).

Corollary. Let $X_1, X_2, \ldots, X_n$ be independent Cauchy RVs, $X_k \sim \mathcal{C}(\mu_k, \theta_k)$, $k = 1, 2, \ldots, n$. Then $S_n = \sum_1^n X_k$ is a $\mathcal{C}(\sum_1^n \mu_k, \sum_1^n \theta_k)$ RV.

In particular, if $X_1, X_2, \ldots, X_n$ are iid $\mathcal{C}(1,0)$ RVs, $n^{-1}S_n$ is also a $\mathcal{C}(1,0)$ RV. This is a remarkable result, the importance of which will become clear in Chapter 7. Actually, this property uniquely characterizes the Cauchy distribution. If F is a nondegenerate DF with the property that $n^{-1}S_n$ also has DF F, then F must be a Cauchy distribution (see Thompson [113, p. 112]).

The proof of the following result is simple.

Theorem 19. Let X be $\mathcal{C}(\mu,0)$. Then λ/X, where λ is a constant, is a $\mathcal{C}(|\lambda|/\mu,0)$ RV.

Corollary. X is $\mathcal{C}(1,0)$ if and only if $1/X$ is $\mathcal{C}(1,0)$.

We emphasize that if X and $1/X$ have the same PDF on $(-\infty,\infty)$, it does not follow* that X is $\mathcal{C}(1,0)$, for let X be an RV with PDF

$$\begin{aligned} f(x) &= \frac{1}{4} \qquad \text{if } |x| \le 1, \\ &= \frac{1}{4x^2} \qquad \text{if } |x| > 1. \end{aligned}$$

Then X and $1/X$ have the same PDF, as can be easily checked.

Theorem 20. Let X be a $U(-\pi/2,\pi/2)$ RV. Then $Y = \tan X$ is a Cauchy RV.

Many important properties of the Cauchy distribution can be derived from this result (see Pitman and Williams [80]).

5.3.5 Normal Distribution (the Gaussian Law)

One of the most important distributions in the study of probability and mathematical statistics is the *normal distribution*, which we will examine presently.

Definition 6. An RV X is said to have a *standard normal distribution* if its PDF is given by

$$\varphi(x) = \frac{1}{\sqrt{2\pi}} e^{-(x^2/2)}, \qquad -\infty < x < \infty. \tag{52}$$

We first check that f defines a PDF. Let

$$I = \int_{-\infty}^{\infty} e^{-x^2/2}\, dx.$$

---

* Menon [73] has shown that we need the condition that both X and $1/X$ be *stable* to conclude that X is Cauchy. A nondegenerate distribution function F is said to be stable if, for two iid RVs X_1, X_2 with common DF F, and given constants $a_1, a_2 > 0$, we can find $\alpha > 0$ and $\beta(a_1,a_2)$ such that the RV

$$X_3 = \alpha^{-1}(a_1X_1 + a_2X_2 - \beta)$$

again has the same distribution F. Examples are the Cauchy (see the corollary to Theorem 18) and normal (discussed in Section 5.3.5) distributions.

Then

$$0 < e^{-x^2/2} < e^{-|x|+1}, \qquad -\infty < x < \infty,$$
$$\int_{-\infty}^{\infty} e^{-|x|+1}\, dx = 2e,$$

and it follows that I exists. We have

$$\begin{aligned} I &= \int_0^\infty y^{-1/2} e^{-y/2}\, dy \\ &= \Gamma\left(\frac{1}{2}\right) 2^{1/2} \\ &= \sqrt{2\pi}. \end{aligned}$$

Thus $\int_{-\infty}^{\infty} \varphi(x)\, dx = 1$, as required.

Let us write $Y = \sigma X + \mu$, where $\sigma > 0$. Then the PDF of Y is given by

$$\begin{aligned} \psi(y) &= \frac{1}{\sigma} \varphi\left(\frac{y-\mu}{\sigma}\right) \\ &= \frac{1}{\sigma\sqrt{2\pi}} e^{-[(y-\mu)^2/2\sigma^2]}, \qquad -\infty < y < \infty;\ \sigma > 0,\ -\infty < \mu < \infty. \end{aligned} \tag{53}$$

Definition 7. An RV X is said to have a normal distribution with parameters μ $(-\infty < \mu < \infty)$ and $\sigma (> 0)$ if its PDF is given by (53).

If X is a normally distributed RV with parameters μ and σ, we will write $X \sim \mathcal{N}(\mu, \sigma^2)$. In this notation φ defined by (53) is the PDF of an $\mathcal{N}(0,1)$ RV. The DF of an $\mathcal{N}(0,1)$ RV will be denoted by $\Phi(x)$, where

$$\Phi(x) = \frac{1}{\sqrt{2\pi}} \int_{-\infty}^{x} e^{-u^2/2}\, du. \tag{54}$$

Clearly, if $X \sim \mathcal{N}(\mu, \sigma^2)$, then $Z = (X - \mu)/\sigma \sim N(0,1)$. Z is called a *standard normal RV*. For the MGF of an $\mathcal{N}(\mu, \sigma^2)$ RV, we have

$$\begin{aligned} M(t) &= \frac{1}{\sqrt{2\pi}\sigma} \int_{-\infty}^{\infty} \exp\left\{\frac{-x^2}{2\sigma^2} + x\frac{(t\sigma^2 + \mu)}{\sigma^2} - \frac{\mu^2}{2\sigma^2}\right\} dx \\ &= \frac{1}{\sqrt{2\pi}\sigma} \int_{-\infty}^{\infty} \exp\left\{\frac{-(x - \mu - \sigma^2 t)^2}{2\sigma^2} + \mu t + \frac{\sigma^2 t^2}{2}\right\} dx \\ &= \exp\left(\mu t + \frac{\sigma^2 t^2}{2}\right), \end{aligned} \tag{55}$$

for all real values of t. Moments of all order exist and may be computed from the MGF. Thus

$$EX = M'(t)|_{t=0} = (\mu + \sigma^2 t) M(t)|_{t=0} = \mu \tag{56}$$

and

$$EX^2 = M''(t)|_{t=0} = [M(t)\sigma^2 + (\mu + \sigma^2 t)^2 M(t)]_{t=0}$$
$$= \sigma^2 + \mu^2. \tag{57}$$

Thus

$$\text{var}(X) = \sigma^2. \tag{58}$$

Clearly, the central moments of odd order are all 0. The central moments of even order are as follows:

$$E\{X-\mu\}^{2n} = \frac{1}{\sigma\sqrt{2\pi}} \int_{-\infty}^{\infty} x^{2n} e^{-x^2/2\sigma^2}\, dx \qquad (n \text{ is a positive integer})$$
$$= \frac{\sigma^{2n}}{\sqrt{2\pi}} 2^{n+1/2} \Gamma\left(n + \frac{1}{2}\right)$$
$$= [(2n-1)(2n-3)\cdots 3\cdot 1]\sigma^{2n}. \tag{59}$$

As for the absolute moment of order α, for a standard normal RV Z we have

$$E|Z|^\alpha = \frac{1}{\sqrt{2\pi}} 2 \int_0^\infty z^\alpha e^{-z^2/2}\, dz$$
$$= \frac{1}{\sqrt{2\pi}} \int_0^\infty y^{[(\alpha+1)/2)]-1} e^{-y/2}\, dy$$
$$= \frac{\Gamma[(\alpha+1)/2] 2^{\alpha/2}}{\sqrt{\pi}}. \tag{60}$$

As remarked earlier, the normal distribution is one of the most important distributions in probability and statistics, and for this reason the standard normal distribution is available in tabular form. Table ST2 at the end of the book gives the probability $P\{Z > z\}$ for various values of $z(> 0)$ in the tail of an $\mathcal{N}(0,1)$ RV. In this book we will write z_α for the value of Z that satisfies $\alpha = P\{Z > z_\alpha\}, 0 \leq \alpha \leq 1$.

Example 4. By Chebychev's inequality, if $E|X|^2 < \infty$, $EX = \mu$, and $\text{var}(X) = \sigma^2$, then

$$P\{|X-\mu| \geq K\sigma\} \leq \frac{1}{K^2}.$$

For $K = 2$, we get $P\{|X-\mu| \geq K\sigma\} \leq 0.25$, and for $K = 3$, we have $P\{|X-\mu| \geq K\sigma\} \leq \frac{1}{9}$. If X is, in particular, $\mathcal{N}(\mu\,\sigma^2)$, then

$$P\{|X-\mu| \geq K\sigma\} = P\{|Z| \geq K\},$$

where Z is $\mathcal{N}(0,1)$. From Table ST2.

$$P\{|Z| \geq 1\} = 0.318, \qquad P\{|Z| \geq 2\} = 0.046, \qquad \text{and} \qquad P\{|Z| \geq 3\} = 0.002.$$

Thus practically all the distribution is concentrated within three standard deviations of the mean.

Example 5. Let $X \sim \mathcal{N}(3,4)$. Then

$$\begin{aligned} P\{2 < X \leq 5\} &= P\left\{\frac{2-3}{2} < \frac{X-3}{2} \leq \frac{5-3}{2}\right\} = P\{-0.5 < Z \leq 1\} \\ &= P\{Z \leq 1\} - P\{Z \leq -0.5\} \\ &= 0.841 - P\{Z \geq 0.5\} \\ &= 0.0841 - 0.309 = 0.532. \end{aligned}$$

Theorem 21. (Feller [25, p. 175]). Let Z be a standard normal RV. Then

$$P\{Z > x\} \approx \frac{1}{\sqrt{2\pi}x} e^{-x^2/2} \qquad \text{as } x \to \infty. \tag{61}$$

More precisely, for every $x > 0$

$$\frac{1}{\sqrt{2\pi}} e^{-x^2/2}\left(\frac{1}{x} - \frac{1}{x^3}\right) < P\{Z > x\} < \frac{1}{x\sqrt{2\pi}} e^{-x^2/2}. \tag{62}$$

Proof. We have

$$\frac{1}{\sqrt{2\pi}} \int_x^\infty e^{-(1/2)y^2}\left(1 - \frac{3}{y^4}\right) dy = \frac{1}{\sqrt{2\pi}} e^{-x^2/2}\left(\frac{1}{x} - \frac{1}{x^3}\right) \tag{63}$$

and

$$\frac{1}{\sqrt{2\pi}} \int_x^\infty e^{-y^2/2}\left(1 + \frac{1}{y^2}\right) dy = \frac{1}{\sqrt{2\pi}} e^{-x^2/2} \frac{1}{x}, \tag{64}$$

as can be checked on differentiation. Approximation (61) follows immediately.

Theorem 22. Let $X_1, X_2, \ldots, X_n$ be independent RVs with $X_k \sim \mathcal{N}(\mu_k, \sigma_k^2)$, $k = 1, 2, \ldots, n$. Then $S_n = \sum_{k=1}^n X_k$ is an $\mathcal{N}(\sum_{k=1}^n \mu_k, \sum_1^n \sigma_k^2)$ RV.

Corollary 3. If $X_1, X_2, \ldots, X_n$ are iid $\mathcal{N}(\mu, \sigma^2)$ RVs, then S_n is an $\mathcal{N}(n\mu, n\sigma^2)$ RV and $n^{-1}S_n$ is an $\mathcal{N}(\mu, \sigma^2/n)$ RV.

Corollary 4. If $X_1, X_2, \ldots, X_n$ are iid $\mathcal{N}(0,1)$ RVs, then $n^{-1/2}S_n$ is also an $\mathcal{N}(0,1)$ RV.

We remark that if $X_1, X_2, \ldots, X_n$ are iid RVs with $EX = 0$, $EX^2 = 1$ such that $n^{-1/2}S_n$ also has the same distribution for each $n = 1, 2, \ldots$, that distribution can only be $\mathcal{N}(0,1)$. This characterization of the normal distribution will become clear when we study the central limit theorem in Chapter 7.

Theorem 23. Let X and Y be independent RVs. Then $X+Y$ is normally distributed if and only if X and Y are both normal.

If X and Y are independent normal RVs, $X+Y$ is normal by Theorem 22. The converse is due to Cramér [16] and will not be proved here.

Theorem 24. Let X and Y be independent RVs with common $\mathcal{N}(0,1)$ distribution. Then $X+Y$ and $X-Y$ are independent.

The converse is due to Bernstein [4] and is stated here without proof.

Theorem 25. If X and Y are independent RVs with the same distribution and if $Z_1 = X+Y$ and $Z_2 = X-Y$ are independent, all RVs X, Y, Z_1, and Z_2 are normally distributed.

The following result generalizes Theorem 24.

Theorem 26. If $X_1, X_2, \ldots, X_n$ are independent normal RVs and $\sum_{i=1}^{n} a_i b_i \operatorname{var}(X_i) = 0$, then $L_1 = \sum_{i=1}^{n} a_i X_i$ and $L_2 = \sum_{i=1}^{n} b_i X_i$ are independent. Here $a_1, a_2, \ldots, a_n$ and $b_1, b_2, \ldots, b_n$ are fixed (nonzero) real numbers.

Proof. Let $\operatorname{var}(X_i) = \sigma_i^2$, and assume without loss of generality that $EX_i = 0$, $i = 1, 2, \ldots, n$. For any real numbers α, β, and t

$$
\begin{aligned}
Ee^{(\alpha L_1 + \beta L_2)t} &= E\exp\left\{ t\sum_1^n (\alpha a_i + \beta b_i) X_i \right\} \\
&= \prod_{i=1}^{n} \exp\left\{ \frac{t^2}{2} (\alpha a_i + \beta b_i)^2 \sigma_i^2 \right\} \\
&= \exp\left\{ \frac{\alpha^2 t^2}{2} \sum_1^n a_i^2 \sigma_i^2 + \frac{\beta^2 t^2}{2} \sum_1^n b_i^2 \sigma_i^2 \right\} \left(\text{since } \sum_i^n a_i b_i \sigma_i^2 = 0 \right) \\
&= \prod_{i=1}^{n} \exp\left\{ \frac{t^2 \alpha^2}{2} a_i^2 \sigma_i^2 \right\} \cdot \prod_{i=1}^{n} \exp\left\{ \frac{t^2 \beta^2}{2} b_i^2 \sigma_i^2 \right\} \\
&= \prod_1^n Ee^{t\alpha a_i X_i} \cdot \prod_1^n Ee^{t\beta b_i X_i} \\
&= E\exp\left(t\alpha \sum_1^n a_i X_i \right) \cdot E\exp\left(t\beta \sum_1^n b_i X_i \right) \\
&= Ee^{\alpha t L_1} Ee^{\beta t L_2}.
\end{aligned}
$$

Thus we have shown that

$$
M(\alpha t, \beta t) = M(\alpha t, 0) M(0, \beta t) \qquad \text{for all } \alpha, \beta, t.
$$

It follows that L_1 and L_2 are independent.

Corollary. If X_1, X_2 are independent $\mathcal{N}(\mu_1,\sigma^2)$ and $\mathcal{N}(\mu_2,\sigma^2)$ RVs, then $X_1 - X_2$ and $X_1 + X_2$ are independent. (This gives Theorem 24.)

Darmois [20] and Skitovitch [106] provided the converse of Theorem 26, which we state without proof.

Theorem 27. If $X_1, X_2, \ldots, X_n$ are independent RVs, $a_1, a_2, \ldots, a_n$, $b_1, b_2, \ldots, b_n$ are real numbers none of which equals 0, and if the linear forms

$$L_1 = \sum_{i=1}^{n} a_i X_i, \qquad L_2 = \sum_{i=1}^{n} b_i X_i$$

are independent, then all the RVs are normally distributed.

Corollary. If X and Y are independent RVs such that $X+Y$ and $X-Y$ are independent, X, Y, $X+Y$, and $X-Y$ are all normal.

Yet another result of this type is the following theorem.

Theorem 28. Let $X_1, X_2, \ldots, X_n$ be iid RVs. Then the common distribution is normal if and only if

$$S_n = \sum_{k=1}^{n} X_k \qquad \text{and} \qquad Y_n = \sum_{i=1}^{n} (X_i - n^{-1} S_n)^2$$

are independent.

In Chapter 6 we will prove the necessity part of this result, which is basic to the theory of t-tests in statistics (Chapter 10; see also Example 4.4.6). The sufficiency part was proved by Lukacs [67], and we will not prove it here.

Theorem 29. $X \sim \mathcal{N}(0,1) \Rightarrow X^2 \sim \chi^2(1)$.

Proof. See Example 2.5.7 for the proof.

Corollary 1. If $X \sim \mathcal{N}(\mu,\sigma^2)$, the RV $Z^2 = (X-\mu)^2/\sigma^2$ is $\chi^2(1)$.

Corollary 2. If $X_1, X_2, \ldots, X_n$ are independent RVs and $X_k \sim \mathcal{N}(\mu_k, \sigma_k^2)$, $k = 1, 2, \ldots, n$, then $\sum_{k=1}^{n} (X_k - \mu_k)^2/\sigma_k^2$ is $\chi^2(n)$.

Theorem 30. Let X and Y be iid $\mathcal{N}(0,\sigma^2)$ RVs. Then X/Y is $\mathcal{C}(1,0)$.

Proof. For the proof see Example 2.5.7.

We remark that the converse of this result does not hold; that is, if $Z = X/Y$ is the quotient of two iid RVs and Z has a $\mathcal{C}(1,0)$ distribution, it does not follow that X and Y

are normal, for take X and Y to be iid with PDF

$$f(x) = \frac{\sqrt{2}}{\pi}\frac{1}{1+x^4}, \qquad -\infty < x < \infty.$$

We leave the reader to verify that $Z = X/Y$ is $\mathcal{C}(1,0)$.

5.3.6 Some Other Continuous Distributions

Several other distributions which are related to distributions studied earlier also arise in practice. We record briefly some of these and their important characteristics. We will use these distributions infrequently. We say that X has a *lognormal distribution* if $Y = \ell\text{n}\, X$ has a normal distribution. The PDF of X is then

$$f(x) = \frac{1}{x\sigma\sqrt{2\pi}} \exp\left\{-\frac{(\log x - \mu)^2}{2\sigma^2}\right\}, \quad x \geq 0, \tag{65}$$

and $f(x) = 0$ for $x < 0$, where $-\infty < \mu < \infty$, $\sigma > 0$. In fact for $x > 0$

$$\begin{aligned} P(X \leq x) &= P(\ell\text{n}\, X \leq \ell\text{n}\, x) \\ &= P(Y \leq \ell\text{n}\, x) = P\left(\frac{Y-\mu}{\sigma} \leq \frac{\ell\text{n}\, x - \mu}{\sigma}\right) \\ &= \Phi\left(\frac{\ell\text{n}\, x - \mu}{\sigma}\right), \end{aligned}$$

where Φ is the DF of a $\mathcal{N}(0,1)$ RV which leads to (65). It is easily seen that for $n \geq 0$

$$\begin{cases} EX^n = \exp\left(\frac{n\mu + n^2\sigma^2}{2}\right) \\ EX = \exp\left(\frac{\mu+\sigma^2}{2}\right), \ \text{var}(X) = \exp(2\mu + 2\sigma^2) - \exp(2\mu + \sigma^2). \end{cases} \tag{66}$$

The MGF of X does not exist.

We say that the RV X has a *Pareto* distribution with parameters $\theta > 0$ and $\alpha > 0$ if its PDF is given by

$$f(x) = \frac{\alpha\theta^\alpha}{(x+\theta)^{\alpha+1}}, \quad x > 0 \tag{67}$$

and 0 otherwise. Here θ is scale parameter and α is a shape parameter. It is easy to check that

$$\begin{cases} F(x) = P(X \leq x) = 1 - \dfrac{\theta^\alpha}{(\theta + x)^\alpha}, \quad x > 0 \\ EX = \frac{\theta}{\alpha - 1}, \alpha > 1, \text{ and } \text{var}(X) = \frac{\alpha\theta^2}{(\alpha-2)(\alpha-1)^2} \end{cases} \tag{68}$$

for $\alpha > 2$. The MGF of X does not exist since all moments of X do not.

Suppose X has a Pareto distribution with parameters θ and α. Writing $Y = \ell\text{n}(X/\theta)$ we see that Y has PDF

$$f_Y(y) = \frac{\alpha e^y}{(1+e^y)^{\alpha+1}}, \quad -\infty < y < \infty, \tag{69}$$

and DF

$$F_Y(y) = 1 - (1+e^y)^{-\alpha}, \quad \text{for all } y.$$

The PDF in (69) is known as a *logistic* distribution. We introduce location and scale parameters μ and σ by writing $Z = \mu + \sigma Y$, taking $\alpha = 1$ and then the PDF of Z is easily seen to be

$$f_Z(z) = \frac{1}{\sigma} \frac{\exp\{(z-\mu)/\sigma\}}{\{1+\exp[(z-\mu)/\sigma]\}^2} \tag{70}$$

for all real z. This is the PDF of a logistic RV with location–scale parameters μ and σ. We leave the reader to check that

$$\begin{cases} F_Z(z) = \exp\left\{\frac{(z-\mu)}{\sigma}\right\}\left\{1+\exp\left[\frac{(z-\mu)}{\sigma}\right]\right\}^{-1} \\ EZ = \mu, \ \text{var}(Z) = \frac{\pi^2\sigma^2}{3} \\ M_Z(t) = \exp(\mu t)\Gamma(1-\sigma t)\Gamma(1+\sigma t), \ t < \frac{1}{\sigma}. \end{cases} \tag{71}$$

Pareto distribution is also related to an exponential distribution. Let X have Pareto PDF of the form

$$f_X(s) = \frac{\alpha\sigma^\alpha}{x^{\alpha+1}}, \quad x > \sigma \tag{72}$$

and 0 otherwise. A simple transformation leads to PDF (72) from (67). Then it is easily seen that $Y = \ell\text{n}(X/\sigma)$ has an exponential distribution with mean $1/\alpha$. Thus some properties of exponential distribution which are preserved under monotone transformations can be derived for Pareto PDF (72) by using the logarithmic transformation.

Some other distributions are related to the gamma distribution. Suppose $X \sim G(1,\beta)$. Let $Y = X^{1/\alpha}$, $\alpha > 0$. Then Y has PDF

$$f_Y(y) = \left(\frac{\alpha}{\beta}\right) y^{\alpha-1} \exp\left\{\frac{-y^\alpha}{\beta}\right\}, \quad y > 0 \tag{73}$$

and 0 otherwise. The RV Y is said to have a *Weibull distribution*. We leave the reader to show that

$$\begin{cases} F_Y(y) = 1 - \exp\left(\frac{-y^\alpha}{\beta}\right), \quad y > 0 \\ EY^n = \beta^{n/\alpha}\Gamma\left(1+\frac{n}{\alpha}\right), \quad EY = \beta^{1/\beta}\Gamma\left(1+\frac{1}{\alpha}\right), \\ \text{var}(Y) = \beta^{2/\alpha}\left[\Gamma\left(1+\frac{2}{\alpha}\right) - \Gamma^2\left(1+\frac{1}{\alpha}\right)\right]. \end{cases} \tag{74}$$

The MGF of Y exists only for $\alpha \geq 1$ but for $\alpha > 1$ it does not have a form useful in applications. The special case $\alpha = 2$ and $\beta = \theta^2$ is known as a *Rayleigh distribution*.

Suppose X has a Weibull distribution with PDF (73). Let $Y = \ell n\, X$. Then Y has DF

$$F_Y(y) = 1 - \exp\left\{-\frac{1}{\beta}e^{\alpha y}\right\}, \quad -\infty < y < \infty.$$

Setting $\theta = (1/\alpha)\ell n\, \beta$ and $\sigma = 1/\alpha$ we get

$$F_Y(y) = 1 - \exp\left\{-\exp\left[\frac{y-\theta}{\sigma}\right]\right\} \tag{75}$$

with PDF

$$f_Y(y) = \frac{1}{\sigma}\exp\left\{\left[\frac{(y-\theta)}{\sigma}\right] - \exp\left[\frac{(y-\theta)}{\sigma}\right]\right\}, \tag{76}$$

for $-\infty < y < \infty$ and $\sigma > 0$. An RV with PDF (76) is called an *extreme value* distribution with location–scale parameters θ and σ. It can be shown that

$$\begin{cases} EY = \theta - \gamma\sigma, \ \mathrm{var}(Y) = \frac{\pi^2\sigma^2}{6}, \ \text{and} \\ M_Y(t) = e^{\theta t}\Gamma(1+\sigma t), \end{cases} \tag{77}$$

where $\gamma \approx 0.577216$ is the Euler constant.

The final distribution we consider is also related to a $G(1,\beta)$ RV. Let f_1 be the PDF of $G(1,\beta)$ and f_2 the PDF

$$f_2(x) = \frac{1}{\beta}\exp\left(\frac{x}{\beta}\right), \quad x < 0, \ = 0 \text{ otherwise.}$$

Clearly f_2 is also an exponential PDF defined on $(-\infty, 0)$. Consider the *mixture* PDF

$$f(x) = \frac{1}{2}[f_1(x) + f_2(x)], \quad -\infty < x < \infty. \tag{78}$$

Clearly,

$$f(x) = \frac{1}{2}\exp\left\{\frac{-|x|}{\beta}\right\}, \quad -\infty < x < \infty, \tag{79}$$

and the PDF f defined in (79) is called a *Laplace or double exponential* pdf. It is convenient to introduce a location parameter μ and consider instead the PDF

$$f(x) = \frac{1}{2}\exp\left\{\frac{-|x-\mu|}{\beta}\right\} \quad -\infty < x < \infty, \tag{80}$$

where $-\infty < \mu < \infty$, $\beta > 0$. It is easy to see that for RV X with PDF (80) we have

$$EX = \mu, \ \mathrm{var}(X) = 2\beta^2, \text{ and } M(t) = e^{\mu t}[1 - (\beta t)^2]^{-1}, \tag{81}$$

for $|t| < 1/\beta$.

For completeness let us define a *mixture* PDF (PMF). Let $g(x|\theta)$ be a PDF and let $h(\theta)$ be a *mixing* PDF. Then the PDF

$$f(x) = \int g(x|\theta)h(\theta)d\theta \tag{82}$$

is called a *mixture density function*. In case h is a PMF with support set $\{\theta_1, \theta_2, \ldots, \theta_k\}$, then (82) reduces to a *finite mixture density function*

$$f(x) = \sum_{i=1}^{k} g(x|\theta_i)h(\theta_i). \tag{83}$$

The quantities $h(\theta_i)$ are called *mixing proportions*. The PDF (78) is an example with $k = 2$, $h(\theta_1) = h(\theta_2) = 1/2$, $g(x|\theta_1) = f_1(x)$, and $g(x|\theta_2) = f_2(x)$.

PROBLEMS 5.3

1. Prove Theorem 1.
2. Let X be an RV with PMF $p_k = P\{X = k\}$ given below. If F is the corresponding DF, find the distribution of $F(X)$, in the following cases:
   (a) $p_k = \binom{n}{k} p^k (1-p)^{n-k}$, $k = 0, 1, 2, \ldots, n$; $0 < p < 1$.
   (b) $p_k = e^{-\lambda}(\lambda^k/k!)$, $k = 0, 1, 2, \ldots$; $\lambda > 0$.
3. Let $Y_1 \sim U[0,1]$, $Y_2 \sim U[0, Y_1], \ldots, Y_n \sim U[0, Y_{n-1}]$. Show that

$$Y_1 \sim X_1, \qquad Y_2 \sim X_1X_2, \ldots, Y_n \sim X_1X_2\cdots X_n,$$

   where $X_1, X_2, \ldots, X_n$ are iid $U[0,1]$ RVs. If U is the number of $Y_1, Y_2, \ldots, Y_n$ in $[t, 1]$, where $0 < t < 1$, show that U has a Poisson distribution with parameter $-\log t$.
4. Let $X_1, X_2, \ldots, X_n$ be iid $U[0,1]$ RVs. Prove by induction or otherwise that $S_n = \sum_{k=1}^{n} X_k$ has the PDF

$$f_n(x) = [(n-1)!]^{-1} \sum_{k=0}^{n} (-1)^k \binom{n}{k} [\varepsilon(x-k)]^{n-1} (x-k)^{n-1},$$

   where $\varepsilon(x) = 1$ if $x \geq 0$, $= 0$ if $x < 0$.
5. (a) Let X be an RV with PMF $p_j = P(X = x_j)$, $j = 0, 1, 2, \ldots$, and let F be the DF of X. Show that

$$EF(X) = \frac{1}{2}\left\{1 + \sum_{j=0}^{\infty} p_j^2\right\}$$

$$\operatorname{var} F(X) = \sum_{j=0}^{\infty} p_j q_{j+1}^2 - \frac{1}{2}\left(1 - \sum_{j=0}^{\infty} p_j^2\right)^2,$$

   where $q_{j+1} = \sum_{i=j+1}^{\infty} p_i$.

(b) Let $p_j > 0$ for $j = 0, 1, \ldots, N$ and $\sum_{j=0}^{N} p_j = 1$. Show that

$$EF(X) \geq \frac{(N+2)}{[2(N+1)]}$$

with equality if and only if $p_j = 1/(N+1)$ for all j.

(Rohatgi [91])

6. Prove (a) Theorem 6 and its corollary, and (b) Theorem 10.

7. Let X be a nonnegative RV of the continuous type, and let $Y \sim U(0,X)$. Also, let $Z = X - Y$. Then the RVs Y and Z are independent if and only if X is $G(2, 1/\lambda)$ for some $\lambda > 0$. (Lamperti [59])

8. Let X and Y be independent RVs with common PDF $f(x) = \beta^{-\alpha}\alpha x^{\alpha-1}$ if $0 < x < \beta$, and $= 0$ otherwise; $\alpha \geq 1$. Let $U = \min(X,Y)$ and $V = \max(X,Y)$. Find the joint PDF of U and V and the PDF of $U + V$. Show that U/V and V are independent.

9. Prove Theorem 14.

10. Prove Theorem 8.

11. Prove Theorems 19 and 20.

12. Let $X_1, X_2, \ldots, X_n$ be independent RVs with $X_i \sim \mathcal{C}(\mu_i, \lambda_i)$, $i = 1, 2, \ldots, n$. Show that the RV $X = 1/\sum_{i=1}^{n} X_i^{-1}$ is also a Cauchy RV with parameters $\mu/(\lambda^2 + \mu^2)$ and $\lambda/(\lambda^2 + \mu^2)$, where

$$\lambda = \sum_{i=1}^{n} \frac{\lambda_i}{\lambda_i^2 + \mu_i^2} \quad \text{and} \quad \mu = \sum_{i=1}^{n} \frac{\mu_i}{\lambda_i^2 + \mu_i^2}.$$

13. Let $X_1, X_2, \ldots, X_n$ be iid $\mathcal{C}(1,0)$ RVs and $a_i \neq 0$, b_i, $i = 1, 2, \ldots, n$, be any real numbers. Find the distribution of $\sum_{i=1}^{n} 1/(a_i X_i + b_i)$.

14. Suppose that the load of an airplane wing is a random variable X with $\mathcal{N}(1000, 14400)$ distribution. The maximum load that the wing can withstand is an RV Y, which is $\mathcal{N}(1260, 2500)$. If X and Y are independent, find the probability that the load encountered by the wing is less than its critical load.

15. Let $X \sim \mathcal{N}(0,1)$. Find the PDF of $Z = 1/X^2$. If X and Y are iid $\mathcal{N}(0,1)$, deduce that $U = XY/\sqrt{X^2 + Y^2}$ is $\mathcal{N}(0, 1/4)$.

16. In Problem 15 let X and Y be independent normal RVs with zero means. Show that $U = XY/\sqrt{(X^2 + Y^2)}$ is normal. If, in addition, $\text{var}(X) = \text{var}(Y)$ show that $V = (X^2 - Y^2)/\sqrt{(X^2 + Y^2)}$ is also normal. Moreover, U and V are independent. (Shepp [104])

17. Let X_1, X_2, X_3, X_4 be independent $\mathcal{N}(0,1)$. Show that $Y = X_1X_2 + X_3X_4$ has the PDF $f(y) = \frac{1}{2}e^{-|y|}$, $-\infty < y < \infty$.

18. Let $X \sim \mathcal{N}(15, 16)$. Find (a) $P\{X \leq 12\}$, (b) $P\{10 \leq X \leq 17\}$, (c) $P\{10 \leq X \leq 19 \mid X \leq 17\}$ and (d) $P\{|X - 15| \geq 0.5\}$.

19. Let $X \sim \mathcal{N}(-1, 9)$. Find x such that $P\{X > x\} = 0.38$. Also find x such that $P\{|X + 1| < x\} = 0.4$.

20. Let X be an RV such that $\log(X-a)$ is $\mathcal{N}(\mu,\sigma^2)$. Show that X has PDF

$$f(x)=\begin{cases}\dfrac{1}{\sigma(x-a)\sqrt{2\pi}}\exp\left\{-\dfrac{[\log(x-a)-\mu]^2}{2\sigma^2}\right\} & \text{if } x>a,\\ 0 & \text{if } x\le a.\end{cases}$$

If m_1, m_2 are the first two moments of this distribution and $\alpha_3=\mu_3/\mu_2^{3/2}$ is the coefficient of skewness, show that a, μ, σ are given by

$$a=m_1-\frac{\sqrt{m_2-m_1^2}}{\eta},\qquad \sigma^2=\log(1+\eta^2),$$

and

$$\mu=\log(m_1-a)-\frac{1}{2}\sigma^2,$$

where η is the real root of the equation $\eta^3+3\eta-\alpha_3=0$.

21. Let $X\sim G(\alpha,\beta)$ and let $Y\sim U(0,X)$.

(a) Find the PDF of Y.

(b) Find the conditional PDF of X given $Y=y$.

(c) Find $P(X+Y\le 2)$.

22. Let X and Y be iid $\mathcal{N}(0,1)$ RVs. Find the PDF of $X/|Y|$. Also, find the PDF of $|X|/|Y|$.

23. It is known that $X\sim B(\alpha,\beta)$ and $P(X<0.2)=0.22$. If $\alpha+\beta=26$, find α and β. [*Hint*: Use Table ST1.]

24. Let $X_1,X_2,\ldots,X_n$ be iid $\mathcal{N}(\mu,\sigma^2)$ RVs. Find the distribution of

$$Y_n=\frac{\sum_{k=1}^n kX_k-\mu\sum_{k=1}^n k}{\left(\sum_{k=1}^n k^2\right)^{1/2}}.$$

25. Let $F_1,F_2,\ldots,F_n$ be n DFs. Show that $\min[F_1(x_1),F_2(x_2),\ldots,F_n(x_n)]$ is an n-dimensional DF with marginal DFs $F_1,F_2,\ldots F_n$. (Kemp [50])

26. Let $X\sim NB(1;p)$ and $Y\sim G(1,1/\lambda)$. Show that X and Y are related by the equation

$$P\{X\le x\}=P\{Y\le[x]\}\qquad \text{for } x>0,\qquad \lambda=\log\left(\frac{1}{1-p}\right),$$

where $[x]$ is the largest integer $\le x$. Equivalently, show that

$$P\{Y\in(n,n+1]\}=P_\theta\{X=n\},$$

where $\theta=1-e^{-\lambda}$ (Prochaska [82]).

27. Let T be an RV with DF F and write $S(t)=1-F(t)=P(T>t)$. The function F is called the *survival (or reliability) function* of X (or DF F). The function $\lambda(t)=\frac{f(t)}{S(t)}$ is called *hazard (or failure-rate) function*. For the following PDF find the hazard function:

(a) Rayleigh: $f(t) = (t/\alpha^2)\exp\{-t^2/(2\alpha^2)\}$, $t > 0$.
(b) Lognormal: $f(t) = 1/(t\sigma\sqrt{2\pi})\exp\{-(\ell n\, t-\mu)^2/2\sigma^2\}$.
(c) Pareto: $f(t) = \alpha\theta^\alpha/t^{\alpha+1}$, $t > \theta$, and $= 0$ otherwise.
(d) Weibull: $f(t) = (\alpha/\beta)t^{\alpha-1}\exp(-t^\alpha/\beta)$, $t > 0$.
(e) Logistic: $f(t) = (1/\beta)\exp\{-(t-\mu)/\beta\}[1+\exp\{-(t-\mu)/\beta\}]^{-2}$, $-\infty < t < \infty$.

28. Consider the PDF

$$f(x) = \left(\frac{\lambda}{2\pi x^3}\right)^{1/2} \exp\left\{-\left[\frac{\lambda(x-\mu)^2}{2\mu^2 x}\right]\right\}, \; x > 0$$

and $= 0$ otherwise. An RV X with PDF f is said to have an *inverse Gaussian distribution* with parameters μ and λ, both positive. Show that

$$EX = \mu, \text{var}(X) = \mu^3/\lambda \quad \text{and}$$

$$M(t) = E\exp(tX) = \exp\left\{\frac{\lambda}{\mu}\left[1-\left(1-\frac{2t\mu^2}{\lambda}\right)^{1/2}\right]\right\}.$$

29. Let f be the PDF of a $\mathcal{N}(\mu, \sigma^2)$ RV:
(a) For what value of c is the function cf^n, $n > 0$, a pdf?
(b) Let Φ be the DF of $Z \sim \mathcal{N}(0,1)$. Find $E\{Z\Phi(Z)\}$ and $E\{Z^2\Phi(Z)\}$.

5.4 BIVARIATE AND MULTIVARIATE NORMAL DISTRIBUTIONS

In this section we introduce the bivariate and multivariate normal distributions and investigate some of their important properties. We note that bivariate analogs of other PDFs are known but they are not always uniquely identified. For example, there are several versions of bivariate exponential PDFs so-called because each has exponential marginals. We will not encounter any of these bivariate PDFs in this book.

Definition 1. A two-dimensional RV (X, Y) is said to have a bivariate normal distribution if the joint PDF is of the form

$$f(x,y) = \frac{1}{2\pi\sigma_1\sigma_2\sqrt{1-\rho^2}} e^{-Q(x,y)/2}, \tag{1}$$
$$-\infty < x < \infty, \quad -\infty < y < \infty,$$

where $\sigma_1 > 0$, $\sigma_2 > 0$, $|\rho| < 1$, and Q is the positive definite quadratic form

$$Q(x,y) = \frac{1}{1-\rho^2}\left[\left(\frac{x-\mu_1}{\sigma_1}\right)^2 - 2\rho\left(\frac{x-\mu_1}{\sigma_1}\right)\left(\frac{y-\mu_2}{\sigma_2}\right) + \left(\frac{y-\mu_2}{\sigma_2}\right)^2\right]. \tag{2}$$

Figure 1 gives graphs of bivariate normal PDF for selected values of ρ.

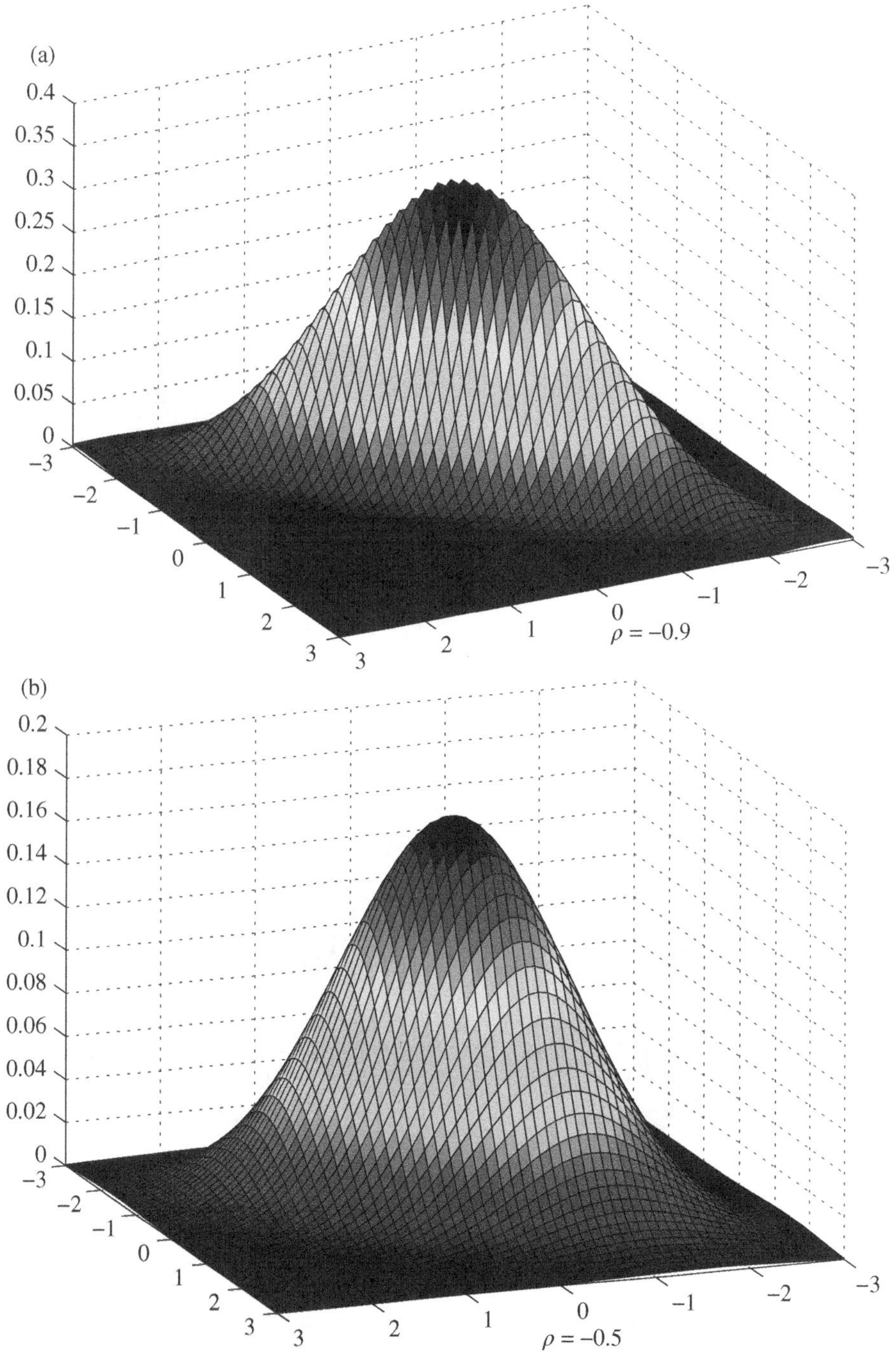

Fig. 1 Bivariate normal with $\mu_1 = \mu_2 = 0$, $\sigma_1 = \sigma_2 = 1$, and $\rho = -0.9, -0.5, 0.5, 0.9$.

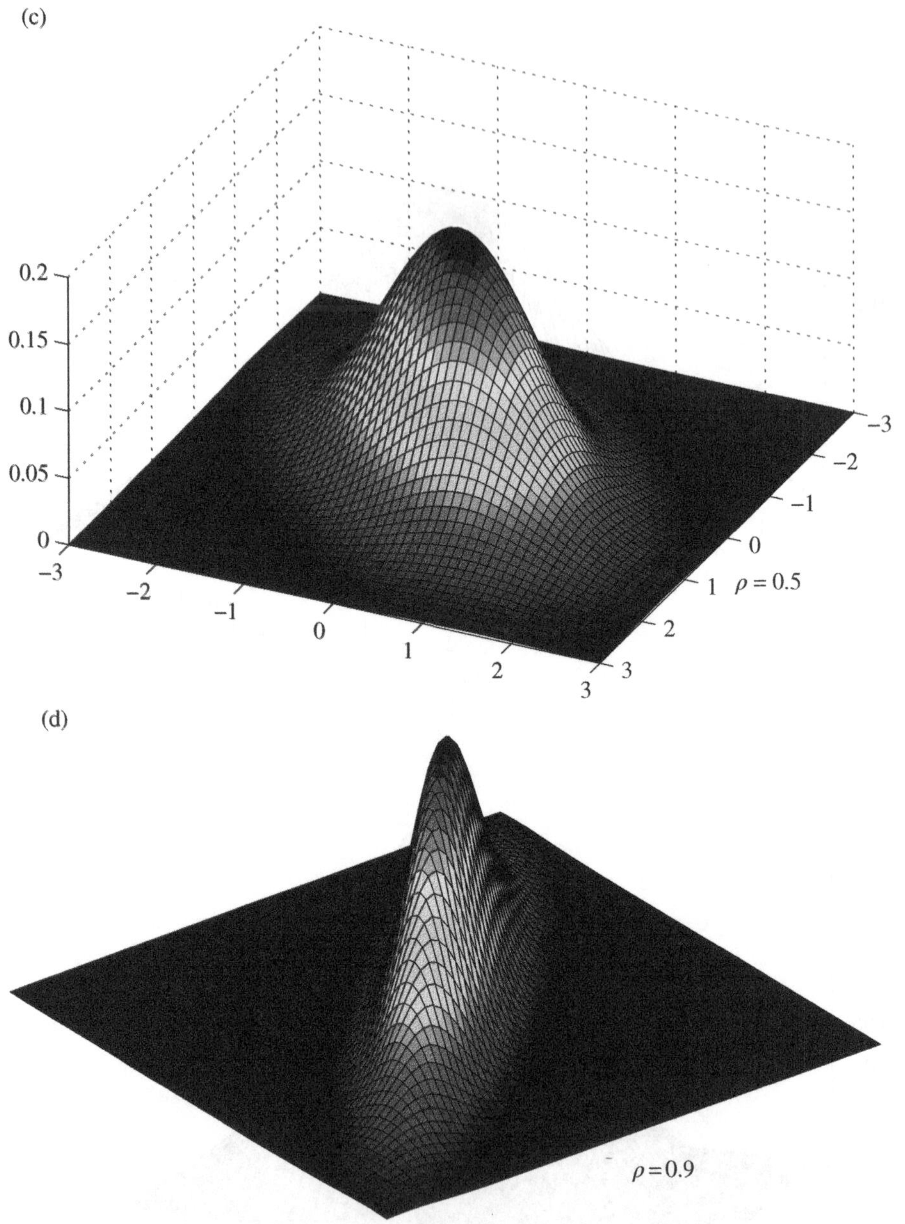

Fig. 1 (continued).

We first show that (1) indeed defines a joint PDF. In fact, we prove the following result.

Theorem 1. The function defined by (1) and (2) with $\sigma_1 > 0$, $\sigma_2 > 0$, $|\rho| < 1$ is a joint PDF. The marginal PDFs of X and Y are, respectively, $\mathcal{N}(\mu_1, \sigma_1^2)$ and $\mathcal{N}(\mu_2, \sigma_2^2)$, and ρ is the correlation coefficient between X and Y.

Proof. Let $f_1(x) = \int_{-\infty}^{\infty} f(x,y)\,dy$. Note that

$$\begin{aligned}(1-\rho^2)Q(x,y) &= \left(\frac{y-\mu_2}{\sigma_2} - \rho\frac{x-\mu_1}{\sigma_1}\right)^2 + (1-\rho^2)\left(\frac{x-\mu_1}{\sigma_1}\right)^2 \\ &= \left\{\frac{y-[\mu_2+\rho(\sigma_2/\sigma_1)(x-\mu_1)]}{\sigma_2}\right\}^2 + (1-\rho^2)\left(\frac{x-\mu_1}{\sigma_1}\right)^2.\end{aligned}$$

It follows that

$$f_1(x) = \frac{1}{\sigma_1\sqrt{2\pi}}\exp\left\{\frac{-(x-\mu_1)^2}{2\sigma_1^2}\right\}\int_{-\infty}^{\infty}\frac{\exp\{-(y-\beta_x)^2/[2\sigma_2^2(1-\rho^2)]\}}{\sigma_2\sqrt{1-\rho^2}\sqrt{2\pi}}\,dy, \tag{3}$$

where we have written

$$\beta_x = \mu_2 + \rho\left(\frac{\sigma_2}{\sigma_1}\right)(x-\mu_1). \tag{4}$$

The integrand is the PDF of an $\mathcal{N}(\beta_x, \sigma_2^2(1-\rho^2))$ RV, so that

$$f_1(x) = \frac{1}{\sigma_1\sqrt{2\pi}}\exp\left\{-\frac{1}{2}\left(\frac{x-\mu_1}{\sigma_1}\right)^2\right\}, \qquad -\infty < x < \infty.$$

Thus

$$\int_{-\infty}^{\infty}\left\{\int_{-\infty}^{\infty} f(x,y)\,dy\right\}dx = \int_{-\infty}^{\infty} f_1(x)\,dx = 1,$$

and $f(x,y)$ is a joint PDF of two RVs of the continuous type. It also follows that f_1 is the marginal PDF of X, so that X is $\mathcal{N}(\mu_1,\sigma_1^2)$. In a similar manner we can show that Y is $\mathcal{N}(\mu_2,\sigma_2^2)$.

Furthermore, we have

$$\frac{f(x,y)}{f_1(x)} = \frac{1}{\sigma_2\sqrt{1-\rho^2}\sqrt{2\pi}}\exp\left\{\frac{-(y-\beta_x)^2}{2\sigma_2^2(1-\rho^2)}\right\}, \tag{5}$$

where β_x is given by (4). It is clear, then, that the conditional PDF $f_{Y|X}(y \mid x)$ given by (5) is also normal, with parameters β_x and $\sigma_2^2(1-\rho^2)$. We have

$$E\{Y \mid x\} = \beta_x = \mu_2 + \rho\frac{\sigma_2}{\sigma_1}(x-\mu_1) \tag{6}$$

and

$$\text{var}\{Y|x\} = \sigma_2^2(1-\rho^2). \tag{7}$$

In order to show that ρ is the correlation coefficient between X and Y, it suffices to show that $\text{cov}(X,Y) = \rho\sigma_1\sigma_2$. We have from (6)

$$\begin{aligned} E(XY) &= E\{E\{XY|X\}\} \\ &= E\left\{X\left[\mu_2 + \rho\frac{\sigma_2}{\sigma_1}(X-\mu_1)\right]\right\} \\ &= \mu_1\mu_2 + \frac{\rho\sigma_2}{\sigma_1}\sigma_1^2. \end{aligned}$$

It follows that

$$\text{cov}(X,Y) = E(XY) - \mu_1\mu_2 = \rho\sigma_1\sigma_2.$$

Remark 1. If $\rho^2 = 1$, then (1) becomes meaningless. But in that case we know (Theorem 4.5.1) that there exist constants a and b such that $P\{Y = aX + b\} = 1$. We thus have a univariate distribution, which is called the *bivariate degenerate* (or *singular*) *normal* distribution. The bivariate degenerate normal distribution does not have a PDF but corresponds to an RV (X,Y) whose marginal distributions are normal or degenerate and are such that (X,Y) falls on a fixed line with probability 1. It is for this reason that degenerate distributions are considered as normal distributions with variance 0.

Next we compute the MGF $M(t_1,t_2)$ of a bivariate normal RV (X,Y). We have, if $f(x,y)$ is the PDF given in (1) and f_1 is the marginal PDF of X,

$$\begin{aligned} M(t_1,t_2) &= \int_{-\infty}^{\infty}\int_{-\infty}^{\infty} e^{t_1x+t_2y} f(x,y)\,dx\,dy, \\ &= \int_{-\infty}^{\infty}\left\{\int_{-\infty}^{\infty} f_{Y|X}(y\mid x)e^{t_2y}\,dy\right\} e^{t_1x} f_1(x)\,dx \\ &= \int_{-\infty}^{\infty} e^{t_1x} f_1(x)\left\{\exp\left[\frac{1}{2}\sigma_2^2t_2^2(1-\rho^2) + t_2\left(\mu_2 + \rho\frac{\sigma_2}{\sigma_1}(x-\mu_1)\right)\right]\right\} dx \\ &= \exp\left[\frac{1}{2}\sigma_2^2t_2^2(1-\rho^2) + t_2\mu_2 - \rho t_2\frac{\sigma_2}{\sigma_1}\mu_1\right]\int_{-\infty}^{\infty} e^{t_1x}e^{(\rho\sigma_2/\sigma_1)xt_2} f_1(x)\,dx. \end{aligned}$$

Now

$$\int_{-\infty}^{\infty} e^{(t_1+\rho t_2\sigma_2/\sigma_1)x} f_1(x)\,dx = \exp\left[\mu_1\left(t_1 + \rho\frac{\sigma_2}{\sigma_1}t_2\right) + \frac{1}{2}\sigma_1^2\left(t_1 + \rho t_2\frac{\sigma_2}{\sigma_1}\right)^2\right].$$

Therefore,

$$M(t_1,t_2) = \exp\left(\mu_1t_1 + \mu_2t_2 + \frac{\sigma_1^2t_1^2 + \sigma_2^2t_2^2 + 2\rho\sigma_1\sigma_2t_1t_2}{2}\right). \tag{8}$$

The following result is an immediate consequence of (8).

Theorem 2. If (X, Y) has a bivariate normal distribution, X and Y are independent if and only if $\rho = 0$.

Remark 2. It is quite possible for an RV (X, Y) to have a bivariate density such that the marginal densities of X and Y are normal and the correlation coefficient is 0, yet X and Y are not independent. Indeed, if the marginal densities of X and Y are normal, it does not follow that the joint density of (X, Y) is a bivariate normal. Let

$$f(x,y) = \frac{1}{2}\left\{\frac{1}{2\pi(1-\rho^2)^{1/2}}\exp\left[\frac{-1}{2(1-\rho^2)}(x^2-2\rho xy+y^2)\right]\right. \tag{9}$$
$$\left.+\frac{1}{2\pi(1-\rho^2)^{1/2}}\exp\left[\frac{-1}{2(1-\rho^2)}(x^2+2\rho xy+y^2)\right]\right\}.$$

Here $f(x,y)$ is a joint PDF such that both marginal densities are normal, $f(x,y)$ is not bivariate normal, and X and Y have zero correlation. But X and Y are not independent. We have

$$f_1(x) = \frac{1}{\sqrt{2\pi}}e^{-x^2/2}, \qquad -\infty < x < \infty,$$
$$f_2(y) = \frac{1}{\sqrt{2\pi}}e^{-y^2/2}, \qquad -\infty < y < \infty,$$
$$EXY = 0.$$

Example 1. (Rosenberg [93]). Let f and g be PDFs with corresponding DFs F and G. Also, let

$$h(x,y) = f(x)g(y)[1+\alpha(2F(x)-1)(2G(y)-1)], \tag{10}$$

where $|\alpha| \leq 1$ is a constant. It was shown in Example 4.3.1 that h is a bivariate density function with given marginal densities f and g.

In particular, take f and g to be the PDF of $\mathcal{N}(0,1)$, that is,

$$f(x) = g(x) = \frac{1}{\sqrt{2\pi}}e^{-x^2/2}, \qquad -\infty < x < \infty, \tag{11}$$

and let (X, Y) have the joint PDF $h(x,y)$. We will show that $X + Y$ is not normal except in the trivial case $\alpha = 0$, when X and Y are independent.

Let $Z = X + Y$. Then

$$EZ = 0, \qquad \text{var}(Z) = \text{var}(X) + \text{var}(Y) + 2\,\text{cov}(X, Y).$$

It is easy to show (Problem 2) that $\text{cov}(X, Y) = \alpha/\pi$, so that $\text{var}(Z) = 2[1 + (\alpha/\pi)]$. If Z is normal, its MGF must be

$$M_z(t) = e^{t^2[1+(\alpha/\pi)]}. \tag{12}$$

Next we compute the MGF of Z directly from the joint PDF (10). We have

$$\begin{aligned} M_1(t) &= E\{e^{tX+tY}\} \\ &= e^{t^2} + \alpha \int_{-\infty}^{\infty}\int_{-\infty}^{\infty} e^{tx+ty}[2F(x)-1][2F(y)-1]f(x)f(y)\,dx\,dy \\ &= e^{t^2} + \alpha \left[\int_{-\infty}^{\infty} e^{tx}[2F(x)-1]f(x)\,dx\right]^2. \end{aligned}$$

Now

$$\begin{aligned} &\int_{-\infty}^{\infty} e^{tx}[2F(x)-1]f(x)\,dx = -2\int_{-\infty}^{\infty} e^{tx}[1-F(x)]f(x)\,dx + e^{t^2/2} \\ &= e^{t^2/2} - 2\int_{-\infty}^{\infty}\int_{x}^{\infty} \frac{1}{2\pi}\exp\left\{-\frac{1}{2}(x^2+u^2-2tx)\right\} du\,dx \\ &= e^{t^2/2} - \int_{-\infty}^{\infty}\int_{0}^{\infty} \frac{\exp\left\{-\frac{1}{2}[x^2+(v+x)^2-2tx]\right\}}{\pi} dv\,dx \\ &= e^{t^2/2} - \int_{0}^{\infty} \frac{\exp\{-v^2/2+(v-t)^2/4\}}{\sqrt{\pi}} \int_{-\infty}^{\infty} \frac{\exp\{-[x+(v-t)/2]^2\}}{\sqrt{\pi}}\,dx\,dv \\ &= e^{t^2/2} - 2e^{t^2/2}\int_{0}^{\infty} \frac{\exp\left\{-\frac{1}{2}[(v+t)^2/2]\right\}}{2\sqrt{\pi}}\,dv \\ &= e^{t^2/2} - 2e^{t^2/2}P\left\{Z_1 > \frac{t}{\sqrt{2}}\right\}, \end{aligned} \tag{13}$$

where Z_1 is an $\mathcal{N}(0,1)$ RV.

It follows that

$$\begin{aligned} M_1(t) &= e^{t^2} + \alpha\left[e^{t^2/2} - 2e^{t^2/2}P\left\{Z_1 > \frac{1}{\sqrt{2}}\right\}\right]^2 \\ &= e^{t^2}\left[1+\alpha\left(1-2P\left\{Z_1 > \frac{t}{\sqrt{2}}\right\}\right)^2\right]. \end{aligned} \tag{14}$$

If Z were normally distributed, we must have $M_z(t) = M_1(t)$ for all t and all $|\alpha| \le 1$, that is,

$$e^{t^2}e^{(\alpha/\pi)t^2} = e^{t^2}\left[1+\alpha\left(1-2P\left\{Z_1 > \frac{t}{\sqrt{2}}\right\}\right)^2\right]. \tag{15}$$

For $\alpha = 0$, the equality clearly holds. The expression within the brackets on the right side of (15) is bounded by $1+\alpha$, whereas the expression $e^{(\alpha/\pi)t^2}$ is unbounded, so the equality cannot hold for all t and α.

Next we investigate the multivariate normal distribution of dimension n, $n \ge 2$. Let $\mathbf{M}$ be an $n \times n$ real, symmetric, and positive definite matrix. Let $\mathbf{x}$ denote the $n \times 1$ column vector of real numbers $(x_1,x_2,\ldots,x_n)'$ and let $\boldsymbol{\mu}$ denote the column vector $(\mu_1,\mu_2,\ldots,\mu_n)'$, where $\mu_i (i=1,2,\ldots,n)$ are real constants.

Theorem 3. The nonnegative function

$$f(\mathbf{x}) = c\exp\left\{-\frac{(\mathbf{x}-\boldsymbol{\mu})'\mathbf{M}(\mathbf{x}-\boldsymbol{\mu})}{2}\right\} \qquad -\infty < x_i < \infty, \quad i=1,2,\ldots,n, \tag{16}$$

defines the joint PDF of some random vector $\mathbf{X} = (X_1,X_2,\ldots,X_n)'$, provided that the constant c is chosen appropriately. The MGF of $\mathbf{X}$ exists and is given by

$$M(t_1,t_2,\ldots,t_n) = \exp\left\{\mathbf{t}'\boldsymbol{\mu} + \frac{\mathbf{t}'\mathbf{M}^{-1}\mathbf{t}}{2}\right\}, \tag{17}$$

where $\mathbf{t} = (t_1,t_2,\ldots,t_n)'$ and $t_1,t_2,\ldots,t_n$ are arbitrary real numbers.

Proof. Let

$$I = c\int_{-\infty}^{\infty}\cdots\int_{-\infty}^{\infty}\exp\left\{\mathbf{t}'\mathbf{x} - \frac{(\mathbf{x}-\boldsymbol{\mu})'\mathbf{M}(\mathbf{x}-\boldsymbol{\mu})}{2}\right\}\prod_{i=1}^{n} dx_i. \tag{18}$$

Changing the variables of integration to $y_1,y_2,\ldots,y_n$ by writing $x_i - \mu_i = y_i$, $i=1,2,\ldots,n$, and $\mathbf{y} = (y_1,y_2,\ldots,y_n)'$, we have $\mathbf{x}-\boldsymbol{\mu} = \mathbf{y}$ and

$$I = c\exp(\mathbf{t}'\boldsymbol{\mu})\int_{-\infty}^{\infty}\cdots\int_{-\infty}^{\infty}\exp\left(\mathbf{t}'\mathbf{y} - \frac{\mathbf{y}'\mathbf{M}\mathbf{y}}{2}\right)\prod_{i=1}^{n} dy_i. \tag{19}$$

Since $\mathbf{M}$ is positive definite, it follows that all the n characteristic roots of $\mathbf{M}$, say $m_1,m_2,\ldots,m_n$, are positive. Moreover, since $\mathbf{M}$ is symmetric there exists an $n\times n$ orthogonal matrix $\mathbf{L}$ such that $\mathbf{L}'\mathbf{M}\mathbf{L}$ is a diagonal matrix with diagonal elements $m_1,m_2,\ldots,m_n$. Let us change the variables to $z_1,z_2,\ldots,z_n$ by writing $\mathbf{y} = \mathbf{L}\mathbf{z}$, where $\mathbf{z}' = (z_1,z_2,\ldots,z_n)$, and note that the Jacobian of this orthogonal transformation is $|\mathbf{L}|$. Since $\mathbf{L}'\mathbf{L} = \mathbf{I}_n$, where $\mathbf{I}_n$ is an $n\times n$ unit matrix, $|\mathbf{L}| = 1$ and we have

$$I = c\exp(\mathbf{t}'\boldsymbol{\mu})\int_{-\infty}^{\infty}\cdots\int_{-\infty}^{\infty}\exp\left(\mathbf{t}'\mathbf{L}\mathbf{z} - \frac{\mathbf{z}'\mathbf{L}'\mathbf{M}\mathbf{L}\mathbf{z}}{2}\right)\prod_{i=1}^{n} dz_i. \tag{20}$$

If we write $\mathbf{t}'\mathbf{L} = \mathbf{u}' = (u_1,u_2,\ldots,u_n)$ then $\mathbf{t}'\mathbf{L}\mathbf{z} = \sum_{i=1}^{n} u_i z_i$. Also $\mathbf{L}'\mathbf{M}\mathbf{L} = \text{diag}(m_1,m_2,\ldots,m_n)$ so that $\mathbf{z}'\mathbf{L}'\mathbf{M}\mathbf{L}\mathbf{z} = \sum_{i=1}^{n} m_i z_i^2$. The integral in (20) can therefore be written as

$$\prod_{i=1}^{n}\left[\int_{-\infty}^{\infty}\exp\left(u_i z_i - \frac{m_1 z_i^2}{2}\right)dz_i\right] = \prod_{i=1}^{n}\left[\sqrt{\frac{2\pi}{m_i}}\exp\left(\frac{u_i^2}{2m_i}\right)\right].$$

If follows that

$$I = c\exp(\mathbf{t}'\mathbf{u})\frac{(2\pi)^{n/2}}{(m_1 m_2\cdots m_n)^{1/2}}\exp\left(\sum_{i=1}^{n}\frac{u_i^2}{2m_i}\right). \tag{21}$$

Setting $t_1 = t_2 = \cdots = t_n = 0$, we see from (18) and (21) that

$$\int_{-\infty}^{\infty}\cdots\int_{-\infty}^{\infty} f(x_1,x_2,\ldots,x_n)dx_1dx_2\cdots dx_n = \frac{c(2\pi)^{n/2}}{(m_1m_2\cdots m_n)^{1/2}}.$$

By choosing

$$c = \frac{(m_1m_2\cdots m_n)^{1/2}}{(2\pi)^{n/2}} \tag{22}$$

we see that f is a joint PDF of some random vector $\mathbf{X}$, as asserted.

Finally, since

$$(\mathbf{L}'\mathbf{M}\mathbf{L})^{-1} = \text{diag}(m_1^{-1}, m_2^{-1},\ldots,m_n^{-1}),$$

we have

$$\sum_{i=1}^{n}\frac{u_i^2}{m_i} = \mathbf{u}'(\mathbf{L}'\mathbf{M}^{-1}\mathbf{L})\mathbf{u} = \mathbf{t}'\mathbf{M}^{-1}\mathbf{t}.$$

Also

$$|\mathbf{M}^{-1}| = |\mathbf{L}'\mathbf{M}^{-1}\mathbf{L}| = (m_1m_2\cdots m_n)^{-1}.$$

It follows from (21) and (22) that the MGF of $\mathbf{X}$ is given by (17), and we may write

$$c = \frac{1}{\{(2\pi)^n|\mathbf{m}^{-1}|\}^{1/2}}. \tag{23}$$

This completes the proof of Theorem 3.

Let us write $\mathbf{M}^{-1} = (\sigma_{ij})_{i,j=1,2,\ldots,n}$. Then

$$M(0,0,\ldots,0,t_i,0,\ldots,0) = \exp\left(t_i\mu_i + \sigma_{ii}\frac{t_i^2}{2}\right)$$

is the MGF of X_i, $i = 1,2,\ldots,n$. Thus each X_i is $\mathcal{N}(\mu_i,\sigma_{ii})$, $i = 1,2,\ldots,n$. For $i \neq j$, we have for the MGF of X_i and X_j

$$\begin{aligned} &M(0,0,\ldots,0,t_i,0,\ldots,0,t_j,0,\ldots,0) \\ &\quad = \exp\left(t_i\mu_i + t_j\mu_j + \frac{\sigma_{ii}t_i^2 + 2\sigma_{ij}t_it_j + t_j^2\sigma_{jj}}{2}\right). \end{aligned}$$

This is the MGF of a bivariate normal distribution with means μ_i, μ_j, variances σ_{ii}, σ_{jj}, and covariance σ_{ij}. Thus we see that

$$\boldsymbol{\mu}' = (\mu_1, \mu_2, \dots, \mu_n) \tag{24}$$

is the mean vector of $\mathbf{X}' = (X_1, \dots, X_n)$,

$$\sigma_{ii} = \sigma_i^2 = \operatorname{var}(X_i), \qquad i = 1, 2, \dots, n, \tag{25}$$

and

$$\sigma_{ij} = \rho_{ij}\sigma_i\sigma_j, \qquad i \neq j;\ i,j = 1, 2, \dots, n. \tag{26}$$

The matrix $\mathbf{M}^{-1}$ is called the *dispersion* (*variance-covariance*) matrix of the multivariate normal distribution.

If $\sigma_{ij} = 0$ for $i \neq j$, the matrix $\mathbf{M}^{-1}$ is a diagonal matrix, and it follows that the RVs $X_1, X_2, \dots, X_n$ are independent. Thus we have the following analog of Theorem 2.

Theorem 4. The components $X_1, X_2, \dots, X_n$ of a jointly normally distributed RV $\mathbf{X}$ are independent if and only if the covariances $\sigma_{ij} = 0$ for all $i \neq j$ $(i,j = 1, 2, \dots, n)$.

The following result is stated without proof. The proof is similar to the two-variate case except that now we consider the quadratic form in n variables: $E\{\sum_{i=1}^{n} t_i(X_i - \mu_i)\}^2 \geq 0$.

Theorem 5. The probability that the RVs $X_1, X_2, \dots, X_n$ with finite variances satisfy at least one linear relationship is 1 if and only if $|\mathbf{M}| = 0$.

Accordingly, if $|\mathbf{M}| = 0$ all the probability mass is concentrated on a hyperplane of dimension $< n$.

Theorem 6. Let $(X_1, X_2, \dots, X_n)$ be an n-dimensional RV with a normal distribution. Let $Y_1, Y_2, \dots, Y_k$, $k \leq n$, be linear functions of X_j $(j = 1, 2, \dots, n)$. Then $(Y_1, Y_2, \dots, Y_k)$ also has a multivariate normal distribution.

Proof. Without loss of generality let us assume that $EX_i = 0$, $i = 1, 2, \dots, n$. Let

$$Y_p = \sum_{j=1}^{n} A_{pj}X_j, \qquad p = 1, 2, \dots, k; \quad k \leq n. \tag{27}$$

Then $EY_p = 0$, $p = 1, 2, \dots, k$, and

$$\operatorname{cov}(Y_p, Y_q) = \sum_{i,j=1}^{n} A_{pi}A_{qj}\sigma_{ij}, \tag{28}$$

where $E(X_iX_j) = \sigma_{ij}, i, j = 1, 2, \dots, n$.

The MGF of $(Y_1, Y_2, \ldots, Y_k)$ is given by

$$M^*(t_1,t_2,\ldots,t_k) = E\left\{\exp\left(t_1\sum_{j=1}^{n} A_{1j}X_j + \cdots + t_k\sum_{j=1}^{n} A_{kj}X_j\right)\right\}.$$

Writing $u_j = \sum_{p=1}^{k} t_p A_{pj}$, $j = 1,2,\ldots,n$, we have

$$\begin{aligned} M^*(t_1,t_2,\ldots,t_k) &= E\left\{\exp\left(\sum_{i=1}^{n} u_i X_i\right)\right\} \\ &= \exp\left(\frac{1}{2}\sum_{i,j=1}^{n} \sigma_{ij} u_i u_j\right) \quad \text{by (17)} \\ &= \exp\left(\frac{1}{2}\sum_{i,j=1}^{n} \sigma_{ij} \sum_{l,m=1}^{k} t_l t_m A_{li} A_{mj}\right) \\ &= \exp\left(\frac{1}{2}\sum_{l,m=1}^{k} t_l t_m \sum_{i,j=1}^{n} A_{li} A_{mj} \sigma_{ij}\right) \\ &= \exp\left\{\frac{1}{2}\sum_{l,m=1}^{k} t_l t_m \operatorname{cov}(Y_l, Y_m)\right\}. \end{aligned} \tag{29}$$

When (17) and (29) are compared, the result follows.

Corollary 1. Every marginal distribution of an n-dimensional normal distribution is univariate normal. Moreover, any linear function of $X_1, X_2, \ldots, X_n$ is univariate normal.

Corollary 2. If $X_1, X_2, \ldots, X_n$ are iid $\mathcal{N}(\mu, \sigma^2)$ and $\mathbf{A}$ is an $n \times n$ orthogonal transformation matrix, the components $Y_1, Y_2, \ldots, Y_n$ of $\mathbf{Y} = \mathbf{AX}'$, where $\mathbf{X} = (X_1, \ldots, X_n)'$, are independent RVs, each normally distributed with the same variance σ^2.

We have from (27) and (28)

$$\begin{aligned} \operatorname{cov}(Y_p, Y_q) &= \sum_{i=1}^{n} A_{pi} A_{qi} \sigma_{ii} + \sum_{i \neq j} A_{pi} A_{qj} \sigma_{ij} \\ &= \begin{cases} 0 & \text{if } p \neq q, \\ \sigma^2 & \text{if } p = q, \end{cases} \end{aligned}$$

since $\sum_{i=1}^{n} A_{pi} A_{qi} = 0$ and $\sum_{j=1}^{n} A_{pj}^2 = 1$. It follows that

$$M^*(t_1,t_2,\ldots,t_n) = \exp\left(\frac{1}{2}\sum_{l=1}^{n} t_l^2 \sigma^2\right)$$

and Corollary 2 follows.

Theorem 7. Let $\mathbf{X} = (X_1, X_2, \ldots, X_n)'$. Then $\mathbf{X}$ has an n-dimensional normal distribution if and only if every linear function of $\mathbf{X}$

$$\mathbf{X}'\mathbf{t} = t_1X_1 + t_2X_2 + \cdots + t_nX_n$$

has a univariate normal distribution.

Proof. Suppose that $\mathbf{X}'\mathbf{t}$ is normal for any $\mathbf{t}$. Then the MGF of $\mathbf{X}'\mathbf{t}$ is given by

$$M(s) = \exp\left(bs + \frac{1}{2}\sigma^2 s^2\right). \tag{30}$$

Here $b = E\{\mathbf{X}'\mathbf{t}\} = \sum_1^n t_{ij}\mu_i = \mathbf{t}'\boldsymbol{\mu}$, where $\boldsymbol{\mu}' = (\mu_1, \ldots, \mu_n)$, and $\sigma^2 = \text{var}(\mathbf{X}'\mathbf{t}) = \text{var}(\sum t_i X_i) = \mathbf{t}'\mathbf{M}^{-1}\mathbf{t}$, where $\mathbf{M}^{-1}$ is the dispersion matrix of $\mathbf{X}$. Thus

$$M(s) = \exp\left(\mathbf{t}'\boldsymbol{\mu}s + \frac{1}{2}\mathbf{t}'\mathbf{M}^{-1}\mathbf{t}s^2\right). \tag{31}$$

Let $s = 1$ then

$$M(1) = \exp\left(\mathbf{t}'\boldsymbol{\mu} + \frac{1}{2}\mathbf{t}'\mathbf{M}^{-1}\mathbf{t}\right), \tag{32}$$

and since the MGF is unique, it follows that $\mathbf{X}$ has a multivariate normal distribution. The converse follows from Corollary 1 to Theorem 6.

Many characterization results for the multivariate normal distribution are now available. We refer the reader to Lukacs and Laha [70, p. 79].

PROBLEMS 5.4

1. Let (X, Y) have joint PDF

$$f(x,y) = \frac{1}{6\pi\sqrt{7}} \exp\left\{-\frac{8}{7}\left(\frac{x^2}{16} - \frac{31}{32}x + \frac{xy}{8} + \frac{y^2}{9} - \frac{4}{3}y + \frac{71}{16}\right)\right\},$$

   for $-\infty < x < \infty$, $-\infty < y < \infty$.
   (a) Find the means and variances of X and Y. Also find ρ.
   (b) Find the conditional PDF of Y given $X = x$ and $E\{Y|x\}$, $\text{var}\{Y|x\}$.
   (c) Find $P\{4 \le Y \le 6 | X = 4\}$.
2. In Example 1 show that $\text{cov}(X, Y) = \alpha/\pi$.
3. Let (X, Y) be a bivariate normal RV with parameters μ_1, μ_2, σ_1^2, σ_2^2, and ρ. What is the distribution of $X + Y$? Compare your result with that of Example 1.
4. Let (X, Y) be a bivariate normal RV with parameters μ_1, μ_2, σ_1^2, σ_2^2, and ρ, and let $U = aX + b$, $a \neq 0$, and $V = cY + d$, $c \neq 0$. Find the joint distribution of (U, V).

5. Let (X,Y) be a bivariate normal RV with parameters $\mu_1=5, \mu_2=8, \sigma_1^2=16, \sigma_2^2=9$, and $\rho=0.6$. Find $P\{5<Y<11 \mid X=2\}$.

6. Let X and Y be jointly normal with means 0. Also, let

$$W=X\cos\theta+Y\sin\theta,\ Z=X\cos\theta-Y\sin\theta.$$

Find θ such that W and Z are independent.

7. Let (X,Y) be a normal RV with parameters μ_1, μ_2, σ_1^2, σ_2^2, and ρ. Find a necessary and sufficient condition for $X+Y$ and $X-Y$ to be independent.

8. For a bivariate normal RV with parameters $\mu_1,\mu_2,\sigma_1,\sigma_2$, and ρ show that

$$P(X>\mu_1,\ Y>\mu_2)=\frac{1}{4}+\frac{1}{2\pi}\tan^{-1}\frac{\rho}{\sqrt{1-\rho^2}}.$$

[*Hint*: The required probability is $P\big((X-\mu_1)/\sigma_1>0,\ (Y-\mu_2)/\sigma_2>0\big)$. Change to polar coordinates and integrate.]

9. Show that every variance–covariance matrix is symmetric positive semidefinite and conversely. If the variance–covariance matrix is not positive definite, then with probability 1 the random (column) vector $\mathbf{X}$ lies in some hyperplane $\mathbf{c}'\mathbf{X}=a$ with $\mathbf{c}\neq\mathbf{0}$.

10. Let (X,Y) be a bivariate normal RV with $EX=EY=0$, $\text{var}(X)=\text{var}(Y)=1$, and $\text{cov}(X,Y)=\rho$. Show that the RV $Z=Y/X$ has a Cauchy distribution.

11. (a) Show that

$$f(x)=\frac{1}{(2\pi)^{n/2}}\exp\left\{-\frac{\sum x_i^2}{2}\right\}\left[1+\prod_1^n\left(x_ie^{-x_i^2/2}\right)\right]$$

is a joint PDF on $\mathcal{R}_n$.

(b) Let $(X_1,X_2,\ldots,X_n)$ have PDF f given in (a). Show that the RVs in any proper subset of $\{X_1,X_2,\ldots,X_n\}$ containing two or more elements are independent standard normal RVs.

5.5 EXPONENTIAL FAMILY OF DISTRIBUTIONS

Most of the distributions that we have so far encountered belong to a general family of distributions that we now study. Let Θ be an interval on the real line, and let $\{f_\theta:\theta\in\Theta\}$ be a family of PDFs (PMFs). Here and in what follows we write $\mathbf{x}=(x_1,x_2,\ldots,x_n)$ unless otherwise specified.

Definition 1. If there exist real-valued functions $Q(\theta)$ and $D(\theta)$ on Θ and Borel-measurable functions $T(x_1,x_2,\ldots,x_n)$ and $S(x_1,x_2,\ldots,x_n)$ on $\mathcal{R}_n$ such that

$$f_\theta(x_1,x_2,\ldots,x_n)=\exp\{Q(\theta)T(\mathbf{x})+D(\theta)+S(\mathbf{x})\}, \tag{1}$$

we say that the family $\{f_\theta,\theta\in\Theta\}$ is a *one-parameter exponential family*.

Let $\mathbf{X}_1, \mathbf{X}_2, \ldots, \mathbf{X}_m$ be iid with PMF (PDF) f_θ. Then the joint distribution of $\mathbf{X} = (\mathbf{X}_1, \mathbf{X}_2, \ldots, \mathbf{X}_m)$ is given by

$$\begin{aligned} g_\theta(\mathbf{x}) &= \prod_{i=1}^{m} f_\theta(\mathbf{x}_i) = \prod_{i=1}^{m} \exp\{Q(\theta)T(\mathbf{x}_i) + D(\theta) + S(\mathbf{x}_i)\} \\ &= \exp\left\{ Q(\theta)\sum_{i=1}^{m} T(\mathbf{x}_i) + mD(\theta) + \sum_{i=1}^{m} S(\mathbf{x}_i) \right\}, \end{aligned}$$

where $\mathbf{x} = (\mathbf{x}_1, \mathbf{x}_2, \ldots, \mathbf{x}_m)$, $\mathbf{x}_j = (x_{j1}, x_{j2}, \ldots, x_{jn})$, $j = 1, 2, \ldots, m$, and it follows that $\{g_\theta : \theta \in \Theta\}$ is again a one-parameter exponential family.

Example 1. Let $X \sim \mathcal{N}(\mu_0, \sigma^2)$, where μ_0 is known and σ^2 unknown. Then

$$\begin{aligned} f_{\sigma^2}(x) &= \frac{1}{\sigma\sqrt{2\pi}} \exp\left\{ -\frac{(x-\mu_0)^2}{2\sigma^2} \right\} \\ &= \exp\left\{ -\log(\sigma\sqrt{2\pi}) - \frac{(x-\mu_0)^2}{2\sigma^2} \right\} \end{aligned}$$

is a one-parameter exponential family with

$$\begin{aligned} &Q(\sigma^2) = -\frac{1}{2\sigma^2}, \qquad T(x) = (x-\mu_0)^2, \qquad S(x) = 0, \quad \text{and} \\ &D(\sigma^2) = -\log(\sigma\sqrt{2\pi}). \end{aligned}$$

If $X \sim \mathcal{N}(\mu, \sigma_0^2)$, where σ_0 is known but μ is unknown, then

$$\begin{aligned} f_\mu(x) &= \frac{1}{\sigma_0\sqrt{2\pi}} \exp\left\{ -\frac{(x-\mu)^2}{2\sigma_0^2} \right\} \\ &= \frac{1}{\sigma_0\sqrt{2\pi}} \exp\left(-\frac{x^2}{2\sigma_0^2} + \frac{\mu x}{\sigma_0^2} - \frac{\mu^2}{2\sigma_0^2} \right) \end{aligned}$$

is a one-parameter exponential family with

$$Q(\mu) = \frac{\mu}{\sigma_0^2}, \qquad D(\mu) = -\frac{\mu}{2\sigma_0^2}, \qquad T(x) = x,$$

and

$$S(x) = -\left[\frac{x^2}{2\sigma_0^2} + \frac{1}{2}\log(2\pi\sigma_0^2) \right].$$

Example 2. Let $X \sim P(\lambda)$, $\lambda > 0$ unknown. Then

$$P_\lambda\{X = x\} = e^{-\lambda}\frac{\lambda^x}{x!} = \exp\{-\lambda + x\log\lambda - \log(x!)\},$$

and we see that the family of Poisson PMFs with parameter λ is a one-parameter exponential family.

Some other important examples of one-parameter exponential families are binomial, $G(\alpha,\beta)$ (provided that one of α, β is fixed), $B(\alpha,\beta)$ (provided that one of α, β is fixed), negative binomial, and geometric. The Cauchy family of densities and the uniform distribution on $[0,\theta]$ do not belong to this class.

Theorem 1. Let $\{f_\theta\colon \theta \in \Theta\}$ be a one-parameter exponential family of PDFs (PMFs) given in (1). Then the family of distributions of $T(\mathbf{X})$ is also a one-parameter exponential family of PDFs (PMFs), given by

$$g_\theta(t) = \exp\{tQ(\theta) + D(\theta) + S^*(t)\}$$

for suitable $S^*(t)$.

Proof. The proof of Theorem 1 is a simple application of the transformation of variables technique studied in Section 4.4 and is left as an exercise, at least for the cases considered in Section 4.4. For the general case we refer to Lehmann [64, p. 58].

Let us now consider the k-parameter exponential family, $k \geq 2$. Let $\Theta \subseteq \mathcal{R}_k$ be a k-dimensional interval.

Definition 2. If there exist real-valued functions $Q_1, Q_2, \ldots, Q_k, D$ defined on Θ, and Borel-measurable functions $T_1, T_2, \ldots, T_k, S$ on $\mathcal{R}_n$ such that

$$f_{\boldsymbol{\theta}}(\mathbf{x}) = \exp\left\{\sum_{i=1}^{k} Q_i(\boldsymbol{\theta})T_i(\mathbf{x}) + D(\boldsymbol{\theta}) + S(\mathbf{x})\right\}, \tag{2}$$

we say that the family $\{f_{\boldsymbol{\theta}},\ \boldsymbol{\theta} \in \Theta\}$ is a *k-parameter exponential family*.

Once again, if $\mathbf{X} = (\mathbf{X}_1, \mathbf{X}_2, \ldots, \mathbf{X}_m)$ and $\mathbf{X}_j$ are iid with common distribution (2), the joint distributions of $\mathbf{X}$ form a k-parameter exponential family. An analog of Theorem 1 also holds for the k-parameter exponential family.

Example 3. The most important example of a k-parameter exponential family is $\mathcal{N}(\mu,\sigma^2)$ when both μ and σ^2 are unknown. We have

$$\boldsymbol{\theta} = (\mu,\sigma^2), \qquad \Theta = \{(\mu,\sigma^2) : -\infty < \mu < \infty, \sigma^2 > 0\}$$

and

$$\begin{aligned} f_{\boldsymbol{\theta}}(x) &= \frac{1}{\sigma\sqrt{2\pi}} \exp\left(-\frac{x^2 - 2\mu x + \mu^2}{2\sigma^2}\right) \\ &= \exp\left\{-\frac{x^2}{2\sigma^2} + \frac{\mu}{\sigma^2}x - \frac{1}{2}\left[\frac{\mu^2}{\sigma^2} + \log(2\pi\sigma^2)\right]\right\}. \end{aligned}$$

It follows that $f_{\boldsymbol{\theta}}$ is a two-parameter exponential family with

$$Q_1(\boldsymbol{\theta}) = -\frac{1}{2\sigma^2}, \qquad Q_2(\boldsymbol{\theta}) = \frac{\mu}{\sigma^2}, \qquad T_1(x) = x^2, \qquad T_2(x) = x,$$
$$D(\boldsymbol{\theta}) = -\frac{1}{2}\left[\frac{\mu^2}{\sigma^2} + \log(2\pi\sigma^2)\right], \quad \text{and} \quad S(x) = 0.$$

Other examples are the $G(\alpha,\beta)$ and $B(\alpha,\beta)$ distributions when both α,β are unknown, and the multinomial distribution. $U[\alpha,\beta]$ does not belong to this family, nor does $\mathcal{C}(\alpha,\beta)$.

Some general properties of exponential families will be studied in Chapter 8, and the importance of these families will then become evident.

Remark 1. The form in (2) is not unique as easily seen by substituting αQ_i for Q_i and $(1/\alpha)T_i$ for T_i. This, however, is not going to be a problem in statistical considerations.

Remark 2. The integer k in Definition 2 is also not unique since the family $\{1, Q_1, \ldots, Q_k\}$ or $\{1, T_1, \ldots, T_k\}$ may be linearly dependent. In general, k need not be the dimension of Θ.

Remark 3. The support $\{\mathbf{x} : f_{\boldsymbol{\theta}}(\mathbf{x}) > 0\}$ does not depend on $\boldsymbol{\theta}$.

Remark 4. In (2), one can change parameters to $\eta_i = Q_i(\boldsymbol{\theta}), i = 1,2,\ldots,k$ so that

$$f_{\boldsymbol{\eta}}(\mathbf{x}) = \exp\left\{\sum_{i=1}^{k} \eta_i T_i(\mathbf{x}) + D(\boldsymbol{\eta}) + S(\mathbf{x})\right\} \tag{3}$$

where the parameters $\boldsymbol{\eta} = (\eta_1, \eta_2, \ldots, \eta_k)$ are called *natural parameters*. Again η_i may be linearly dependent so one of η_i may be eliminated.

PROBLEMS 5.5

1. Show that the following families of distributions are one-parameter exponential families:
   (a) $X \sim b(n,p)$.
   (b) $X \sim G(\alpha,\beta)$, (i) if α is known and (ii) if β is known.
   (c) $X \sim B(\alpha,\beta)$, (i) if α is known and (ii) if β is known.
   (d) $X \sim NB(r;p)$, where r is known, p unknown.
2. Let $X \sim \mathcal{C}(1,\theta)$. Show that the family of distributions of X is not a one-parameter exponential family.
3. Let $X \sim U[0,\theta]$, $\theta \in [0,\infty)$. Show that the family of distributions of X is not an exponential family.
4. Is the family of PDFs

   $$f_\theta(x) = \frac{1}{2}e^{-|x-\theta|}, \qquad -\infty < x < \infty, \theta \in (-\infty,\infty),$$

   an exponential family?

5. Show that the following families of distributions are two-parameter exponential families:

(a) $X \sim G(\alpha, \beta)$, both α and β unknown.

(b) $X \sim B(\alpha, \beta)$, both α and β unknown.

6. Show that the families of distributions $U[\alpha, \beta]$ and $\mathcal{C}(\alpha, \beta)$ do not belong to the exponential families.

7. Show that the multinomial distributions form an exponential family.

6

SAMPLE STATISTICS AND THEIR DISTRIBUTIONS

6.1 INTRODUCTION

In the preceding chapters we discussed fundamental ideas and techniques of probability theory. In this development we created a mathematical model of a random experiment by associating with it a sample space in which random events correspond to sets of a certain σ-field. The notion of probability defined on this σ-field corresponds to the notion of uncertainty in the outcome on any performance of the random experiment.

In this chapter we begin the study of some problems of mathematical statistics. The methods of probability theory learned in preceding chapters will be used extensively in this study.

Suppose that we seek information about some numerical characteristics of a collection of elements called a *population*. For reasons of time or cost we may not wish or be able to study each individual element of the population. Our object is to draw conclusions about the unknown population characteristics on the basis of information on some characteristics of a suitably selected *sample*. Formally, let X be a random variable which describes the population under investigation, and let F be the DF of X. There are two possibilities. Either X has a DF F_θ with a known functional form (except perhaps for the parameter θ, which may be a vector) or X has a DF F about which we know nothing (except perhaps that F is, say, absolutely continuous). In the former case let Θ be the set of possible values of the unknown parameter θ. Then the job of a statistician is to decide, on the basis of a suitably selected sample, which member or members of the family $\{F_\theta, \theta \in \Theta\}$ can represent the DF of X. Problems of this type are called problems of *parametric statistical inference* and will be the subject of investigation in Chapters 8 through 12. The case in which nothing is

An Introduction to Probability and Statistics, Third Edition. Vijay K. Rohatgi and A.K. Md. Ehsanes Saleh.

known about the functional form of the DF F of X is clearly much more difficult. Inference problems of this type fall into the domain of *nonparametric statistics* and will be discussed in Chapter 13.

To be sure, the scope of statistical methods is much wider than the statistical inference problems discussed in this book. Statisticians, for example, deal with problems of planning and designing experiments, of collecting information, and of deciding how best the collected information should be used. However, here we concern ourselves only with the best methods of making inferences about probability distributions.

In Section 6.2 of this chapter we introduce the notions of (*simple*) *random sample* and *sample statistics*. In Section 6.3 we study sample moments and their exact distributions. In Section 6.4 we consider some important distributions that arise in sampling from a normal population. Sections 6.5 and 6.6 are devoted to the study of sampling from univariate and bivariate normal distributions.

6.2 RANDOM SAMPLING

Consider a statistical experiment that culminates in outcomes x, which are the values assumed by an RV X. Let F be the DF of X. In practice, F will not be completely known, that is, one or more parameters associated with F will be unknown. The job of a statistician is to estimate these unknown parameters or to test the validity of certain statements about them. She can obtain n independent observations on X. This means that she observes n values $x_1,x_2,\dots,x_n$ assumed by the RV X. Each x_i can be regarded as the value assumed by an RV X_i, $i=1,2,\dots,n$, where $X_1,X_2,\dots,X_n$ are independent RVs with common DF F. The observed values $(x_1,x_2,\dots,x_n)$ are then values assumed by $(X_1,X_2,\dots,X_n)$. The set $\{X_1,X_2,\dots,X_n\}$ is then a *sample* of size n taken from a *population distribution* F. The set of n values $x_1,x_2,\dots,x_n$ is called a *realization* of the sample. Note that the possible values of the RV $(X_1,X_2,\dots,X_n)$ can be regarded as points in $\mathcal{R}_n$, which may be called the *sample space*. In practice one observes not $x_1,x_2,\dots,x_n$ but some function $f(x_1,x_2,\dots,x_n)$. Then $f(x_1,x_2,\dots,x_n)$ are values assumed by the RV $f(X_1,X_2,\dots,X_n)$.

Let us now formalize these concepts.

Definition 1. Let X be an RV with DF F, and let $X_1,X_2,\dots,X_n$ be iid RVs with common DF F. Then the collection $X_1,X_2,\dots,X_n$ is known as a random sample of size n from the DF F or simply as n independent observations on X.

If $X_1,X_2,\dots,X_n$ is a random sample from F, their joint DF is given by

$$F^*(x_1,x_2,\dots,x_n)=\prod_{i=1}^{n}F(x_i). \tag{1}$$

Definition 2. Let $X_1,X_2,\dots,X_n$ be n independent observations on an RV X, and let $f\colon \mathcal{R}_n \to \mathcal{R}_k$ be a Borel-measurable function. Then the RV $f(X_1,X_2,\dots,X_n)$ is called a (*sample*) *statistic* provided that it is not a function of any unknown parameter(s).

Two of the most commonly used statistics are defined as follows.

Definition 3. Let $X_1, X_2, \ldots, X_n$ be a random sample from a distribution function F. Then the statistic

$$\overline{X} = n^{-1} S_n = \sum_{i=1}^{n} \frac{X_i}{n} \tag{2}$$

is called the *sample mean*, and the statistic

$$S^2 = \sum_{1}^{n} \frac{(X_i - \overline{X})^2}{n-1} = \frac{\sum_{i=1}^{n} X_i^2 - n\overline{X}^2}{n-1} \tag{3}$$

is called the *sample variance* and S is called the *sample standard deviation*.

Remark 1. Whenever the word "sample" is used subsequently, it will mean "random sample."

Remark 2. Sampling from a probability distribution (Definition 1) is sometimes referred to as sampling from an infinite population since one can obtain samples of any size one desires even if the population is finite (by sampling *with replacement*).

Remark 3. In sampling *without replacement* from a finite population, the independence condition of Definition 1 is not satisfied. Suppose a sample of size 2 is taken from a finite population $(a_1, a_2, \ldots, a_N)$ without replacement. Let X_i be the outcome on the ith draw. Then $P\{X_1 = a_1\} = 1/N$, $P\{X_2 = a_2 \mid X_1 = a_1\} = \frac{1}{N-1}$, and $P\{X_2 = a_2 \mid X_1 = a_2\} = 0$. Thus the PMF of X_2 depends on the outcome of the first draw (that is, on the value of X_1), and X_1 and X_2 are not independent. Note, however, that

$$\begin{aligned} P\{X_2 = a_2\} &= \sum_{j=1}^{N} P\{X_1 = a_j\} P\{X_2 = a_2 \mid a_j\} \\ &= \sum_{j \neq 2} P\{X_1 = a_j\} P\{X_2 = a_2 \mid a_j\} = \frac{1}{N}, \end{aligned}$$

and $X_1 \stackrel{d}{=} X_2$. A similar argument can be used to show that $X_1, X_2, \ldots, X_n$ all have the same distribution but they are not independent. In fact, $X_1, X_2, \ldots, X_n$ are exchangeable RVs. Sampling without replacement from a finite population is often referred to as *simple random sampling*.

Remark 4. It should be remembered that sample statistics $\overline{X}$, S^2 (and others that we will define later on) are random variables, while the population parameters μ, σ^2, and so on are fixed constants that may be unknown.

Remark 5. In (3) we divide by $n-1$ rather than n. The reason for this will become clear in the next section.

Remark 6. Other frequently occurring examples of statistics are sample order statistics $X_{(1)}, X_{(2)}, \ldots, X_{(n)}$ and their functions, as well as sample moments, which will be studied in the next section.

Example 1. Let $X \sim b(1, p)$, where p is possibly unknown. The DF of X is given by

$$F(x) = p\varepsilon(x-1) + (1-p)\varepsilon(x), \qquad x \in \mathcal{R}.$$

Suppose that five independent observations on X are 0, 1, 1, 1, 0. Then 0, 1, 1, 1, 0 is a realization of the sample $X_1, X_2, \ldots, X_5$. The sample mean is

$$\overline{x} = \frac{0+1+1+1+0}{5} = 0.6,$$

which is the value assumed by the RV $\overline{X}$. The sample variance is

$$s^2 = \sum_{i=1}^{5} \frac{(x_i - \overline{x})^2}{5-1} = \frac{2(0.6)^2 + 3(0.4)^2}{4} = 0.3,$$

which is the value assumed by the RV S^2. Also $s = \sqrt{0.3} = 0.55$.

Example 2. Let $X \sim \mathcal{N}(\mu, \sigma^2)$, where μ is known but σ^2 is unknown. Let $X_1, X_2, \ldots, X_n$ be a sample from $\mathcal{N}(\mu, \sigma^2)$. Then, according to our definition, $\sum_{i=1}^{n} X_i/\sigma^2$ is not a statistic.

Suppose that five observations on X are $-0.864, 0.561, 2.355, 0.582, -0.774$. Then the sample mean is 0.372, and the sample variance is 1.648.

PROBLEMS 6.2

1. Let X be a $b(1, \frac{1}{2})$ RV, and consider all possible random samples of size 3 on X. Compute $\overline{X}$ and S^2 for each of the eight samples, and also compute the PMFs of $\overline{X}$ and S^2.
2. A fair die is rolled. Let X be the face value that turns up, and X_1, X_2 be two independent observations on X. Compute the PMF of $\overline{X}$.
3. Let $X_1, X_2, \ldots, X_n$ be a sample from some population. Show that

$$\max_{1 \le i \le n} |X_i - \overline{X}| < \frac{(n-1)S}{\sqrt{n}}$$

unless either all the n observations are equal or exactly $n-1$ of the X_j's are equal. (Samuelson [99])

4. Let $x_1, x_2, \ldots, x_n$ be real numbers, and let $x_{(n)} = \max\{x_1, x_2, \ldots, x_n\}$, $x_{(1)} = \min\{x_1, x_2, \ldots, x_n\}$. Show that for any set of real numbers $a_1, a_2, \ldots, a_n$ such that $\sum_{i=1}^{n} a_i = 0$ the following inequality holds:

$$\left| \sum_{i=1}^{n} a_i x_i \right| \le \frac{1}{2}\left(x_{(n)} - x_{(1)}\right) \sum_{i=1}^{n} |a_i|.$$

5. For any set of real numbers $x_1, x_2, \ldots, x_n$ show that the fraction of $x_1, x_2, \ldots, x_n$ included in the interval $(\overline{x} - ks, \overline{x} + ks)$ for $k \geq 1$ is at least $1 - 1/k^2$. Here $\overline{x}$ is the mean and s the standard deviation of x's.

6.3 SAMPLE CHARACTERISTICS AND THEIR DISTRIBUTIONS

Let $X_1, X_2, \ldots, X_n$ be a sample from a population DF F. In this section we consider some commonly used sample characteristics and their distributions.

Definition 1. Let $F_n^*(x) = n^{-1} \sum_{j=1}^{n} \varepsilon(x - X_j)$. Then $nF_n^*(x)$ is the number of X_k's $(1 \leq k \leq n)$ that are $\leq x$. $F_n^*(x)$ is called the *sample (or empirical) distribution function*.

We note that $0 \leq F_n^*(x) \leq 1$ for all x, and, moreover, that F_n^* is right continuous, nondecreasing, and $F_n^*(-\infty) = 0$, $F_n^*(\infty) = 1$. Thus F_n^* is a DF.

If $X_{(1)}, X_{(2)}, \ldots, X_{(n)}$ is the order statistic for $X_1, X_2, \ldots, X_n$, then clearly

$$F_n^*(x) = \begin{cases} 0 & \text{if } x < X_{(1)} \\ \dfrac{k}{n} & \text{if } X_{(k)} \leq x < X_{(k+1)} \\ 1 & \text{if } x \geq X_{(n)}. \end{cases} \qquad (k = 1, 2, \ldots, n-1). \tag{1}$$

For fixed but otherwise arbitrary $x \in \mathcal{R}$, $F_n^*(x)$ itself is an RV of the discrete type. The following result is immediate.

Theorem 1. The RV $F_n^*(x)$ has the probability function

$$P\left\{F_n^*(x) = \frac{j}{n}\right\} = \binom{n}{j} [F(x)]^j [1 - F(x)]^{n-j}, \qquad j = 0, 1, \ldots, n, \tag{2}$$

with mean

$$EF_n^*(x) = F(x) \tag{3}$$

and variance

$$\operatorname{var}(F_n^*(x)) = \frac{F(x)[1 - F(x)]}{n}. \tag{4}$$

Proof. Since $\varepsilon(x - X_j), j = 1, 2, \ldots, n$, are iid RVs, each with PMF

$$P\{\varepsilon(x - X_j) = 1\} = P\{x - X_j \geq 0\} = F(x)$$

and

$$P\{\varepsilon(x - X_j) = 0\} = 1 - F(x),$$

their sum $nF_n^*(x)$ is a $b(n, p)$ RV, where $p = F(x)$. Relations (2), (3), and (4) follow immediately.

We next consider some typical values of the DF $F_n^*(x)$, called *sample statistics*. Since $F_n^*(x)$ has jump points $X_j, j = 1, 2, \ldots, n$, it is clear that all moments of $F_n^*(x)$ exist. Let us write

$$a_k = n^{-1} \sum_{j=1}^{n} X_j^k \tag{5}$$

for the moment of order k about 0. Here a_k will be called the *sample moment of order k*. In this notation

$$a_1 = n^{-1} \sum_{j=1}^{n} X_j = \overline{X}. \tag{6}$$

The *sample central moment* is defined by

$$b_k = n^{-1} \sum_{j=1}^{n} (X_j - a_1)^k = n^{-1} \sum_{j=1}^{n} (X_j - \overline{X})^k. \tag{7}$$

Clearly,

$$b_1 = 0 \quad \text{and} \quad b_2 = \left(\frac{n-1}{n}\right) S^2.$$

As mentioned earlier, we do not call b_2 the sample variance. S^2 will be referred to as the *sample variance* for reasons that will subsequently become clear. We have

$$b_2 = a_2 - a_1^2. \tag{8}$$

For the MGF of DF $F_n^*(x)$, we have

$$M^*(t) = n^{-1} \sum_{j=1}^{n} e^{tX_j}. \tag{9}$$

Similar definitions are made for sample moments of bivariate and multivariate distributions. For example, if $(X_1, Y_1), (X_2, Y_2), \ldots, (X_n, Y_n)$ is a sample from a bivariate distribution, we write

$$\overline{X} = n^{-1} \sum_{j=1}^{n} X_j \quad \text{and} \quad \overline{Y} = n^{-1} \sum_{j=1}^{n} Y_j \tag{10}$$

for the two sample means, and for the second-order sample central moments we write

$$b_{20} = n^{-1} \sum_{j=1}^{n} (X_j - \overline{X})^2, \qquad b_{02} = n^{-1} \sum_{j=1}^{n} (Y_j - \overline{Y})^2, \tag{11}$$

$$b_{11} = n^{-1} \sum_{j=1}^{n} (X_j - \overline{X})(Y_j - \overline{Y}).$$

Once again we write

$$S_1^2 = (n-1)^{-1}\sum_{j=1}^{n}(X_j - \overline{X})^2 \quad \text{and} \quad S_2^2 = (n-1)^{-1}\sum_{j=1}^{n}(Y_j - \overline{Y})^2 \tag{12}$$

for the two *sample variances*, and for the *sample covariance* we use the quantity

$$S_{11} = (n-1)^{-1}\sum_{j=1}^{n}(X_j - \overline{X})(Y_j - \overline{Y}). \tag{13}$$

In particular, the *sample correlation coefficient* is defined by

$$R = \frac{b_{11}}{\sqrt{b_{20}b_{02}}} = \frac{S_{11}}{S_1 S_2}. \tag{14}$$

It can be shown (Problem 4) that $|R| \le 1$, the extreme values ± 1 can occur only when all sample points $(X_1, Y_1), \ldots, (X_n, Y_n)$ lie on a straight line.

The sample quantiles are defined in a similar manner. Thus, if $0 < p < 1$, the *sample quantile of order p*, denoted by Z_p, is the order statistic $X_{(r)}$, where

$$r = \begin{cases} np & \text{if } np \text{ is an integer,} \\ [np+1] & \text{if } np \text{ is not an integer.} \end{cases}$$

As usual, $[x]$ is the largest integer $\le x$. Note that, if np is an integer, we can take any value between $X_{(np)}$ and $X_{(np)+1}$ as the pth sample quantile. Thus, if $p = \frac{1}{2}$ and n is even, we can take any value between $X_{(n/2)}$ and $X_{(n/2)+1}$, the two middle values, as the median. It is customary to take the average. Thus the sample median is defined as

$$Z_{1/2} = \begin{cases} X_{((n+1)/2)} & \text{if } n \text{ is odd,} \\ \dfrac{X_{(n/2)} + X_{((n/2)+1)}}{2} & \text{if } n \text{ is even.} \end{cases} \tag{15}$$

Note that

$$\left[\frac{n}{2} + 1\right] = \left(\frac{n+1}{2}\right)$$

if n is odd.

Example 1. A random sample of 25 observations is taken from the interval (0,1):

0.50	0.24	0.89	0.54	0.34	0.89	0.92	0.17	0.32	0.80
0.06	0.21	0.58	0.07	0.56	0.20	0.31	0.17	0.41	0.38
0.88	0.61	0.35	0.06	0.90					

In order to compute F_{25}^*, the first step is to order the observations from smallest to largest. The ordered sample is

$$0.06,\ 0.06,\ 0.07,\ 0.17,\ 0.17,\ 0.20,\ 0.21,\ 0.24,\ 0.31,\ 0.32,\ 0.34,$$
$$0.35,\ 0.38,\ 0.41,\ 0.50,\ 0.54,\ 0.56,\ 0.58,\ 0.61,\ 0.80,\ 0.88,\ 0.89,$$
$$0.89,\ 0.90,\ 0.92$$

Then the empirical DF is given by

$$F_{25}^*(x) = \begin{cases} 0, & x < 0.06 \\ 2/25, & 0.06 \le x < 0.07 \\ 3/25, & 0.07 \le x < 0.17 \\ 5/25, & 0.17 \le x < 0.20 \\ \vdots & \\ 24/25, & 0.90 \le x < 0.92 \\ 1, & x \ge 0.92 \end{cases}.$$

A plot of F_{25}^* is shown in Fig. 1. The sample mean and variance are

$$\bar{x} = 0.45, \quad s^2 = 0.084, \text{ and } s = 0.29.$$

Also sample median is the 13th observation in the ordered sample, namely, $z_{1/2} = 0.38$, and if $p = 0.2$ then $np = 5$ and $z_{0.2} = 0.17$.

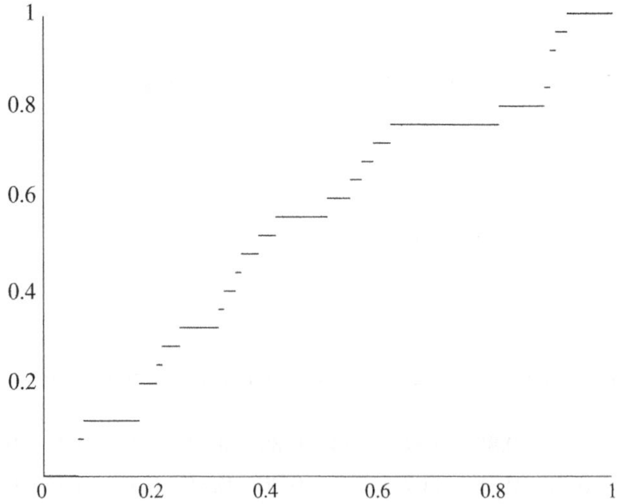

Fig. 1 Empirical DF for data of Example 1.

Next we consider the moments of sample characteristics. In the following we write $EX^k = m_k$ and $E(X-\mu)^k = \mu_k$ for the kth-order population moments. Wherever we use m_k (or μ_k), it will be assumed to exist. Also, σ^2 represents the population variance.

Theorem 2. Let $X_1, X_2, \ldots, X_n$ be a sample from a population with DF F. Then

$$E\overline{X} = \mu, \tag{16}$$

$$\text{var}(\overline{X}) = \frac{\sigma^2}{n}, \tag{17}$$

$$E(\overline{X})^3 = \frac{m_3 + 3(n-1)m_2\mu + (n-1)(n-2)\mu^3}{n^2}, \tag{18}$$

and

$$E(\overline{X})^4 = \frac{m_4 + 4(n-1)m_3\mu + 6(n-1)(n-2)m_2\mu^2 + 3(n-1)m_2^2}{n^3} + \frac{(n-1)(n-2)(n-3)\mu^4}{n^3}. \tag{19}$$

Proof. In view of Theorems 4.5.3 and 4.5.7, it suffices to prove (18) and (19). We have

$$\left(\sum_{j=1}^{n} X_j\right)^3 = \sum_{j=1}^{n} X_j^3 + 3\sum_{j\neq k} X_j^2 X_k + \sum_{j\neq k\neq l} X_j X_k X_l,$$

and (18) follows. Similarly,

$$\begin{aligned}\left(\sum_{i=1}^{n} X_i\right)^4 &= \left(\sum_{i=1}^{n} X_i\right)\left(\sum_{j=1}^{n} X_j^3 + 3\sum_{j\neq k} X_j^2 X_k + \sum_{j\neq k\neq l} X_j X_k X_l\right)\\ &= \sum_{i=1}^{n} X_i^4 + 4\sum_{j\neq k} X_j X_k^3 + 3\sum_{j\neq k} X_j^2 X_k^2 + 6\sum_{i\neq j\neq k} X_i^2 X_j X_k\\ &\quad + \sum_{i\neq j\neq k\neq l} X_i X_j X_k X_l,\end{aligned}$$

and (19) follows.

Theorem 3. For the third and fourth central moments of $\overline{X}$, we have

$$\mu_3(\overline{X}) = \frac{\mu_3}{n^2} \tag{20}$$

and

$$\mu_4(\overline{X}) = \frac{\mu_4}{n^3} + 3\frac{(n-1)\mu_2^2}{n^3}. \tag{21}$$

Proof. We have

$$\begin{aligned}\mu_3(\overline{X}) &= E(\overline{X}-\mu)^3 = \frac{1}{n^3}E\left\{\sum_{i=1}^{n}(X_i-\mu)\right\}^3\\ &= \frac{1}{n^3}\sum_{i=1}^{n}E(X_i-\mu)^3 = \frac{\mu_3}{n^2},\end{aligned}$$

and

$$\begin{aligned}\mu_4(\overline{X}) &= E(\overline{X}-\mu)^4 = \frac{1}{n^4}E\left\{\sum_{i=1}^{n}(X_i-\mu)\right\}^4\\ &= \frac{1}{n^4}\sum_{i=1}^{n}E(X_i-\mu)^4 + \binom{4}{2}\frac{1}{n^4}\sum_{i<j}E\{(X_i-\mu)^2(X_j-\mu)^2\}\\ &= \frac{\mu_4}{n^3} + \frac{3(n-1)}{n^3}\mu_2^2.\end{aligned}$$

Theorem 4. For the moments of b_2, we have

$$E(b_2) = \frac{(n-1)\sigma^2}{n}, \tag{22}$$

$$\mathrm{var}(b_2) = \frac{\mu_4-\mu_2^2}{n} - \frac{2(\mu_4-2\mu_2^2)}{n^2} + \frac{\mu_4-3\mu_2^2}{n^3}, \tag{23}$$

$$E(b_3) = \frac{(n-1)(n-2)}{n^2}\mu_3, \tag{24}$$

and

$$E(b_4) = \frac{(n-1)(n^2-3n+3)}{n^3}\mu_4 + \frac{3(n-1)(2n-3)}{n^3}\mu_2^2. \tag{25}$$

Proof. We have

$$\begin{aligned}Eb_2 &= \frac{1}{n}E\left\{\sum_{1}^{n}(X_i-\mu+\mu-\overline{X})^2\right\}\\ &= \frac{1}{n}E\left\{\sum_{i=1}^{n}(X_i-\mu)^2 - n(\overline{X}-\mu)^2\right\}\\ &= \frac{1}{n}(n\sigma^2-\sigma^2) = \frac{n-1}{n}\sigma^2.\end{aligned}$$

Now

$$n^2b_2^2 = \left[\sum_{i=1}^{n}(X_i-\mu)^2 - n(\overline{X}-\mu)^2\right]^2.$$

Writing $Y_i = X_i - \mu$, we see that $EY_i = 0$, $\text{var}(Y_i) = \sigma^2$, and $EY_i^4 = \mu_4$. We have

$$\begin{aligned}
n^2 Eb_2^2 &= E\left(\sum_1^n Y_i^2 - n\overline{Y}^2\right)^2 \\
&= E\left[\sum_{i=1}^n Y_i^4 + \sum_{i\neq j} Y_i^2 Y_j^2 - \frac{2}{n}\left(\sum_{i\neq j} Y_i^2 Y_j^2 + \sum_{j=1}^n Y_j^4\right)\right. \\
&\quad \left. + \frac{1}{n^2}\left(3\sum_{i\neq j} Y_i^2 Y_j^2 + \sum_1^n Y_j^4\right)\right].
\end{aligned}$$

It follows that

$$\begin{aligned}
n^2 Eb_2^2 &= n\mu_4 + n(n-1)\sigma^4 - \frac{2}{n}[n(n-1)\sigma^4 + n\mu_4] \\
&\quad + \frac{1}{n^2}[3n(n-1)\sigma^4 + n\mu_4] \\
&= \left(n - 2 + \frac{1}{n}\right)\mu_4 + \left(n - 2 + \frac{3}{n}\right)(n-1)\mu_2^2 \quad (\mu_2 = \sigma^2).
\end{aligned}$$

Therefore,

$$\begin{aligned}
\text{var}(b_2) &= Eb_2^2 - (Eb_2)^2 \\
&= \left(n - 2 + \frac{1}{n}\right)\frac{\mu_4}{n^2} + (n-1)\left(n - 2 + \frac{3}{n}\right)\frac{\mu_2^2}{n^2} - \left(\frac{n-1}{n}\right)^2 \mu_2^2 \\
&= \left(n - 2 + \frac{1}{n}\right)\frac{\mu_4}{n^2} + (n-1)(3-n)\frac{\mu_2^2}{n^3},
\end{aligned}$$

as asserted.

Relations (24) and (25) can be proved similarly.

Corollary 1. $ES^2 = \sigma^2$.

This is precisely the reason why we call S^2, and not b_2, the sample variance.

Corollary 2. $\text{var}(S^2) = \dfrac{\mu_4}{n} + \dfrac{3-n}{n(n-1)}\mu_2^2$.

Remark 1. The results of Theorems 3 to 5 can easily be modified and stated for the case when the X_i's are exchangeable RVs. Thus (16) holds and (17) has to be modified to

$$\text{var}(\overline{X}) = \frac{\sigma^2}{n} + \frac{n-1}{n}\rho\sigma^2, \tag{17$'$}$$

where ρ is the correlation coefficient between X_i and X_j. The expressions for $(\Sigma X_j)^3$ and $(\Sigma X_j)^4$ in the proof of Theorem 3 still hold but both (18) and (19) need appropriate modification. For example, (18) changes to

$$E\overline{X}^3 = \frac{m_3 + 3(n-1)E(X_j^2 X_k) + (n-1)(n-2)E(X_j X_k X_l)}{n^2}. \tag{18'}$$

Let us show how Corollary 1 changes for exchangeable RVs. Clearly,

$$(n-1)S^2 = \sum_{i=1}^{n}(X_i - \mu)^2 - n(\overline{X} - \mu)^2$$

so that

$$\begin{aligned}(n-1)ES^2 &= n\sigma^2 - nE(\overline{X} - \mu)^2 \\ &= n\sigma^2 - \left\{\sigma^2 + (n-1)\rho\sigma^2\right\}.\end{aligned}$$

in view of (17′). It follows that

$$ES^2 = \sigma^2(1-\rho).$$

We note that $E(S^2 - \sigma^2) = -\rho\sigma^2$ and, moreover, from Problem 4.5.19 (or from (17′)) we note that $\rho \geq -1/(n-1)$ so that $1 - \rho \leq n/(n-1)$ and hence

$$0 \leq ES^2 \leq \frac{n}{n-1}\sigma^2.$$

Remark 2. In simple random sampling from a (finite) population of size N we note that when $n = N$, $\bar{X} = \mu$, which is a constant so that (17′) reduces to

$$0 = \frac{\sigma^2}{N} + \frac{N-1}{N}\rho\sigma^2,$$

so that $\rho = -1/(N-1)$. It follows that

$$\operatorname{var}(\bar{X}) = \frac{\sigma^2}{n}\left(1 - \frac{n-1}{N-1}\right) = \left(\frac{N-n}{N-1}\right)\frac{\sigma^2}{n}. \tag{17''}$$

The factor $(N-n)/(N-1)$ in (17″) is called the *finite population correction factor*. As $N \to \infty$, with n fixed, $(N-n)/(N-1) \to 1$ so that the expression for $\operatorname{var}(\bar{X})$ in (17″) approaches that in (17).

Remark 3. In view of (17′) if the X_i's are uncorrelated, that is, if $\rho = 0$, then var $(\overline{X}) = \sigma^2/n$, the SD of $\overline{X}$ is $\sigma/\sqrt{n}$. The SD of $\overline{X}$ is sometimes called *standard error* (SE) although if σ is unknown $S/\sqrt{n}$ is most commonly referred to as the SE of $\overline{X}$.

The following result provides a justification for our definition of sample covariance.

Theorem 5. Let $(X_1, Y_1), (X_2, Y_2), \ldots, (X_n, Y_n)$ be a sample from a bivariate population with variances σ_1^2, σ_2^2 and covariance $\rho\sigma_1\sigma_2$. Then,

$$ES_1^2 = \sigma_1^2, \qquad ES_2^2 = \sigma_2^2, \quad \text{and} \quad ES_{11} = \rho\sigma_1\sigma_2, \tag{26}$$

where S_1^2, S_2^2, and S_{11} are defined in (12) and (13).

Proof. It follows from Corollary 1 to Theorem 4 that $ES_1^2 = \sigma_1^2$ and $ES_2^2 = \sigma_2^2$. To prove that $ES_{11} = \rho\sigma_1\sigma_2$ we note that X_i is independent of $X_j (i \neq j)$ and Y_j $(i \neq j)$. We have

$$(n-1)ES_{11} = E\left\{\sum_{j=1}^{n}(X_j - \overline{X})(Y_j - \overline{Y})\right\}.$$

Now

$$\begin{aligned} E\{(X_j - \overline{X})(Y_j - \overline{Y})\} &= E\left\{X_jY_j - X_j\frac{\sum_1^n Y_j}{n} - Y_j\frac{\sum_1^n X_j}{n} + \frac{\sum X_j \sum Y_j}{n^2}\right\} \\ &= EXY - \frac{1}{n}[EXY + (n-1)EX\,EY] - \frac{1}{n}[EXY + (n-1)EX\,EY] \\ &\quad + \frac{1}{n^2}[nEXY + n(n-1)EX\,EY] \\ &= \frac{n-1}{n}(EXY - EX\,EY), \end{aligned}$$

and it follows that

$$(n-1)ES_{11} = n\left(\frac{n-1}{n}\right)(EXY - EX\,EY),$$

that is,

$$ES_{11} = EXY - EX\,EY = \operatorname{cov}(X,Y) = \rho\sigma_1\sigma_2,$$

as asserted.

We next turn our attention to the distributions of sample characteristics. Several possibilities exist. If the exact sampling distribution is required, the method of transformation described in Section 4.4 can be used. Sometimes the technique of MGF or CF can be applied. Thus, if $X_1, X_2, \ldots, X_n$ is a random sample from a population distribution for which the MGF exists, the MGF of the sample mean $\overline{X}$ is given by

$$M_{\overline{X}}(t) = \prod_{i=1}^{n} Ee^{tX_i/n} = \left[M\left(\frac{t}{n}\right)\right]^n, \tag{27}$$

where M is the MGF of the population distribution. If $M_{\overline{X}}(t)$ has one of the known forms, it is possible to write the PDF of $\overline{X}$. Although this method has the obvious drawback that it applies only to distributions for which all moments exist, we will see in Section 6.5 its effectiveness in the important case of sampling from a normal population where this condition is satisfied. An analog of (27) holds for CFs without any condition on existence of moments. Indeed,

$$\phi_{\overline{X}}(t) = \sum_{j=1}^{n} Ee^{itX_j/n} = \left[\phi\left(\frac{t}{n}\right)\right]^n, \tag{28}$$

where ϕ is the CF of X_j.

Example 2. Let $X_1, X_2, \ldots, X_n$ be a sample from a $G(\alpha, 1)$ distribution. We will compute the PDF of $\overline{X}$. We have

$$M_{\overline{X}}(t) = \left[M\left(\frac{t}{n}\right)\right]^n = \frac{1}{(1-t/n)^{\alpha n}}, \qquad \frac{t}{n} < 1,$$

so that $\overline{X}$ is a $G(\alpha n, 1/n)$ variate.

Example 3. Let $X_1, X_2, \ldots, X_n$ be a random sample from a uniform distribution on $(0,1)$. Consider the geometric mean

$$Y_n = \left(\prod_{i=1}^{n} X_i\right)^{1/n}.$$

We have $\log Y_n = (1/n)\sum_{i=1}^{n} \log X_i$, so that $\log Y_n$ is the mean of $\log X_1, \ldots, \log X_n$.

The common PDF of $\log X_1, \ldots, \log X_n$ is

$$f(x) = \begin{cases} e^x & \text{if } x < 0, \\ 0 & \text{otherwise,} \end{cases}$$

which is the negative exponential distribution with parameter $\beta = 1$. We see that the MGF of $\log Y_n$ is given by

$$M(t) = \prod_{i=1}^{n} E e^{t \log X_i / n} = \frac{1}{(1+t/n)^n},$$

and the PDF of $\log Y_n$ is given by

$$f^*(x) = \begin{cases} \dfrac{n^n}{\Gamma(n)}(-x)^{n-1} e^{nx}, & -\infty < x < 0, \\ 0, & \text{otherwise.} \end{cases}$$

It follows that Y_n has PDF

$$f_{Y_n}(y) = \begin{cases} \dfrac{n^n}{\Gamma(n)} y^{n-1}(-\log y)^{n-1}, & 0 < y < 1, \\ 0, & \text{otherwise.} \end{cases}$$

Example 4. (Hogben [46]). Let $X_1, X_2, \ldots, X_n$ be a random sample from a Bernoulli distribution with parameter p, $0 < p < 1$. Let $\overline{X}$ be the sample mean and S^2 the sample variance. We will find the PMF of S^2. Note that $S_n = \sum_{i=1}^{n} X_i = \sum_{i=1}^{n} X_i^2$ and that S_n is $b(n,p)$. Since

$$\begin{aligned}(n-1)S^2 &= \sum_{i=1}^{n} X_i^2 - n(\overline{X})^2 \\ &= \frac{S_n(n-S_n)}{n},\end{aligned}$$

S^2 only assumes values of the form

$$t = \frac{i(n-i)}{n(n-1)}, \qquad i = 0,1,2,\ldots,\left[\frac{n}{2}\right],$$

where $[x]$ is the largest integer $\le x$. Thus

$$\begin{aligned} P\{S^2 = t\} &= P\{nS_n - S_n^2 = i(n-1)\} \\ &= P\left\{\left(S_n - \frac{n}{2}\right)^2 = \left(i - \frac{n}{2}\right)^2\right\} \\ &= P\{S_n = 1 \text{ or } S_n = n-i\} \\ &= \binom{n}{i} p^i (1-p)^{n-i} + \binom{n}{i} p^{n-i}(1-p)^i \\ &= \binom{n}{i} p^i (1-p)^i \{(1-p)^{n-2i} + p^{n-2i}\}, \qquad i < \left[\frac{n}{2}\right]. \end{aligned}$$

If n is even, $n = 2m$, say, where $m \ge 0$ is an integer, and $i = m$, then

$$P\left\{S^2 = \frac{m}{2(2m-1)}\right\} = 2\binom{2m}{m} p^m (1-p)^m.$$

In particular, if $n = 7$, $S^2 = 0$, $\frac{1}{7}$, $\frac{5}{21}$, and $\frac{2}{7}$ with probabilities $\{p^7 + (1-p)^7\}$, $7p(1-p)\{p^5 + (1-p)^5\}$, $21p^2(1-p)^2\{p^3 + (1-p)^3\}$, and $35p^3(1-p)^3$, respectively.

If $n = 6$, then $S^2 = 0$, $\frac{1}{6}$, $\frac{4}{15}$, and $\frac{3}{10}$ with probabilities $\{p^6 + (1-p)^6\}$, $6p(1-p)\{p^4 + (1-p)^4\}$, $15p^2(1-p)^2\{p^2 + (1-p)^2\}$, and $40p^3(1-p)^3$, respectively.

We have already considered the distribution of the sample quantiles in Section 4.7 and the distribution of range $X_{(n)} - X_{(1)}$ in Example 4.7.4. It can be shown, without much difficulty, that the distribution of the sample median is given by

$$f_r(y) = \frac{n!}{(r-1)!\,(n-r)!}[F(y)]^{r-1}[1-F(y)]^{n-r}f(y) \qquad \text{if } r = \frac{n+1}{2}, \tag{29}$$

where F and f are the population DF and PDF, respectively. If $n = 2m$ and the median is taken as the average of $X_{(m)}$ and $X_{(m+1)}$, then

$$f_r(y) = \frac{2(2m)!}{[(m-1)!]^2}\int_y^\infty [F(2y-v)]^{m-1}[1-F(v)]^{m-1}f(2y-v)f(v)\,dv. \tag{30}$$

Example 5. Let $X_1, X_2, \ldots, X_n$ be a random sample from $U(0,1)$. Then the integrand in (30) is positive for the intersection of the regions $0 < 2y - v < 1$ and $0 < v < 1$. This gives $(v/2) < y < (v+1)/2$, $y < v$, and $0 < v < 1$. The shaded area in Fig. 2 gives the limits on the integral as

$$y < v < 2y \qquad \text{if } 0 < y \le \frac{1}{2}$$

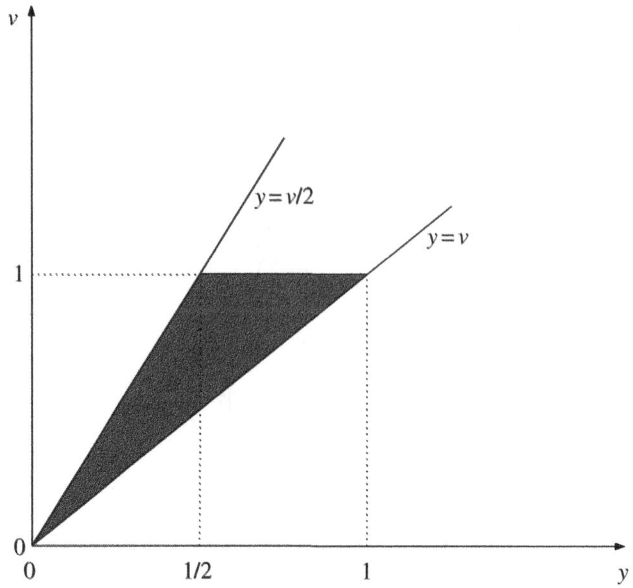

Fig. 2 $\{y < \theta \leq 2y, 0 < y < 1/2, \text{ and } y < \theta < 1, 1/2 < y \leq 1\}$.

and

$$y < v < 1 \qquad \text{if } \frac{1}{2} < y < 1.$$

In particular, if $m = 2$, the PDF of the median, $(X_{(2)} + X_{(3)})/2$, is given by

$$f_r(y) = \begin{cases} 8y^2(3-4y) & \text{if } 0 < y < \frac{1}{2}, \\ 8(4y^3 - 9y^2 + 6y - 1) & \text{if } \frac{1}{2} < y < 1, \\ 0 & \text{otherwise.} \end{cases}$$

The method of MGF (or CF) introduced in this section is particularly effective in computing distributions of commonly used statistics in sampling from a univariate or bivariate normal distribution as we shall see in the next two sections. However, when sampling from nonnormal populations these methods may not be very fruitful in determining the exact distribution of the statistic under consideration. Often the statistic itself may be too intractable. Then we have some of other alternatives at our disposal. One may be able to use the asymptotic distribution of the statistic or one may resort to simulation methods. In Chapter 7 we study some of these procedures.

PROBLEMS 6.3

1. Let $X_1, X_2, \ldots, X_n$ be random sample from a DF F, and let $F_n^*(x)$ be the sample distribution function. Find $\text{cov}(F_n^*(x), F_n^*(y))$ for fixed real numbers x, y.

2. Let F_n^* be the empirical DF of a random sample from DF F. Show that

$$P\left\{|F_n^*(x) - F(x)| \geq \frac{\varepsilon}{2\sqrt{n}}\right\} \leq \frac{1}{\varepsilon^2} \qquad \text{for all } \varepsilon > 0.$$

3. For the data of Example 6.2.2 compute the sample distribution function.
4. (a) Show that the sample correlation coefficient R satisfies $|R| \leq 1$ with equality if and only if all sample points lie on a straight line.
   (b) If we write $U_i = aX_i + b$ $(a \neq 0)$ and $V_i = cY_i + d$ $(c \neq 0)$, what is the sample correlation coefficient between the U's and the V's?
5. (a) A sample of size 2 is taken from the PDF $f(x) = 1, 0 \leq x \leq 1$, and $= 0$ otherwise. Find $P(\bar{X} \geq 0.9)$.
   (b) A sample of size 2 is taken from $b(1,p)$:
   (i) Find $P(\bar{X} \leq p)$. (ii) Find $P(S^2 \geq 0.5)$.
6. Let $X_1, X_2, \ldots, X_n$ be a random sample from $\mathcal{N}(\mu, \sigma^2)$. Compute the first four sample moments of $\overline{X}$ about the origin and about the mean. Also compute the first four sample moments of S^2 about the mean.
7. Derive the PDF of the median given in (29) and (30).
8. Let $U_{(1)}, U_{(2)}, \ldots, U_{(n)}$ be the order statistic of a sample size n from $U(0,1)$. Compute $EU_{(r)}^k$ for any $1 \leq r \leq n$ and integer k (> 0). In particular, show that

$$EU_{(r)} = \frac{r}{n+1} \quad \text{and} \quad \text{var}(U_{(r)}) = \frac{r(n-r+1)}{(n+1)^2(n+2)}.$$

   Show also that the correlation coefficient between $U_{(r)}$ and $U_{(s)}$ for $1 \leq r < s \leq n$ is given by $[r(n-s+1)/s(n-r+1)]^{1/2}$.
9. Let $X_1, X_2, \ldots, X_n$ be n independent observations on X. Find the sampling distribution of $\overline{X}$, the sample mean, if (a) $X \sim P(\lambda)$, (b) $X \sim \mathcal{C}(1,0)$, and (c) $X \sim \chi^2(m)$.
10. Let $X_1, X_2, \ldots, X_n$ be a random sample from $G(\alpha, \beta)$. Let us write $Y_n = (\overline{X} - \alpha\beta)/\beta\sqrt{(\alpha/n)}$, $n = 1, 2, \ldots$.
    (a) Compute the first four moments of Y_n, and compare them with the first four moments of the standard normal distribution.
    (b) Compute the coefficients of skewness α_3 and of kurtosis α_4 for the RVs Y_n. (For definitions of α_3, α_4 see Problem 3.2.10.)
11. Let $X_1, X_2, \ldots, X_n$ be a random sample from $U[0,1]$. Also let $Z_n = (\overline{X} - 0.5)/\sqrt{(1/12n)}$. Repeat Problem 10 for the sequence Z_n.
12. Let $X_1, X_2, \ldots, X_n$ be a random sample from $P(\lambda)$. Find $\text{var}(S^2)$, and compare it with $\text{var}(\overline{X})$. Note that $E\overline{X} = \lambda = ES^2$. [*Hint:* Use Problem 3.2.9.]
13. Prove (24) and (25).
14. Multiple RVs $\mathbf{X}_1, \mathbf{X}_2, \ldots, \mathbf{X}_n$ are exchangeable if the $n!$ permutations $(\mathbf{X}_{i_1}, \mathbf{X}_{i_2}, \ldots, \mathbf{X}_{i_n})$ have the same n-dimensional distribution. Consider the special case when $\mathbf{X}$'s are two dimensional. Find an analog of Theorem 6 for exchangeable bivariate RVs $(X_1, Y_1), (X_2, Y_2), \ldots, (X_n, Y_n)$.

15. Let $X_1, X_2, \ldots, X_n$ be a random sample from a distribution with finite third moment. Show that $\text{cov}(\overline{X}, S^2) = \mu_3/n$.

6.4 CHI-SQUARE, t-, AND F-DISTRIBUTIONS: EXACT SAMPLING DISTRIBUTIONS

In this section we investigate certain distributions that arise in sampling from a normal population. Let $X_1, X_2, \ldots, X_n$ be a sample from $\mathcal{N}(\mu, \sigma^2)$. Then we know that $\overline{X} \sim \mathcal{N}(\mu, \sigma^2/n)$. Also, $\{\sqrt{n}(\overline{X} - \mu)/\sigma\}^2$ is $\chi^2(1)$. We will determine the distribution of S^2 in the next section. Here we mainly define chi-square, t-, and F-distributions and study their properties. Their importance will become evident in the next section and later in the testing of statistical hypotheses (Chapter 10).

The first distribution of interest is the *chi-square distribution*, defined in Chapter 5 as a special case of the gamma distribution. Let $n > 0$ be an integer. Then $G(n/2, 2)$ is a $\chi^2(n)$ RV. In view of Theorem 5.3.29 and Corollary 2 to Theorem 5.3.4, the following result holds.

Theorem 1. Let $X_1, X_2, \ldots, X_n$ be iid RVs, and let $S_n = \sum_{k=1}^n X_k$. Then

(a) $S_n \sim \chi^2(n) \Leftrightarrow X_1 \sim \chi^2(1)$

and

(b) $X_1 \sim \mathcal{N}(0,1) \Rightarrow \sum_{k=1}^n X_k^2 \sim \chi^2(n)$.

If X has a chi-square distribution with n d.f., we write $X \sim \chi^2(n)$. We recall that, if $X \sim \chi^2(n)$, its PDF is given by

$$f(x) = \begin{cases} \dfrac{x^{n/2-1}e^{-x/2}}{2^{n/2}\Gamma(n/2)} & \text{if } x \geq 0, \\ 0 & \text{if } x < 0, \end{cases} \tag{1}$$

the MGF by

$$M(t) = (1-2t)^{-n/2} \quad \text{for } t < \frac{1}{2}, \tag{2}$$

and the mean and the variance by

$$EX = n, \qquad \text{var}(X) = 2n. \tag{3}$$

The $\chi^2(n)$ distribution is tabulated for values of $n = 1, 2, \ldots$. Tables usually go up to $n = 30$, since for $n > 30$ it is possible to use normal approximation. In Fig. 1 we plot the PDF (1) for selected values of n.

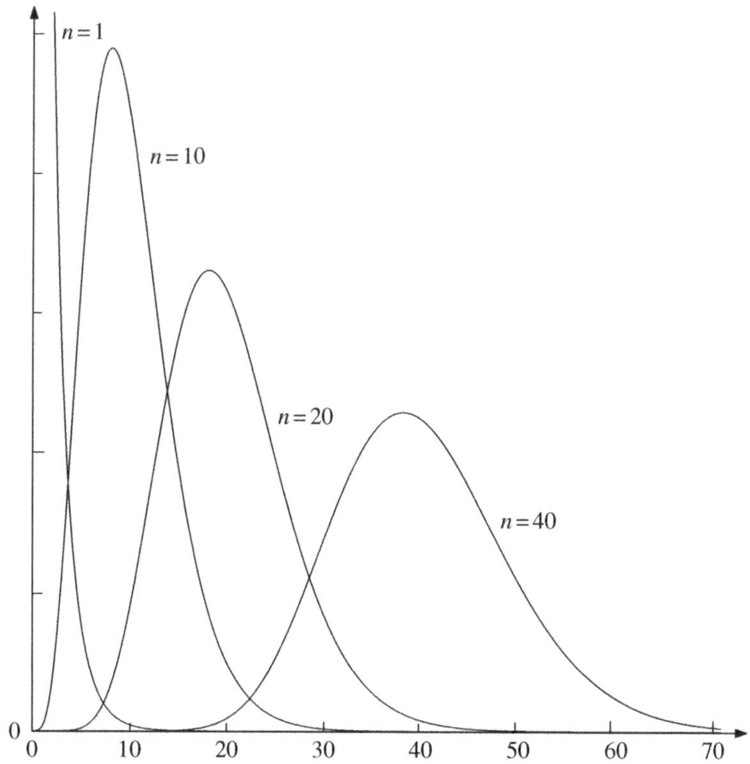

Fig. 1 Chi-square densities.

We will write $\chi^2_{n,\alpha}$ for the upper α percent point of the $\chi^2(n)$ distribution, that is,

$$P\{\chi^2(n) > \chi^2_{n,\alpha}\} = \alpha. \tag{4}$$

Table ST3 at the end of the book gives the values of $\chi^2_{n,\alpha}$ for some selected values of n and α.

Example 1. Let $n = 25$. Then, from Table ST3,

$$P\{\chi^2(25) \le 34.382\} = 0.90.$$

Let us approximate this probability using CLT. We see that $E\chi^2(25) = 25$, $\text{var}\,\chi^2(25) = 50$, so that

$$\begin{aligned} P\{\chi^2(25) \le 34.382\} &= P\left\{\frac{\chi^2(25) - 25}{\sqrt{50}} \le \frac{34.382 - 25}{5\sqrt{2}}\right\} \\ &\approx P\{Z \le 1.32\} \\ &= 0.9066. \end{aligned}$$

Definition 1. Let $X_1, X_2, \ldots, X_n$ be independent normal RVs with $EX_i = \mu_i$ and $\text{var}(X_i) = \sigma^2$, $i = 1, 2, \ldots, n$. Also, let $Y = \sum_{i=1}^n X_i^2/\sigma^2$. The RV Y is said to be a *non-central chi-square RV with noncentrality parameter* $\sum_{i=1}^n \mu_i^2/\sigma^2$ and n d.f. We will write $Y \sim \chi^2(n, \delta)$, where $\delta = \sum_{i=1}^n \mu_i^2/\sigma^2$.

Although the PDF of a $\chi^2(n, \delta)$ RV is hard to compute (see Problem 16), its MGF is easily evaluated. We have

$$M(t) = Ee^{t\sum_1^n X_i^2/\sigma^2} = \prod_1^n Ee^{tX_i^2/\sigma^2},$$

where $X_i \sim \mathcal{N}(\mu_i, \sigma^2)$. Thus

$$Ee^{tX_i^2/\sigma^2} = \int_{-\infty}^{\infty} \frac{1}{\sigma\sqrt{2\pi}} \exp\left\{\frac{tx_i^2}{\sigma^2} - \frac{(x_i - \mu_i)^2}{2\sigma^2}\right\} dx_i,$$

where the integral exists for $t < \frac{1}{2}$. In the integrand we complete squares, and after some simple algebra we obtain

$$Ee^{tX_i^2/\sigma^2} = \frac{1}{\sqrt{1-2t}} \exp\left\{\frac{t\mu_i^2}{\sigma^2(1-2t)}\right\}, \qquad t < \frac{1}{2}.$$

It follows that

$$M(t) = (1-2t)^{-n/2} \exp\left(\frac{t}{1-2t}\frac{\sum \mu_i^2}{\sigma^2}\right), \qquad t < \frac{1}{2}, \tag{5}$$

and the MGF of a $\chi^2(n, \delta)$ RV is therefore

$$M(t) = (1-2t)^{-n/2} \exp\left(\frac{t}{1-2t}\delta\right), \qquad t < \frac{1}{2}. \tag{6}$$

It is immediate that, if $X_1, X_2, \ldots, X_k$ are independent, $X_i \sim \chi^2(n_i, \delta_i)$, $i = 1, 2, \ldots, k$, then $\sum_{i=1}^k X_i$ is $\chi^2(\sum_{i=1}^k n_i, \sum_{i=1}^k \delta_i)$.

The mean and variance of $\chi^2(n, \delta)$ are easy to calculate. We have

$$\begin{aligned} EY &= \frac{\sum_1^n EX_i^2}{\sigma^2} = \frac{\sum_1^n[\text{var}(X_i) + (EX_i)^2]}{\sigma^2} \\ &= \frac{n\sigma^2 + \sum_1^n \mu_i^2}{\sigma^2} = n + \delta, \end{aligned}$$

and

$$\begin{aligned} \text{var}(Y) &= \text{var}\left(\frac{\sum_1^n X_i^2}{\sigma^2}\right) = \frac{1}{\sigma^4}\left[\sum_{i=1}^n \text{var}(X_i^2)\right] \\ &= \frac{1}{\sigma^4}\left\{\sum_{i=1}^n EX_i^4 - \sum_{i=1}^n [E(X_i^2)]^2\right\} \end{aligned}$$

$$= \frac{1}{\sigma^4}\left\{\sum_{i=1}^{n}(3\sigma^4 + 6\sigma^2\mu_i^2 + \mu_i^4) - \sum_{i=1}^{n}(\sigma^2 + \mu_i^2)^2\right\}$$

$$= \frac{1}{\sigma^4}(2n\sigma^4 + 4\sigma^2\sum\mu_i^2) = 2n + 4\delta.$$

We next turn our attention to *Student's t-statistic*, which arises quite naturally in sampling from a normal population.

Definition 2. Let $X \sim \mathcal{N}(0,1)$ and $Y \sim \chi^2(n)$, and let X and Y be independent. Then the statistic

$$T = \frac{X}{\sqrt{Y/n}} \tag{7}$$

is said to have a *t-distribution* with n d.f. and we write $T \sim t(n)$.

Theorem 2. The PDF of T defined in (7) is given by

$$f_n(t) = \frac{\Gamma[(n+1)/2]}{\Gamma(n/2)\sqrt{n\pi}}(1 + t^2/n)^{-(n+1)/2}, \qquad -\infty < t < \infty. \tag{8}$$

Proof. The proof is left as an exercise.

Remark 1. For $n = 1$, T is a Cauchy RV. We will therefore assume that $n > 1$. For each n, we have a different PDF. In Fig. 2 we plot $f_n(t)$ for some selected values of n. Like the

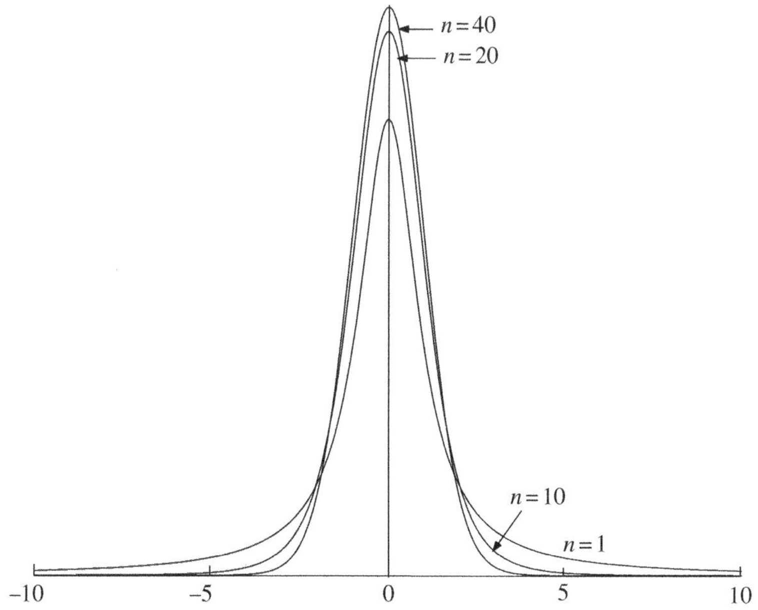

Fig. 2 Student's t-densities.

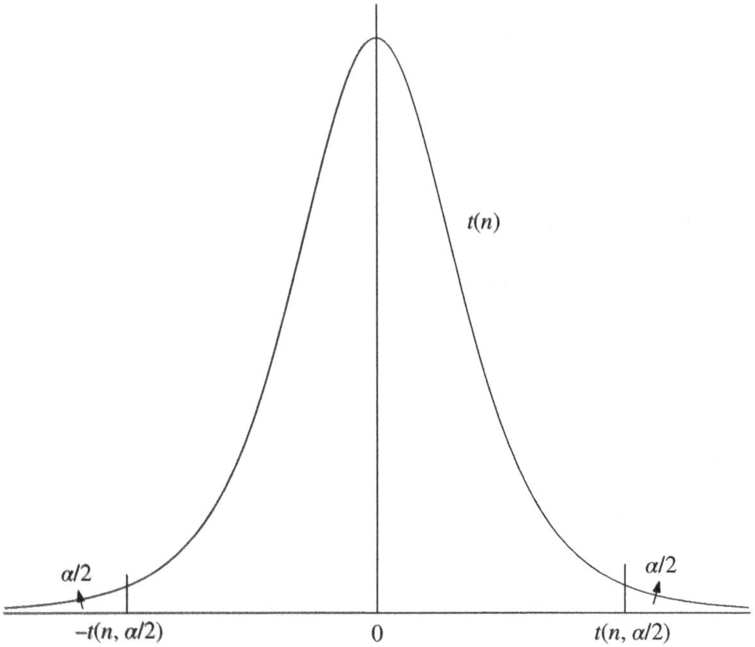

Fig. 3

normal distribution, the t-distribution is important in the theory of statistics and hence is tabulated (Table ST4).

Remark 2. The PDF $f_n(t)$ is symmetric in t, and $f_n(t) \to 0$ as $t \to +\infty$. For large n, the t-distribution is close to the normal distribution. Indeed, $(1+t^2/n)^{-(n+1)/2} \to e^{-t^2/2}$ as $n \to \infty$. Moreover, as $t \to \infty$ or $t \to -\infty$, the tails of $f_n(t) \to 0$ much more slowly than do the tails of the $N(0,1)$ PDF. Thus for small n and large t_0

$$P\{|T| > t_0\} \geq P\{|Z| > t_0\}, \qquad Z \sim \mathcal{N}(0,1),$$

that is, there is more probability in the tail of the t-distribution than in the tail of the standard normal. In what follows we will write $t_{n,\alpha/2}$ for the value (Fig. 3) of T for which

$$P\{|T| > t_{n,\alpha/2}\} = \alpha. \tag{9}$$

In Table ST4 positive values of $t_{n,\alpha}$ are tabulated for some selected values of n and α. Negative values may be obtained from symmetry, $t_{n,1-\alpha} = -t_{n,\alpha}$.

Example 2. Let $n = 5$. Then from Table ST4, we get $t_{5,0.025} = 2.571$ and $t_{5,0.05} = 2.015$. The corresponding values under the $\mathcal{N}(0,1)$ distribution are $z_{0.025} = 1.96$ and $z_{0.05} = 1.65$. For $n = 30$,

$$t_{30,0.05} = 1.697 \quad \text{and} \quad z_{0.05} = 1.65.$$

Theorem 3. Let $X \sim t(n)$, $n > 1$. Then EX^r exists for $r < n$. In particular, if $r < n$ is odd,

$$EX^r = 0, \tag{10}$$

and if $r < n$ is even,

$$EX^r = n^{r/2} \frac{\Gamma[(r+1)/2]\Gamma[(n-r)/2]}{\Gamma(1/2)\Gamma(n/2)}. \tag{11}$$

Corollary. If $n > 2$, $EX = 0$ and $EX^2 = \text{var}(X) = n/(n-2)$.

Remark 3. If in Definition 2 we take $X \sim \mathcal{N}(\mu, \sigma^2)$, $Y/\sigma^2 \sim \chi^2(n)$, and X and Y independent,

$$T = \frac{X}{\sqrt{Y/n}}$$

is said to have a *noncentral t-distribution with parameter* (also called *noncentrality parameter*) $\delta = \mu/\sigma$ and d.f. n. Various moments of noncentral t-distribution may be computed by using the fact that expectation of a product of independent RVs is the product of their expectations.

We leave the reader to show (Problem 3) that, if T has a noncentral t-distribution with n d.f. and noncentrality parameter δ, then

$$ET = \delta \frac{\Gamma[(n-1)/2]}{\Gamma(n/2)} \sqrt{\frac{n}{2}}, \qquad n > 1, \tag{12}$$

and

$$\text{var}(T) = \frac{n(1+\delta^2)}{n-2} - \frac{\delta^2 n}{2} \left(\frac{\Gamma[(n-1)/2]}{\Gamma(n/2)} \right)^2, \qquad n > 2. \tag{13}$$

Definition 3. Let X and Y be independent χ^2 RVs with m and n d.f., respectively. The RV

$$F = \frac{X/m}{Y/n} \tag{14}$$

is said to have an *F-distribution with* (m,n) d.f., and we write $F \sim F(m,n)$.

Theorem 4. The PDF of the F-statistic defined in (14) is given by

$$g(f) = \begin{cases} \dfrac{\Gamma[(m+n)/2]}{\Gamma(m/2)\Gamma(n/2)} \left(\dfrac{m}{n}\right) \left(\dfrac{m}{n}f\right)^{(m/2)-1} \\ \quad \cdot \left(1 + \dfrac{m}{n}f\right)^{-(m+n)/2}, & f > 0, \\ 0, & f \le 0. \end{cases} \tag{15}$$

Proof. The proof is left as an exercise.

Remark 4. If $X \sim F(m,n)$, then $1/X \sim F(n,m)$. If we take $m = 1$, then $F = [t(n)]^2$, so that $F(1,n)$ and $t^2(n)$ have the same distribution. It also follows that, if Z is $\mathcal{C}(1,0)$ [which is the same as $t(1)$], Z^2 is $F(1,1)$.

Remark 5. As usual, we write $F_{m,n,\alpha}$ for the upper α percent point of the $F(m,n)$ distribution, that is,

$$P\{F(m,n) > F_{m,n,\alpha}\} = \alpha. \tag{16}$$

From Remark 4, we have the following relation:

$$F_{m,n,1-\alpha} = \frac{1}{F_{n,m,\alpha}}. \tag{17}$$

It therefore suffices to tabulate values of F that are ≥ 1. This is done in Table ST5, where values of $F_{m,n,\alpha}$ are listed for some selected values of m, n, and α. See Fig. 4 for a plot of $g(f)$.

Theorem 5. Let $X \sim F(m,n)$. Then, for $k > 0$, integral,

$$EX^k = \left(\frac{n}{m}\right)^k \frac{\Gamma[k+(m/2)]\Gamma[(n/2)-k]}{\Gamma[(m/2)\Gamma(n/2)]} \qquad \text{for } n > 2k. \tag{18}$$

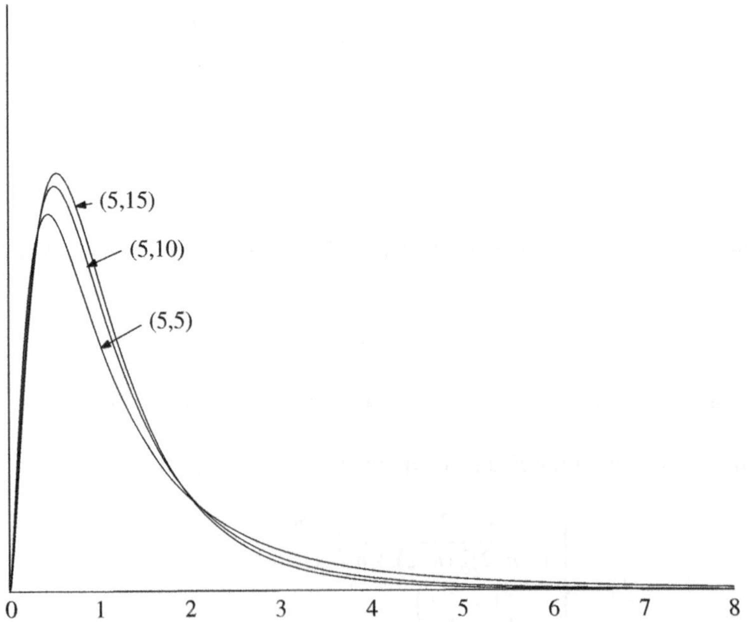

Fig. 4 *F* densities.

In particular,

$$EX = \frac{n}{n-2}, \qquad n > 2, \tag{19}$$

and

$$\text{var}(X) = \frac{n^2(2m+2n-4)}{m(n-2)^2(n-4)}, \qquad n > 4. \tag{20}$$

Proof. We have, for a positive integer k,

$$\int_0^\infty f^k f^{m/2-1}\left(1+\frac{m}{n}f\right)^{-(m+n)/2} df = \left(\frac{m}{n}\right)^{k+(m/2)} \int_0^1 x^{k+(m/2)-1}(1-x)^{(n/2)-k-1}\,dx, \tag{21}$$

where we have changed the variable to $x = (m/n)f[1+(m/n)f]^{-1}$. The integral in the right side of (21) converges for $(n/2)-k > 0$ and diverges for $(n/2)-k \leq 0$. We have

$$EX^k = \frac{\Gamma[(m+n)/2]}{\Gamma(m/2)\Gamma(n/2)}\left(\frac{m}{n}\right)^{m/2}\left(\frac{n}{m}\right)^{k+(m/2)} B\left(k+\frac{m}{2},\frac{n}{2}-k\right),$$

as asserted.

For $k = 1$, we get

$$EX = \frac{n}{m}\frac{m/2}{(n/2)-1} = \frac{n}{n-2}, \qquad n > 2.$$

Also,

$$\begin{aligned} EX^2 &= \left(\frac{n}{m}\right)^2 \frac{(m/2)[(m/2)+1]}{[(n/2)-1][(n/2)-2]}, \qquad n > 4,\\ &= \left(\frac{n}{m}\right)^2 \frac{m(m+2)}{(n-2)(n-4)}, \end{aligned}$$

and

$$\begin{aligned} \text{var}(X) &= \left(\frac{n}{m}\right)^2 \frac{m(m+2)}{(n-2)(n-4)} - \left(\frac{n}{n-2}\right)^2\\ &= \frac{2n^2(m+n-2)}{m(n-2)^2(n-4)}, \qquad n > 4. \end{aligned}$$

Theorem 6. If $X \sim F(m,n)$, then $Y = 1/[1+(m/n)X]$ is $B(n/2,m/2)$. Consequently, for each $x > 0$,

$$F_X(x) = 1 - F_Y\left[\frac{1}{1+(m/n)x}\right].$$

If in Definition 3 we take X to be a noncentral χ^2 RV with n d.f. and noncentrality parameter δ, we get a *noncentral F* RV.

Definition 4. Let $X \sim \chi^2(m,\delta)$ and $Y \sim \chi^2(n)$, and let X and Y be independent. Then the RV

$$F = \frac{X/m}{Y/n} \tag{22}$$

is said to have a noncentral F-distribution with noncentrality parameter δ.

It is shown in Problem 2 that if F has a noncentral F-distribution with (m,n) d.f. and noncentrality parameter δ,

$$EF = \frac{n(m+\delta)}{m(n-2)}, \qquad n > 2,$$

and

$$\operatorname{var}(F) = \frac{2n^2}{m^2(n-4)(n-2)^2}[(m+\delta)^2 + (n-2)(m+2\delta)], \qquad n > 4.$$

PROBLEMS 6.4

1. Let

$$P_x = \left\{\Gamma\left(\frac{n}{2}\right)2^{n/2}\right\}^{-1} \int_0^x \omega^{(n-2)/2} e^{-\omega/2}\, d\omega, \qquad x > 0.$$

Show that

$$x < \frac{n}{1-P_x}.$$

2. Let $X \sim F(m,n,\delta)$. Find EX and $\operatorname{var}(X)$.
3. Let T be a noncentral t-statistic with n d.f. and noncentrality parameter δ. Find ET and $\operatorname{var}(T)$.
4. Let $F \sim F(m,n)$. Then

$$Y = \left(1 + \frac{m}{n}F\right)^{-1} \sim B\left(\frac{n}{2}, \frac{m}{2}\right).$$

Deduce that for $x > 0$

$$P\{F \le x\} = 1 - P\left\{Y \le \left(1 + \frac{m}{n}x\right)^{-1}\right\}.$$

5. Derive the PDF of an F-statistic with (m,n) d.f.
6. Show that the square of a noncentral t-statistic is a noncentral F-statistic.
7. A sample of size 16 showed a variance of 5.76. Find c such that $P\{|\overline{X} - \mu| < c\} = 0.95$, where $\overline{X}$ is the sample mean and μ is the population mean. Assume that the sample comes from a normal population.

8. A sample from a normal population produced variance 4.0. Find the size of the sample if the sample mean deviates from the population mean by no more than 2.0 with a probability of at least 0.95.

9. Let X_1, X_2, X_3, X_4, X_5 be a sample from $\mathcal{N}(0,4)$. Find $P\{\sum_{i=1}^{5} X_i^2 \geq 5.75\}$.

10. Let $X \sim \chi^2(61)$. Find $P\{X > 50\}$.

11. Let $F \sim F(m,n)$. The random variable $Z = \frac{1}{2}\log F$ is known as *Fisher's Z statistic*. Find the PDF of Z.

12. Prove Theorem 1.

13. Prove Theorem 2.

14. Prove Theorem 3.

15. Prove Theorem 4.

16. (a) Let $f_1, f_2, \ldots$ be PDFs with corresponding MGFs $M_1, M_2, \ldots$, respectively. Let α_j $(0 < \alpha_j < 1)$ be constants such that $\sum_{j=1}^{\infty} \alpha_j = 1$. Then $f = \sum_1^{\infty} \alpha_j f_j$ is a PDF with MGF $M = \sum_{j=1}^{\infty} \alpha_j M_j$.

(b) Write the MGF of a $\chi^2(n,\delta)$ RV in (6) as

$$M(t) = \sum_{j=0}^{\infty} \alpha_j M_j(t),$$

where $M_j(t) = (1-2t)^{-(2j+n)/2}$ is the MGF of a $\chi^2(2j+n)$ RV and $\alpha_j = e^{-\delta/2}(\delta/2)^j/j!$ is the PMF of a $P(\delta/2)$ RV. Conclude that PDF of $Y \sim \chi^2(n,\delta)$ is the weighted sum of PDFs of $\chi^2(2j+n)$ RVs, $j = 0,1,2,\ldots$ with Poisson weights and hence

$$f_Y(y) = \sum_{j=0}^{\infty} \frac{e^{-\delta/2}(\delta/2)^j}{j!} \frac{y^{(2j+n)/2-1}\exp(-y/2)}{2^{(2j+n)/2}\,\Gamma\left(\frac{2j+n}{2}\right)}.$$

6.5 DISTRIBUTION OF $(\overline{X}, S^2)$ IN SAMPLING FROM A NORMAL POPULATION

Let $X_1, X_2, \ldots, X_n$ be a sample from $\mathcal{N}(\mu,\sigma^2)$, and write $\overline{X} = n^{-1}\sum_{i=1}^{n} X_i$ and $S^2 = (n-1)^{-1}\sum_{i=1}^{n}(X_i - \overline{X})^2$. In this section we show that $\overline{X}$ and S^2 are independent and derive the distribution of S^2. More precisely, we prove the following important result.

Theorem 1. Let $X_1, X_2, \ldots, X_n$ be iid $\mathcal{N}(\mu,\sigma^2)$ RVs. Then $\overline{X}$ and $(X_1 - \overline{X}, X_2 - \overline{X}, \ldots, X_n - \overline{X})$ are independent.

Proof. We compute the joint MGF of $\overline{X}$ and $X_1 - \overline{X}, X_2 - \overline{X}, \ldots, X_n - \overline{X}$ as follows:

$$\begin{aligned} M(t, t_1, t_2, \ldots, t_n) &= E\exp\{t\overline{X} + t_1(X_1 - \overline{X}) + t_2(X_2 - \overline{X}) + \cdots + t_n(X_n - \overline{X})\} \\ &= E\exp\left\{\sum_{i=1}^{n} t_i X_i - \left(\sum_{i=1}^{n} t_i - t\right)\overline{X}\right\} \end{aligned}$$

$$
\begin{aligned}
&= E\exp\left\{\sum_{i=1}^{n} X_i\left(t_i - \frac{t_1+t_2+\cdots+t_n-t}{n}\right)\right\}\\
&= E\left[\prod_{i=1}^{n}\exp\left\{\frac{X_i(nt_i-n\bar{t}+t)}{n}\right\}\right] \qquad \left(\text{where } \bar{t}=n^{-1}\sum_{i=1}^{n} t_i\right)\\
&= \prod_{i=1}^{n} E\exp\left\{\frac{X_i[t+n(t_i-\bar{t})]}{n}\right\}\\
&= \prod_{1}^{n}\exp\left\{\frac{\mu[t+n(t_i-\bar{t})]}{n}+\frac{\sigma^2}{2}\frac{1}{n^2}[t+n(t_i-\bar{t})]^2\right\}\\
&= \exp\left\{\frac{\mu}{n}[nt+n\sum_{i=1}^{n}(t_i-\bar{t})]+\frac{\sigma^2}{2n^2}\sum_{i=1}^{n}[t+n(t_i-\bar{t})]^2\right\}\\
&= \exp(\mu t)\exp\left\{\frac{\sigma^2}{2n^2}\left(nt^2+n^2\sum_{i=1}^{n}(t_i-\bar{t})^2\right)\right\}\\
&= \exp\left(\mu t+\frac{\sigma^2}{2n}t^2\right)\exp\left\{\frac{\sigma^2}{2}\sum_{i=1}^{n}(t_i-\bar{t})^2\right\}\\
&= M_{\overline{X}}(t)M_{X_1-\overline{X},\ldots,X_n-\overline{X}}(t_1,t_2,\ldots,t_n)\\
&= M(t,0,0,\ldots,0)M(0,t_1,t_2,\ldots,t_n).
\end{aligned}
$$

Corollary 1. $\overline{X}$ and S^2 are independent.

Corollary 2. $(n-1)S^2/\sigma^2$ is $\chi^2(n-1)$.

Since

$$\sum_{i=1}^{n}\frac{(X_i-\mu)^2}{\sigma^2}\sim\chi^2(n), \qquad n\left(\frac{\overline{X}-\mu}{\sigma}\right)^2\sim\chi^2(1),$$

and $\overline{X}$ and S^2 are independent, it follows from

$$\frac{\sum_1^n(X_i-\mu)^2}{\sigma^2}=n\left(\frac{\overline{X}-\mu}{\sigma}\right)^2+(n-1)\frac{S^2}{\sigma^2}$$

that

$$
\begin{aligned}
E\left\{\exp\left[t\sum_{1}^{n}\frac{(X_i-\mu)^2}{\sigma^2}\right]\right\} &= E\left\{\exp\left[tn\left(\frac{\overline{X}-\mu}{\sigma}\right)^2+(n-1)\frac{S^2}{\sigma^2}t\right]\right\}\\
&= E\exp\left[n\left(\frac{\overline{X}-\mu}{\sigma}\right)^2 t\right]E\exp\left[(n-1)\frac{S^2}{\sigma^2}t\right],
\end{aligned}
$$

that is,

$$(1-2t)^{-n/2} = (1-2t)^{-1/2} E\exp\left[(n-1)\frac{S^2}{\sigma^2}t\right], \qquad t < \frac{1}{2},$$

and it follows that

$$E\exp\left[(n-1)\frac{S^2}{\sigma^2}t\right] = (1-2t)^{-(n-1)/2}, \qquad t < \frac{1}{2}.$$

By the uniqueness of the MGF it follows that $(n-1)S^2/\sigma^2$ is $\chi^2(n-1)$.

Corollary 3. The distribution of $\sqrt{n}(\overline{X}-\mu)/S$ is $t(n-1)$.

Proof. Since $\sqrt{n}(\overline{X}-\mu)/\sigma$ is $\mathcal{N}(0,1)$, and $(n-1)S^2/\sigma^2 \sim \chi^2(n-1)$ and since $\overline{X}$ and S^2 are independent,

$$\frac{\sqrt{n}\,(\overline{X}-\mu)/\sigma}{\sqrt{[(n-1)S^2/\sigma^2]/(n-1)}} = \frac{\sqrt{n}\,(\overline{X}-\mu)}{S}$$

is $t(n-1)$.

Corollary 4. If $X_1, X_2, \ldots, X_m$ are iid $\mathcal{N}(\mu_1, \sigma_1^2)$ RVs, $Y_1, Y_2, \ldots, Y_n$ are iid $\mathcal{N}(\mu_2, \sigma_2^2)$ RVs, and the two samples are independently taken, $(S_1^2/\sigma_1^2)/(S_2^2/\sigma_2^2)$ is $F(m-1, n-1)$. If, in particular, $\sigma_1 = \sigma_2$, then S_1^2/S_2^2 is $F(m-1, n-1)$.

Corollary 5. Let $X_1, X_2, \ldots, X_m$ and $Y_1, Y_2, \ldots, Y_n$, respectively, be independent samples from $\mathcal{N}(\mu_1, \sigma_1^2)$ and $\mathcal{N}(\mu_2, \sigma_2^2)$. Then

$$\frac{\overline{X}-\overline{Y}-(\mu_1-\mu_2)}{\{[(m-1)S_1^2/\sigma_1^2]+[(n-1)S_2^2/\sigma_2^2]\}^{1/2}}\sqrt{\frac{m+n-2}{\sigma_1^2/m+\sigma_2^2/n}} \sim t(m+n-2).$$

In particular, if $\sigma_1 = \sigma_2$, then

$$\frac{\overline{X}-\overline{Y}-(\mu_1-\mu_2)}{\sqrt{[(m-1)S_1^2+(n-1)S_2^2]}}\sqrt{\frac{mn(m+n-2)}{m+n}} \sim t(m+n-2).$$

Corollary 5 follows since

$$\overline{X}-\overline{Y} \sim \mathcal{N}\left(\mu_1-\mu_2, \frac{\sigma_1^2}{m}+\frac{\sigma_2^2}{n}\right) \quad \text{and}$$

$$\frac{(m-1)S_1^2}{\sigma_1^2}+\frac{(n-1)S_2^2}{\sigma_2^2} \sim \chi^2(m+n-2)$$

and the two statistics are independent.

Remark 1. The converse of Corollary 1 also holds. See Theorem 5.3.28.

Remark 2. In sampling from a symmetric distribution, $\overline{X}$ and S^2 are uncorrelated. See Problem 4.5.14.

Remark 3. Alternatively, Corollary 1 could have been derived from Corollary 2 to Theorem 5.4.6 by using the Helmert orthogonal matrix:

$$\mathbf{A} = \begin{bmatrix} 1/\sqrt{n} & 1/\sqrt{n} & 1/\sqrt{n} & \cdots & 1/\sqrt{n} \\ -1/\sqrt{2} & 1/\sqrt{2} & 0 & \cdots & 0 \\ -1/\sqrt{6} & -1/\sqrt{6} & 2/\sqrt{6} & \cdots & 0 \\ \cdot & \cdot & \cdot & \cdots & 0 \\ \cdot & \cdot & \cdot & \cdots & \cdot \\ \cdot & \cdot & \cdot & \cdots & 0 \\ -1/\sqrt{n(n-1)} & -1/\sqrt{n(n-1)} & -1/\sqrt{n(n-1)} & \cdots & (n-1)/\sqrt{n(n-1)} \end{bmatrix}.$$

For the case of $n=3$ this was done in Example 4.4.6. In Problem 7 the reader is asked to work out the details in the general case.

Remark 4. An analytic approach to the development of the distribution of $\overline{X}$ and S^2 is as follows. Assuming without loss of generality that X_i is $\mathcal{N}(0,1)$, we have as the joint PDF of $(X_1, X_2, \ldots, X_n)$

$$\begin{aligned} f(x_1, x_2, \ldots, x_n) &= \frac{1}{(2\pi)^{n/2}} \exp\left\{-\frac{1}{2}\sum_{j=1}^{n} x_j^2\right\} \\ &= \frac{1}{(2\pi)^{n/2}} \exp\left\{-\frac{(n-1)s^2 + n\overline{x}^2}{2}\right\}. \end{aligned}$$

Changing the variables to $y_1, y_2, \ldots, y_n$ by using the transformation $y_k = (x_k - \overline{x})/s$, we see that

$$\sum_{k=1}^{n} y_k = 0 \quad \text{and} \quad \sum_{k=1}^{n} y_k^2 = n-1.$$

It follows that two of the y_k's, say y_{n-1} and y_n, are functions of the remaining y_k. Thus either

$$y_{n-1} = \frac{\alpha+\beta}{2}, \qquad y_n = \frac{\alpha-\beta}{2},$$

or

$$y_{n-1} = \frac{\alpha-\beta}{2}, \qquad y_n = \frac{\alpha+\beta}{2},$$

where

$$\alpha = -\sum_{k=1}^{n-2} y_k \quad \text{and} \quad \beta = \sqrt{2(n-1) - 2\sum_{k=1}^{n-2} y_k^2 - \left(\sum_{k=1}^{n-2} y_k\right)^2}.$$

We leave the reader to derive the joint PDF of $(Y_1, Y_2, \ldots, Y_{n-2}, \overline{X}, S^2)$, using the result described in Remark 2, and to show that the RVs $\overline{X}$, S^2, and $(Y_1, Y_2, \ldots, Y_{n-2})$ are independent.

PROBLEMS 6.5

1. Let $X_1, X_2, \ldots, X_n$ be a random sample from $\mathcal{N}(\mu, \sigma^2)$ and $\overline{X}$ and S^2, respectively, be the sample mean and the sample variance. Let $X_{n+1} \sim \mathcal{N}(\mu, \sigma^2)$, and assume that $X_1, X_2, \ldots, X_n, X_{n+1}$ are independent. Find the sampling distribution of $[(X_{n+1} - \overline{X})/S]$
$\sqrt{n/(n+1)}$.
2. Let $X_1, X_2, \ldots, X_m$ and $Y_1, Y_2, \ldots, Y_n$ be independent random samples from $\mathcal{N}(\mu_1, \sigma^2)$ and $\mathcal{N}(\mu_2, \sigma^2)$, respectively. Also, let α, β be two fixed real numbers. If $\overline{X}, \overline{Y}$ denote the corresponding sample means, what is the sampling distribution of
$$\frac{\alpha(\overline{X} - \mu_1) + \beta(\overline{Y} - \mu_2)}{\sqrt{\dfrac{(m-1)S_1^2 + (n-1)S_2^2}{m+n-2}}\sqrt{\dfrac{\alpha^2}{m} + \dfrac{\beta^2}{n}}},$$
where S_1^2 and S_2^2, respectively, denote the sample variances of the X's and the Y's?
3. Let $X_1, X_2, \ldots, X_n$ be a random sample from $\mathcal{N}(\mu, \sigma^2)$, and k be a positive integer. Find $E(S^{2k})$. In particular, find $E(S^2)$ and $\text{var}(S^2)$.
4. A random sample of 5 is taken from a normal population with mean 2.5 and variance $\sigma^2 = 36$.
   (a) Find the probability that the sample variance lies between 30 and 44.
   (b) Find the probability that the sample mean lies between 1.3 and 3.5, while the sample variance lies between 30 and 44.
5. The mean life of a sample of 10 light bulbs was observed to be 1327 hours with a standard deviation of 425 hours. A second sample of 6 bulbs chosen from a different batch showed a mean life of 1215 hours with a standard deviation of 375 hours. If the means of the two batches are assumed to be same, how probable is the observed difference between the two sample means?
6. Let S_1^2 and S_2^2 be the sample variances from two independent samples of sizes $n_1 = 5$ and $n_2 = 4$ from two populations having the same unknown variance σ^2. Find (approximately) the probability that $S_1^2/S_2^2 < 1/5.2$ or > 6.25.
7. Let $X_1, X_2, \ldots, X_n$ be a sample from $\mathcal{N}(\mu, \sigma^2)$. By using the Helmert orthogonal transformation defined in Remark 3, show that $\overline{X}$ and S^2 are independent.
8. Derive the joint PDF of $\overline{X}$ and S^2 by using the transformation described in Remark 4.

6.6 SAMPLING FROM A BIVARIATE NORMAL DISTRIBUTION

Let $(X_1,Y_1),(X_2,Y_2),\dots,(X_n,Y_n)$ be a sample from a bivariate normal population with parameters μ_1, μ_2, ρ, σ_1^2, σ_2^2. Let us write

$$\overline{X}=n^{-1}\sum_{i=1}^{n}X_i, \qquad \overline{Y}=n^{-1}\sum_{i=1}^{n}Y_i,$$

$$S_1^2=(n-1)^{-1}\sum_{i=1}^{n}(X_i-\overline{X})^2, \qquad S_2^2=(n-1)^{-1}\sum_{i=1}^{n}(Y_i-\overline{Y})^2,$$

and

$$S_{11}=(n-1)^{-1}\sum_{i=1}^{n}(X_i-\overline{X})(Y_i-\overline{Y}).$$

In this section we show that $(\overline{X},\overline{Y})$ is independent of (S_1^2,S_{11},S_2^2) and obtain the distribution of the sample correlation coefficient and regression coefficients (at least in the special case where $\rho=0$).

Theorem 1. The random vectors $(\overline{X},\overline{Y})$ and $(X_1-\overline{X},\ X_2-\overline{X},\dots,X_n-\overline{X},\ Y_1-\overline{Y},\ Y_2-\overline{Y},\dots,Y_n-\overline{Y})$ are independent. The joint distribution of $(\overline{X},\overline{Y})$ is bivariate normal with parameters μ_1, μ_2, ρ, σ_1^2/n, σ_2^2/n.

Proof. The proof follows along the lines of the proof of Theorem 1. The MGF of $(\overline{X},\overline{Y},\ X_1-\overline{X},\dots,X_n-\overline{X},Y_1-\overline{Y},\dots,Y_n-\overline{Y})$ is given by

$$\begin{aligned}M^* &= M(u,v,t_1,t_2,\dots,t_n,s_1,s_2,\dots,s_n)\\ &= E\exp\left\{u\overline{X}+v\overline{Y}+\sum_{i=1}^{n}t_i(X_i-\overline{X})+\sum_{i=1}^{n}s_i(Y_i-\overline{Y})\right\}\\ &= E\exp\left\{\sum_{i=1}^{n}X_i\left(\frac{u}{n}+t_i-\overline{t}\right)+\sum_{i=1}^{n}Y_i\left(\frac{v}{n}+s_i-\overline{s}\right)\right\},\end{aligned}$$

where $\overline{t}=n^{-1}\sum_{i=1}^{n}t_i$, $\overline{s}=n^{-1}\sum_{i=1}^{n}s_i$. Therefore,

$$\begin{aligned}M^* &= \prod_{i=1}^{n}E\exp\left\{\left(\frac{u}{n}+t_i-\overline{t}\right)X_i+\left(\frac{v}{n}+s_i-\overline{s}\right)Y_i\right\}\\ &= \prod_{i=1}^{n}\exp\left\{\left(\frac{u}{n}+t_i-\overline{t}\right)\mu_1+\left(\frac{v}{n}+s_i-\overline{s}\right)\mu_2\right.\\ &\quad+\frac{\sigma_1^2[(u/n)+t_i-\overline{t}]^2+2\rho\sigma_1\sigma_2[(u/n)+t_i-\overline{t}][(v/n)+s_i-\overline{s}]}{2}\\ &\quad\left.\frac{+\sigma_2^2[(v/n)+s_i-\overline{s}]^2}{2}\right\}\end{aligned}$$

$$= \exp\left(\mu_1 u + \mu_2 v + \frac{u^2\sigma_1^2 + 2\rho\sigma_1\sigma_2 uv + v^2\sigma_2^2}{2n}\right)$$
$$\cdot \exp\left\{\frac{1}{2}\sigma_1^2 \sum_{i=1}^{n} (t_i - \bar{t})^2 + \rho\sigma_1\sigma_2 \sum_{i=1}^{n} (t_i - \bar{t})(s_i - \bar{s})\right.$$
$$\left. + \frac{1}{2}\sigma_2^2 \sum_{i=1}^{2} (s_i - \bar{s})^2\right\}$$
$$= M_1(u,v) M_2(t_1, t_2, \ldots, t_n, s_1, s_2, \ldots, s_n)$$

for all real u, v, $t_1, t_2, \ldots, t_n$, $s_1, s_2, \ldots, s_n$ where M_1 is the MGF of $(\overline{X}, \overline{Y})$ and M_2 is the MGF of $(X_1 - \overline{X}, \ldots, X_n - \overline{X}, Y_1 - \overline{Y}, \ldots, Y_n - \overline{Y})$. Also, M_1 is the MGF of a bivariate normal distribution. This completes the proof.

Corollary. The sample mean vector $(\overline{X}, \overline{Y})$ is independent of the sample variance–covariance matrix $\begin{pmatrix} s_1^2 & s_{11} \\ s_{11} & s_2^2 \end{pmatrix}$ in sampling from a bivariate normal population.

Remark 1. The result of Theorem 1 can be generalized to the case of sampling from a k-variate normal population. We do not propose to do so here.

Remark 2. Unfortunately the method of proof of Theorem 1 does not lead to the distribution of the variance–covariance matrix. The distribution of $(\overline{X}, \overline{Y}, S_1^2, S_{11}, S_2^2)$ was found by Fisher [30] and Romanovsky [92]. The general case is due to Wishart [119], who determined the distribution of the sample variance–covariance matrix in sampling from a k-dimensional normal distribution. The distribution is named after him.

We will next compute the distribution of the sample correlation coefficient:

$$R = \frac{\sum_{i=1}^{n} (X_i - \overline{X})(Y_i - \overline{Y})}{\left\{\sum_{i=1}^{n} (X_i - \overline{X})^2 \sum_{i=1}^{n} (Y_i - \overline{Y})^2\right\}^{1/2}} = \frac{S_{11}}{S_1 S_2}. \tag{1}$$

It is convenient to introduce the so-called *sample regression coefficient* of Y on X

$$B_{Y|X} = \frac{\sum_{i=1}^{n} (X_i - \overline{X})(Y_i - \overline{Y})}{\sum_{i=1}^{n} (X_i - \overline{X})^2} = \frac{S_{11}}{S_1^2} = R\frac{S_2}{S_1}. \tag{2}$$

Since we will need only the distribution of R and $B_{Y|X}$ whenever $\rho = 0$, we will make this simplifying assumption in what follows. The general case is computationally quite complicated. We refer the reader to Cramér [17] for details.

We note that

$$R = \frac{\sum_{i=1}^{n} Y_i (X_i - \overline{X})}{(n-1) S_1 S_2} \tag{3}$$

and

$$B_{Y|X} = \frac{\sum_{i=1}^{n} Y_i(X_i - \overline{X})}{(n-1)S_1^2}. \tag{4}$$

Moreover,

$$R^2 = \frac{B_{Y|X}^2 S_1^2}{S_2^2}. \tag{5}$$

In the following we write $B = B_{Y|X}$.

Theorem 2. Let $(X_1, Y_1), \ldots, (X_n, Y_n)$, $n \geq 2$, be a sample from a bivariate normal population with parameters $EX = \mu_1$, $EY = \mu_2$, $\text{var}(X) = \sigma_1^2$, $\text{var}(Y) = \sigma_2^2$, and $\text{cov}(X, Y) = 0$. In other words, let $X_1, X_2, \ldots, X_n$ be iid $\mathcal{N}(\mu_1, \sigma_1^2)$ RVs, and $Y_1, Y_2, \ldots, Y_n$ be iid $\mathcal{N}(\mu_2, \sigma_2^2)$ RVs, and suppose that the X's and Y's are independent. Then the PDF of R is given by

$$f_1(r) = \begin{cases} \dfrac{\Gamma[(n-1)/2]}{\Gamma(\frac{1}{2})\Gamma[(n-2)/2]}(1-r^2)^{(n-4)/2}, & -1 \leq r \leq 1, \\ 0, & \text{otherwise}; \end{cases} \tag{6}$$

and the PDF of B is given by

$$h_1(b) = \frac{\Gamma(n/2)}{\Gamma\left(\dfrac{1}{2}\right)\Gamma[(n-1)/2]} \frac{\sigma_1 \sigma_2^{n-1}}{(\sigma_2^2 + \sigma_1^2 b^2)^{n/2}}, \qquad -\infty < b < \infty. \tag{7}$$

Proof. Without any loss of generality, we assume that $\mu_1 = \mu_2 = 0$ and $\sigma_1^2 = \sigma_2^2 = 1$, for we can always define

$$X_i^* = \frac{X_i - \mu_1}{\sigma_1} \quad \text{and} \quad Y_i^* = \frac{Y_i - \mu_2}{\sigma_2}. \tag{8}$$

Now note that the conditional distribution of Y_i, given $X_1, X_2, \ldots, X_n$, is $\mathcal{N}(0, 1)$, and Y_1, $Y_2, \ldots, Y_n$, given $X_1, X_2, \ldots, X_n$, are mutually independent. Let us define the following orthogonal transformation:

$$u_i = \sum_{j=1}^{n} c_{ij} y_j, \qquad i = 1, 2, \ldots, n, \tag{9}$$

where $(c_{ij})_{i,j=1,2,\ldots,n}$ is an orthogonal matrix with the first two rows

$$c_{1j} = \frac{1}{\sqrt{n}}, \quad j = 1, 2, \ldots, n, \tag{10}$$

$$c_{2j} = \frac{x_j - \overline{x}}{\left\{\sum_{i=1}^{n}(x_i - \overline{x})^2\right\}^{1/2}}, \quad j = 1, 2, \ldots, n. \tag{11}$$

It follows from orthogonality that for any $i \geq 2$

$$\sum_{j=1}^{n} c_{ij} = \sqrt{n} \sum_{j=1}^{n} c_{ij} \frac{1}{\sqrt{n}} = \sqrt{n} \sum_{j=1}^{n} c_{ij} c_{1j} = 0 \tag{12}$$

and

$$\begin{aligned}\sum_{i=1}^{n} u_i^2 &= \sum_{i=1}^{n} \left(\sum_{j=1}^{n} c_{ij} y_j \sum_{j'=1}^{n} c_{ij'} y_{j'} \right) \\ &= \sum_{j=1}^{n} \sum_{j'=1}^{n} \left(\sum_{i=1}^{n} c_{ij} c_{ij'} \right) y_j y_{j'} \\ &= \sum_{j=1}^{n} y_{j'}^2 .\end{aligned} \tag{13}$$

Moreover,

$$u_1 = \sqrt{n}\,\overline{y} \tag{14}$$

and

$$u_2 = b\sqrt{\sum (x_i - \overline{x})^2}, \tag{15}$$

where b is a value assumed by RV B. Also $U_1, U_2, \ldots, U_n$, given $X_1, X_2, \ldots, X_n$, are normal RVs (being linear combinations of the Y's). Thus

$$\begin{aligned}E\{U_i \mid X_1, X_2, \ldots, X_n\} &= \sum_{j=1}^{n} c_{ij} E\{Y_j \mid X_1, X_2, \ldots, X_n\} \\ &= 0\end{aligned} \tag{16}$$

and

$$\begin{aligned}\operatorname{cov}\{U_i, U_k \mid X_1, X_2, \ldots, X_n\} &= \operatorname{cov}\left\{ \sum_{j=1}^{n} c_{ij} Y_j, \sum_{p=1}^{n} c_{kp} Y_p \mid X_1, X_2, \ldots, X_n \right\} \\ &= \sum_{j=1}^{n} \sum_{p=1}^{n} c_{ij} c_{kp} \operatorname{cov}\{Y_j, Y_p \mid X_1, X_2, \ldots, X_n\} \\ &= \sum_{j=1}^{n} c_{ij} c_{kj}.\end{aligned}$$

This last equality follows since

$$\operatorname{cov}\{Y_j, Y_p \mid X_1, X_2, \ldots, X_n\} = \begin{cases} 0, & j \neq p, \\ 1, & j = p. \end{cases}$$

From orthogonality, we have

$$\operatorname{cov}\{U_i, U_k \mid X_1, X_2, \ldots, X_n\} = \begin{cases} 0, & i \neq k, \\ 1, & i = k; \end{cases} \tag{17}$$

and it follows that the RVs $U_1, U_2, \ldots, U_n$, given $X_1, X_2, \ldots, X_n$, are mutually independent $\mathcal{N}(0,1)$. Now

$$\begin{aligned} \sum_{j=1}^{n}(y_j - \bar{y}^2) &= \sum_{i=1}^{n} y_i^2 - n\bar{y}^2 \\ &= \sum_{j=1}^{n} u_j^2 - u_1^2 \\ &= \sum_{j=2}^{n} u_j^2. \end{aligned} \tag{18}$$

Thus

$$R^2 = \frac{U_2^2}{\sum_{i=2}^{n} U_i^2} = \frac{U_2^2}{U_2^2 + \sum_{i=3}^{n} U_i^2}. \tag{19}$$

Writing $U = U_2^2$ and $W = \sum_{i=3}^{n} U_i^2$, we see that the conditional distribution of U, given X_1, $X_2, \ldots, X_n$, is $\chi^2(1)$, and that of W, given $X_1, X_2, \ldots, X_n$, is $\chi^2(n-2)$. Moreover U and W are independent. Since these conditional distributions do not involve the X's, we see that U and W are unconditionally independent with $\chi^2(1)$ and $\chi^2(n-2)$ distributions, respectively. The joint PDF of U and W is

$$f(u,w) = \frac{1}{\Gamma(\frac{1}{2})\sqrt{2}} u^{1/2-1} e^{-u/2} \frac{1}{\Gamma[(n-2)/2] 2^{(n-2)/2}} w^{(n-2)/2-1} e^{-w/2}.$$

Let $u + w = z$, then $u = r^2 z$ and $w = z(1-r^2)$. The Jacobian of this transformation is z, so that the joint PDF of R^2 and Z is given by

$$f^*(r^2, z) = \frac{1}{\Gamma(\frac{1}{2})\Gamma[(n-2)/2] 2^{(n-1)/2}} z^{n/2-3/2} e^{-z/2} (r^2)^{-1/2} (1-r^2)^{n/2-2}.$$

The marginal PDF of R^2 is easily computed as

$$f_1^*(r^2) = \frac{\Gamma[(n-1)/2]}{\Gamma(\frac{1}{2})\Gamma[(n-2)/2]} (r^2)^{-1/2} (1-r^2)^{n/2-2}, \qquad 0 \leq r^2 \leq 1. \tag{20}$$

Finally, using Theorem 2.5.4, we get the PDF of R as

$$f_1(r) = \frac{\Gamma[(n-1)/2]}{\Gamma(\frac{1}{2})\Gamma[(n-2)/2]} (1-r^2)^{n/2-2}, \qquad -1 \leq r \leq 1.$$

As for the distribution of B, note that the conditional PDF of $U_2 = \sqrt{n-1}\,BS_1$, given X_1, $X_2, \ldots, X_n$, is $\mathcal{N}(0,1)$, so that the conditional PDF of B, given $X_1, X_2, \ldots, X_n$, is $\mathcal{N}(0, 1/\sum(x_i - \bar{x})^2)$. Let us write $\Lambda = (n-1)S_1^2$. Then the PDF of RV Λ is that of a $\chi^2(n-1)$ RV. Thus the joint PDF of B and Λ is given by

$$h(b,\lambda) = g(b \mid \lambda) h_2(\lambda), \tag{21}$$

where $g(b \mid \lambda)$ is $\mathcal{N}(0, 1/\lambda)$, and $h_2(\lambda)$ is $\chi^2(n-1)$. We have

$$\begin{aligned} h_1(b) &= \int_0^\infty h(b,\lambda)\,d\lambda \\ &= \frac{1}{2^{n/2}\Gamma(\frac{1}{2})\Gamma[(n-1)/2]} \int_0^\infty \lambda^{n/2-1} e^{-\lambda/2(1+b^2)}\,d\lambda \\ &= \frac{\Gamma(n/2)}{\Gamma(\frac{1}{2})\Gamma[(n-1)/2]} \frac{1}{(1+b^2)^{n/2}}, \qquad -\infty < b < \infty. \end{aligned} \tag{22}$$

To complete the proof let us write

$$X_i = \mu_1 + X_i^* \sigma_1 \quad \text{and} \quad Y_i = \mu_2 + Y_i^* \sigma_2,$$

where $X_i^* \sim \mathcal{N}(0,1)$ and $Y_i^* \sim \mathcal{N}(0,1)$. Then $X_i \sim \mathcal{N}(\mu_1, \sigma_1^2)$, $Y_i \sim \mathcal{N}(\mu_2, \sigma_2^2)$, and

$$\begin{aligned} R &= \frac{\sum_{i=1}^n (X_i - \overline{X})(Y_i - \overline{Y})}{\sqrt{\sum_{i=1}^n (X_i - \overline{X})^2 \sum_{i=1}^n (Y_i - \overline{Y})^2}} \\ &= R^*, \end{aligned} \tag{23}$$

so that the PDF of R is the same as derived above. Also

$$\begin{aligned} B &= \frac{\sigma_1\sigma_2 \sum_{i=1}^n (X_i^* - \overline{X}^*)(Y_i^* - \overline{Y}^*)}{\sigma_1^2 \sum_{i=1}^n (X_i^* - \overline{X}^*)^2} \\ &= \frac{\sigma_2}{\sigma_1} B^*, \end{aligned} \tag{24}$$

where the PDF of B^* is given by (22). Relations (22) and (24) are used to find the PDF of B. We leave the reader to carry out these simple details.

Remark 3. In view of (23), namely the invariance of R under translation and (positive) scale changes, we note that for fixed n the sampling distribution of R, under $\rho = 0$, does not depend on μ_1, μ_2, σ_1, and σ_2. In the general case when $\rho \neq 0$, one can show that for fixed n the distribution of R depends only on ρ but not on μ_1, μ_2, σ_1, and σ_2 (see, for example, Cramér [17], p. 398).

Remark 4. Let us change the variable to

$$T = \frac{R}{\sqrt{1-R^2}} \sqrt{n-2}. \tag{25}$$

Then

$$1 - R^2 = \left(1 + \frac{T^2}{n-2}\right)^{-1},$$

and the PDF of T is given by

$$p(t) = \frac{1}{\sqrt{n-2}} \frac{1}{B[(n-2)/2, \frac{1}{2}]} \frac{1}{[1 + t^2/(n-2)]^{(n-1)/2}}, \tag{26}$$

which is the PDF of a t-statistic with $n-2$ d.f. Thus T defined by (25) has a $t(n-2)$ distribution, provided that $\rho = 0$. This result facilitates the computation of probabilities under the PDF of R when $\rho = 0$.

Remark 5. To compute the PDF of $B_{X|Y} = R(S_1/S_2)$, the so-called *sample regression coefficient* of X on Y, all we need to do is to interchange σ_1 and σ_2 in (7).

Remark 6. From (7) we can compute the mean and variance of B. For $n > 2$, clearly

$$EB = 0,$$

and for $n > 3$, we can show that

$$EB^2 = \text{var}(B) = \frac{\sigma_2^2}{\sigma_1^2} \frac{1}{n-3}.$$

Similarly, we can use (6) to compute the mean and variance of R. We have, for $n > 4$, under $\rho = 0$,

$$ER = 0$$

and

$$ER^2 = \text{var}(R) = \frac{1}{n-1}.$$

PROBLEMS 6.6

1. Let $(X_1, Y_1), (X_2, Y_2), \ldots, (X_n, Y_n)$ be a random sample from a bivariate normal population with $EX = \mu_1$, $EY = \mu_2$, $\text{var}(X) = \text{var}(Y) = \sigma^2$, and $\text{cov}(X, Y) = \rho\sigma^2$. Let $\overline{X}, \overline{Y}$ denote the corresponding sample means, S_1^2, S_2^2, the corresponding sample variances, and S_{11}, the sample covariance. Write $R = 2S_{11}/(S_1^2 + S_2^2)$. Show that the PDF of R is given by

$$f(r) = \frac{\Gamma\left(\frac{n}{2}\right)}{\sqrt{\pi}\,\Gamma\left(\frac{n-1}{2}\right)} (1-\rho^2)^{(n-1)/2} (1-\rho r)^{-(n-1)} (1-r^2)^{(n-3)/2}, \quad |r| < 1.$$

(Rastogi [89])

[*Hint:* Let $U = (X+Y)/2$ and $V = (X-Y)/2$, and observe that the random vector (U, V) is also bivariate normal. In fact, U and V are independent.]

2. Let X and Y be independent normal RVs. A sample of $n = 11$ observations on (X, Y) produces sample correlation coefficient $r = 0.40$. Find the probability of obtaining a value of R that exceeds the observed value.

3. Let X_1, X_2 be jointly normally distributed with zero means, unit variances, and correlation coefficient ρ. Let S be a $\chi^2(n)$ RV that is independent of (X_1, X_2). Then the joint distribution of $Y_1 = X_1/\sqrt{S/n}$ and $Y_2 = X_2/\sqrt{S/n}$ is known as a *central bivariate t-distribution*. Find the joint PDF of (Y_1, Y_2) and the marginal PDFs of Y_1 and Y_2, respectively.

4. Let $(X_1, Y_1), \ldots, (X_n, Y_n)$ be a sample from a bivariate normal distribution with parameters $EX_1 = \mu_1$, $EY_i = \mu_2$, $\text{var}(X_i) = \text{var}(Y_i) = \sigma^2$, and $\text{cov}(X_i, Y_i) = \rho\sigma^2$, $i = 1, 2, \ldots, n$. Find the distribution of the statistic

$$T(\mathbf{X}, \mathbf{Y}) = \sqrt{n}\frac{(\overline{X} - \mu_1) - (\overline{Y} - \mu_2)}{\sqrt{\sum_{i=1}^{n}(X_i - Y_i - \overline{X} + \overline{Y})^2}}.$$

7

BASIC ASYMPTOTICS: LARGE SAMPLE THEORY

7.1 INTRODUCTION

In Chapter 6 we described some methods of finding exact distributions of sample statistics and their moments. While these methods are used in some cases such as sampling from a normal population when the sample statistic of interest is $\overline{X}$ or S^2, often either the statistics of interest, say $T_n = T(X_1, \ldots, X_n)$, is either too complicated or its exact distribution is not simple to work with. In such cases we are interested in the convergence properties of T_n. We want to know what happens when the sample size is large. What is the limiting distribution of T_n? When the exact distribution of T_n (and its moments) is unknown or too complicated we will often use their asymptotic approximations when n is large.

In this chapter, we discuss some basic elements of statistical asymptotics. In Section 7.2 we discuss various modes of convergence of a sequence of random variables. In Sections 7.3 and 7.4 the laws of large numbers are discussed. Section 7.5 deals with limiting moment generating functions and in Section 7.6 we discuss one of the most fundamental theorem of classical statistics called the central limit theorem. In Section 7.7 we consider some statistical applications of these methods.

The reader may find some parts of this chapter a bit difficult on first reading. Such a discussion has been indicated with a†.

7.2 MODES OF CONVERGENCE

In this section we consider several modes of convergence and investigate their interrelationships. We begin with the weakest mode of convergence.

Definition 1. Let $\{F_n\}$ be a sequence of distribution functions. If there exists a DF F such that, as $n \to \infty$,

$$F_n(x) \to F(x) \tag{1}$$

at every point x at which F is continuous, we say that F_n *converges in law (or, weakly)*, to F, and we write $F_n \xrightarrow{w} F$.

If $\{X_n\}$ is a sequence of RVs and $\{F_n\}$ is the corresponding sequence of DFs, we say that X_n *converges in distribution (or law)* to X if there exists an RV X with DF F such that $F_n \xrightarrow{w} F$. We write $X_n \xrightarrow{L} X$.

It must be remembered that it is quite possible for a given sequence DFs to converge to a function that is not a DF.

Example 1. Consider the sequence of DFs

$$F_n(x) = \begin{cases} 0, & x < n, \\ 1, & x \geq n. \end{cases}$$

Here $F_n(x)$ is the DF of the RV X_n degenerate at $x = n$. We see that $F_n(x)$ converges to a function F that is identically equal to 0, and hence it is not a DF.

Example 2. Let $X_1, X_2, \ldots, X_n$ be iid RVs with common density function

$$f(x) = \begin{cases} \dfrac{1}{\theta} & 0 < x < \theta,\ (0 < \theta < \infty), \\ 0 & \text{otherwise}. \end{cases}$$

Let $X_{(n)} = \max(X_1, X_2, \ldots, X_n)$. Then the density function of $X_{(n)}$ is

$$f_n(x) = \begin{cases} \dfrac{nx^{n-1}}{\theta^n} & 0 < x < \theta, \\ 0 & \text{otherwise}, \end{cases}$$

and the DF of $X_{(n)}$ is

$$F_n(x) = \begin{cases} 0 & x < 0, \\ (x/\theta)^n & 0 \leq x < \theta, \\ 1, & x \geq \theta. \end{cases}$$

We see that, as $n \to \infty$,

$$F_n(x) \to F(x) = \begin{cases} 0 & x < \theta, \\ 1, & x \geq \theta, \end{cases}$$

which is a DF. Thus $F_n \xrightarrow{w} F$.

Example 3. Let F_n be a sequence of DFs defined by

$$F_n(x) = \begin{cases} 0, & x < 0, \\ 1 - \dfrac{1}{n}, & 0 \le x < n, \\ 1, & n \le x. \end{cases}$$

Clearly $F_n \xrightarrow{w} F$, where F is the DF given by

$$F(x) = \begin{cases} 0, & x < 0, \\ 1, & x \ge 0. \end{cases}$$

Note that F_n is the DF of the RV X_n with PMF

$$P\{X_n = 0\} = 1 - \frac{1}{n}, \qquad P\{X_n = n\} = \frac{1}{n},$$

and F is the DF of the RV X degenerate at 0. We have

$$EX_n^k = n^k \left(\frac{1}{n}\right) = n^{k-1},$$

where k is a positive integer. Also $EX^k = 0$. So that

$$EX_n^k \nrightarrow EX^k \qquad \text{for any } k \ge 1.$$

We next give an example to show that weak convergence of distribution functions does not imply the convergence of corresponding PMF's or PDF's.

Example 4. Let $\{X_n\}$ be a sequence of RVs with PMF

$$f_n(x) = P\{X_n = x\} = \begin{cases} 1 & \text{if } x = 2 + 1/n, \\ 0 & \text{otherwise.} \end{cases}$$

Note that none of the f_n's assigns any probability to the point $x = 2$. It follows that

$$f_n(x) \to f(x) \qquad \text{as} \quad n \to \infty,$$

where $f(x) = 0$ for all x. However, the sequence of DFs $\{F_n\}$ of RVs X_n converges to the function

$$F(x) = \begin{cases} 0, & x < 2, \\ 1 & x \ge 2, \end{cases}$$

at all continuity points of F. Since F is the DF of the RV degenerate at $x=2$, $F_n \xrightarrow{w} F$.

The following result is easy to prove.

Theorem 1. Let X_n be a sequence of integer-valued RVs. Also, let $f_n(k) = P\{X_n = k\}$, $k = 0,1,2,\ldots$, be the PMF of X_n, $n = 1,2,\ldots$, and $f(k) = P\{x = k\}$ be the PMF of X. Then

$$f_n(x) \to f(x) \qquad \text{for all } x \Leftrightarrow X_n \xrightarrow{L} X.$$

In the continuous case we state the following result of Scheffé [100] without proof.

Theorem 2. Let X_n, $n = 1,2,\ldots$, and X be continuous RVs such that

$$f_n(x) \to f(x) \qquad \text{for (almost) all } x \text{ as } n \to \infty.$$

Here, f_n and f are the PDFs of X_n and X, respectively. Then $X_n \xrightarrow{L} X$.

The following result is easy to establish.

Theorem 3. Let $\{X_n\}$ be a sequence of RVs such that $X_n \xrightarrow{L} X$, and let c be a constant. Then

(a) $X_n + c \xrightarrow{L} X + c$,
(b) $cX_n \xrightarrow{L} cX$, $c \neq 0$.

A slightly stronger concept of convergence is defined by *convergence in probability*.

Definition 2. Let $\{X_n\}$ be a sequence of RVs defined on some probability space $(\Omega, \mathcal{S}, P)$. We say that the sequence $\{X_n\}$ *converges in probability* to the RV X if, for every $\varepsilon > 0$.

$$P\{|X_n - X| > \varepsilon\} \to 0 \qquad \text{as } n \to \infty. \tag{2}$$

We write $X_n \xrightarrow{P} X$.

Remark 1. We emphasize that the definition says nothing about the convergence of the RVs X_n to the RV X in the sense in which it is understood in real analysis. Thus $X_n \xrightarrow{P} X$ does not imply that, given $\varepsilon > 0$, we can find an N such that $|X_n - X| < \varepsilon$ for $n \geq N$. Definition 2 speaks only of the convergence of the sequence of probabilities $P\{|X_n - X| > \varepsilon\}$ to 0.

Example 5. Let $\{X_n\}$ be a sequence of RVs with PMF

$$P\{X_n = 1\} = \frac{1}{n}, \qquad P\{X_n = 0\} = 1 - \frac{1}{n}.$$

Then

$$P\{|X_n| > \varepsilon\} = \begin{cases} P\{X_n = 1\} = \dfrac{1}{n} & \text{if } 0 < \varepsilon < 1, \\ 0 & \text{if } \varepsilon \geq 1. \end{cases}$$

It follows that $P\{|X_n| > \varepsilon\} \to 0$ as $n \to \infty$, and we conclude that $X_n \xrightarrow{P} 0$.

The following statements can be verified.

1. $X_n \xrightarrow{P} X \Leftrightarrow X_n - X \xrightarrow{P} 0$.
2. $X_n \xrightarrow{P} X$, $X_n \xrightarrow{P} Y \Rightarrow P\{X = Y\} = 1$, for $P\{|X - Y| > c\} \leq P\{|X_n - X| > \frac{c}{2}\} + P\{|X_n - Y| > \frac{c}{2}\}$, and it follows that $P\{|X - Y| > c\} = 0$ for every $c > 0$.
3. $X_n \xrightarrow{P} X \Rightarrow X_n - X_m \xrightarrow{P} 0$ as $n, m \to \infty$, for

$$P\{|X_n - X_m| > \varepsilon\} \leq P\left\{|X_n - X| > \frac{\varepsilon}{2}\right\} + P\left\{|X_m - X| > \frac{\varepsilon}{2}\right\}.$$

4. $X_n \xrightarrow{P} X$, $Y_n \xrightarrow{P} Y \Rightarrow X_n \pm Y_n \xrightarrow{P} X \pm Y$.
5. $X_n \xrightarrow{P} X$, k constant, $\Rightarrow kX_n \xrightarrow{P} kX$.
6. $X_n \xrightarrow{P} k \Rightarrow X_n^2 \xrightarrow{P} k^2$.
7. $X_n \xrightarrow{P} a$, $Y_n \xrightarrow{P} b$, a, b constants $\Rightarrow X_n Y_n \xrightarrow{P} ab$, for

$$X_n Y_n = \frac{(X_n + Y_n)^2 - (X_n - Y_n)^2}{4} \xrightarrow{P} \frac{(a+b)^2 - (a-b)^2}{4} = ab.$$

8. $X_n \xrightarrow{P} 1 \Rightarrow X_n^{-1} \xrightarrow{P} 1$, for

$$\begin{aligned} P\left\{\left|\frac{1}{X_n} - 1\right| \geq \varepsilon\right\} &= P\left\{\frac{1}{X_n} \geq 1 + \varepsilon\right\} + P\left\{\frac{1}{X_n} \leq 1 - \varepsilon\right\} \\ &= P\left\{\frac{1}{X_n} \geq 1 + \varepsilon\right\} + P\left\{\frac{1}{X_n} \leq 0\right\} \\ &\quad + P\left\{0 < \frac{1}{X_n} \leq 1 - \varepsilon\right\} \end{aligned}$$

   and each of the three terms on the right goes to 0 as $n \to \infty$.
9. $X_n \xrightarrow{P} a$, $Y_n \xrightarrow{P} b$, a, b constants, $b \neq 0 \Rightarrow X_n Y_n^{-1} \xrightarrow{P} ab^{-1}$.
10. $X_n \xrightarrow{P} X$, and Y an RV $\Rightarrow X_n Y \xrightarrow{P} XY$.

    Note that Y is an RV so that, given $\delta > 0$, there exists a $k > 0$ such that $P\{|Y| > k\} < \delta/2$. Thus

$$\begin{aligned} P\{|X_n Y - XY| > \varepsilon\} &= P\{|X_n - X||Y| > \varepsilon, |Y| > k\} \\ &\quad + P\{|X_n - X||Y| > \varepsilon, |Y| \leq k\} \\ &< \frac{\delta}{2} + P\left\{|X_n - X| > \frac{\varepsilon}{k}\right\}. \end{aligned}$$

11. $X_n \xrightarrow{P} X$, $Y_n \xrightarrow{P} Y \Rightarrow X_nY_n \xrightarrow{P} XY$, for

$$(X_n - X)(Y_n - Y) \xrightarrow{P} 0.$$

The result now follows on multiplication, using result 10. It also follows that $X_n \xrightarrow{P} X \Rightarrow X_n^2 \xrightarrow{P} X^2$.

Theorem 4. Let $X_n \xrightarrow{P} X$ and g be a continuous function defined on $\mathcal{R}$. Then $g(X_n) \xrightarrow{P} g(X)$ as $n \to \infty$.

Proof. Since X is an RV, we can, given $\varepsilon > 0$, find a constant $k = k(\varepsilon)$ such that

$$P\{|X| > k\} < \frac{\varepsilon}{2}.$$

Also, g is continuous on $\mathcal{R}$, so that g is uniformly continuous on $[-k,k]$. It follows that there exists a $\delta = \delta(\varepsilon,k)$ such that

$$|g(x_n) - g(x)| < \varepsilon$$

whenever $|x| \le k$ and $|x_n - x| < \delta$. Let

$$A = \{|X| \le k\}, \qquad B = \{|X_n - X| < \delta\}, \qquad C = \{|g(X_n) - g(X)| < \varepsilon\}.$$

Then $\omega \in A \cap B \Rightarrow \omega \in C$, so that

$$A \cap B \subseteq C.$$

It follows that

$$P\{C^c\} \le P\{A^c\} + P\{B^c\},$$

that is,

$$P\{|g(X_n) - g(X)| \ge \varepsilon\} \le P\{|X_n - X| \ge \delta\} + P\{|X| > k\} < \varepsilon$$

for $n \ge N(\varepsilon,\delta,k)$, where $N(\varepsilon,\delta,k)$ is chosen so that

$$P\{|X_n - X| \ge \delta\} < \frac{\varepsilon}{2} \qquad \text{for} \quad n \ge N(\varepsilon,\delta,k).$$

Corollary. $X_n \xrightarrow{P} c$, where c is a constant $\Rightarrow g(X_n) \xrightarrow{P} g(c)$, g being a continuous function.

We remark that a more general result than Theorem 4 is true and state it without proof (see Rao [88, p. 124]): $X_n \xrightarrow{L} X$ and g continuous on $\mathcal{R} \Rightarrow g(X_n) \xrightarrow{L} g(X)$.

The following two theorems explain the relationship between weak convergence and convergence in probability.

Theorem 5. $X_n \xrightarrow{P} X \Rightarrow X_n \xrightarrow{L} X$.

Proof. Let F_n and F, respectively, be the DFs of X_n and X. We have

$$\{\omega\colon X(\omega) \le x'\} = \{\omega\colon X_n(\omega) \le x, X(\omega) \le x'\} \cup \{\omega\colon X_n(\omega) > x, X(\omega) \le x'\} \subseteq \{X_n \le x\} \cup \{X_n > x, X \le x'\}.$$

It follows that

$$F(x') \le F_n(x) + P\{X_n > x, X \le x'\}.$$

Since $X_n - X \xrightarrow{P} 0$, we have for $x' < x$

$$P\{X_n > x, X \le x'\} \le P\{|X_n - X| > x - x'\} \to 0 \qquad \text{as} \quad n \to \infty.$$

Therefore

$$F(x') \le \varliminf_{n\to\infty} F_n(x), \qquad x' < x.$$

Similarly, by interchanging X and X_n, and x and x', we get

$$\varlimsup_{n\to\infty} F_n(x) \le F(x''), \qquad x < x''.$$

Thus, for $x' < x < x''$, we have

$$F(x') \le \varliminf F_n(x) \le \varlimsup F_n(x) \le F(x'').$$

Since F has only a countable number of discontinuity points, we choose x to be a point of continuity of F, and letting $x'' \downarrow x$ and $x' \uparrow x$, we have

$$F(x) = \lim_{n\to\infty} F_n(x)$$

at all points of continuity of F.

Theorem 6. Let k be a constant. Then

$$X_n \xrightarrow{L} k \Rightarrow X_n \xrightarrow{P} k.$$

Proof. The proof is left as an exercise.

Corollary. Let k be a constant. Then

$$X_n \xrightarrow{L} k \Leftrightarrow X_n \xrightarrow{P} k.$$

Remark 2. We emphasize that we cannot improve the above result by replacing k by an RV, that is, $X_n \xrightarrow{L} X$ in general does not imply $X_n \xrightarrow{P} X$, for let $X, X_1, X_2 \ldots$ be identically distributed RVs, and let the joint distribution of (X_n, X) be as follows:

$X \setminus X_n$	0	1	
0	0	$\frac{1}{2}$	$\frac{1}{2}$
1	$\frac{1}{2}$	0	$\frac{1}{2}$
	$\frac{1}{2}$	$\frac{1}{2}$	1

Clearly, $X_n \xrightarrow{L} X$. But

$$\begin{aligned} P\left\{|X_n - X| > \frac{1}{2}\right\} &= P\{|X_n - X| = 1\} \\ &= P\{X_n = 0, X = 1\} + P\{X_n = 1, X = 0\} \\ &= 1 \nrightarrow 0. \end{aligned}$$

Hence, $X_n \stackrel{P}{\nrightarrow} X$, but $X_n \xrightarrow{L} X$.

Remark 3. Example 3 shows that $X_n \xrightarrow{P} X$ does not imply $EX_n^k \to EX^k$ for any $k > 0$, k integral.

Definition 3. Let $\{X_n\}$ be a sequence of RVs such that $E|X_n|^r < \infty$, for some $r > 0$. We say that X_n converges in the rth *mean* to an RV X if $E|X|^r < \infty$ and

$$E|X_n - X|^r \to 0 \qquad \text{as} \quad n \to \infty, \tag{3}$$

and we write $X_n \xrightarrow{r} X$.

Example 6. Let $\{X_n\}$ be a sequence of RVs defined by

$$P\{X_n = 0\} = 1 - \frac{1}{n}, \qquad P\{X_n = 1\} = \frac{1}{n}, \qquad n = 1, 2, \ldots.$$

Then

$$E|X_n|^2 = \frac{1}{n} \to 0 \qquad \text{as} \quad n \to \infty,$$

and we see that $X_n \xrightarrow{2} X$, where RV X is degenerate at 0.

Theorem 7. Let $X_n \xrightarrow{r} X$ for some $r > 0$. Then $X_n \xrightarrow{P} X$.

Proof. The proof is left as an exercise.

Example 7. Let $\{X_n\}$ be a sequence of RVs defined by

$$P\{X_n = 0\} = 1 - \frac{1}{n^r}, \qquad P\{X_n = n\} = \frac{1}{n^r}, \qquad r > 0, \quad n = 1, 2, \ldots.$$

Then $E|X_n|^r = 1$, so that $X_n \stackrel{r}{\nrightarrow} 0$. We show that $X_n \stackrel{P}{\to} 0$.

$$P\{|X_n| > \varepsilon\} = \begin{cases} P\{X_n = n\} & \text{if} \quad \varepsilon < n \\ 0 & \text{if} \quad \varepsilon > n \end{cases} \to 0 \text{ as } n \to \infty.$$

Theorem 8. Let $\{X_n\}$ be a sequence of RVs such that $X_n \stackrel{2}{\to} X$. Then $EX_n \to EX$ and $EX_n^2 \to EX^2$ as $n \to \infty$.

Proof. We have

$$|E(X_n - X)| \le E|X_n - X| \le E^{1/2}|X_n - X|^2 \to 0 \qquad \text{as } n \to \infty.$$

To see that $EX_n^2 \to EX^2$ (see also Theorem 9), we write

$$EX_n^2 = E(X_n - X)^2 + EX^2 + 2E\{X(X_n - X)\}$$

and note that

$$|E\{X(X_n - X)\}| \le \sqrt{EX^2 E(X_n - X)^2}$$

by the Cauchy–Schwarz inequality. The result follows on passing to the limits.

We get, in addition, that $X_n \stackrel{2}{\to} X$ implies $\text{var}(X_n) \to \text{var}(X)$.

Corollary. Let $\{X_m\}$, $\{Y_n\}$ be two sequences of RVs such that $X_m \stackrel{2}{\to} X$, $Y_n \stackrel{2}{\to} Y$. Then $E(X_m Y_n) \to E(XY)$ as $m, n \to \infty$.

Proof. The proof is left to the reader.

As a simple consequence of Theorem 8 and its corollary we see that $X_m \stackrel{2}{\to} X$, $Y_n \stackrel{2}{\to} Y$ together imply $\text{cov}(X_m, Y_n) \to \text{cov}(X, Y)$.

Theorem 9. If $X_n \stackrel{r}{\to} X$, then $E|X_n|^r \to E|X|^r$.

Proof. Let $0 < r \le 1$. Then

$$E|X_n|^r = E|X_n - X + X|^r$$

so that

$$E|X_n|^r - E|X|^r \le E|X_n - X|^r.$$

Interchanging X_n and X, we get

$$E|X|^r - E|X_n|^r \le E|X_n - X|^r.$$

It follows that

$$|E|X|^r - E|X_n|^r| \le E|X_n - X|^r \to 0 \qquad \text{as} \quad n \to \infty.$$

For $r > 1$, we use Minkowski's inequality and obtain

$$[E|X_n|^r]^{1/r} \le [E|X_n - X|^r]^{1/r} + [E|X|^r]^{1/r}$$

and

$$[E|X|^r]^{1/r} \le [E|X_n - X|^r]^{1/r} + [E|X_n|^r]^{1/r}.$$

It follows that

$$|E^{1/r}|X_n|^r - E^{1/r}|X|^r| \le E^{1/r}|X_n - X|^r \to 0 \qquad \text{as} \quad n \to \infty.$$

This completes the proof.

Theorem 10. Let $r > s$. Then $X_n \xrightarrow{r} X \Rightarrow X_n \xrightarrow{s} X$.

Proof. From Theorem 3.4.3 it follows that for $s < r$

$$E|X_n - X|^s \le [E|X_n - X|^r]^{s/r} \to 0 \qquad \text{as } n \to \infty$$

since $X_n \xrightarrow{r} X$.

Remark 4. Clearly the converse to Theorem 10 cannot hold, since $E|X|^s < \infty$ for $s < r$ does not imply $E|X|^r < \infty$.

Remark 5. In view of Theorem 9, it follows that $X_n \xrightarrow{r} X \Rightarrow E|X_n|^s \to E|X|^s$ for $s \le r$.

Definition 4. † Let $\{X_n\}$ be a sequence of RVs. We say that X_n *converges almost surely* (a.s.) to an RV X if and only if

$$P\{\omega\colon X_n(\omega) \to X(\omega) \quad \text{as } n \to \infty\} = 1, \tag{4}$$

and we write $X_n \xrightarrow{\text{a.s.}} X$ or $X_n \to X$ with probability 1.

The following result elucidates Definition 4.

Theorem 11. $X_n \xrightarrow{\text{a.s.}} X$ if and only if $\lim_{n\to\infty} P\{\sup_{m\ge n} |X_m - X| > \varepsilon\} = 0$ for all $\varepsilon > 0$.

Proof. Since $X_n \xrightarrow{\text{a.s.}} X$, $X_n - X \xrightarrow{\text{a.s.}} 0$, and it will be sufficient to show the equivalence of

† May be omitted on the first reading.

(a) $X_n \xrightarrow{\text{a.s.}} 0$ and
(b) $\lim_{n\to\infty} P\{\sup_{m\geq n}|X_m| > \varepsilon\} = 0$.

Let us suppose that (a) holds. Let $\varepsilon > 0$, and write

$$A_n(\varepsilon) = \left\{\sup_{m\geq n}|X_m| > \varepsilon\right\} \qquad \text{and} \quad C = \left\{\lim_{n\to\infty} X_n = 0\right\}.$$

Also write $B_n(\varepsilon) = C \cap A_n(\varepsilon)$, and note that $B_{n+1}(\varepsilon) \subset B_n(\varepsilon)$, and the limit set $\cap_{n=1}^{\infty} B_n(\varepsilon) = \phi$. It follows that

$$\lim_{n\to\infty} PB_n(\varepsilon) = P\left\{\bigcap_{n=1}^{\infty} B_n(\varepsilon)\right\} = 0.$$

Since $PC = 1$, $PC^c = 0$, we have

$$\begin{aligned} PB_n(\varepsilon) - P(A_n \cap C) &= 1 - P(C^c \cup A_n^c) \\ &= 1 - PC^c - PA_n^c + P(C^c \cap A_n^c) \\ &= PA_n + P(C^c \cap A_n^c) \\ &= PA_n. \end{aligned}$$

It follows that (b) holds.

Conversely, let $\lim_{n\to\infty} PA_n(\varepsilon) = 0$, and write

$$D(\varepsilon) = \left\{\overline{\lim_{n\to\infty}}\, |X_n| > \varepsilon > 0\right\}.$$

Since $D(\varepsilon) \subset A_n(\varepsilon)$ for $n = 1, 2, \ldots$, it follows that $PD(\varepsilon) = 0$. Also,

$$C^c = \left\{\lim_{n\to\infty} X_n \neq 0\right\} \subset \bigcup_{k=1}^{\infty}\left\{\overline{\lim}\,|X_n| > \frac{1}{k}\right\},$$

so that

$$1 - PC \leq \sum_{k=1}^{\infty} PD\left(\frac{1}{k}\right) = 0,$$

and (a) holds.

Remark 6. Thus $X_n \xrightarrow{\text{a.s.}} 0$ means that, for $\varepsilon > 0$, $\eta > 0$ arbitrary, we can find an n_0 such that

$$P\left\{\sup_{n\geq n_0}|X_n| > \varepsilon\right\} < \eta. \tag{5}$$

Indeed, we can write, equivalently, that

$$\lim_{n_0\to\infty} P\left[\bigcup_{n\geq n_0}\{|X_n| > \varepsilon\}\right] = 0. \tag{6}$$

Theorem 12. $X_n \xrightarrow{\text{a.s.}} X \Rightarrow X_n \xrightarrow{P} X$.

Proof. By Remark 6, $X_n \xrightarrow{\text{a.s.}} X$ implies that, for arbitrary $\varepsilon > 0$, $\eta > 0$, we can choose an $n_0 = n_0(\varepsilon, \eta)$ such that

$$P\left[\bigcap_{n=n_0}^{\infty} \{|X_n - X| \le \varepsilon\}\right] \ge 1 - \eta.$$

Clearly,

$$\bigcap_{n=n_0}^{\infty} \{|X_n - X| \le \varepsilon\} \subset \{|X_n - X| \le \varepsilon\} \qquad \text{for} \quad n \ge n_0.$$

It follows that for $n \ge n_0$

$$P\{|X_n - X| \le \varepsilon\} \ge P\left[\bigcap_{n=n_0}^{\infty} \{|X_n - X| \le \varepsilon\}\right] \ge 1 - \eta,$$

that is

$$P\{|X_n - X| > \varepsilon\} < \eta \qquad \text{for} \quad n \ge n_0,$$

which is the same as saying $X_n \xrightarrow{P} X$.

That the converse of Theorem 12 does not hold is shown in the following example.

Example 8. For each positive integer n there exist integers m and k (uniquely determined) such that

$$n = 2^k + m, \qquad 0 \le m < 2^k, \quad k = 0, 1, 2, \ldots.$$

Thus, for $n = 1$, $k = 0$ and $m = 0$; for $n = 5$, $k = 2$ and $m = 1$; and so on. Define RVs X_n, for $n = 1, 2, \ldots$, on $\Omega = [0, 1]$ by

$$X_n(\omega) = \begin{cases} 2^k, & \dfrac{m}{2^k} \le \omega < \dfrac{m+1}{2^k}, \\ 0, & \text{otherwise.} \end{cases}$$

Let the probability distribution of X_n be given by $P\{I\}$ = length of the interval $I \subseteq \Omega$. Thus

$$P\{X_n = 2^k\} = \frac{1}{2^k}, \quad P\{X_n = 0\} = 1 - \frac{1}{2^k}.$$

The limit $\lim_{n\to\infty} X_n(\omega)$ does not exist for any $\omega \in \Omega$, so that X_n does not converge almost surely. But

$$P\{X_n| > \varepsilon\} = P\{X_n > \varepsilon\} = \begin{cases} 0 & \text{if} \quad \varepsilon \ge 2^k, \\ \dfrac{1}{2^k} & \text{if} \quad 0 < \varepsilon < 2^k, \end{cases}$$

and we see that

$$P\{|X_n| > \varepsilon\} \to 0 \qquad \text{as } n \text{ (and hence } k) \to \infty.$$

Theorem 13. Let $\{X_n\}$ be a strictly decreasing sequence of positive RVs, and suppose that $X_n \xrightarrow{P} 0$. Then $X_n \xrightarrow{\text{a.s.}} 0$.

Proof. The proof is left as an exercise.

Example 9. Let $\{X_n\}$ be a sequence of independent RVs defined by

$$P\{X_n = 0\} = 1 - \frac{1}{n}, \quad P\{X_n = 1\} = \frac{1}{n}, \qquad n = 1, 2, \ldots.$$

Then

$$E|X_n - 0|^2 = E|X_n|^2 = \frac{1}{n} \to 0 \qquad \text{as } n \to \infty,$$

so that $X_n \xrightarrow{2} 0$. Also

$$\begin{aligned} P\{X_n = 0 \quad &\text{for every } m \le n \le n_0\} \\ &= \prod_{n=m}^{n_0} \left(1 - \frac{1}{n}\right) = \frac{m-1}{n_0}, \end{aligned}$$

which diverges to 0 as $n_0 \to \infty$ for all values of m. Thus X_n does not converge to 0 with probability 1.

Example 10. Let $\{X_n\}$ be independent defined by

$$P\{X_n = 0\} = 1 - \frac{1}{n^r}, \quad P\{X_n = n\} = \frac{1}{n^r}, \quad r \ge 2, \quad n = 1, 2, \ldots.$$

Then

$$P\{X_n = 0 \quad \text{for } m \le n \le n_0\} = \prod_{n=m}^{n_0} \left(1 - \frac{1}{n^r}\right).$$

As $n_0 \to \infty$, the infinite product converges to some nonzero quantity, which itself converges to 1 as $m \to \infty$. Thus $X_n \xrightarrow{\text{a.s.}} 0$. However, $E|X_n|^r = 1$ and $X_n \stackrel{r}{\nrightarrow} 0$ as $n \to \infty$.

Example 11. Let $\{X_n\}$ be a sequence of RVs with $P\{X_n = \pm 1/n\} = \frac{1}{2}$. Then $E|X_n|^r = 1/n^r \to 0$ as $n \to \infty$ and $X_n \xrightarrow{r} 0$. For $j < k$, $|X_j| > |X_k|$, so that $\{|X_k| > \varepsilon\} \subset \{|X_j| > \varepsilon\}$. It follows that

$$\bigcup_{j=n}^{\infty} \{|X_j| > \varepsilon\} = \{|X_n| > \varepsilon\}.$$

Choosing $n > 1/\varepsilon$, we see that

$$P\left[\bigcup_{j=n}^{\infty}\{|X_j| > \varepsilon\}\right] = P\{|X_n| > \varepsilon\} \leq P\left\{|X_n| > \frac{1}{n}\right\} = 0,$$

and (6) implies that $X_n \xrightarrow{\text{a.s.}} 0$.

Remark 7. In Theorem 7.4.3 we prove a result which is sometimes useful in proving a.s. convergence of a sequence of RVs.

Theorem 14. Let $\{X_n, Y_n\}$, $n = 1, 2, \ldots$, be a sequence of RVs. Then

$$|X_n - Y_n| \xrightarrow{P} 0 \quad \text{and} \quad Y_n \xrightarrow{L} Y \Rightarrow X_n \xrightarrow{L} Y.$$

Proof. Let x be a point of continuity of the DF of Y and $\varepsilon > 0$. Then

$$\begin{aligned} P\{X_n \leq x\} &= P\{Y_n \leq x + Y_n - X_n\} \\ &= P\{Y_n \leq x + Y_n - X_n; Y_n - X_n \leq \varepsilon\} \\ &\quad + P\{Y_n \leq x + Y_n - X_n; Y_n - X_n > \varepsilon\} \\ &\leq P\{Y_n \leq x + \varepsilon\} + P\{Y_n - X_n > \varepsilon\}. \end{aligned}$$

It follows that

$$\overline{\lim_{n\to\infty}} P\{X_n \leq x\} \leq \underline{\lim_{n\to\infty}} P\{Y_n \leq x + \varepsilon\}.$$

Similarly

$$\underline{\lim_{n\to\infty}} P\{X_n \leq x\} \geq \overline{\lim_{n\to\infty}} P\{Y_n \leq x - \varepsilon\}.$$

Since $\varepsilon > 0$ is arbitrary and x is a continuity point of $P\{Y \leq x\}$, we get the result by letting $\varepsilon \to 0$.

Corollary. $X_n \xrightarrow{P} X \Rightarrow X_n \xrightarrow{L} X$.

Theorem 15. (Slutsky's Theorem). Let $\{X_n, Y_n\}$, $n = 1, 2, \ldots$, be a sequence of pairs of RVs, and let c be a constant. Then

(a) $X_n \xrightarrow{L} X$, $Y_n \xrightarrow{P} c \Rightarrow X_n + Y_n \xrightarrow{L} X + c$;
(b) $X_n \xrightarrow{L} X$,

$$Y_n \xrightarrow{P} c \Rightarrow \begin{cases} X_nY_n \xrightarrow{L} cX & \text{if } c \neq 0, \\ X_nY_n \xrightarrow{P} 0 & \text{if } c = 0; \end{cases}$$

(c) $X_n \xrightarrow{L} X,\ Y_n \xrightarrow{P} c \Rightarrow \dfrac{X_n}{Y_n} \xrightarrow{L} X/c$ if $c \neq 0$

Proof. (a) $X_n \xrightarrow{L} X \Rightarrow X_n + c \xrightarrow{L} X + c$ (Theorem 3). Also, $Y_n - c = (Y_n + X_n) - (X_n + c) \xrightarrow{P} 0$.

A simple use of Theorem 14 shows that

$$X_n + Y_n \xrightarrow{L} X + c.$$

(b) We first consider the case where $c = 0$. We have, for any fixed number $k > 0$,

$$\begin{aligned} P\{|X_nY_n| > \varepsilon\} &= P\left\{|X_nY_n| > \varepsilon, |Y_n| \leq \frac{\varepsilon}{k}\right\} + P\left\{|X_nY_n| > \varepsilon, |Y_n| > \frac{\varepsilon}{k}\right\} \\ &\leq P\{|X_n| > k\} + P\left\{|Y_n| > \frac{\varepsilon}{k}\right\}. \end{aligned}$$

Since $Y_n \xrightarrow{P} 0$ and $X_n \xrightarrow{L} X$, it follows that, for any fixed $k > 0$,

$$\overline{\lim_{n\to\infty}}\, P\{|X_nY_n| > \varepsilon\} \leq P\{|X| > k\}.$$

Since k is arbitrary, we can make $P\{|X| > k\}$ as small as we please by choosing k large. It follows that

$$X_nY_n \xrightarrow{P} 0.$$

Now, let $c \neq 0$. Then

$$X_nY_n - cX_n = X_n(Y_n - c)$$

and, since $X_n \xrightarrow{L} X$, $Y_n \xrightarrow{P} c$, $X_n(Y_n - c) \xrightarrow{P} 0$. Using Theorem 14, we get the result that

$$X_nY_n \xrightarrow{L} cX.$$

(c) $Y_n \xrightarrow{P} c$, and $c \neq 0 \Rightarrow Y_n^{-1} \xrightarrow{P} c^{-1}$. It follows that $X_n \xrightarrow{L} X$, $Y_n \xrightarrow{P} c \Rightarrow X_nY_n^{-1} \xrightarrow{L} c^{-1}X$, and the proof of the theorem is complete.

As an application of Theorem 15 we present the following example.

Example 12. Let $X_1, X_2, \ldots,$ be iid RVs with common law $\mathcal{N}(0,1)$. We shall determine the limiting distribution of the RV

$$W_n = \sqrt{n}\frac{X_1 + X_2 + \cdots + X_n}{X_1^2 + X_2^2 + \cdots + X_n^2}.$$

Let us write

$$U_n = \frac{1}{\sqrt{n}}(X_1 + X_2 + \cdots + X_n) \quad \text{and } V_n = \frac{X_1^2 + X_2^2 + \cdots + X_n^2}{n}.$$

Then

$$W_n = \frac{U_n}{V_n}.$$

For the MGF of U_n we have

$$M_{U_n}(t) = \prod_{i=1}^{n} E e^{tX_i/\sqrt{n}} = \prod_{i=1}^{n} e^{t^2/2n}$$
$$= e^{t^2/2},$$

so that U_n is an $\mathcal{N}(0,1)$ variate (see also Corollary 2 to Theorem 5.3.22). It follows that $U_n \xrightarrow{L} Z$, where Z is an $\mathcal{N}(0,1)$ RV. As for V_n, we note that each X_i^2 is a chi-square variate with 1 d.f. Thus

$$M_{V_n}(t) = \prod_{i=1}^{n} \left(\frac{1}{1 - 2t/n}\right)^{1/2}, \qquad t < \frac{n}{2},$$
$$= \left(1 - \frac{2t}{n}\right)^{-n/2}, \qquad t < \frac{n}{2},$$

which is the MGF of a gamma variate with parameters $\alpha = n/2$ and $\beta = 2/n$. Thus the density function of V_n is given by

$$f_{V_n}(x) = \begin{cases} \dfrac{1}{\Gamma(n/2)} \dfrac{1}{(2/n)^{n/2}} x^{n/2-1} e^{-nx/2}, & 0 < x < \infty, \\ 0, & \text{otherwise.} \end{cases}$$

We will show that $V_n \xrightarrow{P} 1$. We have, for any $\varepsilon > 0$,

$$P\{|V_n - 1| > \varepsilon\} \le \frac{\text{var}(V_n)}{\varepsilon^2} = \left(\frac{n}{2}\right)\left(\frac{2}{n}\right)^2 \frac{1}{\varepsilon^2} \to 0 \quad \text{as} \quad n \to \infty.$$

We have thus shown that

$$U_n \xrightarrow{L} Z \quad \text{and} \quad V_n \xrightarrow{P} 1.$$

It follows by Theorem 15 (c) that $W_n = U_n/V_n \xrightarrow{L} Z$, where Z is an $\mathcal{N}(0,1)$ RV.

Later on we will see that the condition that the X_i's be $\mathcal{N}(0,1)$ is not needed. All we need is that $E|X_i|^2 < \infty$.

PROBLEMS 7.2

1. Let $X_1, X_2, \ldots$ be a sequence of RVs with corresponding DFs given by $F_n(x) = 0$ if $x < -n$, $= (x+n)/2n$ if $-n \le x < n$, and $= 1$ if $x \ge n$. Does F_n converge to a DF?
2. Let $X_1, X_2 \ldots$ be iid $\mathcal{N}(0,1)$ RVs. Consider the sequence of RVs $\{\overline{X}_n\}$, where $\overline{X}_n = n^{-1}\sum_{i=1}^n X_i$. Let F_n be the DF of $\overline{X}_n$, $n = 1, 2, \ldots$. Find $\lim_{n\to\infty} F_n(x)$. Is this limit a DF?
3. Let $X_1, X_2, \ldots$ be iid $U(0,\theta)$ RVs. Let $X_{(1)} = \min(X_1, X_2, \cdots, X_n)$, and consider the sequence $Y_n = nX_{(1)}$. Does Y_n converge in distribution to some RV Y? If so, find the DF of RV Y.
4. Let $X_1, X_2, \ldots$ be iid RVs with common absolutely continuous DF F. Let $X_{(n)} = \max(X_1, X_2, \ldots, X_n)$, and consider the sequence of RVs $Y_n = n[1 - F(X_{(n)})]$. Find the limiting DF of Y_n.
5. Let $X_1, X_2, \ldots$ be a sequence of iid RVs with common PDF $f(x) = e^{-x+\theta}$ if $x \ge \theta$, and $= 0$ if $x < \theta$. Write $\overline{X}_n = n^{-1}\sum_{i=1}^n X_i$.
   (a) Show that $\overline{X}_n \xrightarrow{P} 1+\theta$.
   (b) Show that $\min\{X_1, X_2, \cdots, X_n\} \xrightarrow{P} \theta$.
6. Let $X_1, X_2, \ldots$ be iid $U[0,\theta]$ RVs. Show that $\max\{X_1, X_2, \ldots, X_n\} \xrightarrow{P} \theta$.
7. Let $\{X_n\}$ be a sequence of RVs such that $X_n \xrightarrow{L} X$. Let a_n be a sequence of positive constants such that $a_n \to \infty$ as $n \to \infty$. Show that $a_n^{-1}X_n \xrightarrow{P} 0$.
8. Let $\{X_n\}$ be a sequence of RVs such that $P\{|X_n| \le k\} = 1$ for all n and some constant $k > 0$. Suppose that $X_n \xrightarrow{P} X$. Show that $X_n \xrightarrow{r} X$ for any $r > 0$.
9. Let $X_1, X_2, \ldots, X_{2n}$ be iid $\mathcal{N}(0,1)$ RVs. Define

$$U_n = \left\{\frac{X_1}{X_2} + \frac{X_3}{X_4} + \cdots + \frac{X_{2n-1}}{X_{2n}}\right\}, \quad V_n = X_1^2 + X_2^2 + \cdots + X_n^2, \quad \text{and}$$

$$Z_n = \frac{U_n}{V_n}.$$

   Find the limiting distribution of Z_n.
10. Let $\{X_n\}$ be a sequence of geometric RVs with parameter λ/n, $n > \lambda > 0$. Also, let $Z_n = X_n/n$. Show that $Z_n \xrightarrow{L} G(1, 1/\lambda)$ as $n \to \infty$ (Prochaska [82]).
11. Let X_n be a sequence of RVs such that $X_n \xrightarrow{\text{a.s.}} 0$, and let c_n be a sequence of real numbers such that $c_n \to 0$ as $n \to \infty$. Show that $X_n + c_n \xrightarrow{\text{a.s.}} 0$.
12. Does convergence almost surely imply convergence of moments?
13. Let $X_1, X_2, \ldots$ be a sequence of iid RVs with common DF F, and write $X_{(n)} = \max\{X_1, X_2, \ldots, X_n\}$, $n = 1, 2, \ldots$.
   (a) For $\alpha > 0$, $\lim_{x\to\infty} x^{\alpha}P\{X_1 > x\} = b > 0$. Find the limiting distribution of $(bn)^{-1/\alpha}X_{(n)}$. Also, find the PDF corresponding to the limiting DF and compute its moments.

(b) If F satisfies

$$\lim_{x\to\infty} e^x[1-F(x)] = b > 0,$$

find the limiting DF of $X_{(n)} - \log(bn)$ and compute the corresponding PDF and the MGF.

(c) If X_i is bounded above by x_0 with probability 1, and for some $\alpha > 0$

$$\lim_{x\to x_0-} (x_0 - x)^{-\alpha}[1-F(x)] = b > 0,$$

find the limiting distribution of $(bn)^{1/\alpha}\{X_{(n)} - x_0\}$, the corresponding PDF, and the moments of the limiting distribution.

(*The above remarkable result, due to Gnedenko* [36], *exhausts all limiting distributions of* $X_{(n)}$ *with suitable norming and centering.*)

14. Let $\{F_n\}$ be a sequence of DFs that converges weakly to a DF F which is continuous everywhere. Show that $F_n(x)$ converges to $F(x)$ uniformly.

15. Prove Theorem 1.

16. Prove Theorem 6.

17. Prove Theorem 13.

18. Prove Corollary 1 to Theorem 8.

19. Let V be the class of all random variables defined on a probability space with finite expectations, and for $X \in V$ define

$$\rho(X) = E\left\{\frac{|X|}{1+|X|}\right\}.$$

Show the following:

(a) $\rho(X+Y) \le \rho(X) + \rho(Y)$; $\rho(\sigma X) \le \max(|\sigma|, 1)\rho(X)$.

(b) $d(X,Y) = \rho(X-Y)$ is a distance function on V (assuming that we identify RVs that are a.s. equal).

(c) $\lim_{n\to\infty} d(X_n, X) = 0 \Leftrightarrow X_n \xrightarrow{P} X$.

20. For the following sequences of RVs $\{X_n\}$, investigate convergence in probability and convergence in rth mean.

(a) $X_n \sim \mathcal{C}(1/n, 0)$.

(b) $P(X_n = e^n) = \frac{1}{n^2}$, $P(X_n = 0) = 1 - \frac{1}{n^2}$.

7.3 WEAK LAW OF LARGE NUMBERS

Let $\{X_n\}$ be a sequence of RVs. Write $S_n = \sum_{k=1}^n X_k$, $n = 1, 2, \ldots$. In this section we answer the following question in the affirmative: Do there exist sequences of constants A_n and $B_n > 0$, $B_n \to \infty$ as $n \to \infty$, such that the sequence of RVs $B_n^{-1}(S_n - A_n)$ converges in probability to 0 as $n \to \infty$?

Definition 1. Let $\{X_n\}$ be a sequence of RVs, and let $S_n = \sum_{k=1}^{n} X_k$, $n = 1, 2, \ldots$. We say that $\{X_n\}$ obeys the *weak law of large numbers* (WLLN) with respect to the sequence of constants $\{B_n\}$, $B_n > 0$, $B_n \uparrow \infty$, if there exists a sequence of real constants A_n such that $B_n^{-1}(S_n - A_n) \xrightarrow{P} 0$ as $n \to \infty$. A_n are called *centering constants* and B_n *norming constants*.

Theorem 1. Let $\{X_n\}$ be a sequence of pairwise uncorrelated RVs with $EX_i = \mu_i$ and $\text{var}(X_i) = \sigma_i^2$, $i = 1, 2, \ldots$. If $\sum_{i=1}^{n} \sigma_i^2 \to \infty$ as $n \to \infty$, we can choose $A_n = \sum_{k=1}^{n} \mu_k$ and $B_n = \sum_{i=1}^{n} \sigma_i^2$, that is,

$$\sum_{i=1}^{n} \frac{X_i - \mu_i}{\sum_{i=1}^{n} \sigma_i^2} \xrightarrow{P} 0 \qquad \text{as } n \to \infty.$$

Proof. We have, by Chebychev's inequality,

$$P\left\{|S_n - \sum_{k=1}^{n} \mu_k| > \varepsilon \sum_{i=1}^{n} \sigma_i^2\right\} \leq \frac{E\{\sum_{i=1}^{n}(X_i - \mu_i)\}^2}{\varepsilon^2(\sum_{i=1}^{n} \sigma_i^2)^2}$$
$$= \frac{1}{\varepsilon^2 \sum_{i=1}^{n} \sigma_i^2} \to 0 \qquad \text{as } n \to \infty.$$

Corollary 1. If the X_n's are identically distributed and pairwise uncorrelated with $EX_i = \mu$ and $\text{var}(X_i) = \sigma^2 < \infty$, we can choose $A_n = n\mu$ and $B_n = n\sigma^2$.

Corollary 2. In Theorem 1 we can choose $B_n = n$, provided that $n^{-2} \sum_{i=1}^{n} \sigma_i^2 \to 0$ as $n \to \infty$.

Corollary 3. In Corollary 1, we can take $A_n = n\mu$ and $B_n = n$, since $n\sigma^2/n^2 \to 0$ as $n \to \infty$. Thus, if $\{X_n\}$ are pairwise-uncorrelated identically distributed RVs with finite variance, $S_n/n \xrightarrow{P} \mu$.

Example 1. Let $X_1, X_2, \ldots$ be iid RVs with common law $b(1, p)$. Then $EX_i = p$, $\text{var}(X_i) = p(1-p)$, and we have

$$\frac{S_n}{n} \xrightarrow{P} p \qquad \text{as } n \to \infty.$$

Note that S_n/n is the proportion of successes in n trials. In particular, recall from Section 6.3 that $n\, F_n^*(x)$ is a $b(x, F(x))$ RV. It follows that for each $x \in \mathcal{R}$,

$$F_n^*(x) \xrightarrow{P} F(x) \qquad \text{as } n \to \infty.$$

Hereafter, we shall be interested mainly in the case where $B_n = n$. When we say that $\{X_n\}$ obeys the WLLN, this is so with respect to the sequence $\{n\}$.

Theorem 2. Let $\{X_n\}$ be any sequence of RVs. Write $Y_n = n^{-1} \sum_{k=1}^{n} X_k$. A necessary and sufficient condition for the sequence $\{X_n\}$ to satisfy the weak law of large numbers is that

$$E\left\{\frac{Y_n^2}{1 + Y_n^2}\right\} \to 0 \qquad \text{as } n \to \infty. \tag{1}$$

Proof. For any two positive numbers a, b, $a \geq b > 0$, we have

$$\left(\frac{a}{1+a}\right)\left(\frac{1+b}{b}\right) \geq 1. \tag{2}$$

Let $A = \{|Y_n| \geq \varepsilon\}$. Then $\omega \in A \Rightarrow |Y_n|^2 \geq \varepsilon^2 > 0$. Using (2), we see that $\omega \in A$ implies

$$\frac{Y_n^2}{1+Y_n^2}\frac{1+\varepsilon^2}{\varepsilon^2} \geq 1.$$

It follows that

$$\begin{aligned} PA &\leq P\left\{\frac{Y_n^2}{1+Y_n^2} \geq \frac{\varepsilon^2}{1+\varepsilon^2}\right\} \\ &\leq E\frac{|Y_n^2/(1+Y_n^2)|}{\varepsilon^2/(1+\varepsilon^2)} \qquad \text{by Markov's inequality} \\ &\to 0 \qquad \text{as } n \to \infty. \end{aligned}$$

That is,

$$Y_n \xrightarrow{P} 0 \qquad \text{as } n \to \infty.$$

Conversely, we will show that for every $\varepsilon > 0$

$$P\{|Y_n| \geq \varepsilon\} \geq E\left\{\frac{Y_n^2}{1+Y_n^2}\right\} - \varepsilon^2. \tag{3}$$

We will prove (3) for the case in which Y_n is of the continuous type. The discrete case being similar, we ask the reader to complete the proof. If Y_n has PDF $f_n(y)$, then

$$\begin{aligned} \int_{-\infty}^{\infty} \frac{y^2}{1+y^2} f_n(y)\,dy &= \left(\int_{|y|>\varepsilon} + \int_{|y|\leq\varepsilon}\right) \frac{y^2}{1+y^2} f_n(y)\,dy \\ &\leq P\{|Y_n| > \varepsilon\} + \int_{-\varepsilon}^{\varepsilon}\left(1 - \frac{1}{1+y^2}\right) f_n(y)\,dy \\ &\leq P\{|Y_n| > \varepsilon\} + \frac{\varepsilon^2}{1+\varepsilon^2} \leq P\{|Y_n| > \varepsilon\} + \varepsilon^2, \end{aligned}$$

which is (3).

Remark 1. Since condition (1) applies not to the individual variables but to their sum, Theorem 2 is of limited use. We note, however, that all weak laws of large numbers obtained as corollaries to Theorem 1 follow easily from Theorem 2 (Problem 6).

Example 2. Let $(X_1, X_2, \ldots, X_n)$ be jointly normal with $EX_i = 0$, $EX_i^2 = 1$ for all i, and $\text{cov}(X_i, X_j) = \rho$ if $|j-i| = 1$, and $= 0$ otherwise. Then $S_n = \sum_{k=1}^n X_k$ is $\mathcal{N}(0, \sigma^2)$, where

$$\sigma^2 = \text{var}(S_n) = n + 2(n-1)\rho,$$

and

$$\begin{aligned}
E\left\{\frac{Y_n^2}{1+Y_n^2}\right\} &= E\left\{\frac{S_n^2}{n^2+S_n^2}\right\} \\
&= \frac{2}{\sigma\sqrt{2\pi}}\int_0^\infty \frac{x^2}{n^2+x^2} e^{-x^2/2\sigma^2}\,dx \\
&= \frac{2}{\sqrt{2\pi}}\int_0^\infty \frac{y^2[n+2(n-1)\rho]}{n^2+y^2[n+2(n-1)\rho]} e^{-y^2/2}\,dy \\
&\le \frac{n+2(n-1)\rho}{n^2}\int_0^\infty \frac{2}{\sqrt{2\pi}} y^2 e^{-y^2/2}\,dy \to 0 \qquad \text{as } n\to\infty.
\end{aligned}$$

It follows from Theorem 2 that $n^{-1}S_n \xrightarrow{P} 0$. We invite the reader to compare this result to that of Problem 7.5.6.

Example 3. Let $X_1, X_2, \ldots$ be iid $\mathcal{C}(1,0)$ RVs. We have seen (corollary to Theorem 5.3.18) that $n^{-1}S_n \sim \mathcal{C}(1,0)$, so that $n^{-1}S_n$ does not converge in probability to 0. It follows that the WLLN does not hold (see also Problem 10).

Let $X_1, X_2, \ldots$ be an arbitrary sequence of RVs, and let $S_n = \sum_{k=1}^n X_k$, $n = 1, 2, \ldots$. Let us truncate each X_i at $c > 0$, that is, let

$$X_i^c = \begin{cases} X_i & \text{if } |X_i| \le c \\ 0 & \text{if } |X_i| > c \end{cases}, \qquad i = 1, 2, \ldots, n.$$

Write

$$S_n^c = \sum_{i=1}^n X_i^c, \text{ and } m_n = \sum_{i=1}^n EX_i^c.$$

Lemma 1. For any $\varepsilon > 0$,

$$P\{|S_n - m_n| > \varepsilon\} \le P\{|S_n^c - m_n| > \varepsilon\} + \sum_{k=1}^n P\{|X_k| > c\}. \tag{4}$$

Proof. We have

$$\begin{aligned}
P\{|S_n - m_n| > \varepsilon\} = {} & P\{|S_n - m_n| > \varepsilon \text{ and } |X_k| \le c \quad \text{for } k = 1, 2, \ldots, n\} \\
& + P\{|S_n - m_n| > \varepsilon \text{ and } |X_k| > c \quad \text{for at least one } k, \\
& \qquad k = 1, 2, \ldots, n\}
\end{aligned}$$

$$\leq P\{|S_n^c - m_n| > \varepsilon\} + P\{|X_k| > c \quad \text{for at least one } k, \; 1 \leq k \leq n\}$$

$$\leq P\{|S_n^c - m_n| > \varepsilon\} + \sum_{k=1}^{n} P\{|X_k| > c\}.$$

Corollary. If $X_1, X_2, \ldots, X_n$ are exchangeable, then

$$P\{|S_n - m_n| > \varepsilon\} \leq P\{|S_n^c - m_n| > \varepsilon\} + nP\{|X_1| > c\}. \tag{5}$$

If, in addition, the RVs $X_1, X_2, \ldots, X_n$ are independent, then

$$P\{|S_n - m_n| > \varepsilon\} \leq \frac{nE(X_1^c)^2}{\varepsilon^2} + nP\{|X_1| > c\}. \tag{6}$$

Inequality (6) yields the following important theorem.

Theorem 3. Let $\{X_n\}$ be a sequence of iid RVs with common finite mean $\mu = EX_1$. Then

$$n^{-1}S_n \xrightarrow{P} \mu \qquad \text{as } n \to \infty.$$

Proof. Let us take $c = n$ in (6) and replace ε by $n\varepsilon$; then we have

$$P\{|S_n - m_n| > n\varepsilon\} \leq \frac{1}{n\varepsilon^2} E(X_1^n)^2 + nP\{|X_1| > n\},$$

where X_1^n is X_1 truncated at n.

First note that $E|X_1| < \infty \Rightarrow nP\{|X_1| > n\} \to 0$ as $n \to \infty$. Now (see remarks following Lemma 3.2.1)

$$\begin{aligned} E(X_1^n)^2 &= 2\int_0^n xP\{|X_1| > x\}\,dx \\ &= 2\left(\int_0^A + \int_A^n\right) xP\{|X_1| > x\}\,dx, \end{aligned}$$

where A is chosen sufficiently large that

$$xP\{|X_1| > x\} < \frac{\delta}{2} \qquad \text{for all } x \geq A, \delta > 0 \text{ arbitrary.}$$

Thus

$$E(X_1^n)^2 \leq c + \delta \int_A^n dx \leq c + n\delta,$$

where c is a constant. It follows that

$$\frac{1}{n\varepsilon^2} E(X_1^n)^2 \leq \frac{c}{n\varepsilon^2} + \frac{\delta}{\varepsilon^2},$$

and since δ is arbitrary, $(1/n\varepsilon^2)E(X_1^n)^2$ can be made arbitrarily small for sufficiently large n. The proof is now completed by the simple observation that, since $EX_j = \mu$,

$$\frac{m_n}{n} \to \mu \qquad \text{as } n \to \infty.$$

We emphasize that in Theorem 3 we require only that $E|X_1| < \infty$; nothing is said about the variance. Theorem 3 is due to Khintchine.

Example 4. Let $X_1, X_2, \ldots$ be iid RVs with $E|X_1|^k < \infty$ for some positive integer k. Then

$$\sum_{j=1}^{n} \frac{X_j^k}{n} \xrightarrow{P} EX_1^k \qquad \text{as } n \to \infty.$$

Thus, if $EX_1^2 < \infty$, then $\sum_1^n X_j^2/n \xrightarrow{P} EX_1^2$, and since $(\sum_{j=1}^n X_j/n)^2 \xrightarrow{P} (EX_1)^2$ it follows that

$$\frac{\Sigma X_j^2}{n} - \left(\frac{\Sigma X_j}{n}\right)^2 \xrightarrow{P} \operatorname{var}(X_1).$$

Example 5. Let $X_1, X_2, \ldots$ be iid RVs with common PDF

$$f(x) = \begin{cases} \dfrac{1+\delta}{x^{2+\delta}}, & x \geq 1 \\ 0, & x < 1 \end{cases}, \quad \delta > 0.$$

Then

$$\begin{aligned} E|X| &= (1+\delta)\int_1^\infty \frac{1}{x^{1+\delta}}\,dx \\ &= \frac{1+\delta}{\delta} < \infty, \end{aligned}$$

and the law of large numbers holds, that is,

$$n^{-1}S_n \xrightarrow{P} \frac{1+\delta}{\delta} \qquad \text{as } n \to \infty.$$

PROBLEMS 7.3

1. Let $X_1, X_2, \ldots$ be a sequence of iid RVs with common uniform distribution on $[0,1]$. Also, let $Z_n = (\prod_{i=1}^n X_i)^{1/n}$ be the geometric mean of $X_1, X_2, \ldots, X_n$, $n = 1, 2, \ldots$. Show that $Z_n \xrightarrow{P} c$, where c is some constant. Find c.

2. Let $X_1, X_2, \ldots$ be iid RVs with finite second moment. Let

$$Y_n = \frac{2}{n(n+1)} \sum_{i=1}^{n} iX_i.$$

Show that $Y_n \xrightarrow{P} EX_1$.

3. Let $X_1, X_2, \ldots$ be a sequence of iid RVs with $EX_i = \mu$ and $\text{var}(X_i) = \sigma^2$. Let $S_k = \sum_{j=1}^{k} X_j$. Does the sequence S_k obey the WLLN in the sense of Definition 1? If so, find the centering and the norming constants.

4. Let $\{X_n\}$ be a sequence of RVs for which $\text{var}(X_n) \leq C$ for all n and $\rho_{ij} = \text{cov}(X_i, X_j) \to 0$ as $|i-j| \to \infty$. Show that the WLLN holds.

5. For the following sequences of independent RVs does the WLLN hold?
   (a) $P\{X_k = \pm 2^k\} = \frac{1}{2}$.
   (b) $P\{X_k = \pm k\} = 1/2\sqrt{k}$, $P\{X_k = 0\} = 1 - (1/\sqrt{k})$.
   (c) $P\{X_k = \pm 2^k\} = 1/2^{2k+1}$, $P\{X_k = 0\} = 1 - (1/2^{2k})$.
   (d) $P\{X_k = \pm 1/k\} = 1/2$.
   (e) $P\{X_k = \pm \sqrt{k}\} = \frac{1}{2}$.

6. Let $X_1, X_2, \ldots$ be a sequence of independent RVs such that var $(X_k) < \infty$ for $k = 1, 2, \ldots$, and $(1/n^2) \sum_{k=1}^{n} \text{var}(X_k) \to 0$ as $n \to \infty$. Prove the WLLN, using Theorem 2.

7. Let X_n be a sequence of RVs with common finite variance σ^2. Suppose that the correlation coefficient between X_i and X_j is < 0 for all $i \neq j$. Show that the WLLN holds for the sequence $\{X_n\}$.

8. Let $\{X_n\}$ be a sequence of RVs such that X_k is independent of X_j for $j \neq k+1$ or $j \neq k-1$. If $\text{var}(X_k) < C$ for all k, where C is some constant, the WLLN holds for $\{X_k\}$.

9. For any sequence of RVs $\{X_n\}$ show that

$$\max_{1 \leq k \leq n} |X_k| \xrightarrow{P} 0 \Rightarrow n^{-1} S_n \xrightarrow{P} 0.$$

10. Let $X_1, X_2, \ldots$ be iid $\mathcal{C}(1,0)$ RVs. Use Theorem 2 to show that the weak law of large numbers does not hold. That is, show that

$$E\frac{S_n^2}{n^2 + S_n^2} \nrightarrow 0 \qquad \text{as } n \to \infty, \text{ where } S_n = \sum_{k=1}^{n} X_k, n = 1, 2, \ldots.$$

11. Let $\{X_n\}$ be a sequence of iid RVs with $P\{X_n \geq 0\} = 1$. Let $S_n = \sum_{j=1}^{n} X_j$, $n = 1, 2, \ldots$. Suppose $\{a_n\}$ is a sequence of constants such that $a_n^{-1} S_n \xrightarrow{P} 1$. Show that
   (a) $a_n \to \infty$ as $n \to \infty$ and (b) $a_{n+1}/a_n \to 1$.

7.4 STRONG LAW OF LARGE NUMBERS[†]

In this section we obtain a stronger form of the law of large numbers discussed in Section 7.3. Let $X_1, X_2, \ldots$ be a sequence of RVs defined on some probability space $(\Omega, \mathcal{S}, P)$.

[†] This section may be omitted on the first reading.

Definition 1. We say that the sequence $\{X_n\}$ obeys the *strong law of large numbers* (SLLN) with respect to the norming constants $\{B_n\}$ if there exists a sequence of (centering) constants $\{A_n\}$ such that

$$B_n^{-1}(S_n - A_n) \xrightarrow{\text{a.s.}} 0 \qquad \text{as } n \to \infty. \tag{1}$$

Here $B_n > 0$ and $B_n \to \infty$ as $n \to \infty$.

We will obtain sufficient conditions for a sequence $\{X_n\}$ to obey the SLLN. In what follows, we will be interested mainly in the case $B_n = n$. Indeed, when we speak of the SLLN we will assume that we are speaking of the norming constants $B_n = n$, unless specified otherwise.

We start with the *Borel–Cantelli lemma*. Let $\{A_j\}$ be any sequence of events in $\mathcal{S}$. We recall that

$$\overline{\lim_{n\to\infty}} A_n = \lim_{n\to\infty} \bigcup_{k=n}^{\infty} A_k = \bigcap_{n=1}^{\infty} \bigcup_{k=n}^{\infty} A_k. \tag{2}$$

We will write $A = \overline{\lim}_{n\to\infty} A_n$. Note that A is the event that *infinitely many* of the A_n occur. We will sometimes write

$$PA = P\left(\overline{\lim_{n\to\infty}} A_n\right) = P(A_n \text{i.o.}),$$

where "i.o." stands for "infinitely often." In view of Theorem 7.2.11 and Remark 7.2.6 we have $X_n \xrightarrow{\text{a.s.}} 0$ if and only if $P\{|X_n| > \varepsilon \text{ i.o.}\} = 0$ for all $\varepsilon > 0$.

Theorem 1 (Borel–Cantelli Lemma).

(a) Let $\{A_n\}$ be a sequence of events such that $\sum_{n=1}^{\infty} PA_n < \infty$. Then $PA = 0$.

(b) If $\{A_n\}$ is an independent sequence of events such that $\sum_{n=1}^{\infty} PA_n = \infty$, then $PA = 1$.

Proof.

(a) $PA = P(\lim_{n\to\infty} \bigcup_{k=n}^{\infty} A_k) = \lim_{n\to\infty} P(\bigcup_{k=n}^{\infty} A_k) \leq \lim_{n\to\infty} \sum_{k=n}^{\infty} PA_k = 0.$

(b) We have $A^c = \bigcup_{n=1}^{\infty} \bigcap_{k=n}^{\infty} A_k^c$, so that

$$PA^c = P\left(\lim_{n\to\infty} \bigcap_{k=n}^{\infty} A_k^c\right) = \lim_{n\to\infty} P\left(\bigcap_{k=n}^{\infty} A_k^c\right).$$

For $n_0 > n$, we see that $\bigcap_{k=n}^{\infty} A_k^c \subset \bigcap_{k=n}^{n_0} A_k^c$, so that

$$P\left(\bigcap_{k=n}^{\infty} A_k^c\right) \leq \lim_{n_0\to\infty} P\left(\bigcap_{k=n}^{n_0} A_k^c\right) = \lim_{n_0\to\infty} \prod_{k=n}^{n_0} (1 - PA_k),$$

because $\{A_n\}$ is an independent sequence of events. Now we use the elementary inequality

$$1-\exp\left(-\sum_{j=n}^{n_0}\alpha_j\right)\leq 1-\prod_{j=n}^{n_0}(1-\alpha_j)\leq\sum_{j=n}^{n_0}\alpha_j,\quad n_0>n, 1\geq\alpha_j\geq 0,$$

to conclude that

$$P\left(\bigcap_{k=n}^{\infty}A_k^c\right)\leq\lim_{n_0\to\infty}\exp\left(-\sum_{k=n}^{n_0}PA_k\right).$$

Since the series $\sum_{n=1}^{\infty}PA_n$ diverges, it follows that $PA^c=0$ or $PA=1$.

Corollary. Let $\{A_n\}$ be a sequence of independent events. Then PA is either 0 or 1.

The corollary follows since $\sum_{n=1}^{\infty}PA_n$ either converges or diverges.

As a simple application of the Borel–Cantelli lemma, we obtain a version of the SLLN.

Theorem 2. If $X_1,X_2,\ldots$ are iid RVs with common mean μ and finite fourth moment, then

$$P\left\{\lim_{n\to\infty}\frac{S_n}{n}=\mu\right\}=1.$$

Proof. We have

$$E\{\Sigma(X_i-\mu)\}^4=nE(X_1-\mu)^4+6\binom{n}{2}\sigma^4\leq Cn^2.$$

By Markov's inequality

$$P\left\{\left|\sum_1^n(X_1-\mu)\right|>n\varepsilon\right\}\leq\frac{E\{\sum_1^n(X_1-\mu)\}^4}{(n\varepsilon)^4}\leq\frac{Cn^2}{(n\varepsilon)^4}=\frac{C'}{n^2}.$$

Therefore,

$$\sum_{n=1}^{\infty}P\{|S_n-\mu n|>n\varepsilon\}<\infty,$$

and it follows by the Borel–Cantelli lemma that with probability 1 only finitely many of the events $\{\omega\colon|(S_n/n)-\mu|>\varepsilon\}$ occur, that is, $PA_\varepsilon=0$, where

$$A_\varepsilon=\limsup_{n\to\infty}\left\{\left|\frac{S_n}{n}-\mu\right|>\varepsilon\right\}.$$

The sets A_ε increase, as $\varepsilon \to 0$, to the ω set on which $S_n/n \nrightarrow \mu$. Letting $\varepsilon \to 0$ through a countable set of values, we have

$$P\left\{\frac{S_n}{n} - \mu \nrightarrow 0\right\} = P\left\{\bigcup_k A_{1/k}\right\} = 0.$$

Corollary. If $X_1, X_2, \ldots$ are iid RVs such that $P\{|X_n| < K\} = 1$ for all n, where K is a positive constant, then $n^{-1}S_n \xrightarrow{\text{a.s.}} \mu$.

Theorem 3. Let $X_1, X_2, \ldots$ be a sequence of independent RVs. Then

$$X_n \xrightarrow{\text{a.s.}} 0 \Leftrightarrow \sum_{n=1}^{\infty} P\{|X_n| > \varepsilon\} < \infty \qquad \text{for all } \varepsilon > 0.$$

Proof. Writing $A_n = \{|X_n| > \varepsilon\}$, we see that $\{A_n\}$ is a sequence of independent events. Since $X_n \xrightarrow{\text{a.s.}} 0$, $X_n \to 0$ on a set E^c with $PE = 0$. A point $\omega \in E^c$ belongs only to a finite number of A_n. It follows that

$$\limsup_{n\to\infty} A_n \subset E,$$

hence, $P(A_n \text{ i.o.}) = 0$. By the Borel–Cantelli lemma (Theorem 1(b)) we must have $\sum_{n=1}^{\infty} PA_n < \infty$. (Otherwise, $\sum_{n=1}^{\infty} PA_n = \infty$, and then $P(A_n \text{ i.o.}) = 1$.)

In the other direction, let

$$A_{1/k} = \limsup_{n\to\infty}\left\{|X_n| > \frac{1}{k}\right\},$$

and use the argument in the proof of Theorem 2.

Example 1. We take an application of Borel–Cantelli Lemma to prove a.s. convergence.

Let $\{X_n\}$ have PMF

$$P(X_n = 0) = 1 - \frac{1}{n^\alpha}, \quad P(X_n = \pm n) = \frac{1}{2n^\alpha}.$$

Then $P(|X_n| > \varepsilon) = \frac{1}{n^\alpha}$ and it follows that

$$\sum_{n=1}^{\infty} P(|X_n| > \varepsilon) = \sum_{n=1}^{\infty} \frac{1}{n^\alpha} < \infty \quad \text{for } \alpha > 1.$$

Thus from Borel–Cantelli lemma $P(A_n \text{ i.o.}) = 0$, where $A_n = \{|X_n| > \varepsilon\}$. Now using the argument in the proof of Theorem 2 we can show that $P(X_n \nrightarrow 0\} = 0$.

We next prove some important lemmas that we will need subsequently.

Lemma 1 (Kolmogorov's Inequality). Let $X_1, X_2, \ldots, X_n$ be independent RVs with common mean 0 and variances σ_k^2, $k = 1, 2, \ldots, n$, respectively. Then for any $\varepsilon > 0$

$$P\left\{ \max_{1 \le k \le n} |S_k| > \varepsilon \right\} \le \sum_1^n \frac{\sigma_i^2}{\varepsilon^2}. \tag{3}$$

Proof. Let $A_0 = \Omega$,

$$A_k = \left\{ \max_{1 \le j \le k} |S_j| \le \varepsilon \right\}, \qquad k = 1, 2, \ldots, n,$$

and

$$\begin{aligned} B_k &= A_{k-1} \cap A_k^c \\ &= \{|S_1| \le \varepsilon, \ldots, |S_{k-1}| \le \varepsilon\} \cap \{\text{at least one of } |S_1|, \ldots, |S_k| \text{ is } > \varepsilon\} \\ &= \{|S_1| \le \varepsilon, \ldots, |S_{k-1}| \le \varepsilon, |S_k| > \varepsilon\}. \end{aligned}$$

It follows that

$$A_n^c = \sum_{k=1}^n B_k$$

and

$$B_k \subset \{|S_{k-1}| \le \varepsilon, |S_k| > \varepsilon\}.$$

As usual, let us write I_{B_k}, for the indicator function of the event B_k. Then

$$\begin{aligned} E(S_n I_{B_k})^2 &= E\{(S_n - S_k) I_{B_k} + S_k I_{B_k}\}^2, \\ &= E\{(S_n - S_k)^2 I_{B_k} + S_k^2 I_{B_k} + 2 S_k (S_n - S_k) I_{B_k}\}. \end{aligned}$$

Since $S_n - S_k = X_{k+1} + \cdots + X_n$ and $S_k I_{B_k}$ are independent, and $EX_k = 0$ for all k, it follows that

$$\begin{aligned} E(S_n I_{B_k})^2 &= E\{(S_n - S_k) I_{B_k}\}^2 + E(S_k I_{B_k})^2 \\ &\ge E(S_k I_{B_k})^2 \ge \varepsilon^2 P B_k. \end{aligned}$$

The last inequality follows from the fact that, in B_k, $|S_k| > \varepsilon$. Moreover,

$$\sum_{k=1}^n E(S_n I_{B_k})^2 = E(S_n^2 I_{A_n^c}) \le E(S_n^2) = \sum_1^n \sigma_k^2$$

so that

$$\sum_1^n \sigma_k^2 \ge \varepsilon^2 \sum_1^n P B_k = \varepsilon^2 P(A_n^c),$$

as asserted.

Corollary. Take $n = 1$ then

$$P\{|X_1| > \varepsilon\} \leq \frac{\sigma_1^2}{\varepsilon^2},$$

which is Chebychev's inequality.

Lemma 2 (Kronecker Lemma). If $\sum_{n=1}^{\infty} x_n$ converges to s (finite) and $b_n \uparrow \infty$, then

$$b_n^{-1} \sum_{k=1}^{n} b_k x_k \to 0.$$

Proof. Writing $b_0 = 0$, $a_k = b_k - b_{k-1}$, and $s_{n+1} = \sum_{k=1}^{n} x_k$, we have

$$\begin{aligned}
\frac{1}{b_n} \sum_{k=1}^{n} b_k x_k &= \frac{1}{b_n} \sum_{k=1}^{n} b_k (s_{k+1} - s_k) \\
&= \frac{1}{b_n} \left(b_n s_{n+1} + \sum_{1}^{n} b_{k-1} s_k \right) - \frac{1}{b_n} \sum_{k=1}^{n} b_k s_k \\
&= s_{n+1} - \frac{1}{b_n} \sum_{k=1}^{n} (b_k - b_{k-1}) s_k \\
&= s_{n+1} - \frac{1}{b_n} \sum_{k=1}^{n} a_k s_k.
\end{aligned}$$

It therefore suffices to show that $b_n^{-1} \sum_{k=1}^{n} a_k s_k \to s$. Since $s_n \to s$, there exists an $n_0 = n_0(\varepsilon)$ such that

$$|s_n - s| < \frac{\varepsilon}{2} \qquad \text{for } n > n_0.$$

Since $b_n \uparrow \infty$, let n_1 be an integer $> n_0$ such that

$$b_n^{-1} \left| \sum_{1}^{n_0} (b_k - b_{k-1})(s_k - s) \right| < \frac{\varepsilon}{2} \qquad \text{for } n > n_1.$$

Writing

$$r_n = b_n^{-1} \sum_{k=1}^{n} (b_k - b_{k-1}) s_k,$$

we see that

$$|r_n - s| = \frac{1}{b_n} \left| \sum_{k=1}^{n} (b_k - b_{k-1})(s_k - s) \right|,$$

and, choosing $n > n_1$, we have

$$|r_n - s| \le \left| \frac{1}{b_n} \sum_{k=1}^{n_0} (b_k - b_{k-1})(s_k - s) \right| + \frac{1}{b_n} \left| \sum_{k=n_0+1}^{n} (b_k - b_{k-1}) \frac{\varepsilon}{2} \right| < \varepsilon.$$

This completes the proof.

Theorem 4. If $\sum_{n=1}^{\infty} \operatorname{var} X_n < \infty$, then $\sum_{n=1}^{\infty} (X_n - EX_n)$ converges almost surely.

Proof. Without loss of generality assume that $EX_n = 0$. By Kolmogorov's inequality

$$P\left\{ \max_{1 \le k \le n} |S_{m+k} - S_m| \ge \varepsilon \right\} \le \frac{1}{\varepsilon^2} \sum_{k=1}^{n} \operatorname{var}(X_{m+k}).$$

Letting $n \to \infty$ we have

$$\begin{aligned} P\left\{ \max_{k \ge 1} |S_{m+k} - S_m| \ge \varepsilon \right\} &= P\left\{ \max_{k \ge m+1} |S_k - S_m| \ge \varepsilon \right\} \\ &\le \frac{1}{\varepsilon^2} \sum_{k=m+1}^{\infty} \operatorname{var}(X_k). \end{aligned}$$

It follows that

$$\lim_{m \to \infty} P\left\{ \max_{k > m} |S_k - S_m| < \varepsilon \right\} = 1,$$

and since $\varepsilon > 0$ is arbitrary we have

$$P\left\{ \lim_{m \to \infty} \left| \sum_{j=m}^{\infty} X_j \right| = 0 \right\} = 1.$$

Consequently, $\sum_{j=1}^{\infty} X_j$ converges a.s.

As a corollary we get a version of the SLLN for nonidentically distributed RVs which subsumes Theorem 2.

Corollary 1. Let $\{X_n\}$ be independent RVs. If

$$\sum_{k=1}^{\infty} \frac{\operatorname{var}(X_k)}{B_k^2} < \infty, \qquad B_n \uparrow \infty,$$

then

$$\frac{S_n - ES_n}{B_n} \xrightarrow{\text{a.s.}} 0.$$

The corollary follows from Theorem 4 and the Kronecker lemma.

Corollary 2. Every sequence $\{X_n\}$ of independent RVs with uniformly bounded variances obeys the SLLN.

If $\operatorname{var}(X_k) \le A$ for all k, and $B_k = k$, then

$$\sum_{k=1}^{\infty} \frac{\sigma_k^2}{k^2} \le A \sum_{1}^{\infty} \frac{1}{k^2} < \infty,$$

and it follows that

$$\frac{S_n - ES_n}{n} \xrightarrow{\text{a.s.}} 0.$$

Corollary 3 (Borel's Strong Law of Large Numbers). For a sequence of Bernoulli trials with (constant) probability p of success, the SLLN holds (with $B_n = n$ and $A_n = np$).

Since

$$EX_k = p, \quad \operatorname{var}(X_k) = p(1-p) \le \frac{1}{4}, \quad 0 < p < 1,$$

the result follows from Corollary 2.

Corollary 4. Let $\{X_n\}$ be iid RVs with common mean μ and finite variance σ^2. Then

$$P\left\{\lim_{n\to\infty} \frac{S_n}{n} = \mu\right\} = 1.$$

Remark 1. Kolmogorov's SLLN is much stronger than Corollaries 1 and 4 to Theorem 4. It states that if $\{X_n\}$ is a sequence of iid RVs then

$$n^{-1} S_n \xrightarrow{a.s.} \mu \Longleftrightarrow E|X_1| < \infty,$$

and then $\mu = EX_1$. The proof requires more work and will not be given here. We refer the reader to Billingsley [6], Chung [15], Feller [26], or Laha and Rohatgi [58].

PROBLEMS 7.4

1. For the following sequences of independent RVs does the SLLN hold?
   (a) $P\{X_k = \pm 2^k\} = \frac{1}{2}$.
   (b) $P\{X_k = \pm k\} = 1/2\sqrt{k}$, $P\{X_k = 0\} = 1 - (1/\sqrt{k})$.
   (c) $P\{X_k = \pm 2^k\} = 1/2^{2k+1}$, $P\{X_k = 0\} = 1 - (1/2^{2k})$.
2. Let $X_1, X_2, \ldots$ be a sequence of independent RVs with $\sum_{k=1}^{\infty} \operatorname{var}(X_k)/k^2 < \infty$. Show that

$$\frac{1}{n^2} \sum_{k=1}^{n} \operatorname{var}(X_k) \to 0 \qquad \text{as } n \to \infty.$$

   Does the converse also hold?

3. For what values of α does the SLLN hold for the sequence

$$P\{X_k = \pm k^\alpha\} = \frac{1}{2}?$$

4. Let $\{\sigma_k^2\}$ be a sequence of real numbers such that $\sum_{k=1}^{\infty} \sigma_k^2/k^2 = \infty$. Show that there exists a sequence of independent RVs $\{X_k\}$ with $\text{var}(X_k) = \sigma_k^2$, $k = 1, 2, \ldots$, such that $n^{-1}\sum_{k=1}^{n}(X_k - EX_k)$ does not converge to 0 almost surely.
[*Hint:* Let $P\{X_k = \pm k\} = \sigma_k^2/2k^2$, $P\{X_k = 0\} = 1 - (\sigma_k^2/k^2)$ if $\sigma_k/k \le 1$, and $P\{X_k = \pm\sigma_k\} = \frac{1}{2}$ if $\sigma_k/k > 1$. Apply the Borel–Cantelli lemma to $\{|X_n| > n\}$.]

5. Let X_n be a sequence of iid RVs with $E|X_n| = +\infty$. Show that, for every positive number A, $P\{|X_n| > nA \text{ i.o.}\} = 1$ and $P\{|S_n| < nA \text{ i.o.}\} = 1$.

6. Construct an example to show that the converse of Theorem 1(a) does not hold.

7. Investigate a.s. convergence of $\{X_n\}$ to 0 in each case.
   (a) $P(X_n = e^n) = 1/n^2$, $P(X_n = 0) = 1 - 1/n^2$.
   (b) $P(X_n = 0) = 1 - 1/n$, $P(X_n = \pm 1) = 1/(2n)$.
   (X_n's are independent in each case.)

7.5 LIMITING MOMENT GENERATING FUNCTIONS

Let $X_1, X_2, \ldots$ be a sequence of RVs. Let F_n be the DF of X_n, $n = 1, 2, \ldots$, and suppose that the MGF $M_n(t)$ of F_n exists. What happens to $M_n(t)$ as $n \to \infty$? If it converges, does it always converge to an MGF?

Example 1. Let $\{X_n\}$ be a sequence of RVs with PMF $P\{X_n = -n\} = 1$, $n = 1, 2, \ldots$. We have

$$M_n(t) = Ee^{tX_n} = e^{-tn} \to 0 \qquad \text{as } n \to \infty \quad \text{for all } t > 0,$$
$$M_n(t) \to +\infty \quad \text{for all } t < 0, \text{ and } M_n(t) \to 1 \quad \text{at } t = 0.$$

Thus

$$M_n(t) \to M(t) = \begin{cases} 0, & t > 0 \\ 1, & t = 0 \\ \infty, & t < 0 \end{cases} \quad \text{as } n \to \infty.$$

But $M(t)$ is not an MGF. Note that if F_n is the DF of X_n then

$$F_n(x) = \begin{cases} 0 & \text{if } x < -n \\ 1 & \text{if } x \ge -n \end{cases} \to F(x) = 1 \quad \text{for all } x,$$

and F is not a DF.

Next suppose that X_n has MGF M_n and $X_n \xrightarrow{L} X$, where X is an RV with MGF M. Does $M_n(t) \to M(t)$ as $n \to \infty$? The answer to this question is in the negative.

Example 2. (Curtiss [19]). Consider the DF

$$F_n(x) = \begin{cases} 0, & x < -n, \\ \frac{1}{2} + c_n \tan^{-1}(nx), & -n \le x < n, \\ 1, & x \ge n, \end{cases}$$

where $c_n = 1/[2\tan^{-1}(n^2)]$. Clearly, as $n \to \infty$,

$$F_n(x) \to F(x) = \begin{cases} 0, & x < 0, \\ 1, & x \ge 0, \end{cases}$$

at all points of continuity of the DF F. The MGF associated with F_n is

$$M_n(t) = \int_{-n}^{n} c_n e^{tx} \frac{n}{1+n^2x^2}\,dx,$$

which exists for all t. The MGF corresponding to F is $M(t) = 1$ for all t. But $M_n(t) \nrightarrow M(t)$, since $M_n(t) \to \infty$ if $t \ne 0$. Indeed

$$M_n(t) > \int_0^n c_n \frac{|t|^3 x^3}{6} \frac{n}{1+n^2x^2}\,dx.$$

The following result is a weaker version of the continuity theorem due to Lévy and Cramér. We refer the reader to Lukacs [69, p. 47], or Curtiss [19], for details of the proof.

Theorem 1 (Continuity Theorem). Let $\{F_n\}$ be a sequence of DFs with corresponding MGFs $\{M_n\}$, and suppose that $M_n(t)$ exists for $|t| \le t_0$ for every n. If there exists a DF F with corresponding MGF M which exists for $|t| \le t_1 < t_0$, such that $M_n(t) \to M(t)$ as $n \to \infty$ for every $t \in [-t_1, t_1]$, then $F_n \xrightarrow{w} F$.

Example 3. Let X_n be an RV with PMF

$$P\{X_n = 1\} = \frac{1}{n} \quad \text{and} \quad P\{X_n = 0\} = 1 - \frac{1}{n}.$$

Then $M_n(t) = (1/n)e^t + [1 - (1/n)]$ exists for all $t \in \mathcal{R}$, and $M_n(t) \to 1$ as $n \to \infty$ for all t. Here $M(t) = 1$ is the MGF of an RV X degenerate at 0. Thus $X_n \xrightarrow{L} X$.

Remark 1. The following notation on orders of magnitude is quite useful. We write $x_n = o(r_n)$ if, given $\varepsilon > 0$, there exists an N such that $|x_n/r_n| < \varepsilon$ for all $n \ge N$ and $x_n = O(r_n)$

if there exists an N and a constant $c > 0$, such that $|x_n/r_n| < c$ for all $n \geq N$. We write $x_n = O(1)$ to express the fact that x_n is bounded for large n, and $x_n = o(1)$ to mean that $x_n \to 0$ as $n \to \infty$.

This notation is extended to RVs in an obvious manner. Thus $X_n = o_p(r_n)$ if, for every $\varepsilon > 0$ and $\delta > 0$, there exists an N such that $P(|X_n/r_n| \leq \delta) \geq 1-\varepsilon$ for $n \geq N$, and $X_n = O_p(r_n)$ if, for $\varepsilon > 0$, there exists a $c > 0$ and an N such that $P(|X_n/r_n| \leq c) \geq 1-\varepsilon$. We write $X_n = o_p(1)$ to mean $X_n \xrightarrow{P} 0$. This notation can be easily extended to the case where r_n itself is an RV.

The following lemma is quite useful in applications of Theorem 1.

Lemma 1. Let us write $f(x) = o(x)$, if $f(x)/x \to 0$ as $x \to 0$. We have

$$\lim_{n\to\infty}\left\{1+\frac{a}{n}+o\left(\frac{1}{n}\right)\right\}^n = e^a \qquad \text{for every real } a.$$

Proof. By Taylor's expansion we have

$$\begin{aligned} f(x) &= f(0)+xf'(\theta x)\\ &= f(0)+xf'(0)+\{f'(\theta x)-f'(0)\}x, \qquad 0<\theta<1. \end{aligned}$$

If $f'(x)$ is continuous at $x=0$, then as $x \to 0$

$$f(x) = f(0)+xf'(0)+o(x).$$

Taking $f(x) = \log(1+x)$, we have $f'(x) = (1+x)^{-1}$, which is continuous at $x=0$, so that

$$\log(1+x) = x+o(x).$$

Then for sufficiently large n

$$\begin{aligned} n\log\left\{1+\frac{a}{n}+o\left(\frac{1}{n}\right)\right\} &= n\left\{\frac{a}{n}+o\left(\frac{1}{n}\right)+o\left[\frac{a}{n}+o\left(\frac{1}{n}\right)\right]\right\}\\ &= a+n\,o\left(\frac{1}{n}\right)\\ &= a+o(1). \end{aligned}$$

It follows that

$$\left\{1+\frac{a}{n}+o\left(\frac{1}{n}\right)\right\}^n = e^{a+o(1)},$$

as asserted.

Example 4. Let $X_1, X_2, \ldots$ be iid $b(1,p)$ RVs. Also, let $S_n = \sum_1^n X_k$, and let $M_n(t)$ be the MGF of S_n. Then

$$M_n(t) = (q+pe^t)^n \qquad \text{for all } t,$$

where $q = 1 - p$. If we let $n \to \infty$ in such a way that np remains constant at λ, say, then, by Lemma 1,

$$M_n(t) = \left(1 - \frac{\lambda}{n} + \frac{\lambda}{n}e^t\right)^n = \left\{1 + \frac{\lambda}{n}(e^t - 1)\right\}^n \to \exp\{\lambda(e^t - 1)\} \quad \text{for all } t,$$

which is the MGF of a $P(\lambda)$ RV. Thus, the binomial distribution function approaches the Poisson df, provided that $n \to \infty$ in such a way that $np = \lambda > 0$.

Example 5. Let $X \sim P(\lambda)$. The MGF of X is given by

$$M(t) = \exp\{\lambda(e^t - 1)\} \qquad \text{for all } t.$$

Let $Y = (X - \lambda)/\sqrt{\lambda}$. Then the MGF of Y is given by

$$M_Y(t) = e^{-t\sqrt{\lambda}} M\left(\frac{t}{\sqrt{\lambda}}\right).$$

Also,

$$\begin{aligned}\log M_Y(t) &= -t\sqrt{\lambda} + \log M\left(\frac{t}{\sqrt{\lambda}}\right) \\ &= -t\sqrt{\lambda} + \lambda(e^{t/\sqrt{\lambda}} - 1) \\ &= -t\sqrt{\lambda} + \lambda\left(\frac{t}{\sqrt{\lambda}} + \frac{t^2}{2\lambda} + \frac{t^3}{3!\lambda^{3/2}} + \cdots\right) \\ &= \frac{t^2}{2} + \frac{t^3}{3!\lambda^{3/2}} + \cdots.\end{aligned}$$

It follows that

$$\log M_Y(t) \to \frac{t^2}{2} \qquad \text{as } \lambda \to \infty,$$

so that $M_Y(t) \to e^{t^2/2}$ as $\lambda \to \infty$, which is the MGF of an $\mathcal{N}(0, 1)$ RV.

For more examples see Section 7.6.

Remark 2. As pointed out earlier working with MGFs has the disadvantage that the existence of MGFs is a very strong condition. Working with CFs which always exist, on the other hand, permits a much wider application of the continuity theorem. Let ϕ_n be the CF of F_n. Then $F_n \xrightarrow{w} F$ if and only if $\phi_n \to \phi$ as $n \to \infty$ on $\mathcal{R}$, where ϕ is continuous at $t = 0$. In this case ϕ, the limit function, is the CF of the limit DF F.

Example 6. Let X be a $\mathcal{C}(0,1)$ RV. Then its CF is given by

$$\begin{aligned} E\exp(itX) &= \frac{1}{\pi}\int_{-\infty}^{\infty}\frac{\cos tx}{1+x^2}dx + i\frac{1}{\pi}\int_{-\infty}^{\infty}\frac{\sin tx}{1+x^2}dx \\ &= \frac{1}{\pi}\int_{-\infty}^{\infty}\frac{\cos tx}{1+x^2}dx = e^{-|t|} \end{aligned}$$

since the second integral on the right side vanishes.

Let $\{X_n\}$ be iid RVs with common law $\mathcal{L}(X)$ and set $Y_n = \sum_{j=1}^{n} X_j/n$. Then the CF of Y_n is given by

$$\begin{aligned} \varphi_n(t) &= E\exp\left\{it\sum_{j=1}^{n} X_j/n\right\} = \prod_{j=1}^{n}\exp\left\{-\frac{|t|}{n}\right\} \\ &= \exp(-|t|) \end{aligned}$$

for all n. It follows φ_n is the CF of a $\mathcal{C}(1,0)$ RV. We could not have derived this result using MGFs. Also if $U_n = \sum_{j=1}^{n} X_j/n^{\alpha}$ for $\alpha > 1$, then

$$\varphi U_n(t) = \exp\left\{-|t|/n^{\alpha-1}\right\} \to 1$$

as $n \to \infty$ for all t. Since $\varphi(t) = 1$ is continuous at $t = 0$, φ is the CF of the limit DF F. Clearly F is the DF of an RV degenerate at 0. Thus $\sum_{j=1}^{n} X_j/n^{\alpha} \xrightarrow{L,P} U$, where $P(U=0)=1$.

PROBLEMS 7.5

1. Let $X \sim NB(r;p)$. Show that

$$2pX \xrightarrow{L} Y \qquad \text{as } p \to 0,$$

where $Y \sim \chi^2(2r)$.

2. Let $X_n \sim NB(r_n; 1-p_n)$, $n = 1,2,\ldots$. Show that $X_n \xrightarrow{L} X$ as $r_n \to \infty$, $p_n \to 0$, in such a way that $r_n p_n \to \lambda$, where $X \sim P(\lambda)$.
3. Let $X_1, X_2, \ldots$ be independent RVs with PMF given by $P\{X_n = \pm 1\} = \frac{1}{2}, n = 1,2,\ldots$. Let $Z_n = \sum_{j=1}^{n} X_j/2^j$. Show that $Z_n \xrightarrow{L} Z$, where $Z \sim U[-1,1]$.
4. Let $\{X_n\}$ be a sequence of RVs with $X_n \sim G(n,\beta)$ where $\beta > 0$ is a constant (independent of n). Find the limiting distribution of X_n/n.
5. Let $X_n \sim \chi^2(n)$, $n = 1,2,\ldots$. Find the limiting distribution of X_n/n^2.
6. Let $X_1, X_2, \ldots, X_n$ be jointly normal with $EX_i = 0$, $EX_i^2 = 1$ for all i and $\text{cov}(X_i,X_j) = \rho$, $i,j = 1,2,\ldots$ $(i \neq j)$. What is the limiting distribution of $n^{-1}S_n$, where $S_n = \sum_{k=1}^{n} X_k$?

7.6 CENTRAL LIMIT THEOREM

Let $X_1, X_2, \ldots$ be a sequence of RVs, and let $S_n = \sum_{k=1}^{n} X_k$, $n = 1, 2, \ldots$. In Sections 7.3 and 7.4 we investigated the convergence of the sequence of RVs $B_n^{-1}(S_n - A_n)$ to the degenerate RV. In this section we examine the convergence of $B_n^{-1}(S_n - A_n)$ to a nondegenerate RV. Suppose that, for a suitable choice of constants A_n and $B_n > 0$, the RVs $B_n^{-1}(S_n - A_n) \xrightarrow{L} Y$. What are the properties of this limit RV Y? The question as posed is far too general and is not of much interest unless the RVs X_i are suitably restricted. For example, if we take X_1 with DF F and $X_2, X_3, \ldots$ to be 0 with probability 1, choosing $A_n = 0$ and $B_n = 1$ leads to F as the limit DF.

We recall (Example 7.5.6) that, if $X_1, X_2, \ldots, X_n$ are iid RVs with common law $\mathcal{C}(1,0)$, then $n^{-1}S_n$ is also $\mathcal{C}(1,0)$. Again, if $X_1, X_2, \ldots, X_n$ are iid $\mathcal{N}(0,1)$ RVs then $n^{-1/2}S_n$ is also $\mathcal{N}(0,1)$ (Corollary 2 to Theorem 5.3.22). We note thus that for certain sequences of RVs there exist sequences A_n and $B_n > 0$, $B_n \to \infty$, such that $B_n^{-1}(S_n - A_n) \xrightarrow{L} Y$. In the Cauchy case $B_n = n$, $A_n = 0$, and in the normal case $B_n = n^{1/2}$, $A_n = 0$. Moreover, we see that Cauchy and normal distributions appear as limiting distributions—in these two cases, because of the reproductive nature of the distributions. Cauchy and normal distributions are examples of *stable distributions*.

Definition 1. Let X_1, X_2, be iid nondegenerate RVs with common DF F. Let a_1, a_2 be any positive constants. We say that F is *stable* if there exist constants A and B (depending on a_1, a_2) such that the RV $B^{-1}(a_1X_1 + a_2X_2 - A)$ also has the DF F.

Let $X_1, X_2, \ldots$ be iid RVs with common DF F. We remark without proof (see Loève [66, p. 339]) that only stable distributions occur as limits. To make this statement more precise we make the following definition.

Definition 2. Let $X_1, X_2, \ldots$ be iid RVs with common DF F. We say that F belongs to the domain of attraction of a distribution V if there exist norming constants $B_n > 0$ and centering constants A_n such that, as $n \to \infty$,

$$P\{B_n^{-1}(S_n - A_n) \le x\} \to V(x), \tag{1}$$

at all continuity points x of V.

In view of the statement after Definition 1, we see that only stable distributions possess domains of attraction. From Definition 1 we also note that each stable law belongs to its own domain of attraction. The study of stable distributions is beyond the scope of this book. We shall restrict ourselves to seeking conditions under which the limit law V is the normal distribution. The importance of the normal distribution in statistics is due largely to the fact that a wide class of distributions F belongs to the domain of attraction of the normal law. Let us consider some examples.

Example 1. Let $X_1, X_2, \ldots, X_n$ be iid $b(1,p)$ RVs. Let

$$S_n = \sum_{k=1}^{n} X_k, \qquad A_n = ES_n = np, \qquad B_n = \sqrt{\text{var}(S_n)} = \sqrt{np(1-p)}.$$

Then

$$\begin{aligned}
M_n(t) &= E\exp\left\{\frac{S_n-np}{\sqrt{np(1-p)}}t\right\}\\
&= \prod_{i=1}^{n} E\exp\left\{\frac{X_i-p}{\sqrt{np(1-p)}}t\right\}\\
&= \exp\left\{-\frac{npt}{\sqrt{np(1-p)}}\right\}\left\{q+p\exp\left[\frac{t}{\sqrt{np(1-p)}}\right]\right\}^n, \quad q=1-p,\\
&= \left\{q\exp\left(-\frac{pt}{\sqrt{npq}}\right)+p\exp\left(\frac{qt}{\sqrt{npq}}\right)\right\}^n\\
&= \left[1+\frac{t^2}{2n}+o\left(\frac{1}{n}\right)\right]^n.
\end{aligned}$$

It follows from Lemma 7.5.1 that

$$M_n(t) \to e^{t^2/2} \qquad \text{as } n\to\infty,$$

and since $e^{t^2/2}$ is the MGF of an $\mathcal{N}(0,1)$ RV, we have by the continuity theorem

$$P\left\{\frac{S_n-np}{\sqrt{npq}} \le x\right\} \to \frac{1}{\sqrt{2\pi}}\int_{-\infty}^{x} e^{t^2/2}\,dt \qquad \text{for all } x\in\mathcal{R}.$$

In particular, we note that for each $x\in\mathcal{R}$, $F_n^*(x) \xrightarrow{P} F(x)$ as $n\to\infty$ and

$$\frac{\sqrt{n}[F_n^*(x)-F(x)]}{\sqrt{F(x)(1-F(x))}} \xrightarrow{L} Z \qquad \text{as } n\to\infty,$$

where Z is $\mathcal{N}(0,1)$. It is possible to make a probability statement simultaneously for all x. This is the so-called Glivenko–Cantelli theorem: $F_n^*(x)$ converges uniformly to $F(x)$. For a proof, we refer to Fisz [31, p. 391].

Example 2. Let $X_1, X_2, \ldots, X_n$ be iid $\chi^2(1)$ RVs. Then $S_n \sim \chi^2(n)$, $ES_n = n$, and $\text{var}(S_n) = 2n$. Also let $Z_n = (S_n-n)/\sqrt{2n}$ then

$$\begin{aligned}
M_n(t) &= Ee^{tZ_n}\\
&= \exp\left(-t\sqrt{\frac{n}{2}}\right)\left(1-\frac{2t}{\sqrt{2n}}\right)^{-n/2}, \qquad 2t<\sqrt{2n},\\
&= \left[\exp\left(t\sqrt{\frac{2}{n}}\right)-t\sqrt{\frac{2}{n}}\exp\left(t\sqrt{\frac{2}{n}}\right)\right]^{-n/2}, \qquad t<\sqrt{\frac{n}{2}}.
\end{aligned}$$

Using Taylor's approximation, we get

$$\exp\left(t\sqrt{\frac{2}{n}}\right) = 1+t\sqrt{\frac{2}{n}}+\frac{t^2}{2}\left(\sqrt{\frac{2}{n}}\right)^2+\frac{1}{6}\exp(\theta_n)\left(t\sqrt{\frac{2}{n}}\right)^3,$$

where $0 < \theta_n < t\sqrt{(2/n)}$. It follows that

$$M_n(t) = \left(1 - \frac{t^2}{n} + \frac{\zeta(n)}{n}\right)^{-n/2},$$

where

$$\zeta(n) = -\sqrt{\frac{2}{n}}t^3 + \left(\frac{t^3}{3}\sqrt{\frac{2}{n}} - \frac{2t^4}{3n}\right)\exp(\theta_n) \to 0 \qquad \text{as } n \to \infty,$$

for every fixed t. We have from Lemma 1 that $M_n(t) \to e^{t^2/2}$ as $n \to \infty$ for all real t, and it follows that $Z_n \xrightarrow{L} Z$, where Z is $\mathcal{N}(0,1)$.

These examples suggest that if we take iid RVs with finite variance and take $A_n = ES_n$, $B_n = \sqrt{\text{var}(S_n)}$, then $B_n^{-1}(S_n - A_n) \xrightarrow{L} Z$, where Z is $\mathcal{N}(0,1)$. This is the central limit result, which we now prove. The reader should note that in both Examples 1 and 2 we used more than just the existence of $E|X|^2$. Indeed, the MGF exists and hence moments of all order exist. The existence of MGF is not a necessary condition.

Theorem 1 (Lindeberg–Lévy Central Limit Theorem). Let $\{X_n\}$ be a sequence of iid RVs with $0 < \text{var}(X_n) = \sigma^2 < \infty$ and common mean μ. Let $S_n = \sum_{j=1}^n X_j$, $n = 1, 2, \ldots$. Then for every $x \in \mathcal{R}$

$$\lim_{n\to\infty} P\left\{\frac{S_n - n\mu}{\sigma\sqrt{n}} \le x\right\} = \lim_{n\to\infty} P\left\{\frac{\overline{X} - \mu}{\sigma/\sqrt{n}} \le x\right\} = \frac{1}{\sqrt{2\pi}}\int_{-\infty}^{x} e^{-u^2/2}du.$$

Proof. The proof we give here assumes that the MGF of X_n exists. Without loss of generality, we also assume that $EX_n = 0$ and $\text{var}(X_n) = 1$. Let M be the MGF of X_n. Then the MGF of $S_n/\sqrt{n}$ is given by

$$M_n(t) = E\exp(tS_n/\sqrt{n}) = [M(t/\sqrt{n})]^n$$

and

$$\ell\text{n}\, M_n(t) = n\,\ell\text{n}\, M(t/\sqrt{n}) = \frac{\ell\text{n}\, M(t/\sqrt{n})}{1/n} = \frac{L(t/\sqrt{n})}{1/n},$$

where $L(t/\sqrt{n}) = \ell\text{n}\, M(t/\sqrt{n})$. Clearly $L(0) = \ell\text{n}(1) = 0$, so that as $n \to \infty$, the conditions for L'Hospital's rule are satisfied. It follows that

$$\lim_{n\to\infty} \ell\text{n}\, M_n(t) = \lim_{n\to\infty} \frac{L'(t/\sqrt{n})t}{2/\sqrt{n}}$$

and since $L'(0) = EX = 0$, we can use L'Hospital's Rule once again to get

$$\lim_{n\to\infty} \ell\text{n}\, M_n(t) = \lim_{n\to\infty} \frac{L''(t/\sqrt{n})t^2}{2} = \frac{t^2}{2}$$

using $L''(0) = \mathrm{var}(X) = 1$. Thus

$$M_n(t) \longrightarrow \exp(t^2/2) = M(t)$$

where $M(t)$ is the MGT of a $\mathcal{N}(0,1)$ RV.

Remark 1. In the proof above we could have used the Taylor series expansion of M to arrive at the same result.

Remark 2. Even though we proved Theorem 1 for the case when the MGF of X_n's exists, we will use the result whenever $0 < EX_n^2 = \sigma^2 < \infty$. The use of CFs would have provided a complete proof of Theorem 1. Let ϕ be the CF of X_n. Assuming again, without loss of generality, that $EX_n = 0$, $\mathrm{var}(X_n) = 1$, we can write

$$\phi(t) = 1 - \frac{1}{2}t^2 + t^2 o(1).$$

Thus the CF of $S_n/\sqrt{n}$ is

$$[\phi(t/\sqrt{n})]^n = \left[1 - \frac{1}{2n}t^2 + \frac{t^2}{n}o(1)\right]^n,$$

which converges to $\exp(-t^2/2)$ which is the CF of a $\mathcal{N}(0,1)$ RV. The devil is in the details of the proof.

The following converse to Theorem 1 holds.

Theorem 2. Let $X_1, X_2, \ldots, X_n$ be iid RVs such that $n^{-1/2}S_n$ has the same distribution for every $n = 1, 2, \ldots$. Then, if $EX_i = 0$, $\mathrm{var}(X_i) = 1$, the distribution of X_i must be $\mathcal{N}(0,1)$.

Proof. Let F be the DF of $n^{-1/2}S_n$. By the central limit theorem,

$$\lim_{n\to\infty} P\{n^{-1/2}S_n \le x\} = \Phi(x).$$

Also, $P\{n^{-1/2}S_n \le x\} = F(x)$ for each n. It follows that we must have $F(x) = \Phi(x)$.

Example 3. Let $X_1, X_2, \ldots$ be iid RVs with common PMF

$$P\{X = k\} = p(1-p)^k, \qquad k = 0, 1, 2, \ldots, \qquad 0 < p < 1, q = 1 - p.$$

Then $EX = q/p$, $\mathrm{var}(X) = q/p^2$. By Theorem 1 we see that

$$P\left\{\frac{S_n - n(q/p)}{\sqrt{nq}}p \le x\right\} \to \Phi(x) \qquad \text{as } n \to \infty \text{ for all } x \in \mathcal{R}.$$

Example 4. Let $X_1, X_2, \ldots$ be iid RVs with common $B(\alpha, \beta)$ distribution. Then

$$EX = \frac{\alpha}{\alpha+\beta} \qquad \text{and} \qquad \mathrm{var}(X) = \frac{\alpha\beta}{(\alpha+\beta)^2(\alpha+\beta+1)}.$$

By the corollary to Theorem 1, it follows that

$$\frac{S_n - n[\alpha/(\alpha+\beta)]}{\sqrt{\alpha\beta n/[(\alpha+\beta+1)(\alpha+\beta)^2]}} \xrightarrow{L} Z,$$

where Z is $\mathcal{N}(0,1)$.

For nonidentically distributed RVs we state, without proof, the following result due to Lindeberg.

Theorem 3. Let $X_1, X_2, \ldots$ be independent RVs with DFs $F_1, F_2, \ldots$, respectively. Let $EX_k = \mu_k$ and $\text{var}(X_k) = \sigma_k^2$, and write

$$s_n^2 = \sum_{j=1}^{n} \sigma_j^2.$$

If the F_k's are absolutely continuous with PDF f_k, assume that the relation

$$\lim_{n\to\infty} \frac{1}{s_n^2} \sum_{k=1}^{n} \int_{|x-\mu_k|>\varepsilon s_n} (x-\mu_k)^2 f_k(x)\,dx = 0 \tag{2}$$

holds for all $\varepsilon > 0$. (A similar condition can be stated for the discrete case.) Then

$$S_n^* = \frac{\sum_{j=1}^{n} X_j - \sum_{j=1}^{n} \mu_j}{s_n} \xrightarrow{L} Z \sim \mathcal{N}(0,1). \tag{3}$$

Condition (2) is known as the *Lindeberg condition*.

Feller [24] has shown that condition (2) is necessary as well in the following sense. For independent RVs $\{X_k\}$ for which (3) holds and

$$P\left\{\max_{1\le k\le n} |X_k - EX_k| > \varepsilon\sqrt{\text{var}(S_n)}\right\} \to 0,$$

(2) holds for every $\varepsilon > 0$.

Example 5. Let $X_1, X_2, \ldots$ be independent RVs such that X_k is $U(-a_k, a_k)$. Then $EX_k = 0$, $\text{var}(X_k) = (1/3)a_k^2$. Suppose that $|a_k| < a$ and $\sum_1^n a_k^2 \to \infty$ as $n \to \infty$. Then

$$\begin{aligned}
\frac{1}{s_n^2}\sum_{k=1}^{n} \int_{|x|>\varepsilon s_n} x^2 f_k(x)\,dx &\le \frac{1}{s_n^2}\sum_{k=1}^{n} \int_{|x|>\varepsilon s_n} a^2 \frac{1}{2a_k}\,dx \\
&\le \frac{a^2}{s_n^2}\sum_{k=1}^{n} P\{|X_k| > \varepsilon s_n\} \le \frac{a^2}{s_n^2}\sum_{k=1}^{n} \frac{\text{var}(X_k)}{\varepsilon^2 s_n^2} \\
&= \frac{a^2}{\varepsilon^2 s_n^2} \to 0 \qquad \text{as } n \to \infty.
\end{aligned}$$

If $\sum_1^\infty a_k^2 < \infty$, then $s_n^2 \uparrow A^2$, say, as $n \to \infty$. For fixed k, we can find ε_k such that $\varepsilon_k A < a_k$ and then $P\{|X_k| > \varepsilon_k s_n\} \geq P\{|X_k| > \varepsilon_k A\} > 0$. For $n \geq k$, we have

$$\begin{aligned}\frac{1}{s_n^2}\sum_{j=1}^n \int_{|x|>\varepsilon_k s_n} x^2 f_j(x)\,dx &\geq \frac{s_n^2\varepsilon_k^2}{s_n^2}\sum_{j=1}^n P\{|X_j| > \varepsilon_k s_n\}\\ &\geq \varepsilon_k^2 P\{|X_k| > \varepsilon_k s_n\}\\ &> 0,\end{aligned}$$

so that the Lindeberg condition does not hold. Indeed, if $X_1, X_2, \ldots$ are independent RVs such that there exists a constant A with $P\{|X_n| \leq A\} = 1$ for all n, the Lindeberg condition (2) is satisfied if $s_n^2 \to \infty$ as $n \to \infty$. To see this, suppose that $s_n^2 \to \infty$. Since the X_k's are uniformly bounded, so are the RVs $X_k - EX_k$. It follows that for every $\varepsilon > 0$ we can find an N_ε such that, for $n \geq N_\varepsilon$, $P\{|X_k - EX_k| < \varepsilon s_n, k = 1, 2, \ldots, n\} = 1$. The Lindeberg condition follows immediately. The converse also holds, for, if $\lim_{n\to\infty} s_n^2 < \infty$ and the Lindeberg condition holds, there exists a constant $A < \infty$ such that $s_n^2 \to A^2$. For any fixed j, we can find an $\varepsilon > 0$ such that $P\{|X_j - \mu_j| > \varepsilon A\} > 0$. Then, for $n \geq j$,

$$\begin{aligned}&\frac{1}{s_n^2}\sum_{k=1}^n \int_{|x-\mu_k|>\varepsilon s_n} (x-\mu_k)^2 f_k(x)\,dx\\ &\geq \varepsilon^2 \sum_{k=1}^n P\{|X_k - \mu_k| > \varepsilon s_n\}\\ &\geq \varepsilon^2 P\{|X_j - \mu_j| > \varepsilon A\}\\ &> 0,\end{aligned}$$

and the Lindeberg condition does not hold. This contradiction shows that $s_n^2 \to \infty$ is also a necessary condition that is, for a sequence of uniformly bounded independent RVs, a necessary and sufficient condition for the central limit theorem to hold is $s_n^2 \to \infty$ as $n \to \infty$.

Example 6. Let $X_1, X_2, \ldots$ be independent RVs such that $\alpha_k = E|X_k|^{2+\delta} < \infty$ for some $\delta > 0$ and $\alpha_1 + \alpha_2 + \cdots + \alpha_n = o(s_n^{2+\delta})$. Then the Lindeberg condition is satisfied, and the central limit theorem holds. This result is due to Lyapunov. We have

$$\begin{aligned}&\frac{1}{s_n^2}\sum_{k=1}^n \int_{|x|>\varepsilon s_n} x^2 f_k(x)\,dx\\ &\leq \frac{1}{\varepsilon^\delta s_n^{2+\delta}}\sum_{k=1}^n \int_{-\infty}^{\infty} |x|^{2+\delta} f_k(x)\,dx\\ &= \frac{\sum_{k=1}^n \alpha_k}{\varepsilon^\delta s_n^{2+\delta}} \to 0 \qquad \text{as } n \to \infty.\end{aligned}$$

A similar argument applies in the discrete case.

Remark 3. Both the central limit theorem (CLT) and the (weak) law of large numbers (WLLN) hold for a large class of sequences of RVs $\{X_n\}$. If the $\{X_n\}$ are independent uniformly bounded RVs, that is, if $P\{|X_n| \le M\} = 1$, the WLLN (Theorem 7.3.1) holds; the CLT holds provided that $s_n^2 \to \infty$ (Example 5).

If the RVs $\{X_n\}$ are iid, then the CLT is a stronger result than the WLLN in that the former provides an estimate of the probability $P\{|S_n - n\mu|/n \ge \varepsilon\}$. Indeed,

$$\begin{aligned} P\{|S_n - n\mu| > n\varepsilon\} &= P\left\{\frac{|S_n - n\mu|}{\sigma\sqrt{n}} > \frac{\varepsilon}{\sigma}\sqrt{n}\right\} \\ &\approx 1 - P\left\{|Z| \le \frac{\varepsilon}{\sigma}\sqrt{n}\right\}, \end{aligned}$$

where Z is $\mathcal{N}(0,1)$, and the law of large number follows. On the other hand, we note that the WLLN does not require the existence of a second moment.

Remark 4. If $\{X_n\}$ are independent RVs, it is quite possible that the CLT may apply to the X_n's, but not the WLLN.

Example 7 (Feller [25, p. 255]). Let $\{X_k\}$ be independent RVs with PMF

$$P\{X_k = k^\lambda\} = P\{X_k = -k^\lambda\} = \frac{1}{2}, \qquad k = 1, 2, \ldots.$$

Then $EX_k = 0$, $\text{var}(X_k) = k^{2\lambda}$. Also let $\lambda > 0$, then

$$s_n^2 = \sum_{k=1}^{n} k^{2\lambda} \le \int_0^{n+1} x^{2\lambda}\,dx = \frac{(n+1)^{2\lambda+1}}{2\lambda+1}$$

It follows that, if $0 < \lambda < \frac{1}{2}$, $s_n/n \to 0$, and by Corollary 2 to Theorem 7.3.1 the WLLN holds. Now $k^\lambda < n^\lambda$, so that the sum $\sum_{k=1}^{n}\sum_{|x_{kl}|>\varepsilon s_n} x_{kl}^2 p_{kl}$ will be nonzero if $n^\lambda > \varepsilon s_n \approx \varepsilon[n^{\lambda+1/2}/\sqrt{(2\lambda+1)}]$. It follows that, as long as $n > (2\lambda+1)\varepsilon^{-2}$,

$$\frac{1}{s_n^2}\sum_{k=1}^{n}\sum_{|x_{kl}|>\varepsilon s_n} x_{kl}^2 p_{kl} = 0$$

and the Lindeberg condition holds. Thus the CLT holds for $\lambda > 0$. This means that

$$P\left\{a < \sqrt{\frac{2\lambda+1}{n^{2\lambda+1}}}S_n < b\right\} \to \int_a^b \frac{e^{-t^2/2}\,dt}{\sqrt{2\pi}}.$$

Thus

$$P\left\{\frac{an^{\lambda+1/2-1}}{\sqrt{2\lambda+1}} < \frac{S_n}{n} < \frac{bn^{\lambda+1/2-1}}{\sqrt{2\lambda+1}}\right\} \to \int_a^b \frac{e^{-t^2/2}}{\sqrt{2\pi}}\,dt$$

and the WLLN cannot hold for $\lambda \ge \frac{1}{2}$.

We conclude this section with some remarks concerning the application of the CLT. Let $X_1, X_2, \ldots$ be iid RVs with common mean μ and variance σ^2. Let us write

$$Z_n = \frac{S_n - n\mu}{\sigma\sqrt{n}},$$

and let z_1, z_2 be two arbitrary real numbers with $z_1 < z_2$. If F_n is the DF of Z_n, then

$$\begin{aligned}\lim_{n\to\infty} P\{z_1 < Z_n \le z_2\} &= \lim_{n\to\infty} [F_n(z_2) - F_n(z_1)] \\ &= \frac{1}{\sqrt{2\pi}} \int_{z_1}^{z_2} e^{-t^2/2}\,dt,\end{aligned}$$

that is,

$$\lim_{n\to\infty} P\{z_1\sigma\sqrt{n} + n\mu < S_n \le z_2\sigma\sqrt{n} + n\mu\} = \frac{1}{\sqrt{2\pi}} \int_{z_1}^{z_2} e^{-t^2/2}\,dt. \tag{4}$$

It follows that the RV $S_n = \sum_{k=1}^{n} X_k$ is *asymptotically normally distributed* with mean $n\mu$ and variance $n\sigma^2$. Equivalently, the RV $n^{-1}S_n$ is asymptotically $\mathcal{N}(\mu, \sigma^2/n)$. This result is of great importance in statistics.

In Fig. 1 we show the distribution of $\overline{X}$ in sampling from $P(\lambda)$ and $G(1,1)$. We have also superimposed, in each case, the graph of the corresponding normal approximation.

How large should n be before we apply approximation (4)? Unfortunately the answer is not simple. Much depends on the underlying distribution, the corresponding speed of convergence, and the accuracy one desires. There is a vast amount of literature on the speed of convergence and error bounds. We will content ourselves with some examples. The reader is referred to Rohatgi [90] for a detailed discussion.

In the discrete case when the underlying distribution is integer-valued, approximation (4) is improved by applying the *continuity correction*. If X is integer-valued, then for integers x_1, x_2

$$P\{x_1 \le X \le x_2\} = P\{x_1 - 1/2 < X < x_2 + 1/2\},$$

which amounts to making the discrete space of values of X continuous by considering intervals of length 1 with midpoints at integers.

Example 8. Let $X_1, X_2, \ldots, X_n$ be iid $b(1,p)$ RVs. Then $ES_n = np$ and $\text{var}(S_n) = np(1-p)$ so $(S_n - np)/\sqrt{np(1-p)}$ is approximately $\mathcal{N}(0,1)$.

Suppose $n = 10, p = 1/2$. Then from binomial tables $P(X \le 4) = 0.3770$. Using normal approximation without continuity correction

$$P(X \le 4) \approx P\left(Z \le \frac{4-5}{\sqrt{2.5}}\right) = P(Z \le -0.63) = 0.2643.$$

Applying continuity correction,

$$P(X \le 4) = P(X < 4.5) \approx P(Z \le -0.32) = 0.3745.$$

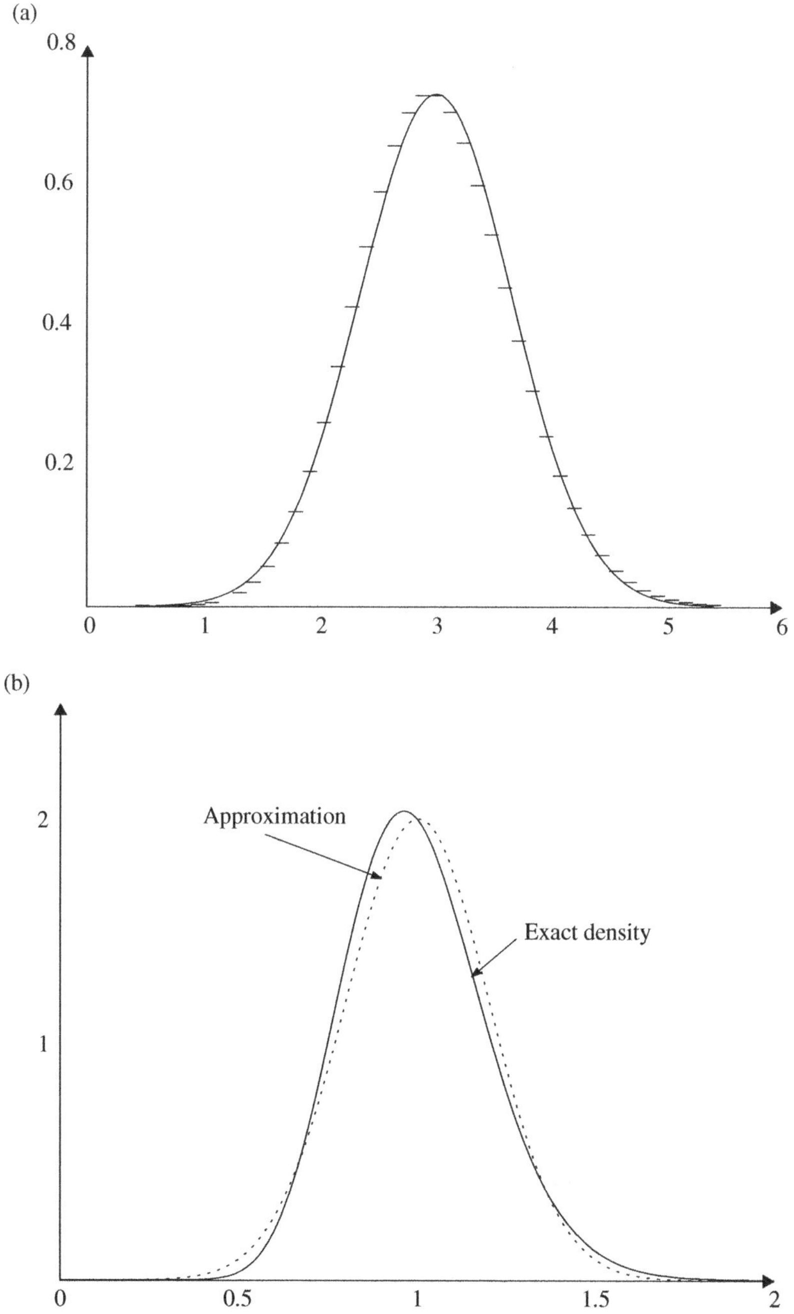

Fig. 1 (*a*) Distribution of $\overline{X}$ for Poisson RV with mean 3 and normal approximation and (*b*) distribution of $\overline{X}$ for exponential RV with mean 1 and normal approximation.

Next suppose that $n = 100$, $p = 0.1$. Then from binomial tables $P(X = 7) = 0.0889$. Using normal approximation, without continuity correction

$$\begin{aligned} P(X=7) &= P(6.0 < X < 8.0) \approx P(-1.33 < Z < -0.67) \\ &= 0.1596 \end{aligned}$$

and with continuity correction

$$\begin{aligned} P(X=7) &= P(6.5 < X < 7.5) \approx P(-1.17 < Z < -0.83) \\ &= 0.0823 \end{aligned}$$

The rule of thumb is to use continuity correction, and use normal approximation whenever $np(1-p) > 10$, and use Poisson approximation with $\lambda = np$ for $p < 0.1$, $\lambda \le 10$.

Example 9. Let $X_1, X_2, \ldots$ be iid $P(\lambda)$ RVs. Then S_n has approximately an $\mathcal{N}(n\lambda, n\lambda)$ distribution for large n. Let $n = 64$, $\lambda = 0.125$. Then $S_n \sim P(8)$ and from Poisson distribution tables $P(S_n = 10) = 0.099$. Using normal approximation

$$\begin{aligned} P(S_n = 10) &= P(9.5 < S_n < 10.5) \approx P(0.53 < Z < 0.88) \\ &= 0.1087. \end{aligned}$$

If $n = 96$, $\lambda = 0.125$, then $S_n \sim P(12)$ and

$$\begin{aligned} P(S_n = 10) &= 0.105, \quad \text{exact}, \\ P(S_n = 10) &\approx 0.1009, \quad \text{normal approximation}. \end{aligned}$$

PROBLEMS 7.6

1. Let $\{X_n\}$ be a sequence of independent RVs with the following distributions. In each case, does the Lindeberg condition hold?
   (a) $P\{X_n = \pm(1/2^n)\} = \frac{1}{2}$.
   (b) $P\{X_n = \pm 2^{n+1}\} = 1/2^{n+3}$, $P\{X_n = 0\} = 1 - (1/2^{n+2})$.
   (c) $P\{X_n = \pm 1\} = (1 - 2^{-n})/2$, $P\{X_n = \pm 2^{-n}\} = 1/2^{n+1}$.
   (d) $\{X_n\}$ is a sequence of independent Poisson RVs with parameter λ_n, $n = 1, 2, \ldots$, such that $\sum_{k=1}^{n} \lambda_k \to \infty$.
   (e) $P\{X_n = \pm 2^n\} = \frac{1}{2}$.
2. Let $X_1, X_2, \ldots$ be iid RVs with mean 0, variance 1, and $EX_i^4 < \infty$. Find the limiting distribution of

$$Z_n = \sqrt{n} \frac{X_1X_2 + X_3X_4 + \cdots + X_{2n-1}X_{2n}}{X_1^2 + X_2^2 + \cdots + X_{2n}^2}.$$

3. Let $X_1, X_2, \ldots$ be iid RVs with mean α and variance σ^2, and let $Y_1, Y_2, \ldots$ be iid RVs with mean β $(\neq 0)$ and variance τ^2. Find the limiting distribution of $Z_n = \sqrt{n}(\overline{X}_n - \alpha)/\overline{Y}_n$, where $\overline{X}_n = n^{-1}\sum_{i=1}^{n} X_i$ and $\overline{Y}_n = n^{-1}\sum_{i=1}^{n} Y_i$.

4. Let $X \sim b(n,\theta)$. Use the CLT to find n such that $P_\theta\{X > n/2\} \ge 1-\alpha$. In particular, let $\alpha = 0.10$ and $\theta = 0.45$. Calculate n, satisfying $P\{X > n/2\} \ge 0.90$.

5. Let $X_1, X_2, \ldots$ be a sequence of iid RVs with common mean μ and variance σ^2. Also, let $\overline{X} = n^{-1}\sum_{k=1}^{n} X_k$ and $S^2 = (n-1)^{-1}\sum_{i=1}^{n}(X_i - \overline{X})^2$. Show that $\sqrt{n}(\overline{X}-\mu)/S \xrightarrow{L} Z$, where $Z \sim \mathcal{N}(0,1)$.

6. Let $X_1, X_2, \ldots, X_{100}$ be iid RVs with mean 75 and variance 225. Use Chebychev's inequality to calculate the probability that the sample mean will not differ from the population mean by more than 6. Then use the CLT to calculate the same probability and compare your results.

7. Let $X_1, X_2, \ldots, X_{100}$ be iid $P(\lambda)$ RVs, where $\lambda = 0.02$. Let $S = S_{100} = \sum_{i=1}^{100} X_i$. Use the central limit result to evaluate $P\{S \ge 3\}$ and compare your result to the exact probability of the event $S \ge 3$.

8. Let $X_1, X_2, \ldots, X_{81}$ be iid RVs with mean 54 and variance 225. Use Chebychev's inequality to find the possible difference between the sample mean and the population mean with a probability of at least 0.75. Also use the CLT to do the same.

9. Use the CLT applied to a Poisson RV to show that $\lim_{n\to\infty} e^{-nt}\sum_{k=1}^{n-1}\frac{(nt)^k}{k!} = 1$ for $0 < t < 1$, $= \frac{1}{2}$ if $t = 1$, and 0 if $t > 1$.

10. Let $X_1, X_2, \ldots$ be a sequence of iid RVs with mean μ and variance σ^2, and assume that $EX_1^4 < \infty$. Write $V_n = \sum_{k=1}^{n}(X_k - \mu)^2$. Find the centering and norming constants A_n and B_n such that $B_n^{-1}(V_n - A_n) \xrightarrow{L} Z$, where Z is $\mathcal{N}(0,1)$.

11. From an urn containing 10 identical balls numbered 0 through 9, n balls are drawn with replacement.

(a) What does the law of large numbers tell you about the appearance of 0's in the n drawings?

(b) How many drawings must be made in order that, with probability at least 0.95, the relative frequency of the occurrence of 0's will be between 0.09 and 0.11?

(c) Use the CLT to find the probability that among the n numbers thus chosen the number 5 will appear between $(n - 3\sqrt{n})/10$ and $(n + 3\sqrt{n})/10$ times (inclusive) if (i) $n = 25$ and (ii) $n = 100$.

12. Let $X_1, X_2, \ldots, X_n$ be iid RVs with $EX_1 = 0$ and $EX_1^2 = \sigma^2 < \infty$. Let $\overline{X} = \sum_{i=1}^{n} X_i/n$, and for any positive real number ε let $P_{n,\varepsilon} = P\{\overline{X} \ge \varepsilon\}$. Show that

$$P_{n,\varepsilon} \approx \frac{\sigma}{\varepsilon\sqrt{n}}\frac{1}{\sqrt{2\pi}}e^{-n\varepsilon^2/2\sigma^2}, \qquad \text{as } n \to \infty.$$

[*Hint:* Use (5.3.61).]

7.7 LARGE SAMPLE THEORY

In many applications of probability one needs the distribution of a statistic or some function of it. The methods of Section 7.3 when applicable lead to the exact distribution of the statistic under consideration. If not, it may be sufficient to approximate this distribution provided the sample size is large enough.

Let $\{X_n\}$ be a sequence of RVs which converges in law to $N(\mu, \sigma^2)$. Then $\{(X_n - \mu)/\sigma)\}$ converges in law to $N(0,1)$, and conversely. We will say alternatively and equivalently that $\{X_n\}$ is *asymptotically normal* with mean μ and variance σ^2. More generally, we say that X_n is *asymptotically normal* with "mean" μ_n and "variance" σ_n^2, and write X_n is $\text{AN}(\mu_n, \sigma_n^2)$, if $\sigma_n > 0$ and as $n \to \infty$.

$$\frac{X_n - \mu_n}{\sigma_n} \xrightarrow{L} \mathcal{N}(0,1). \tag{1}$$

Here μ_n is not necessarily the mean of X_n and σ_n^2, not necessarily its variance. In this case we can approximate, for sufficiently large n, $P(X_n \le t)$ by $P\left(Z \le \frac{t-\mu_n}{\sigma_n}\right)$, where Z is $\mathcal{N}(0,1)$.

The most common method to show that X_n is $\text{AN}(\mu_n, \sigma_n^2)$ is the central limit theorem of Section 6. Thus, according to Theorem 7.6.1 $\sqrt{n}(\overline{X}_n - \mu) \xrightarrow{L} \mathcal{N}(0, \sigma^2)$ as $n \to \infty$, where $\overline{X}_n$ is the sample mean of n iid RVs with mean μ and variance σ^2. The same result applies to kth sample moment, provided $E|X|^{2k} < \infty$. Thus

$$\sum_{j=1}^{n} X_n^k/n \text{ is AN}\left(EX^k, \frac{\text{var}(X^k)}{n}\right).$$

In many large sample approximations an application of the CLT along with Slutsky's theorem suffices.

Example 1. Let $X_1, X_2, \ldots$ be iid $\mathcal{N}(\mu, \sigma^2)$. Consider the RV

$$T_n = \frac{\sqrt{n}(\overline{X} - \mu)}{S}.$$

The statistic T_n is well-known for its applications in statistics and in Section 6.5 we determined its exact distribution. From Example 6.3.4 $(n-1)S^2/n \xrightarrow{P} \sigma^2$ and hence $S/\sigma \xrightarrow{P} 1$. Since $\sqrt{n}(\overline{X} - \mu)/\sigma \xrightarrow{L} Z \sim \mathcal{N}(0,1)$, it follows from Slutsky's theorem that $T_n \xrightarrow{L} Z$. Thus for sufficiently large n ($n \ge 30$) we can approximate $P(T_n \le t)$ by $P(Z \le t)$.

Actually, we do not need X's to be normally distributed (see Problem 7.6.5).

Often we need to approximate the distribution of $g(Y_n)$ given that Y_n is $\text{AN}(\mu, \sigma^2)$.

Theorem 1 (Delta Method). Suppose Y_n is $\text{AN}(\mu, \sigma_n^2)$, with $\sigma_n \to 0$ and μ a fixed real number. Let g be a real-valued function which is differentiable at $x = \mu$, with $g'(\mu) \ne 0$. Then

$$g(Y_n) \text{ is AN}\left(g(\mu), [g'(\mu)]^2 \sigma_n^2\right). \tag{2}$$

Proof. We first show that

$$\frac{[g(Y_n) - g(\mu)]}{g'(\mu)\sigma_n} - \frac{(Y_n - \mu)}{\sigma_n} \xrightarrow{P} 0. \tag{3}$$

Set

$$h(x) = \begin{cases} \frac{g(x)-g(\mu)}{x-\mu} - g'(\mu), & x \neq \mu \\ 0, & x = \mu. \end{cases}$$

Then h is continuous at $x = \mu$. Since

$$Y_n - \mu = \sigma_n \left[\frac{Y_n - \mu}{\sigma_n}\right] \xrightarrow{L} 0$$

by Problem 7.2.7, $Y_n - \mu \xrightarrow{P} 0$, and it follows from Theorem 7.2.4 that $h(Y_n) \xrightarrow{P} h(\mu) = 0$. By Slutsky's theorem, therefore,

$$h(Y_n)\frac{Y_n - \mu}{\sigma_n} \xrightarrow{P} 0.$$

That is,

$$\frac{g(Y_n) - g(\mu)}{\sigma_n g'(\mu)} - \frac{Y_n - \mu}{\sigma_n} \xrightarrow{P} 0.$$

It follows again by Slutsky's theorem that $[g(Y_n) - g(\mu)]/[g'(\mu)\sigma_n]$ has the same limit law as $(Y_n - \mu)/\sigma_n$.

Example 2. We know by CLT theorem that $Y_n = \overline{X}$ is AN$(\mu, \sigma^2/n)$. Suppose $g(\overline{X}) = \overline{X}(1-\overline{X})$ where $\overline{X}$ is the sample mean in random sampling from a population with mean μ and variance σ^2. Since $g'(\mu) = 1 - 2\mu \neq 0$ for $\mu \neq 1/2$, it follows that for $\mu \neq 1/2$, $\sigma^2 < \infty$, $\overline{X}(1-\overline{X})$ is AN$(\mu(1-\mu), (1-2\mu)^2\sigma^2/n)$. Thus

$$\begin{aligned} P(\overline{X}(1-\overline{X}) \leq y) &= P\left(\frac{\overline{X}(1-\overline{X}) - \mu(1-\mu)}{|1-2\mu|\sigma/\sqrt{n})} \leq \frac{y - \mu(1-\mu)}{|1-2\mu|\sigma/\sqrt{n}}\right) \\ &\approx \Phi\left(\frac{y-\mu(1-\mu)}{|1-2\mu|\sigma/\sqrt{n}}\right) \end{aligned}$$

for large n.

Remark 1. Suppose g in Theorem 1 is differentiable k times, $k \geq 1$, at $x = \mu$ and $g^{(i)}(\mu) = 0$ for $1 \leq i \leq k-1$, $g^{(k)}(\mu) \neq 0$. Then a similar argument using Taylor's theorem shows that

$$[g(Y_n) - g(\mu)]/\left\{\frac{1}{k!}g^{(k)}(\mu)\sigma_n^k\right\} \xrightarrow{L} Z^k, \tag{4}$$

where Z is a $\mathcal{N}(0,1)$ RV. Thus in Example 2, when $\mu = 1/2$, $g'(1/2) = 0$ and $g''(1/2) = -2 \neq 0$. It follows that

$$n[\overline{X}(1-\overline{X}) - 1/4] \xrightarrow{L} -\sigma^2\chi^2(1)$$

since $Z^2 \stackrel{d}{=} \chi^2(1)$.

Remark 2. Theorem 1 can be extended to the multivariate case but we will not pursue the development. We refer the reader to Ferguson [29] or Serfling [102].

Remark 3. In general the asymptotic variance $[g'(\mu)]^2\sigma_n^2$ of $g(Y_n)$ will depend on the parameter μ. In problems of inference it will often be desirable to use transformation g such that the approximate variance $\operatorname{var} g(Y_n)$ is free of the parameter. Such transformations are called *variance stabilizing transformations*. Let us write $\sigma_n^2 = \sigma^2(\mu)/n$. Then finding a g such that var $g(Y_n)$ is free of μ is equivalent to finding a g such that

$$g'(\mu) = c/\sigma(\mu)$$

for all μ, where c is a constant independent of μ. It follows that

$$g(x) = c\int \frac{dx}{\sigma(x)}. \tag{5}$$

Example 3. In Example 2, $\sigma^2(\mu) = \mu(1-\mu)$. Suppose $X_1,\ldots,X_n$ are iid $b(1,p)$. Then $\sigma^2(p) = p(1-p)$ and (5) reduces to

$$g(x) = c\int \frac{dx}{\sqrt{x(1-x)}} = 2\arcsin\sqrt{x}.$$

Since $g(0) = 0$, $g(1) = 1$, $c = (2/\pi)$, and $g(x) = (2/\pi)\arcsin\sqrt{x}$.

Remark 4. In Section 6.3 we computed exact moments of some statistics in terms of population parameters. Approximations for moments of $g(\overline{X})$ can also be obtained from series expansions of g. Suppose g is twice differentiable at $x = \mu$. Then

$$Eg(X) \approx g(\mu) + E(X-\mu)g'(\mu) + \frac{1}{2}g''(\mu)E(X-\mu)^2 \tag{6}$$

and

$$E[g(X)-g(\mu)]^2 \approx [g'(\mu)]^2 E(X-\mu)^2, \tag{7}$$

by dropping remainder terms. The case of most interest is to approximate $Eg(\overline{X})$ and $\operatorname{var} g(\overline{X})$. In this case, under suitable conditions, one can show that

$$Eg(\overline{X}) \approx g(\mu) + \frac{\sigma^2}{2n}g''(\mu) \tag{8}$$

and

$$\operatorname{var} g(\overline{X}) \approx \frac{\sigma^2}{n}[g'(\mu)]^2, \tag{9}$$

where $E\overline{X} = \mu$ and $\operatorname{var}(X) = \sigma^2$.

In Example 2, when X_i's are iid $b(1,p)$, and $g(x)=x(1-x)$, $g'(x)=1-2x$, $g''(x)=-2$ so that

$$\begin{aligned} Eg(\overline{X}) &\approx E[\overline{X}(1-\overline{X})] \approx p(1-p)+\frac{\sigma^2}{2n}(-2) \\ &= p(1-p)\frac{n-1}{n} \end{aligned}$$

and

$$\operatorname{var} g(\overline{X}) \approx \frac{p(1-p)}{n}(1-2p)^2.$$

In this case we can compute $Eg(\overline{X})$ and $\operatorname{var} g(\overline{X})$ exactly. We have

$$Eg(\overline{X}) = E\overline{X} - E\overline{X}^2 = p - \left(\frac{p(1-p)}{n}+p^2\right) = p(1-p)\frac{n-1}{n}$$

so that (8) is exact. Also since $X_i^k = X_i$, using Theorem 6.3.4 we have

$$\begin{aligned} \operatorname{var} g(\overline{X}) &= \operatorname{var}(\overline{X}-\overline{X}^2) \\ &= \operatorname{var}\overline{X} - 2\operatorname{cov}(\overline{X},\overline{X}^2) + E\overline{X}^4 - (E\overline{X}^2)^2 \\ &= \frac{p(1-p)}{n}\left\{(1-2p)^2 + \frac{2p(1-p)}{n-1}\right\}\left(\frac{n-1}{n}\right)^2. \end{aligned}$$

Thus the error in approximation (9) is

$$\text{Error} = \frac{2p^2(1-p)^2}{n^3}(n-1).$$

Remark 5. Approximations (6) through (9) do not assert the existence of $Eg(X)$ or $Eg(\overline{X})$, or $\operatorname{var} g(X)$ or $\operatorname{var} g(\overline{X})$.

Remark 6. It is possible to extend (6) through (9) to two (or more) variables by using Taylor series expansion in two (or more) variables.

Finally, we state the following result which gives the asymptotic distribution of the rth order statistic, $1 \le r \le n$, in sampling from a population with an absolutely continuous DF F with PDF f. For a proof see Problem 4.

Theorem 2. If $X_{(r)}$ denotes the rth order statistic of a sample $X_1, X_2, \ldots, X_n$ from an absolutely continuous DF F with PDF f, then

$$\left\{\frac{n}{p(1-p)}\right\}^{1/2} f(\mathfrak{z}_p)\{X_{(r)} - \mathfrak{z}_p\} \xrightarrow{L} Z \quad \text{as } n \to \infty, \tag{10}$$

so that r/n remains fixed, $r/n = p$, where Z is $\mathcal{N}(0,1)$, and $\mathfrak{z}_p$ is the unique solution of $F(\mathfrak{z}_p) = p$ (that is, $\mathfrak{z}_p$ is the population quantile of order p assumed unique).

Remark 7. The sample quantile of order p, Z_p, is

$$\text{AN}\left(\mathfrak{z}_p, \frac{1}{[f(\mathfrak{z}_p)]^2}\frac{p(1-p)}{n}\right),$$

where $\mathfrak{z}_p$ is the corresponding population quantile, and f is the PDF of the population distribution function. It also follows that $Z_p \xrightarrow{P} \mathfrak{z}_p$.

PROBLEMS 7.7

1. In sampling from a distribution with mean μ and variance σ^2 find the asymptotic distribution of
(a) $\overline{X}^2$, (b) $1/\overline{X}$, (c) $\ell n|\overline{X}|^2$, (d) $\exp(\overline{X})$
both when $\mu \neq 0$ and when $\mu = 0$.

2. Let $X \sim P(\lambda)$. Then $(X-\lambda)/\sqrt{\lambda} \xrightarrow{L} \mathcal{N}(0,1)$. Find a transformation g such that $(g(X)-g(\lambda))$ has an asymptotic $\mathcal{N}(0,c)$ distribution for large μ where c is a suitable constant.

3. Let $X_1, X_2, \ldots, X_n$ be a sample from an absolutely continuous DF F with PDF f. Show that

$$EX_{(r)} \approx F^{-1}\left(\frac{r}{n+1}\right) \quad \text{and}$$

$$\text{var}(X_{(r)}) \approx \frac{r(n-r+1)}{(n+1)^2(n+2)} \frac{1}{\{f[F^{-1}(r/n+1)]\}^2}.$$

[*Hint*: Let Y be an RV with mean μ and ϕ be a Borel function such that $E\phi(Y)$ exists. Expand $\phi(Y)$ about the point μ by a Taylor series expansion, and use the fact that $F(X_{(r)}) = U_{(r)}$.]

4. Prove Theorem 2. [*Hint:* For any real μ and σ (>0) compute the PDF of $(U_{(r)}-\mu)/\sigma$ and show that the standardized $U_{(r)}$, $(U_{(r)}-\mu)/\sigma$, is asymptotically $\mathcal{N}(0,1)$ under the conditions of the theorem.]

5. Let $X \sim \chi^2(n)$. Then $(X-n)/\sqrt{2n}$ is $\text{AN}(0,1)$ and X/n is $\text{AN}\left(1,\frac{2}{n}\right)$. Find a transformation g such that the distribution of $g(X)-g(n)$ is $\text{AN}(0,c)$.

6. Suppose X is $G(1,\theta)$. Find g such that $g(\overline{X})-g(\theta)$ is $\text{AN}(0,c)$.

7. Let $X_1, X_2, \ldots, X_n$ be iid RVs with $E|X_1|^4 < \infty$. Let $\text{var}(X) = \sigma^2$ and $\beta_2 = \mu_4/\sigma^4$:
(a) Show, using the CLT for iid RVs, that $\sqrt{n}(S^2-\sigma^2) \xrightarrow{L} \mathcal{N}(0, \mu_4-\sigma^4)$.
(b) Find a transformation g such that $g(S^2)$ has an asymptotic distribution which depends on β_2 alone but not on σ^2.

8

PARAMETRIC POINT ESTIMATION

8.1 INTRODUCTION

In this chapter we study the theory of point estimation. Suppose, for example, that a random variable X is known to have a normal distribution $\mathcal{N}(\mu, \sigma^2)$, but we do not know one of the parameters, say μ. Suppose further that a sample $X_1, X_2, \ldots, X_n$ is taken on X. The problem of point estimation is to pick a (one-dimensional) statistic $T(X_1, X_2, \ldots, X_n)$ that best estimates the parameter μ. The numerical value of T when the realization is $x_1, x_2, \ldots, x_n$ is frequently called an *estimate* of μ, while the statistic T is called an *estimator* of μ. If both μ and σ^2 are unknown, we seek a joint statistic $T = (U, V)$ as an estimator of (μ, σ^2).

In Section 8.2 we formally describe the problem of parametric point estimation. Since the class of all estimators in most problems is too large it is not possible to find the "best" estimator in this class. One narrows the search somewhat by requiring that the estimators have some specified desirable properties. We describe some of these and also outline some criteria for comparing estimators.

Section 8.3 deals, in detail, with some important properties of statistics such as sufficiency, completeness, and ancillarity. We use these properties in later sections to facilitate our search for optimal estimators. Sufficiency, completeness, and ancillarity also have applications in other branches of statistical inference such as testing of hypotheses and nonparametric theory.

In Section 8.4 we investigate the criterion of unbiased estimation and study methods for obtaining optimal estimators in the class of unbiased estimators. In Section 8.5 we derive two lower bounds for variance of an unbiased estimator. These bounds can sometimes help in obtaining the "best" unbiased estimator.

An Introduction to Probability and Statistics, Third Edition. Vijay K. Rohatgi and A.K. Md. Ehsanes Saleh.

In Section 8.6 we describe one of the oldest methods of estimation and in Section 8.7 we study the method of maximum likelihood estimation and its large sample properties. Section 8.8 is devoted to Bayes and minimax estimation, and Section 8.9 deals with equivariant estimation.

8.2 PROBLEM OF POINT ESTIMATION

Let $\mathbf{X}$ be an RV defined on a probability space $(\Omega, \mathcal{S}, P)$. Suppose that the DF F of $\mathbf{X}$ depends on a certain number of parameters, and suppose further that the functional form of F is known except perhaps for a finite number of these parameters. Let $\boldsymbol{\theta} = (\theta_1, \theta_2, \ldots, \theta_k)$ be the unknown parameter associated with F.

Definition 1. The set of all admissible values of the parameters of a DF F is called the parameter space.

Let $\mathbf{X} = (X_1, X_2, \ldots, X_n)$ be an RV with DF $F_{\boldsymbol{\theta}}$, where $\boldsymbol{\theta} = (\theta_1, \theta_2, \ldots, \theta_k)$ is a vector of unknown parameters, $\boldsymbol{\theta} \in \Theta$. Let ψ be a real-valued function on Θ. In this chapter we investigate the problem of approximating $\psi(\boldsymbol{\theta})$ on the basis of the observed value $\mathbf{x}$ of $\mathbf{X}$.

Definition 2. Let $\mathbf{X} = (X_1, X_2, \ldots, X_n) \sim P_{\boldsymbol{\theta}}, \boldsymbol{\theta} \in \Theta$. A statistic $\delta(\mathbf{X})$ is said to be a (point) estimator of ψ if $\delta : \mathfrak{X} \to \Theta$, where $\mathfrak{X}$ is the space of values of $\mathbf{X}$.

The problem of point estimation is to find an estimator δ for the unknown parametric function $\psi(\boldsymbol{\theta})$ that has some nice properties. The value $\delta(\mathbf{x})$ of $\delta(\mathbf{X})$ for the data $\mathbf{x}$ is called the estimate of $\psi(\boldsymbol{\theta})$.

In most problems $X_1, X_2, \ldots, X_n$ are iid RVs with common DF $F_{\boldsymbol{\theta}}$.

Example 1. Let $X_1, X_2, \ldots, X_n$ be iid $G(1, \theta)$, where $\Theta = \{\theta > 0\}$ and θ is to be estimated. Then $\mathfrak{X} = R_n^+$ and any map $\delta : \mathfrak{X} \to (0, \infty)$ is an estimator of θ. Some typical estimators of θ are $\bar{X} = n^{-1} \sum_{j=1}^n X_j$ and $\{2/[n(n+1)]\} \sum_{j=1}^n j X_j$.

Example 2. Let $X_1, X_2, \ldots, X_n$ be iid $b(1, p)$ RVs, where $p \in [0, 1]$. Then $\bar{X}$ is an estimator of p and so also are $\delta_1(\mathbf{X}) = X_1$, $\delta_2(\mathbf{X}) = (X_1 + X_n)/2$, and $\delta_3(\mathbf{X}) = \sum_{j=1}^n a_j X_j$, where $0 \le a_j \le 1$, $\sum_{j=1}^n a_j = 1$.

It is clear that in any given problem of estimation we may have a large, often an infinite, class of appropriate estimators to choose from. Clearly we would like the estimator δ to be close to $\psi(\boldsymbol{\theta})$, and since δ is a statistic, the usual measure of closeness $|\delta(\mathbf{X}) - \psi(\boldsymbol{\theta})|$ is also an RV, we interpret "δ close to ψ" to mean "close on the average." Examples of such measures of closeness are

$$P_{\boldsymbol{\theta}}\{|\delta(\mathbf{X}) - \psi(\boldsymbol{\theta})| < \varepsilon\} \tag{1}$$

for some $\varepsilon > 0$, and

$$E_{\boldsymbol{\theta}}|\delta(\mathbf{X}) - \psi(\boldsymbol{\theta})|^r \tag{2}$$

for some $r > 0$. Obviously we want (1) to be large whereas (2) to be small. For $r = 2$, the quantity defined in (2) is called *mean square error* and we denote it by

$$\text{MSE}_{\boldsymbol{\theta}}(\delta) = E_{\boldsymbol{\theta}}\{\delta(\mathbf{X}) - \psi(\boldsymbol{\theta})\}^2. \tag{3}$$

Among all estimators for ψ we would like to choose one say δ_0 such that

$$P_{\boldsymbol{\theta}}\{|\delta_0(\mathbf{X}) - \psi(\boldsymbol{\theta})| < \varepsilon\} \geq P_{\boldsymbol{\theta}}\{|\delta(\mathbf{X}) - \psi(\boldsymbol{\theta})| < \varepsilon\} \tag{4}$$

for all δ, all $\varepsilon > 0$ and all $\boldsymbol{\theta}$. In case of (2) the requirement is to choose δ_0 such that

$$\text{MSE}_{\boldsymbol{\theta}}(\delta_0) \leq \text{MSE}_{\boldsymbol{\theta}}(\delta) \tag{5}$$

for all δ, and all $\boldsymbol{\theta} \in \Theta$. Estimators satisfying (4) or (5) do not generally exist.

We note that

$$\begin{aligned}\text{MSE}_{\boldsymbol{\theta}}(\delta) &= E_{\boldsymbol{\theta}}\{\delta(\mathbf{X}) - E_{\boldsymbol{\theta}}\delta(\mathbf{X})\}^2 + \{E_{\boldsymbol{\theta}}\delta(\mathbf{X}) - \psi(\boldsymbol{\theta})\}^2 \\ &= \text{var}_{\boldsymbol{\theta}}\,\delta(\mathbf{X}) + \{b(\delta,\psi)\}^2,\end{aligned} \tag{6}$$

where

$$b(\delta,\psi) = E_{\boldsymbol{\theta}}\delta(\mathbf{X}) - \psi(\boldsymbol{\theta}) \tag{7}$$

is called the *bias* of δ. An estimator that has small MSE has small bias and variance. In order to control MSE, we need to control both variance and bias.

One approach is to restrict attention to estimators which have zero bias, that is,

$$E_{\boldsymbol{\theta}}\delta(\mathbf{X}) = \psi(\boldsymbol{\theta}) \qquad \text{for all } \boldsymbol{\theta} \in \Theta. \tag{8}$$

The condition of *unbiasedness* (8) ensures that, on the average the estimator δ has no systematic error; it neither over-nor underestimates ψ on the average. If we restrict attention only to the class of unbiased estimators then we need to find an estimator δ_0 in this class such that δ_0 has the least variance for all $\boldsymbol{\theta} \in \Theta$. The theory of unbiased estimation is developed in Section 8.4.

Another approach is to replace $|\delta - \psi|^r$ in (2) by a more general function. Let $L(\boldsymbol{\theta},\delta)$ measure the loss in estimating ψ by δ. Assume that L, the *loss function*, satisfies $L(\boldsymbol{\theta},\delta) \geq 0$ for all $\boldsymbol{\theta}$ and δ, and $L(\boldsymbol{\theta},\psi(\boldsymbol{\theta})) = 0$ for all $\boldsymbol{\theta}$. Measure average loss by the *risk function*

$$R(\boldsymbol{\theta},\delta) = E_{\theta}L(\boldsymbol{\theta},\delta(\mathbf{X})). \tag{9}$$

Instead of seeking an estimator which minimizes R the risk uniformly in θ, we minimize

$$\int R(\boldsymbol{\theta},\delta)\pi(\boldsymbol{\theta})\,d\boldsymbol{\theta} \tag{10}$$

for some weight function π on Θ and minimize

$$\sup_{\boldsymbol{\theta}\in\Theta} R(\boldsymbol{\theta},\delta). \tag{11}$$

The estimator that minimizes the average risk defined in (10) leads to the Bayes estimator and the estimator that minimizes (11) leads to the minimax estimator. Bayes and minimax estimation are discussed in Section 8.8.

Sometimes there are symmetries in the problem which may be used to restrict attention only to estimators which also exhibit the same symmetry. Consider, for example, an experiment in which the length of life of a light bulb is measured. Then an estimator obtained from the measurements expressed in hours and minutes must agree with an estimator obtained from the measurements expressed in minutes. If $\mathbf{X}$ represents measurements in original units (hours) and $\mathbf{Y}$ represents corresponding measurements in transformed units (minutes) then $\mathbf{Y} = c\mathbf{X}$ (here $c = 60$). If $\delta(\mathbf{X})$ is an estimator of the true mean, then we would expect $\delta(\mathbf{Y})$, the estimator of the true mean to correspond to $\delta(\mathbf{X})$ according to the relation $\delta(\mathbf{Y}) = c\delta(\mathbf{X})$. That is, $\delta(c\mathbf{X}) = c\delta(\mathbf{X})$, for all $c > 0$. This is an example of an *equivariant estimator* which is the topic under extensive discussion in Section 8.9.

Finally, we consider some large sample properties of estimators. As the sample size $n \to \infty$, the data $\mathbf{x}$ are practically the whole population, and we should expect $\delta(\mathbf{X})$ to approach $\psi(\boldsymbol{\theta})$ in some sense. For example, if $\delta(\mathbf{X}) = \overline{X}$, $\psi(\theta) = E_\theta X_1$, and $X_1, X_2, \ldots, X_n$ are iid RVs with finite mean then strong law of large numbers tells us that $\overline{X} \to E_\theta X_1$ with probability 1. This property of a sequence of estimators is called *consistency*.

Definition 3. Let $X_1, X_2, \ldots$ be a sequence of iid RVs with common DF $F_{\boldsymbol{\theta}}$, $\boldsymbol{\theta} \in \Theta$. A sequence of point estimators $T_n(X_1, X_2, \ldots, X_n) = T_n$ will be called consistent for $\psi(\boldsymbol{\theta})$ if

$$T_n \xrightarrow{P} \psi(\boldsymbol{\theta}) \qquad \text{as } n \to \infty$$

for each fixed $\boldsymbol{\theta} \in \Theta$.

Remark 1. Recall that $T_n \xrightarrow{P} \psi(\boldsymbol{\theta})$ if and only if $P\{|T_n - \psi(\boldsymbol{\theta})| > \varepsilon\} \to 0$ as $n \to \infty$ for every $\varepsilon > 0$. One can similarly define *strong consistency* of a sequence of estimators T_n if $T_n \xrightarrow{\text{a.s.}} \psi(\boldsymbol{\theta})$. Sometimes one speaks of *consistency in the rth mean* when $T_n \xrightarrow{r} \psi(\boldsymbol{\theta})$. In what follows, "consistency" will mean *weak consistency* of T_n for $\psi(\boldsymbol{\theta})$, that is, $T_n \xrightarrow{P} \psi(\boldsymbol{\theta})$.

It is important to remember that consistency is a large sample property. Moreover, we speak of consistency of a sequence of estimators rather than one point estimator.

Example 3. Let $X_1, X_2, \ldots$ be iid $b(1,p)$ RVs. Then $EX_1 = p$, and it follows by the WLLN that

$$\frac{\sum_1^n X_i}{n} \xrightarrow{P} p.$$

Thus $\overline{X}$ is consistent for p. Also $(\sum_1^n X_i + 1)/(n+2) \xrightarrow{P} p$, so that a consistent estimator need not be unique. Indeed, if $T_n \xrightarrow{P} p$, and $c_n \to 0$ as $n \to \infty$, then $T_n + c_n \xrightarrow{P} p$ and if $d_n \to 1$ then $d_n T_n \xrightarrow{P} p$.

Theorem 1. If $X_1, X_2 \ldots$ are iid RVs with common law $\mathcal{L}(X)$, and $E|X|^p < \infty$ for some positive integer p, then

$$\frac{\sum_1^n X_i^k}{n} \xrightarrow{P} EX^k \qquad \text{for } 1 \le k \le p,$$

and $n^{-1}\sum_1^n X_i^k$ is consistent for EX^k, $1 \le k \le p$. Moreover, if c_n is any sequence of constants such that $c_n \to 0$ as $n \to \infty$, then $\{n^{-1}\sum X_i^k + c_n\}$ is also consistent for EX^k, $1 \le k \le p$. Also, if $c_n \to 1$ as $n \to \infty$, then $\{c_n n^{-1}\sum X_i^k\}$ is consistent for EX^k. This is simply a restatement of the WLLN for iid RVs.

Example 4. Let $X_1, X_2, \ldots$ be iid $\mathcal{N}(\mu, \sigma^2)$ RVs. If S^2 is the sample variance, we know that $(n-1)S^2/\sigma^2 \sim \chi^2(n-1)$. Thus $E(S^2/\sigma^2) = 1$ and $\text{var}(S^2/\sigma^2) = 2/(n-1)$. It follows that

$$P\{|S^2 - \sigma^2| > \varepsilon\} \le \frac{\text{var}(S^2)}{\varepsilon^2} = \frac{2\sigma^4}{(n-1)\varepsilon^2} \to 0 \qquad \text{as } n \to \infty.$$

Thus $S^2 \xrightarrow{P} \sigma^2$. Actually, this result holds for any sequence of iid RVs with $E|X|^2 < \infty$ and can be obtained from Theorem 1.

Example 4 is a particular case of the following theorem.

Theorem 2. If T_n is a sequence of estimators such that $ET_n \to \psi(\boldsymbol{\theta})$ and $\text{var}(T_n) \to 0$ as $n \to \infty$, then T_n is consistent for $\psi(\boldsymbol{\theta})$.

Proof. We have

$$\begin{aligned} P\{T_n - \psi(\boldsymbol{\theta})| > \varepsilon\} &\le \varepsilon^{-2} E\{T_n - ET_n + ET_n - \psi(\boldsymbol{\theta})\}^2 \\ &= \varepsilon^{-2}\{\text{var}(T_n) + (ET_n - \psi(\boldsymbol{\theta}))^2\} \to 0 \qquad \text{as } n \to \infty. \end{aligned}$$

Other large sample of properties of estimators are asymptotic unbiasedness, asymptotic normality, and asymptotic efficiency. A sequence of estimators $\{T_n\}$ is *asymptotically unbiased* for $\psi(\boldsymbol{\theta})$ if

$$\lim_{n \to \infty} E_{\boldsymbol{\theta}} T_n(\mathbf{X}) = \psi(\boldsymbol{\theta})$$

for all $\boldsymbol{\theta}$. A consistent sequence of estimators $\{T_n\}$ is said to be consistent asymptotically normal (CAN) for $\psi(\boldsymbol{\theta})$ if $T_n \sim \text{AN}(\psi(\boldsymbol{\theta}), v(\boldsymbol{\theta})/n)$ for all $\boldsymbol{\theta} \in \Theta$. If $v(\theta) = 1/I(\theta)$, where $I(\theta)$ is the Fisher information (Section 8.7), then $\{T_n\}$ is known as a *best asymptotically normal* (BAN) estimator.

Example 5. Let $X_1, X_2, \ldots, X_n$ be iid $\mathcal{N}(\theta, 1)$ RVs. Then $T_n = \sum_{i=1}^n X_i/(n+1)$ is asymptotically unbiased for θ and BAN estimator for θ with $v(\theta) = 1$.

In Section 8.7 we consider large sample properties of maximum likelihood estimators and in Section 8.5 asymptotic efficiency is introduced.

PROBLEMS 8.2

1. Suppose that T_n is a sequence of estimators for parameter θ that satisfies the conditions of Theorem 2. Then $T_n \xrightarrow{2} \theta$, that is, T_n is squared error consistent for θ. If T_n is consistent for θ and $|T_n - \theta| \le A < \infty$ for all θ and all $(x_1, x_2, \ldots, x_n) \in \mathcal{R}_n$, show

that $T_n \xrightarrow{2} \theta$. If, however, $|T_n - \theta| \le A_n < \infty$, then show that T_n may not be squared error consistent for θ.

2. Let $X_1, X_2, \ldots, X_n$ be a sample from $U[0,\theta], \theta \in \Theta = (0,\infty)$. Let $X_{(n)} = \max\{X_1, X_2, \ldots, X_n\}$. Show that $X_{(n)} \xrightarrow{P} \theta$. Write $Y_n = 2\overline{X}$. Is Y_n consistent for θ?
3. Let $X_1, X_2, \ldots, X_n$ be iid RVs with $EX_i = \mu$ and $E|X_i|^2 < \infty$. Show that $T(X_1, X_2, \ldots, X_n) = 2[n(n+1)]^{-1} \sum_{i=1}^{n} iX_i$ is a consistent estimator for μ.
4. Let $X_1, X_2, \ldots, X_n$ be a sample from $U[0,\theta]$. Show that $T(X_1, X_2, \ldots, X_n) = (\prod_{i=1}^{n} X_i)^{1/n}$ is a consistent estimator for θe^{-1}.
5. In Problem 2 show that $T(\mathbf{X}) = X_{(n)}$ is asymptotically biased for θ and is not BAN. (Show that $n(\theta - X_{(n)}) \xrightarrow{L} G(1,\theta)$.)
6. In Problem 5 consider the class of estimators $T(\mathbf{X}) = cX_{(n)}$, $c > 0$. Show that the estimator $T_0(\mathbf{X}) = (n+2)X_{(n)}/(n+1)$ in this class has the least MSE.
7. Let $X_1, X_2, \ldots, X_n$ be iid with PDF $f_\theta(x) = \exp\{-(x-\theta)\}, x > \theta$. Consider the class of estimators $T(\mathbf{X}) = X_{(1)} + b$, $b \in \mathcal{R}$. Show that the estimator that has the smallest MSE in this class is given by $T(\mathbf{X}) = X_{(1)} - 1/n$.

8.3 SUFFICIENCY, COMPLETENESS AND ANCILLARITY

After the completion of any experiment, the job of a statistician is to interpret the data she has collected and to draw some statistically valid conclusions about the population under investigation. The raw data by themselves, besides being costly to store, are not suitable for this purpose. Therefore the statistician would like to condense the data by computing some statistics from them and to base her analysis on these statistics, provided that there is "no loss of information" in doing so. In many problems of statistical inference a function of the observations contains as much information about the unknown parameter as do all the observed values. The following example illustrates this point.

Example 1. Let $X_1, X_2, \ldots, X_n$ be a sample from $\mathcal{N}(\mu, 1)$, where μ is unknown. Suppose that we transform variables $X_1, X_2, \ldots, X_n$ to $Y_1, Y_2, \ldots, Y_n$ with the help of an orthogonal transformation so that Y_1 is $\mathcal{N}(\sqrt{n}\mu, 1)$, $Y_2, \ldots, Y_n$ are iid $\mathcal{N}(0,1)$, and $Y_1, Y_2, \ldots, Y_n$ are independent. (Take $y_1 = \sqrt{n}\bar{x}$, and, for $k = 2, \ldots, n$, $y_k = [(k-1)x_k - (x_1 + \cdots + x_{k-1})]/\sqrt{k(k-1)}$). To estimate μ we can use either the observed values of $X_1, X_2, \ldots, X_n$ or simply the observed value of $Y_1 = \sqrt{n}\overline{X}$. The RVs $Y_2, Y_3, \ldots, Y_n$ provide no information about μ. Clearly, Y_1 is preferable since one need not keep a record of all the observations; it suffices to cumulate the observations and compute y_1. Any analysis of the data based on y_1 is just as effective as any analysis that could be based on x_i's. We note that Y_1 takes values in $\mathcal{R}_1$ whereas $(X_1, X_2, \ldots, X_n)$ takes values in $\mathcal{R}_n$.

A rigorous definition of the concept involved in the above discussion requires the notion of a conditional distribution and is beyond the scope of this book. In view of the discussion of conditional probability distributions in Section 4.2, the following definition will suffice for our purposes.

Definition 1. Let $\mathbf{X} = (X_1, X_2, \ldots, X_n)$ be a sample from $\{F_\theta : \theta \in \Theta\}$. A statistic $T = T(\mathbf{X})$ is sufficient for θ or for the family of distributions $\{F_\theta : \theta \in \Theta\}$ if and only if the conditional distribution of X, given $T = t$, does not depend on θ (except perhaps for a null set $A, P_\theta\{T \in A\} = 0$ for all θ).

Remark 1. The outcome $X_1, X_2, \ldots, X_n$ is always sufficient, but we will exclude this trivial statistic from consideration. According to Definition 1, if T is sufficient for θ, we need only concentrate on T since it exhausts all the information that the sample has about θ. In practice, there will be several sufficient statistics for a family of distributions, and the question arises as to which of these should be used in a given problem. We will return to this topic in more detail later in this section.

Example 2. We show that the statistic Y_1 in Example 1 is sufficient for μ. By construction $Y_2, \ldots, Y_n$ are iid $\mathcal{N}(0,1)$ RVs that are independent of Y_1. Hence the conditional distribution of $Y_2, \ldots, Y_n$, given $Y_1 = \sqrt{n}\overline{X}$, is the same as the unconditional distribution of $(Y_2, \ldots, Y_n)$, which is multivariate normal with mean $(0, 0, \ldots, 0)$ and dispersion matrix $\mathbf{I}_{n-1}$. Since this distribution is independent of μ, the conditional distribution of $(Y_1, Y_2, \ldots, Y_n)$, and hence $(X_1, X_2, \ldots, X_n)$, given $Y_1 = y_1$, is also independent of μ and Y_1 is sufficient.

Example 3. Let $X_1, X_2, \ldots, X_n$ be iid $b(1, p)$ RVs. Intuitively, if a loaded coin is tossed with probability p of heads n times, it seems unnecessary to know which toss resulted in a head. To estimate p, it should be sufficient to know the number of heads in n trials. We show that this is consistent with our definition. Let $T(X_1, X_2, \ldots, X_n) = \sum_{i=1}^n X_i$. Then

$$P\left\{X_1 = x_1, \ldots, X_n = x_n \,\middle|\, \sum_{i=1}^n X_i = t\right\} = \frac{P\{X_1 = x_1, \ldots, X_n = x_n, T = t\}}{\binom{n}{t} p^t (1-p)^{n-t}},$$

if $\sum_1^n x_i = t$, and $= 0$ otherwise. Thus, for $\sum_1^n x_i = t$, we have

$$\begin{aligned} P\{X_1 = x_1, \ldots, X_n = x_n \mid T = t\} &= \frac{p^{\sum_1^n x_i}(1-p)^{n - \sum x_i}}{\binom{n}{t} p^t (1-p)^{n-t}} \\ &= \frac{1}{\binom{n}{t}}, \end{aligned}$$

which is independent of p. It is therefore sufficient to concentrate on $\sum_1^n X_i$.

Example 4. Let X_1, X_2 be iid $P(\lambda)$ RVs. Then $X_1 + X_2$ is sufficient for λ, for

$$\begin{aligned} &P\{X_1 = x_1, X_2 = x_2 \mid X_1 + X_2 = t\} \\ &\quad = \begin{cases} \dfrac{P\{X_1 = x_1, X_2 = t - x_1\}}{P\{X_1 + X_2 = t\}} & \text{if } t = x_1 + x_2, x_i = 0, 1, 2, \ldots, \\ 0 & \text{otherwise.} \end{cases} \end{aligned}$$

Thus, for $x_i = 0, 1, 2, \ldots$, $i = 1, 2, x_1 + x_2 = t$, we have

$$P\{X_1 = x_1, X_2 = x_2 \mid X_1 + X_2 = t\} = \binom{t}{x_1}\left(\frac{1}{2}\right)^t,$$

which is independent of λ.

Not every statistic is sufficient.

Example 5. Let X_1, X_2 be iid $P(\lambda)$ RVs, and consider the statistic $T = X_1 + 2X_2$. We have

$$\begin{aligned}
P\{X_1 = 0, X_2 = 1 \mid X_1 + 2X_2 = 2\} &= \frac{P\{X_1 = 0, X_2 = 1\}}{P\{X_1 + 2X_2 = 2\}} \\
&= \frac{e^{-\lambda}(\lambda e^{-\lambda})}{P\{X_1 = 0, X_2 = 1\} + P\{X_1 = 2, X_2 = 0\}} \\
&= \frac{\lambda e^{-2\lambda}}{\lambda e^{-2\lambda} + (\lambda^2/2)e^{-2\lambda}} = \frac{1}{1 + (\lambda/2)},
\end{aligned}$$

and we see that $X_1 + 2X_2$ is not sufficient for λ.

Definition 1 is not a constructive definition since it requires that we first guess a statistic T and then check to see whether T is sufficient. Moreover, the procedure for checking that T is sufficient is quite time-consuming. We now give a criterion for determining sufficient statistics.

Theorem 1 (The Factorization Criterion). Let $X_1, X_2, \ldots, X_n$ be discrete RVs with PMF $p_\theta(x_1, x_2, \ldots, x_n)$, $\theta \in \Theta$. Then $T(X_1, X_2, \ldots, X_n)$ is sufficient for θ if and only if we can write

$$p_\theta(x_1, x_2, \ldots, x_n) = h(x_1, x_2, \ldots, x_n)\, g_\theta(T(x_1, x_2, \ldots, x_n)), \tag{1}$$

where h is a nonnegative function of $x_1, x_2, \ldots, x_n$ only and does not depend on θ, and g_θ is a nonnegative nonconstant function of θ and $T(x_1, x_2, \ldots, x_n)$ only. The statistic $T(X_1, \ldots, X_n)$ and parameter θ may be multidimensional.

Proof. Let T be sufficient for θ. Then $P\{\mathbf{X} = \mathbf{x} \mid T = t\}$ is independent of θ, and we may write

$$\begin{aligned}
P_\theta\{\mathbf{X} = \mathbf{x}\} &= P_\theta\{\mathbf{X} = \mathbf{x}, T(X_1, X_2, \ldots, X_n) = t\} \\
&= P_\theta\{T = t\} P\{\mathbf{X} = \mathbf{x} \mid T = t\},
\end{aligned}$$

provided that $P\{\mathbf{X} = \mathbf{x} \mid T = t\}$ is well defined.

For values of $\mathbf{x}$ for which $P_\theta\{\mathbf{X} = \mathbf{x}\} = 0$ for all θ, let us define $h(x_1, x_2, \ldots, x_n) = 0$, and for $\mathbf{x}$ for which $P_\theta\{\mathbf{X} = \mathbf{x}\} > 0$ for some θ, we define

$$h(x_1, x_2, \ldots, x_n) = P\{X_1 = x_1, \ldots, X_n = x_n \mid T = t\}$$

and

$$g_\theta(T(x_1, x_2, \ldots, x_n)) = P_\theta\{T(x_1, \ldots, x_n) = t\}.$$

Thus we see that (1) holds.

Conversely, suppose that (1) holds. Then for fixed t_0 we have

$$\begin{aligned}
P_\theta\{T = t_0\} &= \sum_{(\mathbf{x}:\, T(\mathbf{x}) = t_0)} P_\theta\{\mathbf{X} = \mathbf{x}\} \\
&= \sum_{(\mathbf{x}:\, T(\mathbf{x}) = t_0)} g_\theta(T(\mathbf{x})) h(\mathbf{x}) \\
&= g_\theta(t_0) \sum_{T(\mathbf{x}) = t_0} h(\mathbf{x}).
\end{aligned}$$

Suppose that $P_\theta\{T=t_0\}>0$ for some $\theta>0$. Then

$$P_\theta\{\mathbf{X}=\mathbf{x}\mid T=t_0\}=\frac{P_\theta\{\mathbf{X}=\mathbf{x},T(\mathbf{x})=t_0\}}{P_\theta\{T(\mathbf{x})=t_0\}}=\begin{cases}0 & \text{if } T(\mathbf{x})\neq t_0,\\ \dfrac{P_\theta\{\mathbf{X}=\mathbf{x}\}}{P_\theta\{T(\mathbf{x})=t_0\}} & \text{if } T(\mathbf{x})=t_0.\end{cases}$$

Thus, if $T(\mathbf{x})=t_0$, then

$$\frac{P_\theta\{\mathbf{X}=\mathbf{x}\}}{P_\theta\{T(\mathbf{x})=t_0\}}=\frac{g_\theta(t_0)h(\mathbf{x})}{g_\theta(t_0)\sum_{T(\mathbf{x})=t_0}h(\mathbf{x})},$$

which is free of θ, as asserted. This completes the proof.

Remark 2. Theorem 1 holds also for the continuous case and, indeed, for quite arbitrary families of distributions. The general proof is beyond the scope of this book, and we refer the reader to Halmos and Savage [41] or to Lehmann [64, pp. 53–56]. We will assume that the result holds for the absolutely continuous case. We leave the reader to write the analog of (1) and to prove it, at least under the regularity conditions assumed in Theorem 4.4.2.

Remark 3. Theorem 1 (and its analog for the continuous case) holds if θ is a vector of parameters and T is a multiple RV, and we say that T is *jointly sufficient* for θ. We emphasize that, even if θ is scalar, T may be multidimensional (Example 9). If θ and T are of the same dimension, and if T is sufficient for θ, it does not follow that the jth component of T is sufficient for the jth component of θ (Example 8). The converse is true under mild conditions (see Fraser [32, p. 21]).

Remark 4. If T is sufficient for θ, any one-to-one function of T is also sufficient. This follows from Theorem 1, if $U=k(T)$ is a one-to-one function of T, then $t=k^{-1}(u)$ and we can write

$$f_\theta(\mathbf{x})=g_\theta(t)h(\mathbf{x})=g_\theta(k^{-1}(u))h(\mathbf{x})=g_\theta^*(u)h(\mathbf{x}).$$

If T_1,T_2 are two distinct sufficient statistics, then

$$f_\theta(\mathbf{x})=g_\theta(t_1)h_1(\mathbf{x})=g_\theta(t_2)h_2(\mathbf{x}),$$

and it follows that T_1 is a function of T_2. It does not follow, however, that every function of a sufficient statistic is itself sufficient. For example, in sampling from a normal population, $\overline{X}$ is sufficient for the mean μ but $\overline{X}^2$ is not. Note that $\overline{X}$ is sufficient for μ^2.

Remark 5. As a rule, Theorem 1 cannot be used to show that a given statistic T is not sufficient. To do this, one would normally have to use the definition of sufficiency. In most cases Theorem 1 will lead to a sufficient statistic if it exists.

Remark 6. If $T(\mathbf{X})$ is sufficient for $\{F_\theta\colon \theta\in\Theta\}$, then T is sufficient for $\{F_\theta\colon \theta\in\omega\}$, where $\omega\subseteq\Theta$. This follows trivially from the definition.

Example 6. Let $X_1, X_2, \ldots, X_n$ be iid $b(1,p)$ RVs. Then $T = \sum_{i=1}^{n} X_i$ is sufficient. We have

$$P_p\{X_1 = x_1, X_2 = x_2, \ldots, X_n = x_n\} = p^{\sum_1^n x_i}(1-p)^{n-\sum_1^n x_i},$$

and, taking

$$h(x_1, x_2, \ldots, x_n) = 1 \qquad \text{and} \qquad g_p(x_1, x_2, \ldots, x_n) = (1-p)^n \left(\frac{p}{1-p}\right)^{\sum_{i=1}^n x_i},$$

we see that T is sufficient. We note that $T_1(\mathbf{X}) = (X_1, X_2 + X_3 + \cdots + X_n)$ and $T_2(\mathbf{X}) = (X_1 + X_2, X_3, X_4 + X_5 + \cdots + X_n)$ are also sufficient for p although T is preferable to T_1 or T_2.

Example 7. Let $X_1, X_2, \ldots, X_n$ be iid RVs with common PMF

$$P\{X_i = k\} = \frac{1}{N}, \qquad k = 1, 2, \ldots, N; \quad i = 1, 2, \ldots, n.$$

Then

$$\begin{aligned} P_N\{X_1 = k_1, X_2 = k_2, \ldots, X_n = k_n\} &= \frac{1}{N^n} \qquad \text{if } 1 \le k_1, \ldots, k_n \le N, \\ &= \frac{1}{N^n}\varphi(1, \min_{1\le i\le n} k_i)\varphi(\max_{1\le i\le n} k_i, N), \end{aligned}$$

where $\varphi(a,b) = 1$ if $b \ge a$, and $= 0$ if $b < a$. It follows, by taking $g_N[\max(k_1, \ldots k_n)] = (1/N^n)\varphi(\max_{1\le i\le n} k_i, N)$ and $h = \varphi(1, \min k_i)$, that $\max(X_1, X_2, \ldots, X_n)$ is sufficient for the family of joint PMFs P_N.

Example 8. Let $X_1, X_2, \ldots, X_n$ be a sample from $\mathcal{N}(\mu, \sigma^2)$, where both μ and σ^2 are unknown. The joint PDF of $(X_1, X_2, \ldots, X_n)$ is

$$\begin{aligned} f_{\mu,\sigma^2}(\mathbf{x}) &= \frac{1}{(\sigma\sqrt{2\pi})^n} \exp\left\{-\frac{\sum(x_i - \mu)^2}{2\sigma^2}\right\} \\ &= \frac{1}{(\sigma\sqrt{2\pi})^n} \exp\left(-\frac{\sum_1^n x_i^2}{2\sigma^2} + \frac{\mu\sum_1^n x_i}{\sigma^2} - \frac{n\mu^2}{2\sigma^2}\right). \end{aligned}$$

It follows that the statistic

$$T(X_1, \ldots, X_n) = \left(\sum_1^n X_i, \sum_1^n X_i^2\right)$$

is jointly sufficient for the parameter (μ, σ^2). An equivalent sufficient statistic that is frequently used is $T_1(X_1, \ldots, X_n) = (\overline{X}, S^2)$. Note that $\overline{X}$ is not sufficient for μ if σ^2 is unknown, and S^2 is not sufficient for σ^2 if μ is unknown. If, however, σ^2 is known, $\overline{X}$ is sufficient for μ. If $\mu = \mu_0$ is known, $\sum_1^n (X_i - \mu_0)^2$ is sufficient for σ^2.

Example 9. Let $X_1,X_2,\ldots,X_n$ be a sample from PDF

$$f_\theta(x)=\begin{cases}\dfrac{1}{\theta}, & x\in\left[-\dfrac{\theta}{2},\dfrac{\theta}{2}\right], \quad \theta>0,\\ 0, & \text{otherwise.}\end{cases}$$

The joint PDF of $X_1,X_2,\ldots,X_n$ is given by

$$f_\theta(x_1,x_2,\ldots,x_n)=\frac{1}{\theta^n}I_A(x_1,\ldots,x_n),$$

where

$$A=\left\{(x_1,x_2,\ldots,x_n):-\frac{\theta}{2}\le \min x_i\le \max x_i\le\frac{\theta}{2}\right\}.$$

It follows that $(X_{(1)},X_{(n)})$ is sufficient for θ.

We note that the order statistic $(X_{(1)},X_{(2)},\ldots,X_{(n)})$ is also sufficient. Note also that the parameter is one-dimensional, the statistics $(X_{(1)},X_{(n)})$ is two-dimensional, whereas the order statistic is n-dimensional.

In Example 9 we saw that order statistic is sufficient. This is not a mere coincidence. In fact, if $\mathbf{X}=(X_1,X_2,\ldots,X_n)$ are exchangeable then the joint PDF of $\mathbf{X}$ is a symmetric function of its arguments. Thus

$$f_\theta(x_1,x_2,\ldots,x_n)=f_\theta(x_{(1)},x_{(2)},\ldots,x_{(n)}),$$

and it follows that the order statistic is sufficient for f_θ.

The concept of sufficiency is frequently used with another concept, called *completeness*, which we now define.

Definition 2. Let $\{f_\theta(x),\theta\in\Theta\}$ be a family of PDFs (or PMFs). We say that this family is complete if

$$E_\theta g(X)=0 \qquad \text{for all } \theta\in\Theta,$$

which implies

$$P_\theta\{g(X)=0\}=1 \qquad \text{for all } \theta\in\Theta.$$

Definition 3. A statistic $T(X)$ is said to be complete if the family of distributions of T is complete.

In Definition 3 X will usually be a multiple RV. The family of distributions of T is obtained from the family of distributions of $X_1,X_2,\ldots,X_n$ by the usual transformation technique discussed in Section 4.4.

Example 10. Let $X_1, X_2, \ldots, X_n$ be iid $b(1,p)$ RVs. Then $T = \sum_1^n X_i$ is a sufficient statistic. We show that T is also complete, that is, the family of distributions of T, $\{b(n,p), 0 < p < 1\}$, is complete.

$$E_p g(T) = \sum_{t=0}^{n} g(t)\binom{n}{t} p^t (1-p)^{n-t} = 0 \qquad \text{for all } p \in (0,1)$$

may be rewritten as

$$(1-p)^n \sum_{t=0}^{n} g(t)\binom{n}{t}\left(\frac{p}{1-p}\right)^t = 0 \qquad \text{for all } p \in (0,1).$$

This is a polynomial in $p/(1-p)$. Hence the coefficients must vanish, and it follows that $g(t) = 0$ for $t = 0, 1, 2, \ldots, n$, as required.

Example 11. Let X be $\mathcal{N}(0,\theta)$. Then the family of PDFs $\{\mathcal{N}(0,\theta), \theta > 0\}$ is not complete since $EX = 0$ and $g(x) = x$ is not identically 0. Note that $T(X) = X^2$ is complete, for the PDF of $X^2 \sim \theta\chi^2(1)$ is given by

$$f(t) = \begin{cases} \dfrac{e^{-t/2\theta}}{\sqrt{2\pi\theta t}}, & t > 0, \\ 0, & \text{otherwise.} \end{cases}$$

$$E_\theta g(T) = \frac{1}{\sqrt{2\pi\theta}} \int_0^\infty g(t) t^{-1/2} e^{-t/2\theta}\, dt = 0 \qquad \text{for all } \theta > 0,$$

which holds if and only if $\int_0^\infty g(t) t^{-1/2} e^{-t/2\theta}\, dt = 0$, and using the uniqueness property of Laplace transforms, it follows that

$$g(t) t^{-1/2} = 0 \qquad \text{for all } t > 0,$$

that is, $g(t) = 0$.

The next example illustrates the existence of a sufficient statistic which is not complete.

Example 12. Let $X_1, X_2, \ldots, X_n$ be a sample from $\mathcal{N}(\theta, \theta^2)$. Then $T = (\sum_1^n X_i, \sum_1^n X_i^2)$ is sufficient for θ. However, T is not complete since

$$E_\theta\left\{2\left(\sum_1^n X_i\right)^2 - (n+1)\sum_1^n X_i^2\right\} = 0 \qquad \text{for all } \theta,$$

and the function $g(x_1, \ldots, x_n) = 2(\sum_1^n x_i)^2 - (n+1)\sum_1^n x_j^2$ is not identically 0.

Example 13. Let $X \sim U(0,\theta)$, $\theta \in (0,\infty)$. We show that the family of PDFs of X is complete. We need to show that

$$E_\theta g(X) = \int_0^\theta \frac{1}{\theta} g(x)\, dx = 0 \qquad \text{for all } \theta > 0$$

if and only if $g(x) = 0$ for all x. In general, this result follows from Lebesgue integration theory. If g is continuous, we differentiate both sides in

$$\int_0^\theta g(x)\,dx = 0$$

to get $g(\theta) = 0$ for all $\theta > 0$.

Now let $X_1, X_2, \ldots, X_n$ be iid $U(0,\theta)$ RVs. Then the PDF of $X_{(n)}$ is given by

$$f_n(x \mid \theta) = \begin{cases} n\theta^{-n}x^{n-1}, & 0 < x < \theta, \\ 0, & \text{otherwise.} \end{cases}$$

We see by a similar argument that $X_{(n)}$ is complete, which is the same as saying that $\{f_n(x \mid \theta); \theta > 0\}$ is a complete family of densities. Clearly, $X_{(n)}$ is sufficient.

Example 14. Let $X_1, X_2, \ldots, X_n$ be a sample from PMF

$$P_N(x) = \begin{cases} \dfrac{1}{N}, & x = 1, 2, \ldots, N, \\ 0, & \text{otherwise.} \end{cases}$$

We first show that the family of PMFs $\{P_N, N \geq 1\}$ is complete. We have

$$E_N g(X) = \frac{1}{N}\sum_{k=1}^{N} g(k) = 0 \qquad \text{for all } N \geq 1,$$

and this happens if and only if $g(k) = 0$, $k = 1, 2, \ldots, N$. Next we consider the family of PMFs of $X_{(n)} = \max(X_1, \ldots, X_n)$. The PMF of $X_{(n)}$ is given by

$$P_N^{(n)}(x) = \frac{x^n}{N^n} - \frac{(x-1)^n}{N^n}, \qquad x = 1, 2, \ldots, N.$$

Also

$$E_N g(X_{(n)}) = \sum_{k=1}^{N} g(k)\left[\frac{k^n}{N^n} - \frac{(k-1)^n}{N^n}\right] = 0 \qquad \text{for all } N \geq 1.$$

$$E_1 g(X_{(n)}) = g(1) = 0$$

implies $g(1) = 0$. Again,

$$E_2 g(X_{(n)}) = \frac{g(1)}{2^n} + g(2)\left(1 - \frac{1}{2^n}\right) = 0$$

so that $g(2) = 0$.

Using an induction argument, we conclude that $g(1) = g(2) = \cdots = g(N) = 0$ and hence $g(x) = 0$. It follows that $P_N^{(n)}$ is a complete family of distributions, and $X_{(n)}$ is a complete sufficient statistic.

Now suppose that we exclude the value $N = n_0$ for some fixed $n_0 \geq 1$ from the family $\{P_N : N \geq 1\}$. Let us write $\mathcal{P} = \{P_N : N \geq 1, N \neq n_0\}$. Then $\mathcal{P}$ is not complete. We ask the reader to show that the class of all functions g such that $E_P g(X) = 0$ for all $P \in \mathcal{P}$ consists of functions of the form

$$g(k) = \begin{cases} 0, & k = 1, 2, \ldots, n_0 - 1, n_0 + 2, n_0 + 3, \ldots, \\ c, & k = n_0, \\ -c, & k = n_0 + 1, \end{cases}$$

where c is a constant, $c \neq 0$.

Remark 7. Completeness is a property of a family of distributions. In Remark 6 we saw that if a statistic is sufficient for a class of distributions it is sufficient for any subclass of those distributions. Completeness works in the opposite direction. Example 14 shows that the exclusion of even one member from the family $\{P_N : N \geq 1\}$ destroys completeness.

The following result covers a large class of probability distributions for which a complete sufficient statistic exists.

Theorem 2. Let $\{f_{\boldsymbol{\theta}} : \boldsymbol{\theta} \in \Theta\}$ be a k-parameter exponential family given by

$$f_{\boldsymbol{\theta}}(\mathbf{x}) = \exp\left\{\sum_{j=1}^{k} Q_j(\boldsymbol{\theta}) T_j(\mathbf{x}) + D(\boldsymbol{\theta}) + S(\mathbf{x})\right\}, \tag{2}$$

where $\boldsymbol{\theta} = (\theta_1, \theta_2, \ldots, \theta_k) \in \Theta$, an interval in $\mathcal{R}_k, T_1, T_2, \ldots, T_k$, and S are defined on $\mathcal{R}_n, \mathbf{T} = (T_1, T_2, \ldots, T_k)$, and $\mathbf{x} = (x_1, x_2, \ldots, x_n)$, $k \leq n$. Let $\mathbf{Q} = (Q_1, Q_2, \ldots, Q_k)$, and suppose that the range of $\mathbf{Q}$ contains an open set in $\mathcal{R}_k$. Then

$$\mathbf{T} = (T_1(\mathbf{X}), T_2(\mathbf{X}), \ldots, T_k(\mathbf{X}))$$

is a complete sufficient statistic.

Proof. For a complete proof in a general setting we refer the reader to Lehmann [64, pp. 142–143]. Essentially, the unicity of the Laplace transform is used on the probability distribution induced by $\mathbf{T}$. We will content ourselves here by proving the result for the $k = 1$ case when f_θ is a PMF.

Let us write $Q(\theta) = \theta$ in (2), and let $(\alpha, \beta) \subseteq \Theta$. We wish to show that

$$\begin{aligned} E_\theta g(T(\mathbf{X})) &= \sum_t g(t) P_\theta\{T(\mathbf{X}) = t\} \\ &= \sum_t g(t) \exp\{\theta t + D(\theta) + S^*(t)\} = 0 \qquad \text{for all } \theta \end{aligned} \tag{3}$$

implies that $g(t) = 0$.

Let us write $x^+ = x$ if $x \geq 0$, $= 0$ if $x < 0$, and $x^- = -x$ if $x < 0$, $= 0$ if $x \geq 0$. Then $g(t) = g^+(t) - g^-(t)$, and both g^+ and g^- are nonnegative functions. In terms of g^+ and g^-, (3) is the same as

$$\sum_t g^+(t)e^{\theta t + S^*(t)} = \sum_t g^-(t)e^{\theta t + S^*(t)} \tag{4}$$

for all θ.

Let $\theta_0 \in (\alpha, \beta)$ be fixed, and write

$$p^+(t) = \frac{g^+(t)e^{\theta_0 t + S^*(t)}}{\sum_t g^+(t)e^{\theta_0 t + S^*(t)}} \quad \text{and} \quad p^-(t) = \frac{g^-(t)e^{\theta_0 t + S^*(t)}}{\sum_t g^-(t)e^{\theta_0 t + S^*(t)}} \tag{5}$$

Then both p^+ and p^- are PMFs, and it follows from (4) that

$$\sum_t e^{\delta t}p^+(t) = \sum_t e^{\delta t}p^-(t) \tag{6}$$

for all $\delta \in (\alpha - \theta_0, \beta - \theta_0)$. By the uniqueness of MGFs (6) implies that

$$p^+(t) = p^-(t) \qquad \text{for all } t$$

and hence that $g^+(t) = g^-(t)$ for all t, which is equivalent to $g(t) = 0$ for all t. Since T is clearly sufficient (by the factorization criterion), it is proved that T is a complete sufficient statistic.

Example 15. Let $X_1, X_2, \ldots, X_n$ be iid $\mathcal{N}(\mu, \sigma^2)$ RVs where both μ and σ^2 are unknown. We know that the family of distributions of $\mathbf{X} = (X_1, \ldots, X_n)$ is a two-parameter exponential family with $T(X_1, \ldots, X_n) = (\sum_1^n X_i, \sum_1^n X_i^2)$. From Theorem 2 it follows that T is a complete sufficient statistic. Examples 10 and 11 fall in the domain of Theorem 2.

In Example 6, 8, and 9 we have shown that a given family of probability distributions that admits a nontrivial sufficient statistic usually admits several sufficient statistics. Clearly we would like to be able to choose the sufficient statistic that results in the greatest reduction of data collection. We next study the notion of a *minimal sufficient statistic*. For this purpose it is convenient to introduce the notion of a *sufficient partition*. The reader will recall that a *partition* of a space $\mathfrak{X}$ is just a collection of disjoint sets E_α such that $\sum_\alpha E_\alpha = \mathfrak{X}$. Any statistic $T(X_1, X_2, \ldots, X_n)$ induces a partition of the space of values of $(X_1, X_2, \ldots, X_n)$, that is, T induces a covering of $\mathfrak{X}$ by a family $\mathfrak{U}$ of disjoint sets $A_t = \{(x_1, x_2, \ldots, x_n) \in \mathfrak{X} : T(x_1, x_2, \ldots, x_n) = t\}$, where t belongs to the range of T. The sets A_t are called *partition sets*. Conversely, given a partition, any assignment of a number to each set so that no two partition sets have the same number assigned defines a statistic. Clearly this function is not, in general, unique.

Definition 4. Let $\{F_\theta \colon \theta \in \Theta\}$ be a family of DFs, and $\mathbf{X} = (X_1, X_2, \ldots, X_n)$ be a sample from F_θ. Let $\mathfrak{U}$ be a partition of the sample space induced by a statistic

$T = T(X_1, X_2, \ldots, X_n)$. We say that $\mathfrak{U} = \{A_t : t \text{ is in the range of } T\}$ is a sufficient partition for θ (or the family $\{F_\theta \colon \theta \in \Theta\}$) if the conditional distribution of $\mathbf{X}$, given $T = t$, does not depend on θ for any A_t, provided that the conditional probability is well defined.

Example 16. Let $X_1, X_2, \ldots, X_n$ be iid $b(1, p)$ RVs. The sample space of values of $(X_1, X_2, \ldots, X_n)$ is the set of n-tuples $(x_1, x_2, \ldots, x_n)$, where each $x_i = 0$ or $= 1$ and consists of 2^n points. Let $T(X_1, X_2, \ldots, X_n) = \sum_1^n X_i$, and consider the partition $\mathfrak{U} = \{A_0, A_1, \ldots, A_n\}$, where $\mathbf{x} \in A_j$ if and only if $\sum_1^n x_i = j$, $0 \le j \le n$. Each A_j contains $\binom{n}{j}$ sample points. The conditional probability

$$P_p\{\mathbf{x} \mid A_j\} = \frac{P_p\{\mathbf{x}\}}{P_p(A_j)} = \binom{n}{j}^{-1} \qquad \text{if } \mathbf{x} \in A_j,$$

and we see that $\mathfrak{U}$ is a sufficient partition.

Example 17. Let $X_1, X_2, \ldots, X_n$ be iid $U[0, \theta]$ RVs. Consider the statistic $T(\mathbf{X}) = \max_{1 \le i \le n} X_i$. The space of values of $X_1, X_2, \ldots, X_n$ is the set of points $\{\mathbf{x} : 0 \le x_i \le \theta, i = 1, 2, \ldots, n\}$. T induces a partition $\mathfrak{U}$ on this set. The sets of this partition are $A_t = \{(x_1, x_2, \ldots, x_n) : \max(x_1, \ldots, x_n) = t\}$, $t \in [0, \theta]$.

We have

$$f_\theta(\mathbf{x} \mid t) = \frac{f_\theta(\mathbf{x})}{f_\theta^T(t)} \qquad \text{if} \qquad \mathbf{x} \in A_t,$$

where $f_\theta^T(t)$ is the PDF of T. We have

$$f_\theta(\mathbf{x} \mid t) = \frac{1/\theta^n}{nt^{n-1}/\theta^n} = \frac{1}{nt^{n-1}} \qquad \text{if} \qquad \mathbf{x} \in A_t.$$

It follows that $\mathfrak{U} = \{A_t\}$ defines a sufficient partition.

Remark 8. Clearly a sufficient statistic T for a family of DFs $\{F_\theta \colon \theta \in \Theta\}$ induces a sufficient partition and, conversely, given a sufficient partition, we can define a sufficient statistic (not necessarily uniquely) for the family.

Remark 9. Two statistics T_1, T_2 that define the same partition must be in one-to-one correspondence, that is, there exists a function h such that $T_1 = h(T_2)$ with a unique inverse, $T_2 = h^{-1}(T_1)$. It follows that if T_1 is sufficient every one-to-one function of T_1 is also sufficient.

Let $\mathfrak{U}_1, \mathfrak{U}_2$ be two partitions of a space $\mathfrak{X}$. We say that $\mathfrak{U}_1$ is a *subpartition* of $\mathfrak{U}_2$ if every partition set in $\mathfrak{U}_2$ is a union of sets of $\mathfrak{U}_1$. We sometimes say also that $\mathfrak{U}_1$ is *finer* than $\mathfrak{U}_2$ ($\mathfrak{U}_2$ is *coarser* than $\mathfrak{U}_1$) or that $\mathfrak{U}_2$ is a *reduction* of $\mathfrak{U}_1$. In this case, a statistic T_2 that defines $\mathfrak{U}_2$ must be a function of any statistic T_1 that defines $\mathfrak{U}_1$. Clearly, this function need not have a unique inverse unless the two partitions have exactly the same partition sets.

Given a family of distributions $\{F_\theta \colon \theta \in \Theta\}$ for which a sufficient partition exists, we seek to find a sufficient partition $\mathfrak{U}$ that is as coarse as possible, that is, any reduction of $\mathfrak{U}$ leads to a partition that is not sufficient.

Definition 5. A partition $\mathfrak{U}$ is said to be minimal sufficient if

(i) $\mathfrak{U}$ is a sufficient partition, and
(ii) if $\mathfrak{C}$ is any sufficient partition, $\mathfrak{C}$ is a subpartition of $\mathfrak{U}$.

The question of the existence of the minimal partition was settled by Lehmann and Scheffé [65] and, in general, involves measure-theoretic considerations. However, in the cases that we consider where the sample space is either discrete or a finite-dimensional Euclidean space and the family of distributions of $\mathbf{X}$ is defined by a family of PDFs (PMFs) $\{f_\theta, \theta \in \Theta\}$ such difficulties do not arise. The construction may be described as follows.

Two points $\mathbf{x}$ and $\mathbf{y}$ in the sample space are said to be *likelihood equivalent*, and we write $\mathbf{x} \sim \mathbf{y}$, if and only if there exists a $k(\mathbf{y},\mathbf{x}) \neq 0$ which does not depend on θ such that $f_\theta(\mathbf{y}) = k(\mathbf{y},\mathbf{x}) f_\theta(\mathbf{x})$. We leave the reader to check that "$\sim$" is an equivalence relation (that is, it is reflexive, symmetric, and transitive) and hence "$\sim$" defines a partition of the sample space. This partition defines the minimal sufficient partition.

Example 18. Consider again Example 16. Then

$$\frac{f_p(\mathbf{x})}{f_p(\mathbf{y})} = p^{\sum x_i - \sum y_i}(1-p)^{-\sum x_i + \sum y_i},$$

and this ratio is independent of p if and only if

$$\sum_1^n x_i = \sum_1^n y_i,$$

so that $\mathbf{x} \sim \mathbf{y}$ if and only if $\sum_1^n x_i = \sum_1^n y_i$. It follows that the partition $\mathfrak{U} = \{A_0, A_1, \ldots, A_n\}$, where $\mathbf{x} \in A_j$ if and only if $\sum_1^n x_i = j$, introduced in Example 16 is minimal sufficient.

A rigorous proof of the above assertion is beyond the scope of this book. The basic ideas are outlined in the following theorem.

Theorem 3. The relation "$\sim$" defined above induces a minimal sufficient partition.

Proof. If T is a sufficient statistic, we have to show that $\mathbf{x} \sim \mathbf{y}$ whenever $T(\mathbf{x}) = T(\mathbf{y})$. This will imply that every set of the minimal sufficient partition is a union of sets of the form $A_t = \{T = t\}$, proving condition (ii) of Definition 5.

Sufficiency of T means that whenever $\mathbf{x} \in A_t$, then

$$f_\theta\{\mathbf{x} \mid T = t\} = \frac{f_\theta(\mathbf{x})}{f_\theta^T(t)} \qquad \text{if} \qquad \mathbf{x} \in A_t$$

is free of θ. It follows that if both $\mathbf{x}$ and $\mathbf{y} \in A_t$, then

$$\frac{f_\theta(\mathbf{x} \mid t)}{f_\theta(\mathbf{y} \mid t)} = \frac{f_\theta(\mathbf{x})}{f_\theta(\mathbf{y})}$$

is independent of θ, and hence $\mathbf{x} \sim \mathbf{y}$.

To prove the sufficiency of the minimal sufficient partition $\mathfrak{U}$, let T_1 be an RV that induces $\mathfrak{U}$. Then T_1 takes on distinct values over distinct sets of $\mathfrak{U}$ but remains constant on the same set. If $\mathbf{x} \in \{T_1 = t_1\}$, then

$$f_\theta(\mathbf{x} \mid T_1 = t_1) = \frac{f_\theta(\mathbf{x})}{P_\theta\{T_1 = t_1\}}. \tag{7}$$

Now

$$P_\theta\{T_1 = t_1\} = \int_{(\mathbf{y}:T_1(\mathbf{y})=t_1)} f_\theta(\mathbf{y})\,d\mathbf{y} \qquad \text{or} \qquad \sum_{(\mathbf{y}:T_1(\mathbf{y})=t_1)} f_\theta(\mathbf{y}),$$

depending on whether the joint distribution of $\mathbf{X}$ is absolutely continuous or discrete. Since $f_\theta(\mathbf{x})/f_\theta(\mathbf{y})$ is independent of θ whenever $\mathbf{x} \sim \mathbf{y}$, it follows that the ratio on the right-hand side of (7) does not depend on θ. Thus T_1 is sufficient.

Definition 6. A statistic that induces the minimal sufficient partition is called a minimal sufficient statistic.

In view of Theorem 3 a minimal sufficient statistic is a function of every sufficient statistic. It follows that if T_1 and T_2 are both minimal sufficient, then both must induce the same minimal sufficient partition and hence T_1 and T_2 must be equivalent in the sense that each must be a function of the other (with probability 1).

How does one show that a statistic T is not sufficient for a family of distributions $\mathcal{P}$? Other than using the definition of sufficiency one can sometimes use a result of Lehmann and Scheffé [65] according to which if $T_1(\mathbf{X})$ is sufficient for $\boldsymbol{\theta}$, $\boldsymbol{\theta} \in \Theta$, then $T_2(\mathbf{X})$ is also sufficient if and only if $T_1(\mathbf{X}) = g(T_2(\mathbf{X}))$ for some Borel-measurable function g and all $\mathbf{x} \in B$, where B is a Borel set with $P_{\boldsymbol{\theta}}B = 1$.

Another way to prove T nonsufficient is to show that there exist $\mathbf{x}$ for which $T(\mathbf{x}) = T(\mathbf{y})$ but $\mathbf{x}$ and $\mathbf{y}$ are not likelihood equivalent. We refer to Sampson and Spencer [98] for this and other similar results.

The following important result will be proved in the next section.

Theorem 4. A complete sufficient statistic is minimal sufficient.

We emphasize that the converse is not true. A minimal sufficient statistic may not be complete.

Example 19. Suppose $X \sim U(\theta, \theta+1)$. Then X is a minimal sufficient statistic. However, X is not complete. Take for example $g(x) = \sin 2\pi x$. Then

$$Eg(X) = \int_\theta^{\theta+1} \sin 2\pi x\,dx = \int_0^1 \sin 2\pi x\,dx = 0.$$

for all θ and it follows that X is not complete.

If $X_1, X_2, \ldots, X_n$ is a sample from $U(\theta, \theta+1)$, then $(X_{(1)}, X_{(n)})$ is minimal sufficient for θ but not complete since

$$E_\theta(X_{(n)} - X_{(1)}) = \frac{n-1}{n+1}$$

for all θ.

Finally, we consider statistics that have distributions free of the parameter(s) $\boldsymbol{\theta}$ and seem to contain no information about $\boldsymbol{\theta}$. We will see (Example 23) that such statistics can sometimes provide useful information about $\boldsymbol{\theta}$.

Definition 7. A statistic $A(\mathbf{x})$ is said to be ancillary if its distribution does not depend on the underlying model parameter θ.

Example 20. Let $X_1, X_2, \ldots, X_n$ be a random sample from $\mathcal{N}(\mu, 1)$. Then the statistic $A(\mathbf{X}) = (n-1)S^2 = \sum_{i=1}^n (X_i - \overline{X})^2$ is ancillary since $(n-1)S^2 \sim \chi^2(n-1)$ which is free of μ. Some other ancillary statistics are

$$X_1 - \overline{X}, X_{(n)} - X_{(1)}, \sum_{i=1}^n |X_i - \overline{X}|.$$

Also, $\overline{X}$, a complete sufficient statistic (hence minimal sufficient) for μ is independent of $A(\mathbf{X})$.

Example 21. Let $X_1, X_2, \ldots, X_n$ be a random sample from $\mathcal{N}(0, \sigma^2)$. Then, $A(\mathbf{X}) = \overline{X}$ follows a $\mathcal{N}(0, n^{-1}\sigma^2)$ and not ancillary with respect to the parameter σ^2.

Example 22. Let $X_{(1)}, X_{(2)}, \ldots, X_{(n)}$ be the order statistics of a random sample from the PDF $f(x-\theta)$, where $\theta \in \mathcal{R}$. Then the statistic $A(\mathbf{X}) = (X_{(2)} - X_{(1)}, \ldots X_{(n)} - X_{(1)})$ is ancillary for θ.

In Example 20 we saw that S^2 was independent of the minimal sufficient statistic $\overline{X}$. The following result due to Basu shows that it is not a mere coincidence.

Theorem 5. If $S(\mathbf{X})$ is a complete sufficient statistic for θ, then any ancillary statistic $A(\mathbf{X})$ is independent of S.

Proof. If A is ancillary, then $P_\theta\{A(\mathbf{X}) \le a\}$ is free of θ for all a. Consider the conditional probability $g_a(s) = P\{A(\mathbf{X}) \le a \mid S(\mathbf{X}) = s\}$. Clearly

$$E_\theta\{g_a(S(\mathbf{X}))\} = P_\theta\{A(\mathbf{X}) \le a\}.$$

Thus

$$E_\theta(g_a(S) - P\{A(\mathbf{X}) \le a\}) = 0$$

for all θ. By completeness of S it follows that

$$P_\theta\{g_a(S) - P\{A \le a\} = 0\} = 1,$$

that is ,

$$P_\theta\{A(\mathbf{X}) \le a \mid S(\mathbf{X}) = s\} = P\{A(\mathbf{X}) \le a\},$$

with probability 1. Hence A and S are independent.

The converse of Basu's Theorem is not true. A statistic S that is independent of every ancillary statistic need not be complete (see, for example, Lehmann [62]).

The following example due to R.A. Fisher shows that if there is no sufficient statistic for θ, but there exists a reasonable statistic not independent of an ancillary statistic $A(\mathbf{X})$, then the recovery of information is sometimes helped by the ancillary statistic via a conditional analysis. Unfortunately, the lack of uniqueness of ancillary statistics creates problems with this conditional analysis.

Example 23. Let $X_1, X_2, \ldots, X_n$ be a random sample from an exponential distribution with mean θ, and let $Y_1, Y_2, \ldots, Y_n$ be another random sample from an exponential distribution and mean $1/\theta$. Assume X's and Y's are independent and consider the problem of estimation of θ based on the observations $(X_1, X_2, \ldots, X_n;\ Y_1, Y_2, \ldots, Y_n)$. Let $S_1(\mathbf{x}) = \sum_{i=1}^n x_i$ and $S_2(\mathbf{y}) = \sum_{i=1}^n y_i$. Then $\big(S_1(\mathbf{X}), S_2(\mathbf{Y})\big)$ is jointly sufficient for θ. It is easily seen that (S_1, S_2) is a minimal sufficient statistic for θ.

Consider the statistics

$$S(\mathbf{X}, \mathbf{Y}) = \big(S_1(\mathbf{X})/S_2(\mathbf{Y})\big)^{1/2}$$

and

$$A(\mathbf{X}, \mathbf{Y}) = S_1(\mathbf{X})S_2(\mathbf{Y}).$$

Then the joint PDF of S and A is given by

$$\frac{2}{[\Gamma(n)]^2} \exp\left\{-A(\mathbf{x}, \mathbf{y})\left(\frac{S(\mathbf{x}, \mathbf{y})}{\theta} + \frac{\theta}{S(\mathbf{x}, \mathbf{y})}\right)\right\} \frac{[A(\mathbf{x}, \mathbf{y})]^{2n-1}}{S(\mathbf{x}, \mathbf{y})},$$

and it is clear that S and A are not independent. The marginal distribution of A is given by the PDF

$$C(\mathbf{x}, \mathbf{y})[A(\mathbf{x}, \mathbf{y})]^{2n-1},$$

where $C(\mathbf{x}, \mathbf{y})$ is the constant of integration which depends only on $\mathbf{x}, \mathbf{y}$, and n but not on θ. In fact, $C(\mathbf{x}, \mathbf{y}) = 4K_0[2A(\mathbf{x}, \mathbf{y})]/[\Gamma(n)]^2$, where K_0 is the standard form of a Bessel function (Watson [116]). Consequently A is ancillary for θ.

Clearly, the conditional PDF of S given $A = a$ is of the form

$$\frac{1}{2K_0[2a]S(\mathbf{x},\mathbf{y})}\exp\left\{-a\left(\frac{S(\mathbf{x},\mathbf{y})}{\theta}+\frac{\theta}{S(\mathbf{x},\mathbf{y})}\right)\right\}.$$

The amount of information lost by using $S(\mathbf{X},\mathbf{Y})$ alone is $(\frac{1}{2n+1})$th part of the total and this loss of information is gained by the knowledge of the ancillary statistic $A(\mathbf{X},\mathbf{Y})$. These calculations will be discussed in Example 8.5.9.

PROBLEMS 8.3

1. Find a sufficient statistic in each of the following cases based on a random sample of size n:

(a) $X \sim B(\alpha,\beta)$ when (i) α is unknown, β known; (ii) β; is unknown, α known; and (iii) α,β are both unknown.

(b) $X \sim G(\alpha,\beta)$ when (i) α is unknown, β known; (ii) β is unknown, α known; and (iii) α,β are both unknown.

(c) $X \sim P_{N_1,N_2}(x)$, where

$$P_{N_1,N_2}(x) = \frac{1}{N_2 - N_1}, \qquad x = N_1+1, N_1+2,\ldots,N_2,$$

and $N_1, N_2 (N_1 < N_2)$ are integers, when (i) N_1 is known, N_2 unknown; (ii) N_2 known, N_1 unknown; and (iii) N_1, N_2 are both unknown.

(d) $X \sim f_\theta(x)$, where

$$f_\theta(x) = \begin{cases} e^{-x+\theta} & \text{if } < x < \infty, \\ 0 & \text{otherwise.} \end{cases}$$

(e) $X \sim f(x;\mu,\sigma)$, where

$$f(x;\mu,\sigma) = \frac{1}{x\sigma\sqrt{2\pi}}\exp\left\{-\frac{1}{2\sigma^2}(\log x - \mu)^2\right\}, x > 0$$

(f) $X \sim f_\theta(x)$, where

$$f_\theta(x) = P_\theta\{X = x\} = c(\theta)2^{-x/\theta}, \quad x = \theta, \theta+1, \ldots, \theta > 0$$

and

$$c(\theta) = 2^{1-1/\theta}(2^{1/\theta} - 1).$$

(g) $X \sim P_{\theta,p}(x)$, where

$$P_{\theta,p}(x) = (1-p)p^{x-\theta}, \qquad x = \theta, \theta+1, \ldots, \quad 0 < p < 1,$$

when (i) p is known, θ unknown; (ii) p is unknown, θ known; and (iii) p, θ are both unknown.

2. Let $\mathbf{X} = (X_1, X_2, \ldots, X_n)$ be a sample from $\mathcal{N}(\alpha\sigma, \sigma^2)$, where α is a known real number. Show that the statistic $T(\mathbf{X}) = (\sum_{i=1}^n X_i, \sum_{i=1}^n X_i^2)$ is sufficient for σ but that the family of distributions of $T(\mathbf{X})$ is not complete.

3. Let $X_1, X_2, \ldots, X_n$ be a sample from $\mathcal{N}(\mu, \sigma^2)$. Then $\mathbf{X} = (X_1, X_2, \ldots, X_n)$ is clearly sufficient for the family $\mathcal{N}(\mu, \sigma^2), \mu \in \mathcal{R}, \sigma > 0$. Is the family of distributions of $\mathbf{X}$ complete?

4. Let $X_1, X_2, \ldots, X_n$ be a sample from $U(\theta - \frac{1}{2}, \theta + \frac{1}{2})$, $\theta \in \mathcal{R}$. Show that the statistic $T(X_1, \ldots, X_n) = (\min X_i, \max X_i)$ is sufficient for θ but not complete.

5. If $T = g(U)$ and T is sufficient, then so also is U.

6. In Example 14 show that the class of all functions g for which $E_P g(X) = 0$ for all $P \in \mathcal{P}$ consists of functions of the form

$$g(k) = \begin{cases} 0, & k = 1, 2, \ldots, n_0 - 1, n_0 + 2, n_0 + 3, \ldots, \\ c, & k = n_0, \\ -c, & k = n_0 + 1, \end{cases}$$

where c is a constant.

7. For the class $\{F_{\theta_1}, F_{\theta_2}\}$ of two DFs where F_{θ_1} is $\mathcal{N}(0,1)$ and F_{θ_1} is $\mathcal{C}(1,0)$, find a sufficient statistic.

8. Consider the class of hypergeometric probability distributions $\{P_D : D = 0, 1, 2, \ldots, N\}$, where

$$P_D\{X = x\} = \binom{N}{n}^{-1} \binom{D}{x} \binom{N-D}{n-x}, \qquad x = 0, 1, \ldots, \min\{n, D\}.$$

Show that it is a complete class. If $\mathcal{P} = \{P_D : D = 0, 1, 2, \ldots, N, D \neq d, d \text{ integral } 0 \leq d \leq N\}$, is $\mathcal{P}$ complete?

9. Is the family of distributions of the order statistic in sampling from a Poisson distribution complete?

10. Let $(X_1, X_2, \ldots, X_n)$ be a random vector of the discrete type. Is the statistic $T(X_1, \ldots, X_n) = (X_1, \ldots, X_{n-1})$ sufficient?

11. Let $X_1, X_2, \ldots, X_n$ be a random sample from a population with law $\mathcal{L}(X)$. Find a minimal sufficient statistic in each of the following cases:

(a) $X \sim P(\lambda)$.

(b) $X \sim U[0, \theta]$.

(c) $X \sim NB(1; p)$.

(d) $X \sim P_N$, where $P_N\{X = k\} = 1/N$ if $k = 1, 2, \ldots, N$, and $= 0$ otherwise.

(e) $X \sim \mathcal{N}(\mu, \sigma^2)$.

(f) $X \sim G(\alpha, \beta)$.

(g) $X \sim B(\alpha, \beta)$.

(h) $X \sim f_\theta(x)$, where $f_\theta(x) = (2/\theta^2)(\theta - x), 0 < x < \theta$.

12. Let X_1, X_2 be a sample of size 2 from $P(\lambda)$. Show that the statistic $X_1 + \alpha X_2$, where $\alpha > 1$ is an integer, is not sufficient for λ.

13. Let $X_1, X_2, \ldots, X_n$ be a sample from the PDF

$$f_\theta(x) = \begin{cases} \frac{x}{\theta} e^{-x^2/2\theta} & \text{if } x > 0 \\ 0 & \text{if } x \le 0 \end{cases} \qquad \theta > 0.$$

Show that $\sum_{i=1}^n X_i^2$ is a minimal sufficient statistic for θ, but $\sum_{i=1}^n X_i$ is not sufficient.

14. Let $X_1, X_2, \ldots, X_n$ be a sample from $\mathcal{N}(0, \sigma^2)$. Show that $\sum_{i=1}^n X_i^2$ is a minimal sufficient statistic but $\sum_{i=1}^n X_i$ is not sufficient for σ^2.

15. Let $X_1, X_2, \ldots, X_n$ be a sample from PDF $f_{\alpha,\beta}(x) = \beta e^{-\beta(x-\alpha)}$ if $x > \alpha$, and $= 0$ if $x \le \alpha$. Find a minimal sufficient statistic for (α, β).

16. Let T be a minimal sufficient statistic. Show that a necessary condition for a sufficient statistic U to be complete is that U be minimal.

17. Let $X_1, X_2, \ldots, X_n$ be iid $\mathcal{N}(\mu, \sigma^2)$. Show that $(\bar{X}, S^2)$ is independent of each of $(X_{(n)} - X_{(1)})/S$, $(X_{(n)} - \bar{X})/S$, and $\sum_{i=1}^{n-1} (X_{i+1} - X_i)^2/S^2$.

18. Let $X_1, X_2, \ldots, X_n$ be iid $\mathcal{N}(\theta, 1)$. Show that a necessary and sufficient condition for $\sum_{i=1}^n a_i X_i$ and $\sum_{i=1}^n X_i$ to be independent is $\sum_{i=1}^n a_i = 0$.

19. Let $X_1, X_2, \ldots, X_n$ be a random sample from $f_\theta(x) = \exp\{-(x-\theta)\}$, $x > \theta$. Show that $X_{(1)}$ is a complete sufficient statistic which is independent of S^2.

20. Let $X_1, X_2, \ldots, X_n$ be iid RVs with common PDF $f_\theta(x) = (1/\theta)\exp(-x/\theta)$, $x > 0$, $\theta > 0$. Show that X must be independent of every scale-invariant statistic such as $X_1 / \sum_{j=1}^n X_j$.

21. Let T_1, T_2 be two statistics with common domain D. Then T_1 is a function of T_2 if and only if

$$\text{for all } x, y \in D, \quad T_1(x) = T_1(y) \Longrightarrow T_2(x) = T_2(y).$$

22. Let S be the support of f_θ, $\theta \in \Theta$ and let T be a statistic such that for some $\theta_1, \theta_2 \in \Theta$, and $x, y \in S$, $x \ne y$, $T(x) = T(y)$ but $f_{\theta_1}(x) f_{\theta_2}(y) \ne f_{\theta_2}(x) f_{\theta_1}(y)$. Then show that T is not sufficient for θ.

23. Let $X_1, X_2, \ldots, X_n$ be iid $\mathcal{N}(\theta, 1)$. Use the result in Problem 22 to show that $\left(\sum_1^n X_i\right)^2$ is not sufficient for θ.

24. (a) If T is complete then show that any one-to-one mapping of T is also complete.

(b) Show with the help of an example that a complete statistic is not unique for a family of distributions.

8.4 UNBIASED ESTIMATION

In this section we focus attention on the class of unbiased estimators. We develop a criterion to check if an unbiased estimator is optimal in this class. Using sufficiency and completeness, we describe a method of constructing uniformly minimum variance unbiased estimators.

Definition 1. Let $\{F_{\boldsymbol{\theta}},\ \boldsymbol{\theta} \in \Theta\}$, $\Theta \subseteq \mathcal{R}_k$, be a nonempty set of probability distributions. Let $\mathbf{X} = (X_1, X_2, \ldots, X_n)$ be a multiple RV with DF $F_{\boldsymbol{\theta}}$ and sample space $\mathfrak{X}$. Let $\psi : \Theta \to \mathcal{R}$ be a real-valued parametric function. A Borel-measurable function $T : \mathfrak{X} \to \Theta$ is said to be unbiased for ψ if

$$E_{\boldsymbol{\theta}} T(\mathbf{X}) = \psi(\boldsymbol{\theta}) \qquad \text{for all } \boldsymbol{\theta} \in \Theta. \tag{1}$$

Any parametric function ψ for which there exists a T satisfying (1) is called an *estimable function*. An estimator that is not unbiased is called *biased*, and the function $b(T, \psi)$, defined by

$$b(T, \psi) = E_{\boldsymbol{\theta}} T(\mathbf{X}) - \psi(\boldsymbol{\theta}), \tag{2}$$

is called the *bias* of T.

Remark 1. Definition 1, in particular, requires that $E_{\boldsymbol{\theta}}|T| < \infty$ for all $\boldsymbol{\theta} \in \Theta$ and can be extended to the case when both ψ and T are multidimensional. In most applications we consider $\Theta \subseteq \mathcal{R}_1$, $\psi(\theta) = \theta$, and $X_1, X_2, \ldots, X_n$ are iid RVs.

Example 1. Let $X_1, X_2, \ldots, X_n$ be a random sample from some population with finite mean. Then $\overline{X}$ is unbiased for the population mean. If the population variance is finite, the sample variance S^2 is unbiased for the population variance. In general, if the kth population moment m_k exists, the kth sample moment is unbiased for m_k.

Note that S is not, in general, unbiased for σ. If $X_1, X_2, \ldots, X_n$ are iid $\mathcal{N}(\mu, \sigma^2)$ RVs we know that $(n-1)S^2/\sigma^2$ is $\chi^2(n-1)$. Therefore,

$$\begin{aligned} E(S\sqrt{n-1}/\sigma) &= \int_0^\infty \sqrt{x}\, \frac{1}{2^{(n-1)/2}\Gamma[(n-1)/2]} x^{(n-1)/2-1} e^{-x/2}\, dx \\ &= \sqrt{2}\,\Gamma\left(\frac{n}{2}\right)\left[\Gamma\left(\frac{n-1}{2}\right)\right]^{-1}, \\ E_\sigma(S) &= \sigma\left\{\sqrt{\frac{2}{n-1}}\,\Gamma\left(\frac{n}{2}\right)\left[\Gamma\left(\frac{n-1}{2}\right)\right]^{-1}\right\}. \end{aligned}$$

The bias of S is given by

$$b(S, \sigma) = \sigma\left\{\sqrt{\frac{2}{n-1}}\,\Gamma\left(\frac{n}{2}\right)\left[\Gamma\left(\frac{n-1}{2}\right)\right]^{-1} - 1\right\}.$$

We note that $b(s, \sigma) \to 0$ as $n \to \infty$ so that S is asymptotically unbiased for σ.

If T is unbiased for θ, $g(T)$ is not, in general, an unbiased estimator of $g(\theta)$ unless g is a linear function.

Example 2. Unbiased estimators do not always exist. Consider an RV with PMF $b(1, p)$. Suppose that we wish to estimate $\psi(p) = p^2$. Then, in order that T be unbiased for p^2, we must have

$$p^2 = E_p T = pT(1) + (1-p)T(0), \qquad 0 \le p \le 1,$$

that is,

$$p^2 = p\{T(1) - T(0)\} + T(0)$$

must hold for all p in the interval $[0, 1]$, which is impossible. (If a convergent power series vanishes in an open interval, each of the coefficients must be 0. See also Problem 1.)

Example 3. Sometimes an unbiased estimator may be absurd. Let X be $P(\lambda)$, and $\psi(\lambda) = e^{-3\lambda}$. We show that $T(X) = (-2)^X$ is unbiased for $\psi(\lambda)$. We have

$$E_\lambda T(X) = e^{-\lambda} \sum_{x=0}^{\infty} (-2)^x \frac{\lambda^x}{x!} = e^{-\lambda} \sum_{x=0}^{\infty} \frac{(-2\lambda)^x}{x!} = e^{-\lambda} e^{-2\lambda} = \psi(\lambda).$$

However, $T(x) = (-2)^x > 0$ if x is even, and < 0 if x is odd, which is absurd since $\psi(\lambda) > 0$.

Example 4. Let $X_1, X_2, \ldots, X_n$ be a sample from $P(\lambda)$. Then $\overline{X}$ is unbiased for λ and so also is S^2, since both the mean and the variance are equal to λ. Indeed, $\alpha\overline{X} + (1-\alpha)S^2$, $0 \le \alpha \le 1$, is unbiased for λ.

Let θ be estimable, and let T be an unbiased estimator of θ. Let T_1 be another unbiased estimator of θ, different from T. This means that there exists at least one θ such that $P_\theta\{T \ne T_1\} > 0$. In this case there exist infinitely many unbiased estimators of θ of the form $\alpha T + (1-\alpha)T_1$, $0 < \alpha < 1$. It is therefore desirable to find a procedure to differentiate among these estimators.

Definition 2. Let $\theta_0 \in \Theta$ and $\mathcal{U}(\theta_0)$ be the class of all unbiased estimators T of θ_0 such that $E_{\theta_0} T^2 < \infty$. Then $T_0 \in \mathcal{U}(\theta_0)$ is called a locally minimum variance unbiased estimator (LMVUE) at θ_0 if

$$E_{\theta_0}(T_0 - \theta_0)^2 \le E_{\theta_0}(T - \theta_0)^2 \tag{3}$$

holds for all $T \in \mathcal{U}(\theta_0)$.

Definition 3. Let $\mathcal{U}$ be the set of all unbiased estimators T of $\theta \in \Theta$ such that $E_\theta T^2 < \infty$ for all $\theta \in \Theta$. An estimator $T_0 \in \mathcal{U}$ is called a uniformly minimum variance unbiased estimator (UMVUE) of θ if

$$E_\theta(T_0 - \theta)^2 \le E_\theta(T - \theta)^2 \tag{4}$$

for all $\theta \in \Theta$ and every $T \in \mathcal{U}$.

Remark 2. Let $a_1, a_2, \ldots, a_n$ be any set of real numbers with $\sum_{i=1}^n a_i = 1$. Let $X_1, X_2, \ldots, X_n$ be independent RVs with common mean μ and variances σ_k^2, $k = 1, 2, \ldots, n$. Then $T = \sum_{i=1}^n a_i X_i$ is an unbiased estimator of μ with variance $\sum_{i=1}^n a_i^2 \sigma_i^2$ (see Theorem 4.5.6). T is called a *linear unbiased estimator* of μ. Linear unbiased estimators of μ that have minimum variance (among all linear unbiased estimators) are called *best linear unbiased estimators* (BLUEs). In Theorem 4.5.6 (Corollary 2) we have shown that, if X_i are iid RVs with common variance σ^2, the BLUE of μ is $\overline{X} = n^{-1}\sum_{i=1}^{n} X_i$. If X_i are independent with common mean μ but different variance σ_i^2, the BLUE of μ is obtained if we choose a_i proportional to $1/\sigma_i^2$, then the minimum variance is H/n, where H is the harmonic mean of $\sigma_1^2, \ldots, \sigma_n^2$ (see Example 4.5.4).

Remark 3. Sometimes the precision of an estimator T of parameter θ is measured by the so-called *mean square error* (MSE). We say that an estimator T_0 is at least as good as any other estimator T in the sense of the MSE if

$$E_\theta(T_0 - \theta)^2 \leq E_\theta(T - \theta)^2 \qquad \text{for all } \theta \in \Theta. \tag{5}$$

In general, a particular estimator will be better than another for some values of θ and worse for others. Definitions 2 and 3 are special cases of this concept if we restrict attention only to unbiased estimators.

The following result gives a necessary and sufficient condition for an unbiased estimator to be a UMVUE.

Theorem 1. Let $\mathcal{U}$ be the class of all unbiased estimators T of a parameter $\theta \in \Theta$ with $E_\theta T^2 < \infty$ for all θ, and suppose that $\mathcal{U}$ is nonempty. Let $\mathcal{U}_0$ be the class of all unbiased estimators v of 0, that is,

$$\mathcal{U}_0 = \{v : E_\theta v = 0, E_\theta v^2 < \infty \qquad \text{for all } \theta \in \Theta\}.$$

Then $T_0 \in \mathcal{U}$ is a UMVUE if and only if

$$E_\theta(vT_0) = 0 \qquad \text{for all } \theta \text{ and all } v \in \mathcal{U}_0. \tag{6}$$

Proof. The conditions of the theorem guarantee the existence of $E_\theta(vT_0)$ for all θ and $v \in \mathcal{U}_0$. Suppose that $T_0 \in \mathcal{U}$ is a UMVUE and $E_{\theta_0}(v_0T_0) \neq 0$ for some θ_0 and some $v_0 \in \mathcal{U}_0$. Then $T_0 + \lambda v_0 \in \mathcal{U}$ for all real λ. If $E_{\theta_0} v_0^2 = 0$, then $E_{\theta_0}(v_0T_0) = 0$ must hold since $P_{\theta_0}\{v_0 = 0\} = 1$. Let $E_{\theta_0} v_0^2 > 0$. Choose $\lambda_0 = -E_{\theta_0}(T_0v_0)/E_{\theta_0}v_0^2$. Then

$$E_{\theta_0}(T_0 + \lambda_0 v_0)^2 = E_{\theta_0}T_0^2 - \frac{E_{\theta_0}^2(v_0T_0)}{E_{\theta_0}v_0^2} < E_{\theta_0}T_0^2. \tag{7}$$

Since $T_0 + \lambda_0 v_0 \in \mathcal{U}$ and $T_0 \in \mathcal{U}$, it follows from (7) that

$$\text{var}_{\theta_0}(T_0 + \lambda_0 v_0) < \text{var}_{\theta_0}(T_0), \tag{8}$$

which is a contradiction. It follows that (6) holds.

Conversely, let (6) hold for some $T_0 \in \mathcal{U}$, all $\theta \in \Theta$ and all $v \in \mathcal{U}_0$, and let $T \in \mathcal{U}$. Then $T_0 - T \in \mathcal{U}_0$, and for every θ

$$E_\theta\{T_0(T_0 - T)\} = 0.$$

We have

$$E_\theta T_0^2 = E_\theta(TT_0) \le (E_\theta T_0^2)^{1/2}(E_\theta T^2)^{1/2}$$

by the Cauchy–Schwarz inequality. If $E_\theta T_0^2 = 0$, then $P(T_0 = 0) = 1$ and there is nothing to prove. Otherwise

$$(E_\theta T_0^2)^{1/2} \le (E_\theta T^2)^{1/2}$$

or $\text{var}_\theta(T_0) \le \text{var}_\theta(T)$. Since T is arbitrary, the proof is complete.

Theorem 2. Let $\mathcal{U}$ be the nonempty class of unbiased estimators as defined in Theorem 1. Then there exists at most one UMVUE for θ.

Proof. If T and $T_0 \in \mathcal{U}$ are both UMVUEs, then $T - T_0 \in U_0$ and

$$E_\theta\{T_0(T - T_0)\} = 0 \qquad \text{for all } \theta \in \Theta,$$

that is, $E_\theta T_0^2 = E_\theta(TT_0)$, and it follows that

$$\text{cov}(T, T_0) = \text{var}_\theta(T_0) \qquad \text{for all } \theta.$$

Since T_0 and T are both UMVUEs $\text{var}_\theta(T) = \text{var}_\theta(T_0)$, and it follows that the correlation coefficient between T and T_0 is 1. This implies that $P_\theta\{aT + bT_0 = 0\} = 1$ for some a, b and all $\theta \in \Theta$. Since T and T_0 are both unbiased for θ, we must have $P_\theta\{T = T_0\} = 1$ for all θ.

Remark 4. Both Theorems 1 and 2 have analogs for LMVUE's at $\theta_0 \in \Theta$, θ_0 fixed.

Theorem 3. If UMVUEs T_i exist for real functions ψ_i, $i = 1, 2$, of θ, they also exist for $\lambda\psi_i$ (λ real), as well as for $\psi_1 + \psi_2$, and are given by λT_i and $T_1 + T_2$, respectively.

Theorem 4. Let $\{T_n\}$ be a sequence of UMVUEs and T be a statistic with $E_\theta T^2 < \infty$ such that $E_\theta\{T_n - T\}^2 \to 0$ as $n \to \infty$ for all $\theta \in \Theta$. Then T is also the UMVUE.

Proof. That T is unbiased follows from $|E_\theta T - \theta| \le E_\theta|T - T_n| \le E_0^{1/2}\{T_n - T\}^2$. For all $v \in \mathcal{U}_0$, all θ, and every $n = 1, 2, \ldots$,

$$E_\theta(T_n v) = 0$$

by Theorem 1. Therefore,

$$\begin{aligned} E_\theta(vT) &= E_\theta(vT) - E_\theta(vT_n) \\ &= E_\theta[v(T - T_n)] \end{aligned}$$

and

$$|E_\theta(vT)| \le (E_\theta v^2)^{1/2}[E_\theta(T - T_n)^2]^{1/2} \to 0 \qquad \text{as } n \to \infty$$

for all θ and all $v \in \mathcal{U}$. Thus

$$E_\theta(vT) = 0 \qquad \text{for all } v \in \mathcal{U}_0, \quad \text{all } \theta \in \Theta,$$

and, by Theorem 1, T must be the UMVUE.

Example 5. Let $X_1, X_2, \ldots, X_n$ be iid $P(\lambda)$. Then $\overline{X}$ is the UMVUE of λ. Surely $\overline{X}$ is unbiased. Let g be an unbiased estimator of 0. Then $T(\mathbf{X}) = \overline{X} + g(\overline{X})$ is unbiased for θ. But $\overline{X}$ is complete. It follows that

$$E_\lambda g(\overline{X}) = 0 \qquad \text{for all } \lambda > 0 \Rightarrow g(x) = 0 \quad \text{for } x = 0, 1, 2, \ldots.$$

Hence $\overline{X}$ must be the UMVUE of λ.

Example 6. Sometimes an estimator with larger variance may be preferable.

Let X be a $G(1, 1/\beta)$ RV. X is usually taken as a good model to describe the time to failure of a piece of equipment. Let $X_1, X_2, \ldots, X_n$ be a sample of n observations on X. Then $\overline{X}$ is unbiased for $EX = 1/\beta$ with variance $1/(n\beta^2)$. ($\overline{X}$ is actually the UMVUE for $1/\beta$.) Now consider $X_{(n)} = \min(X_1, X_2, \ldots, X_n)$. Then $nX_{(n)}$ is unbiased for $1/\beta$ with variance $1/\beta^2$, and it has a larger variance than $\overline{X}$. However, if the length of time is of importance, $nX_{(n)}$ may be preferable to $\overline{X}$, since to observe $nX_{(n)}$ one needs to wait only until the first piece of equipment fails, whereas to compute $\overline{X}$ one would have to wait until all the n observations $X_1, X_2, \ldots, X_n$ are available.

Theorem 5. If a sample consists of n independent observations $X_1, X_2, \ldots, X_n$ from the same distribution, the UMVUE, if it exists, is a symmetric function of the X_i's.

Proof. The proof is left as an exercise.

The converse of Theorem 5 is not true. If $X_1, X_2, \ldots, X_n$ are iid $P(\lambda)$ RVs, $\lambda > 0$, both $\overline{X}$ and S^2 are unbiased for θ. But $\overline{X}$ is the UMVUE, whereas S^2 is not.

We now turn our attention to some methods for finding UMVUE's.

Theorem 6. (Blackwell [10], Rao [87]). Let $\{F_\theta: \theta \in \Theta\}$ be a family of probability DFs and h be any statistic in $\mathcal{U}$, where $\mathcal{U}$ is the (nonempty) class of all unbiased estimators of θ with $E_\theta h^2 < \infty$. Let T be a sufficient statistic for $\{F_\theta, \theta \in \Theta\}$. Then the conditional expectation $E_\theta\{h \mid T\}$ is independent of θ and is an unbiased estimator of θ. Moreover,

$$E_\theta(E\{h \mid T\} - \theta)^2 \le E_\theta(h - \theta)^2 \qquad \text{for all } \theta \in \Theta. \tag{9}$$

The equality in (9) holds if and only if $h = E\{h \mid T\}$ (that is, $P_\theta\{h = E\{h \mid T\}\} = 1$ for all θ).

Proof. We have

$$E_\theta\{E\{h \mid T\}\} = E_\theta h = \theta.$$

It is therefore sufficient to show that

$$E_\theta\{E\{h \mid T\}\}^2 \leq E_\theta h^2 \qquad \text{for all } \theta \in \Theta. \tag{10}$$

But $E_\theta h^2 = E_\theta\{E\{h^2 \mid T\}\}$, so that it will be sufficient to show that

$$[E\{h \mid T\}]^2 \leq E\{h^2 \mid T\}. \tag{11}$$

By the Cauchy–Schwarz inequality

$$E^2\{h \mid T\} \leq E\{h^2 \mid T\}E\{1 \mid T\},$$

and (11) follows. The equality holds in (9) if and only if

$$E_\theta[E\{h \mid T\}]^2 = E_\theta h^2, \tag{12}$$

that is,

$$E_\theta[E\{h^2 \mid T\} - E^2\{h \mid T\}] = 0,$$

which is the same as

$$E_\theta\{\text{var}\{h \mid T\}\} = 0.$$

This happens if and only if $\text{var}\{h \mid T\} = 0$, that is, if and only if

$$E\{h^2 \mid T\} = E^2\{h \mid T\},$$

as will be the case if and only if h is a function of T. Thus $h = E\{h \mid T\}$ with probability 1.

Theorem 6 is applied along with completeness to yield the following result.

Theorem 7. (Lehmann-Scheffé [65]). If T is a complete sufficient statistic and there exists an unbiased estimator h of θ, there exists a unique UMVUE of θ, which is given by $E\{h \mid T\}$.

Proof. If $h_1, h_2 \in \mathfrak{U}$, then $E\{h_1 \mid T\}$ and $E\{h_2 \mid T\}$ are both unbiased and

$$E_\theta[E\{h_1 \mid T\} - E\{h_2 \mid T\}] = 0, \qquad \text{for all } \theta \in \Theta.$$

Since T is a complete sufficient statistic, it follows that $E\{h_1 \mid T\} = E\{h_2 \mid T\}$. By Theorem 6 $E\{h \mid T\}$ is the UMVUE.

Remark 5. According to Theorem 6, we should restrict our search to Borel-measurable functions of a sufficient statistic (whenever it exists). According to Theorem 7, if a complete sufficient statistic T exists, all we need to do is to find a Borel-measurable function of T that is unbiased. If a complete sufficient statistic does not exist, an UMVUE may still exist (see Example 11).

Example 7. Let $X_1, X_2, \ldots, X_n$ be $\mathcal{N}(\theta, 1)$. X_1 is unbiased for θ. However, $\overline{X} = n^{-1}\sum_1^n X_i$ is a complete sufficient statistic, so that $E\{X_1 \mid \overline{X}\}$ is the UMVUE.

We will show that $E\{X_1 \mid \overline{X}\} = \overline{X}$. Let $Y = n\overline{X}$. Then Y is $\mathcal{N}(n\theta, n)$, X_1 is $\mathcal{N}(\theta, 1)$, and (X_1, Y) is a bivariate normal RV with variance covariance matrix $\begin{pmatrix} 1 & 1 \\ 1 & n \end{pmatrix}$. Therefore,

$$\begin{aligned} E\{X_1 \mid y\} &= EX_1 + \frac{\operatorname{cov}(X_1, Y)}{\operatorname{var}(Y)}(y - EY) \\ &= \theta + \frac{1}{n}(y - n\theta) = \frac{y}{n}, \end{aligned}$$

as asserted.

If we let $\psi(\theta) = \theta^2$, we can show similarly that $\overline{X}^2 - 1/n$ is the UMVUE for $\psi(\theta)$. Note that $\overline{X}^2 - 1/n$ may occasionally be negative, so that an UMVUE for θ^2 is not very sensible in this case.

Example 8. Let $X_1, X_2, \ldots, X_n$ be iid $b(1, p)$ RVs. Then $T = \sum_1^n X_i$ is a complete sufficient statistic. The UMVUE for p is clearly $\overline{X}$. To find the UMVUE for $\psi(p) = p(1-p)$, we have $E(nT) = n^2 p$, $ET^2 = np + n(n-1)p^2$, so that $E\{nT - T^2\} = n(n-1)p(1-p)$, and it follows that $(nT - T^2)/n(n-1)$ is the UMVUE for $\psi(p) = p(1-p)$.

Example 9. Let $X_1, X_2, \ldots, X_n$ be a sample from $\mathcal{N}(\mu, \sigma^2)$. Then $(\overline{X}, S^2)$ is a complete sufficient statistic for (μ, σ^2). $\overline{X}$ is the UMVUE for μ, and S^2 is the UMVUE for σ^2. Also $k(n)S$ is the UMVUE for σ, where $k(n) = \sqrt{[(n-1)/2]}\,\Gamma[(n-1)/2]/\Gamma(n/2)$. We wish to find the UMVUE for the pth quantile $\mathfrak{z}_p$. We have

$$p = P\{X \leq \mathfrak{z}_p\} = P\left\{Z \leq \frac{\mathfrak{z}_p - \mu}{\sigma}\right\},$$

where Z is $\mathcal{N}(0, 1)$. Thus $\mathfrak{z}_p = \sigma z_{1-p} + \mu$, and the UMVUE is

$$T(X_1, X_2, \ldots, X_n) = z_{1-p} k(n) S + \overline{X}.$$

Example 10. (Stigler [110]). We return to Example 14. We have seen that the family $\{P_N^{(n)} : N \geq 1\}$ of PMFs of $X_{(n)} = \max_{1 \leq i \leq n} X_i$ is complete and $X_{(n)}$ is sufficient for N. Now $EX_1 = (N+1)/2$, so that $T(X_1) = 2X_1 - 1$ is unbiased for N. It follows from Theorem 7 that $E\{T(X_1) \mid X_{(n)}\}$ is the UMVUE of N. We have

$$P\{X_1 = x_1 \mid X_{(n)} = y\} = \begin{cases} \dfrac{y^{n-1} - (y-1)^{n-1}}{y^n - (y-1)^n} & \text{if } x_1 = 1, 2, \ldots, y-1, \\[2ex] \dfrac{y^{n-1}}{y^n - (y-1)^n} & x_1 = y. \end{cases}$$

Thus

$$E\{T(X_1) \mid X_{(n)} = y\} = \frac{y^{n-1} - (y-1)^{n-1}}{y^n - (y-1)^n} \sum_{x_1=1}^{y-1} (2x_1 - 1) + (2y-1)\frac{y^{n-1}}{y^n - (y-1)^n}$$
$$= \frac{y^{n+1} - (y-1)^{n+1}}{y^n - (y-1)^n}$$

is the UMVUE of N.

If we consider the family $\mathcal{P}$ instead, we have seen (Example 8.3.14 and Problem 8.3.6) that $\mathcal{P}$ is not complete. The UMVUE for the family $\{P_N : N \geq 1\}$ is $T(X_1) = 2X_1 - 1$, which is not the UMVUE for $\mathcal{P}$. The UMVUE for $\mathcal{P}$ is in fact, given by

$$T_1(k) = \begin{cases} 2k - 1, & k \neq n_0, \quad k \neq n_0 + 1, \\ 2n_0, & k = n_0, \quad k = n_0 + 1. \end{cases}$$

The reader is asked to check that T_1 has covariance 0 with all unbiased estimators g of 0 that are of the form described in Example 8.3.14 and Problem 8.3.6, and hence Theorem 1 implies that T_1 is the UMVUE. Actually $T_1(X_1)$ is a complete sufficient statistic for $\mathcal{P}$. Since $E_{n_0} T_1(X_1) = n_0 + 1/n_0$, T_1 is not even unbiased for the family $\{P_N : N \geq 1\}$. The minimum variance is given by

$$\text{var}_N(T_1(X_1)) = \begin{cases} \text{var}_N(T(X_1)) & \text{if } N < n_0, \\ \text{var}_N(T(X_1)) - \dfrac{2}{N} & \text{if } N > n_0. \end{cases}$$

The following example shows that UMVUE may exist while minimal sufficient statistic may not.

Example 11. Let X be an RV with PMF

$$P_\theta(X = -1) = \theta \text{ and } P_\theta(X = x) = (1-\theta)^2 \theta^x,$$

$x = 0, 1, 2, \ldots$, where $0 < \theta < 1$. Let $\psi(\theta) = P_\theta(X = 0) = (1-\theta)^2$. Then X is clearly sufficient, in fact minimal sufficient, for θ but since

$$E_\theta X = (-1)\theta + \sum_{x=0}^{\infty} x(1-\theta)^2 \theta^x$$
$$= -\theta + \theta(1-\theta)^2 \frac{d}{d\theta} \sum_{x=1}^{\infty} \theta^x = 0,$$

it follows that X is not complete for $\{P_\theta : 0 < \theta < 1\}$. We will use Theorem 1 to check if a UMVUE for $\psi(\theta)$ exists. Suppose

$$E_\theta h(X) = h(-1)\theta + \sum_{x=0}^{\infty} (1-\theta)^2 \theta^x h(x) = 0$$

for all $0 < \theta < 1$. Then, for $0 < \theta < 1$,

$$0 = \theta h(-1) + \sum_{x=0}^{\infty} \theta^x h(x) - 2\sum_{x=0}^{\infty} \theta^{x+1} h(x) + \sum_{x=0}^{\infty} \theta^{x+2} h(x)$$

$$= h(0) + \sum_{x=0}^{\infty} \theta^{x+1} [h(x+1) - 2h(x) + h(x-1)]$$

which is a power series in θ.

It follows that $h(0) = 0$, and for $x \geq 1$, $h(x+1) - 2h(x) + h(x-1) = 0$. Thus

$$h(1) = h(-1),\ h(2) = 2h(1) - h(0) = 2h(-1),$$

$$h(3) = 2h(2) - h(1) = 4h(-1) - h(-1) = 3h(-1),$$

and so on. Consequently, all unbiased estimators of 0 are of the form $h(X) = cX$. Clearly, $T(X) = 1$ if $X = 0$, and $= 0$ otherwise is unbiased for $\psi(\theta)$. Moreover, for all θ

$$E\{cX \cdot T(X)\} = 0$$

so that T is UMVUE of $\psi(\theta)$.

We conclude this section with a proof of Theorem 8.3.4.

Theorem 8. (Theorem 8.3.4) A complete sufficient statistic is minimal sufficient statistic.

Proof. Let $S(\mathbf{X})$ be a complete sufficient statistic for $\{f_\theta : \theta \in \Theta\}$ and let T be any statistic for which $E_\theta|T^2| < \infty$. Writing $h(S) = E_\theta\{T|S\}$ we see that h is UMVUE of $E_\theta T$. Let $S_1(\mathbf{X})$ be another sufficient statistic. We show that $h(S)$ is a function of S_1. If not, then $h_1(S_1) = E_\theta\{h(S)|S_1\}$ is unbiased for $E_\theta T$ and by Rao–Blackwell theorem

$$\text{var}_\theta\, h_1(S_1) \leq \text{var}_\theta\, h(S),$$

contradicting the fact that $h(S)$ is UMVUE for $E_\theta T$. It follows that $h(S)$ is a function of S_1. Since h and S_1 are arbitrary, S must be a function of every sufficient statistic and hence, minimal sufficient.

PROBLEMS 8.4

1. Let $X_1, X_2, \ldots, X_n (n \geq 2)$ be a sample from $b(1, p)$. Find an unbiased estimator for $\psi(p) = p^2$.
2. Let $X_1, X_2, \ldots, X_n (n \geq 2)$ be a sample from $\mathcal{N}(\mu, \sigma^2)$. Find an unbiased estimator for σ^p, where $p + n > 1$. Find a minimum MSE estimator of σ^p.
3. Let $X_1, X_2, \ldots, X_n$ be iid $\mathcal{N}(\mu, \sigma^2)$ RVs. Find a minimum MSE estimator of the form αS^2 for the parameter σ^2. Compare the variances of the minimum MSE estimator and the obvious estimator S^2.

4. Let $X \sim b(1,\theta^2)$. Does there exist an unbiased estimator of θ?

5. Let $X \sim P(\lambda)$. Does there exist an unbiased estimator of $\psi(\lambda) = \lambda^{-1}$?

6. Let $X_1, X_2, \ldots, X_n$ be a sample from $b(1,p)$, $0 < p < 1$, and $0 < s < n$ be an integer. Find the UMVUE for (a) $\psi(p) = p^s$ and (b) $\psi(p) = p^s + (1-p)^{n-s}$.

7. Let $X_1, X_2, \ldots, X_n$ be a sample from a population with mean θ and finite variance, and T be an estimator of θ of the form $T(X_1, X_2, \ldots, X_n) = \sum_{i=1}^{n} \alpha_i X_i$. If T is an unbiased estimator of θ that has minimum variance and T' is another linear unbiased estimator of θ, then

$$\text{cov}_\theta(T, T') = \text{var}_\theta(T).$$

8. Let T_1, T_2 be two unbiased estimators having common variance $\alpha\sigma^2 (\alpha > 1)$, where σ^2 is the variance of the UMVUE. Show that the correlation coefficient between T_1 and T_2 is $\geq (2-\alpha)/\alpha$.

9. Let $X \sim NB(1;\theta)$ and $d(\theta) = P_\theta\{X = 0\}$. Let $X_1, X_2, \ldots, X_n$ be a sample on X. Find the UMVUE of $d(\theta)$.

10. This example covers most discrete distributions. Let $X_1, X_2, \ldots, X_n$ be a sample from PMF

$$P_\theta\{X = x\} = \frac{\alpha(x)\theta^x}{f(\theta)}, \qquad x = 0, 1, 2, \ldots,$$

where $\theta > 0, \alpha(x) > 0, f(\theta) = \sum_{x=0}^{\infty} \alpha(x)\theta^x, \alpha(0) = 1$, and let $T = X_1 + X_2 + \cdots + X_n$. Write

$$c(t,n) = \sum_{x_1, x_2, \ldots, x_n} \prod_{i=1}^{n} \alpha(x_i).$$

$$\text{with} \quad \sum_{i=1}^{n} x_i = t$$

Show that T is a complete sufficient statistic for θ and that the UMVUE for $d(\theta) = \theta^r$ ($r > 0$ is an integer) is given by

$$Y_r(t) = \begin{cases} 0 & \text{if } t < r \\ \dfrac{c(t-r,n)}{c(t,n)} & \text{if } t \geq r. \end{cases}$$

(Roy and Mitra [94])

11. Let X be a hypergeometric RV with PMF

$$P_M\{X = x\} = \binom{N}{n}^{-1} \binom{M}{x} \binom{N-M}{n-x},$$

where $\max(0, M+n-N) \leq x \leq \min(M,n)$.

(a) Find the UMVUE for M when N is assumed to be known.

(b) Does there exist an unbiased estimator of N (M known)?

12. Let $X_1, X_2, \ldots, X_n$ be iid $G(1, 1/\lambda)$ RVs $\lambda > 0$. Find the UMVUE of $P_\lambda\{X_1 \le t_0\}$, where $t_0 > 0$ is a fixed real number.

13. Let $X_1, X_2, \ldots, X_n$ be a random sample from $P(\lambda)$. Let $\psi(\lambda) = \sum_{k=0}^{\infty} c_k \lambda^k$ be a parametric function. Find the UMVUE for $\psi(\lambda)$. In particular, find the UMVUE for (a) $\psi(\lambda) = 1/(1-\lambda)$, (b) $\psi(\lambda) = \lambda^s$ for some fixed integer $s > 0$, (c) $\psi(\lambda) = P_\lambda\{X=0\}$, and (d) $\psi(\lambda) = P_\lambda\{X = 0 \text{ or } 1\}$.

14. Let $X_1, X_2, \ldots, X_n$ be a sample from PMF

$$P_N(x) = \frac{1}{N}, \qquad x = 1, 2, \ldots, N.$$

Let $\psi(N)$ be some function of N. Find the UMVUE of $\psi(N)$.

15. Let $X_1, X_2, \ldots, X_n$ be a random sample from $P(\lambda)$. Find the UMVUE of $\psi(\lambda) = P_\lambda\{X = k\}$, where k is a fixed positive integer.

16. Let $(X_1, Y_1), (X_2, Y_2), \ldots, (X_n, Y_n)$ be a sample from a bivariate normal population with parameters μ_1, μ_2, σ_1^2, σ_2^2, and ρ. Assume that $\mu_1 = \mu_2 = \mu$, and it is required to find an unbiased estimator of μ. Since a complete sufficient statistic does not exist, consider the class of all linear unbiased estimators

$$\hat{\mu}(\alpha) = \alpha\overline{X} + (1-\alpha)\overline{Y}.$$

(a) Find the variance of $\hat{\mu}$.

(b) Choose $\alpha = \alpha_0$ to minimize $\text{var}(\hat{\mu})$ and consider the estimator

$$\hat{\mu}_0 = \alpha_0\overline{X} + (1-\alpha_0)\overline{Y}.$$

Compute $\text{var}(\hat{\mu}_0)$. If $\sigma_1 = \sigma_2$, the BLUE of μ (in the sense of minimum variance) is

$$\hat{\mu}_1 = \frac{\overline{X} + \overline{Y}}{2}$$

irrespective of whether σ_1 and ρ are known or unknown.

(c) If $\sigma_1 \ne \sigma_2$ and ρ, σ_1, σ_2 are unknown, replace these values in α_0 by their corresponding estimators. Let

$$\hat{\alpha} = \frac{S_2^2 - S_{11}}{S_1^2 + S_2^2 - 2S_{11}}.$$

Show that

$$\hat{\mu}_2 = \overline{Y} + (\overline{X} - \overline{Y})\hat{\alpha}$$

is an unbiased estimator of μ.

17. Let $X_1, X_2, \ldots, X_n$ be iid $\mathcal{N}(\theta, 1)$. Let $p = \Phi(x - \theta)$, where Φ is the DF of a $\mathcal{N}(0,1)$ RV. Show that the UMVUE of p is given by $\Phi\left((x-\overline{x})\sqrt{\frac{n}{n-1}}\right)$.

18. Prove Theorem 5.

19. In Example 10 show that T_1 is the UMVUE for N (restricted to the family $\mathcal{P}$), and compute the minimum variance.

20. Let $(X_1, Y_1), \ldots, (X_n, Y_n)$ be a sample from a bivariate population with finite variances σ_1^2 and σ_2^2, respectively, and covariance γ. Show that

$$\operatorname{var}(S_{11}) = \frac{1}{n}\left(\mu_{22} - \frac{n-2}{n-1}\gamma^2 + \frac{\sigma_1^2\sigma_2^2}{n-1}\right),$$

where $\mu_{22} = E[(X - EX)^2(Y - EY)^2]$. It is assumed that appropriate order moments exist.

21. Suppose that a random sample is taken on (X, Y) and it is desired to estimate γ, the unknown covariance between X and Y. Suppose that for some reason a set S of n observations is available on both X and Y, an additional $n_1 - n$ observations are available on X but the corresponding Y values are missing, and an additional $n_2 - n$ observations of Y are available for which the X values are missing. Let S_1 be the set of all $n_1 (\geq n)$ X values, and S_2, the set of all $n_2 (\geq n)$ Y values, and write

$$\hat{X} = \frac{\sum_{j \in S_1} X_j}{n_1}, \qquad \hat{Y} = \frac{\sum_{j \in S_2} Y_j}{n_2}, \qquad \overline{X} = \frac{\sum_{i \in S} X_i}{n}, \qquad \overline{Y} = \frac{\sum_{i \in S} Y_i}{n}.$$

Show that

$$\hat{\gamma} = \frac{n_1 n_2}{n(n_1 n_2 - n_1 - n_2 + n)} \sum_{i \in S} (X_i - \hat{X})(Y_i - \hat{Y})$$

is an unbiased estimator of γ. Find the variance of $\hat{\gamma}$, and show that $\operatorname{var}(\hat{\gamma}) \leq \operatorname{var}(S_{11})$, where S_{11} is the usual unbiased estimator of γ based on the n observations in S (Boas [11]).

22. Let $X_1, X_2, \ldots, X_n$ be iid with common PDF $f_\theta(x) = \exp(-x + \theta)$, $x > \theta$. Let x_0 be a fixed real number. Find the UMVUE of $f_\theta(x_0)$.

23. Let $X_1, X_2, \ldots, X_n$ be iid $\mathcal{N}(\mu, 1)$ RVs. Let $T(\mathbf{X}) = \sum_{i=1}^n X_i$. Show that $\varphi(x; t/n, n - 1/n)$ is UMVUE of $\varphi(x; \mu, 1)$ where $\varphi(x; \mu, \sigma^2)$ is the PDF of a $\mathcal{N}(\mu, \sigma^2)$ RV.

24. Let $X_1, X_2, \ldots, X_n$ be iid $G(1, \theta)$ RVs. Show that the UMVUE of $f(x; \theta) = (1/\theta)\exp(-x/\theta)$, $x > 0$, is given by $h(x|t)$ the conditional PDF of X_1 given $T(X) = \sum_{i=1}^n X_i = t$, where

$$h(x|t) = (n-1)(t-x)^{n-2}/t^{n-1} \text{ for } x < t \text{ and } = 0 \text{ for } x > t.$$

25. Let $X_1, X_2, \ldots, X_n$ be iid RVs with common PDF $f_\theta(x) = 1/(2\theta)$, $|x| < \theta$, and $= 0$ elsewhere. Show that $T(\mathbf{X}) = \max\{-X_{(1)}, X_{(1)}\}$ is a complete sufficient statistic for θ. Find the UMVU estimator of θ^r.

26. Let $X_1, X_2, \ldots, X_n$ be a random sample from PDF

$$f_\theta(x) = (1/\sigma)\exp\{-(x - \mu)/\sigma\}, \; x > \mu, \sigma > 0,$$

where $\theta = (\mu, \sigma)$.

(a) $\left(X_{(1)}, \sum_{j=1}^{n}\left(X_j - X_{(1)}\right)\right)$ is a complete sufficient statistic for θ.

(b) Show that the UMVUEs of μ and σ are given by

$$\hat{\mu} = X_{(1)} - \frac{1}{n(n-1)}\sum_{j=1}^{n}\left(X_j - X_{(1)}\right), \; \hat{\sigma} = \frac{1}{n-1}\sum_{j=1}^{n}\left(X_j - X_{(1)}\right).$$

(c) Find the UMVUE of $\psi(\mu, \sigma) = E_{\mu,\sigma}X_1$.

(d) Show that the UMVUE of $P_\theta(X_1 \geq t)$ is given by

$$\hat{P}(X_1 \geq t) = \frac{n-1}{n}\left\{\left(1 - \frac{t - X_{(1)}}{\sum_1^n (X_j - X_{(1)})}\right)^+\right\}^{n-2},$$

where $x^+ = \max(x, 0)$.

8.5 UNBIASED ESTIMATION (*CONTINUED*): A LOWER BOUND FOR THE VARIANCE OF AN ESTIMATOR

In this section we consider two inequalities, each of which provides a lower bound for the variance of an estimator. These inequalities can sometimes be used to show that an unbiased estimator is the UMVUE. We first consider an inequality due to Fréchet, Cramér, and Rao (the FCR inequality).

Theorem 1. (Cramér [18], Fréchet [34], Rao [86]). Let $\Theta \subseteq \mathcal{R}$ be an open interval and suppose the family $\{f_\theta : \theta \in \Theta\}$ satisfies the following regularity conditions:

(i) It has common support set S. Thus $S = \{x : f_\theta(\mathbf{x}) > 0\}$ does not depend on θ.

(ii) For $x \in S$ and $\theta \in \Theta$, the derivative $\frac{\theta}{\partial\theta}\log f_\theta(x)$ exists and is finite.

(iii) For any statistic h with $E_\theta|h(\mathbf{X})| < \infty$ for all θ, the operations of integration (summation) and differentiation with respect to θ can be interchanged in $E_\theta h(\mathbf{X})$. That is,

$$\frac{\partial}{\partial\theta}\int h(\mathbf{x}) f_\theta(\mathbf{x})d\mathbf{x} = \int h(\mathbf{x})\frac{\partial}{\partial\theta}f_\theta(\mathbf{x})d\mathbf{x} \tag{1}$$

whenever the right-hand side of (1) is finite.

Let $T(\mathbf{X})$ be such that $\mathrm{var}_\theta T(\mathbf{X}) < \infty$ for all θ and set $\psi(\theta) = E_\theta T(\mathbf{X})$. If $I(\theta) = E_\theta\left\{\frac{\partial}{\partial\theta}\log f_\theta(\mathbf{X})\right\}^2$ satisfies $0 < I(\theta) < \infty$ then

$$\mathrm{var}_\theta T(\mathbf{X}) \geq \frac{[\psi'(\theta)]^2}{I(\theta)}. \tag{2}$$

Proof. Since (iii) holds for $h \equiv 1$, we get

$$\begin{aligned}
0 &= \int_S \frac{\partial}{\partial\theta} f_\theta(\mathbf{x})dx \\
&= \int_S \left\{ \frac{\partial}{\partial\theta} \log f_\theta(\mathbf{x}) \right\} f_\theta(\mathbf{x})d\mathbf{x} \\
&= E\left\{ \frac{\partial}{\partial\theta} \log f_\theta(\mathbf{X}) \right\}. \qquad (3)
\end{aligned}$$

Differentiating $\psi(\theta) = E_\theta T(\mathbf{X})$ and using (1) we get

$$\begin{aligned}
\psi'(\theta) &= \int_S T(\mathbf{x}) \frac{\partial}{\partial\theta} f_\theta(\mathbf{x})d\mathbf{x} \\
&= \int_S \left\{ T(\mathbf{x}) \frac{\partial}{\partial\theta} \log f_\theta(\mathbf{x}) \right\} f_\theta(\mathbf{x})d\mathbf{x} \\
&= \operatorname{cov}\left(T(\mathbf{X}), \frac{\partial}{\partial\theta} \log f_\theta(\mathbf{X}) \right). \qquad (4)
\end{aligned}$$

Also, in view of (3) we have

$$\operatorname{var}_\theta\left(\frac{\partial}{\partial\theta} \log f_\theta(\mathbf{X}) \right) = E_\theta \left\{ \frac{\partial}{\partial\theta} \log f_\theta(\mathbf{X}) \right\}^2$$

and using Cauchy–Schwarz inequality in (4) we get

$$[\psi'(\theta)]^2 \le \operatorname{var}_\theta T(\mathbf{X}) E_\theta \left\{ \frac{\partial}{\partial\theta} \log f_\theta(\mathbf{X}) \right\}^2$$

which proves (2). Practically the same proof may be given when f_θ is a PMF by replacing $\int$ by Σ.

Remark 1. If, in particular, $\psi(\theta) = \theta$, then (2) reduces to

$$\operatorname{var}_\theta(T(\mathbf{X})) \ge \frac{1}{I(\theta)}. \qquad (5)$$

Remark 2. Let $X_1, X_2, \ldots, X_n$ be iid RVs with common PDF (PMF) $f_\theta(x)$. Then

$$\begin{aligned}
I(\theta) = E_\theta \left\{ \frac{\partial \log f_\theta(\mathbf{X})}{\partial\theta} \right\}^2 &= \sum_{i=1}^n E_\theta \left\{ \frac{\partial \log f_\theta(X_i)}{\partial\theta} \right\}^2 \\
= nE_\theta \left\{ \frac{\partial \log f_\theta(X_1)}{\partial\theta} \right\}^2 &= nI_1(\theta),
\end{aligned}$$

where $I_1(\theta) = E_\theta \left\{ \frac{\partial \log f_\theta(X_1)}{\partial\theta} \right\}^2$. In this case the inequality (2) reduces to

$$\operatorname{var}_\theta(T(\mathbf{X})) \ge \frac{[\psi'(\theta)]^2}{nI_1(\theta)}.$$

Definition 1. The quantity

$$I_1(\theta) = E_\theta\left\{\frac{\partial \log f_\theta(X_1)}{\partial\theta}\right\}^2 \tag{6}$$

is called Fisher's information in X_1 and

$$I_n(\theta) = E_\theta\left\{\frac{\partial \log f_\theta(\mathbf{X})}{\partial\theta}\right\}^2 = nI_1(\theta) \tag{7}$$

is known as Fisher information in the random sample $X_1, X_2, \ldots, X_n$.

Remark 3. As n gets larger, the lower bound for $\text{var}_\theta(T(\mathbf{X}))$ gets smaller. Thus, as the Fisher information increases, the lower bound decreases and the "best" estimator (one for which equality holds in (2)) will have smaller variance, consequently more information about θ.

Remark 4. Regularity condition (i) is unnecessarily restrictive. An examination of the proof shows that it is only necessary that (ii) and (iii) hold for (2) to hold. Condition (i) excludes distributions such as $f_\theta(x) = (1/\theta), 0 < x < \theta$, for which (3) fails to hold. It also excludes densities such as $f_\theta(x) = 1, \theta < x < \theta+1$, or $f_\theta(x) = \frac{2}{\pi}\sin^2(x+\pi), \theta \le x \le \theta+\pi$, each of which satisfies (iii) for $h \equiv 1$ so that (3) holds but not (1) for all h with $E_\theta|h| < \infty$.

Remark 5. Sufficient conditions for regularity condition (iii) may be found in most calculus textbooks. For example if (i) and (ii) hold then (iii) holds provided that for all h with $E_\theta|h| < \infty$ for all $\theta \in \Theta$, both $E_\theta\left\{h(\mathbf{X})\frac{\partial \log f_\theta(\mathbf{X})}{\partial\theta}\right\}$ and $E_\theta\left|h(\mathbf{X})\frac{\partial f_\theta(\mathbf{X})}{\partial\theta}\right|$ are continuous functions of θ. Regularity conditions (i) to (iii) are satisfied for a one-parameter exponential family.

Remark 6. The inequality (2) holds trivially if $I(\theta) = \infty$ (and $\psi'(\theta)$ is finite) or if $\text{var}_\theta(T(\mathbf{X})) = \infty$.

Example 1. Let $X \sim b(n,p)$; $\Theta = (0,1) \subset \mathcal{R}$. Here the Fisher Information may be obtained as follows:

$$\log f_p(x) = \log\binom{n}{x} + x\log p + (n-x)\log(1-p),$$

$$\frac{\partial \log f_p(x)}{\partial p} = \frac{x}{p} - \frac{n-x}{1-p},$$

and

$$E_p\left(\frac{\partial \log f_p(x)}{\partial p}\right)^2 = \frac{n}{p(1-p)} = I(p).$$

Let $\psi(p)$ be a function of p and $T(X)$ be an unbiased estimator of $\psi(p)$. The only condition that need be checked is differentiability under the summation sign. We have

$$\psi(p) = E_pT(X) = \sum_{x=0}^{n}\binom{n}{x}T(x)p^x(1-p)^{n-x},$$

which is a polynomial in p and hence can be differentiated with respect to p. For any unbiased estimator $T(X)$ of p we have

$$\text{var}_p(T(X)) \geq \frac{1}{n}p(1-p) = \frac{1}{I(p)},$$

and since

$$\text{var}\left(\frac{X}{n}\right) = \frac{np(1-p)}{n^2} = \frac{p(1-p)}{n},$$

it follows that the variance of the estimator X/n attains the lower bound of the FCR inequality, and hence $T(\mathbf{X})$ has least variance among all unbiased estimators of p. Thus $T(\mathbf{X})$ is UMVUE for p.

Example 2. Let $X \sim P(\lambda)$. We leave the reader to check that the regularity conditions are satisfied and

$$\text{var}_\lambda(T(X)) \geq \lambda.$$

Since $T(X) = X$ has variance λ, X is the UMVUE of λ. Similarly, if we take a sample of size n from $P(\lambda)$, we can show that

$$I_n(\lambda) = \frac{n}{\lambda} \quad \text{and} \quad \text{var}_\lambda(T(X_1,\ldots,X_n)) \geq \frac{\lambda}{n}$$

and $\overline{X}$ is the UMVUE.

Let us next consider the problem of unbiased estimation of $\psi(\lambda) = e^{-\lambda}$ based on a sample of size 1. The estimator

$$\partial(X) = \begin{cases} 1 & \text{if } X = 0 \\ 0 & \text{if } X \geq 1 \end{cases}$$

is unbiased for $\psi(\lambda)$ since

$$E_\lambda \partial(X) = E_\lambda[\partial(X)]^2 = P_\lambda\{X = 0\} = e^{-\lambda}.$$

Also,

$$\text{var}_\lambda(\partial(X)) = e^{-\lambda}(1 - e^{-\lambda}).$$

To compute the FCR lower bound we have

$$\log f_\lambda(x) = x \log \lambda - \lambda - \log x!.$$

This has to be differentiated with respect to $e^{-\lambda}$, since we want a lower bound for an estimator of the parameter $e^{-\lambda}$. Let $\theta = e^{-\lambda}$. Then

$$\log f_\theta(x) = x \log \log \frac{1}{\theta} + \log \theta - \log x!,$$

$$\frac{\partial}{\partial \theta} \log f_\theta(x) = x \frac{1}{\theta \log \theta} + \frac{1}{\theta},$$

and

$$E_\theta\left\{\frac{\partial}{\partial\theta}\log f_\theta(X)\right\}^2 = \frac{1}{\theta^2}\left\{1+\frac{2}{\log\theta}\log\frac{1}{\theta}+\frac{1}{(\log\theta)^2}\left(\log\frac{1}{\theta}+\left(\log\frac{1}{\theta}\right)^2\right)\right\}$$
$$= e^{2\lambda}\left\{1-2+\frac{1}{\lambda^2}(\lambda+\lambda^2)\right\}$$
$$= \frac{e^{2\lambda}}{\lambda} = I(e^{-\lambda}),$$

so that

$$\text{var}_\theta\, T(X) \geq \frac{\lambda}{e^{2\lambda}} = \frac{1}{I(e^{-\lambda})},$$

where $\theta = e^{-\lambda}$.

Since $e^{-\lambda}(1-e^{-\lambda}) > \lambda e^{-2\lambda}$ for $\lambda > 0$, we see that $\text{var}(\delta(X))$ is greater than the lower bound obtained from the FCR inequality. We show next that $\delta(X)$ is the only unbiased estimator of θ and hence is the UMVUE.

If h is any unbiased estimator of θ, it must satisfy $E_\theta h(X) = \theta$. That is, for all $\lambda > 0$

$$e^{-\lambda} = \sum_{k=0}^{\infty} h(k)e^{-\lambda}\frac{\lambda^k}{k!}.$$

Equating coefficients of powers of λ we see immediately that $h(0) = 1$ and $h(k) = 0$ for $k = 1,2,\ldots$. It follows that $h(X) = \partial(X)$.

The same computation can be carried out when $X_1, X_2, \ldots, X_n$ is random sample from $P(\lambda)$. We leave the reader to show that the FCR lower bound for any unbiased estimator of $\theta = e^{-\lambda}$ is $\lambda e^{-2\lambda}/n$. The estimator $\sum_{i=1}^{n}\partial(X_i)/n$ is clearly unbiased for $e^{-\lambda}$ with variance $e^{-\lambda}(1-e^{-\lambda})/n > (\lambda e^{-2\lambda})/n$. The UMVUE of $e^{-\lambda}$ is given by $T_0 = \left(\frac{n-1}{n}\right)^{\sum_{i=1}^{n}X_i}$ with $\text{var}_\lambda(T_0) = e^{-2\lambda}(e^{\lambda/n}-1) > (\lambda e^{-2\lambda})/n$ for all $\lambda > 0$.

Corollary. Let $X_1, X_2, \ldots, X_n$ be iid with common PDF $f_\theta(x)$. Suppose the family $\{f_\theta : \theta \in \Theta\}$ satisfies the conditions of Theorem 1. Then equality holds in (2) if and only if, for all $\theta \in \Theta$,

$$T(\mathbf{x}) - \psi(\theta) = k(\theta)\frac{\partial}{\partial\theta}\log f_\theta(\mathbf{x}) \tag{8}$$

for some function $k(\theta)$.

Proof. Recall that we derived (2) by an application of Cauchy–Schwatz inequality where equality holds if and only if (8) holds.

Remark 7. Integrating (8) with respect to θ we get

$$\log f_\theta(\mathbf{x}) = Q(\theta)T(\mathbf{x}) + S(\theta) + A(\mathbf{x})$$

for some functions Q, S, and A. It follows that f_θ is a one-parameter exponential family and the statistic T is sufficient for θ.

Remark 8. A result that simplifies computations is the following. If f_θ is twice differentiable and $E_\theta\left\{\frac{\partial}{\partial\theta}\log f_\theta(\mathbf{X})\right\}$ can be differentiated under the expectation sign, then

$$I(\theta) = E_\theta\left\{\frac{\partial}{\partial\theta}\log f_\theta(\mathbf{X})\right\}^2 = -E_\theta\left\{\frac{\partial^2}{\partial\theta^2}\log f_\theta(\mathbf{X})\right\}. \tag{9}$$

For the proof of (9), it is straightforward to check that

$$\frac{\partial^2}{\partial\theta^2}\log f_\theta(\mathbf{x}) = \frac{f_\theta''(\mathbf{x})}{f_\theta(\mathbf{x})} - \left\{\frac{\partial}{\partial\theta}\log f_\theta(\mathbf{x})\right\}^2.$$

Taking expectations on both side we get (9).

Example 3. Let $X_1, X_2, \ldots, X_n$ be iid $\mathcal{N}(\mu, 1)$. Then

$$\begin{aligned}
\log f_\mu(x) &= -\frac{1}{2}\log(2\pi) - \frac{(x-\mu)^2}{2},\\
\frac{\partial}{\partial\mu}\log f_\mu(x) &= x - \mu,\\
\frac{\partial^2}{\partial\mu^2}\log f_\mu(x) &= -1.
\end{aligned}$$

Hence $I(\mu) = 1$ and $I_n(\mu) = n$.

We next consider an inequality due to Chapman, Robbins, and Kiefer (the CRK inequality) that gives a lower bound for the variance of an estimator but does not require regularity conditions of the Fréchet–Cramér–Rao type.

Theorem 2 (Chapman and Robbins [12], Kiefer [52]). Let $\Theta \subset \mathcal{R}$ and $\{f_\theta(\mathbf{x}) : \theta \in \Theta\}$ be a class of PDFs (PMFs). Let ψ be defined on Θ, and let T be an unbiased estimator of $\psi(\theta)$ with $E_\theta T^2 < \infty$ for all $\theta \in \Theta$. If $\theta \neq \varphi$, assume that f_θ and f_φ are different and assume further that there exists a $\varphi \in \Theta$ such that $\theta \neq \varphi$ and

$$S(\theta) = \{f_\theta(\mathbf{x}) > 0\} \supset S(\varphi) = \{f_\varphi(\mathbf{x}) > 0\}. \tag{10}$$

Then

$$\operatorname{var}_\theta(T(\mathbf{X})) \geq \sup_{\{\varphi : S(\varphi) \subset S(\theta), \varphi \neq \theta\}} \frac{[\psi(\varphi) - \psi(\theta)]^2}{\operatorname{var}_\theta\{f_\varphi(\mathbf{X})/f_\theta(\mathbf{X})\}} \tag{11}$$

for all $\theta \in \Omega$.

Proof. Since T is unbiased for ψ, $E_\varphi T(\mathbf{X}) = \psi(\varphi)$ for all $\varphi \in \Theta$. Hence, for $\varphi \neq \theta$,

$$\int_{S(\theta)} T(\mathbf{x}) \frac{f_\varphi(\mathbf{x}) - f_\theta(\mathbf{x})}{f_\theta(\mathbf{x})} f_\theta(\mathbf{x})\, d\mathbf{x} = \psi(\varphi) - \psi(\theta), \tag{12}$$

which yields

$$\operatorname{cov}_\theta \left\{ T(\mathbf{X}), \frac{f_\varphi(\mathbf{X})}{f_\theta(\mathbf{X})} - 1 \right\} = \psi(\varphi) - \psi(\theta).$$

Using the Cauchy–Schwarz inequality, we get

$$\begin{aligned}\operatorname{cov}^2_\theta \left\{ T(\mathbf{X}), \frac{f_\varphi(\mathbf{X})}{f_\theta(\mathbf{X})} - 1 \right\} &\leq \operatorname{var}_\theta(T(\mathbf{X}))\operatorname{var}_\theta \left\{ \frac{f_\varphi(\mathbf{X})}{f_\theta(\mathbf{X})} - 1 \right\} \\ &= \operatorname{var}_\theta(T(\mathbf{X}))\operatorname{var}_\theta \left(\frac{f_\varphi(\mathbf{X})}{f_\theta(\mathbf{X})} \right).\end{aligned}$$

Thus

$$\operatorname{var}_\theta(T(\mathbf{X})) \geq \frac{[\psi(\varphi) - \psi(\theta)]^2}{\operatorname{var}_\theta\{f_\varphi(\mathbf{X})/f_\theta(\mathbf{X})\}},$$

and the result follows. In the discrete case it is necessary only to replace the integral in the left side of (12) by a sum. The rest of the proof needs no change.

Remark 9. Inequality (11) holds without any regularity conditions on f_θ or $\psi(\theta)$. We will show that it covers some nonregular cases of the FCR inequality. Sometimes (11) is available in an alternative form. Let θ and $\theta + \delta(\delta \neq 0)$ be any two distinct values in Θ such that $S(\theta + \delta) \subset S(\theta)$, and take $\psi(\theta) = \theta$. Write

$$J = J(\theta, \delta) = \frac{1}{\delta^2} \left\{ \left(\frac{f_{\theta+\delta}(\mathbf{X})}{f_\theta(\mathbf{X})} \right)^2 - 1 \right\}.$$

Then (11) can be written as

$$\operatorname{var}_\theta(T(\mathbf{X})) \geq \frac{1}{\inf\limits_{\delta} E_\theta J}, \tag{13}$$

where the infimum is taken over all $\delta \neq 0$ such that $S(\theta + \delta) \subset S(\theta)$.

Remark 10. Inequality (11) applies if the parameter space is discrete, but the Fréchet–Cramér–Rao regularity conditions do not hold in that case.

Example 4. Let X be $U[0,\theta]$. The regularity conditions of FCR inequality do not hold in this case. Let $\psi(\theta)=\theta$. If $\varphi<\theta$, then $S(\varphi)\subset S(\theta)$. Also,

$$E_\theta\left\{\frac{f_\varphi(X)}{f_\theta(X)}\right\}^2=\int_0^\varphi\left(\frac{\theta}{\varphi}\right)^2\frac{1}{\theta}\,dx=\frac{\theta}{\varphi}.$$

Thus

$$\operatorname{var}_\theta(T(X))\ge\sup_{(\varphi:\varphi<\theta)}\frac{(\varphi-\theta)^2}{(\theta/\varphi)-1}=\sup_{(\varphi:\varphi<\theta)}\{\varphi(\theta-\varphi)\}=\frac{\theta^2}{4}$$

for any unbiased estimator $T(X)$ of θ. X is a complete sufficient statistic, and $2X$ is unbiased for θ so that $T(X)=2X$ is the UMVUE. Also

$$\operatorname{var}_\theta(2X)=4\operatorname{var}X=\frac{\theta^2}{3}>\frac{\theta^2}{4}.$$

Thus the lower bound of $\theta^2/4$ of the CRK inequality is not achieved by any unbiased estimator of θ.

Example 5. Let X have PMF

$$P_N\{X=k\}=\begin{cases}\dfrac{1}{N}, & k=1,2,\ldots,N\\ 0, & \text{otherwise.}\end{cases}$$

Let $\Theta=\{N:N\ge M, M>1\text{ given}\}$. Take $\psi(N)=N$. Although the FCR regularity conditions do not hold, (11) is applicable since, for $N\ne N'\in\Theta\subset\mathcal{R}$,

$$S(N)=\{1,2,\ldots,N\}\supset S(N')=\{1,2,\ldots,N'\}\qquad\text{if } N'<N.$$

Also, P_N and $P_{N'}$ are different for $N\ne N'$. Thus

$$\operatorname{var}_N(T)\ge\sup_{N'<N}\frac{(N-N')^2}{\operatorname{var}_N\{P_{N'}/P_N\}}.$$

Now

$$\frac{P_{N'}}{P_N}(x)=\frac{P_{N'}(x)}{P_N(x)}=\begin{cases}\dfrac{N}{N'}, & x=1,2,\ldots,N',\ N'<N,\\ 0, & \text{otherwise,}\end{cases}$$

$$E_N\left\{\frac{P_{N'}(X)}{P_N(X)}\right\}^2=\frac{1}{N}\sum_1^{N'}\left(\frac{N}{N'}\right)^2=\frac{N}{N'},$$

and

$$\operatorname{var}_N\left\{\frac{P_{N'}(X)}{P_N(X)}\right\} = \frac{N}{N'} - 1 > 0 \qquad \text{for } N > N'.$$

It follows that

$$\operatorname{var}_N(T(X)) \geq \sup_{N'<N} \frac{(N-N')^2}{(N-N')/N'} = \sup_{N'<N} N'(N-N').$$

Now

$$\frac{k(N-k)}{(k-1)(N-k+1)} > 1 \qquad \text{if and only if } k < \frac{N+1}{2},$$

so that $N'(N-N')$ increases as long as $N' < (N+1)/2$ and decreases if $N' > (N+1)/2$. The maximum is achieved at $N' = [(N+1)/2]$ if $M \leq (N+1)/2$ and at $N' = M$ if $M > (N+1)/2$, where $[x]$ is the largest integer $\leq x$. Therefore,

$$\operatorname{var}_N(T(X)) \geq \left[\frac{N+1}{2}\right]\left\{N - \left[\frac{N+1}{2}\right]\right\} \quad \text{if } M \leq (N+1)/2$$

and

$$\operatorname{var}_N(T(X)) \geq M(N-M) \quad \text{if } M > (N+1)/2,$$

Example 6. Let $X \sim \mathcal{N}(0, \sigma^2)$. Let us compute J (see Remark 9) for $\delta \neq 0$.

$$J = \frac{1}{\delta^2}\left\{\left(\frac{f_{\sigma+\delta}(\mathbf{X})}{f_\sigma(\mathbf{X})}\right)^2 - 1\right\} = \frac{1}{\delta^2}\left[\frac{\sigma^{2n}}{(\sigma+\delta)^{2n}} \exp\left\{-\frac{\sum X_i^2}{(\sigma+\delta)^2} + \frac{\sum X_i^2}{\sigma^2}\right\} - 1\right]$$

$$= \frac{1}{\delta^2}\left[\left(\frac{\sigma}{\sigma+\delta}\right)^{2n} \exp\left\{\frac{\sum X_i^2(\delta^2+2\sigma\delta)}{\sigma^2(\sigma+\delta)^2}\right\} - 1\right],$$

$$E_\sigma J = \frac{1}{\delta^2}\left(\frac{\sigma}{\sigma+\delta}\right)^{2n} E_\sigma\left\{\exp\left(c\frac{\sum X_i^2}{\sigma^2}\right)\right\} - \frac{1}{\delta^2},$$

where $c = (\delta^2 + 2\sigma\delta)/(\sigma+\delta)^2$.

Since $\sum X_i^2/\sigma^2 \sim \chi^2(n)$

$$E_\sigma J = \frac{1}{\delta^2}\left\{\left(\frac{\sigma}{\sigma+\delta}\right)^{2n} \frac{1}{(1-2c)^{n/2}} - 1\right\} \qquad \text{for } c < \frac{1}{2}.$$

Let $k = \delta/\sigma$ then

$$c = \frac{2k+k^2}{(1+k)^2} \qquad \text{and} \qquad 1 - 2c = \frac{1-2k-k^2}{(1+k)^2},$$

$$E_\sigma J = \frac{1}{k^2\sigma^2}[(1+k)^{-n}(1-2k-k^2)^{-n/2} - 1].$$

Here $1+k>0$ and $1-2c>0$, so that $1-2k-k^2>0$, implying $-\sqrt{2}<k+1<\sqrt{2}$ and also $k>-1$. Thus $-1<k<\sqrt{2}-1$ and $k\neq 0$. Also,

$$\lim_{k\to 0} E_\sigma J = \lim_{k\to 0} \frac{(1+k)^{-n}(1-2k-k^2)^{-n/2}-1}{k^2\sigma^2} = \frac{2n}{\sigma^2}$$

by L'Hospital's rule. We leave the reader to check that this is the FCR lower bound for $\text{var}_\sigma(T(\mathbf{X}))$. But the minimum value of $E_\sigma J$ is not achieved in the neighborhood of $k=0$ so that the CRK inequality is sharper than the FCR inequality. Next, we show that for $n=2$ we can do better with the CRK inequality. We have

$$\begin{aligned} E_\sigma J &= \frac{1}{k^2\sigma^2}\left\{\frac{1}{(1-2k-k^2)(1+k)^2}-1\right\} \\ &= \frac{(k+2)^2}{\sigma^2(1+k)^2(1-2k-k^2)}, \qquad -1<k<\sqrt{2}-1, \quad k\neq 0. \end{aligned}$$

For $k=-0.1607$ we achieve the lower bound as $(E_\sigma J)^{-1}=0.2698\sigma^2$, so that $\text{var}_\sigma(T(X))\geq 0.2698\sigma^2>\sigma^2/4$. Finally, we show that this bound is by no means the best available; it is possible to improve on the Chapman–Robbins–Kiefer bounds too in some cases. Take

$$T(X_1,X_2,\ldots,X_n) = \frac{\Gamma(n/2)}{\Gamma[(n+1)/2]}\frac{\sigma}{\sqrt{2}}\sqrt{\frac{\sum_1^n X_i^2}{\sigma^2}}$$

to be an estimate of σ. Now $E_\sigma T=\sigma$ and

$$\begin{aligned} E_\sigma T^2 &= \frac{\sigma^2}{2}\left(\frac{\Gamma(n/2)}{\Gamma[(n+1)/2]}\right)^2 E\left\{\frac{\sum_1^n X_i^2}{\sigma^2}\right\} \\ &= \frac{n\sigma^2}{2}\left\{\frac{\Gamma(n/2)}{\Gamma[(n+1)/2]}\right\}^2 \end{aligned}$$

so that

$$\text{var}_\sigma(T) = \sigma^2\left\{\frac{n}{2}\left(\frac{\Gamma(n/2)}{\Gamma[(n+1)/2]}\right)^2-1\right\}.$$

For $n=2$,

$$\text{var}_\sigma(T) = \sigma^2\left[\frac{4}{\pi}-1\right] = 0.2732\sigma^2,$$

which is $>0.2698\sigma^2$, the CRK bound. Note that T is the UMVUE.

Remark 11. In general the CRK inequality is as sharp as the FCR inequality. See Chapman and Robbins [12, pp. 584–585], for details.

We next introduce the concept of *efficiency*.

Definition 2. Let T_1, T_2 be two unbiased estimators for a parameter θ. Suppose that $E_\theta T_1^2 < \infty$, $E_\theta T_2^2 < \infty$. We define the efficiency of T_1 relative to T_2 by

$$\text{eff}_\theta(T_1 \mid T_2) = \frac{\text{var}_\theta(T_2)}{\text{var}_\theta(T_1)} \tag{14}$$

and say that T_1 is more efficient than T_2 if

$$\text{eff}_\theta(T_1 \mid T_2) > 1. \tag{15}$$

It is usual to consider the performance of an unbiased estimator by comparing its variance with the lower bound given by the FCR inequality.

Definition 3. Assume that the regularity conditions of the FCR inequality are satisfied by the family of DFs $\{F_\theta, \theta \in \Theta\}$, $\Theta \subseteq \mathcal{R}$. We say that an unbiased estimator T for parameter θ is most efficient for the family $\{F_\theta\}$ if

$$\text{var}_\theta(T) = \left[E_\theta\left\{\frac{\partial \log f_\theta(\mathbf{X})}{\partial\theta}\right\}^2\right]^{-1} = 1/I_n(\theta). \tag{16}$$

Definition 4. Let T be the most efficient estimator for the regular family of DFs $\{F_\theta, \theta \in \Theta\}$. Then the efficiency of any unbiased estimator T_1 of θ is defined as

$$\text{eff}_\theta(T_1) = \text{eff}_\theta(T_1 \mid T) = \frac{\text{var}_\theta(T)}{\text{var}_\theta(T_1)} = \frac{1}{I_n(\theta)\,\text{var}_\theta(T_1)}. \tag{17}$$

Clearly, the efficiency of the most efficient estimator is 1, and the efficiency of any unbiased estimator T_1 is < 1.

Definition 5. We say that an estimator T_1 is asymptotically (most) efficient if

$$\lim_{n\to\infty} \text{eff}_\theta(T_1) = 1 \tag{18}$$

and T_1 is at least asymptotically unbiased in the sense that $\lim_{n\to\infty} E_\theta T_1 = \theta$. Here n is the sample size.

Remark 12. Definition 3, although in common usage, has many drawbacks. We have already seen cases in which the regularity conditions are not satisfied and yet UMVUEs exist. The definition does not cover such cases. Moreover, in many cases where the regularity conditions are satisfied and UMVUEs exist, the UMVUE is not most efficient since the variance of the best estimator (the UMVUE) does not achieve the lower bound of the FCR inequality.

Example 7. Let $X \sim b(n,p)$. Then we have seen in Example 1 that X/n is the UMVUE since its variance achieves the lower bound of the FCR inequality. It follows that X/n is most efficient.

Example 8. Let $X_1, X_2, \ldots, X_n$ be iid $P(\lambda)$ RVs and suppose $\psi(\lambda) = P_\lambda(X=0) = e^{-\lambda}$. From Example 2, the UMVUE of ψ is given by $T_0 = \left(\frac{n-1}{n}\right)^{\sum_{i=1}^n X_i}$ with

$$\text{var}_\lambda(T_0) = e^{-2\lambda}(e^{\lambda/n} - 1).$$

Also $I_n(\lambda) = n/(\lambda e^{-2\lambda})$. It follows that

$$\text{eff}_\lambda(T_0) = \frac{(\lambda e^{-2\lambda})/n}{e^{-2\lambda}(e^{\lambda/n} - 1)} < \frac{\lambda e^{-2\lambda}/n}{e^{-2\lambda}(\lambda/n)} = 1$$

since $e^x - 1 > x$ for $x > 0$. Thus T_0 is not most efficient. However, since $\text{eff}_\lambda(T_0) \to 1$ as $n \to \infty$, T_0 is asymptotically efficient.

In view of Remarks 6 and 7, the following result describes the relationship between most efficient unbiased estimators and UMVUEs.

Theorem 3. A necessary and sufficient condition for an unbiased estimator T of ψ to be most efficient is that T be sufficient and the relation (8) holds for some function $k(\theta)$.

Clearly, an estimator T satisfying the conditions of Theorem 3 will be the UMVUE, and two estimators coincide. We emphasize that we have assumed the regularity conditions of FCR inequality in making this statement.

Example 9. Let (X, Y) be jointly distributed with PDF

$$f_\theta(x,y) = \exp\left\{-\left(\frac{x}{\theta} + \theta y\right)\right\}, \quad x > 0,\ y > 0.$$

For a sample (x, y) of size 1, we have

$$-\frac{\partial}{\partial\theta}\log f_\theta(x,y) = \frac{\partial}{\partial\theta}\left(\frac{x}{\theta} + \theta y\right) = -\frac{x}{\theta^2} + y.$$

Hence, information for this sample is

$$I(\theta) = E_\theta\left(Y - \frac{X}{\theta^2}\right)^2 = E_\theta(Y^2) + \frac{E(X^2)}{\theta^4} - \frac{2E(XY)}{\theta^2}.$$

Now

$$E_\theta(Y^2) = \frac{2}{\theta^2}, \quad E_\theta(X^2) = 2\theta^2$$

$$\text{and} \quad E(XY) = 1,$$

so that

$$I(\theta) = \frac{2}{\theta^2} + \frac{2}{\theta^2} - \frac{2}{\theta^2} = \frac{2}{\theta^2}.$$

Therefore, amount of Fisher's Information in a sample of n pairs is $\frac{2n}{\theta^2}$.

We return to Example 8.3.23 where $X_1, X_2, \ldots, X_n$ are iid $G(1,\theta)$ and $Y_1, Y_2, \ldots, Y_n$ are iid $G(1, 1/\theta)$, and X's and Y's are independent. Then (X_1, Y_1) has common PDF $f_\theta(x,y)$ given above. We will compute Fisher's Information for θ in the family of PDFs of $S(\mathbf{X},\mathbf{Y}) = (\sum X_i / \sum Y_i)^{1/2}$. Using the PDFs of $\sum X_i \sim G(n,\theta)$ and $\sum Y_i \sim G(n, 1/\theta)$ and the transformation technique, it is easy to see that $S(\mathbf{X},\mathbf{Y})$ has PDF

$$g_\theta(s) = \frac{2\Gamma(2n)}{[\Gamma(n)]^2} s^{-1} \left(\frac{s}{\theta} + \frac{\theta}{s}\right)^{-2n}, \quad s > 0.$$

Thus

$$\frac{\partial \log g_\theta(s)}{\partial \theta} = -2n \left(-\frac{s}{\theta^2} + \frac{1}{s}\right) \left(\frac{s}{\theta} + \frac{\theta}{s}\right)^{-1}.$$

It follows that

$$\begin{aligned} E_\theta \left\{ \frac{\partial}{\partial \theta} \log g_\theta(S) \right\}^2 &= \frac{4n^2}{\theta^2} E_\theta \left\{ 1 - 4\left(\frac{S}{\theta} + \frac{\theta}{S}\right)^{-2} \right\} \\ &= \frac{4n^2}{\theta^2} \left\{ 1 - 4 \frac{n}{2(2n+1)} \right\} = \frac{2n}{\theta^2} \left(\frac{2n}{2n+1}\right) \\ &< \frac{2n}{\theta^2}. \end{aligned}$$

That is, the information about θ in S is smaller than that in the sample.

The Fisher Information in the conditional PDF of S given $A = a$, where $A(\mathbf{X},\mathbf{Y}) = S_1(\mathbf{X}) S_2(\mathbf{Y})$, can be shown (Problem 12) to equal

$$\frac{2a}{\theta^2} \frac{K_1(2a)}{K_0(2a)},$$

where K_0 and K_1 are Bessel functions of order 0 and 1, respectively. Averaging over all values of A, one can show that the information is $2n/\theta^2$ which is the total Fisher information in the sample of n pairs (x_j, y_j)'s.

PROBLEMS 8.5

1. Are the following families of distributions regular in the sense of Fréchet, Cramér, and Rao? If so, find the lower bound for the variance of an unbiased estimator based on a sample size n.
   (a) $f_\theta(x) = \theta^{-1} e^{-x/\theta}$ if $x > 0$, and $= 0$ otherwise; $\theta > 0$.
   (b) $f_\theta(x) = e^{-(x-\theta)}$ if $\theta < x < \infty$, and $= 0$ otherwise.
   (c) $f_\theta(x) = \theta(1-\theta)^x$, $x = 0, 1, 2, \ldots$; $0 < \theta < 1$.
   (d) $f(x;\sigma^2) = (1/\sigma\sqrt{2\pi}) e^{-x^2/2\sigma^2}$, $-\infty < x < \infty$; $\sigma^2 > 0$.
2. Find the CRK lower bound for the variance of an unbiased estimator of θ, based on a sample of size n from the PDF of Problem 1(b).
3. Find the CRK bound for the variance of an unbiased estimator of θ in sampling from $\mathcal{N}(\theta, 1)$.

4. In Problem 1 check to see whether there exists a most efficient estimator in each case.

5. Let $X_1, X_2, \ldots, X_n$ be a sample from a three-point distribution:

$$P\{X = y_1\} = \frac{1-\theta}{2}, \qquad P\{X = y_2\} = \frac{1}{2}, \qquad P\{X = y_3\} = \frac{\theta}{2},$$

where $0 < \theta < 1$. Does the FCR inequality apply in this case? If so, what is the lower bound for the variance of an unbiased estimator of θ?

6. Let $X_1, X_2, \ldots, X_n$ be iid RVs with mean μ and finite variance. What is the efficiency of the unbiased (and consistent) estimator $[2/n(n+1)]\sum_{i=1}^{n} iX_i$ relative to $\overline{X}$?

7. When does the equality hold in the CRK inequality?

8. Let $X_1, X_2, \ldots, X_n$ be a sample from $\mathcal{N}(\mu, 1)$, and let $d(\mu) = \mu^2$:

(a) Show that the minimum variance of any estimator of μ^2 from the FCR inequality is $4\mu^2/n$:

(b) Show that $T(X_1, X_2, \ldots, X_n) = \overline{X}^2 - (1/n)$ is the UMVUE of μ^2 with variance $(4\mu^2/n + 2/n^2)$.

9. Let $X_1, X_2, \ldots, X_n$ be iid $G(1, 1/\alpha)$ RVs:

(a) Show that the estimator $T(X_1, X_2, \ldots, X_n) = (n-1)/n\overline{X}$ is the UMVUE for α with variance $a^2/(n-2)$.

(b) Show that the minimum variance from FCR inequality is α^2/n.

10. In Problem 8.4.16 compute the relative efficiency of $\hat{\mu}_0$ with respect to $\hat{\mu}_1$.

11. Let $X_1, X_2, \ldots, X_n$ and $Y_1, Y_2, \ldots, Y_m$ be independent samples from $\mathcal{N}(\mu, \sigma_1^2)$ and $\mathcal{N}(\mu, \sigma_2^2)$, respectively, where $\mu, \sigma_1^2, \sigma_2^2$ are unknown. Let $\rho = \sigma_2^2/\sigma_1^2$ and $\theta = m/n$, and consider the problem of unbiased estimation of μ:

(a) If ρ is known, show that

$$\hat{\mu}_0 = \alpha\overline{X} + (1-\alpha)\overline{Y},$$

where $\alpha = \rho/(\rho + \theta)$ is the BLUE of μ. Compute $\text{var}(\hat{\mu}_0)$.

(b) If ρ is unknown, the unbiased estimator

$$\bar{\mu} = \frac{\overline{X} + \theta\overline{Y}}{1+\theta}$$

is optimum in the neighborhood of $\rho = 1$. Find the variance of $\bar{\mu}$.

(c) Compute the efficiency of $\bar{\mu}$ relative to $\hat{\mu}_0$.

(d) Another unbiased estimator of μ is

$$\hat{\mu} = \frac{\rho F\overline{X} + \theta\overline{Y}}{\theta + \rho F},$$

where $F = S_2^2/\rho S_1^2$ is an $F(m-1, n-1)$ RV.

12. Show that the Fisher Information on θ based on the PDF

$$\frac{1}{2K_0(2a)} s \exp\left\{-a\left(\frac{s}{\theta} + \frac{\theta}{s}\right)\right\}$$

for fixed a equals $\frac{2a}{\theta^2}\frac{K_1(2a)}{K_0(2a)}$, where $K_0(2a)$ and $K_1(2a)$ are Bessel functions of order 0 and 1 respectively.

8.6 SUBSTITUTION PRINCIPLE (METHOD OF MOMENTS)

One of the simplest and oldest methods of estimation is the *substitution principle*: Let $\psi(\theta)$, $\theta \in \Theta$ be a parametric function to be estimated on the basis of a random sample $X_1, X_2, \ldots, X_n$ from a population DF F. Suppose we can write $\psi(\theta) = h(F)$ for some known function h. Then the substitution principle estimator of $\psi(\theta)$ is $h(F_n^*)$, where F_n^* is the sample distribution function. Accordingly we estimate $\mu = \mu(F)$ by $\mu(F_n^*) = \overline{X}$, $m_k = E_F X^k$ by $\sum_{j=1}^n X_j/n$, and so on. The *method of moments* is a special case when we need to estimate some known function of a finite number of unknown moments. Let us suppose that we are interested in estimating

$$\theta = h(m_1, m_2, \ldots, m_k), \tag{1}$$

where h is some known numerical function and m_j is the jth-order moment of the population distribution that is known to exist for $1 \le j \le k$.

Definition 1. The method of moments consists in estimating θ by the statistic

$$T(X_1, \ldots, X_n) = h\left(n^{-1}\sum_1^n x_i, n^{-1}\sum_1^n X_i^2, \ldots, n^{-1}\sum_1^n X_i^k\right). \tag{2}$$

To make sure that T is a statistic, we will assume that $h : \mathcal{R}_k \to \mathcal{R}$ is a Borel-measurable function.

Remark 1. It is easy to extend the method to the estimation of joint moments. Thus we use $n^{-1}\sum_1^n X_i Y_i$ to estimate $E(XY)$ and so on.

Remark 2. From the WLLN, $n^{-1}\sum_{i=1}^n X_i^j \xrightarrow{P} EX^j$. Thus, if one is interested in estimating the population moments, the method of moments leads to consistent and unbiased estimators. Moreover, the method of moments estimators in this case are asymptotically normally distributed (see Section 7.5).

Again, if one estimates parameters of the type θ defined in (1) and h is a continuous function, the estimators $T(X_1, X_2, \ldots, X_n)$ defined in (2) are consistent for θ (see Problem 1). Under some mild conditions on h, the estimator T is also asymptotically normal (see Cramér [17, pp. 386–387]).

Example 1. Let $X_1, X_2, \ldots, X_n$ be iid RVs with common mean μ and variance σ^2. Then $\sigma = \sqrt{(m_2 - m_1^2)}$, and the method of moments estimator for σ is given by

$$T(X_1, \ldots, X_n) = \sqrt{\frac{1}{n}\sum_1^n X_i^2 - \frac{(\sum X_i)^2}{n^2}}.$$

Although T is consistent and asymptotically normal for σ, it is not unbiased.

In particular, if $X_1, X_2, \ldots, X_n$ are iid $P(\lambda)$ RVs, we know that $EX_1 = \lambda$ and $\text{var}(X_1) = \lambda$. The method of moments leads to using either $\overline{X}$ or $\sum_1^n (X_i - \overline{X})^2/n$ as an estimator of λ. To avoid this kind of ambiguity we take the estimator involving the lowest-order sample moment.

Example 2. Let $X_1, X_2, \ldots, X_n$ be a sample from

$$f(x) = \begin{cases} \dfrac{1}{b-a}, & a \le x \le b, \\ 0, & \text{otherwise.} \end{cases}$$

Then

$$EX = \frac{a+b}{2} \qquad \text{and} \qquad \text{var}(X) = \frac{(b-a)^2}{12}.$$

The method of moments leads to estimating EX by $\overline{X}$ and $\text{var}(X)$ by $\sum_1^n (X_i - \overline{X})^2/n$ so that the estimators for a and b, respectively, are

$$T_1(X_1, \ldots, X_n) = \overline{X} - \sqrt{\frac{3\sum_1^n (X_i - \overline{X})^2}{n}}$$

and

$$T_2(X_1, \ldots, X_n) = \overline{X} + \sqrt{\frac{3\sum_1^n (X_i - \overline{X})^2}{n}}.$$

Example 3. Let $X_1, X_2, \ldots, X_N$ be iid $b(n,p)$ RVs, where both n and p are unknown. The method of moments estimators of p and n are given by

$$\overline{X} = EX = np$$

and

$$\frac{1}{N} \sum_1^N X_i^2 = EX^2 = np(1-p) + n^2p^2.$$

Solving for n and p, we get the estimator for p as

$$T_1(X_1, \ldots, X_N) = \frac{\overline{X}}{T_2(X_1, \ldots, X_N)},$$

where $T_2(X_1, \ldots, X_N)$ is the estimator for n, given by

$$T_2(X_1, X_2, \ldots, X_N) = \frac{(\overline{X})^2}{\overline{X} + \overline{X}^2 - \left(\sum_1^N X_i^2/N\right)}.$$

Note that $\overline{X} \xrightarrow{P} np, \sum_1^N X_i^2/N \xrightarrow{P} np(1-p) + n^2p^2$, so that both T_1 and T_2 are consistent estimators.

Method of moments may lead to absurd estimators. The reader is asked to compute estimators of θ in $\mathcal{N}(\theta, \theta)$ or $\mathcal{N}(\theta, \theta^2)$ by the method of moments and verify this assertion.

PROBLEMS 8.6

1. Let $X_n \xrightarrow{P} a$, and $Y_n \xrightarrow{P} b$, where a and b are constants. Let $h: \mathcal{R}_2 \to \mathcal{R}$ be a continuous function. Show that $h(X_n, Y_n) \xrightarrow{P} h(a,b)$.
2. Let $X_1, X_2, \ldots, X_n$ be a sample from $G(\alpha, \beta)$. Find the method of moments estimator for (α, β).
3. Let $X_1, X_2, \ldots, X_n$ be a sample from $\mathcal{N}(\mu, \sigma^2)$. Find the method of moments estimator for (μ, σ^2).
4. Let $X_1, X_2, \ldots, X_n$ be a sample from $B(\alpha, \beta)$. Find the method of moments estimator for (α, β).
5. A random sample of size n is taken from the lognormal PDF

$$f(x; \mu, \sigma) = (\sigma\sqrt{2\pi})^{-1} x^{-1} \exp\left\{-\frac{1}{2\sigma^2}(\log x - \mu)^2\right\}, \qquad x > 0.$$

Find the method of moments estimators for μ and σ^2.

8.7 MAXIMUM LIKELIHOOD ESTIMATORS

In this section we study a frequently used method of estimation, namely, the *method of maximum likelihood estimation*. Consider the following example.

Example 1. Let $X \sim b(n,p)$. One observation on X is available, and it is known that n is either 2 or 3 and $p = \frac{1}{2}$ or $\frac{1}{3}$. Our objective is to estimate the pair (n,p). The following table gives the probability that $X = x$ for each possible pair (n,p):

x	$(2,\frac{1}{2})$	$(2,\frac{1}{3})$	$(3,\frac{1}{2})$	$(3,\frac{1}{3})$	Maximum Probability
0	$\frac{1}{4}$	$\frac{4}{9}$	$\frac{1}{8}$	$\frac{8}{27}$	$\frac{4}{9}$
1	$\frac{1}{2}$	$\frac{4}{9}$	$\frac{3}{8}$	$\frac{12}{27}$	$\frac{1}{2}$
2	$\frac{1}{4}$	$\frac{1}{9}$	$\frac{3}{8}$	$\frac{6}{27}$	$\frac{3}{8}$
3	0	0	$\frac{1}{8}$	$\frac{1}{27}$	$\frac{1}{8}$

The last column gives the maximum probability in each row, that is, for each value that X assumes. If the value $x = 1$, say, is observed, it is more probable that it came from the distribution $b(2, \frac{1}{2})$ than from any of the other distributions and so on. The following estimator is, therefore, reasonable in that it maximizes the probability of the observed value:

$$(\hat{n}, \hat{p})(x) = \begin{cases} (2, \frac{1}{3}) & \text{if } x = 0, \\ (2, \frac{1}{2}) & \text{if } x = 1, \\ (3, \frac{1}{2}) & \text{if } x = 2, \\ (3, \frac{1}{2}) & \text{if } x = 3. \end{cases}$$

The *principle of maximum likelihood* essentially assumes that the sample is representative of the population and chooses as the estimator that value of the parameter which maximizes the PDF (PMF) $f_\theta(x)$.

Definition 1. Let $(X_1, X_2, \ldots, X_n)$ be a random vector with PDF (PMF) $f_\theta(x_1, x_2, \ldots, x_n)$, $\theta \in \Theta$. The function

$$L(\theta; x_1, x_2, \ldots, x_n) = f_\theta(x_1, x_2, \ldots, x_n), \tag{1}$$

considered as a function of θ, is called the likelihood function.

Usually θ will be a multiple parameter. If $X_1, X_2, \ldots, X_n$ are iid with PDF (PMF) $f_\theta(x)$, the likelihood function is

$$L(\theta; x_1, x_2, \ldots, x_n) = \prod_{i=1}^{n} f_\theta(x_i). \tag{2}$$

Let $\Theta \subseteq \mathcal{R}_k$ and $\mathbf{X} = (X_1, X_2, \ldots, X_n)$.

Definition 2. The principle of maximum likelihood estimation consists of choosing as an estimator of $\boldsymbol{\theta}$ a $\hat{\boldsymbol{\theta}}(\mathbf{X})$ that maximizes $L(\boldsymbol{\theta}; x_1, x_2, \ldots, x_n)$, that is, to find a mapping $\hat{\boldsymbol{\theta}}$ of $\mathcal{R}_n \to \mathcal{R}_k$ that satisfies

$$L(\hat{\boldsymbol{\theta}}; x_1, x_2, \ldots, x_n) = \sup_{\boldsymbol{\theta} \in \boldsymbol{\theta}} L(\boldsymbol{\theta}; x_1, x_2, \ldots, x_n). \tag{3}$$

(Constants are not admissible as estimators.)

If a $\hat{\boldsymbol{\theta}}$ satisfying (3) exists, we call it a *maximum likelihood estimator* (MLE).

It is convenient to work with the logarithm of the likelihood function. Since log is a monotone function,

$$\log L(\hat{\boldsymbol{\theta}}; x_1, \ldots, x_n) = \sup_{\boldsymbol{\theta} \in \boldsymbol{\theta}} \log L(\boldsymbol{\theta}; x_1, \ldots, x_n). \tag{4}$$

Let Θ be an open subset of $\mathcal{R}_k$, and suppose that $f_\theta(\mathbf{x})$ is a positive, differentiable function of $\boldsymbol{\theta}$ (that is, the first-order partial derivatives exist in the components of $\boldsymbol{\theta}$). If a supremum $\hat{\boldsymbol{\theta}}$ exists, it must satisfy the *likelihood equations*

$$\frac{\partial \log L(\hat{\boldsymbol{\theta}}; x_1, \ldots, x_n)}{\partial \theta_j} = 0, \qquad j = 1, 2, \ldots, k, \quad \boldsymbol{\theta} = (\theta_1, \ldots, \theta_k). \tag{5}$$

Any nontrivial root of the likelihood equations (5) is called an MLE in the *loose sense*. A parameter value that provides the absolute maximum of the likelihood function is called an MLE in the *strict* sense or, simply, an MLE.

Remark 1. If $\Theta \subseteq \mathcal{R}$, there may still be many problems. Often the likelihood equation $\partial L / \partial \theta = 0$ has more than one root, or the likelihood function is not differentiable everywhere in Θ, or $\hat{\theta}$ may be a terminal value. Sometimes the likelihood equation may be quite complicated and difficult to solve explicitly. In that case one may have to resort to some numerical procedure to obtain the estimator. Similar remarks apply to the multiparameter case.

Example 2. Let $X_1, X_2, \ldots, X_n$ be a sample from $\mathcal{N}(\mu, \sigma^2)$, where both μ and σ^2 are unknown. Here $\Theta = \{(\mu, \sigma^2), -\infty < \mu < \infty, \sigma^2 > 0\}$. The likelihood function is

$$L(\mu, \sigma^2; x_1, \ldots, x_n) = \frac{1}{\sigma^n (2\pi)^{n/2}} \exp\left\{-\sum_{i=1}^{n} \frac{(x_i - \mu)^2}{2\sigma^2}\right\}$$

and

$$\log L(\mu, \sigma^2; \mathbf{x}) = -\frac{n}{2}\log \sigma^2 - \frac{n}{2}\log(2\pi) - \frac{\sum_1^n (x_i - \mu)^2}{2\sigma^2}.$$

The likelihood equations are

$$\frac{1}{\sigma^2}\sum_{i=1}^{n}(x_i - \mu) = 0$$

and

$$-\frac{n}{2}\frac{1}{\sigma^2} + \frac{1}{2\sigma^4}\sum_{i=1}^{n}(x_i - \mu)^2 = 0.$$

Solving the first of these equations for μ, we get $\hat{\mu} = \overline{X}$ and, substituting in the second, $\hat{\sigma}^2 = \sum_{i=1}^n [(X_i - \overline{X})^2/n]$. We see that $(\hat{\mu}, \hat{\sigma}^2) \in \Theta$ with probability 1. We show that $(\hat{\mu}, \hat{\sigma}^2)$ maximizes the likelihood function. First note that $\overline{X}$ maximizes $L(\mu, \sigma^2; \mathbf{x})$ whatever σ^2 is, since $L(\mu, \sigma^2; \mathbf{x}) \to 0$ as $|\mu| \to \infty$, and in that case $L(\hat{\mu}, \sigma^2; \mathbf{x}) \to 0$ as $\sigma^2 \to 0$ or ∞ whenever $\boldsymbol{\theta} \in \Theta$, $\hat{\boldsymbol{\theta}} = (\hat{\mu}, \hat{\sigma}^2)$.

Note that $\hat{\sigma}^2$ is not unbiased for σ^2. Indeed, $E\hat{\sigma}^2 = [(n-1)/n]\sigma^2$. But $n\hat{\sigma}^2/(n-1) = S^2$ is unbiased, as we already know. Also, $\hat{\mu}$ is unbiased, and both $\hat{\mu}$ and $\hat{\sigma}^2$ are consistent. In addition, $\hat{\mu}$ and $\hat{\sigma}^2$ are method of moments estimators for μ and σ^2, and $(\hat{\mu}, \hat{\sigma}^2)$ is jointly sufficient.

Finally, note that $\hat{\mu}$ is the MLE of μ if σ^2 is known; but if μ is known, the MLE of σ^2 is not $\hat{\sigma}^2$ but $\sum_1^n (X_i - \mu)^2/n$.

Example 3. Let $X_1, X_2, \ldots, X_n$ be a sample from PMF

$$P_N(k) = \begin{cases} \dfrac{1}{N}, & k = 1, 2, \ldots, N, \\ 0 & \text{otherwise.} \end{cases}$$

The likelihood function is

$$L(N; k_1, k_2, \ldots, k_n) = \begin{cases} \dfrac{1}{N^n}, & 1 \le \max(k_1, \ldots, k_n) \le N, \\ 0, & \text{otherwise.} \end{cases}$$

Clearly the MLE of N is given by

$$\hat{N}(X_1, X_2, \ldots, X_n) = \max(X_1, X_2, \ldots, X_n),$$

if we take any $\hat{\alpha} < \hat{N}$ as the MLE, then $P_{\hat{\alpha}}(k_1,k_2,\ldots,k_n) = 0$; and if we take any $\hat{\beta} > \hat{N}$ as the MLE, then $P_{\hat{\beta}}(k_1,k_2,\ldots,k_n) = 1/(\hat{\beta})^n < 1/(\hat{N})^n = P_{\hat{N}}(k_1,k_2,\ldots,k_n)$.

We see that the MLE $\hat{N}$ is consistent, sufficient, and complete, but not unbiased.

Example 4. Consider the hypergeometric PMF

$$P_N(x) = \begin{cases} \dfrac{\binom{M}{x}\binom{N-M}{n-x}}{\binom{N}{n}}, & \max(0, n-N+M) \leq x \leq \min(n,M), \\ 0, & \text{otherwise.} \end{cases}$$

To find the MLE $\hat{N} = \hat{N}(X)$ of N consider the ratio

$$R(N) = \frac{P_N(x)}{P_{N-1}(x)} = \frac{N-n}{N}\,\frac{N-M}{N-M-n+x}.$$

For values of N for which $R(N) > 1$, $P_N(x)$ increases with N, and for values of N for which $R(N) < 1$, $P_N(x)$ is a decreasing function of N:

$$R(N) > 1 \qquad \text{if and only if} \qquad N < \frac{nM}{x}$$

and

$$R(N) < 1 \qquad \text{if and only if} \qquad N > \frac{nM}{x}.$$

It follows that $P_N(x)$ reaches its maximum value where $N \approx nM/x$. Thus $\hat{N}(X) = [nM/X]$, where $[x]$ denotes the largest integer $\leq x$.

Example 5. Let $X_1, X_2, \ldots, X_n$ be a sample from $U[\theta - \frac{1}{2}, \theta + \frac{1}{2}]$. The likelihood function is

$$L(\theta; x_1, x_2, \ldots, x_n) = \begin{cases} 1 & \text{if } \theta - \frac{1}{2} \leq \min(x_1,\ldots,x_n) \\ & \qquad \leq \max(x_1,\ldots,x_n) \leq \theta + \frac{1}{2}, \\ 0 & \text{otherwise.} \end{cases}$$

Thus $L(\theta; \mathbf{x})$ attains its maximum provided that

$$\theta - \frac{1}{2} \leq \min(x_1,\ldots,x_n) \qquad \text{and} \qquad \theta + \frac{1}{2} \geq \max(x_1,\ldots,x_n),$$

or when

$$\theta < \min(x_1,\ldots,x_n) + \frac{1}{2} \qquad \text{and} \qquad \theta \geq \max(x_1,\ldots,x_n) - \frac{1}{2}.$$

It follows that every statistic $T(X_1, X_2, \ldots, X_n)$ such that

$$\max_{1 \leq i \leq n} X_i - \frac{1}{2} \leq T(X_1, X_2, \ldots, X_n) \leq \min_{1 \leq i \leq n} X_i + \frac{1}{2} \tag{6}$$

is an MLE of θ. Indeed, for $0 < \alpha < 1$,

$$T_\alpha(X_1,\ldots,X_n) = \max_{1\le i\le n} X_i - \frac{1}{2} + \alpha(1 + \min_{1\le i\le n} X_i - \max_{1\le i\le n} X_i)$$

lies in interval (6), and hence for each α, $0 < \alpha < 1$, $T_\alpha(X_1,\ldots,X_n)$ is an MLE of θ. In particular, if $\alpha = \frac{1}{2}$,

$$T_{1/2}(X_1,\ldots,X_n) = \frac{\min X_i + \max X_i}{2}$$

is an MLE of θ.

Example 6. Let $X \sim b(1,p)$, $p \in [\frac{1}{4}, \frac{3}{4}]$. In this case $L(p;x) = p^x(1-p)^{1-x}$, $x = 0, 1$, and we cannot differentiate $L(p;x)$ to get the MLE of p, since that would lead to $\hat{p} = x$, a value that does not lie in $\Theta = [\frac{1}{4}, \frac{3}{4}]$. We have

$$L(p;x) = \begin{cases} p, & x = 1, \\ 1-p, & x = 0, \end{cases}$$

which is maximized if we choose $\hat{p}(x) = \frac{1}{4}$ if $x = 0$, and $= \frac{3}{4}$ if $x = 1$. Thus the MLE of p is given by

$$\hat{p}(X) = \frac{2X+1}{4}.$$

Note that $E_p\hat{p}(X) = (2p+1)/4$, so that $\hat{p}$ is biased. Also, the mean square error for $\hat{p}$ is

$$E_p(\hat{p}(X) - p)^2 = \frac{1}{16} E_p(2X + 1 - 4p)^2 = \frac{1}{16}.$$

In the sense of the MSE, the MLE is worse than the trivial estimator $\delta(X) = \frac{1}{2}$, for $E_p(\frac{1}{2} - p)^2 = (\frac{1}{2} - p)^2 \le \frac{1}{16}$ for $p \in [\frac{1}{4}, \frac{3}{4}]$.

Example 7. Let $X_1, X_2, \ldots, X_n$ be iid $b(1,p)$ RVs, and suppose that $p \in (0,1)$. If $(0,0,\ldots,0)((1,1,\ldots,1))$ is observed, $\overline{X} = 0(\overline{X} = 1)$ is the MLE, which is not an admissible value of p. Hence an MLE does not exist.

Example 8. (Oliver [78]). This example illustrates a distribution for which an MLE is necessarily an actual observation, but not necessarily any particular observation. Let $X_1, X_2, \ldots, X_n$ be a sample from PDF

$$f_\theta(x) = \begin{cases} \dfrac{2}{\alpha}\dfrac{x}{\theta}, & 0 \le x \le \theta, \\ \dfrac{2}{\alpha}\dfrac{\alpha - x}{\alpha - \theta}, & \theta \le x \le \alpha, \\ 0, & \text{otherwise}, \end{cases}$$

where $\alpha > 0$ is a (known) constant. The likelihood function is

$$L(\theta; x_1, x_2, \ldots, x_n) = \left(\frac{2}{\alpha}\right)^n \prod_{x_i \le \theta} \left(\frac{x_i}{\theta}\right) \prod_{x_i > 0} \left(\frac{\alpha - x_i}{\alpha - \theta}\right),$$

where we have assumed that observations are arranged in increasing order of magnitude, $0 \le x_1 < x_2 < \cdots < x_n \le \alpha$. Clearly L is continuous in θ (even for $\theta =$ some x_i) and differentiable for values of θ between any two x_i's. Thus, for $x_j < \theta < x_{j+1}$, we have

$$L(\theta) = \left(\frac{2}{\alpha}\right)^n \theta^{-j} (\alpha - \theta)^{-(n-j)} \prod_{i=1}^{j} x_i \prod_{i=j+1}^{n} (\alpha - x_i),$$

$$\frac{\partial \log L}{\partial \theta} = -\frac{j}{\theta} + \frac{n-j}{\alpha - \theta}, \qquad \text{and} \qquad \frac{\partial^2 \log L}{\partial \theta^2} = \frac{j}{\theta^2} + \frac{n-j}{(\alpha - \theta)^2} > 0.$$

It follows that any stationary value that exists must be a minimum, so that there can be no maximum in any range $x_j < \theta < x_{j+1}$. Moreover, there can be no maximum in $0 \le \theta < x_1$ or $x_n < \theta \le \alpha$. This follows since, for $0 \le \theta < x_1$,

$$L(\theta) = \left(\frac{2}{\alpha}\right)^n (\alpha - \theta)^{-n} \prod_{i=1}^{n} (\alpha - x_i)$$

is a strictly increasing function of θ. By symmetry, $L(\theta)$ is a strictly decreasing function of θ in $x_n < \theta \le \alpha$. We conclude that an MLE has to be one of the observations.

In particular, let $\alpha = 5$ and $n = 3$, and suppose that the observations, arranged in increasing order of magnitude, are $1, 2, 4$. In this case the MLE can be shown to be $\hat{\theta} = 1$, which corresponds to the first-order statistic. If the sample values are $2, 3, 4$, the third-order statistic is the MLE.

Example 9. Let $X_1, X_2, \ldots, X_n$ be a sample from $G(r, 1/\beta)$; $\beta > 0$ and $r > 0$ are both unknown. The likelihood function is

$$L(\beta, r; x_1, x_2, \ldots, x_n) = \begin{cases} \dfrac{\beta^{nr}}{\{\Gamma(r)\}^n} \prod_{i=1}^{n} x_i^{r-1} \exp\left(-\beta \sum_{i=1}^{n} x_i\right), & x_i \ge 0, \\ 0, & \text{otherwise}. \end{cases}$$

Then

$$\log L(\beta, r) = nr \log \beta - n \log \Gamma(r) + (r-1) \sum_{i=1}^{n} \log x_i - \beta \sum_{i=1}^{n} x_i,$$

$$\frac{\partial \log L(\beta, r)}{\partial \beta} = \frac{nr}{\beta} - \sum_{i=1}^{n} x_i = 0,$$

$$\frac{\partial \log L(\beta, r)}{\partial r} = n \log \beta - n \frac{\Gamma'(r)}{\Gamma(r)} + \sum_{i=1}^{n} \log x_i = 0.$$

The first of the likelihood equations yields $\hat{\beta}(x_1, x_2, \ldots, x_n) = \hat{r}/\bar{x}$, while the second gives

$$n \log \frac{r}{\bar{x}} + \sum_{i=1}^{n} \log x_i - n \frac{\Gamma'(r)}{\Gamma(r)} = 0,$$

that is,

$$\log r - \frac{\Gamma'(r)}{\Gamma(r)} = \log \overline{x} - \frac{1}{n}\sum_{i=1}^{n} \log x_i,$$

which is to be solved for $\hat{r}$. In this case, the likelihood equation is not easily solvable and it is necessary to resort to numerical methods, using tables for $\Gamma'(r)/\Gamma(r)$.

Remark 2. We have seen that MLEs may not be unique, although frequently they are. Also, they are not necessarily unbiased even if a unique MLE exists. In terms of MSE, an MLE may be worthless. Moreover, MLEs may not even exist. We have also seen that MLEs are functions of sufficient statistics. This is a general result, which we now prove.

Theorem 1. Let T be a sufficient statistic for the family of PDFs (PMFs) $\{f_\theta : \theta \in \Theta\}$. If a unique MLE of θ exists, then it is a (nonconstant) function of T. If a MLE of θ exists but is not unique, then one can find a MLE that is a function of T.

Proof. Since T is sufficient, we can write

$$L(\theta) = f_\theta(\mathbf{x}) = h(\mathbf{x})g_\theta(T(\mathbf{x})),$$

for all $\mathbf{x}$, all θ, and some h and g_θ. If a unique MLE $\hat{\theta}$ exists that maximizes $L(\theta)$, it also maximizes $g_\theta(T(\mathbf{x}))$ and hence $\hat{\theta}$ is a function of T. If a MLE of θ exists but is not unique, we choose a particular MLE $\hat{\theta}$ from the set of all MLE's which is a function of T.

Example 10. Let $X_1, X_2, \ldots, X_n$ be a random sample from $U[\theta, \theta+1]$, $\theta \in R$. Then the likelihood function is given by

$$L(\theta; \mathbf{x}) = \left(\frac{1}{2}\right)^n I_{[\theta-1 \le x_{(1)} \le x_{(n)} \le \theta+1]}(\mathbf{x}).$$

We note that $T(\mathbf{X}) = (X_{(1)}, X_{(n)})$ is jointly sufficient for θ and any θ satisfying

$$\theta - 1 \le x_{(1)} \le x_{(n)} \le \theta + 1,$$

or, equivalently,

$$x_{(n)} - 1 \le \theta \le x_{(1)} + 1$$

maximizes the likelihood and hence is an MLE for θ. Thus, for $0 \le \alpha \le 1$,

$$\hat{\theta}_\alpha = \alpha(X_{(n)} - 1) + (1-\alpha)(X_{(1)} + 1)$$

is an MLE of θ. If α is a constant independent of the X's, then $\hat{\theta}_\alpha$ is a function of T. If, on the other hand, α depends on the X's, then $\hat{\theta}_\alpha$ may not be a function of T alone. For example

$$\hat{\theta}_\alpha = (\sin^2 X_1)(X_{(n)} - 1) + (\cos^2 X_1)(X_{(1)} + 1)$$

is an MLE of θ but not a function of T alone.

Theorem 2. Suppose that the regularity conditions of the FCR inequality are satisfied and θ belongs to an open interval on the real line. If an estimator $\hat{\theta}$ of θ attains the FCR lower bound for the variance, the likelihood equation has a unique solution $\hat{\theta}$ that maximizes the likelihood.

Proof. If $\hat{\theta}$ attains the FCR lower bound, we have [see (8.5.8)]

$$\frac{\partial \log f_\theta(\mathbf{X})}{\partial \theta} = [k(\theta)]^{-1}[\hat{\theta}(\mathbf{X}) - \theta]$$

with probability 1, and the likelihood equation has a unique solution $\theta = \hat{\theta}$.

Let us write $A(\theta) = [k(\theta)]^{-1}$. Then

$$\frac{\partial^2 \log f_\theta(\mathbf{X})}{\partial \theta^2} = A'(\theta)(\hat{\theta} - \theta) - A(\theta),$$

so that

$$\left.\frac{\partial^2 \log f_\theta(\mathbf{X})}{\partial \theta^2}\right|_{\theta=\hat{\theta}} = -A(\theta).$$

We need only to show that $A(\theta) > 0$.

Recall from (8.5.4) with $\psi(\theta) = \theta$ that

$$E_\theta\left\{[T(\mathbf{X}) - \theta]\,\frac{\partial \log f_\theta(\mathbf{X})}{\partial \theta}\right\} = 1,$$

and substituting $T(\mathbf{X}) - \theta = k(\theta)\,\frac{\partial \log f_\theta(\mathbf{X})}{\partial \theta}$ we get

$$k(\theta)E_\theta\left\{\frac{\partial \log f_\theta(\mathbf{X})}{\partial \theta}\right\}^2 = 1.$$

That is,

$$A(\theta) = E\left\{\frac{\partial \log f_\theta(\mathbf{X})}{\partial \theta}\right\}^2 > 0$$

and the proof is complete.

Remark 3. In Theorem 2 we assumed the differentiability of $A(\theta)$ and the existence of the second-order partial derivative $\partial^2 \log f_\theta / \partial \theta^2$. If the conditions of Theorem 2 are satisfied, the most efficient estimator is necessarily the MLE. It does not follow, however, that every MLE is most efficient. For example, in sampling from a normal population, $\hat{\sigma}^2 = \sum_1^n (X_i - \overline{X})^2/n$ is the MLE of σ^2, but it is not most efficient. Since $\sum(X_i - \overline{X})^2/\sigma^2$ is $\chi^2(n-1)$, we see that $\text{var}(\hat{\sigma}^2) = 2(n-1)\sigma^4/n^2$, which is not equal to the FCR lower bound, $2\sigma^4/n$. Note that $\hat{\sigma}^2$ is not even an unbiased estimator of σ^2.

We next consider an important property of MLEs that is not shared by other methods of estimation. Often the parameter of interest is not θ but some function $h(\theta)$. If $\hat{\theta}$ is MLE of θ what is the MLE of $h(\theta)$? If $\lambda = h(\theta)$ is a one to one function of θ, then the inverse function $h^{-1}(\lambda) = \theta$ is well defined and we can write the likelihood function as a function of λ. We have

$$L^*(\lambda;\mathbf{x}) = L(h^{-1}(\lambda);\mathbf{x})$$

so that

$$\sup_{\lambda} L^*(\lambda;\mathbf{x}) = \sup_{\lambda} L(h^{-1}(\lambda);\mathbf{x}) = \sup_{\theta} L(\theta;\mathbf{x}).$$

It follows that the supremum of L^* is achieved at $\lambda = h(\hat{\theta})$. Thus $h(\hat{\theta})$ is the MLE of $h(\theta)$.

In many applications $\lambda = h(\theta)$ is not one-to-one. It is still tempting to take $\hat{\lambda} = h(\hat{\theta})$ as the MLE of λ. The following result provides a justification.

Theorem 3 (Zehna [122]). Let $\{f_{\boldsymbol{\theta}} : \boldsymbol{\theta} \in \Theta\}$ be a family of PDFs (PMFs), and let $L(\boldsymbol{\theta})$ be the likelihood function. Suppose that $\Theta \subseteq \mathcal{R}_k$, $k \geq 1$. Let $h : \Theta \to \Lambda$ be a mapping of Θ onto Λ, where Λ is an interval in $\mathcal{R}_p (1 \leq p \leq k)$. If $\hat{\boldsymbol{\theta}}$ is an MLE of $\boldsymbol{\theta}$, then $h(\hat{\boldsymbol{\theta}})$ is an MLE of $h(\boldsymbol{\theta})$.

Proof. For each $\lambda \in \Lambda$, let us define

$$\Theta_\lambda = \{\boldsymbol{\theta} : \boldsymbol{\theta} \in \Theta, h(\boldsymbol{\theta}) = \lambda\}$$

and

$$M(\lambda;\mathbf{x}) = \sup_{\boldsymbol{\theta} \in \Theta_\lambda} L(\boldsymbol{\theta};\mathbf{x}).$$

Then M defined on Λ is called the likelihood function induced by h. If $\hat{\boldsymbol{\theta}}$ is any MLE of $\boldsymbol{\theta}$, then $\hat{\boldsymbol{\theta}}$ belongs to one and only one set, $\Theta_{\hat{\lambda}}$. Since $\hat{\boldsymbol{\theta}} \in \Theta_{\hat{\lambda}}$, $\hat{\lambda} = h(\hat{\boldsymbol{\theta}})$. Now

$$M(\hat{\lambda};\mathbf{x}) = \sup_{\boldsymbol{\theta} \in \Theta_\lambda} L(\boldsymbol{\theta};\mathbf{x}) \geq L(\hat{\boldsymbol{\theta}};\mathbf{x})$$

and $\hat{\lambda}$ maximizes M, since

$$M(\hat{\lambda};\mathbf{x}) \leq \sup_{\lambda \in \Lambda} M(\lambda;\mathbf{x}) = \sup_{\boldsymbol{\theta} \in \Theta_\lambda} L(\boldsymbol{\theta};\mathbf{x}) = L(\hat{\boldsymbol{\theta}};\mathbf{x}),$$

so that $M(\hat{\lambda};\mathbf{x}) = \sup_{\lambda \in \Lambda} M(\lambda;\mathbf{x})$. It follows that $\hat{\lambda}$ is an MLE of $h(\boldsymbol{\theta})$, where $\hat{\lambda} = h(\hat{\boldsymbol{\theta}})$.

Example 11. Let $X \sim b(1,p)$, $0 \leq p \leq 1$, and let $h(p) = \text{var}(X) = p(1-p)$. We wish to find the MLE of $h(p)$. Note that $\Lambda = [0, \frac{1}{4}]$. The function h is not one-to-one. The MLE of p based on a sample of size n is $\hat{p}(X_1,\ldots,X_n) = \overline{X}$. Hence the MLE of parameter $h(p)$ is $h(\overline{X}) = \overline{X}(1-\overline{X})$.

Example 12. Consider a random sample from $G(1,\beta)$. It is required to find the MLE of β in the following manner. A sample of size n is taken, and it is known only that k, $0 \leq k \leq n$, of these observations are $\leq M$, where M is a fixed positive number.

Let $p = P\{X_i \le M\} = 1 - e^{-M/\beta}$, so that $-M/\beta = \log(1-p)$ and $\beta = M/\log[1/(1-p)]$. Therefore, the MLE of β is $M/\log[1/(1-\hat{p})]$, where $\hat{p}$ is the MLE of p. To compute the MLE of p we have

$$L(p; x_1, x_2, \ldots, x_n) = p^k(1-p)^{n-k},$$

so that the MLE of p is $\hat{p} = k/n$. Thus the MLE of β is

$$\hat{\beta} = \frac{M}{\log[n/(n-k)]}.$$

Finally we consider some important large-sample properties of MLE's. In the following we assume that $\{f_\theta, \theta \in \Theta\}$ is a family of PDFs (PMFs), where Θ is an open interval on $\mathcal{R}$. The conditions listed below are stated when f_θ is a PDF. Modifications for the case where f_θ is a PMF are obvious and will be left to the reader.

(i) $\partial \log f_\theta/\partial\theta, \partial^2 \log f_\theta/\partial\theta^2, \partial^3 \log f_\theta/\partial\theta^3$ exist for all $\theta \in \Theta$ and every x. Also,

$$\int_{-\infty}^{\infty} \frac{\partial f_\theta(x)}{\partial\theta}\,dx = E_\theta \frac{\partial \log f_\theta(X)}{\partial\theta} = 0 \qquad \text{for all } \theta \in \Theta.$$

(ii) $\int_{-\infty}^{\infty} \frac{\partial^2 f_\theta(x)}{\partial\theta^2}\,dx = 0 \qquad$ for all $\theta \in \Theta$.

(iii) $-\infty < \int_{-\infty}^{\infty} \frac{\partial^2 \log f_\theta(x)}{\partial\theta^2} f_\theta(x)\,dx < 0 \qquad$ for all θ.

(iv) There exists a function $H(x)$ such that for all $\theta \in \Theta$

$$\left|\frac{\partial^3 \log f_\theta(x)}{\partial\theta^3}\right| < H(x) \qquad \text{and} \qquad \int_{-\infty}^{\infty} H(x) f_\theta(x)\,dx = M(\theta) < \infty.$$

(v) There exists a function $g(\theta)$ which is positive and twice differentiable for every $\theta \in \Theta$, and a function $H(x)$ such that for all θ

$$\left|\frac{\partial^2}{\partial\theta^2}\left[g(\theta)\frac{\partial \log f_\theta}{\partial\theta}\right]\right| < H(x) \qquad \text{and} \qquad \int_{-\infty}^{\infty} H(x) f_\theta(x)\,dx < \infty.$$

Note that the condition (v) is equivalent to condition (iv) with the added qualification that $g(\theta) = 1$.

We state the following results without proof.

Theorem 4 (Cramér [17]).

(a) Conditions (i), (iii), and (iv) imply that, with probability approaching 1, as $n \to \infty$, the likelihood equation has a consistent solution.

(b) Conditions (i) through (iv) imply that a consistent solution $\hat{\theta}_n$ of the likelihood equation is asymptotically normal, that is,

$$\sigma^{-1}\sqrt{n}(\hat{\theta}_n - \theta) \xrightarrow{L} Z,$$

where Z is $\mathcal{N}(0,1)$ and

$$\sigma^2 = \left[E_\theta\left\{\frac{\partial \log f_\theta(X)}{\partial\theta}\right\}^2\right]^{-1}.$$

On occasions one encounters examples where the conditions of Theorem 4 are not satisfied and yet a solution of the likelihood equation is consistent and asymptotically normal.

Example 13 **(Kulldorf [57]).** Let $X \sim \mathcal{N}(0,\theta)$, $\theta > 0$. Let $X_1, X_2, \ldots, X_n$ be n independent observations on X. The solution of the likelihood equation is $\hat{\theta}_n = \sum_{i=1}^n X_i^2/n$. Also, $EX^2 = \theta$, $\text{var}(X^2) = 2\theta^2$, and

$$E_\theta\left\{\frac{\partial \log f_\theta(X)}{\partial\theta}\right\}^2 = \frac{1}{2\theta^2}.$$

We note that

$$\hat{\theta}_n \xrightarrow{\text{a.s.}} \theta$$

and

$$\sqrt{n}(\hat{\theta}_n - \theta) = \theta\sqrt{2}\,\frac{\sum_1^n X_i^2 - n\theta}{\sqrt{2n}\,\theta} \xrightarrow{L} \mathcal{N}(0, 2\theta^2).$$

However,

$$\frac{\partial^3 \log f_\theta}{\partial^3\theta} = -\frac{1}{\theta^3} + \frac{3x^2}{\theta^4} \to \infty \qquad \text{as } \theta \to 0$$

and is not bounded in $0 < \theta < \infty$. Thus condition (iv) does not hold.

The following theorem covers such cases also.

Theorem 5 (Kulldorf [57]).

(a) Conditions (i), (iii), and (v) imply that, with probability approaching 1 as $n \to \infty$, the likelihood equation has a solution.
(b) Conditions (i), (ii), (iii), and (v) imply that a consistent solution of the likelihood equation is asymptotically normal.

Proof of Theorems 4 and 5. For proofs we refer to Cramér [17, p. 500], and Kulldorf [57].

Remark 4. It is important to note that the results in Theorems 4 and 5 establish the consistency of some root of the likelihood equation but not necessarily that of the MLE when the likelihood equation has several roots. Huzurbazar [47] has shown that under certain conditions the likelihood equation has at most one consistent solution and that the likelihood function has a relative maximum for such a solution. Since there may be several solutions for which the likelihood function has relative maxima, Cramér's and Huzurbazar's results still do not imply that a solution of the likelihood equation that makes the likelihood function an absolute maximum is necessarily consistent.

Wald [115] has shown that under certain conditions the MLE is strongly consistent. It is important to note that Wald does not make any differentiability assumptions.

In any event, if the MLE is a unique solution of the likelihood equation, we can use Theorems 4 and 5 to conclude that it is consistent and asymptotically normal. Note that the asymptotic variance is the same as the lower bound of the FCR inequality.

Example 14. Consider $X_1, X_2, \ldots, X_n$ iid $P(\lambda)$ RVs, $\lambda \in \Theta = (0, \infty)$. The likelihood equation has a unique solution, $\hat{\lambda}(x_1, \ldots, x_n) = \overline{X}$, which maximizes the likelihood function. We leave the reader to check that the conditions of Theorem 4 hold and that MLE $\overline{X}$ is consistent and asymptotically normal with mean λ and variance λ/n, a result that is immediate otherwise.

We leave the reader to check that in Example 13 conditions of Theorem 5 are satisfied.

Remark 5. The invariance and the large sample properties of MLEs permit us to find MLEs of parametric functions and their limiting distributions. The delta method introduced in Section 7.5 (Theorem 1) comes in handy in these applications. Suppose in Example 13 we wish to estimate $\psi(\theta) = \theta^2$. By invariance of MLEs, the MLE of $\psi(\theta)$ is $\psi(\hat{\theta}_n)$ where $\hat{\theta}_n = \sum_1^n X_i^2/n$ is the MLE of θ. Applying Theorem 7.5.1 we see that $\psi(\hat{\theta}_n)$ is AN$(\theta^2, 8\theta^4/n)$.

In Example 14, suppose we wish to estimate $\psi(\lambda) = P_\lambda(X = 0) = e^{-\lambda}$. Then $\psi(\hat{\lambda}) = e^{-\overline{X}}$ is the MLE of $\psi(\lambda)$ and, in view of Theorem 7.5.1, $\psi(\hat{\lambda}) \sim$ AN$(e^{-\lambda}, \lambda e^{-2\lambda}/n)$.

Remark 6. Neither Theorem 4 nor Theorem 5 guarantee asymptotic normality for a unique MLE. Consider, for example, a random sample from $U(0, \theta]$. Then $X_{(n)}$ is the unique MLE for θ and in Problem 8.2.5 we asked the reader to show that $n(\theta - X_{(n)}) \xrightarrow{L} G(1, \theta)$.

PROBLEMS 8.7

1. Let $X_1, X_2, \ldots, X_n$ be iid RVs with common PMF (pdf) $f_\theta(x)$. Find an MLE for θ in each of the following cases:
  (a) $f_\theta(x) = \frac{1}{2} e^{-|x-\theta|}$, $-\infty < x < \infty$.
  (b) $f_\theta(x) = e^{-x+\theta}$, $\theta \le x < \infty$.
  (c) $f_\theta(x) = (\theta\alpha) x^{\alpha-1} e^{-\theta x^\alpha}$, $x > 0$, and α known.
  (d) $f_\theta(x) = \theta(1-x)^{\theta-1}$, $0 \le x \le 1$, $\theta > 1$.

2. Find an MLE, if it exists, in each of the following cases:
  (a) $X \sim b(n, \theta)$: both n and $\theta \in [0, 1]$ are unknown, and one observation is available.
  (b) $X_1, X_2, \ldots, X_n \sim b(1, \theta)$, $\theta \in [\frac{1}{2}, \frac{3}{4}]$.
  (c) $X_1, X_2, \ldots, X_n \sim \mathcal{N}(\theta, \theta^2)$, $\theta \in \mathcal{R}$.
  (d) $X_1, X_2, \ldots, X_n$ is a sample from

$$P\{X = y_1\} = \frac{1-\theta}{2}, \qquad P\{X = y_2\} = \frac{1}{2}, \qquad P\{X = y_3\} = \frac{\theta}{2} (0 < \theta < 1).$$

(e) $X_1, X_2, \ldots, X_n \sim \mathcal{N}(\theta, \theta)$, $0 < \theta < \infty$.
(f) $X \sim \mathcal{C}(\theta, 0)$.

3. Suppose that n observations are taken on an RV X with distribution $\mathcal{N}(\mu, 1)$, but instead of recording all the observations one notes only whether or not the observation is less than 0. If $\{X < 0\}$ occurs $m(< n)$ times, find the MLE of μ.

4. Let $X_1, X_2, \ldots, X_n$ be a random sample from PDF

$$f(x; \alpha, \beta) = \beta^{-1} e^{-\beta^{-1}(x-\alpha)}, \qquad \alpha < x < \infty, \qquad -\infty < \alpha < \infty, \quad \beta > 0.$$

(a) Find the MLE of (α, β).
(b) Find the MLE of $P_{\alpha,\beta}\{X_1 \geq 1\}$.

5. Let $X_1, X_2, \ldots, X_n$ be a sample from exponential density $f_\theta(x) = \theta e^{-\theta x}$, $x \geq 0$, $\theta > 0$. Find the MLE of θ, and show that it is consistent and asymptotically normal.

6. For Problem 8.6.5 find the MLE for (μ, σ^2).

7. For a sample of size 1 taken from $\mathcal{N}(\mu, \sigma^2)$, show that no MLE of (μ, σ^2) exists.

8. For Problem 8.6.5 suppose that we wish to estimate N on the basis of observations $X_1, X_2, \ldots, X_M$:
(a) Find the UMVUE of N.
(b) Find the MLE of N.
(c) Compare the MSEs of the UMVUE and the MLE.

9. Let $X_{ij}(1 = 1, 2, \ldots, s; j = 1, 2, \ldots, n)$ be independent RVs where $X_{ij} \sim \mathcal{N}(\mu_i, \sigma^2)$, $i = 1, 2, \ldots, s$. Find MLEs for $\mu_1, \mu_2, \ldots, \mu_s$, and σ^2. Show that the MLE for σ^2 is not consistent as $s \to \infty$ (n fixed) (Neyman and Scott [77]).

10. Let (X, Y) have a bivariate normal distribution with parameters $\mu_1, \mu_2, \sigma_1^2, \sigma_2^2$, and ρ. Suppose that n observations are made on the pair (X, Y), and $N - n$ observations on X that is, $N - n$ observations on Y are missing. Find the MLE's of $\mu_1, \mu_2, \sigma_1^2, \sigma_2^2$, and ρ (Anderson [2]).

[*Hint:* If $f(x, y; \mu_1, \mu_2, \sigma_1^2, \sigma_2^2, \rho)$ is the joint PDF of (X, Y) write

$$f(x, y; \mu_1, \mu_2, \sigma_1^2, \sigma_2^2, \rho) = f_1(x; \mu_1, \sigma_1^2) f_{Y|X}(y \mid \beta_x, \sigma_2^2(1 - \rho^2)),$$

where f_1 is the marginal (normal) PDF of X, and $f_{Y|X}$ is the conditional (normal) PDF of Y, given x with mean

$$\beta_x = \left(\mu_2 - \rho \frac{\sigma_2}{\sigma_1} \mu_1\right) + \rho \frac{\sigma_2}{\sigma_1} x$$

and variance $\sigma_2^2(1 - \rho^2)$. Maximize the likelihood function first with respect to μ_1 and σ_1^2 and then with respect to $\mu_2 - \rho(\sigma_2/\sigma_1)\mu_1$, $\rho\sigma_2/\sigma_1$, and $\sigma_2^2(1 - \rho^2)$.]

11. In Problem 5, let $\hat{\theta}$ denote the MLE of θ. Find the MLE of $\mu = EX_1 = 1/\theta$ and its asymptotic distribution.

12. In Problem 1(d), find the asymptotic distribution of the MLE of θ.

13. In Problem 2(a), find MLE of $d(\theta) = \theta^2$ and its asymptotic distribution.

14. Let $X_1, X_2, \ldots, X_n$ be a random sample from some DF F on the real line. Suppose we observe $x_1, x_2, \ldots, x_n$ which are all different. Show that the MLE of F is F_n^*, the empirical DF of the sample.

15. Let $X_1, X_2, \ldots, X_n$ be iid $\mathcal{N}(\mu, 1)$. Suppose $\Theta = \{\mu \geq 0\}$. Find the MLE of μ.

16. Let $(X_1, X_2, \ldots, X_{k-1})$ have a multinomial distribution with parameters $n, p_1, \ldots, p_{k-1}$, $0 \leq p_1, p_2, \ldots, p_{k-1} \leq 1$, $\sum_1^{k-1} p_j \leq 1$, where n is known. Find the MLE of $(p_1, p_2, \ldots, p_{k-1})$.

17. Consider the one parameter exponential density introduced in Section 5.5 in its natural form with PDF

$$f_\theta(x) = \exp\{\eta T(x) + D(\eta) + S(x)\}.$$

(a) Show that the MGF of $T(X)$ is given by

$$M(t) = \exp\{D(\eta) - D(\eta + t)\}$$

for t in some neighborhood of the origin. Moreover, $E_\eta T(X) = -D'(\eta)$ and $\text{var}(T(X)) = -D''(\eta)$.

(b) If the equation $E_\eta T(X) = T(x)$ has a solution, it must be the unique MLE of η.

18. In Problem 1(b) show that the unique MLE of θ is consistent. Is it asymptotically normal?

8.8 BAYES AND MINIMAX ESTIMATION

In this section we consider the problem of point estimation in a decision-theoretic setting. We will consider here Bayes and minimax estimation.

Let $\{f_\theta : \theta \in \Theta\}$ be a family of PDFs (PMFs) and $X_1, X_2, \ldots, X_n$ be a sample from this distribution. Once the sample point $(x_1, x_2, \ldots, x_n)$ is observed, the statistician takes an *action* on the basis of these data. Let us denote by $\mathcal{A}$ the set of all *actions* or *decisions* open to the statistician.

Definition 1. A decision function δ is a statistic that takes values in $\mathcal{A}$, that is, δ is a Borel-measurable function that maps $\mathcal{R}_n$ into $\mathcal{A}$.

If $\mathbf{X} = \mathbf{x}$ is observed, the statistician takes action $\delta(\mathbf{X}) \in \mathcal{A}$.

Example 1. Let $\mathcal{A} = \{a_1, a_2\}$. Then any decision function δ partitions the space of values of $(X_1, \ldots, X_n)$, namely, $\mathcal{R}_n$, into a set C and its complement C^c, such that if $\mathbf{x} \in C$ we take action a_1, and if $\mathbf{x} \in C^c$ action a_2 is taken. This is the problem of testing hypotheses, which we will discuss in Chapter 9.

Example 2. Let $\mathcal{A} = \Theta$. In this case we face the problem of estimation.

Another element of decision theory is the specification of a *loss function*, which measures the loss incurred when we take a decision.

Definition 2. Let $\mathcal{A}$ be an arbitrary space of actions. A nonnegative function L that maps $\Theta \times \mathcal{A}$ into $\mathcal{R}$ is called a loss function.

The value $L(\theta, a)$ is the loss to the statistician if he takes action a when θ is the true parameter value. If we use the decision function $\delta(\mathbf{X})$ and loss function L and θ is the true parameter value, then the loss is the RV $L(\theta, \delta(\mathbf{X}))$. (As always, we will assume that L is a Borel-measurable function.)

Definition 3. Let $\mathcal{D}$ be a class of decision functions that map $\mathcal{R}_n$ into $\mathcal{A}$, and let L be a loss function on $\Theta \times \mathcal{A}$. The function R defined on $\Theta \times \mathcal{D}$ by

$$R(\theta, \delta) = E_\theta L(\theta, \delta(\mathbf{X})) \tag{1}$$

is known as the risk function associated with δ at θ.

Example 3. Let $\mathcal{A} = \Theta \subseteq \mathcal{R}$, $L(\theta, a) = |\theta - a|^2$. Then

$$R(\theta, \delta) = E_\theta L(\theta, \delta(X)) = E_\theta\{\delta(X) - \theta\}^2,$$

which is just the MSE. If we restrict attention to estimators that are unbiased, the risk is just the variance of the estimator.

The basic problem of decision theory is the following: Given a space of actions A, and a loss function $L(\theta, a)$, find a decision function δ in D such that the risk $R(\theta, \delta)$ is "minimum" in some sense for all $\theta \in \Theta$. We need first to specify some criterion for comparing the decision functions δ.

Definition 4. The principle of minimax is to choose $\delta^* \in \mathcal{D}$ so that

$$\max_\theta R(\theta, \delta^*) \le \max_\theta R(\theta, \delta) \tag{2}$$

for all δ in $\mathcal{D}$. Such a rule δ^*, if it exists, is called a minimax (decision) rule.

If the problem is one of estimation, that is, if $\mathcal{A} = \Theta$, we call δ^* satisfying (2) a *minimax estimator* of θ.

Example 4. Let $X \sim b(1, p)$, $p \in \Theta = \{\frac{1}{4}, \frac{1}{2}\}$, and $\mathcal{A} = \{a_1, a_2\}$. Let the loss function be defined as follows.

	a_1	a_2
$p_1 = \frac{1}{4}$	1	4
$p_2 = \frac{1}{2}$	3	2

The set of decision rules includes four functions: $\delta_1, \delta_2, \delta_3, \delta_4$, defined by $\delta_1(0) = \delta_1(1) = a_1$; $\delta_2(0) = a_1$, $\delta_2(1) = a_2$; $\delta_3(0) = a_2$, $\delta_3(1) = a_1$; and $\delta_4(0) = \delta_4(1) = a_2$. The risk function takes the following values

i	$R(p_1, \delta_i)$	$R(p_2, \delta_i)$	$\underset{p_1, p_2}{\text{Max}}\, R(p, \delta_i)$	$\underset{i}{\text{Min}}\, \underset{p_1, p_2}{\text{Max}}\, R(p, \delta_i)$
1	1	3	3	
2	$\frac{7}{4}$	$\frac{5}{2}$	$\frac{5}{2}$	$\frac{5}{2}$
3	$\frac{13}{4}$	$\frac{5}{2}$	$\frac{13}{4}$	
4	4	2	4	

Thus the minimax solution is $\delta_2(x) = a_1$ if $x = 0$ and $= a_2$ if $x = 1$.

The computation of minimax estimators is facilitated by the use of the *Bayes estimation method*. So far, we have considered θ as a fixed constant and $f_\theta(\mathbf{x})$ has represented the PDF (PMF) of the RV $\mathbf{X}$. In Bayesian estimation we treat θ as a random variable distributed according to PDF (PMF) $\pi(\theta)$ on Θ. Also, π is called the *a priori distribution*. Now $f(\mathbf{x} \mid \theta)$ represents the conditional probability density (or mass) function of RV $\mathbf{X}$, given that $\theta \in \Theta$ is held fixed. Since π is the distribution of θ, it follows that the joint density (PMF) of θ and $\mathbf{X}$ is given by

$$f(\mathbf{x}, \theta) = \pi(\theta) f(\mathbf{x} \mid \theta). \tag{3}$$

In this framework $R(\theta, \delta)$ is the conditional average loss, $E\{L(\theta, \delta(\mathbf{X})) \mid \theta\}$, given that θ is held fixed. (Note that we are using the same symbol to denote the RV θ and a value assumed by it.)

Definition 5. The Bayes risk of a decision function δ is defined by

$$R(\pi, \delta) = E_\pi R(\theta, \delta). \tag{4}$$

If θ is a continuous RV and $\mathbf{X}$ is of the continuous type, then

$$\begin{aligned} R(\pi, \delta) &= \int R(\theta, \delta) \pi(\theta)\, d\theta \\ &= \iint L(\theta, \delta(\mathbf{x})) f(\mathbf{x} \mid \theta) \pi(\theta)\, d\mathbf{x}\, d\theta \\ &= \iint L(\theta, \delta(\mathbf{x})) f(\mathbf{x}, \theta)\, d\mathbf{x}\, d\theta. \end{aligned} \tag{5}$$

If θ is discrete with PMF π and $\mathbf{X}$ is of the discrete type, then

$$R(\pi, \delta) = \sum_\theta \sum_{\mathbf{x}} L(\theta, \delta(\mathbf{x})) f(\mathbf{x}, \theta). \tag{6}$$

Similar expressions may be written in the other two cases.

Definition 6. A decision function δ^* is known as a Bayes rule (procedure) if it minimizes the Bayes risk, that is, if

$$R(\pi,\delta^*) = \inf_{\delta} R(\pi,\delta). \tag{7}$$

Definition 7. The conditional distribution of RV θ, given $\mathbf{X} = \mathbf{x}$, is called the a posteriori probability distribution of θ, given the sample.

Let the joint PDF (PMF) be expressed in the form

$$f(\mathbf{x},\theta) = g(\mathbf{x})h(\theta \mid \mathbf{x}), \tag{8}$$

where g denotes the joint marginal density (PMF) of $\mathbf{X}$. The a priori PDF (PMF) $\pi(\theta)$ gives the distribution of θ before the sample is taken, and the a posteriori PDF (PMF) $h(\theta \mid \mathbf{x})$ gives the distribution of θ after sampling. In terms of $h(\theta \mid \mathbf{x})$ we may write

$$R(\pi,\delta) = \int g(\mathbf{x})\left\{\int L(\theta,\, \delta(\mathbf{x}))h(\theta \mid \mathbf{x})\, d\theta\right\} d\mathbf{x} \tag{9}$$

or

$$R(\pi,\delta) = \sum_{\mathbf{x}} g(\mathbf{x})\left\{\sum_{\theta} L(\theta,\delta(\mathbf{x}))h(\theta \mid \mathbf{x})\right\}, \tag{10}$$

depending on whether f and π are both continuous or both discrete. Similar expressions may be written if only one of f and π is discrete.

Theorem 1. Consider the problem of estimation of a parameter $\theta \in \Theta \subseteq \mathcal{R}$ with respect to the quadratic loss function $L(\theta,\delta) = (\theta - \delta)^2$. A Bayes solution is given by

$$\delta(\mathbf{x}) = E\{\theta \mid \mathbf{X} = \mathbf{x}\} \tag{11}$$

($\delta(x)$ defined by (11) is called the *Bayes estimator*).

Proof. In the continuous case, if π is the prior PDF of θ, then

$$R(\pi,\delta) = \int g(\mathbf{x})\left\{\int [\theta - \delta(\mathbf{x})]^2 h(\theta \mid \mathbf{x}) d\theta\right\} d\mathbf{x},$$

where g is the marginal PDF of $\mathbf{X}$, and h is the conditional PDF of θ, given $\mathbf{x}$. The Bayes rule is a function δ that minimizes $R(\pi,\delta)$. Minimization of $R(\pi,\delta)$ is the same as minimization of

$$\int [\theta - \delta(\mathbf{x})]^2 h(\theta \mid \mathbf{x})\, d\theta,$$

which is minimum if and only if

$$\delta(\mathbf{x}) = E\{\theta \mid \mathbf{x}\}.$$

The proof for the remaining cases is similar.

Remark 1. The argument used in Theorem 1 shows that a Bayes estimator is one which minimizes $E\{L(\theta,\delta(\mathbf{X}))\mid\mathbf{X}\}$. Theorem 1 is a special case which says that if $L(\theta,\delta(\mathbf{X}))=[\theta-\delta(\mathbf{X})]^2$ the function

$$\delta(\mathbf{x})=\int\theta h(\theta\mid\mathbf{x})\,d\theta$$

is the Bayes estimator for θ with respect to π, the a priori distribution on Θ.

Remark 2. Suppose $T(\mathbf{X})$ is sufficient for the parameter θ. Then it is easily seen that the posterior distribution of θ given $\mathbf{x}$ depends on $\mathbf{x}$ only through T and it follows that the Bayes estimator of θ is a function of T.

Example 5. Let $X\sim b(n,p)$ and $L(p,\delta(x))=[p-\delta(x)]^2$. Let $\pi(p)=1$ for $0<p<1$ be the a priori PDF of p. Then

$$h(p\mid x)=\frac{\binom{n}{x}p^x(1-p)^{n-x}}{\int_0^1\binom{n}{x}p^x(1-p)^{n-x}dp}$$

It follows that

$$\begin{aligned}E\{p\mid x\}&=\int_0^1 ph\{p\mid x\}\,dp\\&=\frac{x+1}{n+2}.\end{aligned}$$

Hence the Bayes estimator is

$$\delta^*(X)=\frac{X+1}{n+2}.$$

The Bayes risk is

$$\begin{aligned}R(\pi,\delta^*)&=\int\pi(p)\sum_{x=0}^{n}[\delta^*(x)-p]^2f(x\mid p)\,dp\\&=\int_0^1 E\left\{\left(\frac{X+1}{n+2}-p\right)^2\middle|p\right\}dp\\&=\frac{1}{(n+2)^2}\int_0^1[np(1-p)+(1-2p)^2]\,dp\\&=\frac{1}{6(n+2)}.\end{aligned}$$

Example 6. Let $X\sim\mathcal{N}(\mu,1)$, and let the a priori PDF of μ be $\mathcal{N}(0,1)$. Also, let $L(\mu,\delta)=[\mu-\delta(\mathbf{X})]^2$. Then

$$h(\mu\mid\mathbf{x})=\frac{f(\mathbf{x},\mu)}{g(\mathbf{x})}=\frac{\pi(\mu)f(\mathbf{x}\mid\mu)}{g(\mathbf{x})},$$

where

$$
\begin{aligned}
f(\mathbf{x}) &= \int f(\mathbf{x},\mu)\,d\mu \\
&= \frac{1}{(2\pi)^{(n+1)/2}} \exp\left(-\frac{1}{2}\sum_1^n x_i^2\right) \\
&\quad \cdot \int_{-\infty}^{\infty} \exp\left\{-\frac{n+1}{2}\left(\mu^2 - 2\mu\frac{n\bar{x}}{n+1}\right)\right\} d\mu \\
&= \frac{(n+1)^{-1/2}}{(2\pi)^{n/2}} \exp\left\{-\frac{1}{2}\sum x_i^2 + \frac{n^2\bar{x}^2}{2(n+1)}\right\}.
\end{aligned}
$$

It follows that

$$
h(\mu \mid \mathbf{x}) = \frac{1}{\sqrt{2\pi/(n+1)}} \exp\left\{-\frac{n+1}{2}\left(\mu - \frac{n\bar{x}}{n+1}\right)^2\right\},
$$

and the Bayes estimator is

$$
\delta^*(\mathbf{x}) = E\{\mu \mid \mathbf{x}\} = \frac{n\bar{x}}{n+1} = \frac{\sum_1^n x_i}{n+1}.
$$

The Bayes risk is

$$
\begin{aligned}
R(\pi,\delta^*) &= \int \pi(\mu) \int [\delta^*(\mathbf{x}) - \mu]^2 f(\mathbf{x} \mid \mu)\,d\mathbf{x}\,d\mu \\
&= \int_{-\infty}^{\infty} E_\mu\left\{\frac{n\overline{X}}{n+1} - \mu\right\}^2 \pi(\mu)\,d\mu \\
&= \int_{-\infty}^{\infty} (n+1)^{-2}(n+\mu^2)\pi(\mu)\,d\mu \\
&= \frac{1}{n+1}.
\end{aligned}
$$

The quadratic loss function used in Theorem 1 is but one example of a loss function in frequent use. Some of many other loss functions that may be used are

$$
|\theta - \delta(\mathbf{X})|, \qquad \frac{|\theta - \delta(\mathbf{X})|^2}{|\theta|}, \qquad |\theta - \delta|^4, \qquad \text{and} \qquad \left(\frac{|\theta - \delta(\mathbf{X})|}{|\theta|+1}\right)^{1/2}.
$$

Example 7. Let $X_1, X_2, \ldots, X_n$ be iid $\mathcal{N}(\mu, \sigma^2)$ RVs. It is required to find a Bayes estimator of μ of the form $\delta(x_1,\ldots,x_n) = \delta(\bar{x})$, where $\bar{x} = \sum_1^n x_i/n$, using the loss function $L(\mu,\delta) = |\mu - \delta(\bar{x})|$. From the argument used in the proof of Theorem 1 (or by Remark 1), the Bayes estimator is one that minimizes the integral $\int |\mu - \delta(\bar{x})| h(\mu|\bar{x})\,d\mu$. This will be the case if we choose δ to be the median of the conditional distribution (see Problem 3.2.5).

Let the a priori distribution of μ be $\mathcal{N}(\theta, \tau^2)$. Since $\overline{X} \sim \mathcal{N}(\mu, \sigma^2/n)$, we have

$$
f(\bar{x},\mu) = \frac{\sqrt{n}}{2\pi\sigma\tau} \exp\left\{-\frac{(\mu-\theta)^2}{2\tau^2} - \frac{n(\bar{x}-\mu)^2}{2\sigma^2}\right\}.
$$

Writing

$$(\bar{x}-\mu)^2 = (\bar{x}-\theta+\theta-\mu)^2 = (\bar{x}-\theta)^2 - 2(\bar{x}-\theta)(\mu-\theta) + (\mu-\theta)^2,$$

we see that the exponent in $f(\bar{x}, \mu)$ is

$$-\frac{1}{2}\left\{(\mu-\theta)^2\left(\frac{1}{\tau^2}+\frac{n}{\sigma^2}\right) - \frac{2n(\bar{x}-\theta)(\mu-\theta)}{\sigma^2} + \frac{n}{\sigma^2}(\bar{x}-\theta)^2\right\}.$$

It follows that the joint PDF of μ and $\overline{X}$ is bivariate normal with means θ, θ, variances τ^2, $\tau^2+(\sigma^2/n)$, and correlation coefficient $\tau/\sqrt{[\tau^2+(\sigma^2/n)]}$. The marginal of $\overline{X}$ is $\mathcal{N}(\theta, \tau^2 + (\sigma^2/n))$, and the conditional distribution of μ, given $\overline{X}$, is normal with mean

$$\theta + \frac{\tau}{\sqrt{\tau^2+(\sigma^2/n)}}\frac{\tau}{\sqrt{\tau^2+(\sigma^2/n)}}(\bar{x}-\theta) = \frac{\theta(\sigma^2/n)+\bar{x}\tau^2}{\tau^2+(\sigma^2/n)}$$

and variance

$$\tau^2\left[1-\frac{\tau^2}{\tau^2+(\sigma^2/n)}\right] = \frac{\tau^2\sigma^2/n}{\tau^2+(\sigma^2/n)}$$

(see the proof of Theorem 1). The Bayes estimator is therefore the median of this conditional distribution, and since the distribution is symmetric about the mean,

$$\delta^*(\bar{x}) = \frac{\theta(\sigma^2/n)+\bar{x}\tau^2}{\tau^2+(\sigma^2/n)}$$

is the Bayes estimator of μ.

Clearly δ^* is also the Bayes estimator under the quadratic loss function $L(\mu, \delta) = [\mu - \delta(\mathbf{X})]^2$.

Key to the derivation of Bayes estimator is the posteriori distribution, $h(\theta \mid \mathbf{x})$. The derivation of the posteriori distribution $h(\theta \mid \mathbf{x})$, however, is a three-step process:

1. Find the joint distribution of $\mathbf{X}$ and $\boldsymbol{\theta}$ given by $\pi(\theta)f(\mathbf{x} \mid \theta)$.
2. Find the marginal distribution with PDF (PMF) $g(\mathbf{x})$ by integrating (summing) over $\theta \in \Omega$.
3. Divide the joint PDF (PMF) by $g(\mathbf{x})$.

It is not always easy to go through these steps in practice. It may not be possible to obtain $h(\theta \mid \mathbf{x})$ in a closed form.

Example 8. Let $X \sim \mathcal{N}(\mu, 1)$ and the prior PDF of μ be given by

$$\pi(\mu) = \frac{e^{-(\mu-\theta)}}{[1+e^{-(\mu-\theta)}]^2},$$

where θ is a location parameter. Then the joint PDF of X and μ is given by

$$f(x,\mu) = \frac{1}{\sqrt{2\pi}} e^{-(x-\mu)^2/2} \frac{e^{-(\mu-\theta)}}{[1+e^{-(\mu-\theta)}]^2}$$

so that the marginal PDF of X is

$$g(x) = \frac{e^{\theta}}{\sqrt{2\pi}} \int_{-\infty}^{\infty} \frac{e^{-(x-\mu)^2/2} e^{-\mu}}{[1+e^{-(\mu-\theta)}]^2} d\mu.$$

A closed form for g is not known.

To avoid problem of integration such as that in Example 8, statisticians use the so-called *conjugate prior* distributions. Often there is a natural parameter family of distributions such that the posterior distributions also belong to the same family. These priors make the computations much easier.

Definition 8. Let $X \sim f(x|\theta)$ and $\pi(\theta)$ be the prior distribution on Θ. Then π is said to be a conjugate prior family if the corresponding posterior distribution $h(\theta \mid x)$ also belongs to the same family as $\pi(\theta)$.

Example 9. Consider Example 6 where $\pi(\mu)$ is $\mathcal{N}(0,1)$ and $h(\mu \mid \mathbf{x})$ is $\mathcal{N}\left(\frac{n\bar{x}}{n+1}, \frac{1}{n+1}\right)$ so that both h and π belong to the same family. Hence $\mathcal{N}(0,1)$ is a conjugate prior for μ.

Example 10. Let $X \sim b(n,p), 0<p<1$, and $\pi(p)$ be the beta PDF with parameters (α,β). Then

$$h(p \mid x) = \frac{p^{x+\alpha-1}(1-p)^{\beta-1}}{\int_0^1 p^{x+\alpha-1}(1-p)^{\beta-1} dp} = \frac{p^{x+\alpha-1}(1-p)^{\beta-1}}{B(x+\alpha,\beta)}$$

which is also a beta density. Thus the family of beta distributions is a conjugate family of priors for p.

Conjugate priors are popular because whenever the prior family is parametric the posterior distributions are always computable, $h(\theta|x)$ being an updated parametric version of $\pi(\theta)$. One no longer needs to go through a computation of g, the marginal PDF (PMF) of $\mathbf{X}$. Once $h(\theta|x)$ is known g, if needed, is easily determined from

$$g(\mathbf{x}) = \frac{\pi(\theta) f(\mathbf{x}|\theta)}{h(\theta|\mathbf{x})}.$$

Thus in Example 10, we see easily that $g(x)$ is beta $(x+\alpha,\beta)$, while in Example 6 g is given by

$$g(\mathbf{x}) = \frac{1}{(n+1)^{1/2}(2\pi)^{n/2}} \exp\left\{-\frac{1}{2}\sum_{i=1}^{n} x_i^2 + \frac{n^2\bar{x}^2}{2(n-1)}\right\}.$$

Conjugate priors are usually associated with a wide class of sampling distributions, namely, the exponential family of distributions.

Natural Conjugate Priors		
Sampling PDF(PMF), $f(x\|\theta)$	Prior $\pi(\theta)$	Posterior $h(\theta\|x)$
$N(\theta,\sigma^2)$	$N(\mu,\tau^2)$	$N(\frac{\sigma^2\mu+x\tau^2}{\sigma^2+\tau^2},\frac{\sigma^2\tau^2}{\sigma^2+\tau^2})$
$G(\nu,\beta)$	$G(\alpha,\beta)$	$G(\alpha+\nu,\beta+x)$
$b(n,p)$	$B(\alpha,\beta)$	$B(\alpha+x,\beta+n-x)$
$P(\lambda)$	$G(\alpha,\beta)$	$G(\alpha+x,\beta+1)$
$NB(r;p)$	$B(\alpha,\beta)$	$B(\alpha+r,\beta+x)$
$G(\gamma,1/\theta)$	$G(\alpha,\beta)$	$G(\alpha+\nu,\beta+x)$

Another easy way is to use a *noninformative prior* $\pi(\theta)$ though one needs some integration to obtain $g(\mathbf{x})$.

Definition 9. A PDF $\pi(\theta)$ is said to be a noninformative prior if it contains no information about θ, that is, the distribution does not favor any value of θ over others.

Example 11. Some simple examples of noninformative priors are $\pi(\theta)=1$, $\pi(\theta)=\frac{1}{\theta}$ and $\pi(\theta)=\sqrt{I(\theta)}$. These may quite often lead to infinite mass and the PDF may be improper (that is, does not integrate to 1).

Calculation of $h(\theta|x)$ becomes easier by-passing the calculation of $g(\mathbf{x})$ when $f(x|\theta)$ is invariant under a group $\mathcal{G}$ of transformations following Fraser's [33] structural theory.

Let $\mathcal{G}$ be a group of Borel-measurable functions on $\mathcal{R}_n$ onto itself. The group operation is composition, that is, if g_1 and g_2 are mappings from $\mathcal{R}_n$ onto $\mathcal{R}_n$, g_2g_1 is defined by $g_2g_1(\mathbf{x})=g_2(g_1(\mathbf{x}))$. Also, $\mathcal{G}$ is closed under composition and inverse, so that all maps in $\mathcal{G}$ are one-to-one. We define the group G of affine linear transformations $g=\{a,b\}$ by

$$gx=a+bx,\ a\in\mathcal{R},\ b>0.$$

The inverse of $\{a,b\}$ is

$$\{a,b\}^{-1}=\left\{-\frac{a}{b},\frac{1}{b}\right\},$$

and the composition $\{a,b\}$ and $\{c,d\}\in\mathcal{G}$ is given by

$$\begin{aligned}\{a,b\}\{c,d\}(x)&=\{a,b\}(c+dx)=a+b(c+dx)\\&=(a+bc)+bdx=\{a+bc,bd\}(x).\end{aligned}$$

In particular,

$$\{a,b\}\{a,b\}^{-1}=\{a,b\}\left\{-\frac{a}{b},\frac{1}{b}\right\}=\{0,1\}=e.$$

Example 12. Let $X \sim \mathcal{N}(\mu, 1)$ and let $\mathcal{G}$ be the group of translations $\mathcal{G} = \{\{b, 1\}, -\infty < b < \infty\}$. Let $X_1, \ldots, X_n$ be a sample from $\mathcal{N}(\mu, 1)$. Then, we may write

$$X_i = \{\mu, 1\}Z_i, \quad i = 1, \ldots, n,$$

where $Z_1, \ldots, Z_n$ are iid $\mathcal{N}(0, 1)$.

It is clear that $\overline{Z} \sim \mathcal{N}(0, 1/n)$ with PDF

$$\sqrt{\frac{n}{2\pi}} \exp\left\{-\frac{n}{2}\overline{z}^2\right\}$$

and there is a one-to-one correspondence between values of $\{\overline{z}, 1\}$ and $\{\mu, 1\}$ given by

$$\{\overline{x}, 1\} = \{\mu, 1\}\{\overline{z}, 1\} = \{\mu + \overline{z}, 1\}.$$

Thus $\overline{x} = \mu + \overline{z}$ with inverse map $\overline{z} = \overline{x} - \mu$. We fix $\overline{x}$ and consider the variation in $\overline{z}$ as a function of μ. Changing the PDF element of $\overline{z}$ to μ we get

$$\sqrt{\frac{n}{2\pi}} \exp\left\{-\frac{n}{2}(\mu - \overline{x})^2\right\}$$

as the posterior of μ given $\overline{x}$ with prior $\pi(\mu) = 1$.

Example 13. Let $X \sim \mathcal{N}(0, \sigma^2)$ and consider the scale group $\mathcal{G} = \{\{0, c\}, c > 0\}$. Let $X_1, X_2, \ldots, X_n$ be iid $\mathcal{N}(0, \sigma^2)$. Write

$$X_i = \{0, \sigma\}Z_i, \quad i = 1, 2, \ldots, n,$$

where Z_i are iid $\mathcal{N}(0, 1)$ RVs. Then the RV $nS_z^2 = \sum_{i=1}^n Z_i^2 \sim \chi^2(n)$ with PDF

$$\frac{1}{2^{\frac{n}{2}}\Gamma(\frac{n}{2})} \exp\left\{-\frac{ns_z^2}{2}\right\}(ns_z^2)^{\frac{n}{2}-1}.$$

The values of $\{0, s_z\}$ are in one-to-one correspondence with those of $\{0, \sigma\}$ through

$$\{0, s_x\} = \{0, \sigma\}\{0, s_z\},$$

where $nS_x^2 = \sum_{i=1}^n X_i^2$, so that $s_x = \sigma s_z$. Considering the variation in s_z as a function of σ for fixed s_x we see that $ds_z = s_x \frac{d\sigma}{\sigma^2}$. Changing the PDF element of s_z to σ we get the PDF of σ as

$$\frac{1}{2^{n/2}\Gamma(\frac{n}{2})} \exp\left\{-\frac{ns_x^2}{2\sigma^2}\right\}\left(\frac{ns_x^2}{\sigma^2}\right)^{(\frac{n}{2}-1)}$$

which is the same as the posterior of σ given s_x with prior $\pi(\sigma) = 1/\sigma$.

Example 14. Let $X_1 \ldots X_n$ be a sample from $\mathcal{N}(\mu, \sigma^2)$ and consider the affine linear group $\mathcal{G} = \{\{a, b\}, -\infty < a < \infty, b > 0\}$. Then

$$X_i = \{\mu, \sigma\}Z_i, \quad i = 1, \ldots, n$$

where Z_i's are iid $\mathcal{N}(0,1)$. We know that the joint distribution of $(\overline{Z}, \mathcal{S}_z^2)$ is given by

$$\sqrt{\frac{n}{2\pi}}\exp\left\{-\frac{n\bar{z}^2}{2}\right\}d\bar{z}\frac{1}{\sqrt{\left(\frac{n-1}{2}\right)}}\left(\frac{(n-1)s_z^2}{2}\right)^{\frac{n-1}{2}-1}$$
$$\exp\left\{-\frac{(n-1)s_z^2}{2}\right\}d\left[\frac{(n-1)s_z^2}{2}\right].$$

Further,the values of $\{\bar{z}, s_z\}$ are in one-to-one correspondence with the values of $\{\mu,\sigma\}$ through

$$\begin{aligned}\{\bar{x}, s_x\} &= \{\mu,\sigma\}\{\bar{z}, s_z\} = \{\mu+\sigma\bar{z}, \sigma s_z\}\\ \Rightarrow \bar{z} &= \frac{\bar{x}-\mu}{\sigma} \text{ and } s_z = \frac{s_x}{\sigma}.\end{aligned}$$

Consider the variation of $(\bar{z}, s_z)$ as a function of (μ,σ) for fixed $(\bar{x}, s_x)$. The Jacobian of the transformation from $\{\bar{z}, s_z\}$ to $\{\mu,\sigma\}$ is given by

$$J = \begin{vmatrix} -\frac{1}{\sigma} & -\frac{(\bar{x}-\mu)}{\sigma^2} \\ 0 & -\frac{s_x}{\sigma^2}\end{vmatrix} = \frac{s_x}{\sigma^3}.$$

Hence, the joint PDF of (μ,σ) given $(\bar{x}, s_x)$ is given by

$$\sqrt{\frac{n}{2\pi}}\exp\left\{-\frac{n(\mu-\bar{x})^2}{2\sigma^2}\right\}\frac{1}{\sqrt{\left(\frac{n-1}{2}\right)}}\left(\frac{(n-1)s_x^2}{2\sigma^2}\right)^{\frac{n-1}{2}-1}$$
$$\times\exp\left\{-\frac{(n-1)s_x^2}{2\sigma^2}\right\}\left(\frac{(n-1)s_x^2}{2\sigma^2}\right)^{\frac{n-1}{2}-1}\left(\frac{(n-1)s_x^2}{\sigma^3}\right).$$

This is the PDF one obtains if $\pi(\mu) = 1$ and $\pi(\sigma) = \frac{1}{\sigma}$ and μ and σ are independent RV.

The following theorem provides a method for determining minimax estimators.

Theorem 2. Let $\{f_\theta \colon \theta \in \Theta\}$ be a family of PDFs (PMFs), and suppose that an estimator δ^* of θ is a Bayes estimator corresponding to an a priori distribution π on Θ. If the risk function $R(\theta,\delta^*)$ is constant on Θ, then δ^* is a minimax estimator for θ.

Proof. Since δ^* is the Bayes estimator of θ with constant risk r^* (free of θ), we have

$$\begin{aligned} r^* = R(\pi,\delta^*) &= \int_{-\infty}^{\infty} R(\theta,\delta^*)\pi(\theta)d\theta \\ &= \inf_{\delta\in\mathcal{D}}\int R(\theta,\delta)\pi(\theta)d\theta \\ &\le \sup_{\theta\in\Theta}\inf_{\delta\in\mathcal{D}} R(\theta,\delta) \le \inf_{\delta\in\mathcal{D}}\sup_{\theta\in\Theta} R(\theta,\delta).\end{aligned}$$

Similarly, since $r^* = R(\theta, \delta^*)$ for all $\theta \in \Theta$, we have

$$r^* = \sup_{\theta\in\Theta} R(\theta, \delta^*) \geq \inf_{\delta\in\mathcal{D}} \sup_{\theta\in\Theta} R(\theta, \delta).$$

Together we then have

$$\sup_{\theta\in\Theta} R(\theta, \delta^*) = \inf_{\delta\in\mathcal{D}} \sup_{\theta\in\Theta} R(\theta, \delta),$$

which means δ^* is minimax.

The following examples show how to obtain constant risk estimators and the suitable prior distribution.

Example 15. (Hodges and Lehmann [43]). Let $X \sim b(n,p), 0 \leq p \leq 1$. We seek a minimax estimator of p of the form $\alpha X + \beta$, using the squared error loss function. We have

$$\begin{aligned} R(p, \delta) &= E_p\{\alpha X + \beta - p\}^2 = E_p\{\alpha(X - np) + \beta + (\alpha n - 1)p\}^2 \\ &= [(\alpha n - 1)^2 - \alpha^2 n]p^2 + [\alpha^2 n + 2\beta(\alpha n - 1)]p + \beta^2, \end{aligned}$$

which is a quadratic equation in p. To find α and β such that $R(p, \delta)$ is constant for all $p \in \Theta$, we set the coefficients of p^2 and p equal to 0 to get

$$(\alpha n - 1)^2 - \alpha^2 n = 0 \qquad \text{and} \qquad \alpha^2 n + 2\beta(\alpha n - 1) = 0.$$

It follows that

$$\alpha = \frac{1}{\sqrt{n}(1 + \sqrt{n})} \qquad \text{or} \qquad \frac{1}{\sqrt{n}(\sqrt{n} - 1)}$$

and

$$\beta = \frac{1}{2(1 + \sqrt{n})} \qquad \text{or} \qquad -\frac{1}{2(\sqrt{n} - 1)}.$$

Since $0 \leq p \leq 1$, we discard the second set of roots for both α and β, and then the estimator is of the form

$$\delta^*(x) = \frac{X}{\sqrt{n}(1 + \sqrt{n})} + \frac{1}{2(1 + \sqrt{n})}$$

It remains to show that δ^* is Bayes against some a priori PDF π.

Consider the natural conjugate priori PDF

$$\pi(p) = [\beta(\alpha', \beta')]^{-1} p^{\alpha'-1}(1 - p)^{\beta'-1}, \qquad 0 \leq p \leq 1, \quad \alpha', \beta' > 0.$$

The a posteriori PDF of p, given x, is expressed by

$$h(p \mid x) = \frac{p^{x+\alpha'-1}(1 - p)^{n-x+\beta'-1}}{B(x + \alpha', n - x + \beta')}$$

It follows that

$$E\{p \mid x\} = \frac{B(x+\alpha'+1, n-x+\beta')}{B(x+\alpha', n-x+\beta')} = \frac{x+\alpha'}{n+\alpha'+\beta'},$$

which is the Bayes estimator for a squared error loss. For this to be of the form δ^*, we must have

$$\frac{1}{\sqrt{n}(1+\sqrt{n})} = \frac{1}{n+\alpha'+\beta'} \quad \text{and} \quad \frac{1}{2(1+\sqrt{n})} = \frac{\alpha'}{n+\alpha'+\beta'},$$

giving $\alpha' = \beta' = \sqrt{n}/2$. It follows that the estimator $\delta^*(x)$ is minimax with constant risk

$$R(p,\delta^*) = \frac{1}{4(1+\sqrt{n})^2} \quad \text{for all} \quad p \in [0,1].$$

Note that the UMVUE (which is also the MLE) is $\delta(X) = X/n$ with risk $R(p,d) = p(1-p)/n$. Comparing the two risks (Figs. 1 and 2), we see that

$$\frac{p(1-p)}{n} \le \frac{1}{4(1+\sqrt{n})^2} \quad \text{if and only if} \quad |p-\frac{1}{2}| \ge \frac{\sqrt{1+2\sqrt{n}}}{2(1+\sqrt{n})},$$

so that

$$R(p,\delta^*) < R(p,\delta)$$

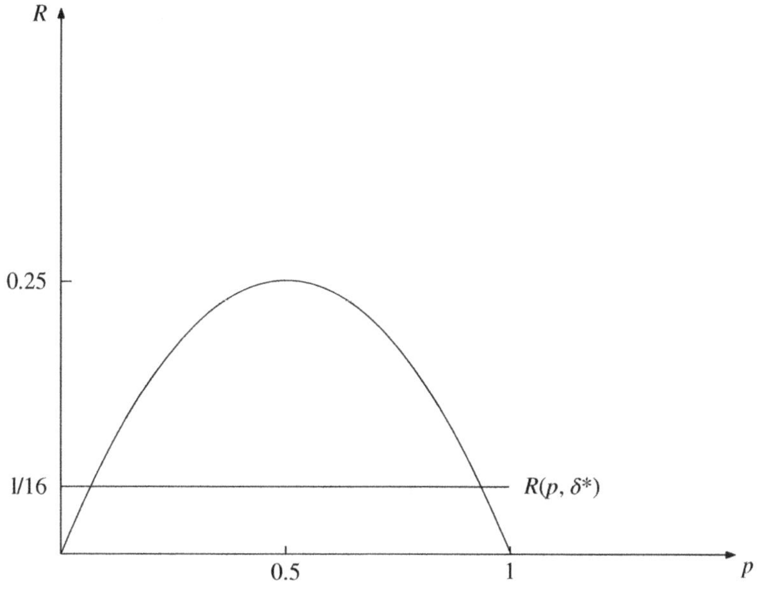

Fig. 1 Comparison of $R(p,\delta)$ and $R(p,\delta^*)$, $n=1$.

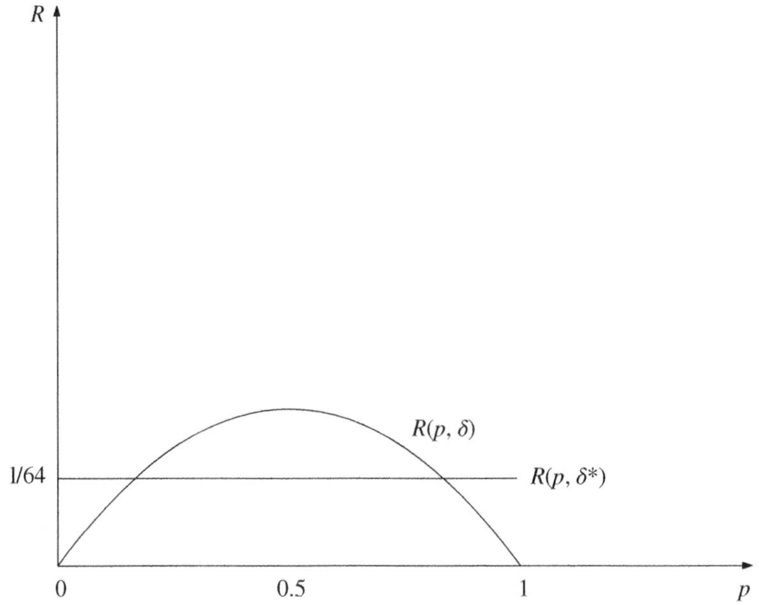

Fig. 2 Comparison of $R(p,\delta)$ and $R(p,\delta^*)$, $n = 9$.

in the interval $(\frac{1}{2} - a_n, \frac{1}{2} + a_n)$, where $a_n \to 0$ as $n \to \infty$. Moreover,

$$\frac{\sup_p R(p,\delta)}{\sup_p R(p,\delta^*)} = \frac{1/4n}{1/[4(1+\sqrt{n})^2]} = \frac{n+2\sqrt{n}+1}{n} \to 1 \qquad \text{as } n \to \infty.$$

Clearly, we would prefer the minimax estimator if n is small and would prefer the UMVUE because of its simplicity if n is large.

Example 16. (Hodges and Lehmann [43]). A lot contains N elements, of which D are defective. A random sample of size n produces X defectives. We wish to estimate D. Clearly,

$$P_D\{X=k\} = \binom{D}{k}\binom{N-D}{n-k}\binom{N}{n}^{-1},$$

$$E_D X = n\frac{D}{N}, \qquad \text{and} \qquad \sigma_D^2 = \frac{nD(N-n)(N-D)}{N^2(N-1)}.$$

Proceeding as in Example 8, we find a linear function of X with constant risk. Indeed, $E_D(\alpha X + \beta - D)^2 = \beta^2$ when

$$\alpha = \frac{N}{n+\sqrt{n(N-n)/(N-1)}} \qquad \text{and} \qquad \beta = \frac{N}{2}\left(1 - \frac{\alpha n}{N}\right).$$

We show that $\alpha X+\beta$ is the Bayes estimator corresponding to a priori PMF

$$P\{D=d\}=c\int_0^1\binom{N}{d}p^d(1-p)^{N-d}p^{a-1}(1-p)^{b-1}\,dp,$$

where $a,b>0$, and $c=\Gamma(a+b)/\Gamma(a)\Gamma(b)$. First note that $\sum_{d=0}^{N}P\{D=d\}=1$ so that

$$\sum_{d=0}^{N}\binom{N}{d}\frac{\Gamma(a+b)}{\Gamma(a)\Gamma(b)}\frac{\Gamma(a+d)\Gamma(N+b-d)}{\Gamma(N+a+b)}=1.$$

The Bayes estimator is given by

$$\delta^*(k)=\frac{\sum_{d=k}^{N-n+k}d\binom{d}{k}\binom{N-d}{n-k}\binom{N}{d}\Gamma(a+d)\Gamma(N+b-d)}{\sum_{d=k}^{N-n+k}\binom{d}{k}\binom{N-d}{n-k}\binom{N}{d}\Gamma(a+d)\Gamma(N+b-d)}.$$

A little simplification, writing $d=(d-a)+a$ and using

$$\binom{d}{k}\binom{N-d}{n-k}\binom{N}{d}=\binom{N-n}{d-k}\binom{N}{n}\binom{n}{k},$$

yields

$$\begin{aligned}\delta^*(k)&=\frac{\sum_{i=0}^{N-n}\binom{N-n}{i}\Gamma(d+a+1)\Gamma(N+b-d)}{\sum_{0}^{N-n}\binom{N-n}{i}\Gamma(d+a)\Gamma(N+b-d)}-a\\&=k\frac{a+b+N}{a+b+n}+\frac{a(N-n)}{a+b+n}.\end{aligned}$$

Now putting

$$\alpha=\frac{a+b+N}{a+b+n}\qquad\text{and}\qquad\beta=\frac{a(N-n)}{a+b+n}$$

and solving for a and b, we get

$$a=\frac{\beta}{\alpha-1}\qquad\text{and}\qquad b=\frac{N-\alpha n-\beta}{\alpha-1}.$$

Since $a>0$, $\beta>0$, and since $b>0$, $N>\alpha n+\beta$. Moreover, $\alpha>1$ if $N>n+1$. If $N=n+1$, the result is obtained if we give D a binomial distribution with parameter $p=\frac{1}{2}$. If $N=n$, the result is immediate.

The following theorem which is an extension of Theorem 2 is of considerable help to prove minimaxity of various estimators.

Theorem 3. Let $\{\pi_k(\theta);\ k \geq 1\}$ be a sequence of prior distributions on Θ and let $\{\delta_k^*\}$ be the corresponding sequence of Bayes estimators with Bayes risks $R(\pi_k;\delta_k^*)$. If $\limsup_{k\to\infty} R(\pi_k;\delta_k^*) = r^*$ and there exists an estimator δ^* for which

$$\sup_{\theta\in\Theta} R(\theta,\delta^*) \leq r^*$$

then δ^* is minimax.

Proof. Suppose δ^* is not minimax. Then there exists an estimator $\tilde{\delta}$ such that

$$\sup_{\theta\in\Theta} R(\theta,\tilde{\delta}) \leq \sup_{\theta\in\Theta} R(\theta,\delta^*).$$

On the other hand, consider the Bayes estimators $\{\delta_k^*\}$ corresponding to the priors $\{\pi_k(\theta)\}$. We obtain

$$R(\pi_k,\delta_k^*) = \int R(\theta,\delta_k^*)\pi_k(\theta)d\theta \tag{12}$$

$$\leq \int R(\theta,\tilde{\delta})\pi_k(\theta)d\theta \tag{13}$$

$$\leq \sup_{\theta\in\Theta} R(\theta,\tilde{\delta}), \tag{14}$$

which contradicts $\sup_{\theta\in\Theta} R(\theta,\delta^*) \leq r^*$. Hence δ^* is minimax.

Example 17. Let $X_1,\ldots,X_n$ be a sample of size n from $N(\mu,1)$. Then, the MLE of μ is $\overline{X}$ with variance $\frac{1}{n}$. We show that $\overline{X}$ is minimax. Let $\mu \sim N(0,\tau^2)$. Then the Bayes estimator of μ is $\overline{X}\left(\frac{n\tau^2}{1+n\tau^2}\right)$. The Bayes risk of this estimator is $R(\pi,\delta_{\tau^2}) = \frac{1}{n}\left(\frac{n\tau^2}{1+\tau^2}\right)$. Now, as $\tau^2 \to \infty$ $R(\pi,\delta_{\tau^2}^*) \to \frac{1}{n}$ which is the risk of $\overline{X}$. Hence $\overline{X}$ is minimax.

Definition 10. A decision rule δ is *inadmissible* if there exists a $\delta^* \in \mathcal{D}$ such that $R(\theta,\delta^*) \leq R(\theta,\delta)$ where the inequality is strict for some $\theta \in \Theta$; otherwise δ is admissible.

Theorem 4. If $X_1,\ldots,X_n$ is a sample from $N(\theta,1)$, then $\overline{X}$ is an admissible estimator of θ under square error loss $L(\theta,a) = (\theta - a)^2$.

Proof. Clearly, $\overline{X} \sim N(\theta,\frac{1}{n})$. Suppose $\overline{X}$ is not admissible, then there exists another rule $\delta^*(\mathbf{x})$ such that $R(\theta,\delta^*) \leq R(\theta,\overline{X})$ while the inequality is strict for some $\theta = \theta_0$ (say). Now, the risk $R(\theta,\delta)$ is a continuous function of θ and hence there exists an $\varepsilon > 0$ such that $R(\theta,\delta^*) < R(\theta,\overline{X}) - \varepsilon$ for $|\theta - \theta_0| < \varepsilon$.

Now consider the prior $N(0,\tau^2)$. Then the Bayes estimator is $\delta(\mathbf{X}) = \overline{X}\left(1+\frac{1}{n\tau^2}\right)^{-1}$ with risk $\frac{1}{n}\left(\frac{n\tau^2}{1+n\tau^2}\right)$. Thus,

$$R(\pi,\overline{X}) - R(\pi,\delta_{\tau^2}) = \frac{1}{n}\left(\frac{1}{1+n\tau^2}\right).$$

However,

$$\begin{aligned}
&\tau[R(\pi,\delta^*)-R(\pi,\overline{X})]\\
&=\tau\int[R(\theta,\delta^*)-R(\theta,\overline{X})]\frac{1}{\sqrt{2\pi}\tau}\exp\left\{-\frac{1}{2\tau^2}\theta^2\right\}d\theta\\
&\le-\frac{\varepsilon}{\sqrt{2\pi}}\int_{\theta_0-\varepsilon}^{\theta_0+\varepsilon}\exp\left\{-\frac{1}{2\tau^2}\theta^2\right\}d\theta.
\end{aligned}$$

We get

$$\begin{aligned}
0&\le\tau[R(\pi,\delta^*)-R(\pi,\overline{X})]+\tau\left[R(\pi,\overline{X})-R(\pi,\delta_{\tau^2})\right]\\
&\le-\frac{\varepsilon}{\sqrt{2\pi}}\int_{\theta_0+\varepsilon}^{\theta_0+\varepsilon}\exp\left\{-\frac{1}{2\tau^2}\theta^2\right\}d\theta+\frac{\tau}{n}\frac{1}{(1+n\tau^2)}.
\end{aligned}$$

The right-hand side goes to $-\frac{2\varepsilon^2}{\sqrt{2\pi}}$ as $\tau\to\infty$. This result leads to a contradiction that δ^* is admissible. Hence $\overline{X}$ is admissible under squared loss.

Thus we have proved that $\overline{X}$ is an admissible minimax estimator of the mean of a normal distribution $\mathcal{N}(\theta,1)$.

PROBLEMS 8.8

1. It rains quite often in Bowling Green, Ohio. On a rainy day a teacher has essentially three choices: (1) to take an umbrella and face the possible prospect of carrying it around in the sunshine; (2) to leave the umbrella at home and perhaps get drenched; or (3) to just give up the lecture and stay at home. Let $\Theta=\{\theta_1,\theta_2\}$, where θ_1 corresponds to rain, and θ_2, to no rain. Let $\mathcal{A}=\{a_1,a_2,a_3\}$, where a_i corresponds to the choice i, $i=1,2,3$. Suppose that the following table gives the losses for the decision problem:

	θ_1	θ_2
a_1	1	2
a_2	4	0
a_3	5	5

The teacher has to make a decision on the basis of a weather report that depends on θ as follows.

	θ_1	θ_2
W_1 (Rain)	0.7	0.2
W_2 (No rain)	0.3	0.8

Find the minimax rule to help the teacher reach a decision.

2. Let $X_1, X_2, \ldots, X_n$ be a random sample from $P(\lambda)$. For estimating λ, using the quadratic error loss function, an a priori distribution over Θ, given by PDF

$$\begin{aligned} \pi(\lambda) &= e^{-\lambda} && \text{if } \lambda > 0, \\ &= 0 && \text{otherwise,} \end{aligned}$$

is used:

(a) Find the Bayes estimator for λ.

(b) If it is required to estimate $\varphi(\lambda) = e^{-\lambda}$ with the same loss function and same a priori PDF, find the Bayes estimator for $\varphi(\lambda)$.

3. Let $X_1, X_2, \ldots, X_n$ be a sample from $b(1, \theta)$. Consider the class of decision rules δ of the form $\delta(x_1, x_2, \ldots, x_n) = n^{-1} \sum_{i=1}^{n} x_i + \alpha$, where α is a constant to be determined. Find α according to the minimax principle, using the loss function $(\theta - \delta)^2$, where δ is an estimator for θ.

4. Let δ^* be a minimax estimator for $a\psi(\theta)$ with respect to the squared error loss function. Show that $a\delta^* + b(a, b$ constants) is a minimax estimator for $a\psi(\theta) + b$.

5. Let $X \sim b(n, \theta)$, and suppose that the a priori PDF of θ is $U(0, 1)$. Find the Bayes estimator of θ, using loss function $L(\theta, \delta) = (\theta - \delta)^2 / [\theta(1 - \theta)]$. Find a minimax estimator for θ.

6. In Example 5 find the Bayes estimator for p^2.

7. Let $X_1, X_2, \ldots, X_n$ be a random sample from $G(1, 1/\lambda)$. To estimate λ, let the a priori PDF on λ be $\pi(\lambda) = e^{-\lambda}$, $\lambda > 0$, and let the loss function be squared error. Find the Bayes estimator of λ.

8. Let $X_1, X_2, \ldots, X_n$ be iid $U(0, \theta)$ RVs. Suppose the prior distribution of θ is a Pareto PDF $\pi(\theta) = \frac{\alpha a^{\alpha}}{\theta^{\alpha+1}}$ for $\theta \geq a$, $= 0$ for $\theta < a$. Using the quadratic loss function find the Bayes estimator of θ.

9. Let T be the unique Bayes estimator of θ with respect to the prior density π. Then T is admissible.

10. Let $X_1, X_2, \ldots, X_n$ be iid with PDF $f_\theta(x) = \exp\{-(x - \theta)\}$, $x > \theta$. Take $\pi(\theta) = e^{-\theta}$, $\theta > 0$. Find the Bayes estimator of θ under quadratic loss.

11. For the PDF of Problem 10 consider the estimation of θ under quadratic loss. Consider the class of estimators $a\left(X_{(1)} - \frac{1}{n}\right)$ for all $a > 0$. Show that $X_{(1)} - 1/n$ is minimax in this class.

8.9 PRINCIPLE OF EQUIVARIANCE

Let $\mathcal{P} = \{P_{\boldsymbol{\theta}} : \boldsymbol{\theta} \in \Theta\}$ be a family of distributions of some RV $\mathbf{X}$. Let $\mathfrak{X} \subseteq \mathcal{R}_n$ be sample space of values of $\mathbf{X}$. In Section 8.8 we saw that the statistical decision theory revolves around the following four basic elements: the parameter space Θ, the action space $\mathcal{A}$, the sample space $\mathfrak{X}$, and the loss function $L(\theta, a)$.

Let $\mathcal{G}$ be a group of transformations which map $\mathfrak{X}$ onto itself. We say that $\mathcal{P}$ is *invariant* under $\mathcal{G}$ if for each $g \in \mathcal{G}$ and every $\boldsymbol{\theta} \in \Theta$, there is a unique $\boldsymbol{\theta}' = \bar{g}\boldsymbol{\theta} \in \Theta$ such that $g(\mathbf{X}) \sim P_{\bar{g}\boldsymbol{\theta}}$ whenever $\mathbf{X} \sim P_{\boldsymbol{\theta}}$. Accordingly,

$$P_{\boldsymbol{\theta}}\{g(\mathbf{X}) \in A\} = P_{\bar{g}\boldsymbol{\theta}}\{\mathbf{X} \in A\} \tag{1}$$

for all Borel subsets in $\mathcal{R}_n$. We note that the invariance of $\mathcal{P}$ under $\mathcal{G}$ does not change the class of distributions we begin with; it only changes the parameter or index $\boldsymbol{\theta}$ to $\bar{g}\boldsymbol{\theta}$. The group $\mathcal{G}$ induces $\bar{\mathcal{G}}$, a group of transformations $\bar{g}$ on Θ onto itself.

Example 1. Let $X \sim b(n,p), 0 \le p \le 1$. Let $\mathcal{G} = \{g, e\}$, where $g(x) = n - x$, and $e(x) = x$. Then $gg^{-1} = e$. Clearly, $g(X) \sim b(n, 1-p)$ so that $\bar{g}p = 1-p$ and $\bar{e}p = e$. The group $\mathcal{G}$ leaves $\{b(n,p);\ 0 \le p \le 1\}$ invariant.

Example 2. Let $X_1, X_2, \ldots, X_n$ be iid $\mathcal{N}(\mu, \sigma^2)$ RVs. Consider the affine group of transformations $\mathcal{G} = \{\{a,b\},\ a \in \mathcal{R},\ b > 0\}$ on $\mathfrak{X}$. The joint PDF of $\{a,b\}X = (a + bX_1, \ldots, a + bX_n)$ is given by

$$f(x_1, x_2, \ldots, x_n) = \frac{1}{(b\sigma\sqrt{2\pi})^n} \exp\left\{-\frac{1}{2b^2\sigma^2}\sum_{i=1}^{n}(x_i - a - b\mu)^2\right\}$$

and we see that

$$\bar{g}(\mu, \sigma) = (a + \mu\sigma, b\sigma) = \{a,b\}\{\mu, \sigma\}.$$

Clearly $\mathcal{G}$ leaves the family of joint PDFs of $\mathbf{X}$ invariant.

In order to apply invariance considerations to a decision problem we need also to ensure that the loss function is invariant.

Definition 1. A decision problem is said to be invariant under a group $\mathcal{G}$ if

(i) $\mathcal{P}$ is invariant under $\mathcal{G}$ and

(ii) the loss function L is invariant in the sense that for every $g \in \mathcal{G}$ and $a \in \mathcal{A}$ there is a unique $a' \in \mathcal{A}$ such that

$$L(\boldsymbol{\theta}, a) = L(\bar{g}\boldsymbol{\theta}, a') \qquad \text{for all } \boldsymbol{\theta}.$$

The $a' \in \mathcal{A}$ in Definition 1 is uniquely determined by g and may be denoted by $\tilde{g}(a)$. One can show that $\tilde{\mathcal{G}} = \{\tilde{g} : g \in \mathcal{G}\}$ is a group of transformations of $\mathcal{A}$ into itself.

Example 3. Consider the estimation of μ in sampling from $\mathcal{N}(\mu, 1)$. In Example 8.9.2 we have shown that the normal family is invariant under the location group $\mathcal{G} = \{\{b, 1\}, -\infty < b < \infty\}$. Consider the quadratic loss function

$$L(\mu, a) = (\mu - a)^2.$$

Then, $\{b,1\}a = b + a$ and $\{b,1\}\{\mu,1\} = \{b + \mu, 1\}$. Hence,

$$L(\{b,1\}\mu, \{b,1\}a) = L[(b+\mu) - (b+a)]^2 = (\mu - a)^2 = L(\mu, a).$$

Thus $L(\mu, a)$ is invariant under $\mathcal{G}$ and the problem of estimation of μ is invariant under group $\mathcal{G}$.

Example 4. Consider the normal family $\mathcal{N}(0,\sigma^2)$ which is invariant under the scale group $\mathcal{G} = \{\{0,c\}, c > 0\}$. Let the loss function be

$$L(\sigma^2,a) = \frac{1}{\sigma^4}(\sigma^2 - a)^2.$$

Now, $\{0,c\}a = ca$ and $\{0,c\}\{0,\sigma^2\} = \{0,c\sigma^2\}$ and

$$L[\{0,c\}\sigma^2, \{0,c\}a] = \frac{1}{c^2\sigma^4}(c\sigma^2 - ca)^2 = \frac{1}{\sigma^4}(\sigma^2 - a)^2 = L(\sigma^2,a).$$

Thus, the loss function $L(\sigma^2,a)$ is invariant under $\mathcal{G} = \{\{0,c\}, c > 0\}$ and the problem of estimation of σ^2 is invariant.

Example 5. Consider the loss function

$$L(\sigma^2,a) = \frac{a}{\sigma^2} - 1 - \log\frac{a}{\sigma^2}$$

for the estimation of σ^2 from the normal family $\mathcal{N}(0,\sigma^2)$. We show that this loss-function is invariant under the scale group. Since

$$\{0,c\}\sigma^2 = \{0,c\sigma^2\} \text{ and } \{0,c\}\{0,a\} = \{0,ca\},$$

we have

$$\begin{aligned} L[\{0,c\}\sigma^2, \{0,c\}a] &= \frac{ca}{c\sigma^2} - 1 - \log\frac{ca}{c\sigma^2} \\ &= L(\sigma^2,a). \end{aligned}$$

Let us now return to the problem of estimation of a parametric function $\psi : \Theta \to \mathcal{R}$. For convenience let us take $\Theta \subseteq \mathcal{R}$ and $\psi(\theta) = \theta$. Then $\mathcal{A} = \Theta$ and $\tilde{\mathcal{G}} = \bar{\mathcal{G}}$.

Suppose θ is the mean of PDF f_θ, $\mathcal{G} = \{\{b,1\},\ b \in \mathcal{R}\}$, and $\{f_\theta\}$ is invariant under $\mathcal{G}$. Consider the estimator $\partial(\mathbf{X}) = \overline{X}$. What we want in an estimator ∂^* of θ is that it changes in the same prescribed way as the data are changed. In our case, since $\mathbf{X}$ changes to $\{b,1\}\mathbf{X} = \mathbf{X} + b$ we would like $\overline{X}$ to transform to $\{b,1\}\overline{X} = \overline{X} + b$.

Definition 2. An estimator $\delta(\mathbf{X})$ of θ is said to be equivariant, under $\mathcal{G}$, if

$$\delta(g\mathbf{X}) = \bar{g}\delta(\mathbf{X}) \qquad \text{for all } g \in \mathcal{G}, \tag{2}$$

where we have written $g\mathbf{X}$ for $g(\mathbf{X})$ for convenience.

Indeed g on S induces $\bar{g}$ on Θ. Thus if $\mathbf{X} \sim f_\theta$, then $g\mathbf{X} \sim f_{\bar{g}\theta}$ so if $\delta(\mathbf{X})$ estimates θ then $\delta(g\mathbf{X})$ should estimate $\bar{g}\theta$. The *principle of equivariance* requires that we restrict attention to equivariant estimators and select the "best" estimator in this class in a sense to be described later in this section.

Example 6. In Example 3, consider the estimators $\partial_1(\mathbf{X}) = \overline{X}$, $\partial_2(\mathbf{X}) = (X_{(1)} + X_{(n)})/2$, and $\partial_3(\mathbf{X}) = \alpha\overline{X}$, α a fixed real number. Then $\mathcal{G} = \{(b,1),\ -\infty < b < \infty\}$ induces $\bar{\mathcal{G}} = \mathcal{G}$ on Θ and both ∂_1, ∂_2 are equivariant under $\mathcal{G}$. The estimator δ_3 is not equivariant unless $\alpha = 1$. In Example 8.9.1 $\partial(X) = X/n$ is an equivariant estimator of p.

In Example 6 consider the statistic $\partial(\mathbf{X}) = S^2$. Note that under the translation group $\{b,1\}\mathbf{X} = \mathbf{X} + b$ and $\partial(\{b,1\}\mathbf{X}) = \partial(\mathbf{X})$. That is, for every $g \in \mathcal{G}$, $\partial(g\mathbf{X}) = \partial(\mathbf{X})$. A statistic ∂ is said to be *invariant* under a group of transformations $\mathcal{G}$ if $\partial(g\mathbf{X}) = \partial(\mathbf{X})$ for all $g \in \mathcal{G}$. When $\mathcal{G}$ is the translation group, an invariant statistic (function) under $\mathcal{G}$ is called *location-invariant*. Similarly if $\mathcal{G}$ is the scale group, we call ∂ *scale-invariant* and if $\mathcal{G}$ is the location-scale group, we call ∂ *location-scale invariant*. In Example 6 $\partial_4(\mathbf{X}) = S^2$ is location-invariant but not equivariant, and $\partial_2(\mathbf{X})$ and $\partial_3(\mathbf{X})$ are not location-invariant.

A very important property of equivariant estimators is that their risk function is constant on orbits of θ.

Theorem 1. Suppose ∂ is an equivariant estimator of θ in a problem which is invariant under $\mathcal{G}$. Then the risk function of ∂ satisfies

$$R(\bar{g}\theta, \partial) = R(\theta, \partial) \tag{3}$$

for all $\theta \in \Theta$ and $g \in \mathcal{G}$. If, in particular, $\bar{\mathcal{G}}$ is transitive over Θ, then $R(\theta, \partial)$ is independent of θ.

Proof. We have for $\theta \in \Theta$ and $g \in \mathcal{G}$

$$\begin{aligned}
R(\theta, \partial(\mathbf{X})) &= E_\theta L(\theta, \partial(\mathbf{X})) \\
&= E_\theta L(\bar{g}\theta, \bar{g}\partial(\mathbf{X})) \quad \text{(Invariance of } L) \\
&= E_\theta L(\bar{g}\theta, \partial(g(\mathbf{X})) \quad \text{(Equivariance of } \delta) \\
&= E_{\bar{g}\theta} L(\bar{g}\theta, \partial(\mathbf{X})) \quad \text{(Invariance of } \{P_\theta\}) \\
&= R(\bar{g}\theta, \partial(\mathbf{X})).
\end{aligned}$$

In the special case when $\bar{\mathcal{G}}$ is transitive over Θ then for any $\theta_1, \theta_2 \in \Theta$, there exists a $\bar{g} \in \bar{\mathcal{G}}$ such that $\theta_2 = \bar{g}\theta_1$. It follows that

$$R(\theta_2, \partial) = R(\bar{g}\theta_1, \partial) = R(\theta_1, \partial)$$

so that R is independent of θ.

Remark 1. When the risk function of every equivariant estimator is constant, an estimator (in the class equivariant estimators) which is obtained by minimizing the constant is called the *minimum risk equivariant* (MRE) estimator.

Example 7. Let $X_1, X_2, \ldots, X_n$ iid RVs with common PDF

$$f(x, \theta) = \exp\{-(x - \theta)\},\ x \geq \theta, \text{ and } = 0, \text{ if } x < 0.$$

Consider the location group $\mathcal{G} = \{\{b,1\}, -\infty < b < \infty\}$ which induces $\bar{\mathcal{G}}$ on Θ where $\bar{\mathcal{G}} = \mathcal{G}$. Clearly $\bar{\mathcal{G}}$ is transitive. Let $L(\theta, \partial) = (\theta - \partial)^2$. Then the problem of estimation of θ is invariant and according to Theorem 1 the risk of every equivariant estimator is free of θ. The estimator $\delta_0(\mathbf{X}) = X_{(1)} - \frac{1}{n}$ is equivariant under $\mathcal{G}$ since

$$\delta_0(\{b,1\}\mathbf{X}) = \min_{1 \le i \le n} (X_i + b) - \frac{1}{n} = b + X_{(1)} - \frac{1}{n} = b + \delta_0(\mathbf{X}).$$

We leave the reader to check that

$$R(\theta, \partial_0) = E_\theta \left(X_{(1)} - \frac{1}{n} - \theta \right)^2 = \frac{1}{n^2},$$

and it will be seen later that ∂_0 is the MRE estimator of θ.

Example 8. In this example we consider sampling from a normal PDF. Let us first consider estimation of μ when $\sigma = 1$. Let $\mathcal{G} = \{\{b,1\},\ -a < b < \infty\}$. Then $\partial(\mathbf{X}) = \overline{X}$ is equivariant under $\mathcal{G}$ and it has the smallest risk $1/n$. Note that $\{\bar{x},1\}^{-1} = \{-\bar{x},1\}$ may be used to designate $\mathbf{x}$ on its orbits

$$\{\bar{x},1\}^{-1}\mathbf{x} = (x_1 - \bar{x}, \ldots, x_n - \bar{x}) = A(\mathbf{x}).$$

Clearly $A(\mathbf{x})$ is invariant under $\mathcal{G}$ and $A(\mathbf{X})$ is ancillary to μ. By Basu's theorem $A(\mathbf{X})$ and $\overline{X}$ are independent.

Next consider estimation of σ^2 with $\mu = 0$ and $\mathcal{G} = \{\{0,c\}, c > 0\}$. Then $S_x^2 = \sum_1^n X_i^2$ is an equivariant estimator of σ^2. Note that $\{0, s_x\}^{-1}$ may be used to designate $\mathbf{x}$ on its orbits

$$\{0, s_x\}^{-1}\mathbf{x} = \left(\frac{x_1}{s_x}, \ldots, \frac{x_n}{s_x} \right) = A(\mathbf{x}).$$

Again $A(\mathbf{x})$ is invariant under $\mathcal{G}$ and $A(\mathbf{X})$ is ancillary to σ^2. Moreover, S_x^2 and $A(\mathbf{X})$ are independent.

Finally, we consider estimation of (μ, σ^2) when $\mathcal{G} = \{\{b,c\},\ -a < b < \infty,\ c > 0\}$. Then $(\overline{X}, S_x^2)$, where $S_x^2 = \sum_1^n (X_i - \overline{X})^2$ is an equivariant estimator of (μ, σ^2). Also $\{\bar{x}, s_x\}^{-1}$ may be used to designate $\mathbf{x}$ on its orbits

$$\{\bar{x}, s_x\}^{-1}\mathbf{x} = \left(\frac{x_1 - \bar{x}}{s_x}, \ldots, \frac{x_n - \bar{x}}{s_x} \right) = A(\mathbf{x}).$$

Note that the statistic $A(\mathbf{X})$ defined in each of the three cases considered in Example 8 is constant on its orbits. A statistic A is said to be *maximal invariant* if

(i) A is invariant, and

(ii) A is maximal, that is, $A(\mathbf{x}_1) = A(\mathbf{x}_2) \Rightarrow \mathbf{x}_1 = g(\mathbf{x}_2)$ for some $g \in \mathcal{G}$.

We now derive an explicit expression for MRE estimator for a location parameter. Let $X_1, X_2, \ldots, X_n$ be iid with common PDF $f_\theta(x) = f(x - \theta)$, $-\infty < \theta < \infty$. Then $\{f_\theta : \theta \in \Theta\}$ is invariant under $\mathcal{G} = \{\{b,1\},\ -\infty < b < \infty\}$ and an estimator of θ is equivariant if

$$\partial(\{b,1\}\mathbf{X} = \partial(\mathbf{X}) + b$$

for all real b.

Lemma 1. An estimator ∂ is equivariant for θ if and only if

$$\partial(\mathbf{X}) = X_1 + q(X_2 - X_1, \ldots, X_n - X_1), \tag{4}$$

for some function q.

Proof. If (4) holds then

$$\begin{aligned}\partial(\{b,1\}\mathbf{x}) &= b + x_1 + q(x_2 - x_1, \ldots, x_n - x_1)\\ &= b + \partial(\mathbf{x}).\end{aligned}$$

Conversely,

$$\begin{aligned}\partial(\mathbf{x}) &= \partial(x_1 + x_1 - x_1, x_1 + x_2 - x_1, \ldots, x_1 + x_n - x_1)\\ &= x_1 + \partial(0, x_2 - x_1, x \cdots, x_n - x_1),\end{aligned}$$

which is (4) with $q(x_2 - x_1, \ldots, x_n - x_1) = \partial(0, x_2 - x_1, \ldots, x_n - x_1)$.

From Theorem 1 the risk function of an equivariant estimator ∂ is constant with risk

$$R(\theta, \partial) = R(0, \partial) = E_0[\partial(\mathbf{X})]^2, \quad \text{for all } \theta,$$

where the expectation is with respect to PDF $f_0(\mathbf{x}) = f(\mathbf{x})$. Consequently, among all equivariant estimators ∂ for θ, the MRE estimator is ∂_0 satisfying

$$R(0, \partial_0) = \min_{\partial} R(0, \partial).$$

Thus we only need to choose the function q in (4).

Let $L(\theta, \partial)$ be the loss function. Invariance considerations require that

$$L(\theta, \partial) = L(\bar{g}\theta, \bar{g}\partial) = L(\theta + b, \partial + b)$$

for all real b so that $L(\theta, \partial)$ must be some function w of $\partial - \theta$.

Let $Y_i = X_i - X_1$, $i = 2, \ldots, n$, $\mathbf{Y} = (Y_2, \ldots, Y_n)$, and $g(\mathbf{y})$ be the joint PDF of $\mathbf{Y}$ under $\theta = 0$. Let $h(x_1|\mathbf{y})$ be the conditional density, under $\theta = 0$, of X_1 given $\mathbf{Y} = y$. Then

$$\begin{aligned}R(0, \partial) &= E_0[w(X_1 - q(\mathbf{Y}))]\\ &= \int \left\{ \int w(x_1 - q(\mathbf{y}))h(x_1|\mathbf{y})dx \right\} g(\mathbf{y})d\mathbf{y}.\end{aligned} \tag{5}$$

Then $R(0, \partial)$ will be minimized by choosing, for each fixed $\mathbf{y}$, $q(\mathbf{y})$ to be that value of c which minimizes

$$\int w(u - c)h(u|\mathbf{y})du. \tag{6}$$

Necessarily q depends on $\mathbf{y}$. In the special case $w(d-\theta)=(d-\theta)^2$, the integral in (6) is minimum when c is chosen to be the mean of the conditional distribution. Thus the unique MRE estimator of θ is given by

$$\partial_0(\mathbf{x}) = x_1 - E_\theta\{X_1|\mathbf{Y}=\mathbf{y}\}. \tag{7}$$

This is the so-called *Pitman estimator*. Let us simplify it a little more by computing $E_0\{x_1 - X_1|\mathbf{Y}=\mathbf{y}\}$.

First we need to compute $h(u|\mathbf{y})$. When $\theta=0$, the joint PDF of $X_1, Y_2, \ldots, Y_n$ is easily seen to be

$$f(x_1)f(x_1+y_2)\ldots f(x_1+y_n)$$

so the joint PDF of $(Y_2,\ldots,Y_n)$ is given by

$$\int_{-\infty}^{\infty} f(u)f(u+y_2)\ldots f(u+y_n)du.$$

It follows that

$$h(u|\mathbf{y}) = \frac{f(u)f(u+y_2)\cdots f(u+y_n)}{\int_{-\infty}^{\infty} f(u)f(u+y_2)\cdots f(u+y_n)du}. \tag{8}$$

Now let $Z = x_1 - X_1$. Then the conditional PDF of Z given $\mathbf{y}$ is $h(x_1 - z \mid \mathbf{y})$. It follows from (8) that

$$\begin{aligned}\partial_0(\mathbf{x}) = E_0\{Z|\mathbf{y}\} &= \int_{-\infty}^{\infty} z h(x_1 - z)dz \\ &= \frac{\int_{-\infty}^{\infty} z\prod_{j=1}^{n} f(x_j - z)dz}{\int_{-\infty}^{\infty} \prod_{j=1}^{n} f(x_j - z)dz}. \end{aligned}\tag{9}$$

Remark 2. Since the joint PDF of $X_1, X_2, \ldots, X_n$ is $\prod_{j=1}^{n} f_\theta(x_j) = \prod_{j=1}^{n} f(x_j-\theta)$, the joint PDF of θ and $\mathbf{X}$ when θ has prior $\pi(\theta)$ is $\pi(\theta)\prod_{j=1}^{n} f(x_j-\theta)$. The joint marginal of $\mathbf{X}$ is $\int_{-\infty}^{\infty}\pi(\theta)\prod_{j=1}^{n} f(x_j-\theta)d\theta$. It follows that the conditional pdf of θ given $\mathbf{X}=\mathbf{x}$ is given by

$$\frac{\pi(\theta)\prod_{j=1}^{n} f(x_j-\theta)}{\int_{-\infty}^{\infty}\pi(\theta)\prod_{j=1}^{n} f(x_j-\theta)d\theta}.$$

Taking $\pi(\theta)=1$, the improper uniform prior on Θ, we see from (9) that $\partial_0(\mathbf{x})$ is the Bayes estimator of θ under squared error loss and prior $\pi(\theta)=1$. Since the risk of ∂_0 is constant, it follows that ∂_0 is also minimax estimator of θ.

Remark 3. Suppose S is sufficient for θ. Then $\prod_{j=1}^{n} f_\theta(x_j) = g_\theta(s)h(\mathbf{x})$ so that the Pitman estimator of θ can be rewritten as

$$\begin{aligned}\partial_0(\mathbf{x}) &= \frac{\int_{-\infty}^{\infty} \theta \prod_{j=1}^{n} f_\theta(x_j)d\theta}{\int_{-\infty}^{\infty} \prod_{j=1}^{n} f_\theta(x_j)d\theta} \\ &= \frac{\int_{-\infty}^{\infty} \theta g_\theta(s)h(\mathbf{x})d\theta}{\int_{-\infty}^{\infty} g_\theta(s)h(\mathbf{x})d\theta} \\ &= \frac{\int_{-\infty}^{\infty} \theta g_\theta(s)d\theta}{\int_{-\infty}^{\infty} g_\theta(s)d\theta},\end{aligned}$$

which is a function of s alone.

Examples 7 and 8 (continued). A direct computation using (9) shows that $X_{(1)} - 1/n$ is the Pitman MRE estimator of θ in Example 7 and $\overline{X}$ is the MRE estimator of μ in Example 8 (when $\sigma = 1$). The results can be obtained by using sufficiency reduction. In Example 7, $X_{(1)}$ is the minimal sufficient statistic for θ. Every (translation) equivariant function based on $X_{(1)}$ must be of the form $\partial_c(\mathbf{X}) = X_{(1)} + c$ where c is a real number. Then

$$\begin{aligned}R(\theta, \partial_c) &= E_\theta\{X_{(1)} + c - \theta\}^2 \\ &= E_\theta\{X_{(1)} - 1/n - \theta + (c + 1/n)\}2 \\ &= R(\theta, \partial_0) + (c + 1/n)^2 = (1/n)^2 + (c + 1/n)^2\end{aligned}$$

which is minimized for $c = -1/n$. In Example 8, $\overline{X}$ is the minimal sufficient statistic so every equivariant function of $\overline{X}$ must be of the form $\partial_c(\mathbf{X}) = \overline{X} + c$, where c is a real constant. Then

$$R(\mu, \partial_c) = E_\mu(\overline{X} + c - \mu)^2 = \frac{1}{n} + c^2,$$

which is minimized for $c = 0$.

Example 9. Let $X_1, X_2, \ldots, X_n$ be iid $U(\theta - 1/2, \theta + 1/2)$. Then $(X_{(1)}, X_{(n)})$ is jointly sufficient for θ. Clearly,

$$f(x_1 - \theta, \ldots, x_n - \theta) = \begin{cases} 1 & x_{(1)} < \theta < x_{(n)} \\ 0 & \text{otherwise} \end{cases}$$

so that Pitman estimator of θ is given by

$$\partial_0(\mathbf{x}) = \frac{\displaystyle\int_{x_{(1)}}^{x_{(n)}} \theta d\theta}{\displaystyle\int_{x_{(1)}}^{x_{(n)}} d\theta} = \frac{(x_{(n)} + x_{(1)})}{2}.$$

We now consider, briefly, Pitman estimator of a scale parameter. Let $\mathbf{X}$ have a joint PDF

$$f_\sigma(\mathbf{x}) = \frac{1}{\sigma^n} f\left(\frac{x_1}{\sigma}, \ldots, \frac{x_n}{\sigma}\right),$$

where f is known and $\sigma > 0$ is a scale parameter. The family $\{f_\sigma : \sigma > 0\}$ remains invariant under $\mathcal{G} = \{\{0,c\}, c > 0\}$ which induces $\bar{\mathcal{G}} = \mathcal{G}$ on Θ. Then for estimation of σ^k loss function $L(\sigma, a)$ is invariant under these transformations if and only if $L(\sigma, a) = w\left(\frac{a}{\sigma^k}\right)$. An estimator ∂ of σ^k is equivariant under $\mathcal{G}$ if

$$\partial(\{0,c\}\mathbf{X}) = c^k \partial(\mathbf{X}) \quad \text{for all } c > 0.$$

Some simple examples of scale-equivariant estimators of σ are the mean deviation $\sum_1^n |X_i - \overline{X}|/n$ and the standard deviation $\sqrt{\sum_1^n (X_i - \overline{X})^2/(n-1)}$. We note that the group $\bar{\mathcal{G}}$ over Θ is transitive so according to Theorem 1, the risk of any equivariant estimator of σ^k is free of σ and an MRE estimator minimizes this risk over the class of all equivariant estimators of σ^k. Using the loss function $L(\sigma, a) = w(a/\sigma^k) = (a - \sigma^k)^2/\sigma^{2k}$ it can be shown that the MRE estimator of σ^k, also known as the *Pitman estimate* of σ^k, is given by

$$\partial_0(\mathbf{x}) = \frac{\int_0^\infty v^{n+k-1} f(vx_1, \ldots, vx_n)dv}{\int_0^\infty v^{n+2k-1} f(vx_1, \ldots, vx_n)dv}.$$

Just as in the location case one can show that ∂_0 is a function of the minimal sufficient statistic and ∂_0 is the Bayes estimator of σ^k with improper prior $\pi(\sigma) = 1/\sigma^{2k+1}$. Consequently, ∂_0 is minimax.

Example 8 (continued). In Example 8, the Pitman estimator of σ^k is easily shown to be

$$\partial_0(\mathbf{X}) = \frac{\Gamma\left(\frac{n+k}{2}\right)}{\Gamma\left(\frac{n+2k}{2}\right)} \left(\sum_1^n X_i^2\right)^{k/2}.$$

Thus the MRE estimator of σ is given by $\left\{\Gamma\left(\frac{n+1}{2}\right)\sqrt{\sum_1^n X_i^2} \Big/ \Gamma\left(\frac{n+2}{2}\right)\right\}$ and that of σ^2 by $\sum_1^n X_i^2/(n+2)$.

Example 10. Let $X_1, X_2, \ldots, X_n$ be iid $U(0, \theta)$. The Pitman estimator of θ is given by

$$\partial_0(\mathbf{X}) = \frac{\int_{X_{(n)}}^\infty v^n dv}{\int_{X_{(n)}}^\infty v^{n+1} dv} = \frac{n+2}{n+1} X_{(n)}.$$

PROBLEMS 8.9

In all problems assume that $X_1, X_2, \ldots, X_n$ is a random sample from the distribution under consideration.

1. Show that the following statistics are equivariant under translation group:
   (a) Median (X_i).
   (b) $(X_{(1)} + X_{(n)})/2$.

(c) $X_{[np]+1}$, the quantile of order p, $0 < p < 1$.
(d) $\left(X_{(r)} + X_{(r+1)} + \cdots + X_{(n-r)}\right) / (n-2r)$.
(e) $\overline{X} + \overline{Y}$, where $\overline{Y}$ is the mean of a sample of size m, $m \neq n$.

2. Show that the following statistics are invariant under location or scale or location-scale group:
(a) $\overline{X} - \text{median}(X_i)$.
(b) $X_{(n+1-k)} - X_{(k)}$.
(c) $\sum_{i=1}^{n} |X_i - \overline{X}|/n$.
(d) $\frac{\sum_{i=1}^{n}(X_i-\overline{X})(Y_i-\overline{Y})}{\left\{\sum_{i=1}^{n}(X_i-\overline{X})^2 \sum_{i=1}^{n}(Y_i-\overline{Y})^2\right\}^{1/2}}$, where $(X_1, Y_1, \ldots, (X_n, Y_n)$ is a random sample from a bivariate distribution.

3. Let the common distribution be $G(\alpha, \sigma)$ where α (>0) is known and $\sigma > 0$ is unknown. Find the MRE estimator of σ under loss $L(\sigma, a) = (1 - a/\sigma)^2$.

4. Let the common PDF be the folded normal distribution

$$\sqrt{\frac{2}{\pi}} \exp\left\{-\frac{1}{2}(x-\mu)^2\right\} I_{[\mu,\infty)}(x).$$

Verify that the best equivariant estimator of μ under quadratic loss is given by

$$\hat{\mu} = \overline{X} - \frac{\exp\{-\frac{n}{2}(X_{(1)} - \overline{X})^2\}}{\sqrt{2n\pi}\left\{\int_0^{\sqrt{n}(X_{(1)}-\overline{X})} \frac{1}{\sqrt{2\pi}} \exp(-z^2/2)dz\right\}}.$$

5. Let $X \sim U(\theta, 2\theta)$.
(a) Show that $(X_{(1)}, X_{(n)})$ is jointly sufficient statistic for θ.
(b) Verify whether or not $(X_{(n)} - X_{(1)})$ is an unbiased estimator of θ. Find an ancillary statistic.
(c) Determine the best invariant estimator of θ under the loss function $L(\theta, a) = \left(1 - \frac{a}{\theta}\right)^2$.

6. Let

$$f_\theta(x) = \frac{1}{2}\exp\{-|x-\theta|\}.$$

Find the Pitman estimator of θ.

7. Let $f_\theta(x) = \exp\{-(x-\theta)\} \cdot [1 + \exp\{-(x-\theta)\}]^{-2}$, for $x \in \mathcal{R}$, $\theta \in \mathcal{R}$. Find the Pitman estimator of θ.

8. Show that an estimator ∂ is (location) equivariant if and only if

$$\partial(\mathbf{x}) = \partial_0(\mathbf{x}) + \phi(\mathbf{x}),$$

where ∂_0 is any equivariant estimator and ϕ is an invariant function.

9. Let X_1,X_2 be iid with PDF

$$f_\sigma(x) = \frac{2}{\sigma}\left(1 - \frac{x}{\sigma}\right),\ 0 < x < \sigma, \text{ and } = 0 \text{ otherwise.}$$

Find, explicitly, the Pitman estimator of σ^r.

10. Let $X_1,X_2,\ldots,X_n$ be iid with PDF

$$f_\theta(x) = \frac{1}{\theta}\exp(-x/\theta),\ x > 0, \text{ and } = 0, \text{ otherwise.}$$

Find the Pitman estimator of θ^k.

9

NEYMAN–PEARSON THEORY OF TESTING OF HYPOTHESES

9.1 INTRODUCTION

Let $X_1, X_2, \ldots, X_n$ be a random sample from a population distribution $F_{\boldsymbol{\theta}}$, $\boldsymbol{\theta} \in \Theta$, where the functional form of $F_{\boldsymbol{\theta}}$ is known except, perhaps, for the parameter $\boldsymbol{\theta}$. Thus, for example, the X_i's may be a random sample from $\mathcal{N}(\theta, 1)$, where $\theta \in \mathcal{R}$ is not known. In many practical problems the experimenter is interested in testing the validity of an assertion about the unknown parameter θ. For example, in a coin-tossing experiment it is of interest to test, in some sense, whether the (unknown) probability of heads p equals a given number p_0, $0 < p_0 < 1$. Similarly, it is of interest to check the claim of a car manufacturer about the average mileage per gallon of gasoline achieved by a particular model. A problem of this type is usually referred to as a problem of *testing of hypotheses* and is the subject of discussion in this chapter. We will develop the fundamentals of Neyman–Pearson theory. In Section 9.2 we introduce the various concepts involved. In Section 9.3 the fundamental Neyman–Pearson lemma is proved, and Sections 9.4 and 9.5 deal with some basic results in the testing of composite hypotheses. Section 9.6 deals with locally optimal tests.

9.2 SOME FUNDAMENTAL NOTIONS OF HYPOTHESES TESTING

In Chapter 8 we discussed the problem of point estimation in sampling from a population whose distribution is known except for a finite number of unknown parameters. Here we consider another important problem in statistical inference, the testing of statistical hypotheses. We begin by considering the following examples.

An Introduction to Probability and Statistics, Third Edition. Vijay K. Rohatgi and A.K. Md. Ehsanes Saleh.

Example 1. In coin-tossing experiments one frequently assumes that the coin is fair, that is, the probability of getting heads or tails is the same: $\frac{1}{2}$. How does one test whether the coin is fair (unbiased) or loaded (biased)? If one is guided by intuition, a reasonable procedure would be to toss the coin n times and count the number of heads. If the proportion of heads observed does not deviate "too much" from $p = \frac{1}{2}$, one would tend to conclude that the coin is fair.

Example 2. It is usual for manufacturers to make quantitative assertions about their products. For example, a manufacturer of 12-volt batteries may claim that a certain brand of his batteries lasts for N hours. How does one go about checking the truth of this assertion? A reasonable procedure suggests itself: Take a random sample of n batteries of the brand in question and note their length of life under more or less identical conditions. If the average length of life is "much smaller" than N, one would tend to doubt the manufacturer's claim.

To fix ideas, let us define formally the concepts involved. As usual, $\mathbf{X} = (X_1, X_2, \ldots, X_n)$ and let $\mathbf{X} \sim F_{\boldsymbol{\theta}}$, $\boldsymbol{\theta} \in \Theta \subseteq \mathcal{R}_k$. It will be assumed that the functional form of $F_{\boldsymbol{\theta}}$ is known except for the parameter $\boldsymbol{\theta}$. Also, we assume that Θ contains at least two points.

Definition 1. A parametric hypothesis is an assertion about the unknown parameter $\boldsymbol{\theta}$. It is usually referred to as the *null hypothesis*, $H_0\colon \boldsymbol{\theta} \in \Theta_0 \subset \Theta$. The statement $H_1 : \boldsymbol{\theta} \in \Theta_1 = \Theta - \Theta_0$ is usually referred to as the alternative hypothesis.

Usually the null hypothesis is chosen to correspond to the smaller or simpler subset Θ_0 of Θ and is a statement of "no difference," whereas the alternative represents change.

Definition 2. If $\Theta_0(\Theta_1)$ contains only one point, we say that $\Theta_0(\Theta_1)$ is *simple*; otherwise, *composite*. Thus, if a hypothesis is simple, the probability distribution of $\mathbf{X}$ is completely specified under that hypothesis.

Example 3. Let $X \sim \mathcal{N}(\mu, \sigma^2)$. If both μ and σ^2 are unknown, $\Theta = \{(\mu, \sigma^2)\colon -\infty < \mu < \infty, \sigma^2 > 0\}$. The hypothesis $H_0\colon \mu \le \mu_0$, $\sigma^2 > 0$, where μ_0 is a known constant, is a composite null hypothesis. The alternative hypothesis is $H_1 : \mu > \mu_0$, $\sigma^2 > 0$, which is also composite. Similarly, the null hypothesis $\mu = \mu_0$, $\sigma^2 > 0$ is also composite.

If $\sigma^2 = \sigma_0^2$ is known, the hypothesis $H_0\colon \mu = \mu_0$ is a simple hypothesis.

Example 4. Let $X_1, X_2, \ldots, X_n$ be iid $b(1, p)$ RVs. Some hypotheses of interest are $p = \frac{1}{2}$, $p \le \frac{1}{2}$, $p \ge \frac{1}{2}$ or, quite generally, $p = p_0$, $p \le p_0$, $p \ge p_0$, where p_0 is a known number, $0 < p_0 < 1$.

The problem of testing of hypotheses may be described as follows: Given the sample point $\mathbf{x} = (x_1, x_2, \ldots, x_n)$, find a decision rule (function) that will lead to a decision to reject or fail to reject the null hypothesis. In other words, partition the sample space into two disjoint sets C and C^c such that, if $\mathbf{x} \in C$, we reject H_0, and if $\mathbf{x} \in C^c$, we fail to reject H_0. In the following we will write accept H_0 when we fail to reject H_0. We emphasize that when the sample point $\mathbf{x} \in C^c$ and we fail to reject H_0, it does not mean that H_0 gets our stamp of approval. It simply means that the sample does not have enough evidence against H_0.

Definition 3. Let $\mathbf{X} \sim F_{\boldsymbol{\theta}}, \boldsymbol{\theta} \in \Theta$. A subset C of $\mathcal{R}_n$ such that if $\mathbf{x} \in C$, then H_0 is rejected (with probability 1) and is called the *critical region* (set):

$$C = \{\mathbf{x} \in \mathcal{R}_n : H_0 \text{ is rejected if } \mathbf{x} \in C\}.$$

There are two types of errors that can be made if one uses such a procedure. One may reject H_0 when in fact it is true, called a *type I error*, or accept H_0 when it is false, called a *type II error*,

		True	
		H_0	H_1
Accept	H_0	Correct	Type II Error
	H_1	Type I Error	Correct

If C is the critical region of a rule, $P_{\boldsymbol{\theta}}C$, $\boldsymbol{\theta} \in \Theta_0$, is a *probability of type* I *error*, and $P_{\boldsymbol{\theta}}C^c$, $\boldsymbol{\theta} \in \Theta_1$, is a *probability of type* II *error*. Ideally, one would like to find a critical region for which both these probabilities are 0. This will be the case if we can find a subset $S \subseteq \mathcal{R}_n$ such that $P_{\boldsymbol{\theta}}S = 1$ for every $\boldsymbol{\theta} \in \Theta_0$ and $P_{\boldsymbol{\theta}}S = 0$ for every $\boldsymbol{\theta} \in \Theta_1$. Unfortunately, situations such as this do not arise in practice, although they are conceivable. For example, let $X \sim \mathcal{C}(1, \theta)$ under H_0 and $X \sim P(\theta)$ under H_1. Usually, if a critical region is such that the probability of type I error is 0, it will be of the form "do not reject H_0" and the probability of type II error will then be 1.

The procedure used in practice is to limit the probability of type I error to some preassigned level α (usually 0.01 or 0.05) that is small and to minimize the probability of type II error. To restate our problem in terms of this requirement, let us formulate these notions.

Definition 4. Every Borel-measurable mapping φ of $\mathcal{R}_n \to [0, 1]$ is known as a *test function*.

Some simple examples of test functions are $\varphi(\mathbf{x}) = 1$ for all $\mathbf{x} \in \mathcal{R}_n$, $\varphi(\mathbf{x}) = 0$ for all $\mathbf{x} \in \mathcal{R}_n$, or $\varphi(\mathbf{x}) = \alpha$, $0 \leq \alpha \leq 1$, for all $\mathbf{x} \in \mathcal{R}_n$. In fact, Definition 4 includes Definition 3 in the sense that, whenever φ is the indicator function of some Borel subset A of $\mathcal{R}_n$, A is called the *critical region* (of the test φ).

Definition 5. The mapping φ is said to be a *test* of hypothesis $H_0 : \boldsymbol{\theta} \in \Theta_0$ against the alternatives $H_1 : \boldsymbol{\theta} \in \Theta_1$ with error probability α (also called *level of significance* or, simply, *level*) if

$$E_{\boldsymbol{\theta}}\varphi(\mathbf{X}) \leq \alpha \qquad \text{for all } \boldsymbol{\theta} \in \Theta_0. \tag{1}$$

We shall say, in short, that φ is a test for the problem $(\alpha, \Theta_0, \Theta_1)$.

Let us write $\beta_\varphi(\boldsymbol{\theta}) = E_{\boldsymbol{\theta}}\varphi(\mathbf{X})$. Our objective, in practice, will be to seek a test φ for a given α, $0 \le \alpha \le 1$, such that

$$\sup_{\boldsymbol{\theta}\in\boldsymbol{\Theta}_0} \beta_\varphi(\boldsymbol{\theta}) \le \alpha. \tag{2}$$

The left-hand side of (2) is usually known as the *size* of the test φ. Condition (1) therefore restricts attention to tests whose size does not exceed a given level of significance α.

The following interpretation may be given to all tests φ satisfying $\beta_\varphi(\boldsymbol{\theta}) \le \alpha$ for all $\boldsymbol{\theta} \in \Theta_0$. To every $\mathbf{x} \in \mathcal{R}_n$ we assign a number $\varphi(\mathbf{x}), 0 \le \varphi(\mathbf{x}) \le 1$, which is the probability of rejecting H_0 that $\mathbf{X} \sim f_{\boldsymbol{\theta}}$, $\boldsymbol{\theta} \in \Theta_0$, if $\mathbf{x}$ is observed. The restriction $\beta_\varphi(\boldsymbol{\theta}) \le \alpha$ for $\boldsymbol{\theta} \in \Theta_0$ then says that, if H_0 were true, φ rejects it with a probability $\le \alpha$. We will call such a test a *randomized* test function. If $\varphi(\mathbf{x}) = I_A(\mathbf{x})$, φ will be called a *nonrandomized* test. If $\mathbf{x} \in A$, we reject H_0 with probability 1; and if $\mathbf{x} \notin A$, this probability is 0. Needless to say, $A \in \mathfrak{B}_n$.

We next turn our attention to the type II error.

Definition 6. Let φ be a test function for the problem $(\alpha, \Theta_0, \Theta_1)$. For every $\boldsymbol{\theta} \in \Theta$ define

$$\beta_\varphi(\boldsymbol{\theta}) = E_{\boldsymbol{\theta}}\varphi(\mathbf{X}) = P_{\boldsymbol{\theta}}\{\text{Reject } H_0\}. \tag{3}$$

As a function of $\boldsymbol{\theta}$, $\beta_\varphi(\boldsymbol{\theta})$ is called the *power function* of the test φ. For any $\boldsymbol{\theta} \in \Theta_1$, $\beta_\varphi(\boldsymbol{\theta})$ is called the power of φ against the alternative $\boldsymbol{\theta}$.

In view of Definitions 5 and 6 the problem of testing of hypotheses may now be reformulated. Let $\mathbf{X} \sim f_{\boldsymbol{\theta}}$, $\boldsymbol{\theta} \in \Theta \subseteq \mathcal{R}_k$, $\Theta = \Theta_0 + \Theta_1$. Also, let $0 \le \alpha \le 1$ be given. Given a sample point $\mathbf{x}$, find a test $\varphi(\mathbf{x})$ such that $\beta_\varphi(\boldsymbol{\theta}) \le \alpha$ for $\boldsymbol{\theta} \in \Theta_0$, and $\beta_\varphi(\boldsymbol{\theta})$ is a maximum for $\boldsymbol{\theta} \in \Theta_1$.

Definition 7. Let Φ_α be the class of all tests for the problem $(\alpha, \Theta_0, \Theta_1)$. A test $\varphi_0 \in \Phi_\alpha$ is said to be a *most powerful* (MP) test against an alternative $\boldsymbol{\theta} \in \Theta_1$ if

$$\beta_{\varphi_0}(\boldsymbol{\theta}) \ge \beta_\varphi(\boldsymbol{\theta}) \qquad \text{for all} \qquad \varphi \in \Phi_\alpha. \tag{4}$$

If Θ_1 contains only one point, this definition suffices. If, on the other hand, Θ_1 contains at least two points, as will usually be the case, we will have an MP test corresponding to each $\boldsymbol{\theta} \in \Theta_1$.

Definition 8. A test $\varphi_0 \in \Phi_\alpha$ for the problem $(\alpha, \Theta_0, \Theta_1)$ is said to be a *uniformly most powerful* (UMP) test if

$$\beta_{\varphi_0}(\boldsymbol{\theta}) \ge \beta_\varphi(\boldsymbol{\theta}) \qquad \text{for all } \varphi \in \Phi_\alpha, \qquad \text{uniformly in } \boldsymbol{\theta} \in \Theta_1. \tag{5}$$

Thus, if Θ_0 and Θ_1 are both composite, the problem is to find a UMP test φ for the problem $(\alpha, \Theta_0, \Theta_1)$. We will see that UMP tests very frequently do not exist, and we will have to place further restrictions on the class of all tests, Φ_α.

Note that if φ_1, φ_2 are two tests and λ is a real number, $0 < \lambda < 1$, then $\lambda\varphi_1 + (1-\lambda)\varphi_2$ is also a test function, and it follows that the class of all test functions Φ_α is convex.

Example 5. Let $X_1, X_2, \ldots, X_n$ be iid $\mathcal{N}(\mu, 1)$ RVs, where μ is unknown but it is known that $\mu \in \Theta = \{\mu_0, \mu_1\}$, $\mu_0 < \mu_1$. Let $H_0\colon X_i \sim \mathcal{N}(\mu_0, 1)$, $H_1: X_i \sim \mathcal{N}(\mu_1, 1)$. Both H_0 and H_1 are simple hypotheses. Intuitively, one would accept H_0 if the sample mean $\overline{X}$ is "closer" to μ_0 than to μ_1; that is to say, one would reject H_0 if $\overline{X} > k$, and accept H_0 otherwise. The constant k is determined from the level requirements. Note that, under $H_0, \overline{X} \sim \mathcal{N}(\mu_0, 1/n)$, and, under $H_1, \overline{X} \sim \mathcal{N}(\mu_1, 1/n)$. Given $0 < \alpha < 1$, we have

$$\begin{aligned} P_{\mu_0}\{\overline{X} > k\} &= P\left\{\frac{\overline{X} - \mu_0}{1/\sqrt{n}} > \frac{k - \mu_0}{1/\sqrt{n}}\right\} \\ &= P\{\text{Type I error}\} = \alpha, \end{aligned}$$

so that $k = \mu_0 + z_\alpha/\sqrt{n}$. The test, therefore, is (Fig. 1)

$$\varphi(\mathbf{x}) = \begin{cases} 1 & \text{if } \overline{x} > \mu_0 + z_\alpha/\sqrt{n}, \\ 0 & \text{otherwise.} \end{cases}$$

Here $\overline{X}$ is known as a *test statistic*, and the test φ is nonrandomized with critical region $C = \{\mathbf{x}\colon \overline{x} > \mu_0 + z_\alpha/\sqrt{n}\}$. Note that in this case the continuity of $\mathbf{X}$ (that is, the absolute continuity of the DF of $\mathbf{X}$) allows us to achieve any size α, $0 < \alpha < 1$.

The power of the test at μ_1 is given by

$$\begin{aligned} E_{\mu_1}\varphi(\mathbf{X}) &= P_{\mu_1}\left\{\overline{X} > \mu_0 + \frac{z_\alpha}{\sqrt{n}}\right\} \\ &= P\left\{\frac{\overline{X} - \mu_1}{1/\sqrt{n}} > (\mu_0 - \mu_1)\sqrt{n} + z_\alpha\right\} \\ &= P\{Z > z_\alpha - \sqrt{n}(\mu_1 - \mu_0)\}, \end{aligned}$$

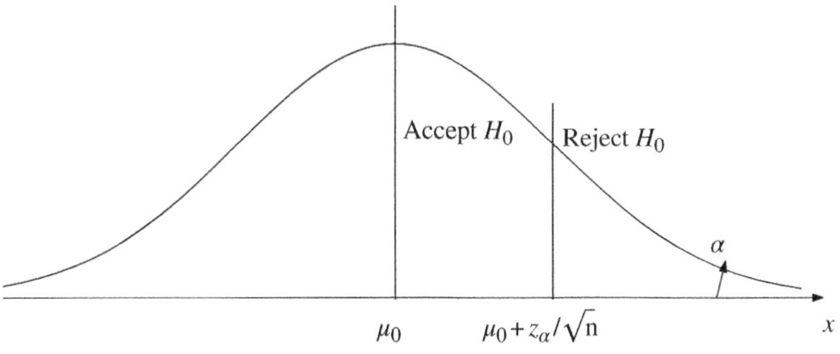

Fig. 1 Rejection region of H_0 in Example 5.

where $Z \sim \mathcal{N}(0,1)$. In particular, $E_{\mu_1}\varphi(\mathbf{X}) > \alpha$ since $\mu_1 > \mu_0$. The probability of type II error is given by

$$\begin{aligned} P\{\text{Type II error}\} &= 1 - E_{\mu_1}\varphi(X) \\ &= P\{Z \leq z_\alpha - \sqrt{n}(\mu_1 - \mu_0)\}. \end{aligned}$$

Figure 2 gives a graph of the power function $\beta_\varphi(\mu)$ of φ for $\mu > 0$ when $\mu_0 = 0$, and $H_1 : \mu > 0$.

Example 6. Let X_1, X_2, X_3, X_4, X_5, be a sample from $b(1,p)$, where p is unknown and $0 \leq p \leq 1$. Consider the simple null hypothesis H_0: $X_i \sim b(1, \frac{1}{2})$, that is, under H_0, $p = \frac{1}{2}$. Then $H_1 : X_i \sim b(1,p), p \neq 1/2$. A reasonable procedure would be to compute the average number of 1's, namely, $\overline{X} = \sum_1^5 X_i/5$, and to accept H_0 if $|\overline{X} - \frac{1}{2}| \leq c$, where c is to be determined. Let $\alpha = 0.10$. Then we would like to choose c such that the size of our test is α, that is,

$$0.10 = P_{p=1/2}\left\{|\overline{X} - \frac{1}{2}| > c\right\},$$

or

$$\begin{aligned} 0.90 &= P_{p=1/2}\left\{-5c \leq \sum_1^5 X_i - \frac{5}{2} \leq 5c\right\} \\ &= P_{p=1/2}\left\{-k \leq \sum_1^5 X_i - \frac{5}{2} \leq k\right\}, \end{aligned} \tag{6}$$

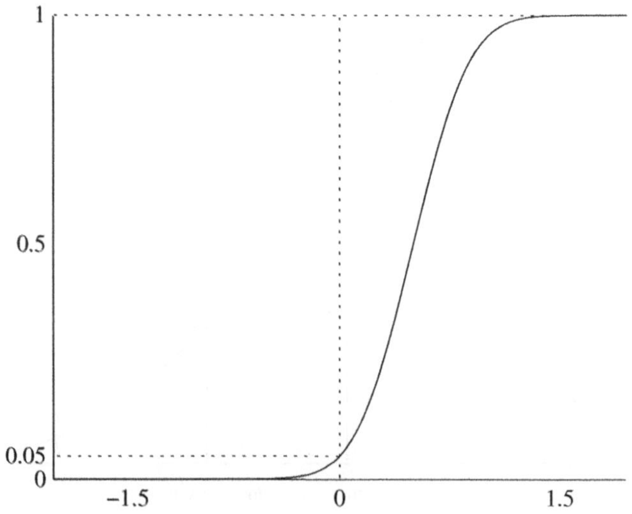

Fig. 2 Power function of φ in Example 5.

where $k = 5c$. Now $\sum_1^5 X_i \sim b(5, \frac{1}{2})$ under H_0, so that the PMF of $\sum_1^5 X_i - \frac{5}{2}$ is given in the following table.

$\sum_1^5 x_i$	$\sum_1^5 x_i - \frac{5}{2}$	$P_{p=1/2}\left\{\sum_1^5 X_i = \sum_1^5 x_i\right\}$
0	−2.5	0.03125
1	−1.5	0.15625
2	−0.5	0.31250
3	0.5	0.31250
4	1.5	0.15625
5	2.5	0.03125

Note that we cannot choose any k to satisfy (6) exactly. It is clear that we have to reject H_0 when $k = \pm 2.5$, that is, when we observe $\sum X_i = 0$ or 5. The resulting size if we use this test is $\alpha = 0.03125 + 0.03125 = 0.0625 < 0.10$. A second procedure would be to reject H_0 if $k = \pm 1.5$ or ± 2.5 ($\sum X_i = 0, 1, 4, 5$), in which case the resulting size is $\alpha = 0.0625 + 2(0.15625) = 0.375$, which is considerably larger than 0.10. A third alternative, if we insist on achieving $\alpha = 0.10$, is to randomize on the boundary. Instead of accepting or rejecting H_0 with probability 1 when $\sum X_i = 1$ or 4, we reject H_0 with probability γ where

$$0.10 = P_{p=1/2}\left\{\sum_1^5 X_i = 0 \text{ or } 5\right\} + \gamma P_{p=1/2}\left\{\sum_1^5 X_i = 1 \text{ or } 4\right\}.$$

Thus

$$\gamma = \frac{0.0375}{0.3125} = 0.114.$$

A randomized test of size $\alpha = 0.10$ is therefore given by

$$\varphi(\mathbf{x}) = \begin{cases} 1 & \text{if } \sum_1^5 x_i = 0 \text{ or } 5, \\ 0.114 & \text{if } \sum_1^5 x_i = 1 \text{ or } 4, \\ 0 & \text{otherwise.} \end{cases}$$

The power of this test is

$$E_p\varphi(\mathbf{X}) = P_p\left\{\sum_1^5 X_i = 0 \text{ or } 5\right\} + 0.114 P_p\left\{\sum_1^5 X_i = 1 \text{ or } 4\right\},$$

where $p \neq \frac{1}{2}$ and can be computed for any value of p. Figure 3 gives a graph of $\beta_\varphi(p)$.

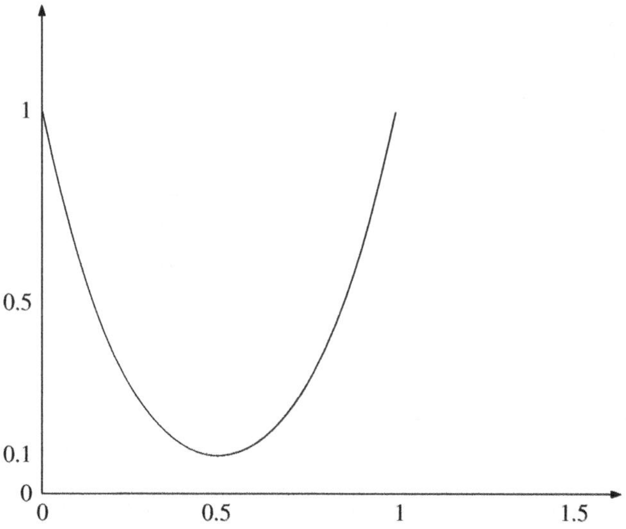

Fig. 3 Power function of φ in Example 6.

We conclude this section with the following remarks.

Remark 1. The problem of testing of hypotheses may be considered as a special case of the general decision problem described in Section 8.8. Let $\mathcal{A} = \{a_0, a_1\}$, where a_0 represents the decision to accept $H_0: \boldsymbol{\theta} \in \Theta_0$ and a_1 represents the decision to reject H_0. A decision function δ is a mapping of $\mathcal{R}_n$ into $\mathcal{A}$. Let us introduce the following loss functions:

$$L_1(\theta, a_1) = \begin{cases} 1 & \text{if } \boldsymbol{\theta} \in \Theta_0 \\ 0 & \text{if } \boldsymbol{\theta} \in \Theta_1 \end{cases} \qquad \text{and } L_1(\boldsymbol{\theta}, a_0) = 0 \text{ for all } \boldsymbol{\theta},$$

and

$$L_2(\boldsymbol{\theta}, a_0) = \begin{cases} 0 & \text{if } \boldsymbol{\theta} \in \Theta_0 \\ 1 & \text{if } \boldsymbol{\theta} \in \Theta_1 \end{cases} \qquad \text{and } L_2(\boldsymbol{\theta}, a_1) = 0 \text{ for all } \boldsymbol{\theta}.$$

Then the minimization of $E_{\boldsymbol{\theta}} L_2(\boldsymbol{\theta}, \delta(\mathbf{X}))$ subject to $E_{\boldsymbol{\theta}} L_1(\boldsymbol{\theta}, \delta(\mathbf{X})) \le \alpha$ is the hypotheses testing problem discussed above. We have

$$\begin{aligned} E_{\boldsymbol{\theta}} L_2(\boldsymbol{\theta}, \delta(\mathbf{X})) &= P_{\boldsymbol{\theta}}\{\delta(\mathbf{X}) = a_0\}, \qquad \boldsymbol{\theta} \in \Theta_1, \\ &= P_{\boldsymbol{\theta}}\{\text{Accept } H_0 \mid H_1 \text{ true}\}, \end{aligned}$$

and

$$\begin{aligned} E_{\boldsymbol{\theta}} L_1(\boldsymbol{\theta}, \delta(\mathbf{X})) &= P_{\boldsymbol{\theta}}\{\delta(\mathbf{X}) = a_1\}, \qquad \boldsymbol{\theta} \in \Theta_0, \\ &= P_{\boldsymbol{\theta}}\{\text{Reject } H_0 \mid \boldsymbol{\theta} \in \Theta_0 \text{ true}\}. \end{aligned}$$

Remark 2. In Example 6 we saw that the chosen size α is often unattainable. The choice of a specific value of α is completely arbitrary and is determined by nonstatistical considerations such as the possible consequences of rejecting H_0 falsely, and the economic and practical implications of the decision to reject H_0. An alternative, and somewhat subjective, approach wherever possible is to report the so-called P-*value* of the observed test statistic. This is the smallest level α at which the observed sample statistic is significant. In Example 6, let $S=\sum_{i=1}^{5}X_i$. If $S=0$ is observed, then $P_{H_0}(S=0)=P_0(S=0)=0.03125$. By symmetry, if we reject H_0 for $S=0$ we should do so also for $S=5$ so the probability of interest is $P_0(S=0 \text{ or } 5)=0.0625$ which is the P-value. If $S=1$ is observed and we decide to reject H_0, then we would do so also for $S=0$ because $S=0$ is *more extreme* than $S=1$. By symmetry considerations

$$P\text{-value} = P_0(S \le 1 \text{ or } S \ge 4) = 2(0.03125+0.15625) = 0.375.$$

This discussion motivates Definition 9 below. Suppose the appropriate critical region for testing H_0 against H_1 is one-sided. That is, suppose C is either of the form $\{T \ge c_1\}$ or $\{T \le c_2\}$, where T is the test statistic.

Definition 9. The probability of observing under H_0 a sample outcome at least as extreme as the one observed is called the *P-value*. The smaller the P-value, the more extreme the outcome and the stronger the evidence against H_0.

If α is given, then we reject H_0 if $P \le \alpha$ and do not reject H_0 if $P > \alpha$. In the two-sided case when the critical region is of the form $C=\{|T(\mathbf{X})|>k\}$, the one-sided P-value is doubled to obtain the P-value. If the distribution of T is not symmetric then the P-value is not well-defined in the two-sided case although many authors recommend doubling the one-sided P-value.

PROBLEMS 9.2

1. A sample of size 1 is taken from a population distribution $P(\lambda)$. To test $H_0: \lambda=1$ against $H_1: \lambda=2$, consider the nonrandomized test $\varphi(x)=1$ if $x>3$, and $=0$ if $x \le 3$. Find the probabilities of type I and type II errors and the power of the test against $\lambda=2$. If it is required to achieve a size equal to 0.05, how should one modify the test φ?
2. Let $X_1, X_2, \ldots, X_n$ be a sample from a population with finite mean μ and finite variance σ^2. Suppose that μ is not known, but σ is known, and it is required to test $\mu=\mu_0$ against $\mu=\mu_1$ ($\mu_1>\mu_0$). Let n be sufficiently large so that the central limit theorem holds, and consider the test

$$\begin{aligned}\varphi(x_1,x_2,\ldots,x_n) &= 1 \qquad \text{if } \overline{x}>k,\\ &= 0 \qquad \text{if } \overline{x}\le k,\end{aligned}$$

where $\overline{x}=n^{-1}\sum_{i=1}^{n}x_i$. Find k such that the test has (approximately) size α. What is the power of this test at $\mu=\mu_1$? If the probabilities of type I and type II errors are fixed at α and β, respectively, find the smallest sample size needed.
3. In Problem 2, if σ is not known, find k such that the test φ has size α.

4. Let $X_1, X_2, \ldots, X_n$ be a sample from $\mathcal{N}(\mu, 1)$. For testing $\mu \le \mu_0$ against $\mu > \mu_0$ consider the test function

$$\varphi(x_1, x_2, \ldots, x_n) = \begin{cases} 1 & \text{if } \overline{x} > \mu_0 + \dfrac{z_\alpha}{\sqrt{n}}, \\ 0 & \text{if } \overline{x} \le \mu_0 + \dfrac{z_\alpha}{\sqrt{n}}. \end{cases}$$

Show that the power function of φ is a nondecreasing function of μ. What is the size of the test?

5. A sample of size 1 is taken from an exponential PDF with parameter θ, that is, $X \sim G(1, \theta)$. To test $H_0\colon \theta = 1$ against $H_1\colon \theta > 1$, the test to be used is the nonrandomized test

$$\begin{aligned} \varphi(x) &= 1 \qquad \text{if } x > 2, \\ &= 0 \qquad \text{if } x \le 2. \end{aligned}$$

Find the size of the test. What is the power function?

6. Let $X_1, X_2, \ldots, X_n$ be a sample from $\mathcal{N}(0, \sigma^2)$. To test $H_0\colon \sigma = \sigma_0$ against $H_1 = \sigma \neq \sigma_0$, it is suggested that the test

$$\varphi(x_1, x_2, \ldots, x_n) = \begin{cases} 1 & \text{if } \sum x_i^2 > c_1 \text{ or } \sum x_i^2 < c_2, \\ 0 & \text{if } c_2 \le \sum x_i^2 \le c_1, \end{cases}$$

be used. How will you find c_1 and c_2 such that the size of φ is a preassigned number α, $0 < \alpha < 1$? What is the power function of this test?

7. An urn contains 10 marbles, of which M are white and $10 - M$ are black. To test that $M = 5$ against the alternative hypothesis that $M = 6$, one draws 3 marbles from the urn without replacement. The null hypothesis is rejected if the sample contains 2 or 3 white marbles; otherwise it is accepted. Find the size of the test and its power.

9.3 NEYMAN–PEARSON LEMMA

In this section we prove the fundamental lemma due to Neyman and Pearson [76], which gives a general method for finding a best (most powerful) test of a simple hypothesis against a simple alternative. Let $\{f_\theta, \theta \in \Theta\}$, where $\Theta = \{\theta_0, \theta_1\}$, be a family of possible distributions of $\mathbf{X}$. Also, f_θ represents the PDF of $\mathbf{X}$ if $\mathbf{X}$ is a continuous type rv, and the PMF of $\mathbf{X}$ if $\mathbf{X}$ is of the discrete type. Let us write $f_0(\mathbf{x}) = f_{\theta_0}(\mathbf{x})$ and $f_1(\mathbf{x}) = f_{\theta_1}(\mathbf{x})$ for convenience.

Theorem 1 (The Neyman–Pearson Fundamental Lemma).

(a) Any test φ of the form

$$\varphi(\mathbf{x}) = \begin{cases} 1 & \text{if } f_1(\mathbf{x}) > k f_0(\mathbf{x}), \\ \gamma(\mathbf{x}) & \text{if } f_1(\mathbf{x}) = k f_0(\mathbf{x}), \\ 0 & \text{if } f_1(\mathbf{x}) < k f_0(\mathbf{x}), \end{cases} \tag{1}$$

for some $k \geq 0$ and $0 \leq \gamma(\mathbf{x}) \leq 1$, is most powerful of its size for testing $H_0: \theta = \theta_0$ against $H_1: \theta = \theta_1$. If $k = \infty$, the test

$$\begin{aligned} \varphi(\mathbf{x}) &= 1 \qquad \text{if } f_0(\mathbf{x}) = 0, \\ &= 0 \qquad \text{if } f_0(\mathbf{x}) > 0, \end{aligned} \tag{2}$$

is most powerful of size 0 for testing H_0 against H_1.

(b) Given α, $0 \leq \alpha \leq 1$, there exists a test of form (1) or (2) with $\gamma(\mathbf{x}) = \gamma$ (a constant), for which $E_{\theta_0}\varphi(\mathbf{X}) = \alpha$.

Proof. Let φ be a test satisfying (1), and φ^* be any test with $E_{\theta_0}\varphi^*(\mathbf{X}) \leq E_{\theta_0}\varphi(\mathbf{X})$. In the continuous case

$$\begin{aligned} &\int (\varphi(\mathbf{x}) - \varphi^*(\mathbf{x}))(f_1(\mathbf{x}) - k f_0(\mathbf{x}))\,d\mathbf{x} \\ &= \left(\int_{f_1 > kf_0} + \int_{f_1 < kf_0} \right) (\varphi(\mathbf{x}) - \varphi^*(\mathbf{x}))(f_1(\mathbf{x}) - k f_0(\mathbf{x}))\,d\mathbf{x}. \end{aligned}$$

For any $\mathbf{x} \in \{f_1(\mathbf{x}) > kf_0(\mathbf{x})\}$, $\varphi(\mathbf{x}) - \varphi^*(\mathbf{x}) = 1 - \varphi^*(\mathbf{x}) \geq 0$, so that the integrand is ≥ 0. For $\mathbf{x} \in \{f_1(\mathbf{x}) < kf_0(\mathbf{x})\}$, $\varphi(\mathbf{x}) - \varphi^*(\mathbf{x}) = -\varphi^*(\mathbf{x}) \leq 0$, so that the integrand is again ≥ 0. It follows that

$$\begin{aligned} &\int (\varphi(\mathbf{x}) - \varphi^*(\mathbf{x}))(f_1(\mathbf{x}) - k f_0(\mathbf{x}))d\mathbf{x} \\ &= E_{\theta_1}\varphi(\mathbf{X}) - E_{\theta_1}\varphi^*(\mathbf{X}) - k(E_{\theta_0}\varphi(\mathbf{X}) - E_{\theta_0}\varphi^*(\mathbf{X})) \geq 0, \end{aligned}$$

which implies

$$E_{\theta_1}\varphi(\mathbf{X}) - E_{\theta_1}\varphi^*(\mathbf{X}) \geq k(E_{\theta_0}\varphi(\mathbf{X}) - E_{\theta_0}\varphi^*(\mathbf{X})) \geq 0$$

since $E_{\theta_0}\varphi^*(\mathbf{X}) \leq E_{\theta_0}\varphi(\mathbf{X})$.

If $k = \infty$, any test φ^* of size 0 must vanish on the set $\{f_0(\mathbf{x}) > 0\}$. We have

$$E_{\theta_1}\varphi(\mathbf{X}) - E_{\theta_1}\varphi^*(\mathbf{X}) = \int_{\{f_0(\mathbf{x})=0\}} (1 - \varphi^*(\mathbf{x}))f_1(\mathbf{x})\,d\mathbf{x} \geq 0.$$

The proof for the discrete case requires the usual change of integral by a sum throughout.

To prove (b) we need to restrict ourselves to the case where $0 < \alpha \leq 1$, since the MP size 0 test is given by (2). Let $\gamma(\mathbf{x}) = \gamma$, and let us compute the size of a test of form (1). We have

$$\begin{aligned} E_{\theta_0}\varphi(\mathbf{X}) &= P_{\theta_0}\{f_1(\mathbf{X}) > k f_0(\mathbf{X})\} + \gamma P_{\theta_0}\{f_1(\mathbf{X}) = k f_0(\mathbf{X})\} \\ &= 1 - P_{\theta_0}\{f_1(\mathbf{X}) \leq k f_0(\mathbf{X})\} + \gamma\, P_{\theta_0}\{f_1(\mathbf{X}) = k f_0(\mathbf{X})\}. \end{aligned}$$

Since $P_{\theta_0}\{f_0(\mathbf{X})=0\}=0$, we may rewrite $E_{\theta_0}\varphi(\mathbf{X})$ as

$$E_{\theta_0}\varphi(\mathbf{X}) = 1 - P_{\theta_0}\left\{\frac{f_1(\mathbf{X})}{f_0(\mathbf{X})} \le k\right\} + \gamma P_{\theta_0}\left\{\frac{f_1(\mathbf{X})}{f_0(\mathbf{X})} = k\right\}. \tag{3}$$

Given $0 < \alpha \le 1$, we wish to find k and γ such that $E_{\theta_0}\varphi(\mathbf{X}) = \alpha$, that is,

$$P_{\theta_0}\left\{\frac{f_1(\mathbf{X})}{f_0(\mathbf{X})} \le k\right\} - \gamma P_{\theta_0}\left\{\frac{f_1(\mathbf{X})}{f_0(\mathbf{X})} \le k\right\} = 1-\alpha. \tag{4}$$

Note that

$$\left\{\frac{f_1(\mathbf{X})}{f_0(\mathbf{X})} \le k\right\}$$

is a DF so that it is a nondecreasing and right continuous function of k. If there exists a k_0 such that

$$P_{\theta_0}\left\{\frac{f_1(\mathbf{X})}{f_0(\mathbf{X})} \le k_0\right\} = 1-\alpha,$$

we choose $\gamma = 0$ and $k = k_0$. Otherwise there exists a k_0 such that

$$P_{\theta_0}\left\{\frac{f_1(\mathbf{X})}{f_0(\mathbf{X})} < k_0\right\} \le 1-\alpha < P_{\theta_0}\left\{\frac{f_1(\mathbf{X})}{f_0(\mathbf{X})} \le k_0\right\}, \tag{5}$$

that is, there is a jump at k_0 (see Fig. 1). In this case we choose $k = k_0$ and

$$\gamma = \frac{P_{\theta_0}\{f_1(\mathbf{X})/f_0(\mathbf{X}) \le k_0\} - (1-\alpha)}{P_{\theta_0}\{f_1(\mathbf{X})/f_0(\mathbf{X}) = k_0\}.} \tag{6}$$

Since γ given by (6) satisfies (4), and $0 \le \gamma \le 1$, the proof is complete.

Remark 1. It is possible to show (see Problem 6) that the test given by (1) or (2) is unique (except on a null set), that is, if φ is an MP test of size α of H_0 against H_1, it must have form (1) or (2), except perhaps for a set A with $P_{\theta_0}(A) = P_{\theta_1}(A) = 0$.

Remark 2. An analysis of proof of part (a) of Theorem 1 shows that test (1) is MP even if f_1 and f_0 are not necessarily densities.

Theorem 2. If a sufficient statistic T exists for the family $\{f_\theta : \theta \in \Theta\}$, $\Theta = \{\theta_0, \theta_1\}$, the Neyman–Pearson MP test is a function of T.

Proof. The proof of this result is left as an exercise.

Remark 3. If the family $\{f_\theta : \theta \in \Theta\}$ admits a sufficient statistic, one can restrict attention to tests based on the sufficient statistic, that is, to tests that are functions of the sufficient

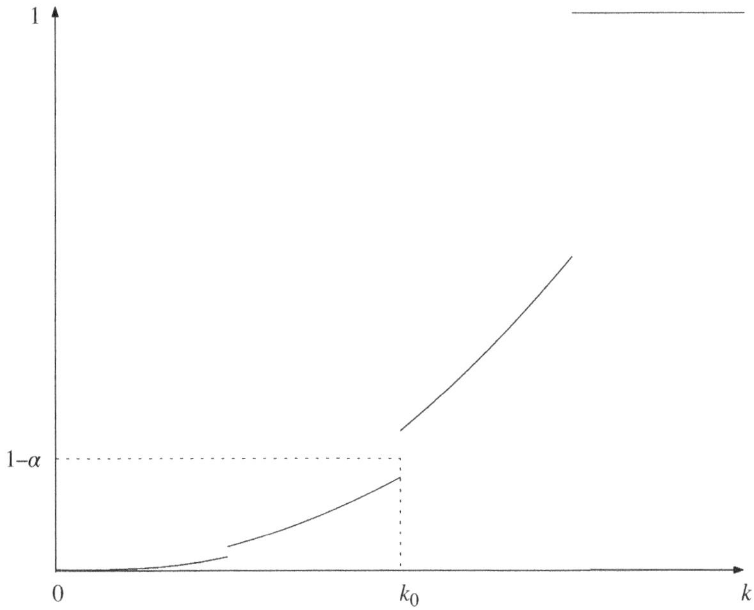

Fig. 1

statistic. If φ is a test function and T is a sufficient statistic, $E\{\varphi(\mathbf{X}) \mid T\}$ is itself a test function, $0 \le E\{\varphi(\mathbf{X}) \mid T\} \le 1$, and

$$E_\theta\{\varphi(\mathbf{X}) \mid T\}\} = E_\theta \varphi(\mathbf{X}),$$

so that φ and $E\{\varphi \mid T\}$ have the same power function.

Example 1. Let X be an RV with PMF under H_0 and H_1 given by

x	1	2	3	4	5	6
$f_0(x)$	0.01	0.01	0.01	0.01	0.01	0.95
$f_1(x)$	0.05	0.04	0.03	0.02	0.01	0.85

Then $\lambda(x) = f_1(x)/f_0(x)$ is given by

x	1	2	3	4	5	6
$\lambda(x)$	5	4	3	2	1	0.89

If $\alpha = 0.03$, for example, then Neyman–Pearson MP size 0.03 test rejects H_0 if $\lambda(X) \ge 3$, that is, if $X \le 3$ and has power

$$P_1(X \le 3) = 0.05 + 0.04 + 0.03 = 0.12$$

with P(Type II error) $= 1 - 0.12 = 0.88$.

Example 2. Let $X \sim \mathcal{N}(0,1)$ under H_0 and $X \sim \mathcal{C}(1,0)$ under H_1. To find an MP size α test of H_0 against H_1,

$$\lambda(x) = \frac{f_1(x)}{f_0(x)} = \frac{(1/\pi)[1/(1+x^2)]}{(1/\sqrt{2\pi})e^{-x^2/2}} = \sqrt{\frac{2}{\pi}}\frac{e^{x^2/2}}{1+x^2}.$$

Figure 2 gives a graph of $\lambda(x)$ and we note that λ has a maximum at $x=0$ and two minimas at $x=\pm 1$. Note that $\lambda(0)=0.7979$ and $\lambda(\pm 1)=0.6578$ so for $k \in (0.6578, 0.7989)$, $\lambda(x)=k$ intersects the graph at four points and the critical region is of the form $|X| \le k_1$ or $|X| \ge k_2$, where k_1 and k_2 are solutions of $\lambda(x)=k$. For $k=0.7979$, the critical region is of the form $|X| \ge k_0$, where k_0 is the positive solution of $e^{-k_0^2/2} = 1+k_0^2$ so that $k_0 \approx 1.59$ with $\alpha = 0.1118$. For $k < 0.6578$, $\alpha = 1$ and for $k=0.6578$, the critical region is $|X| \ge 1$ with $\alpha = 0.3413$. For the traditional level $\alpha = 0.05$, the critical region is of the form $|X| \ge 1.96$.

Example 3. Let $X_1, X_2, \ldots, X_n$ be iid $b(1,p)$ RVs, and let $H_0: p = p_0$, $H_1: p = p_1, p_1 > p_0$. The MP size α test of H_0 against H_1 is of the form

$$\varphi(x_1,x_2,\ldots,x_n) = \begin{cases} 1, & \lambda(\mathbf{x}) = \dfrac{p_1^{\sum x_i}(1-p_1)^{n-\sum x_i}}{p_0^{\sum x_i}(1-p_0)^{n-\sum x_i}} > k, \\ \gamma, & \lambda(\mathbf{x}) = k, \\ 0, & \lambda(\mathbf{x}) < k, \end{cases}$$

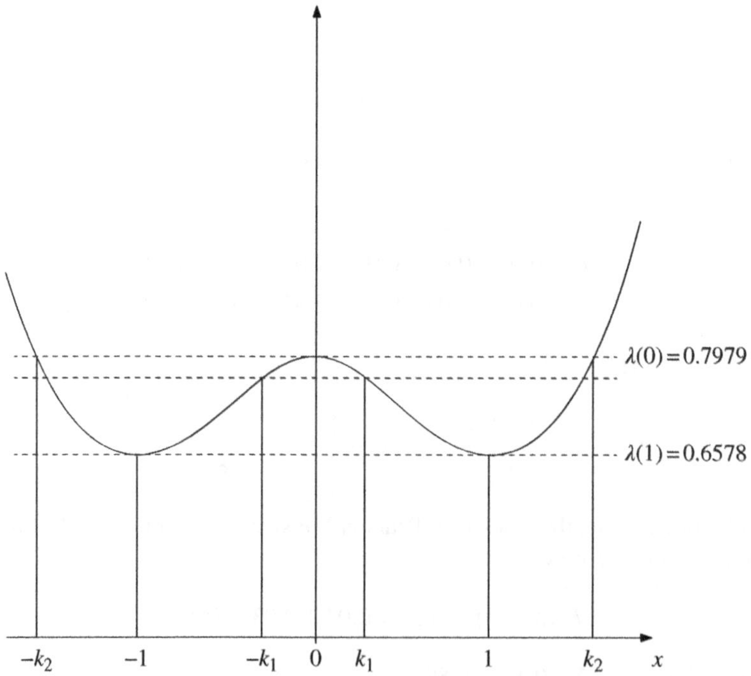

Fig. 2 Graph of $\lambda(x) = (2/\pi)^{1/2}\frac{\exp(x^2/2)}{(1+x^2)}$.

where k and γ are determined from

$$E_{p_0}\varphi(\mathbf{X})=\alpha.$$

Now

$$\lambda(\mathbf{x})=\left(\frac{p_1}{p_0}\right)^{\sum x_i}\left(\frac{1-p_1}{1-p_0}\right)^{n-\sum x_i},$$

and since $p_1>p_0$, $\lambda(\mathbf{x})$ is an increasing function of $\sum x_i$. It follows that $\lambda(\mathbf{x})>k$ if and only if $\sum x_i>k_1$, where k_1 is some constant. Thus the MP size α test is of the form

$$\varphi(\mathbf{x})=\begin{cases}1 & \text{if } \sum x_i>k_1,\\ \gamma & \text{if } \sum x_i=k_1,\\ 0 & \text{otherwise.}\end{cases}$$

Also, k_1 and γ are determined from

$$\begin{aligned}\alpha=E_{p_0}\varphi(\mathbf{X}) &= P_{p_0}\left\{\sum_1^n X_i>k_1\right\}+\gamma P_{p_0}\left\{\sum_1^n X_i=k_1\right\}\\ &=\sum_{r=k_1+1}^{n}\binom{n}{r}p_0^r(1-p_0)^{n-r}+\gamma\binom{n}{k_1}p_0^{k_1}(1-p_0)^{n-k_1}.\end{aligned}$$

Note that the MP size α test is independent of p_1 as long as $p_1>p_0$, that is, it remains an MP size α test against any $p>p_0$ and is therefore a UMP test of $p=p_0$ against $p>p_0$.

In particular, let $n=5$, $p_0=\frac{1}{2}$, $p_1=\frac{3}{4}$, and $\alpha=0.05$. Then the MP test is given by

$$\varphi(\mathbf{x})=\begin{cases}1 & \sum x_i>k,\\ \gamma & \sum x_i=k,\\ 0 & \sum x_i<k,\end{cases}$$

where k and γ are determined from

$$0.05=\alpha=\sum_{k+1}^{5}\binom{5}{r}\left(\frac{1}{2}\right)^5+\gamma\binom{5}{k}\left(\frac{1}{2}\right)^5.$$

It follows that $k=4$ and $\gamma=0.122$. Thus the MP size $\alpha=0.05$ test is to reject $p=\frac{1}{2}$ in favor of $p=\frac{3}{4}$ if $\sum_1^n X_i=5$ and reject $p=\frac{1}{2}$ with probability 0.122 if $\sum_1^n X_i=4$.

It is simply a matter of reversing inequalities to see that the MP size α test of $H_0: p=p_0$ against $H_1: p=p_1$ $(p_1<p_0)$ is given by

$$\varphi(\mathbf{x})=\begin{cases}1 & \text{if } \sum x_i<k,\\ \gamma & \text{if } \sum x_i=k,\\ 0 & \text{if } \sum x_i>k,\end{cases}$$

where γ and k are determined from $E_{p_0}\varphi(\mathbf{X})=\alpha$.

We note that $T(X) = \sum X_i$ is minimal sufficient for p so that, in view of Remark 3, we could have considered tests based only on T. Since $T \sim b(n,p)$,

$$\lambda(t) = \frac{f_1(t)}{f_0(t)} = \frac{\binom{n}{t} p_1^t (1-p_1)^{n-t}}{\binom{n}{t} p_0^t (1-p_0)^{n-t}} = \left(\frac{p_1}{p_0}\right)^t \left(\frac{1-p_1}{1-p_0}\right)^{n-t}$$

so that an MP Test is of the same form as above but the computation is somewhat simpler.

We remark that in both cases $(p_1 > p_0, p_1 < p_0)$ the MP test is quite intuitive. We would tend to accept the larger probability if a larger number of "successes" showed up, and the smaller probability if a smaller number of "successes" were observed. See, however, Example 2.

Example 4. Let $X_1, X_2, \ldots, X_n$ be iid $\mathcal{N}(\mu, \sigma^2)$ RVs where both μ and σ^2 are unknown. We wish to test the null hypothesis $H_0\colon \mu = \mu_0, \sigma^2 = \sigma_0^2$ against the alternative $H_1 : \mu = \mu_1$, $\sigma^2 = \sigma_0^2$. The fundamental lemma leads to the following MP test:

$$\varphi(\mathbf{x}) = \begin{cases} 1 & \text{if } \lambda(\mathbf{x}) > k, \\ 0 & \text{if } \lambda(\mathbf{x}) < k, \end{cases}$$

where

$$\lambda(\mathbf{x}) = \frac{(1/\sigma_0\sqrt{2\pi})^n \exp\{-[\sum(x_i - \mu_1)^2/2\sigma_0^2]\}}{(1/\sigma_0\sqrt{2\pi})^n \exp\{-[\sum(x_i - \mu_0)^2/2\sigma_0^2]\}},$$

and k is determined from $E_{\mu_0,\sigma_0}\varphi(\mathbf{X}) = \alpha$. We have

$$\lambda(\mathbf{x}) = \exp\left\{\sum x_i \left(\frac{\mu_1}{\sigma_0^2} - \frac{\mu_0}{\sigma_0^2}\right) + n\left(\frac{\mu_0^2}{2\sigma_0^2} - \frac{\mu_1^2}{2\sigma_1^2}\right)\right\}.$$

If $\mu_1 > \mu_0$, then

$$\lambda(\mathbf{x}) > k \qquad \text{if and only if } \sum_{i=1}^n x_i > k',$$

where k' is determined from

$$\alpha = P_{\mu_0,\sigma_0}\left\{\sum_{i=1}^n X_i > k'\right\} = P\left\{\frac{\sum X_i - n\mu_0}{\sqrt{n}\sigma_0} > \frac{k' - n\mu_0}{\sqrt{n}\,\sigma_0}\right\},$$

giving $k' = z_\alpha\sqrt{n}\sigma_0 + n\mu_0$. The case $\mu_1 < \mu_0$ is treated similarly. If σ_0 is known, the test determined above is independent of μ_1 as long as $\mu_1 > \mu_0$, and it follows that the test is UMP against $H_1'\colon \mu > \mu_1$, $\sigma^2 = \sigma_0^2$. If, however, σ_0 is not known, that is, the null hypothesis is a composite hypothesis $H_0''\colon \mu = \mu_0$, $\sigma^2 > 0$ to be tested against the alternatives $H_1''\colon \mu = \mu_1$, $\sigma^2 > 0$ $(\mu_1 > \mu_0)$, then the MP test determined above depends on σ^2. In other words, an MP test against the alternative μ_1, σ_0^2 will not be MP against μ_1, σ_1^2, where $\sigma_1^2 \neq \sigma_0^2$.

PROBLEMS 9.3

1. A sample of size 1 is taken from PDF

$$f_\theta(x) = \begin{cases} \frac{2}{\theta^2}(\theta - x) & \text{if } 0 < x < \theta, \\ 0 & \text{otherwise.} \end{cases}$$

Find an MP test of $H_0\colon \theta = \theta_0$ against $H_1: \theta_1$ $(\theta_1 < \theta_0)$.

2. Find the Neyman–Pearson size α test of $H_0\colon \theta = \theta_0$ against $H_1: \theta = \theta_1$ $(\theta_1 < \theta_0)$ based on a sample of size 1 from the PDF

$$f_\theta(x) = 2\theta x + 2(1-\theta)(1-x), \qquad 0 < x < 1, \qquad \theta \in [0,1].$$

3. Find the Neyman–Pearson size α test of $H_0\colon \beta = 1$ against $H_1: \beta = \beta_1$ (> 1) based on a sample of size 1 from

$$f(x;\beta) = \begin{cases} \beta x^{\beta-1}, & 0 < x < 1, \\ 0, & \text{otherwise.} \end{cases}$$

4. Find an MP size α test of $H_0\colon X \sim f_0(x)$, where $f_0(x) = (2\pi)^{-1/2}e^{-x^2/2}$, $-\infty < x < \infty$, against $H_1: X \sim f_1(x)$ where $f_1(x) = 2^{-1}e^{-|x|}$, $-\infty < x < \infty$, based on a sample of size 1.

5. For the PDF $f_\theta(x) = e^{-(x-\theta)}$, $x \geq \theta$, find an MP size α test of $\theta = \theta_0$ against $\theta = \theta_1$ $(> \theta_0)$, based on a sample of size n.

6. If φ^* is an MP size α test of $H_0\colon \mathbf{X} \sim f_0(\mathbf{x})$ against $H_1: \mathbf{X} \sim f_1(\mathbf{x})$ show that it has to be either of form (1) or form (2) (except for a set of $\mathbf{x}$ that has probability 0 under H_0 and H_1).

7. Let φ^* be an MP size α $(0 < \alpha \leq 1)$ test of H_0 against H_1, and let $k(\alpha)$ denote the value of k in (1). Show that if $\alpha_1 < \alpha_2$, then $k(\alpha_2) \leq k(\alpha_1)$.

8. For the family of Neyman–Pearson tests show that the larger the α, the smaller the β (= P[Type II error]).

9. Let $1 - \beta$ be the power of an MP size α test, where $0 < \alpha < 1$. Show that $\alpha < 1 - \beta$ unless $P_{\theta_0} = P_{\theta_1}$.

10. Let α be a real number, $0 < \alpha < 1$, and φ^* be an MP size α test of H_0 against H_1. Also, let $\beta = E_{H_1}\varphi^*(\mathbf{X}) < 1$. Show that $1 - \varphi^*$ is an MP test for testing H_1 against H_0 at level $1 - \beta$.

11. Let $X_1, X_2, \ldots, X_n$ be a random sample from PDF

$$f_\theta(x) = \frac{\theta}{x^2} \qquad \text{if} \qquad 0 < \theta \leq x < \infty.$$

Find an MP test of $\theta = \theta_0$ against $\theta = \theta_1 (\neq \theta_0)$.

12. Let X be an observation in $(0,1)$. Find an MP size α test of $H_0\colon X \sim f(x) = 4x$ if $0 < x < \frac{1}{2}$, and $= 4 - 4x$ if $\frac{1}{2} \leq x < 1$, against $H_1: X \sim f(x) = 1$ if $0 < x < 1$. Find the power of your test.

13. In each of the following cases of simple versus simple hypotheses $H_0 : X \sim f_0$, $H_1 : X \sim f_1$, draw a graph of the ratio $\lambda(x) = f_1(x)/f_0(x)$ and find the form of the Neyman–Pearson test:

(a) $f_0(x) = (1/2)\exp\{-|x+1|\}$; $f_1(x) = (1/2)\exp\{-|x-1|\}$.

(b) $f_0(x) = (1/2)\exp\{-|x|\}$; $f_1(x) = \{1/[\pi(1+x^2)]\}$.

(c) $f_0(x) = (1/\pi)\{1+(1+x)^2\}^{-1}$; $f_1(x) = (1/\pi)\{1+(1-x)^2\}^{-1}$.

14. Let $X_1, X_2, \ldots, X_n$ be a random sample with common PDF

$$f_\theta(x) = \frac{1}{2\theta}\exp\{-|x|/\theta\}, \quad x \in \mathcal{R},\ \theta > 0.$$

Find a size α MP test for testing $H_0 : \theta = \theta_0$ versus $H_1 : \theta = \theta_1\ (> \theta_0)$.

15. Let $X \sim f_j$, $j = 0, 1$, where

x	1	2	3	4	5
$f_0(x)$	1/5	1/5	1/5	1/5	1/5
$f_1(x)$	1/6	1/4	1/6	1/4	1/6

(a) Find the form of the MP test of its size.

(b) Find the size and the power of your test for various values of the cutoff point.

(c) Consider now a random sample of size n from f_0 under H_0 or f_1 under H_1. Find the form of the MP test of its size.

9.4 FAMILIES WITH MONOTONE LIKELIHOOD RATIO

In this section we consider the problem of testing one-sided hypotheses on a single real-valued parameter. Let $\{f_\theta, \theta \in \Theta\}$ be a family of PDFs (PMFs), $\Theta \subseteq \mathcal{R}$, and suppose that we wish to test $H_0\colon \theta \le \theta_0$ against the alternatives $H_1 : \theta > \theta_0$ or its dual, $H_0'\colon \theta \ge \theta_0$, against $H_1' : \theta < \theta_0$. In general, it is not possible to find a UMP test for this problem. The MP test of $H_0\colon \theta \le \theta_0$, say, against the alternative $\theta = \theta_1\ (> \theta_0)$ depends on θ_1 and cannot be UMP. Here we consider a special class of distributions that is large enough to include the one-parameter exponential family, for which a UMP test of a one-sided hypothesis exists.

Definition 1. Let $\{f_\theta, \theta \in \Theta\}$ be a family of PDFs (PMFs), $\theta \subseteq \mathcal{R}$. We say that $\{f_\theta\}$ has a *monotone likelihood ratio* (MLR) in statistic $T(\mathbf{x})$ if for $\theta_1 < \theta_2$, whenever $f_{\theta_1}, f_{\theta_2}$ are distinct, the ratio $f_{\theta_2}(\mathbf{x})/f_{\theta_1}(\mathbf{x})$ is a nondecreasing function of $T(\mathbf{x})$ for the set of values $\mathbf{x}$ for which at least one of f_{θ_1} and f_{θ_2} is > 0.

It is also possible to define families of densities with nonincreasing MLR in $T(\mathbf{x})$, but such families can be treated by symmetry.

Example 1. Let $X_1, X_2, \ldots, X_n \sim U[0, \theta]$, $\theta > 0$. The joint PDF of $X_1, \ldots, X_n$ is

$$f_\theta(\mathbf{x}) = \begin{cases} \dfrac{1}{\theta^n}, & 0 \le \max x_i \le \theta, \\ 0, & \text{otherwise.} \end{cases}$$

Let $\theta_2 > \theta_1$ and consider the ratio

$$\begin{aligned}\frac{f_{\theta_2}(\mathbf{x})}{f_{\theta_1}(\mathbf{x})} &= \frac{(1/\theta_2^n)I_{[\max x_i \le \theta_2]}}{(1/\theta_1^n)I_{[\max x_i \le \theta_1]}} \\ &= \left(\frac{\theta_1}{\theta_2}\right)^n I_{[\max x_i \le \theta_2]}/I_{[\max x_i \le \theta_1]}.\end{aligned}$$

Let

$$\begin{aligned}R(\mathbf{x}) &= \frac{I_{[\max x_i \le \theta_2]}}{I_{[\max x_i \le \theta_1]}} \\ &= \begin{cases} 1, & \max x_i \in [0, \theta_1], \\ \infty, & \max x_i \in [\theta_1, \theta_2]. \end{cases}\end{aligned}$$

Define $R(\mathbf{x}) = \infty$ if $\max x_i > \theta_2$. It follows that $f_{\theta_2}/f_{\theta_1}$ is a nondecreasing function of $\max_{1 \le i \le n} x_i$, and the family of uniform densities on $[0, \theta]$ has an MLR in $\max_{1 \le i \le n} x_i$.

Theorem 1. The one-parameter exponential family

$$f_\theta(\mathbf{x}) = \exp\{Q(\theta)T(\mathbf{x}) + S(\mathbf{x}) + D(\theta)\}, \tag{1}$$

where $Q(\theta)$ is nondecreasing, has an MLR in $T(\mathbf{x})$.

Proof. The proof is left as an exercise.

Remark 1. The nondecreasingness of $Q(\theta)$ can be obtained by a reparametrization, putting $\vartheta = Q(\theta)$, if necessary.

Theorem 1 includes normal, binomial, Poisson, gamma (one parameter fixed), beta (one parameter fixed), and so on. In Example 1 we have already seen that $U[0, \theta]$, which is not an exponential family, has an MLR.

Example 2. Let $X \sim \mathcal{C}(1, \theta)$. Then

$$\frac{f_{\theta_2}(x)}{f_{\theta_1}(x)} = \frac{1 + (x - \theta_1)^2}{1 + (x - \theta_2)^2} \to 1 \qquad \text{as } x \to \pm\infty,$$

and we see that $\mathcal{C}(1, \theta)$ does not have an MLR.

Theorem 2. Let $\mathbf{X} \sim f_\theta$, $\theta \in \Theta$, where $\{f_\theta\}$ has an MLR in $T(\mathbf{x})$. For testing $H_0\colon \theta \le \theta_0$ against $H_1: \theta > \theta_0$, $\theta_0 \in \Theta$, any test of the form

$$\varphi(\mathbf{x}) = \begin{cases} 1 & \text{if } T(\mathbf{x}) > t_0, \\ \gamma & \text{if } T(\mathbf{x}) = t_0, \\ 1 & \text{if } T(\mathbf{x}) < t_0, \end{cases} \tag{2}$$

has a nondecreasing power function and is UMP of its size $E_{\theta_0}\varphi(\mathbf{X}) = \alpha$ (provided that the size is not 0).

Moreover, for every $0 \le \alpha \le 1$ and every $\theta_0 \in \Theta$, there exists a t_0, $-\infty \le t_0 \le \infty$, and $0 \le \gamma \le 1$ such that the test described in (2) is the UMP size α test of H_0 against H_1.

Proof. Let $\theta_1, \theta_2 \in \Theta$, $\theta_1 < \theta_2$. By the fundamental lemma any test of the form

$$\varphi(\mathbf{x}) = \begin{cases} 1, & \lambda(\mathbf{x}) > k, \\ \gamma(\mathbf{x}), & \lambda(\mathbf{x}) = k, \\ 0, & \lambda(\mathbf{x}) < k, \end{cases} \tag{3}$$

where $\lambda(\mathbf{x}) = f_{\theta_2}(\mathbf{x})/f_{\theta_1}(\mathbf{x})$ is MP of its size for testing $\theta = \theta_1$ against $\theta = \theta_2$, provided that $0 \le k < \infty$ and if $k = \infty$, the test

$$\varphi(\mathbf{x}) = \begin{cases} 1, & \text{if } f_{\theta_1}(\mathbf{x}) = 0, \\ 0 & \text{if } f_{\theta_1}(\mathbf{x}) > 0, \end{cases} \tag{4}$$

is MP of size 0. Since f_θ has an MLR in T, it follows that any test of form (2) is also of form (3), provided that $E_{\theta_1}\varphi(\mathbf{X}) > 0$, that is, provided that its size is > 0. The trivial test $\varphi'(\mathbf{x}) \equiv \alpha$ has size α and power α, so that the power of any test (2) is at least α, that is,

$$E_{\theta_2}\varphi(\mathbf{X}) \ge E_{\theta_2}\varphi'(\mathbf{X}) = \alpha = E_{\theta_1}\varphi(\mathbf{X}).$$

It follows that, if $\theta_1 < \theta_2$ and $E_{\theta_1}\varphi(\mathbf{X}) > 0$, then $E_{\theta_1}\varphi(\mathbf{X}) \le E_{\theta_2}\varphi(\mathbf{X})$, as asserted.

Let $\theta_1 = \theta_0$ and $\theta_2 > \theta_0$, as above. We know that (2) is an MP test of its size $E_{\theta_0}\varphi(\mathbf{X})$ for testing $\theta = \theta_0$ against $\theta = \theta_2$ $(\theta_2 > \theta_0)$, provided that $E_{\theta_0}\varphi(\mathbf{X}) > 0$. Since the power function of φ is nondecreasing,

$$E_\theta\varphi(\mathbf{X}) \le E_{\theta_0}\varphi(\mathbf{X}) = \alpha_0 \qquad \text{for all } \theta \le \theta_0. \tag{5}$$

Since, however, φ does not depend on θ_2 (it depends only on constants k and γ), it follows that φ is the UMP size α_0 test for testing $\theta = \theta_0$ against $\theta > \theta_0$. Thus φ is UMP among the class of tests φ'' for which

$$E_{\theta_0}\varphi''(\mathbf{X}) \le E_{\theta_0}\varphi(\mathbf{X}) = \alpha_0. \tag{6}$$

Now the class of tests satisfying (5) is contained in the class of tests satisfying (6) [there are more restrictions in (5)]. It follows that φ, which is UMP in the larger class satisfying (6), must also be UMP in the smaller class satisfying (5). Thus, provided that $\alpha_0 > 0$, φ is the UMP size α_0 test for $\theta \le \theta_0$ against $\theta > \theta_0$.

We ask the reader to complete the proof of the final part of the theorem, using the fundamental lemma.

Remark 2. By interchanging inequalities throughout in Theorem 2, we see that this theorem also provides a solution of the dual problem $H_0'\colon \theta \geq \theta_0$ against $H_1'\colon \theta < \theta_0$.

Example 3. Let X have the hypergeometric PMF

$$P_M\{X = x\} = \frac{\binom{M}{x}\binom{N-M}{n-x}}{\binom{N}{n}}, \qquad x = 0, 1, 2, \ldots, M.$$

Since

$$\frac{P_{M+1}\{X = x\}}{P_M\{X = x\}} = \frac{M+1}{N-M}\,\frac{N-M-n+x}{M+1-x},$$

we see that $\{P_M\}$ has an MLR in $x(P_{M_2}/P_{M_1}$ where $M_2 > M_1$ is just a product of such ratios). It follows that there exists a UMP test of $H_0\colon M \leq M_0$ against $H_1\colon M > M_0$, which rejects H_0 when X is too large, that is, the UMP size α test is given by

$$\varphi(x) = \begin{cases} 1, & x > k, \\ \gamma, & x = k, \\ 0, & x < k, \end{cases}$$

where (integer) k and γ are determined from

$$E_{M_0}\varphi(X) = \alpha.$$

For the one-parameter exponential family UMP tests exist also for some two-sided hypotheses of the form

$$H_0\colon \theta \leq \theta_1 \text{ or } \theta \geq \theta_2 (\theta_1 < \theta_2). \tag{7}$$

We state the following result without proof.

Theorem 3. For the one-parameter exponential family (1), there exists a UMP test of the hypothesis $H_0\colon \theta \leq \theta_1$ or $\theta \geq \theta_2$ $(\theta_1 < \theta_2)$ against $H_1\colon \theta_1 < \theta < \theta_2$ that is of the form

$$\varphi(\mathbf{x}) = \begin{cases} 1 & \text{if } c_1 < T(\mathbf{x}) < c_2, \\ \gamma_i & \text{if } T(\mathbf{x}) = c_i, \qquad i = 1, 2 \qquad (c_1 < c_2), \\ 0 & \text{if } T(\mathbf{x}) < c_1 \text{ or } > c_2, \end{cases} \tag{8}$$

where the c's and the γ's are given by

$$E_{\theta_1}\varphi(\mathbf{X}) = E_{\theta_2}\varphi(\mathbf{X}) = \alpha. \tag{9}$$

See Lehmann [64, pp. 101–103], for proof.

Example 4. Let $X_1, X_2, \ldots, X_n$ be iid $\mathcal{N}(\mu, 1)$ RVs. To test $H_0: \mu \le \mu_0$ or $\mu \ge \mu_1$ $(\mu_1 > \mu_0)$ against $H_1: \mu_0 < \mu < \mu_1$, the UMP test is given by

$$\varphi(\mathbf{x}) = \begin{cases} 1 & \text{if } c_1 < \sum_1^n x_i < c_2, \\ \gamma_i & \text{if } \sum x_i = c_1 \text{ or } c_2, \\ 0 & \text{if } \sum x_i < c_1 \text{ or } > c_2, \end{cases}$$

where we determine c_1, c_2 from

$$\alpha = P_{\mu_0}\{c_1 < \sum X_i < c_2\} = P_{\mu_1}\{c_1 < \sum X_i < c_2\}$$

and $\gamma_1 = \gamma_2 = 0$. Thus

$$\begin{aligned} \alpha &= P\left\{\frac{c_1 - n\mu_0}{\sqrt{n}} < \frac{\sum X_i - n\mu_0}{\sqrt{n}} < \frac{c_2 - n\mu_0}{\sqrt{n}}\right\} \\ &= P\left\{\frac{c_1 - n\mu_1}{\sqrt{n}} < \frac{\sum X_i - n\mu_1}{\sqrt{n}} < \frac{c_2 - n\mu_1}{\sqrt{n}}\right\} \\ &= P\left\{\frac{c_1 - n\mu_0}{\sqrt{n}} < Z < \frac{c_2 - n\mu_0}{\sqrt{n}}\right\} \\ &= P\left\{\frac{c_1 - n\mu_1}{\sqrt{n}} < Z < \frac{c_2 - n\mu_1}{\sqrt{n}}\right\}, \end{aligned}$$

where Z is $\mathcal{N}(0, 1)$. Given α, n, μ_0, and μ_1, we can solve for c_1 and c_2 from the simultaneous equations

$$\Phi\left(\frac{c_2 - n\mu_0}{\sqrt{n}}\right) - \Phi\left(\frac{c_1 - n\mu_0}{\sqrt{n}}\right) = \alpha,$$
$$\Phi\left(\frac{c_2 - n\mu_1}{\sqrt{n}}\right) - \Phi\left(\frac{c_1 - n\mu_1}{\sqrt{n}}\right) = \alpha,$$

where Φ is the DF of Z.

Remark 3. We caution the reader that UMP tests for testing $H_0: \theta_1 \le \theta \le \theta_2$ and $H_0': \theta = \theta_0$ for the one-parameter exponential family do not exist. An example will suffice.

Example 5. Let $X_1, X_2, \ldots, X_n$ be a sample from $\mathcal{N}(0, \sigma^2)$. Since the family of joint PDFs of $\mathbf{X} = (X_1, \ldots, X_n)$ has an MLR in $T(\mathbf{X}) = \sum_1^n X_i^2$, it follows that UMP tests exist for one-sided hypotheses $\sigma \ge \sigma_0$ and $\sigma \le \sigma_0$.

Consider now the null hypotheses $H_0: \sigma = \sigma_0$ against the alternative $H_1: \sigma \ne \sigma_0$. We will show that a UMP test of H_0 does not exist. For testing $\sigma = \sigma_0$ against $\sigma > \sigma_0$, a test of the form

$$\varphi_1(\mathbf{x}) = \begin{cases} 1, & \sum x_i^2 > c_1 \\ 0, & \text{otherwise} \end{cases}$$

is UMP, and for testing $\sigma = \sigma_0$ against $\sigma < \sigma_0$, a test of the form

$$\varphi_2(\mathbf{x}) = \begin{cases} 1 & \sum x_i^2 < c_2 \\ 0 & \text{otherwise} \end{cases}$$

is UMP. If the size is chosen as α, then $c_1 = \sigma_0^2 \chi^2_{n,\alpha}$ and $c_2 = \sigma_0^2 \chi^2_{n,1-\alpha}$. Clearly, neither φ_1 nor φ_2 is UMP for H_0 against $H_1 : \sigma \neq \sigma_0$. The power of any test of H_0 for values $\sigma > \sigma_0$ cannot exceed that of φ_1, and for values of $\sigma < \sigma_0$ it cannot exceed the power of test φ_2. Hence no test of H_0 can be UMP (see Fig. 1).

PROBLEMS 9.4

1. For the following families of PMFs (PDFs) $f_\theta(x)$, $\theta \in \Theta \subseteq \mathcal{R}$, find a UMP size α test of $H_0 \colon \theta \leq \theta_0$ against $H_1 : \theta > \theta_0$, based on a sample of n observations.
   (a) $f_\theta(x) = \theta^x(1-\theta)^{1-x}$, $x = 0, 1$; $0 < \theta < 1$.
   (b) $f_\theta(x) = (1/\sqrt{2\pi})\exp\{-(x-\theta)^2/2\}$, $-\infty < x < \infty$, $-\infty < \theta < \infty$.
   (c) $f_\theta(x) = e^{-\theta}(\theta^x/x!)$, $x = 0, 1, 2, \ldots$; $\theta > 0$.
   (d) $f_\theta(x) = (1/\theta)e^{-x/\theta}$, $x > 0$, $\theta > 0$.
   (e) $f_\theta(x) = [1/\Gamma(\theta)]x^{\theta-1}e^{-x}$, $x > 0$, $\theta > 0$.
   (f) $f_\theta(x) = \theta x^{\theta-1}$, $0 < x < 1$, $\theta > 0$.

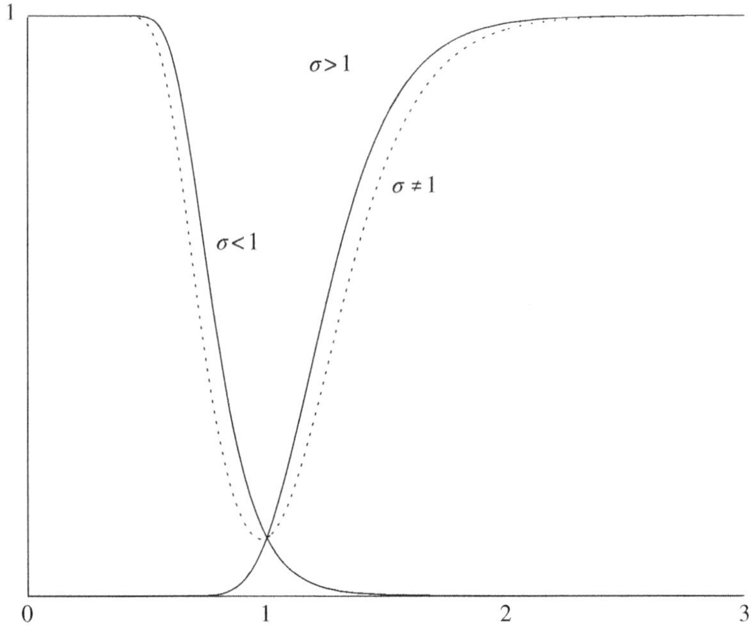

Fig. 1 Power functions of chi-square tests of $H_0 : \sigma = \sigma_0$ against H_1.

2. Let $X_1, X_2, \ldots, X_n$ be a sample of size n from the PMF

$$P_N(x) = \frac{1}{N}, \qquad x = 1, 2, \ldots, N; N \in \{1, 2, \ldots\}.$$

(a) Show that the test

$$\varphi(x_1, x_2, \ldots, x_n) = \begin{cases} 1 & \text{if } \max(x_1, x_2, \ldots, x_n) > N_0 \\ \alpha & \text{if } \max(x_1, x_2, \ldots, x_n) \le N_0 \end{cases}$$

is UMP size α for testing $H_0 \colon N \le N_0$ against $H_1 : N > N_0$.

(b) Show that

$$\varphi(x_1, x_2, \ldots, x_n) = \begin{cases} 1 & \text{if } \max(x_1, x_2, \ldots, x_n) > N_0 \text{ or} \\ & \quad \max(x_1, x_2, \ldots, x_n) \le \alpha^{1/n} N_0 \\ 0 & \text{otherwise,} \end{cases}$$

is a UMP size α test of $H_0' \colon N = N_0$ against $H_1' : N \neq N_0$.

3. Let $X_1, X_2, \ldots, X_n$ be a sample of size n from $U(0, \theta)$, $\theta > 0$. Show that the test

$$\varphi_1(x_1, x_2, \ldots, x_n) = \begin{cases} 1 & \text{if } \max(x_1, \ldots, x_n) > \theta_0 \\ \alpha & \text{if } \max(x_1, x_2, \ldots, x_n) \le \theta_0 \end{cases}$$

is UMP size α for testing $H_0 \colon \theta \le \theta_0$ against $H_1 : \theta > \theta_0$ and that the test

$$\varphi_2(x_1, x_2, \ldots, x_n) = \begin{cases} 1 & \text{if } \max(x_1, \ldots, x_n) > \theta_0 \text{ or} \\ & \quad \max(x_1, x_2, \ldots, x_n) \le \theta_0 \alpha^{1/n} \\ 0 & \text{otherwise} \end{cases}$$

is UMP size α for $H_0' \colon \theta = \theta_0$ against $H_1' : \theta \neq \theta_0$.

4. Does the Laplace family of PDFs

$$f_\theta(x) = \frac{1}{2} \exp\{-|x - \theta|\}, \qquad -\infty < x < \infty, \qquad \theta \in \mathcal{R},$$

possess an MLR?

5. Let X have logistic distribution with PDF

$$f_\theta(x) = e^{-x-\theta}\{1 + e^{-x-\theta}\}^{-2}, \quad x \in \mathcal{R}.$$

Does $\{f_\theta\}$ belong to the exponential family? Does $\{f_\theta\}$ have MLR?

6. (a) Let f_θ be the PDF of a $\mathcal{N}(\theta, \theta)$ RV. Does $\{f_\theta\}$ have MLR?

(b) Do the same as in (a) if $X \sim \mathcal{N}(\theta, \theta^2)$.

9.5 UNBIASED AND INVARIANT TESTS

We have seen that, if we restrict ourselves to the class Φ_α of all size α tests, there do not exist UMP tests for many important hypotheses. This suggests that we reduce the class of tests under consideration by imposing certain restrictions.

Definition 1. A size α test φ of $H_0: \theta \in \Theta_0$ against the alternatives $H_1: \theta \in \Theta_1$ is said to be *unbiased* if

$$E_\theta \varphi(\mathbf{X}) \geq \alpha \qquad \text{for all } \theta \in \Theta_1. \tag{1}$$

It follows that a test φ is unbiased if and only if its power function $\beta_\varphi(\theta)$ satisfies

$$\beta_\varphi(\theta) \leq \alpha \qquad \text{for } \theta \in \Theta_0 \tag{2}$$

and

$$\beta_\varphi(\theta) \geq \alpha \qquad \text{for } \theta \in \Theta_1. \tag{3}$$

This seems to be a reasonable requirement to place on a test. An unbiased test rejects a false H_0 more often than a true H_0.

Definition 2. Let U_α be the class of all unbiased size α tests of H_0. If there exists a test $\varphi \in U_\alpha$ that has maximum power at each $\theta \in \Theta_1$, we call φ a UMP *unbiased* size α test.

Clearly $U_\alpha \subset \Phi_\alpha$. If a UMP test exists in Φ_α, it is UMP in U_α. This follows by comparing the power of the UMP test with that of the trivial test $\varphi(\mathbf{x}) = \alpha$. It is convenient to introduce another class of tests.

Definition 3. A test φ is said to be α-*similar* on a subset Θ^* of Θ if

$$\beta_\varphi(\theta) = E_\theta \varphi(\mathbf{X}) = \alpha \qquad \text{for } \theta \in \Theta^*. \tag{4}$$

A test is said to be *similar* on a set $\Theta^* \subseteq \Theta$ if it is α-similar on Θ^* for some α, $0 \leq \alpha \leq 1$.

It is clear that there exists at least one similar test on every Θ^*, namely, $\varphi(x) \equiv \alpha$, $0 \leq \alpha \leq 1$.

Theorem 1. Let $\beta_\varphi(\theta)$ be continuous in θ for any φ. If φ is an unbiased size α test of $H_0: \theta \in \Theta_0$ against $H_1: \theta \in \Theta_1$, it is α-similar on the boundary $\Lambda = \overline{\Theta}_0 \cap \overline{\Theta}_1$. (Here $\overline{A}$ is the closure of set A.)

Proof. Let $\theta \in \Lambda$. Then there exists a sequence $\{\theta_n\}$, $\theta_n \in \Theta_0$, such that $\theta_n \to \theta$. Since $\beta_\varphi(\theta)$ is continuous, $\beta_\varphi(\theta_n) \to \beta_\varphi(\theta)$; and since $\beta_\varphi(\theta_n) \leq \alpha$, for $\theta_n \in \Theta_0$, $\beta_\varphi(\theta) \leq \alpha$. Similarly, there exists a sequence $\{\theta'_n\}$, $\theta'_n \in \Theta_1$, such that $\beta_\varphi(\theta'_n) \geq \alpha$ (φ is unbiased) and $\theta'_n \to \theta$. Thus $\beta_\varphi(\theta'_n) \to \beta_\varphi(\theta)$, and it follows that $\beta_\varphi(\theta) \geq \alpha$. Hence $\beta_\varphi(\theta) = \alpha$ for $\theta \in \Lambda$, and φ is α-similar on Λ.

Remark 1. Thus, if $\beta_\varphi(\theta)$ is continuous in θ for any φ, an unbiased size α test of H_0 against H_1 is also α-similar for the PDFs (PMFs) of Λ, that is, for $\{f_\theta, \theta \in \Lambda\}$. If we can find an MP similar test of $H_0\colon \theta \in \Lambda$ against H_1, and if this test is unbiased size α, then necessarily it is MP in the smaller class.

Definition 4. A test φ that is UMP among all α-similar tests on the boundary $\Lambda = \overline{\Theta}_0 \cap \overline{\Theta}_1$ is said to be a UMP *α-similar* test.

It is frequently easier to find a UMP α-similar test. Moreover, tests that are UMP similar on the boundary are often UMP unbiased.

Theorem 2. Let the power function of every test φ of $H_0\colon \theta \in \Theta_0$ against $H_1 : \theta \in \Theta_1$ be continuous in θ. Then a UMP α-similar test is UMP unbiased, provided that its size is α for testing H_0 against H_1.

Proof. Let φ_0 be UMP α-similar. Then $E_\theta \varphi_0(\mathbf{X}) \le \alpha$ for $\theta \in \Theta_0$. Comparing its power with that of the trivial similar test $\varphi(\mathbf{x}) \equiv \alpha$, we see that φ_0 is unbiased also. By the continuity of $\beta_\varphi(\theta)$ we see that the class of all unbiased size α tests is a subclass of the class of all α-similar tests. It follows that φ_0 is a UMP unbiased size α test.

Remark 2. The continuity of power function $\beta_\varphi(\theta)$ is not always easy to check but sufficient conditions may be found in most advanced calculus texts. See, for example, Widder [117, p. 356]. If the family of PDF (PMF) f_θ is an exponential family then a proof is given in Lehman [64, p. 59].

Example 1. Let $X_1, X_2, \ldots, X_n$ be a sample from $\mathcal{N}(\mu, 1)$. We wish to test $H_0\colon \mu \le 0$ against $H_1 : \mu > 0$. Since the family of densities has an MLR in $\sum_1^n X_i$, we can use Theorem 2 to conclude that a UMP test rejects H_0 if $\sum_1^n X_i > c$. This test is also UMP unbiased. Nevertheless we use this example to illustrate the concepts introduced above.

Here $\Theta_0 = \{\mu \le 0\}$, $\Theta_1 = \{\mu > 0\}$, and $\Lambda = \overline{\Theta}_0 \cap \overline{\Theta}_1 = \{\mu = 0\}$. Since $T(\mathbf{X}) = \sum_{i=1}^n X_i$ is sufficient, we focus attention to tests based on T alone. Note that $T \sim \mathcal{N}(n\mu, n)$ which is one-parameter exponential. Thus the power function of any test φ based on T is continuous in μ. It follows that any unbiased size α test of H_0 has the property $\beta_\varphi(0) = \alpha$ of similarity over Λ. In order to use Theorem 2, we find a UMP test of $H_0' : \mu \in \Lambda$ against H_1. Let $\mu_1 > 0$. By the fundamental lemma an MP test of $\mu = 0$ against $\mu = \mu_1 > 0$ is given by

$$\varphi(t) = \begin{cases} 1 & \text{if } \exp\left\{\frac{t^2}{2n} - \frac{(t-n\mu)^2}{2n}\right\} > k' \\ 0 & \text{otherwise,} \end{cases}$$

$$= \begin{cases} 1 & \text{if } t > k \\ 0 & \text{if } t \le k \end{cases}$$

where k is determined from

$$\alpha = P_0\{T > k\} = P\left\{Z > \frac{k}{\sqrt{n}}\right\}.$$

Thus $k = \sqrt{n}z_\alpha$. Since φ is independent of μ_1 as long as $\mu_1 > 0$, we see that the test

$$\varphi(t) = \begin{cases} 1, & t > \sqrt{n}z_\alpha \\ 0, & \text{otherwise}, \end{cases},$$

is UMP α-similar. We need only check that φ is of the right size for testing H_0 against H_1. We have, for $\mu \le 0$,

$$\begin{aligned} E_\mu\varphi(T) &= P_\mu\{T > \sqrt{n}z_\alpha\} \\ &= P\left\{\frac{T-n\mu}{\sqrt{n}} > z_\alpha - \sqrt{n}\mu\right\} \\ &\le P\{Z > z_\alpha\}, \end{aligned}$$

since $-\sqrt{n}\mu \ge 0$. Here Z is $\mathcal{N}(0,1)$. It follows that

$$E_\mu\varphi(T) \le \alpha \qquad \text{for } \mu \le 0,$$

hence φ is UMP unbiased.

Theorem 2 can be used only if it is possible to find a UMP α-similar test. Unfortunately this requires heavy use of conditional expectation, and we will not pursue the subject any further. We refer to Lehmann [64, chapters 4 and 5] and Ferguson [28, pp. 224–233] for further details.

Yet another reduction is obtained if we apply the principle of invariance to hypothesis testing problems. We recall that a class of distributions is invariant under a group of transformations $\mathcal{G}$ if for every $g \in \mathcal{G}$ and every $\boldsymbol{\theta} \in \Theta$ there exists a unique $\boldsymbol{\theta}' \in \Theta$ such that $g(\mathbf{X})$ has distribution $P_{\boldsymbol{\theta}'}$, whenever $\mathbf{X} \sim P_{\boldsymbol{\theta}}$. We rewrite $\boldsymbol{\theta}' = \bar{g}\boldsymbol{\theta}$.

In a hypothesis testing problem we need to reformulate the principle of invariance. First, we need to ensure that under transformations $\mathcal{G}$ not only does $\mathcal{P} = \{P_{\boldsymbol{\theta}} \colon \boldsymbol{\theta} \in \Theta\}$ remain invariant but also the problem of testing $H_0 \colon \boldsymbol{\theta} \in \Theta_0$ against $H_1 : \boldsymbol{\theta} \in \Theta_1$ remain invariant. Second, since the problem has not changed by application of $\mathcal{G}$, the decision also must not change.

Definition 5. A group $\mathcal{G}$ of transformations on the space of values of X leaves a hypothesis testing problem *invariant* if $\mathcal{G}$ leaves both $\{P_{\boldsymbol{\theta}} \colon \boldsymbol{\theta} \in \Theta_0\}$ and $\{P_{\boldsymbol{\theta}} \colon \boldsymbol{\theta} \in \Theta_1\}$ invariant.

Definition 6. We say that φ is *invariant* under $\mathcal{G}$ if

$$\varphi(g(\mathbf{x})) = \varphi(\mathbf{x}) \qquad \text{for all } \mathbf{x} \text{ and all } g \in \mathcal{G}.$$

Definition 7. Let $\mathcal{G}$ be a group of transformations on the space of values of the RV $\mathbf{X}$. We say that a statistic $T(\mathbf{x})$ is *maximal invariant* under $\mathcal{G}$ if (a) T is invariant; (b) T is maximal, that is $T(\mathbf{x}_1) = T(\mathbf{x}_2) \Rightarrow \mathbf{x}_1 = g(\mathbf{x}_2)$ for some $g \in \mathcal{G}$.

Example 2. Let $\mathbf{x} = (x_1, x_2, \ldots, x_n)$, and $\mathcal{G}$ be the group of translations

$$g_c(\mathbf{x}) = (x_1 + c, \ldots, x_n + c), \qquad -\infty < c < \infty.$$

Here the space of values of $\mathbf{X}$ is $\mathcal{R}_n$. Consider the statistic

$$T(\mathbf{x}) = (x_n - x_1, \ldots, x_n - x_{n-1}).$$

Clearly,

$$T(g_c(\mathbf{x})) = (x_n - x_1, \ldots, x_n - x_{n-1}) = T(\mathbf{x}).$$

If $T(\mathbf{x}) = T(\mathbf{x}')$, then $x_n - x_i = x'_n - x'_i$, $i = 1, 2, \ldots, n-1$, and we have $x_i - x'_i = x_n - x'_n = c$ $(i = 1, 2, \ldots, n-1)$, that is, $g_c(\mathbf{x}') = (x'_1 + c, \ldots, x'_n + c) = \mathbf{x}$ and T is maximal invariant.

Next consider the group of scale changes

$$g_c(\mathbf{x}) = (cx_1, \ldots, cx_n), \qquad c > 0.$$

Then

$$T(\mathbf{x}) = \begin{cases} 0 & \text{if all } x_i = 0, \\ \left(\dfrac{x_1}{z}, \ldots, \dfrac{x_n}{z}\right) & \text{if at least one } x_i \neq 0, \end{cases} \qquad z = \left(\sum_1^n x_i^2\right)^{1/2},$$

is maximal invariant; for

$$T(g_c(\mathbf{x})) = T(cx_1, \ldots, cx_n) = T(\mathbf{x}),$$

and if $T(\mathbf{x}) = T(\mathbf{x}')$, then either $T(\mathbf{x}) = T(\mathbf{x}') = 0$ in which case $x_i = x'_i = 0$, or $T(\mathbf{x}) = T(\mathbf{x}') \neq 0$, in which case $x_i/z = x'_i/z'$, implying $x'_i = (z'/z)x_i = cx_i$, and T is maximal.

Finally, if we consider the group of translation and scale changes,

$$g(\mathbf{x}) = (ax_1 + b, \ldots, ax_n + b), \qquad a > 0, \qquad -\infty < b < \infty,$$

a maximal invariant is

$$T(\mathbf{x}) = \begin{cases} 0 & \text{if } \beta = 0, \\ \left(\dfrac{x_1 - \overline{x}}{\beta}, \dfrac{x_2 - \overline{x}}{\beta}, \ldots, \dfrac{x_n - \overline{x}}{\beta}\right) & \text{if } \beta \neq 0, \end{cases}$$

where $\overline{x} = n^{-1}\sum_1^n x_i$ and $\beta = n^{-1}\sum_1^n (x_i - \overline{x})^2$.

Definition 8. Let I_α denote the class of all invariant size α tests of $H_0\colon \boldsymbol{\theta} \in \Theta_0$ against $H_1\colon \boldsymbol{\theta} \in \Theta_1$. If there exists a UMP member in I_α, we call the test a UMP *invariant test* of H_0 against H_1.

The search for UMP invariant tests is greatly facilitated by the use of the following result.

Theorem 3. Let $T(\mathbf{x})$ be maximal invariant with respect to $\mathcal{G}$. Then φ is invariant under $\mathcal{G}$ if and only if φ is a function of T.

Proof. Let φ be invariant. We have to show that $T(\mathbf{x}_1) = T(\mathbf{x}_2) \Rightarrow \varphi(\mathbf{x}_1) = \varphi(\mathbf{x}_2)$. If $T(\mathbf{x}_1) = T(\mathbf{x}_2)$, there is a $g \in \mathcal{G}$ such that $\mathbf{x}_1 = g(\mathbf{x}_2)$, so that $\varphi(\mathbf{x}_1) = \varphi(g(\mathbf{x}_2)) = \varphi(\mathbf{x}_2)$.

Conversely, if φ is a function of T, $\varphi(\mathbf{x}) = h[T(\mathbf{x})]$, then

$$\varphi(g(\mathbf{x})) = h[T(g(\mathbf{x}))] = h[T(\mathbf{x})] = \varphi(\mathbf{x}),$$

and φ is invariant.

Remark 3. The use of Theorem 3 is obvious. If a hypothesis testing problem is invariant under a group $\mathcal{G}$, the *principle of invariance* restricts attention to invariant tests. According to Theorem 3, it suffices to restrict attention to test functions that are functions of maximal invariant T.

Example 3. Let $X_1, X_2, \ldots, X_n$ be a sample from $\mathcal{N}(\mu, \sigma^2)$, where both μ and σ^2 are unknown. We wish to test $H_0\colon \sigma \geq \sigma_0$, $-\infty < \mu < \infty$, against $H_1\colon \sigma < \sigma_0$, $-\infty < \mu < \infty$. The family $\{\mathcal{N}(\mu, \sigma^2)\}$ remains invariant under translations $x_i' = x_i + c$, $-\infty < c < \infty$. Moreover, since $\text{var}(X + c) = \text{var}(X)$, the hypothesis testing problem remains invariant under the group of translations, that is, both $\{\mathcal{N}(\mu, \sigma^2)\colon \sigma^2 \geq \sigma_0^2\}$ and $\{\mathcal{N}(\mu, \sigma^2)\colon \sigma^2 < \sigma_0^2\}$ remain invariant. The joint sufficient statistic is $(\overline{X}, \sum(X_i - \overline{X})^2)$, which is transformed to $(\overline{X} + c, \sum(X_i - \overline{X})^2)$ under translations. A maximal invariant is $\sum(X_i - \overline{X})^2$. It follows that the class of invariant tests consists of tests that are functions of $\sum(X_i - \overline{X})^2$.

Now $\sum(X_i - \overline{X})^2/\sigma^2 \sim \chi^2(n-1)$, so that the PDF of $Z = \sum(X_i - \overline{X})^2$ is given by

$$f_{\sigma^2}(z) = \frac{\sigma^{-(n-1)}}{\Gamma[(n-1)/2]2^{(n-1)/2}} z^{(n-3)/2} e^{-z/2\sigma^2}, \qquad z > 0.$$

The family of densities $\{f_{\sigma^2} : \sigma^2 > 0\}$ has an MLR in z, and it follows that a UMP test is to reject $H_0 : \sigma^2 \geq \sigma_0^2$ if $z \leq k$, that is, a UMP invariant test is given by

$$\varphi(\mathbf{x}) = \begin{cases} 1 & \text{if } \sum(x_i - \overline{x})^2 \leq k, \\ 0 & \text{if } \sum(x_i - \overline{x})^2 > k, \end{cases}$$

where k is determined from the size restriction

$$\alpha = P_{\sigma_0}\left\{\sum(X_i - \overline{X})^2 \leq k\right\} = P\left\{\frac{\sum(X_i - \overline{X})^2}{\sigma_0^2} \leq \frac{k}{\sigma_0^2}\right\},$$

that is,

$$k = \sigma_0^2 \chi^2_{n-1,1-\alpha}.$$

Example 4. Let $\mathbf{X}$ have PDF $f_i(x_1-\theta,\ldots,x_n-\theta)$ under H_i $(i=0,1)$, $-\infty<\theta<\infty$. Let $\mathcal{G}$ be the group of translations

$$\mathbf{g}_c(\mathbf{x})=(x_1+c,\ldots,x_n+c),\qquad -\infty<c<\infty,\quad n\geq 2.$$

Clearly, g induces $\overline{g}$ on Θ, where $\overline{g}\theta=\theta+c$. The hypothesis testing problem remains invariant under $\mathcal{G}$. A maximal invariant under $\mathcal{G}$ is $T(\mathbf{X})=(X_1-X_n,\ldots,X_{n-1}-X_n)=(T_1,T_2,\ldots,T_{n-1})$. The class of invariant tests coincides with the class of tests that are functions of T. The PDF of T under H_i is independent of θ and is given by $\int_{-\infty}^{\infty}f_i(t_1+z,\ldots,t_{n-1}+z,z)\,dz$. The problem is thus reduced to testing a simple hypothesis against a simple alternative. By the fundamental lemma the MP test

$$\varphi(t_1,t_2,\ldots,t_{n-1})=\begin{cases}1 & \text{if } \lambda(\mathbf{t})>c,\\ 0 & \text{if } \lambda(\mathbf{t})<c,\end{cases}$$

where $\mathbf{t}=(t_1,t_2,\ldots,t_{n-1})$ and

$$\lambda(\mathbf{t})=\frac{\displaystyle\int_{-\infty}^{\infty}f_1(t_1+z,\ldots,t_{n-1}+z,z)\,dz}{\displaystyle\int_{-\infty}^{\infty}f_0(t_1+z,\ldots,t_{n-1}+z,z)\,dz},$$

is UMP invariant.

A particular case of Example 4 will be, for instance, to test $H_0\colon X\sim\mathcal{N}(\theta,1)$ against $H_1\colon X\sim\mathcal{C}(1,\theta),\theta\in\mathcal{R}$. See Problem 1.

Example 5. Suppose (X,Y) has joint PDF

$$f_{\boldsymbol{\theta}}(x,y)=\lambda\mu\exp\{-\lambda x-\mu y\},\ x>0,\ y>0,$$

and $=0$ elsewhere, where $\boldsymbol{\theta}=(\lambda,\mu)$, $\lambda>0$, $\mu>0$. Consider scale group $\mathcal{G}=\{\{0,c\},\ c>0\}$ which leaves $\{f_{\boldsymbol{\theta}}\}$ invariant. Suppose we wish to test $H_0:\mu\geq\lambda$ against $H_1:\mu<\lambda$. It is easy to see that $\overline{\mathcal{G}}\Theta_0=\Theta_0$ so that $\mathcal{G}$ leaves $(\alpha,\Theta_0,\Theta_1)$ invariant and $T=Y/X$ is maximal invariant. The PDF of T is given by

$$f_{\boldsymbol{\theta}}^{T}(t)=\frac{\lambda\mu}{(\lambda+\mu t)^2},\quad t>0,\ =0\text{ for }t<0.$$

The family $\{f_\theta^T\}$ has MLR in T and hence a UMP invariant test of H_0 is of the form

$$\varphi(t)=\begin{cases}1, & t>c(\alpha),\\ \gamma, & t=c(\alpha),\\ 0, & t<c(\alpha),\end{cases}$$

where

$$\alpha=\int_{c(\alpha)}^{\infty}\frac{1}{(1+t)^2}dt\Rightarrow c(\alpha)=\frac{1-\alpha}{\alpha}.$$

PROBLEMS 9.5

1. To test $H_0: X \sim \mathcal{N}(\theta, 1)$ against $H_1: X \sim \mathcal{C}(1, \theta)$ a sample of size 2 is available on X. Find a UMP invariant test of H_0 against H_1.
2. Let $X_1, X_2, \ldots, X_n$ be a sample from $P(\lambda)$. Find a UMP unbiased size α test for the null hypothesis $H_0: \lambda \leq \lambda_0$ against alternatives $\lambda > \lambda_0$ by the methods of this section.
3. Let $X \sim NB(1; \theta)$. By the methods of this section find a UMP unbiased size α test of $H_0: \theta \geq \theta_0$ against $H_1: \theta < \theta_0$.
4. Let $X_1, X_2, \ldots, X_n$ iid $\mathcal{N}(\mu, \sigma^2)$ RVs. Consider the problem of testing $H_0: \mu \leq 0$ against $H_1: \mu > 0$:
   (a) It suffices to restrict attention to sufficient statistic (U, V) where $U = \overline{X}$ and $V = S^2$. Show that the problem of testing H_0 is invariant under $\mathcal{G} = \{\{a, 1\}, a \in \mathcal{R}\}$ and a maximal invariant is $T = U/\sqrt{V}$.
   (b) Show that the distribution of T has MLR and a UMP invariant test rejects H_0 when $T > c$.
5. Let $X_1, X_2, \ldots, X_n$ be iid RVs and let H_0 be that $X_i \sim \mathcal{N}(\theta, 1)$, and H_1 be that the common PDF is $f_\theta(x) = (1/2)\exp\{-|x - \theta|\}$. Find the form of the UMP invariant test of H_0 against H_1.
6. Let $X_1, X_2, \ldots, X_n$ be iid RVs and suppose $H_0: X_i \sim \mathcal{N}(0, 1)$ and $H_1: X_i \sim f_1(x) = \exp\{-|x|\}/2$:
   (a) Show that the problem of testing H_0 against H_1 is invariant under scale changes $g_c(\mathbf{x}) = c\mathbf{x}$, $c > 0$ and a maximal invariant is $T(\mathbf{X}) = (X_1/X_n, \ldots, X_{n-1}/X_n)$.
   (b) Show that the MP invariant test rejects H_0 when $\sqrt{1 + \sum_{i=1}^{n-1} Y_i^2} \Big/ \left[1 + \sum_{i=1}^{n+1} |Y_i|\right]$ $< k$ where $Y_j = X_j/X_n, j = 1, 2, \ldots, n-1$, or equivalently when

$$\frac{\left(\sum_{j=1}^{n} X_j^2\right)^{1/2}}{\sum_{j=1}^{n} |X_j|} < k.$$

9.6 LOCALLY MOST POWERFUL TESTS

In the previous section we argued that whenever a UMP test does not exist, we restrict the class of tests under consideration and then find a UMP test in the subclass. Yet another approach when no UMP test exists is to restrict the parameter set to a subset of Θ_1. In most problems, the parameter values that are close to the null hypothesis are the hardest to detect. Tests that have good power properties for "local alternatives" may also retain good power properties for "nonlocal" alternatives.

Definition 1. Let $\Theta \subseteq \mathcal{R}$. Then a test φ_0 with power function $\beta_{\varphi_0}(\theta) = E_\theta \varphi_0(\mathbf{X})$ is said to be a *locally most powerful* (LMP) test of $H_0 : \theta \le \theta_0$ against $H_1 : \theta > \theta_0$ if there exists a $\Delta > 0$ such that for any other test φ with

$$\beta_\varphi(\theta_0) = \beta_{\varphi_0}(\theta_0) = \int \varphi(\mathbf{x}) f_{\theta_0}(\mathbf{x}) d\mathbf{x} \tag{1}$$

$$\beta_{\varphi_0}(\theta) \ge \beta_\varphi(\theta) \text{ for every } \theta \in (\theta_0, \theta_0 + \Delta]. \tag{2}$$

We assume that the tests under consideration have continuously differentiable power function at $\theta = \theta_0$ and the derivative may be taken under the integral sign. In that case, an LMP test maximizes

$$\frac{\partial}{\partial\theta}\beta_\varphi(\theta)\Big|_{\theta=\theta_0} = \beta'_\varphi(\theta)\Big|_{\theta=\theta_0} = \int \varphi(\mathbf{x}) \frac{\partial}{\partial\theta} f_\theta(\mathbf{x})\Big|_{\theta=\theta_0} d\mathbf{x} \tag{3}$$

subject to the size constraint (1). A slight extension of the Neyman–Pearson lemma (Remark 9.3.2) implies that a test satisfying (1) and given by

$$\varphi_0(\mathbf{x}) = \begin{cases} 1 & \text{if } \frac{\partial}{\partial\theta} f_\theta(\mathbf{x})\Big|_{\theta_0} > k f_{\theta_0}(\mathbf{x}), \\ \gamma & \text{if } \frac{\partial}{\partial\theta} f_\theta(\mathbf{x})\Big|_{\theta_0} = k f_{\theta_0}(\mathbf{x}), \\ 0 & \text{if } \frac{\partial}{\partial\theta} f_\theta(\mathbf{x})\Big|_{\theta_0} < k f_{\theta_0}(\mathbf{x}) \end{cases} \tag{4}$$

will maximize $\beta'_\varphi(\theta_0)$. It is possible that a test that maximizes $\beta'_\varphi(\theta_0)$ is not LMP, but if the test maximizes $\beta'(\theta_0)$ and is unique then it must be LMP test (see Kallenberg et al. [49, p. 290] and Lehmann [64, p. 528]).

Note that for x for which $f_{\theta_0}(\mathbf{x}) \ne 0$ we can write

$$\frac{\frac{\partial}{\partial\theta} f_\theta(\mathbf{x})\Big|_{\theta_0}}{f_{\theta_0}(\mathbf{x})} = \frac{\partial}{\partial\theta} \log f_\theta(\mathbf{x})\big|_{\theta_0},$$

and then

$$\varphi_0(\mathbf{x}) = \begin{cases} 1 & \text{if } \frac{\partial}{\partial\theta} \log f_\theta(\mathbf{x})\Big|_{\theta_0} > k, \\ \gamma & \text{if } \frac{\partial}{\partial\theta} \log f_\theta(\mathbf{x})\Big|_{\theta_0} = k, \\ 0 & \text{if } \frac{\partial}{\partial\theta} \log f_\theta(\mathbf{x})\Big|_{\theta_0} < k. \end{cases} \tag{5}$$

Example 1. Let $X_1, X_2, \ldots, X_n$ be iid with common normal PDF with mean μ and variance σ^2. If one of these parameters is unknown while the other is known, the family of PDFs has MLR and UMP tests exist for one-sided hypotheses for the unknown parameter. Let us derive the LMP test in each case.

First consider the case when σ^2 is known, say $\sigma^2 = 1$ and $H_0 : \mu \le 0$, $H_1 : \mu > 0$. An easy computation shows that an LMP test is of the form

$$\varphi_0(\mathbf{x}) = \begin{cases} 1 & \text{if } \overline{x} > k \\ 0 & \text{if } \overline{x} \le k \end{cases}$$

which, of course, is the form of the UMP test obtained in Problem 9.4.1 by an application of Theorem 9.4.2.

Next consider the case when μ is known, say $\mu = 0$ and $H_0 : \sigma \le \sigma_0$, $H_1 : \sigma > \sigma_0$. Using (5) we see that an LMP test is of the form

$$\varphi_1(\mathbf{x}) = \begin{cases} 1 & \text{if } \sum_{i=1}^n x_i^2 > k \\ 0 & \text{if } \sum_{i=1}^n x_i^2 \le k \end{cases}$$

which coincides with the UMP test.

In each case the power function is differentiable and the derivatives may be taken inside the integral sign because the PDF is a one–parameter exponential type PDF.

Example 2. Let $X_1, X_2, \ldots, X_n$ be iid RVs with common PDF

$$f_\theta(x) = \frac{1}{\pi}\frac{1}{1+(x-\theta)^2}, \quad x \in \mathcal{R},$$

and consider the problem of testing $H_0 : \theta \le 0$ against $H_1 : \theta > 0$.

In this case $\{f_\theta\}$ does not have MLR. A direct computation using the Neyman–Pearson lemma shows that an MP test of $\theta = 0$ against $\theta = \theta_1$, $\theta_1 > 0$ depends on θ_1 and hence cannot be MP for testing $\theta = 0$ against $\theta = \theta_2$, $\theta_2 \ne \theta_1$. Hence a UMP test of H_0 against H_1 does not exist. An LMP test of H_0 against H_1 is of the form

$$\varphi_0(\mathbf{x}) = \begin{cases} 1 & \text{if } \sum_{i=1}^n \frac{2x_i}{1+x_i^2} > k \\ 0 & \text{otherwise,} \end{cases}$$

where k is chosen so that the size of φ_0 is α. For small n it is hard to compute k but for large n it is easy to compute k using the central limit theorem. Indeed $\{\frac{X_i}{1+X_i^2}\}$ are iid RVs with mean 0 and finite variance $(= 3/8)$ so that $k = z_\alpha\sqrt{n/2}$ will give an (approximate) level α test for large n.

The test φ_0 is good at detecting small departures from $\theta \le 0$ but it is quite unsatisfactory in detecting values of θ away from 0. In fact, for $\alpha < 1/2$, $\beta_{\varphi_0}(\theta) \to 0$ as $\theta \to \infty$.

This procedure for finding locally best tests has applications in nonparametric statistics. We refer the reader to Randles and Wolfe [85, section 9.1] for details.

PROBLEMS 9.6

1. Let $X_1, X_2, \ldots, X_n$ be iid $\mathcal{C}(1,\theta)$ RVs. Show that $E_0(1+X_1^2)^{-k} = (1/\pi)B(k+1/2, 1/2)$. Hence or otherwise show that $E_0\left[\frac{X_1^2}{(1+X_1^2)^2}\right] = \text{var}\left(\frac{X_1}{1+X_1^2}\right) = 1/8$.
2. Let $X_1, X_2, \ldots, X_n$ be a random sample from logistic PDF

$$f_\theta(x) = \frac{1}{2[1+\cosh(x-\theta)]} = \frac{e^{x-\theta}}{\{1+e^{x-\theta}\}^2}.$$

Show that the LMP test of $H_0: \theta = 0$ against $H_1: \theta > 0$ rejects H_0 if $\sum_{i=1}^n \tanh\left(\frac{x_i}{2}\right) > k$.

3. Let $X_1, X_2, \ldots, X_n$ be iid RVs with common Laplace PDF

$$f_\theta(x) = (1/2)\exp\{-|x-\theta|\}.$$

For $n \geq 2$ show that UMP size α $(0 < \alpha < 1)$ test of $H_0: \theta \leq 0$ against $H_1: \theta > 0$ does not exist. Find the form of the LMP test.

10

SOME FURTHER RESULTS ON HYPOTHESES TESTING

10.1 INTRODUCTION

In this chapter we study some commonly used procedures in the theory of testing of hypotheses. In Section 10.2 we describe the classical procedure for constructing tests based on likelihood ratios. This method is sufficiently general to apply to multi-parameter problems and is specially useful in the presence of *nuisance parameters*. These are unknown parameters in the model which are of no inferential interest. Most of the normal theory tests described in Sections 10.3 to 10.5 and those in Chapter 12 can be derived by using methods of Section 10.2. In Sections 10.3 to 10.5 we list some commonly used normal theory-based tests. In Section 10.3 we also deal with goodness-of-fit tests. In Section 10.6 we look at the hypothesis testing problem from a decision-theoretic viewpoint and describe Bayes and minimax tests.

10.2 GENERALIZED LIKELIHOOD RATIO TESTS

In Chapter 9 we saw that UMP tests do not exist for some problems of hypothesis testing. It was suggested that we restrict attention to smaller classes of tests and seek UMP tests in these subclasses or, alternatively, seek tests which are optimal against local alternatives. Unfortunately, some of the reductions suggested in Chapter 9, such as invariance, do not apply to all families of distributions.

In this section we consider a classical procedure for constructing tests that has some intuitive appeal and that frequently, though not necessarily, leads to optimal

An Introduction to Probability and Statistics, Third Edition. Vijay K. Rohatgi and A.K. Md. Ehsanes Saleh.

tests. Also, the procedure leads to tests that have some desirable large-sample properties.

Recall that for testing $H_0 : \mathbf{X} \sim f_0$ against $H_1 : \mathbf{X} \sim f_1$, Neyman–Pearson MP test is based on the ratio $f_1(\mathbf{x})/f_0(\mathbf{x})$. If we interpret the numerator as the best possible explanation of $\mathbf{x}$ under H_1, and the denominator as the best possible explanation of $\mathbf{X}$ under H_0, then it is reasonable to consider the ratio

$$r(\mathbf{x}) = \frac{\sup_{\boldsymbol{\theta}\in\Theta_1} L(\theta;\mathbf{x})}{\sup_{\boldsymbol{\theta}\in\Theta_0} L(\theta;\mathbf{x})} = \frac{\sup_{\boldsymbol{\theta}\in\Theta_1} f_{\theta}(\mathbf{x})}{\sup_{\boldsymbol{\theta}\in\Theta_0} f_{\theta}(\mathbf{x})}$$

as a test statistic for testing $H_0 : \boldsymbol{\theta} \in \Theta_0$ against $H_1 : \boldsymbol{\theta} \in \Theta_1$. Here $L(\boldsymbol{\theta};\mathbf{x})$ is the likelihood function of $\mathbf{x}$. Note that for each $\mathbf{x}$ for which the MLEs of $\boldsymbol{\theta}$ under Θ_1 and Θ_0 exist the ratio is well defined and free of $\boldsymbol{\theta}$ and can be used as a test statistic. Clearly we should reject H_0 if $r(\mathbf{x}) > c$.

The statistic r is hard to compute; only one of the two supremas in the ratio may be attained.

Let $\boldsymbol{\theta} \in \Theta \subseteq \mathcal{R}_k$ be a vector of parameters, and let $\mathbf{X}$ be a random vector with PDF (PMF) $f_{\boldsymbol{\theta}}$. Consider the problem of testing the null hypothesis $H_0 : \mathbf{X} \sim f_{\boldsymbol{\theta}}, \boldsymbol{\theta} \in \Theta_0$ against the alternative $H_1 : \mathbf{X} \sim f_{\boldsymbol{\theta}}, \boldsymbol{\theta} \in \Theta_1$.

Definition 1. For testing H_0 against H_1, a test of the form, reject H_0 if and only if $\lambda(\mathbf{x}) < c$, where c is some constant, and

$$\lambda(\mathbf{x}) = \frac{\sup\limits_{\boldsymbol{\theta}\in\Theta_0} f_{\boldsymbol{\theta}}(x_1,x_2,\ldots,x_n)}{\sup\limits_{\boldsymbol{\theta}\in\Theta} f_{\boldsymbol{\theta}}(x_1,x_2,\ldots,x_n)},$$

is called a *generalized likelihood ratio* (GLR) test.

We leave the reader to show that the statistics $\lambda(\mathbf{X})$ and $r(\mathbf{X})$ lead to the same criterion for rejecting H_0.

The numerator of the likelihood ratio λ is the best *explanation* of $\mathbf{X}$ (in the sense of maximum likelihood) that the null hypothesis H_0 can provide, and the denominator is the best possible explanation of $\mathbf{X}$. H_0 is rejected if there is a much better explanation of $\mathbf{X}$ than the best one provided by H_0.

It is clear that $0 \leq \lambda \leq 1$. The constant c is determined from the size restriction

$$\sup_{\boldsymbol{\theta}\in\Theta_0} P_{\boldsymbol{\theta}}\{\lambda(\mathbf{X}) < c\} = \alpha.$$

If the distribution of λ is continuous (that is, the DF is absolutely continuous), any size α is attainable. If, however, $\lambda(\mathbf{X})$ is a discrete RV, it may not be possible to find a likelihood ratio test whose size exactly equals α. This problem arises because of the nonrandomized nature of the likelihood ratio test and can be handled by randomization. The following result holds.

Theorem 1. If for given α, $0 \leq \alpha \leq 1$, nonrandomized Neyman–Pearson and likelihood ratio tests of a simple hypothesis against a simple alternative exist, they are equivalent.

Proof. The proof is left as an exercise.

Theorem 2. For testing $\boldsymbol{\theta} \in \Theta_0$ against $\boldsymbol{\theta} \in \Theta_1$, the likelihood ratio test is a function of every sufficient statistic for $\boldsymbol{\theta}$.

Theorem 2 follows from the factorization theorem for sufficient statistics.

Example 1. Let $X \sim b(n,p)$, and we seek a level α likelihood ratio test of $H_0\colon p \leq p_0$ against $H_1 : p > p_0$:

$$\lambda(x) = \frac{\sup\limits_{p \leq p_0} \binom{n}{x} p^x (1-p)^{n-x}}{\sup\limits_{0 \leq p \leq 1} \binom{n}{x} p^x (1-p)^{n-x}}.$$

Now

$$\sup_{0 \leq p \leq 1} p^x (1-p)^{n-x} = \left(\frac{x}{n}\right)^x \left(1 - \frac{x}{n}\right)^{n-x}.$$

The function $p^x(1-p)^{n-x}$ first increases, then achieves its maximum at $p = x/n$, and finally decreases, so that

$$\sup_{p \leq p_0} p^x (1-p)^{n-x} = \begin{cases} p_0^x (1-p_0)^{n-x} & \text{if } p_0 < \dfrac{x}{n}, \\ \left(\dfrac{x}{n}\right)^x \left(1 - \dfrac{x}{n}\right)^{n-x} & \text{if } \dfrac{x}{n} \leq p_0. \end{cases}$$

It follows that

$$\lambda(x) = \begin{cases} \dfrac{p_0^x (1-p_0)^{n-x}}{(x/n)^x [1-(x/n)]^{n-x}} & \text{if } p_0 < \dfrac{x}{n}, \\ 1 & \text{if } \dfrac{x}{n} \leq p_0. \end{cases}$$

Note that $\lambda(x) \leq 1$ for $np_0 < x$ and $\lambda(x) = 1$ if $x \leq np_0$, and it follows that $\lambda(x)$ is a decreasing function of x. Thus $\lambda(x) < c$ if and only if $x > c'$, and the GLR test rejects H_0 if $x > c'$.

The GLR test is of the type obtained in Section 9.4 for families with an MLR except for the boundary $\lambda(x) = c$. In other words, if the size of the test happens to be exactly α, the likelihood ratio test is a UMP level α test. Since X is a discrete RV, however, to obtain size α may not be possible. We have

$$\alpha = \sup_{p \leq p_0} P_p\{X > c'\} = P_{p_0}\{X > c'\}.$$

If such a c' does not exist, we choose an integer c' such that

$$P_{p_0}\{X > c'\} \leq \alpha \qquad \text{and} \qquad P_{p_0}\{> c' - 1\} > \alpha.$$

The situation in Example 1 is not unique. For one-parameter exponential family it can be shown (Birkes [7]) that a GLR test of $H_0 : \theta \leq \theta_0$ against $H_1 : \theta > \theta_0$ is UMP of its size. The result holds also for the dual $H_0': \theta \geq \theta_0$ and, in fact, for a much wider class of one-parameter family of distributions.

The GLR test is specially useful when $\boldsymbol{\theta}$ is a multiparameter and we wish to test hypothesis concerning one of the parameters. The remaining parameters act as nuisance parameters.

Example 2. Consider the problem of testing $\mu = \mu_0$ against $\mu \neq \mu_0$ in sampling from $\mathcal{N}(\mu, \sigma^2)$, where both μ and σ^2 are unknown. In this case $\Theta_0 = \{(\mu_0, \sigma^2)\colon \sigma^2 > 0\}$ and $\Theta = \{(\mu, \sigma^2)\colon -\infty < \mu < \infty, \sigma^2 > 0\}$. We write $\boldsymbol{\theta} = (\mu, \sigma^2)$:

$$\begin{aligned}\sup_{\boldsymbol{\theta} \in \Theta_0} f_{\boldsymbol{\theta}}(\mathbf{x}) &= \sup_{\sigma^2 > 0} \left[\frac{1}{(\sigma\sqrt{2\pi})^n} \exp\left\{-\frac{\sum_1^n (x_i - \mu_0)^2}{2\sigma^2}\right\}\right] \\ &= f_{\hat{\sigma}_0^2}(\mathbf{x}),\end{aligned}$$

where $\hat{\sigma}_0^2$ is the MLE, $\hat{\sigma}_0^2 = (1/n)\sum_{i=1}^n (x_i - \mu_0)^2$. Thus

$$\sup_{\boldsymbol{\theta}\Theta_0} f_{\boldsymbol{\theta}}(\mathbf{x}) = \frac{1}{(2\pi/n)^{n/2} \left\{\sum_1^n (x_i - \mu_0)^2\right\}^{n/2}} e^{-n/2}.$$

The MLE of $\boldsymbol{\theta} = (\mu, \sigma^2)$ when both μ and σ^2 are unknown is $(\sum_1^n x_i/n, \sum_1^n (x_i - \bar{x})^2/n)$. It follows that

$$\begin{aligned}\sup_{\boldsymbol{\theta} \in \Theta} f_{\boldsymbol{\theta}}(\mathbf{x}) &= \sup_{\mu, \sigma^2} \left[\frac{1}{(\sigma\sqrt{2\pi})^n} \exp\left\{-\frac{\sum_1^n (x_i - \mu)^2}{2\sigma^2}\right\}\right] \\ &= \frac{1}{(2\pi/n)^{n/2} \left\{\sum_1^n (x_i - \bar{x})^2\right\}^{n/2}} e^{-n/2}.\end{aligned}$$

Thus

$$\begin{aligned}\lambda(\mathbf{x}) &= \left\{\frac{\sum_1^n (x_i - \bar{x})^2}{\sum_1^n (x_i - \mu_0)^2}\right\}^{n/2} \\ &= \left\{\frac{1}{1 + [n(\bar{x} - \mu_0)^2 / \sum_1^n (x_i - \bar{x})^2]}\right\}^{n/2}.\end{aligned}$$

The GLR test rejects H_0 if

$$\lambda(\mathbf{x}) < c,$$

and since $\lambda(\mathbf{x})$ is a decreasing function of $n(\bar{x} - \mu_0)^2 / \sum_1^n n(x_i - \bar{x})^2$, we reject H_0 if

$$\left|\frac{\bar{x} - \mu_0}{\sqrt{\sum_1^n (x_i - \bar{x})^2}}\right| > c',$$

that is, if

$$\left|\frac{\sqrt{n}(\bar{x} - \mu_0)}{s}\right| > c'',$$

where $s^2 = (n-1)^{-1}\sum_1^n (x_i - \bar{x})^2$. The statistic

$$t(\mathbf{X}) = \frac{\sqrt{n}(\overline{X} - \mu_0)}{S}$$

has a t-distribution with $n-1$ d.f. Under $H_0\colon \mu = \mu_0$, $t(\mathbf{X})$ has a central $t(n-1)$ distribution, but under $H_1 : \mu \neq \mu_0$, $t(\mathbf{X})$ has a noncentral t-distribution with $n-1$ d.f. and noncentrality parameter $\delta = (\mu - \mu_0)/\sigma$. We choose $c'' = t_{n-1,\alpha/2}$ in accordance with the distribution of $t(\mathbf{X})$ under H_0. Note that the two-sided t-test obtained here is UMP unbiased. Similarly one can obtain one-sided t-tests also as likelihood ratio tests.

The computations in Example 2 could be slightly simplified by using Theorem 2. Indeed $T(X) = (\overline{X}, S^2)$ is a minimal sufficient statistic for θ and since $\overline{X}$ and S^2 are independent the likelihood is the product of the PDFs of $\overline{X}$ and S^2. We note that $\overline{X} \sim \mathcal{N}(\mu, \sigma^2/n)$ and $S^2 \sim \frac{\sigma^2}{n-1}\chi^2_{n-1}$. We leave it to the reader to carry out the details.

Example 3. Let $X_1, X_2, \ldots, X_m$ and $Y_1, Y_2, \ldots, Y_n$ be independent random samples from $\mathcal{N}(\mu_1, \sigma_1^2)$ and $\mathcal{N}(\mu_2, \sigma_2^2)$, respectively. We wish to test the null hypothesis $H_0\colon \sigma_1^2 = \sigma_2^2$ against $H_1 : \sigma_1^2 \neq \sigma_2^2$. Here

$$\Theta = \{(\mu_1, \sigma_1^2, \mu_2, \sigma_2^2)\colon -\infty < \mu_i < \infty, \sigma_i^2 > 0, i = 1, 2\}$$

and

$$\Theta_0 = \{(\mu_1, \sigma_1^2, \mu_2, \sigma_2^2)\colon -\infty < \mu_i < \infty, i = 1, 2, \sigma_1^2 = \sigma_2^2 > 0\}.$$

Let $\boldsymbol{\theta} = (\mu_1, \sigma_1^2, \mu_2, \sigma_2^2)$. Then the joint PDF is

$$f_{\boldsymbol{\theta}}(\mathbf{x}, \mathbf{y}) = \frac{1}{(2\pi)^{(m+n)/2}\sigma_1^m \sigma_2^n} \exp\left\{-\frac{1}{2\sigma_1^2}\sum_1^m (x_i - \mu_1)^2 - \frac{1}{2\sigma_2^2}\sum_1^n (y_i - \mu_2)^2\right\}.$$

Also,

$$\log f_{\boldsymbol{\theta}}(\mathbf{x}, \mathbf{y}) = -\frac{m+n}{2}\log 2\pi - \frac{m}{2}\log \sigma_1^2 - \frac{n}{2}\log \sigma_2^2 - \frac{\sum_1^m (x_i - \mu_1)^2}{2\sigma_1^2} - \frac{1}{2\sigma_2^2}\sum_1^n (y_i - \mu_2)^2.$$

Differentiating with respect to μ_1 and μ_2, we obtain the MLEs

$$\hat{\mu}_1 = \bar{x} \qquad \text{and} \qquad \hat{\mu}_2 = \bar{y}.$$

Differentiating with respect to σ_1^2 and σ_2^2, we obtain the MLEs

$$\hat{\sigma}_1^2 = \frac{1}{m}\sum_1^m (x_i - \bar{x})^2 \qquad \text{and} \qquad \hat{\sigma}_2^2 = \frac{1}{n}\sum_1^n (y_i - \bar{y})^2.$$

If, however, $\sigma_1^2 = \sigma_2^2 = \sigma^2$, the MLE of σ^2 is

$$\hat{\sigma}^2 = \frac{\sum_1^m (x_i - \overline{x})^2 + \sum_1^n (y_i - \overline{y})^2}{m+n}.$$

Thus

$$\sup_{\boldsymbol{\theta} \in \Theta_0} f_{\boldsymbol{\theta}}(\mathbf{x}, \mathbf{y}) = \frac{e^{-(m+n)/2}}{[2\pi/(m+n)]^{(m+n)/2} \left\{\sum_1^m (x_i - \overline{x})^2 + \sum_1^n (y_i - \overline{y})^2\right\}^{(m+n)/2}}$$

and

$$\sup_{\boldsymbol{\theta} \in \Theta} f_{\boldsymbol{\theta}}(\mathbf{x}, \mathbf{y}) = \frac{e^{-(m+n)/2}}{(2\pi/m)^{m/2}(2\pi/n)^{n/2} \left\{\sum_1^m (x_i - \overline{x})^2\right\}^{m/2} \left\{\sum_1^n (y_i - \overline{y})^2\right\}^{n/2}},$$

so that

$$\lambda(\mathbf{x}, \mathbf{y}) = \left(\frac{m}{m+n}\right)^{m/2} \left(\frac{n}{m+n}\right)^{n/2} \frac{\left\{\sum_1^m (x_i - \overline{x})^2\right\}^{m/2} \left\{\sum_1^n (y_i - \overline{y})^2\right\}^{n/2}}{\left\{\sum_1^m (x_i - \overline{x})^2 + \sum_1^n (y_i - \overline{y})^2\right\}^{(m+n)/2}}.$$

Now

$$\frac{\left\{\sum_1^m (x_i - \overline{x})^2\right\}^{m/2} \left\{\sum_1^n (y_i - \overline{y})^2\right\}^{n/2}}{\left\{\sum_1^m (x_i - \overline{x})^2 + \sum_1^n (y_i - \overline{y})^2\right\}^{(m+n)/2}}$$

$$= \frac{1}{\left\{1 + \sum_1^m (x_i - \overline{x})^2 / \sum_1^n (y_i - \overline{y})^2\right\}^{n/2} \left\{1 + \sum_1^n (y_i - \overline{y})^2 / \sum_1^m (x_i - \overline{x})^2\right\}^{m/2}}.$$

Writing

$$f = \frac{\sum_1^m (x_i - \overline{x})^2/(m-1)}{\sum_1^n (y_i - \overline{y})^2/(n-1)},$$

we have

$$\lambda(\mathbf{x}, \mathbf{y}) = \left(\frac{m}{m+n}\right)^{m/2} \left(\frac{n}{m+n}\right)^{n/2} \cdot \frac{1}{\{1 + [(m-1)/(n-1)]f\}^{n/2} \{1 + [(n-1)/(m-1)](1/f)\}^{m/2}}.$$

We leave the reader to check that $\lambda(\mathbf{x}, \mathbf{y}) < c$ is equivalent to $f < c_1$ or $f > c_2$. (Take logarithms and use properties of convex functions. Alternatively, differentiate $\log \lambda$.)

Under H_0, the statistic

$$F = \frac{\sum_1^m (X_i - \overline{X})^2/(m-1)}{\sum_1^n (Y_i - \overline{Y})^2/(n-1)}$$

has an $F(m-1,n-1)$ distribution, so that c_1, c_2 can be selected. It is usual to take

$$P\{F \le c_1\} = P\{F \ge c_2\} = \frac{\alpha}{2}.$$

Under H_1, $(\sigma_2^2/\sigma_1^2)F$ has an $F(m-1,n-1)$ distribution.

In Example 3 we can obtain the same GLR test by focusing attention on the joint sufficient statistic $(\overline{X},\overline{Y},S_X^2,S_Y^2)$ where S_X^2 and S_Y^2 are sample variances of the X's and the Y's, respectively. In order to write down the likelihood function we note that $\overline{X}$, $\overline{Y}$, S_X^2, S_Y^2 are independent RVs. The distributions $\overline{X}$ and S_X^2 are the same as in Example 2 except that m is the sample size. Distributions of $\overline{Y}$ and S_Y^2 require appropriate modifications. We leave the reader to carry out the details. It turns out that the GLR test coincides with the UMP unbiased test in this case.

In certain situations the GLR test does not perform well. We reproduce here an example due to Stein and Rubin.

Example 4. Let X be a discrete RV with PMF

$$P_{p=0}\{X=x\} = \begin{cases} \dfrac{\alpha}{2} & \text{if } x = \pm 2, \\ \dfrac{1-2\alpha}{2} & \text{if } x = \pm 1, \\ \alpha & \text{if } x = 0, \end{cases}$$

under the null hypothesis $H_0\colon p=0$, and

$$P_p\{X=x\} = \begin{cases} pc & \text{if } x = -2, \\ \dfrac{1-c}{1-\alpha}\left(\dfrac{1}{2}-\alpha\right) & \text{if } x = \pm 1, \\ \alpha\left(\dfrac{1-c}{1-\alpha}\right) & \text{if } x = 0, \\ (1-p)c & \text{if } x = 2, \end{cases}$$

under the alternative $H_1: p \in (0,1)$, where α and c are constants with

$$0 < \alpha < \frac{1}{2} \qquad \text{and} \qquad \frac{\alpha}{2-\alpha} < c < \alpha.$$

To test the simple null hypothesis against the composite alternative at the level of significance α, let us compute the likelihood ratio λ. We have

$$\lambda(2) = \frac{P_0\{X=2\}}{\sup\limits_{0\le p<1} P_p\{X=2\}} = \frac{\alpha/2}{c} = \frac{\alpha}{2c}$$

since $\alpha/2 < c$. Similarly $\lambda(-2) = \alpha/(2c)$. Also

$$\lambda(1) = \lambda(-1) = \frac{\frac{1}{2}-\alpha}{[(1-c)/(1-\alpha)]\left(\frac{1}{2}-\alpha\right)} = \frac{1-\alpha}{1-c}, \qquad \alpha < \frac{1}{2},$$

and

$$\lambda(0) = \frac{1-\alpha}{1-c}.$$

The GLR test rejects H_0 if $\lambda(x) < k$, where k is to be determined so that the level is α. We see that

$$P_0\left\{\lambda(X) < \frac{1-\alpha}{1-c}\right\} = P_0\{X = \pm 2\} = \alpha,$$

provided that $\alpha/2c < [(1-\alpha)/(1-c)]$. But $\alpha/(2-\alpha) < c < \alpha$ implies $\alpha < 2c - c\alpha$, so that $\alpha - c\alpha < 2c - 2c\alpha$, or $\alpha(1-c) < 2c(1-\alpha)$, as required. Thus the GLR size α test is to reject H_0 if $X = \pm 2$. The power of the GLR test is

$$P_p\left\{\lambda(X) < \frac{1-\alpha}{1-c}\right\} = P_p\{X = \pm 2\} = pc + (1-p)c = c < \alpha$$

for all $p \in (0,1)$. The test is not unbiased and is even worse than the trivial test $\varphi(x) \equiv \alpha$.

Another test that is better than the trivial test is to reject H_0 whenever $x = 0$ (this is opposite to what the likelihood ratio test says). Then

$$P_0\{X=0\} = \alpha,$$

$$P_p\{X=0\} = \alpha\frac{1-c}{1-\alpha} > \alpha \qquad \text{(since } c < \alpha\text{)},$$

for all $p \in (0,1)$, and the test is unbiased.

We will use the generalized likelihood ratio procedure quite frequently hereafter because of its simplicity and wide applicability. The exact distribution of the test statistic under H_0 is generally difficult to obtain (despite what we saw in Examples 1 to 3 above) and evaluation of power function is also not possible in many problems. Recall, however, that under certain conditions the asymptotic distribution of the MLE is normal. This result can be used to prove the following large-sample property of the GLR under H_0, which solves the problem of computation of the cut-off point c at least when the sample size is large.

Theorem 3. Under some regularity conditions on $f_{\boldsymbol{\theta}}(\mathbf{x})$, the random variable $-2\log\lambda(\mathbf{X})$ under H_0 is asymptotically distributed as a chi-square RV with degrees of freedom equal to the difference between the number of independent parameters in Θ and the number in Θ_0.

We will not prove this result here; the reader is referred to Wilks [118, p. 419]. The regularity conditions are essentially the ones associated with Theorem 8.7.4. In Example 2 the number of parameters unspecified under H_0 is one (namely, σ^2), and under H_1 two parameters are unspecified (μ and σ^2), so that the asymptotic chi-square distribution will have 1 d.f. Similarly, in Example 3, the d.f. $= 4 - 3 = 1$.

Example 5. In Example 2 we showed that, in sampling from a normal population with unknown mean μ and unknown variance σ^2, the likelihood ratio for testing $H_0: \mu = \mu_0$ against $H_1: \mu \neq \mu_0$ is

$$\lambda(\mathbf{x}) = \left\{1 + \frac{n(\overline{x}-\mu_0)^2}{\sum_{i=1}^{n}(x_i-\overline{x})^2}\right\}^{-n/2}.$$

Thus

$$-2\log\lambda(\mathbf{X}) = n\log\left\{1 + n\frac{(\overline{X}-\mu_0)^2}{\sum_1^n(X_i-\overline{X})^2}\right\}.$$

Under H_0, $\sqrt{n}(\overline{X}-\mu_0)/\sigma \sim \mathcal{N}(0,1)$ and $\sum_1^n(X_i-\overline{X})^2/\sigma^2 \sim \chi^2(n-1)$. Also $\sum_{i=1}^{n}(X_i-\overline{X})^2/[(n-1)\sigma^2] \xrightarrow{P} 1$. It follows that if $Z \sim \mathcal{N}(0,1)$ then $-2\log\lambda(\mathbf{X})$ has the same limiting distribution as $n\log\left\{1+\frac{Z^2}{n-1}\right\}$. Moreover,

$$\left\{1+\frac{Z^2}{n-1}\right\}^n \xrightarrow{L} \exp\{Z^2\}$$

and since logarithm is a continuous function we see that

$$n\log\left\{1+\frac{Z^2}{n-1}\right\} \xrightarrow{L} Z^2.$$

Thus $-2\log\lambda(X) \xrightarrow{L} Y$, where $Y \sim \chi^2(1)$. This result is consistent with Theorem 3.

PROBLEMS 10.2

1. Prove Theorems 1 and 2.
2. A random sample of size n is taken from PMF $P(X_j = x_j) = p_j$, $j = 1,2,3,4$, $0 < p_j < 1$, $\sum_{j=1}^{4} p_j = 1$. Find the form of the GLR test of $H_0 : p_1 = p_2 = p_3 = p_4 = 1/4$ against $H_1 : p_1 = p_2 = p/2$, $p_3 = p_4 = (1-p)/2$, $0 < p < 1$.
3. Find the GLR test of $H_0 : p = p_0$ against $H_1 : p \neq p_0$, based on a sample of size 1 from $b(n,p)$.
4. Let $X_1, X_2, \ldots, X_n$ be a sample from $\mathcal{N}(\mu,\sigma^2)$, where both μ and σ^2 are unknown. Find the GLR test of $H_0 : \sigma = \sigma_0$ against $H_1 : \sigma \neq \sigma_0$.
5. Let $X_1, X_2, \ldots, X_k$ be a sample from PMF

$$P_N\{X=j\} = \frac{1}{N}, \qquad j = 1,2,\ldots,N,\ N \geq 1 \text{ is an integer.}$$

   (a) Find the GLR test of $H_0 : N \leq N_0$ against $H_1 : N > N_0$.
   (b) Find the GLR test of $H_0 : N = N_0$ against $H_1 : N \neq N_0$.
6. For a sample of size 1 from PDF

$$f_\theta(x) = \frac{2}{\theta^2}(\theta - x), \qquad 0 < x < \theta,$$

   find the GLR test of $\theta = \theta_0$ against $\theta \neq \theta_0$.

7. Let $X_1,X_2,\ldots,X_n$ be a sample from $G(1,\beta)$:
(a) Find the GLR test of $\beta = \beta_0$ against $\beta \neq \beta_0$.
(b) Find the GLR test of $\beta \le \beta_0$ against $\beta > \beta_0$.

8. Let $(X_1,Y_1),(X_2,Y_2),\ldots,(X_n,Y_n)$ be a random sample from a bivariate normal population with $EX_i = \mu_1, EY_i = \mu_2$, $\text{var}(X_i) = \sigma^2$, $\text{var}(Y_i) = \sigma^2$, and $\text{cov}(X_i,Y_i) = \rho\sigma^2$. Show that the likelihood ratio test of the null hypothesis $H_0: \rho = 0$ against $H_1: \rho \neq 0$ reduces to rejecting H_0 if $|R| > c$, where $R = 2S_{11}/(S_1^2 + S_2^2)$, S_{11}, S_1^2, and S_2^2 being the sample covariance and the sample variances, respectively. (For the PDF of the test statistic R, see Problem 7.7.1.)

9. Let $X_1,X_2,\ldots,X_m$ be iid $G(1,\theta)$ RVs and let $Y_1,Y_2,\ldots,Y_n$ be iid $G(1,\mu)$ RVs, where θ and μ are unknown positive real numbers. Assume that the X's and the Y's are independent. Develop an α-level GLR test for testing $H_0: \theta = \mu$ against $H_1: \theta \neq \mu$.

10. A die is tossed 60 times in order to test $H_0: P\{j\} = 1/6, j = 1,2,\ldots,6$ (die is fair) against $H_1: P\{2\} = P\{4\} = P\{6\} = 2/9$, $P\{1\} = P\{3\} = P\{5\} = 1/9$. Find the GLR test.

11. Let $X_1,X_2,\ldots,X_n$ be iid with common PDF $f_\theta(x) = \exp\{-(x-\theta)\}, x > 0$, and $= 0$ otherwise. Find the level α GLR test for testing $H_0: \theta \le \theta_0$ against $H_1: \theta > \theta_0$.

12. Let $X_1,X_2,\ldots,X_n$ be iid RVs with common Pareto PDF $f_\theta(x) = \theta/x^2$ for $x > \theta$, and $= 0$ elsewhere. Show that the family of joint PDFs has MLR in $X_{(1)}$ and find a size α test of $H_0: \theta = \theta_0$ against $H_1: \theta > \theta_0$. Show that the GLR test coincides with the UMP test.

10.3 CHI-SQUARE TESTS

In this section we consider a variety of tests where the test statistic has an exact or a limiting chi-square distribution. Chi-square tests are also used for testing some nonparametric hypotheses and will be taken up again in Chapter 13.

We begin with tests concerning variances in sampling from a normal population. Let $X_1,X_2,\ldots,X_n$ be iid $\mathcal{N}(\mu,\sigma^2)$ RVs where σ^2 is unknown. We wish to test a hypothesis of the type $\sigma^2 \ge \sigma_0^2$, $\sigma^2 \le \sigma_0^2$, or $\sigma^2 = \sigma_0^2$, where σ_0 is some given positive number. We summarize the tests in the following table.

	H_0	H_1	Reject H_0 at level α if: μ Known	μ Unknown
I.	$\sigma \ge \sigma_0$	$\sigma < \sigma_0$	$\sum_1^n (x_i-\mu)^2 \le \chi^2_{n,1-\alpha}\sigma_0^2$	$s^2 \le \frac{\sigma_0^2}{n-1}\chi^2_{n-1,1-\alpha}$
II.	$\sigma \le \sigma_0$	$\sigma > \sigma_0$	$\sum_1^n (x_i-\mu)^2 \ge \chi^2_{n,\alpha}\sigma_0^2$	$s^2 \ge \frac{\sigma_0^2}{n-1}\chi^2_{n-1,\alpha}$
III.	$\sigma = \sigma_0$	$\sigma \neq \sigma_0$	$\sum_1^n (x_i-\mu)^2 \le \chi^2_{n,1-\alpha/2}\sigma_0^2$ or $\sum_1^n (x_i-\mu)^2 \ge \chi^2_{n,\alpha/2}\sigma_0^2$	$s^2 \le \frac{\sigma_0^2}{n-1}\chi^2_{n-1,1-\alpha/2}$ or $s^2 \ge \frac{\sigma_0^2}{n-1}\chi^2_{n-1,\alpha/2}$

Remark 1. All these tests can be derived by the standard likelihood ratio procedure. If μ is unknown, tests I and II are UMP unbiased (and UMP invariant). If μ is known, tests I and II are UMP (see Example 9.4.5). For tests III we have chosen constants c_1, c_2 so that each tail has probability $\alpha/2$. This is the customary procedure, even though it destroys the unbiasedness property of the tests, at least for small samples.

Example 1. A manufacturer claims that the lifetime of a certain brand of batteries produced by his factory has a variance of 5000 (hours)^2. A sample of size 26 has a variance of 7200 (hours)^2. Assuming that it is reasonable to treat these data as a random sample from a normal population, let us test the manufacturer's claim at the $\alpha = 0.02$ level. Here $H_0: \sigma^2 = 5000$ is to be tested against $H_1: \sigma^2 \neq 5000$. We reject H_0 if either

$$s^2 = 7200 \leq \frac{\sigma_0^2}{n-1}\chi^2_{n-1,1-\alpha/2} \qquad \text{or} \qquad s^2 > \frac{\sigma_0^2}{n-1}\chi^2_{n-1,\alpha/2}.$$

We have

$$\frac{\sigma_0^2}{n-1}\chi^2_{n-1,1-\alpha/2} = \frac{5000}{25} \times 11.524 = 2304.8$$
$$\frac{\sigma_0^2}{n-1}\chi^2_{n-1,\alpha/2} = \frac{5000}{25} \times 44.314 = 8862.8.$$

Since s^2 is neither ≤ 2304.8 nor ≥ 8862.8, we cannot reject the manufacturer's claim at level 0.02.

A test based on a chi-square statistic is also used for testing the equality of several proportions. Let $X_1, X_2, \ldots, X_k$ be independent RVs with $X_i \sim b(n_i, p_i)$, $i = 1, 2, \ldots, k$, $k \geq 2$.

Theorem 1. The RV $\sum_{i=1}^{k}\{(X_i - n_i p_i)/\sqrt{n_i p_i(1-p_i)}\}^2$ converges in distribution to the $\chi^2(k)$ RV as $n_1, n_2, \ldots, n_k \to \infty$.

Proof. The proof is left as an exercise.

If $n_1, n_2, \ldots, n_k$ are large, we can use Theorem 1 to test $H_0: p_1 = p_2 = \cdots = p_k = p$ against all alternatives. If p is known, we compute

$$y = \sum_{1}^{k}\left\{\frac{x_i - n_i p}{\sqrt{n_i p(1-p)}}\right\}^2,$$

and if $y \geq \chi^2_{k,\alpha}$, we reject H_0. In practice p will be unknown. Let $\mathbf{p} = (p_1, p_2, \ldots, p_k)$. Then the likelihood function is

$$L(\mathbf{p}; x_1, \ldots, x_k) = \prod_{1}^{k}\left\{\binom{n_i}{x_i} p_i^{x_i}(1-p_i)^{n_i - x_i}\right\}$$

so that

$$\log L(\mathbf{p}; x) = \sum_{i=1}^{k} \log \binom{n_i}{x_i} + \sum_{i=1}^{k} x_i \log p_i + \sum_{i=1}^{k} (n_i - x_i) \log(1 - p_i).$$

The MLE $\hat{p}$ of p under H_0 is therefore given by

$$\frac{\sum_1^k x_i}{p} - \frac{\sum_1^k (n_i - x_i)}{1 - p} = 0,$$

that is,

$$\hat{p} = \frac{x_1 + x_2 + \cdots + x_k}{n_1 + n_2 + \cdots + n_k}.$$

Under certain regularity assumptions (see Cramér [17, pp. 426–427]) it can be shown that the statistic

$$Y_1 = \sum_1^k \frac{(X_i - n_i\hat{p})^2}{n_i\hat{p}(1 - \hat{p})} \tag{1}$$

is asymptotically $\chi^2(k-1)$. Thus the test rejects $H_0\colon p_1 = p_2 = \cdots = p_k = p$, p unknown, at level α if $y_1 \geq \chi^2_{k-1,\alpha}$.

It should be remembered that the tests based on Theorem 1 are all large-sample tests and hence not exact, in contrast to the tests concerning the variance discussed above, which are all exact tests. In the case $k = 1$, UMP tests of $p \geq p_0$ and $p \leq p_0$ exist and can be obtained by the MLR method described in Section 9.4. For testing $p = p_0$, the usual test is UMP unbiased.

In the case $k = 2$, if n_1 and n_2 are large, a test based on the normal distribution can be used instead of Theorem 1. In this case the statistic

$$Z = \frac{X_1/n_1 - X_2/n_2}{\sqrt{\hat{p}(1 - \hat{p})(1/n_1 + 1/n_2)}}, \tag{2}$$

where $\hat{p} = (X_1 + X_2)/(n_1 + n_2)$, is asymptotically $\mathcal{N}(0,1)$ under $H_0\colon p_1 = p_2 = p$. If p is known, one uses p instead of $\hat{p}$. It is not too difficult to show that Z^2 is equal to Y_1, so that the two tests are equivalent.

For small samples the so-called *Fisher–Irwin test* is commonly used and is based on the conditional distribution of X_1 given $T = X_1 + X_2$. Let $\rho = [p_1(1 - p_2)]/[p_2(1 - p_1)]$. Then

$$\begin{aligned} P(X_1 + X_2 = t) &= \sum_{j=0}^{t} \binom{n_1}{j} p_1^j (1 - p_1)^{n_1 - j} \binom{n_2}{t - j} p_2^{t-j} (1 - p_2)^{n_1 - t + j} \\ &= \sum_{j=0}^{t} \binom{n_1}{j} \binom{n_2}{t - j} \rho^j a(n_1, n_2), \end{aligned}$$

where

$$a(n_1,n_2) = (1-p_1)^{n_1}(1-p_2)^{n_2}\{p_2/(1-p_2)\}^t.$$

It follows that

$$\begin{aligned} P\{X_1 = x|X_1+X_2 = t\} &= \frac{\binom{n_1}{x}p_1^x(1-p_1)^{n_1-x}\binom{n_2}{t-x}p_2^{t-x}(1-p_2)^{n_2-t+x}}{a(n_1,n_2)\sum_{j=0}^{t}\binom{n_1}{j}\binom{n_2}{t-j}\rho^j} \\ &= \frac{\binom{n_1}{j}\binom{n_2}{t-j}\rho^x}{\sum_{j=0}^{t}\binom{n_1}{j}\binom{n_2}{t-j}\rho^j}. \end{aligned}$$

On the boundary of any of the hypotheses $p_1 = p_2$, $p_1 \leq p_2$, or $p_1 \geq p_2$ we note that $\rho = 1$ so that

$$P\{X_1 = x|X_1+X_2 = t\} = \frac{\binom{n_1}{x}\binom{n_2}{t-x}}{\binom{n_1+n_2}{t}},$$

which is a hypergeometric distribution. For testing $H_0 : p_1 \leq p_2$ this conditional test rejects if $X_1 \leq k(t)$, where $k(t)$ is the largest integer for which $P\{X_1 \leq k(T)|T = t\} \leq \alpha$. Obvious modifications yield critical regions for testing $p_1 = p_2$, and $p_1 \geq p_2$ against corresponding alternatives.

In applications a wide variety of problems can be reduced to the multinomial distribution model. We therefore consider the problem of testing the parameters of a multinomial distribution. Let $(X_1, X_2, \ldots, X_{k-1})$ be a sample from a multinomial distribution with parameters n, $p_1, p_2, \ldots, p_{k-1}$, and let us write $X_k = n - X_1 - \cdots - X_{k-1}$, and $p_k = 1 - p_1 - \cdots - p_{k-1}$. The difference between the model of Theorem 1 and the multinomial model is the independence of the X_i's.

Theorem 2. Let $(X_1, X_2, \ldots, X_{k-1})$ be a multinomial RV with parameters n, $p_1, p_2, \ldots, p_{k-1}$. Then the RV

$$U_k = \sum_{i=1}^{k}\left\{\frac{(X_i - np_i)^2}{np_i}\right\} \tag{3}$$

is asymptotically distributed as a $\chi^2(k-1)$ RV (as $n \to \infty$).

Proof. For the general proof we refer the reader to Cramér [17, pp. 417–419] or Ferguson [29, p. 61]. We will consider here the $k = 2$ case to make the result a little more plausible. We have

$$\begin{aligned} U_2 &= \frac{(X_1 - np_1)^2}{np_1} + \frac{(X_2 - np_2)^2}{np_2} = \frac{(X_1 - np_1)^2}{np_1} + \frac{[n - X_1 - n(1-p_1)]^2}{n(1-p_1)} \\ &= (X_1 - np_1)^2\left[\frac{1}{np_1} + \frac{1}{n(1-p_1)}\right] \\ &= \frac{(X_1 - np_1)^2}{np_1(1-p_1)}. \end{aligned}$$

It follows from Theorem 1 that $U_2 \xrightarrow{L} Y$ as $n \to \infty$, where $Y \sim \chi^2(1)$.

To use Theorem 2 to test $H_0\colon p_1 = p_1', \ldots, p_k = p_k'$, we need only to compute the quantity

$$u = \sum_1^k \left\{ \frac{(x_i - np_i')^2}{np_i'} \right\}$$

from the sample; if n is large, we reject H_0 if $u > \chi^2_{k-1,\alpha}$.

Example 2. A die is rolled 120 times with the following results:

	1	2	3	4	5	6
Frequency:	20	30	20	25	15	10

Let us test the hypothesis that the die is fair at level $\alpha = 0.05$. The null hypothesis is $H_0\colon p_i = \frac{1}{6}$, $i = 1, 2, \ldots, 6$, where p_i is the probability that the face value is i, $1 \le i \le 6$. By Theorem 2 we reject H_0 if

$$u = \sum_1^6 \frac{[x_i - 120\left(\frac{1}{6}\right)]^2}{120\left(\frac{1}{6}\right)} > \chi^2_{5,0.05}.$$

We have

$$u = 0 + \frac{10^2}{20} + 0 + \frac{5^2}{20} + \frac{5^2}{20} + \frac{10^2}{20} = 12.5$$

Since $\chi_{5,0.05} = 11.07$, we reject H_0. Note that, if we choose $\alpha = 0.025$, then $\chi_{5,0.025} = 12.8$, and we cannot reject at this level.

Theorem 2 has much wider applicability, and we will later study its application to contingency tables. Here we consider the application of Theorem 2 to testing the null hypothesis that the DF of an RV X has a specified form.

Theorem 3. Let $X_1, X_2, \ldots, X_n$ be a random sample on X. Also, let $H_0\colon X \sim F$, where the functional form of the DF F is known completely. Consider a collection of disjoint Borel sets $A_1, A_2, \ldots, A_k$ that form a partition of the real line. Let $P\{X \in A_i\} = p_i$, $i = 1, 2, \ldots, k$, and assume that $p_i > 0$ for each i. Let $Y_j =$ number of X_i's in A_j, $j = 1, 2, \ldots, k$, $i = 1, 2, \ldots, n$. Then the joint distribution of $(Y_1, Y_2, \ldots, Y_{k-1})$ is multinomial with parameters n, $p_1, p_2, \ldots, p_{k-1}$. Clearly, $Y_k = n - Y_1 - \cdots - Y_{k-1}$ and $p_k = 1 - p_1 - \cdots - p_{k-1}$.

The proof of Theorem 3 is obvious. One frequently selects $A_1, A_2, \ldots, A_k$ as disjoint intervals. Theorem 3 is especially useful when one or more of the parameters associated with the DF F are unknown. In that case the following result is useful.

Theorem 4. Let $H_0\colon X \sim F_{\boldsymbol{\theta}}$, where $\boldsymbol{\theta} = (\theta_1, \theta_2, \ldots, \theta_r)$ is unknown. Let $X_1, X_2, \ldots, X_n$ be independent observations on X, and suppose that the MLEs of $\theta_1, \theta_2, \ldots, \theta_r$ exist and are, respectively, $\hat{\theta}_1, \hat{\theta}_2, \ldots, \hat{\theta}_r$. Let $A_1, A_2, \ldots, A_k$ be a collection of disjoint Borel sets that cover the real line, and let

$$\hat{p}_i = P_{\hat{\theta}}\{X \in A_i\} > 0 \qquad i = 1, 2, \ldots, k,$$

where $\hat{\boldsymbol{\theta}} = (\hat{\theta}_1, \ldots, \hat{\theta}_r)$, and $P_{\boldsymbol{\theta}}$ is the probability distribution associated with $F_{\boldsymbol{\theta}}$. Let $Y_1, Y_2, \ldots, Y_k$ be the RVs, defined as follows: Y_i = number of $X_1, X_2, \ldots, X_n$ in A_i, $i = 1, 2, \ldots, k$.

Then the RV

$$V_k = \sum_{n=1}^{k} \left\{ \frac{(Y_i - n\hat{p}_i)^2}{n\hat{p}_i} \right\}$$

is asymptotically distributed as a $\chi^2(k - r - 1)$ RV (as $n \to \infty$).

The proof of Theorem 4 and some regularity conditions required on $F_{\boldsymbol{\theta}}$ are given in Rao [88, pp. 391–392].

To test $H_0: X \sim F$, where F is completely specified, we reject H_0 if

$$u = \sum_{1}^{k} \left\{ \frac{(y_i - np_i)^2}{np_i} \right\} > \chi^2_{k-1,\alpha},$$

provided that n is sufficiently large. If the null hypothesis is $H_0: X \sim F_{\boldsymbol{\theta}}$, where $F_{\boldsymbol{\theta}}$ is known except for the parameter $\boldsymbol{\theta}$, we use Theorem 4 and reject H_0 if

$$v = \sum_{i=1}^{k} \left\{ \frac{(y_i - n\hat{p}_i)^2}{n\hat{p}_i} \right\} > \chi^2_{k-r-1,\alpha},$$

where r is the number of parameters estimated.

Example 3. The following data were obtained from a table of random numbers of normal distribution with mean 0 and variance 1.

0.464	0.137	2.455	−0.323	−0.068
0.906	−0.513	−0.525	0.595	0.881
−0.482	1.678	−0.057	−1.229	−0.486
−1.787	−0.261	1.237	1.046	−0.508

We want to test the null hypothesis that the DF F from which the data came is normal with mean 0 and variance 1. Here F is completely specified. Let us choose three intervals $(-\infty, -0.5]$, $(-0.5, 0.5]$, and $(0.5, \infty)$. We see that $Y_1 = 5$, $Y_2 = 8$, and $Y_3 = 7$.

Also, if Z is $\mathcal{N}(0, 1)$, then $p_1 = 0.3085$, $p_2 = 0.3830$, and $p_3 = 0.3085$. Thus

$$u = \sum_{i=1}^{3} \left\{ \frac{(y_i - np_i)^2}{np_i} \right\}$$

$$= \frac{(5-20\times 0.3085)^2}{6.17} + \frac{(8-20\times 0.383)^2}{7.66} + \frac{(7-20\times 0.3085)^2}{6.17}$$
$$< 1.$$

Also, $\chi^2_{2,0.05} = 5.99$, so we cannot reject H_0 at level 0.05.

Example 4. In a 72-hour period on a long holiday weekend there was a total of 306 fatal automobile accidents. The data are as follows:

Number of Fatal Accidents per Hour	Numbers of Hours
0 or 1	4
2	10
3	15
4	12
5	12
6	6
7	6
8 or more	7

Let us test the hypothesis that the number of accidents per hour is a Poisson RV.

Since the mean of the Poisson RV is not given, we estimate it by

$$\hat{\lambda} = \bar{x} = \frac{306}{72} = 4.25.$$

Let us now estimate $\hat{p}_i = P_{\hat{\lambda}}\{X = i\}$, $i = 0, 1, 2, \ldots$, $\hat{p}_0 = e^{-\hat{\lambda}} = 0.0143$. Note that

$$\frac{P_{\hat{\lambda}}\{X = x+1\}}{P_{\hat{\lambda}}\{X = x\}} = \frac{\hat{\lambda}}{x+1},$$

so that $\hat{p}_{i+1} = [\hat{\lambda}/(i+1)]\hat{p}_i$. Thus

$$\hat{p}_1 = 0.0606,\ \hat{p}_2 = 0.1288,\ \hat{p}_3 = 0.1825,\ \hat{p}_4 = 0.1939,$$
$$\hat{p}_5 = 0.1648,\ \hat{p}_6 = 0.1167,\ \hat{p}_7 = 0.0709,\ \hat{p}_8 = 1 - 0.9325 = 0.0675.$$

The observed and expected frequencies are as follows:

	i							
	0 or 1	2	3	4	5	6	7	8 or more
Observed Frequency, o_i	4	10	15	12	12	6	6	7
Expected Frequency $= 72\hat{p}_i = e_i$	5.38	9.28	13.14	13.96	11.87	8.41	5.10	4.86

Thus

$$u = \sum_{i=1}^{8} \frac{(o_i - e_i)^2}{e_i}$$
$$= 2.74.$$

Since we estimated one parameter, the number of degrees of freedom is $k-r-1 = 8-1-1=6$. From Table ST3, $\chi^2_{6,0.05} = 12.6$, and since $2.74 < 12.6$, we cannot reject the null hypothesis.

Remark 2. Any application of Theorem 3 or 4 requires that we choose sets $A_1, A_2, \ldots, A_k$, and frequently these are chosen to be disjoint intervals. As a rule of thumb, we choose the length of each interval in such a way that the probability $P\{X \in A_i\}$ under H_0 is approximately $1/k$. Moreover, it is desirable to have $n/k \geq 5$ or, rather, $e_i \geq 5$ for each i. If any of the e_i's is < 5, the corresponding interval is pooled with one or more adjoining intervals to make the cell frequency at least 5. The number of degrees of freedom, if any pooling is done, is the number of classes after pooling, minus 1, minus the number of parameters estimated.

Finally, we consider a test of *homogeneity* of several multinomial distributions. Suppose we have c samples of sizes $n_1, n_2, \ldots, n_c$ from c multinomial distributions. Let the associated probabilities with the jth population be $(p_{1j}, p_{2j}, \ldots, p_{rj})$, where $\sum_{i=1}^{r} p_{ij} = 1$, $j = 1, 2, \ldots, c$. Given observations N_{ij}, $i = 1, 2, \ldots, r$, $j = 1, 2, \ldots, c$ with $\sum_{i=1}^{r} N_{ij} = n_j$, $j = 1, 2, \ldots, c$ we wish to test $H_0 : p_{ij} = p_i$, for $j = 1, 2, \ldots, c$, $i = 1, 2, \ldots, r-1$. The case $c = 1$ is covered by Theorem 2. By Theorem 2 for each j

$$U_r = \sum_{i=1}^{r} \left\{ \frac{(N_{ij} - n_j p_i)^2}{n_j p_i} \right\}$$

has a limiting χ^2_{r-1} distribution. Since samples are independent, the statistic

$$U_{rc} = \sum_{j=1}^{c} \sum_{i=1}^{r} \frac{(N_{ij} - n_j p_i)^2}{n_j p_i}$$

has a limiting $\chi^2_{c(r-1)}$ distribution. If p_i's are unknown we use the MLEs

$$\hat{p}_i = \frac{\sum_{j=1}^{c} N_{ij}}{\sum_{j=1}^{c} n_j}, \quad i = 1, 2, \ldots, r-1$$

for p_i and we see that the statistic

$$V_{rc} = \sum_{j=1}^{c} \sum_{i=1}^{r} \frac{(N_{ij} - n_j \hat{p}_i)^2}{n_j \hat{p}_i}$$

has a chi-square distribution with $c(r-1) - (r-1) = (c-1)(r-1)$ d.f. We reject H_0 at (approximate) level α is $V_{rc} > \chi^2_{(r-1)(c-1),\alpha}$.

Example 5. A market analyst believes that there is no difference in preferences of television viewers among the four Ohio cities of Toledo, Columbus, Cleveland, and Cincinnati. In order to test this belief, independent random samples of 150, 200, 250, and 200 persons were selected from the four cities and asked, "What type of program do you prefer most: Mystery, Soap, Comedy, or News Documentary?" The following responses were recorded.

	City			
Program Type	Toledo	Columbus	Cleveland	Cincinnati
Mystery	50	70	85	60
Soap	45	50	58	40
Comedy	35	50	72	67
News	20	30	35	33
Sample Size	150	200	250	200

Under the null hypothesis that the proportions of viewers who prefer the four types of programs are the same in each city, the maximum likelihood estimates of p_i, $i = 1, 2, 3, 4$ are given by

$$\hat{p}_1 = \frac{50+70+85+60}{150+200+250+200} = \frac{265}{800} = 0.33, \quad \hat{p}_3 = \frac{35+50+72+67}{800} = \frac{224}{800} = 0.28,$$

$$\hat{p}_2 = \frac{45+50+58+40}{800} = \frac{193}{800} = 0.24, \qquad \hat{p}_4 = \frac{20+30+35+33}{800} = \frac{118}{800} = 0.15.$$

Here p_1 = proportion of people who prefer mystery, and so on. The following table gives the expected frequencies under H_0.

	Expected Number of Responses Under H_0			
Program Type	Toledo	Columbus	Cleveland	Cincinnati
Mystery	$150 \times 0.33 = 49.5$	$200 \times 0.33 = 66$	$250 \times 0.33 = 82.5$	$200 \times 0.33 = 66$
Soap	$150 \times 0.24 = 36$	$200 \times 0.24 = 48$	$250 \times 0.24 = 60$	$200 \times 0.24 = 48$
Comedy	$150 \times 0.28 = 42$	$200 \times 0.28 = 56$	$250 \times 0.28 = 70$	$200 \times 0.28 = 56$
News	$150 \times 0.15 = 22.5$	$200 \times 0.15 = 30$	$250 \times 0.15 = 37.5$	$200 \times 0.15 = 30$
Sample Size	150	200	250	200

It follows that

$$
\begin{aligned}
u_{44} &= \frac{(50-49.5)^2}{49.5} + \frac{(45-36)^2}{36} + \frac{(35-42)^2}{42} + \frac{(20-22.5)^2}{22.5} \\
&+ \frac{(70-66)^2}{66} + \frac{(50-48)^2}{48} + \frac{(50-56)^2}{56} + \frac{(30-30)^2}{30} \\
&+ \frac{(85-82.5)^2}{82.5} + \frac{(58-60)^2}{60} + \frac{(72-70)^2}{70} + \frac{(35-37.5)^2}{37.5} \\
&+ \frac{(60-66)^2}{66} + \frac{(40-48)^2}{48} + \frac{(67-56)^2}{56} + \frac{(33-30)^2}{30} \\
&= 9.37.
\end{aligned}
$$

Since $c = 4$ and $r = 4$, the number of degrees of freedom is $(4-1)(4-1) = 9$ and we note that under H_0

$$0.30 < P(U_{44} \geq 9.37) < 0.50.$$

With such a large P-value we can hardly reject H_0. The data do not offer any evidence to conclude that the proportions in the four cities are different.

PROBLEMS 10.3

1. The standard deviation of capacity for batteries of a standard type is known to be 1.66 ampere-hours. The following capacities (ampere-hours) were recorded for 10 batteries of a new type: 146, 141, 135, 142, 140, 143, 138, 137, 142, 136. Does the new battery differ from the standard type with respect to variability of capacity (Natrella [75, p. 4-1])?
2. A manufacturer recorded the cut-off bias (volts) of a sample of 10 tubes as follows: 12.1, 12.3, 11.8, 12.0, 12.4, 12.0, 12.1, 11.9, 12.2, 12.2. The variability of cut-off bias for tubes of a standard type as measured by the standard deviation is 0.208 volts. Is the variability of the new tube, with respect to cut-off bias less than that of the standard type (Natrella [75, p. 4-5])?
3. Approximately equal numbers of four different types of meters are in service and all types are believed to be equally likely to break down. The actual numbers of breakdowns reported are as follows:

Type of Meter	1	2	3	4
Number of Breakdowns Reported	30	40	33	47

   Is there evidence to conclude that the chances of failure of the four types are not equal (Natrella [75, p. 9-4])?
4. Every clinical thermometer is classified into one of four categories, A, B, C, D, on the basis of inspection and test. From past experience it is known that thermometers produced by a certain manufacturer are distributed among the four categories in the following proportions:

Category	A	B	C	D
Proportion	0.87	0.09	0.03	0.01

A new lot of 1336 thermometers is submitted by the manufacturer for inspection and test, and the following distribution into the four categories results:

Category	A	B	C	D
Number of Thermometers Reported	1188	91	47	10

Does this new lot of thermometers differ from the previous experience with regard to proportion of thermometers in each category (Natrella [75, p. 9-2])?

5. A computer program is written to generate random numbers, X, uniformly in the interval $0 \leq X < 10$. From 250 consecutive values the following data are obtained:

X-value	0–1.99	2–3.99	4–5.99	6–7.99	8–9.99
Frequency	38	55	54	41	62

Do these data offer any evidence that the program is not written properly?

6. A machine working correctly cuts pieces of wire to a mean length of 10.5 cm with a standard deviation of 0.15 cm. Sixteen samples of wire were drawn at random from a production batch and measured with the following results (centimeters): 10.4, 10.6, 10.1, 10.3, 10.2, 10.9, 10.5, 10.8, 10.6, 10.5, 10.7, 10.2, 10.7, 10.3, 10.4, 10.5. Test the hypothesis that the machine is working correctly.

7. An experiment consists in tossing a coin until the first head shows up. One hundred repetitions of this experiment are performed. The frequency distribution of the number of trials required for the first head is as follows:

Number of trials	1	2	3	4	5 or more
Frequency	40	32	15	7	6

Can we conclude that the coin is fair?

8. Fit a binomial distribution to the following data:

x	0	1	2	3	4
Frequency:	8	46	55	40	11

9. Prove Theorem 1.

10. Three dice are rolled independently 360 times each with the following results.

Face Value	Die 1	Die 2	Die 3
1	50	62	38
2	48	55	60
3	69	61	64
4	45	54	58
5	71	78	73
6	77	50	67
Sample Size	360	360	360

Are all the dice equally loaded? That is, test the hypothesis $H_0 : p_{i1} = p_{i2} = p_{i3}$, $i = 1,2,\ldots,6$, where p_{i1} is the probability of getting an i with die 1, and so on.

11. Independent random samples of 250 Democrats, 150 Republicans, and 100 Independent voters were selected 1 week before a nonpartisan election for mayor of a large city. Their preference for candidates Albert, Basu, and Chatfield were recorded as follows.

	Party Affiliation		
Preference	Democrat	Republican	Independent
Albert	160	70	90
Basu	32	45	25
Chatfield	30	23	15
Undecided	28	12	20
Sample Size	250	150	150

Are the proportions of voters in favor of Albert, Basu, and Chatfield the same within each political affiliation?

12. Of 25 income tax returns audited in a small town, 10 were from low- and middle-income families and 15 from high-income families. Two of the low-income families and four of the high-income families were found to have underpaid their taxes. Are the two proportions of families who underpaid taxes the same?

13. A candidate for a congressional seat checks her progress by taking a random sample of 20 voters each week. Last week, six reported to be in her favor. This week nine reported to be in her favor. Is there evidence to suggest that her campaign is working?

14. Let $\{X_{11},X_{21},\ldots,X_{r1}\},\ldots,\{X_{1c},X_{2c},\ldots,X_{rc}\}$ be independent multinomial RVs with parameters $(n_1,p_{11},p_{21},\ldots,p_{r1}),\ldots,(n_c,p_{1c},p_{2c},\ldots,p_{rc})$ respectively. Let

$X_{i\cdot} = \sum_{j=1}^{c} X_{ij}$ and $\sum_{j=1}^{c} n_j = n$. Show that the GLR test for testing $H_0: p_{ij} = p_j$, for $j = 1, 2, \ldots, c$, $i = 1, 2, \ldots, r-1$, where p_j's are unknown against all alternatives can be based on the statistic

$$\lambda(\mathbf{X}) = \frac{\prod_{i=1}^{r} \left(\frac{X_{i\cdot}}{n}\right)^{X_{i\cdot}}}{\prod_{i=1}^{r}\prod_{j=1}^{c} \left(\frac{X_{ij}}{n_j}\right)^{X_{ij}}}.$$

10.4 *t*-TESTS

In this section we investigate one of the most frequently used types of tests in statistics, the tests based on a t-statistic. Let $X_1, X_2, \ldots, X_n$ be a random sample from $\mathcal{N}(\mu, \sigma^2)$, and, as usual, let us write

$$\overline{X} = n^{-1}\sum_{1}^{n} X_i, \qquad S^2 = (n-1)^{-1}\sum_{1}^{n}(X_i - \overline{X})^2.$$

The tests for usual null hypotheses about the mean can be derived using the GLR method. In the following table we summarize the results.

			Reject H_0 at Level α if	
	H_0	H_1	σ^2 Known	σ^2 Unknown
I.	$\mu \leq \mu_0$	$\mu > \mu_0$	$\overline{X} \geq \mu_0 + \frac{\sigma}{\sqrt{n}} z_\alpha$	$\overline{x} \geq \mu_0 + \frac{s}{\sqrt{n}} t_{n-1,\alpha}$
II.	$\mu \geq \mu_0$	$\mu < \mu_0$	$\overline{X} \leq \mu_0 + \frac{\sigma}{\sqrt{n}} z_{1-\alpha}$	$\overline{x} \leq \mu_0 + \frac{s}{\sqrt{n}} t_{n-1,1-\alpha}$
III.	$\mu = \mu_0$	$\mu \neq \mu_0$	$\lvert\overline{x} - \mu_0\rvert \geq \frac{\sigma}{\sqrt{n}} z_{\alpha/2}$	$\lvert\overline{x} - \mu_0\rvert \geq \frac{s}{\sqrt{n}} t_{n-1,\alpha/2}$

Remark 1. A test based on a t-statistic is called a *t-test*. The t-tests in I and II are called *one-tailed tests*; the t-test in III, a *two-tailed test*.

Remark 2. If σ^2 is known, tests I and II are UMP and test III is UMP unbiased. If σ^2 is unknown, the t-tests are UMP unbiased and UMP invariant.

Remark 3. If n is large we may use normal tables instead of t-tables. The assumption of normality may also be dropped because of the central limit theorem. For small samples care is required in applying the proper test, since the tail probabilities under normal distribution and t-distribution differ significantly for small n (see Remark 6.4.2).

Example 1. Nine determinations of copper in a certain solution yielded a sample mean of 8.3 percent with a standard deviation of 0.025 percent. Let μ be the mean of the population of such determinations. Let us test $H_0: \mu = 8.42$ against $H_1: \mu < 8.42$ at level $\alpha = 0.05$.

Here $n = 9$, $\bar{x} = 8.3$, $s = 0.025$, $\mu_0 = 8.42$, and $t_{n-1,1-\alpha} = -t_{8,0.05} = -1.860$. Thus

$$\mu_0 + \frac{s}{\sqrt{n}} t_{n-1,1-\alpha} = 8.42 - \frac{0.025}{3} 1.86 = 8.4045.$$

We reject H_0 since $8.3 < 8.4045$.

We next consider the two-sample case. Let $X_1, X_2, \ldots, X_m$ and $Y_1, Y_2, \ldots, Y_n$ be independent random samples from $\mathcal{N}(\mu_1, \sigma_1^2)$ and $\mathcal{N}(\mu_2, \sigma_2^2)$, respectively. Let us write

$$\overline{X} = m^{-1} \sum_1^m X_i, \qquad \overline{Y} = n^{-1} \sum_1^n Y_i,$$
$$S_1^2 = (m-1)^{-1} \sum_1^m (X_i - \overline{X})^2, \qquad S_2^2 = (n-1)^{-1} \sum_1^n (Y_i - \overline{Y})^2,$$

and

$$S_p^2 = \frac{(m-1)S_1^2 + (n-1)S_2^2}{m+n-2}.$$

S_p^2 is sometimes called the *pooled sample variance*. The following table summarizes the two sample tests comparing μ_1 and μ_2:

H_0	H_1	Reject H_0 at Level α if	
(δ = Known Constant)		σ_1^2, σ_2^2 Known	σ_1^2, σ_2^2 Unknown, $\sigma_1 = \sigma_2$
I. $\mu_1 - \mu_2 \le \delta$	$\mu_1 - \mu_2 > \delta$	$\bar{x} - \bar{y} \ge \delta + z_\alpha \sqrt{\frac{\sigma_1^2}{m} + \frac{\sigma_2^2}{n}}$	$\bar{x} - \bar{y} \ge \delta + t_{m+n-2,\alpha} \cdot s_p \sqrt{\frac{1}{m} + \frac{1}{n}}$
II. $\mu_1 - \mu_2 \ge \delta$	$\mu_1 - \mu_2 < \delta$	$\bar{x} - \bar{y} \le \delta - z_\alpha \sqrt{\frac{\sigma_1^2}{m} + \frac{\sigma_2^2}{n}}$	$\bar{x} - \bar{y} \le \delta - t_{m+n-2,\alpha} \cdot s_p \sqrt{\frac{1}{m} + \frac{1}{n}}$
III. $\mu_1 - \mu_2 = \delta$	$\mu_1 - \mu_2 \ne \delta$	$\lvert\bar{x} - \bar{y} - \delta\rvert \ge z_{\alpha/2} \sqrt{\frac{\sigma_1^2}{m} + \frac{\sigma_2^2}{n}}$	$\lvert\bar{x} - \bar{y} - \delta\rvert \ge t_{m+n-2,\alpha/2} \cdot s_p \sqrt{\frac{1}{m} + \frac{1}{n}}$

Remark 4. The case of most interest is that in which $\delta = 0$. If σ_1^2, σ_2^2 are unknown and $\sigma_2^2 = \sigma_2^2 = \sigma^2$, σ^2 unknown, then S_p^2 is an unbiased estimate of σ^2. In this case all the two-sample t-tests are UMP unbiased and UMP invariant. Before applying the t-test, one should first make sure that $\sigma_1^2 = \sigma_2^2 = \sigma^2$, σ^2 unknown. This means applying another test on the data. We will consider this test in the next section.

Remark 5. If $m+n$ is large, we use normal tables; if both m and n are large, we can drop the assumption of normality, using the CLT.

Remark 6. The problem of equality of means in sampling from several populations will be considered in Chapter 12.

Remark 7. The two sample problem when $\sigma_1 \neq \sigma_2$, both unknown, is commonly referred to as *Behrens–Fisher problem*. The *Welch approximate t-test* of $H_0 : \mu_1 = \mu_2$ is based on a random number of d.f. f given by

$$f = \left\{ \left(\frac{R}{1+R} \right)^2 \frac{1}{m-1} + \frac{1}{(1+R)^2} \frac{1}{n-1} \right\}^{-1},$$

where

$$R = \frac{S_1^2/m}{S_2^2/n},$$

and the t-statistic

$$T = \frac{(\overline{X}-\overline{Y})-(\mu_1-\mu_2)}{\sqrt{S_1^2/m+S_2^2/n}}$$

with f d.f. This approximation has been found to be quite good even for small samples. The formula for f generally leads to noninteger d.f. Linear interpolation in t-table can be used to obtain the required percentiles for f d.f.

Example 2. The mean life of a sample of 9 light bulbs was observed to be 1309 hours with a standard deviation of 420 hours. A second sample of 16 bulbs chosen from a different batch showed a mean life of 1205 hours with a standard deviation of 390 hours. Let us test to see whether there is a significant difference between the means of the two batches, assuming that the population variances are the same (see also Example 10.5.1).

Here $H_0\colon \mu_1 = \mu_2, H_1\colon \mu_1 \neq \mu_2, m = 9, n = 16, \overline{x} = 1309, s_1 = 420, \overline{y} = 1205, s_2 = 390$, and let us take $\alpha = 0.05$. We have

$$s_p = \sqrt{\frac{8(420)^2+15(390)^2}{23}}$$

so that

$$t_{m+n-2,\alpha/2}s_p\sqrt{\frac{1}{m}+\frac{1}{n}} = t_{23,0.025}\sqrt{\frac{8(420)^2+15(390)^2}{23}}\sqrt{\frac{1}{9}+\frac{1}{16}} = 345.44.$$

Since $|\overline{x}-\overline{y}| = |1309-1205| = 104 \not> 345.44$, we cannot reject H_0 at level $\alpha = 0.05$.

Quite frequently one samples from a bivariate normal population with means μ_1, μ_2, variances σ_1^2, σ_2^2, and correlation coefficient ρ, the hypothesis of interest being $\mu_1 = \mu_2$. Let $(X_1,Y_1),(X_2,Y_2),\ldots,(X_n,Y_n)$ be a sample from a bivariate normal distribution with parameters μ_1, μ_2, σ_1^2, σ_2^2, and ρ. Then $X_j - Y_j$ is $\mathcal{N}(\mu_1 - \mu_2, \sigma^2)$, where $\sigma^2 = \sigma_1^2 + \sigma_2^2 - 2\rho\sigma_1\sigma_2$. We can therefore treat $D_j = (X_j - Y_j)$, $j = 1,2,\ldots,n$, as a sample from a normal population. Let us write

$$\overline{d} = \frac{\sum_1^n d_i}{n} \quad \text{and} \quad s_d^2 = \frac{\sum_1^n (d_i - \overline{d})^2}{n-1}.$$

The following table summarizes the resulting tests:

H_0	H_1	
$d_0 =$ Known Constant		Reject H_0 at Level α if
I. $\mu_1 - \mu_2 \geq d_0$	$\mu_1 - \mu_2 < d_0$	$\overline{d} \leq d_0 + \frac{s_d}{\sqrt{n}} t_{n-1,1-\alpha}$
II. $\mu_1 - \mu_2 \leq d_0$	$\mu_1 - \mu_2 > d_0$	$\overline{d} \geq d_0 + \frac{s_d}{\sqrt{n}} t_{n-1,\alpha}$
III. $\mu_1 - \mu_2 = d_0$	$\mu_1 - \mu_2 \neq d_0$	$\lvert\overline{d} - d_0\rvert \geq \frac{s_d}{\sqrt{n}} t_{n-1,\alpha/2}$

Remark 8. The case of most importance is that in which $d_0 = 0$. All the t-tests, based on D_j's, are UMP unbiased and UMP invariant. If σ is known, one can base the test on a standardized normal RV, but in practice such an assumption is quite unrealistic. If n is large one can replace t-values by the corresponding critical values under the normal distribution.

Remark 9. Clearly, it is not necessary to assume that $(X_1, Y_1), \ldots, (X_n, Y_n)$ is a sample from a bivariate normal population. It suffices to assume that the differences D_i form a sample from a normal population.

Example 3. Nine adults agreed to test the efficacy of a new diet program. Their weights (pounds) were measured before and after the program and found to be as follows:

	Participant								
	1	2	3	4	5	6	7	8	9
Before	132	139	126	114	122	132	142	119	126
After	124	141	118	116	114	132	145	123	121

Let us test the null hypothesis that the diet is not effective, $H_0\colon \mu_1 - \mu_2 = 0$, against the alternative, $H_1 : \mu_1 - \mu_2 > 0$, that it is effective at level $\alpha = 0.01$. We compute

$$\overline{d} = \frac{8-2+8-2+8+0-3-4+5}{9} = \frac{18}{9} = 2,$$

$$s_d^2 = 26.75, \qquad s_d = 5.17.$$

Thus

$$d_0 + \frac{s_d}{\sqrt{n}} t_{n-1,\alpha} = 0 + \frac{5.17}{\sqrt{9}} t_{8,0.01} = \frac{5.17}{3} \times 2.896 = 4.99.$$

Since $\overline{d} \not> 4.99$, we cannot reject hypothesis H_0 that the diet is not very effective.

PROBLEMS 10.4

1. The manufacturer of a certain subcompact car claims that the average mileage of this model is 30 miles per gallon of regular gasoline. For nine cars of this model driven in an identical manner, using 1 gallon of regular gasoline, the mean distance traveled was 26 miles with a standard deviation of 2.8 miles. Test the manufacturer's claim if you are willing to reject a true claim no more than twice in 100.
2. The nicotine contents of five cigarettes of a certain brand showed a mean of 21.2 milligrams with a standard deviation of 2.05 milligrams. Test the hypothesis that the average nicotine content of this brand of cigarettes does not exceed 19.7 milligrams. Use $\alpha = 0.05$.
3. The additional hours of sleep gained by eight patients in an experiment with a certain drug were recorded as follows:

Patient	1	2	3	4	5	6	7	8
Hours Gained	0.7	−1.1	3.4	0.8	2.0	0.1	−0.2	3.0

   Assuming that these patients form a random sample from a population of such patients and that the number of additional hours gained from the drug is a normal random variable, test the hypothesis that the drug has no effect at level $\alpha = 0.10$.
4. The mean life of a sample of 8 light bulbs was found to be 1432 hours with a standard deviation of 436 hours. A second sample of 19 bulbs chosen from a different batch produced a mean life of 1310 hours with a standard deviation of 382 hours. Making appropriate assumptions, test the hypothesis that the two samples came from the same population of light bulbs at level $\alpha = 0.05$.
5. A sample of 25 observations has a mean of 57.6 and a variance of 1.8. A further sample of 20 values has a mean of 55.4 and a variance of 2.5. Test the hypothesis that the two samples came from the same normal population.
6. Two methods were used in a study of the latent heat of fusion of ice. Both method A and method B were conducted with the specimens cooled to −0.72°C. The following data represent the change in total heat from −0.72°C to water, 0°C, in calories per gram of mass:

   Method A: 79.98, 80.04, 80.02, 80.04, 80.03, 80.03, 80.04, 79.97, 80.05, 80.03, 80.02, 80.00, 80.02

   Method B: 80.02, 79.74, 79.98, 79.97, 79.97, 80.03, 79.95, 79.97

Perform a test at level 0.05 to see whether the two methods differ with regard to their average performance (Natrella [75, p. 3-23]).

7. In Problem 6, if it is known from past experience that the standard deviations of the two methods are $\sigma_A = 0.024$ and $\sigma_B = 0.033$, test the hypothesis that the methods are same with regard to their average performance at level $\alpha = 0.05$.

8. During World War II bacterial polysaccharides were investigated as blood plasma extenders. Sixteen samples of hydrolyzed polysaccharides supplied by various manufacturers in order to assess two chemical methods for determining the average molecular weight yielded the following results:

Method *A*:	$62,700; 29,100; 44,400; 47,800; 36,300; 40,000; 43,400; 35,800;$
	$33,900; 44,200; 34,300; 31,300; 38,400; 47,100; 42,100; 42,200$
Method *B*:	$56,400; 27,500; 42,200; 46,800; 33,300; 37,100; 37,300; 36,200;$
	$35,200; 38,000; 32,200; 27,300; 36,100; 43,100; 38,400; 39,900$

Perform an appropriate test of the hypothesis that the two averages are the same against a one-sided alternative that the average of Method *A* exceeds that of Method *B*. Use $\alpha = 0.05$. (Natrella [75, p. 3-38]).

9. The following grade-point averages were collected over a period of 7 years to determine whether membership in a fraternity is beneficial or detrimental to grades:

	Year						
	1	2	3	4	5	6	7
Fraternity	2.4	2.0	2.3	2.1	2.1	2.0	2.0
Nonfraternity	2.4	2.2	2.5	2.4	2.3	1.8	1.9

Assuming that the populations were normal, test at the 0.025 level of significance whether membership in a fraternity is detrimental to grades.

10. Consider the two sample t-statistic $T = (\overline{X} - \overline{Y})/[S_p\sqrt{1/m+1/n}]$, where $S_p^2 = [(m-1)S_1^2 + (n-1)S_2^2]/(m+n-2)$. Suppose $\sigma_1 \neq \sigma_2$. Let $m,n \to \infty$ such that $m/(m+n) \to \rho$. Show that, under $\mu_1 = \mu_2$, $T \xrightarrow{L} U$, where $U \sim \mathcal{N}(0,\tau^2)$, where $\tau^2 = [(1-\rho)\sigma_1^2 + \rho\sigma_2^2]/[\rho\sigma_1^2 + (1-\rho)\sigma_2^2]$. Thus when $m \approx n$, $\rho \approx 1/2$ and $\tau^2 \approx 1$ and T is approximately $\mathcal{N}(0,1)$ as $m(\approx n) \to \infty$. In this case, a t-test based on T will have approximately the right level.

10.5 *F*-TESTS

The term *F-tests* refers to tests based on an F-statistic. Let $X_1, X_2, \ldots, X_m$ and $Y_1, Y_2, \ldots, Y_n$ be independent samples from $\mathcal{N}(\mu_1, \sigma_1^2)$ and $\mathcal{N}(\mu_2, \sigma_2^2)$, respectively. We recall that $\sum_1^m (X_i - \overline{X})/\sigma_1^2 \sim \chi^2(m-1)$ and $\sum_1^n (Y_i - \overline{Y})^2/\sigma_2^2 \sim \chi^2(n-1)$ are independent RVs, so that the RV

$$F(\mathbf{X},\mathbf{Y}) = \frac{\sum_1^m (X_i-\overline{X})^2}{\sum_1^n (Y_i-\overline{Y})^2}\frac{\sigma_2^2(n-1)}{\sigma_1^2(m-1)} = \frac{\sigma_2^2}{\sigma_1^2}\frac{S_1^2}{S_2^2}$$

is distributed as $F(m-1,n-1)$.

The following table summarizes the F-tests:

		Reject H_0 at Level α if	
H_0	H_1	μ_1,μ_2 Known	μ_1,μ_2 Unknown
I. $\sigma_1^2 \le \sigma_2^2$	$\sigma_1^2 > \sigma_2^2$	$\frac{\sum_1^m (x_i-\mu_1)^2}{\sum_1^n (y_i-\mu_2)^2} \ge \frac{m}{n}F_{m,n,\alpha}$	$\frac{s_1^2}{s_2^2} \ge F_{m-1,n-1,\alpha}$
II. $\sigma_1^2 \ge \sigma_2^2$	$\sigma_1^2 < \sigma_2^2$	$\frac{\sum_1^n (y_i-\mu_2)^2}{\sum_1^m (x_i-\mu_1)^2} \ge \frac{n}{m}F_{n,m,\alpha}$	$\frac{s_2^2}{s_1^2} \ge F_{n-1,m-1,\alpha}$
III. $\sigma_1^2 = \sigma_2^2$	$\sigma_1^2 \ne \sigma_2^2$	$\frac{\sum_1^m (x_i-\mu_1)^2}{\sum_1^n (y_i-\mu_2)^2} \ge \frac{m}{n}F_{m,n,\alpha/2}$ or $\le \frac{m}{n}F_{m,n,1-\alpha/2}$	$\frac{s_1^2}{s_2^2} \ge F_{m-1,n-1,\alpha/2}$ or $\le F_{m-1,n-1,1-\alpha/2}$

Remark 1. Recall (Remark 6.4.5) that

$$F_{m,n,1-\alpha} = \{F_{n,m,\alpha}\}^{-1}.$$

Remark 2. The tests described above can be easily obtained from the likelihood ratio procedure. Moreover, in the important case where μ_1,μ_2 are unknown, tests I and II are UMP unbiased and UMP invariant. For test III we have chosen equal tails, as is customarily done for convenience even though the unbiasedness property of the test is thereby destroyed.

Example 1 (Example 10.4.2 continued). In Example 10.4.2 let us test the validity of the assumption on which the t-test was based, namely, that the two populations have the same variance at level 0.05. We compute $s_1^2/s_2^2 = (420/390)^2 = 196/169 = 1.16$. Since $F_{m-1,n-1,\alpha/2} = F_{8,15,0.025} = 3.20$, we cannot reject $H_0: \sigma_1 = \sigma_2$.

An important application of the F-test involves the case where one is testing the equality of means of two normal populations under the assumption that the variances are the same, that is, testing whether the two samples come from the same population. Let $X_1,X_2,\ldots,X_m$ and $Y_1,Y_2,\ldots,Y_n$ be independent samples from $\mathcal{N}(\mu_1,\sigma_1^2)$ and $\mathcal{N}(\mu_2,\sigma_2^2)$, respectively. If $\sigma_1^2 = \sigma_2^2$ but is unknown, the t-test rejects $H_0: \mu_1 = \mu_2$ if $|T| > c$, where c is selected so that $\alpha_2 = P\{|T| > c \mid \mu_1 = \mu_2, \sigma_1 = \sigma_2\}$, that is, $c = t_{m+n-2,\alpha_2/2}s_p\sqrt{(1/m+1/n)}$, where

$$s_p^2 = \frac{(m-1)s_1^2+(n-1)s_2^2}{m+n-2},$$

s_1, s_2 being the sample variances. If first an F-test is performed to test $\sigma_1 = \sigma_2$, and then a t-test to test $\mu_1 = \mu_2$ at levels α_1 and α_2, respectively, the probability of accepting both hypotheses when they are true is

$$P\{|T| \le c, c_1 < F < c_2 | \mu_1 = \mu_2, \sigma_1 = \sigma_2\};$$

and if F is independent of T, this probability is $(1-\alpha_1)(1-\alpha_2)$. It follows that the combined test has a significance level $\alpha = 1-(1-\alpha_1)(1-\alpha_2)$. We see that

$$\alpha = \alpha_1 + \alpha_2 - \alpha_1\alpha_2 \le \alpha_1 + \alpha_2$$

and $\alpha \ge \max(\alpha_1, \alpha_2)$. In fact, α will be closer to $\alpha_1 + \alpha_2$, since for small α_1 and α_2, $\alpha_1\alpha_2$ will be closer to 0.

We show that F is independent of T whenever $\sigma_1 = \sigma_2$. The statistic $V = (\overline{X}, \overline{Y}, \sum_1^m (X_i - \overline{X})^2 + \sum_1^n (Y_i - \overline{Y})^2)$ is a complete sufficient statistic for the parameter $(\mu_1, \mu_2, \sigma_1 = \sigma_2)$ (see Theorem 8.3.2). Since the distribution of F does not depend on μ_1, μ_2, and $\sigma_1 = \sigma_2$, it follows (Problem 5) that F is independent of V whenever $\sigma_1 = \sigma_2$. But T is a function of V alone, so that F must be independent of T also.

In Example 1, the combined test has a significance level of

$$\alpha = 1-(0.95)(0.95) = 1-0.9025 = 0.0975.$$

PROBLEMS 10.5

1. For the data of Problem 10.4.4 is the assumption of equality of variances, on which the t-test is based, valid?
2. Answer the same question for Problems 10.4.5 and 10.4.6.
3. The performance of each of two different dive bombing methods is measured a dozen times. The sample variances for the two methods are computed to be 5545 and 4073, respectively. Do the two methods differ in variability?
4. In Problem 3 does the variability of the first method exceed that of the second method?
5. Let $\mathbf{X} = (X_1, X_2, \ldots, X_n)$ be a random sample from a distribution with PDF (PMF) $f(x, \boldsymbol{\theta})$, $\boldsymbol{\theta} \in \Theta$ where Θ is an interval in $\mathcal{R}_k$. Let $T(\mathbf{X})$ be a complete sufficient statistic for the family $\{f(\mathrm{x}; \boldsymbol{\theta}) \colon \boldsymbol{\theta} \in \Theta\}$. If $U(\mathbf{X})$ is a statistic (not a function of T alone) whose distribution does not depend on $\boldsymbol{\theta}$, show that U is independent of T.

10.6 BAYES AND MINIMAX PROCEDURES

Let $X_1, X_2, \ldots, X_n$ be a sample from a probability distribution with PDF (PMF) f_θ, $\theta \in \Theta$. In Section 8.8 we described the general decision problem, namely, once the statistician observes $\mathbf{x}$, she has a set $\mathcal{A}$ of options available. The problem is to find a decision function d that minimizes the risk $R(\theta, \delta) = E_\theta L(\theta, \delta)$ in some sense. Thus a minimax solution requires the minimization of $\max R(\theta, \delta)$, while a Bayes solution requires the minimization of $R(\pi, \delta) = ER(\theta, \delta)$, where π is the a priori distribution on Θ. In Remark 9.2.1

we considered the problem of hypothesis testing as a special case of the general decision problem. The set $\mathcal{A}$ contains two points, a_0 and a_1; a_0 corresponds to the acceptance of $H_0\colon \theta \in \Theta_0$, and a_1 corresponds to the rejection of H_0. Suppose that the loss function is defined by

$$\begin{cases} L(\theta, a_0) = a(\theta) & \text{if } \theta \in \Theta_1, \quad a(\theta) > 0, \\ L(\theta, a_1) = b(\theta) & \text{if } \theta \in \Theta_0, \quad b(\theta) > 0, \\ L(\theta, a_0) = 0 & \text{if } \theta \in \Theta_0, \\ L(\theta, a_1) = 0 & \text{if } \theta \in \Theta_1. \end{cases} \tag{1}$$

Then

$$R(\theta, \delta(\mathbf{X})) = L(\theta, a_0) P_\theta\{\delta(\mathbf{X}) = a_0\} + L(\theta, a_1) P_\theta\{\delta(\mathbf{X}) = a_1\} \tag{2}$$

$$= \begin{cases} a(\theta) P_\theta\{\delta(\mathbf{X}) = a_0\} & \text{if } \theta \in \Theta_1 \\ b(\theta) P_\theta\{\delta(\mathbf{X}) = a_1\} & \text{if } \theta \in \Theta_0. \end{cases} \tag{3}$$

A *minimax solution* to the problem of testing $H_0\colon \theta \in \Theta_0$ against $H_1: \theta \in \Theta_1$, where $\Theta = \Theta_0 + \Theta_1$, is to find a rule δ that minimizes

$$\max_\theta [a(\theta) P_\theta\{\delta(\mathbf{X}) = a_0\}, \quad b(\theta) P_\theta\{\delta(\mathbf{X}) = a_1\}].$$

We will consider here only the special case of testing $H_0\colon \theta = \theta_0$ against $H_1: \theta = \theta_1$. In that case we want to find a rule δ which minimizes

$$\max[a P_{\theta_1}\{\delta(\mathbf{X}) = a_0\}, \qquad b P_{\theta_0}\{\delta(\mathbf{X}) = a_1\}]. \tag{4}$$

We will show that the solution is to reject H_0 if

$$\frac{f_{\theta_1}(\mathbf{x})}{f_{\theta_0}(\mathbf{x})} \geq k, \tag{5}$$

provided that the constant k is chosen so that

$$R(\theta_0, \delta(\mathbf{X})) = R(\theta_1, \delta(\mathbf{X})), \tag{6}$$

where δ is the rule defined in (5); that is, the minimax rule δ is obtained if we choose k in (5) so that

$$a P_{\theta_1}\{\delta(\mathbf{X}) = a_0\} = b P_{\theta_0}\{\delta(\mathbf{X}) = a_1\}, \tag{7}$$

or, equivalently, we choose k so that

$$a P_{\theta_1}\left\{\frac{f_{\theta_1}(\mathbf{X})}{f_{\theta_0}(\mathbf{X})} < k\right\} = b P_{\theta_0}\left\{\frac{f_{\theta_1}(\mathbf{X})}{f_{\theta_0}(\mathbf{X})} \geq k\right\}. \tag{8}$$

Let δ^* be any other rule. If $R(\theta_0,\delta) < R(\theta_0,\delta^*)$, then $R(\theta_0,\delta) = R(\theta_1,\delta) < \max[R(\theta_0,\delta^*),\ R(\theta_1,\delta^*)]$ and δ^* cannot be minimax. Thus, $R(\theta_0,\delta) \geq R(\theta_0,\delta^*)$, which means that

$$P_{\theta_0}\{\delta^*(\mathbf{X}) = a_1\} \leq P_{\theta_0}\{\delta(\mathbf{X}) = a_1\} = P\{\text{Reject } H_0 \mid H_0 \text{ true}\}. \tag{9}$$

By the Neyman–Pearson lemma, rule δ is the most powerful of its size, so that its power must be at least that of δ^*, that is,

$$P_{\theta_1}\{\delta(\mathbf{X}) = a_1\} \geq P_{\theta_1}\{\delta^*(\mathbf{X}) = a_1\}$$

so that

$$P_{\theta_1}\{\delta(\mathbf{X}) = a_0\} \leq P_{\theta_1}\{\delta^*(\mathbf{X}) = a_0\}.$$

It follows that

$$aP_{\theta_1}\{\delta(\mathbf{X}) = a_0\} \leq aP_{\theta_1}\{\delta^*(\mathbf{X}) = a_0\}$$

and hence that

$$R(\theta_1,d) \leq R(\theta_1,\delta^*). \tag{10}$$

This means that

$$\max[R(\theta_0,\delta),\ R(\theta_1,\delta)] = R(\theta_1,\delta) \leq R(\theta_1,\delta^*)$$

and thus

$$\max[R(\theta_0,\delta),\ R(\theta_1,\delta)] \leq \max[R(\theta_0,\delta^*),\ R(\theta_1,\delta^*)].$$

Note that in the discrete case one may need some randomization procedure in order to achieve equality in (8).

Example 1. Let $X_1,X_2,\ldots,X_n$ be iid $\mathcal{N}(\mu,1)$ RVs. To test $H_0\colon \mu = \mu_0$ against $H_1: \mu = \mu_1$ $(> \mu_0)$, we should choose k so that (8) is satisfied. This is the same as choosing c, and thus k, so that

$$aP_{\mu_1}\{\overline{X} < c\} = bP_{\mu_0}\{\overline{X} \geq c\}$$

or

$$aP_{\mu_1}\left\{\frac{\overline{X}-\mu_1}{1/\sqrt{n}} < \frac{c-\mu_1}{1/\sqrt{n}}\right\} = bP\left\{\frac{\overline{X}-\mu_0}{1/\sqrt{n}} \geq \frac{c-\mu_0}{1/\sqrt{n}}\right\}.$$

Thus

$$a\Phi[\sqrt{n}(c-\mu_1)] = b\{1-\Phi[\sqrt{n}(c-\mu_0)]\},$$

where Φ is the DF of an $\mathcal{N}(0,1)$ RV. This can easily be accomplished with the help of normal tables once we know a, b, μ_0, μ_1, and n.

We next consider the problem of testing $H_0: \theta \in \Theta_0$ against $H_1: \theta \in \Theta_1$ from a Bayesian point of view. Let $\pi(\theta)$ be the a priori probability distribution on Θ.

Then

$$\begin{aligned} R(\pi,d) &= E_\theta R(\theta,\delta(\mathbf{X})) \\ &= \begin{cases} \int_\Theta R(\theta,\delta)\pi(\theta)d\theta & \text{if } \pi \text{ is a pdf,} \\ \sum_\Theta R(\theta,\delta)\pi(\theta) & \text{if } \pi \text{ is a pmf,} \end{cases} \\ &= \begin{cases} \int_{\Theta_0} b(\theta)\pi(\theta)P_\theta\{\delta(\mathbf{X})=a_1\}d\theta + \\ \qquad \int_{\Theta_1} a(\theta)\pi(\theta)P_\theta\{\delta(\mathbf{X})=a_0\}d\theta & \text{if } \pi \text{ is a PDF,} \\ \sum_{\Theta_0} b(\theta)\pi(\theta)P_\theta\{\delta(\mathbf{X})=a_1\} + \\ \qquad \sum_{\Theta_1} a(\theta)\pi(\theta)P_\theta\{\delta(\mathbf{X})=a_0\} & \text{if } \pi \text{ is a PMF.} \end{cases} \end{aligned} \tag{11}$$

The Bayes solution is a decision rule that minimizes $R(\pi,\delta)$. In what follows we restrict our attention to the case where both H_0 and H_1 have exactly one point each, that is, $\Theta_0 = \{\theta_0\}$, $\Theta_1 = \{\theta_1\}$. Let $\pi(\theta_0) = \pi_0$ and $\pi(\theta_1) = 1-\pi_0 = \pi_1$. Then

$$R(\pi,\delta) = b\pi_0 P_{\theta_0}\{\delta(\mathbf{X})=a_1\} + a\pi_1 P_{\theta_1}\{\delta(\mathbf{X})=a_0\}, \tag{12}$$

where $b(\theta_0)=b$, $a(\theta_1)=a$; $(a,b>0)$.

Theorem 1. Let $\mathbf{X}=(X_1,X_2,\ldots,X_n)$ be an RV of the discrete (continuous) type with PMF (PDF) f_θ, $\theta \in \Theta = \{\theta_0,\theta_1\}$. Let $\pi(\theta_0)=\pi_0$, $\pi(\theta_1)=1-\pi_0=\pi_1$ be the a priori probability mass function on Θ. A Bayes solution for testing $H_0: \mathbf{X} \sim f_{\theta_0}$ against $H_1: \mathbf{X} \sim f_{\theta_1}$, using the loss function (1), is to reject H_0 if

$$\frac{f_{\theta_1}(\mathbf{x})}{f_{\theta_0}(\mathbf{x})} \geq \frac{b\pi_0}{a\pi_1}. \tag{13}$$

Proof. We wish to find δ which minimizes

$$R(\pi,\delta) = b\pi_0 P_{\theta_0}\{\delta(\mathbf{X})=a_1\} + a\pi_1 P_{\theta_1}\{\delta(\mathbf{X})=a_0\}.$$

Now

$$\begin{aligned} R(\pi,\delta) &= E_\theta R(\theta,\delta) \\ &= E\{E_\theta\{L(\theta,\delta)|\mathbf{X}\}\} \end{aligned}$$

so it suffices to minimize $\{E_\theta\{L(\theta,\delta)|\mathbf{X}\}$.

The a posteriori distribution of θ is given by

$$\begin{aligned} h(\theta|\mathbf{x}) &= \frac{\pi(\theta)f_\theta(\mathbf{x})}{\sum_\theta f_\theta(\mathbf{x})\pi(\theta)} \\ &= \frac{\pi(\theta)f_\theta(\mathbf{x})}{\pi_0 f_{\theta_0}(\mathbf{x}) + \pi_1 f_{\theta_1}(\mathbf{x})} \end{aligned}$$

$$= \begin{cases} \dfrac{\pi_0 f_{\theta_0}(\mathbf{x})}{\pi_0 f_{\theta_0}(\mathbf{x}) + \pi_1 f_{\theta_1}(\mathbf{x})} & \text{if } \theta = \theta_0, \\ \dfrac{\pi_1 f_{\theta_1}(\mathbf{x})}{\pi_0 f_{\theta_0}(\mathbf{x}) + \pi_1 f_{\theta_1}(\mathbf{x})} & \text{if } \theta = \theta_1. \end{cases} \tag{14}$$

Thus

$$E_\theta\{L(\theta, \delta(\mathbf{X}))|\mathbf{X} = \mathbf{x}\} = \begin{cases} bh(\theta_0|\mathbf{x}), & \theta = \theta_0, \delta(\mathbf{X}) = a_1, \\ ah(\theta_1|\mathbf{x}), & \theta = \theta_1, \delta(\mathbf{X}) = a_0, \end{cases}$$

It follows that we reject H_0, that is, $\delta(\mathbf{X}) = a_1$ if

$$bh(\theta_0|\mathbf{x}) \le ah(\theta_1|\mathbf{x}),$$

which is the case if and only if

$$b\pi_0 f_{\theta_0}(\mathbf{x}) \le a\pi_1 f_{\theta_1}(\mathbf{x}),$$

as asserted.

Remark 1. In the Neyman–Pearson lemma we fixed $P_{\theta_0}\{\delta(\mathbf{X}) = a_1\}$, the probability of rejecting H_0 when it is true, and minimized $P_{\theta_1}\{\delta(\mathbf{X}) = a_0\}$, the probability of accepting H_0 when it is false. Here we no longer have a fixed level α for $P_{\theta_0}\{\delta(\mathbf{X}) = a_1\}$. Instead we allow it to assume any value as long as $R(\pi, \delta)$, defined in (12), is minimum.

Remark 2. It is easy to generalize Theorem 1 to the case of multiple decisions. Let $\mathbf{X}$ be an RV with PDF (PMF) f_θ, where θ can take any of the k values $\theta_1, \theta_2, \ldots, \theta_k$. The problem is to observe $\mathbf{x}$ and decide which of the θ_i's is the correct value of θ. Let us write $H_i: \theta = \theta_i$, $i = 1, 2, \ldots, k$, and assume that $\pi(\theta_i) = \pi_i$, $i = 1, 2, \ldots k$, $\sum_1^k \pi_i = 1$, is the prior probability distribution on $\Theta = \{\theta_1, \theta_2, \ldots, \theta_k\}$. Let

$$L(\theta_i, \delta) = \begin{cases} 1 & \text{if } \delta \text{ chooses } \theta_j, j \ne i. \\ 0 & \text{if } \delta \text{ chooses } \theta_i. \end{cases}$$

The problem is to find a rule δ that minimizes $R(\pi, \delta)$. We leave the reader to show that a Bayes solution is to accept $H_i: \theta = \theta_i$ $(i = 1, 2, \ldots, k)$ if

$$\pi_i f_{\theta_i}(\mathbf{x}) \ge \pi_j f_{\theta_j}(\mathbf{x}) \qquad \text{for all } j \ne i, j = 1, 2, \ldots, k, \tag{15}$$

where any point lying in more than one such region is assigned to any one of them.

Example 2. Let $X_1, X_2, \ldots, X_n$ be iid $\mathcal{N}(\mu, 1)$ RVs. To test $H_0: \mu = \mu_0$ against $H_1: \mu = \mu_1$ $(> \mu_0)$, let us take $a = b$ in the loss function (1). Then Theorem 1 says that the Bayes rule is one that rejects H_0 if

$$\frac{f_{\theta_1}(\mathbf{x})}{f_{\theta_0}(\mathbf{x})} \ge \frac{\pi_0}{1 - \pi_0},$$

that is,

$$\exp\left\{-\frac{\sum_1^n(x_i-\mu_1)^2}{2}+\frac{\sum_1^n(x_i-\mu_0)^2}{2}\right\}\geq\frac{\pi_0}{1-\pi_0},$$

$$\exp\left\{(\mu_1-\mu_0)\sum_1^n x_i+\frac{n(\mu_0^2-\mu_1^2)}{2}\right\}\geq\frac{\pi_0}{1-\pi_0}.$$

This happens if and only if

$$\frac{1}{n}\sum_1^n x_i\geq\frac{1}{n}\frac{\log[\pi_0/(1-\pi_0)]}{\mu_1-\mu_0}+\frac{\mu_0+\mu_1}{2},$$

where the logarithm is to the base e. It follows that, if $\pi_0=\frac{1}{2}$, the rejection region consists of

$$\bar{x}\geq\frac{\mu_0+\mu_1}{2}.$$

Example 3. This example illustrates the result described in Remark 2. Let $X_1,X_2,\ldots,X_n$ be a sample from $\mathcal{N}(\mu,1)$ and suppose that μ can take any one of the three values μ_1, μ_2, or μ_3. Let $\mu_1<\mu_2<\mu_3$. Assume, for simplicity, that $\pi_1=\pi_2=\pi_3$. Then we accept H_i: $\mu=\mu_i$, $i=1,2,3$, if

$$\pi_i\exp\left\{-\sum_{k=1}^n\frac{(x_k-\mu_i)^2}{2}\right\}\geq\pi_j\exp\left\{-\sum_{k=1}^n\frac{(x_k-\mu_j)^2}{2}\right\}$$
$$\text{for each } j\neq i, j=1,2,3.$$

It follows that we accept H_i if

$$(\mu_i-\mu_j)\bar{x}+\frac{\mu_j^2-\mu_i^2}{2}\geq 0,\qquad j=1,2,3,(j\neq i),$$

that is,

$$\bar{x}(\mu_i-\mu_j)\geq\frac{(\mu_i-\mu_j)(\mu_i+\mu_j)}{2},\qquad j=1,2,3,(j\neq i).$$

Thus, the acceptance region of H_1 is given by

$$\bar{x}\leq\frac{\mu_1+\mu_2}{2}\quad\text{and}\quad\bar{x}\leq\frac{\mu_1+\mu_3}{2}.$$

Also, the acceptance region of H_2 is given by

$$\bar{x}\geq\frac{\mu_1+\mu_2}{2}\quad\text{and}\quad\bar{x}\leq\frac{\mu_2+\mu_3}{2},$$

and that of H_3 by

$$\overline{x} \geq \frac{\mu_1 + \mu_3}{2} \quad \text{and} \quad \overline{x} \geq \frac{\mu_2 + \mu_3}{2}.$$

In particular, if $\mu_1 = 0$, $\mu_2 = 2$, $\mu_3 = 4$, we accept H_1 if $\overline{x} \leq 1$, H_2 if $1 \leq \overline{x} \leq 3$, and H_3 if $\overline{x} \geq 3$. In this case, boundary points 1 and 3 have zero probability, and it does not matter where we include them.

PROBLEMS 10.6

1. In Example 1 let $n = 15$, $\mu_0 = 4.7$, and $\mu_1 = 5.2$, and choose $a = b > 0$. Find the minimax test, and compute its power at $\mu = 4.7$ and $\mu = 5.2$.
2. A sample of five observations is taken on a $b(1,\theta)$ RV to test $H_0\colon \theta = \frac{1}{2}$ against $H_1 : \theta = \frac{3}{4}$.
   (a) Find the most powerful test of size $\alpha = 0.05$.
   (b) If $L(\frac{1}{2}, \frac{1}{2}) = L(\frac{3}{4}, \frac{3}{4}) = 0$, $L(\frac{1}{2}, \frac{3}{4}) = 1$, and $L(\frac{3}{4}, \frac{1}{2}) = 2$, find the minimax rule.
   (c) If the prior probabilities of $\theta = \frac{1}{2}$ and $\theta = \frac{3}{4}$ are $\pi_0 = \frac{1}{3}$ and $\pi_1 = \frac{2}{3}$, respectively, find the Bayes rule.
3. A sample of size n is to be used from the PDF

$$f_\theta(x) = \theta e^{-\theta x}, \qquad x > 0,$$

   to test $H_0\colon \theta = 1$ against $H_1 : \theta = 2$. If the a priori distribution on θ is $\pi_0 = \frac{2}{3}$, $\pi_1 = \frac{1}{3}$, and $a = b$, find the Bayes solution. Find the power of the test at $\theta = 1$ and $\theta = 2$.
4. Given two normal densities with variances 1 and with means -1 and 1, respectively, find the Bayes solution based on a single observation when $a = b$ and (a) $\pi_0 = \pi_1 = \frac{1}{2}$, and (b) $\pi_0 = \frac{1}{4}$, $\pi_1 = \frac{3}{4}$.
5. Given three normal densities with variances 1 and with means -1, 0, 1, respectively, find the Bayes solution to the multiple decision problem based on a single observation when $\pi_1 = \frac{2}{5}$, $\pi_2 = \frac{2}{5}$, $\pi_3 = \frac{1}{5}$.
6. For the multiple decision problem described in Remark 2 show that a Bayes solution is to accept $H_i\colon \theta = \theta_i$ $(i = 1,2,\ldots.k)$ if (15) holds.

11

CONFIDENCE ESTIMATION

11.1 INTRODUCTION

In many problems of statistical inference the experimenter is interested in constructing a family of sets that contain the true (unknown) parameter value with a specified (high) probability. If X, for example, represents the length of life of a piece of equipment, the experimenter is interested in a lower bound $\underline{\theta}$ for the mean θ of X. Since $\underline{\theta} = \underline{\theta}(X)$ will be a function of the observations, one cannot ensure with probability 1 that $\underline{\theta}(X) \leq \theta$. All that one can do is to choose a number $1-\alpha$ that is close to 1 so that $P_\theta\{\underline{\theta}(X) \leq \theta\} \geq 1-\alpha$ for all θ. Problems of this type are called problems of *confidence estimation*. In this chapter we restrict ourselves mostly to the case where $\Theta \subseteq \mathcal{R}$ and consider the problem of setting confidence limits for the parameter θ.

In Section 11.2 we introduce the basic ideas of confidence estimation. Section 11.3 deals with various methods of finding confidence intervals, and Section 11.4 deals with shortest-length confidence intervals. In Section 11.5 we study unbiased and equivariant confidence intervals.

11.2 SOME FUNDAMENTAL NOTIONS OF CONFIDENCE ESTIMATION

So far we have considered a random variable or some function of it as the basic observable quantity. Let X be an RV, and a, b, be two given positive real numbers. Then

An Introduction to Probability and Statistics, Third Edition. Vijay K. Rohatgi and A.K. Md. Ehsanes Saleh.

$$P\{a < X < b\} = P\{a < X \text{ and } X < b\}$$
$$= P\left\{X < b < \frac{bX}{a}\right\},$$

and if we know the distribution of X and a, b, we can determine the probability $P\{a < X < b\}$. Consider the interval $I(X) = (X, bX/a)$. This is an interval with end points that are functions of the RV X, and hence it takes the value $(x, bx/a)$ when X takes the value x. In other words, $I(X)$ assumes the value $I(x)$ whenever X assumes the value x. Thus $I(X)$ is a random quantity and is an example of a *random interval*. Note that $I(X)$ includes the value b with a certain fixed probability. For example, if $b = 1$, $a = \frac{1}{2}$, and X is $U(0,1)$, the interval $(X, 2X)$ includes point 1 with probability $\frac{1}{2}$. We note that $I(X)$ is a family of intervals with associated *coverage probability* $P(I(X) \ni 1) = \frac{1}{2}$. It has (random) length $\ell(I(X)) = 2X - X = X$. In general the larger the length of the interval the larger the coverage probability. Let us formalize these notions.

Definition 1. Let $\mathcal{P}_{\boldsymbol{\theta}}$, $\boldsymbol{\theta} \in \Theta \subseteq \mathcal{R}_k$, be the set of probability distributions of an RV $\mathbf{X}$. A family of subsets $S(\mathbf{x})$ of Θ, where $S(\mathbf{x})$ depends on the observation $\mathbf{x}$ but not on $\boldsymbol{\theta}$, is called a *family of random sets*. If, in particular, $\Theta \subseteq \mathcal{R}$ and $S(\mathbf{x})$ is an interval $(\underline{\theta}(\mathbf{x}), \overline{\theta}(\mathbf{x}))$, where $\underline{\theta}(\mathbf{x})$ and $\overline{\theta}(\mathbf{x})$ are functions of $\mathbf{x}$ alone (and not θ), we call $S(\mathbf{X})$ a *random interval* with $\underline{\theta}(\mathbf{X})$ and $\overline{\theta}(\mathbf{X})$ as lower and upper bounds, respectively. $\overline{\theta}(\mathbf{X})$ may be $-\infty$, and $\overline{\theta}(\mathbf{X})$ may be $+\infty$.

In a wide variety of inference problems one is not interested in estimating the parameter or testing some hypothesis concerning it. Rather, one wishes to establish a lower or an upper bound, or both, for the real-valued parameter. For example, if X is the time to failure of a piece of equipment, one may be interested in a lower bound for the mean of X. If the RV X measures the toxicity of a drug, the concern is to find an upper bound for the mean. Similarly, if the RV X measures the nicotine content of a certain brand of cigarettes, one may be interested in determining an upper and a lower bound for the average nicotine content of these cigarettes.

In this chapter we are interested in the *problem of confidence estimation*, namely, that of finding a family of random sets $S(\mathbf{x})$ for a parameter $\boldsymbol{\theta}$ such that, for a given α, $0 < \alpha < 1$ (usually small),

$$P_{\boldsymbol{\theta}}\{S(\mathbf{X}) \ni \boldsymbol{\theta}\} \geq 1 - \alpha \qquad \text{for all } \boldsymbol{\theta} \in \Theta. \tag{1}$$

We restrict our attention mainly to the case where $\theta \in \Theta \subseteq \mathcal{R}$.

Definition 2. Let $\theta \in \Theta \subseteq \mathcal{R}$ and $0 < \alpha < 1$. A *statistic* $\underline{\theta}(\mathbf{X})$ satisfying

$$P_\theta\{\underline{\theta}(\mathbf{X}) \leq \theta\} \geq 1 - \alpha \qquad \text{for all } \theta \tag{2}$$

is called a *lower confidence bound* for θ at confidence level $1 - \alpha$. The quantity

$$\inf_{\theta \in \Theta} P_\theta\{\underline{\theta}(\mathbf{X}) \leq \theta\} \tag{3}$$

is called the *confidence coefficient*.

Definition 3. A statistic $\underline{\theta}$ that minimizes

$$P_\theta\{\underline{\theta}(\mathbf{X}) \le \theta'\} \qquad \text{for all } \theta' < \theta \tag{4}$$

subject to (2) is known as a *uniformly most accurate* (UMA) lower confidence bound for θ at confidence level $1-\alpha$.

Remark 1. Suppose $\mathbf{X} \sim P_\theta$ and (2) holds. Then the smallest probability of *true coverage*, $P_\theta\{\underline{\theta}(\mathbf{X}) \le \theta\} = P_\theta\{[\underline{\theta}(\mathbf{X}),\infty) \ni \theta\}$ is $1-\alpha$. Then the probability of *false (or incorrect) coverage* is $P_\theta\{[\underline{\theta}(\mathbf{X}),\infty) \ni \theta'\} = P_\theta\{\underline{\theta}(\mathbf{X}) \le \theta'\}$ for $\theta' < \theta$. According to Definition 3 among the class of all lower confidence bounds satisfying (2), a UMA lower confidence bound has the smallest probability of false coverage.

Similar definitions are given for an upper confidence bound for θ and a UMA upper confidence bound.

Definition 4. A family of subsets $S(\mathbf{x})$ of $\Theta \subseteq \mathcal{R}_k$ is said to constitute a family of confidence sets at confidence level $1-\alpha$ if

$$P_{\boldsymbol{\theta}}\{S(\mathbf{X}) \ni \boldsymbol{\theta}\} \ge 1-\alpha \qquad \text{for all } \boldsymbol{\theta} \in \Theta, \tag{5}$$

that is, the random set $S(\mathbf{X})$ covers the true parameter value $\boldsymbol{\theta}$ with probability $\ge 1-\alpha$. A lower confidence bound corresponds to the special case where $k=1$ and

$$S(\mathbf{X}) = \{\theta\colon \underline{\theta}(\mathbf{x}) \le \theta < \infty\}; \tag{6}$$

and an upper confidence bound, to the case where

$$S(\mathbf{x}) = \{\theta\colon \overline{\theta}(\mathbf{x}) \ge \theta > -\infty\}. \tag{7}$$

If $S(\mathbf{x})$ is of the form

$$S(\mathbf{x}) = (\underline{\theta}(\mathbf{x}), \overline{\theta}(\mathbf{x})) \tag{8}$$

we will call it a *confidence interval* at confidence level $1-\alpha$, provided that

$$P_\theta\{\underline{\theta}(\mathbf{X}) < \theta < \overline{\theta}(\mathbf{X})\} \ge 1-\alpha \qquad \text{for all } \theta, \tag{9}$$

and the quantity

$$\inf_\theta P_\theta\{\underline{\theta}(\mathbf{X}) < \theta < \overline{\theta}(\mathbf{X})\} \tag{10}$$

will be referred to as the confidence coefficient associated with the random interval.

Remark 2. We write $S(\mathbf{X}) \ni \boldsymbol{\theta}$ to indicate that $\mathbf{X}$, and hence $S(\mathbf{X})$, is random here and not $\boldsymbol{\theta}$ so the probability distribution referred to is that of $\mathbf{X}$.

Remark 3. When $\mathbf{X} = \mathbf{x}$ is the realization the confidence interval (set) $S(\mathbf{x})$ is a fixed subset of $\mathcal{R}_k$. No probability is attached to $S(\mathbf{x})$ itself since neither $\boldsymbol{\theta}$ nor $S(\mathbf{x})$ has a probability distribution. In fact either $S(\mathbf{x})$ covers $\boldsymbol{\theta}$ or it does not and we will never know which since $\boldsymbol{\theta}$ is unknown. One can give a relative frequency interpretation. If $(1-\alpha)$-level confidence sets for $\boldsymbol{\theta}$ were computed a large number of times, then a fraction (approximately) $1-\alpha$ of these would contain the true (but unknown) parameter value.

Definition 5. A family of $(1-\alpha)$-level confidence sets $\{S(\mathbf{x})\}$ is said to be a UMA family of confidence sets at level $1-\alpha$ if

$$P_{\boldsymbol{\theta}}\{S(\mathbf{X}) \text{ contains } \boldsymbol{\theta}'\} \le P_{\boldsymbol{\theta}}\{S'(\mathbf{X}) \text{ contains } \boldsymbol{\theta}'\}$$

for all $\boldsymbol{\theta} \neq \boldsymbol{\theta}'$ and any $(1-\alpha)$-level family of confidence sets $S'(\mathbf{X})$.

Example 1. Let $X_1, X_2, \ldots, X_n$ be iid RVs, $X_i \sim \mathcal{N}(\mu, \sigma^2)$. Consider the interval $(\overline{X} - c_1, \overline{X} + c_2)$. In order for this to be a $(1-\alpha)$-level confidence interval, we must have

$$P\{\overline{X} - c_1 < \mu < \overline{X} + c_2\} \ge 1-\alpha,$$

which is the same as

$$P\{\mu - c_2 < \overline{X} < \mu + c_1\} \ge 1-\alpha.$$

Thus

$$P\left\{-\frac{c_2}{\sigma}\sqrt{n} < \frac{\overline{X}-\mu}{\sigma}\sqrt{n} < \frac{c_1}{\sigma}\sqrt{n}\right\} \ge 1-\alpha.$$

Since $\sqrt{n}(\overline{X}-\mu)/\sigma \sim \mathcal{N}(0,1)$, we can choose c_1 and c_2 to have equality, namely,

$$P\left\{-\frac{c_2}{\sigma}\sqrt{n} < \frac{\overline{X}-\mu}{\sigma}\sqrt{n} < \frac{c_1}{\sigma}\sqrt{n}\right\} = 1-\alpha,$$

provided that σ is known. There are infinitely many such pairs of values (c_1, c_2). In particular, an intuitively reasonable choice is $c_1 = -c_2 = c$, say. In that case

$$\frac{c\sqrt{n}}{\sigma} = z_{\alpha/2},$$

and the confidence interval is $(\overline{X} - (\sigma/\sqrt{n})z_{\alpha/2}, \overline{X} + (\sigma/\sqrt{n})z_{\alpha/2})$. The length of this interval is $(2\sigma/\sqrt{n})z_{\alpha/2}$. Given σ and α, we can choose n to get a confidence interval of a fixed length.

If σ is not known, we have from

$$P\{-c_2 < \overline{X} - \mu < c_1\} \ge 1-\alpha$$

that

$$P\left\{-\frac{c_2}{S}\sqrt{n} < \frac{\overline{X}-\mu}{S/\sqrt{n}} < \frac{c_1}{S}\sqrt{n}\right\} \geq 1-\alpha,$$

and once again we can choose pairs of values (c_1, c_2) using a t-distribution with $n-1$ d.f. such that

$$P\left\{-\frac{c_2\sqrt{n}}{S} < \frac{\overline{X}-\mu}{S}\sqrt{n} < \frac{c_1\sqrt{n}}{S}\right\} = 1-\alpha.$$

In particular, if we take $c_1 = -c_2 = c$, say, then

$$c\frac{\sqrt{n}}{S} = t_{n-1,\alpha/2},$$

and $(\overline{X}-(S/\sqrt{n})t_{n-1,\alpha/2}, \overline{X}+(S/\sqrt{n})t_{n-1,\alpha/2})$, is a $(1-\alpha)$-level confidence interval for μ. The length of this interval is $(2S/\sqrt{n})t_{n-1,\alpha/2}$, which is no longer constant. Therefore we cannot choose n to get a fixed-width confidence interval of level $1-\alpha$. Indeed, the length of this interval can be quite large if σ is large. Its expected length is

$$\frac{2}{\sqrt{n}}t_{n-1,\alpha/2}E_\sigma S = \frac{2}{\sqrt{n}}t_{n-1,\alpha/2}\sqrt{\frac{2}{n-1}}\frac{\Gamma(n/2)}{\Gamma[(n-1)/2]}\sigma,$$

which can be made as small as we please by choosing n large enough.

Example 2. In Example 1, suppose that we wish to find a confidence interval for σ^2 instead when μ is unknown. Consider the interval (c_1S^2, c_2S^2), $c_1, c_2 > 0$. We have

$$P\{c_1S^2 < \sigma^2 < c_2S^2\} \geq 1-\alpha,$$

so that

$$P\left\{c_2^{-1} < \frac{S^2}{\sigma^2} < c_1^{-1}\right\} \geq 1-\alpha.$$

Since $(n-1)S^2/\sigma^2$ is $\chi^2(n-1)$, we can choose pairs of values (c_1, c_2) from the tables of the chi-square distribution. In particular, we can choose c_1, c_2 so that

$$P\left\{\frac{S^2}{\sigma^2} \geq \frac{1}{c_1}\right\} = \frac{\alpha}{2} = P\left\{\frac{S^2}{\sigma^2} \leq \frac{1}{c_2}\right\}.$$

Then

$$\frac{n-1}{c_1} = \chi^2_{n-1,\alpha/2} \quad \text{and} \quad \frac{n-1}{c_2} = \chi^2_{n-1,1-\alpha/2}.$$

Thus

$$\left(\frac{(n-1)S^2}{\chi^2_{n-1,\alpha/2}}, \quad \frac{(n-1)S^2}{\chi^2_{n-1,1-\alpha/2}}\right)$$

is a $(1-\alpha)$-level confidence interval for σ^2 whenever μ is unknown. If μ is known, then

$$\sum_1^n \frac{(X_i-\mu)^2}{\sigma^2} \sim \chi^2(n).$$

Thus we can base the confidence interval on $\sum_1^n (X_i-\mu)^2$. Proceeding similarly, we get a $(1-\alpha)$-level confidence interval as

$$\left(\frac{\sum_1^n (X_i-\mu)^2}{\chi^2_{n,\alpha/2}}, \quad \frac{\sum_1^n (X_i-\mu)^2}{\chi^2_{n,1-\alpha/2}} \right).$$

Next suppose that both μ and σ^2 are unknown and that we want a confidence set for (μ,σ^2). We have from Boole's inequality

$$\begin{aligned}
&P\left\{\overline{X} - \frac{S}{\sqrt{n}} t_{n-1,\alpha_1/2} < \mu < \overline{X} + \frac{S}{\sqrt{n}} t_{n-1,\alpha_1/2}, \frac{(n-1)S^2}{\chi^2_{n-1,\alpha_2/2}} < \sigma^2 < \frac{(n-1)S^2}{\chi^2_{n-1,1-\alpha_2/2}}\right\} \\
&\geq 1 - P\left\{\overline{X} + \frac{S}{\sqrt{n}} t_{n-1,\alpha_1/2} \leq \mu \text{ or } \overline{X} - \frac{S}{\sqrt{n}} t_{n-1,\alpha_1/2} \geq \mu\right\} \\
&\quad - P\left\{\frac{(n-1)S^2}{\chi^2_{n-1,1-\alpha_2/2}} \leq \sigma^2 \text{ or } \frac{(n-1)S^2}{\chi^2_{n-1,\alpha_2/2}} \geq \sigma^2\right\} \\
&= 1 - \alpha_1 - \alpha_2,
\end{aligned}$$

so that the Cartesian product,

$$S(\mathbf{X}) = \left(\overline{X} - \frac{S}{\sqrt{n}} t_{n-1,\alpha_1/2}, \overline{X} + \frac{S}{\sqrt{n}} t_{n-1,\alpha_1/2}\right) \times \left(\frac{(n-1)S^2}{\chi^2_{n-1,\alpha_2/2}}, \frac{(n-1)S^2}{\chi^2_{n-1,1-\alpha_2/2}}\right)$$

is a $(1-\alpha_1-\alpha_2)$-level confidence set for (μ,σ^2).

11.3 METHODS OF FINDING CONFIDENCE INTERVALS

We now consider some common methods of constructing confidence sets. The most common of these is the method of *pivots*.

Definition 1. Let $\mathbf{X} \sim P_{\boldsymbol{\theta}}$. A random variable $T(\mathbf{X},\boldsymbol{\theta})$ is known as a *pivot* if the distribution of $T(\mathbf{X},\boldsymbol{\theta})$ does not depend on $\boldsymbol{\theta}$.

In many problems, especially in location and scale problems, pivots are easily found. For example, in sampling from $f(x-\theta)$, $X_{(n)}-\theta$ is a pivot and so is $\overline{X}-\theta$. In sampling from $(1/\sigma)f(x/\sigma)$, a scale family, $X_{(n)}/\sigma$ is a pivot and so is $X_{(1)}/\sigma$, and in sampling from $(1/\sigma)f((x-\theta)/\sigma)$, a location-scale family, $(\overline{X}-\theta)/S$ is a pivot, and so is $(X_{(2)}+X_{(1)}-2\theta)/S$.

If the DF F_θ of X_i is continuous, then $F_\theta(X_i) \sim U[0,1]$ and, in case of random sampling, we can take

$$T(X,\theta) = \prod_{i=1}^{n} F_\theta(X_i),$$

or,

$$-\log T(X,\theta) = -\sum_{i=1}^{n} \log F_\theta(X_i)$$

as a pivot. Since $F_\theta(X_i) \sim U[0,1]$, $-\log F_\theta(X_i) \sim G(1,1)$ and $-\sum_{i=1}^{n} \log F_\theta(X_i) \sim G(n,1)$. It follows that $-\sum_{i=1}^{n} \log F_\theta(X_i)$ is a pivot.

The following result gives a simple sufficient condition for a pivot to yield a confidence interval for a real-valued parameter θ.

Theorem 1. Let $T(\mathbf{X},\theta)$ be a pivot such that for each θ, $T(\mathbf{X},\theta)$ is a statistic, and as a function of θ, T is either strictly increasing or decreasing at each $\mathbf{x} \in \mathcal{R}_n$. Let $\Lambda \subseteq \mathcal{R}$ be the range of T, and for every $\lambda \in \Lambda$ and $\mathbf{x} \in \mathcal{R}_n$ let the equation $\lambda = T(\mathbf{x},\theta)$ be solvable. Then one can construct a confidence interval for θ at any level.

Proof. Let $0 < \alpha < 1$. Then we can choose a pair of numbers $\lambda_1(\alpha)$ and $\lambda_2(\alpha)$ in Λ not necessarily unique, such that

$$P_\theta\{\lambda_1(\alpha) < T(\mathbf{X},\theta) < \lambda_2(\alpha)\} \geq 1-\alpha \qquad \text{for all } \theta. \tag{1}$$

Since the distribution of T is independent of θ, it is clear that λ_1 and λ_2 are independent of θ. Since, moreover, T is monotone in θ, we can solve the equations

$$T(\mathbf{x},\theta) = \lambda_1(\alpha) \qquad \text{and} \qquad T(\mathbf{x},\theta) = \lambda_2(\alpha) \tag{2}$$

for every $\mathbf{x}$ uniquely for θ. We have

$$P_\theta\{\underline{\theta}(\mathbf{X}) < \theta < \overline{\theta}(\mathbf{X})\} \geq 1-\alpha \qquad \text{for all } \theta, \tag{3}$$

where $\underline{\theta}(\mathbf{X}) < \overline{\theta}(\mathbf{X})$ are RVs. This completes the proof.

Remark 1. The condition that $\lambda = T(\mathbf{x},\theta)$ be solvable will be satisfied if, for example, T is continuous and strictly increasing or decreasing as a function of θ in Θ.

Note that in the continuous case (that is, when the DF of T is continuous) we can find a confidence interval with equality on the right side of (1). In the discrete case, however, this is usually not possible.

Remark 2. Relation (1) is valid even when the assumption of monotonicity of T in the theorem is dropped. In that case inversion of the inequalities may yield a set of intervals (random set) $S(\mathbf{X})$ in Θ instead of a confidence interval.

Remark 3. The argument used in Theorem 1 can be extended to cover the multiparameter case, and the method will determine a confidence set for all the parameters of a distribution.

Example 1. Let $X_1, X_2, \ldots, X_n \sim \mathcal{N}(\mu, \sigma^2)$, where σ is unknown and we seek a $(1-\alpha)$-level confidence interval for μ. Let us choose

$$T(\mathbf{X}, \mu) = \frac{\overline{X} - \mu}{S}\sqrt{n},$$

where $\overline{X}, S^2$ are the usual sample statistics. The RV $T(\mathbf{X}, \mu)$ has Student's t-distribution with $n-1$ d.f., which is independent of μ and $T(\mathbf{X}, \mu)$, as a function of μ is monotone. We can clearly choose $\lambda_1(\alpha), \lambda_2(\alpha)$ (not necessarily uniquely) so that

$$P\{\lambda_1(\alpha) < T(\mathbf{X}, \mu) < \lambda_2(\alpha)\} = 1 - \alpha \qquad \text{for all } \mu.$$

Solving

$$\lambda_1(\alpha) = \frac{\overline{X} - \mu}{S}\sqrt{n},$$

we get

$$\underline{\mu}(\mathbf{X}) = \overline{X} - \frac{S}{\sqrt{n}}\lambda_2(\alpha), \qquad \overline{\mu}(\mathbf{X}) = \overline{X} - \frac{S}{\sqrt{n}}\lambda_1(\alpha),$$

and the $(1-\alpha)$-level confidence interval is

$$\left(\overline{X} - \frac{S}{\sqrt{n}}\lambda_2(\alpha), \overline{X} - \frac{S}{\sqrt{n}}\lambda_1(\alpha)\right).$$

In practice, one chooses $\lambda_2(\alpha) = -\lambda_1(\alpha) = t_{n-1,\alpha/2}$.

Example 2. Let $X_1, X_2, \ldots, X_n$ be iid with common PDF

$$f_\theta(x) = \exp\{-(x-\theta)\},\ x > \theta, \quad \text{and 0 elsewhere.}$$

Then the joint PDF of $\mathbf{X}$ is

$$f(\mathbf{x};\theta) = \exp\left\{-\sum_{i=1}^{n} x_i + n\theta\right\} I_{[x_{(1)} > \theta]}.$$

Clearly, $T(\mathbf{X}, \theta) = X_{(1)} - \theta$ is a pivot. We can choose $\lambda_1(\alpha)$, $\lambda_2(\alpha)$ such that

$$P_\theta\left\{\lambda_1(\alpha) < X_{(1)} - \theta < \lambda_2(\alpha)\right\} = 1 - \alpha \ \text{ for all } \theta,$$

which yields $(X_{(1)} - \lambda_2(\alpha), X_{(1)} - \lambda_1(\alpha))$ as a $(1-\alpha)$-level confidence interval for θ.

Remark 4. In Example 1 we chose $\lambda_2 = -\lambda_1$ whereas in Example 2 we did not indicate how to choose the pair (λ_1, λ_2) from an infinite set of solutions to $P_\theta\{\lambda_1(\alpha) < T(\mathbf{X}, \theta) < \lambda_2(\alpha)\} = 1-\alpha$. One choice is the *equal-tails* confidence interval which is arrived at by assigning probability $\alpha/2$ to each tail of the distribution of T. This means that we solve

$$\alpha/2 = P_\theta\{T(\mathbf{X}, \theta) < \lambda_1\} = P\{T(\mathbf{X}, \theta) > \lambda_2\}.$$

In Example 1 symmetry of the distribution leads to the indicated choice. In Example 2, $Y = X_{(1)} - \theta$ has PDF

$$g(y) = n\exp(-ny) \quad \text{for } y > 0$$

so we choose (λ_1, λ_2) from

$$P_\theta\left\{X_{(1)} - \theta < \lambda_1\right\} = \alpha/2 = P_\theta\left\{X_{(1)} - \theta > \lambda_2\right\},$$

giving $\lambda_2(\alpha) = (1/n)\ell\text{n}(\alpha/2)$ and $\lambda_1(\alpha) = -(1/n)\ell\text{n}(1-\alpha/2)$. Yet another method is to choose λ_1, λ_2 in such a way that the resulting confidence interval has smallest length. We will discuss this method in Section 11.4.

We next consider the method of *test inversion* and explore the relationship between a test of hypothesis for a parameter θ and confidence interval for θ. Consider the following example.

Example 3. Let $X_1, X_2, \ldots, X_n$ be a sample from $\mathcal{N}(\mu, \sigma_0^2)$, where σ_0 is known. In Example 11.2.1 we showed that

$$\left(\overline{X} - \frac{1}{\sqrt{n}} z_{\alpha/2}\sigma_0, \overline{X} + \frac{1}{\sqrt{n}} z_{\alpha/2}\sigma_0\right)$$

is a $(1-\alpha)$-level confidence interval for μ. If we define a test φ that rejects a value of $\mu = \mu_0$ if and only if μ_0 lies outside this interval, that is, if and only if

$$\frac{\sqrt{n}\,|\overline{X} - \mu_0|}{\sigma_0} \geq z_{\alpha/2},$$

then

$$P_{\mu_0}\left\{\sqrt{n}\frac{|\overline{X} - \mu_0|}{\sigma_0} \geq z_{\alpha/2}\right\} = \alpha,$$

and the test φ is a size α test of $\mu = \mu_0$ against the alternatives $\mu \neq \mu_0$.

Conversely, a family of α-level tests for the hypothesis $\mu = \mu_0$ generates a family of confidence intervals for μ by simply taking, as the confidence interval for μ_0, the set of those μ for which one cannot reject $\mu = \mu_0$.

Similarly, we can generate a family of α-level tests from a $(1-\alpha)$-level lower (or upper) confidence bound. Suppose that we start with the $(1-\alpha)$-level lower confidence bound

$\overline{X} - z_\alpha(\sigma_0/\sqrt{n})$ for μ. Then, by defining a test $\varphi(\mathbf{X})$ that rejects $\mu \le \mu_0$ if and only if $\mu_0 < \overline{X} - z_\alpha(\sigma_0/\sqrt{n})$, we get an α-level test for a hypothesis of the form $\mu \le \mu_0$.

Example 1 is a special case of the duality principle proved in Theorem 2 below. In the following we restrict attention to the case in which the rejection (acceptance) region of the test is the indicator function of a (Borel-measurable) set, that is, we consider only nonrandomized tests (and confidence intervals). For notational convenience we write $H_0(\theta_0)$ for the hypothesis $H_0\colon \theta = \theta_0$ and $H_1(\theta_0)$ for the alternative hypothesis, which may be one- or two-sided.

Theorem 2. Let $A(\theta_0)$, $\theta_0 \in \Theta$, denote the region of acceptance of an α-level test of $H_0(\theta_0)$. For each observation $\mathbf{x} = (x_1, x_2, \ldots, x_n)$ let $S(\mathbf{x})$ denote the set

$$S(\mathbf{x}) = \{\theta\colon \mathbf{x} \in A(\theta), \theta \in \Theta\}. \tag{4}$$

Then $S(\mathbf{x})$ is a family of confidence sets for θ at confidence level $1-\alpha$. If, moreover, $A(\theta_0)$ is UMP for the problem $(\alpha, H_0(\theta_0), H_1(\theta_0))$, then $S(\mathbf{X})$ minimizes

$$P_\theta\{S(\mathbf{X}) \ni \theta'\} \qquad \text{for all } \theta \in H_1(\theta') \tag{5}$$

among all $(1-\alpha)$-level families of confidence sets. That is, $S(\mathbf{X})$ is UMA.

Proof. We have

$$S(\mathbf{x}) \ni \theta \qquad \text{if and only } \mathbf{x} \in A(\theta), \tag{6}$$

so that

$$P_\theta\{S(\mathbf{X}) \ni \theta\} = P_\theta\{\mathbf{X} \in A(\theta)\} \ge 1 - \alpha,$$

as asserted.

If $S^*(\mathbf{X})$ is any other family of $(1-\alpha)$-level confidence sets, let $A^*(\theta) = \{\mathbf{x}\colon S^*(\mathbf{x}) \ni \theta\}$. Then

$$P_\theta\{\mathbf{X} \in A^*(\theta)\} = P_\theta\{S^*(\mathbf{X}) \ni \theta\} \ge 1 - \alpha$$

and since $A(\theta_0)$ is UMP for $(\alpha, H_0(\theta_0), H_1(\theta_0))$, it follows that

$$P_\theta\{\mathbf{X} \in A^*(\theta_0)\} \ge P_\theta\{\mathbf{X} \in A(\theta_0)\} \qquad \text{for any } \theta \in H_1(\theta_0).$$

Hence

$$P_\theta\{S^*(\mathbf{X}) \ni \theta_0\} \ge P_\theta\{\mathbf{X} \in A(\theta_0)\} = P_\theta\{S(\mathbf{X}) \ni \theta_0\}$$

for all $\theta \in H_1(\theta_0)$. This completes the proof.

Example 4. Let $\mathbf{X}$ be an RV of the continuous type with one-parameter exponential PDF, given by

$$f_\theta(\mathbf{x}) = \exp\{Q(\theta)T(\mathbf{x}) + S'(\mathbf{x}) + D(\theta)\},$$

where $Q(\theta)$ is a nondecreasing function of θ. Let $H_0\colon \theta = \theta_0$ and $H_1 : \theta < \theta_0$. Then the acceptance region of a UMP size α test of H_0 is of the form

$$A(\theta_0) = \{\mathbf{x}\colon T(\mathbf{x}) > c(\theta_0)\}.$$

Since, for $\theta \geq \theta'$,

$$P_{\theta'}\{T(\mathbf{X}) \leq c(\theta')\} = \alpha = P_\theta\{T(\mathbf{X}) \leq c(\theta)\} \leq P_{\theta'}\{T(\mathbf{X}) \leq c(\theta)\},$$

$c(\theta)$ may be chosen to be nondecreasing. (The last inequality follows because the power of the UMP test is at least α, the size.) We have

$$S(\mathbf{x}) = \{\theta\colon \mathbf{x} \in A(\theta)\},$$

so that $S(\mathbf{x})$ is of the form $(-\infty, c^{-1}(T(\mathbf{x})))$, or $(-\infty, c^{-1}(T(\mathbf{x}))]$, where c^{-1} is defined by

$$c^{-1}(T(\mathbf{x})) = \sup_\theta\{\theta\colon c(\theta) \leq T(\mathbf{x})\}.$$

In particular, if $X_1, X_2, \ldots, X_n$ is a sample from

$$f_\theta(x) = \begin{cases} \dfrac{1}{\theta}e^{-x/\theta}, & x > 0, \\ 0, & \text{otherwise}, \end{cases}$$

then $T(\mathbf{x}) = \sum_{i=1}^n x_i$ and for testing $H_0\colon \theta = \theta_0$ against $H_1 : \theta < \theta_0$, the UMP acceptance region is of the form

$$A(\theta_0) = \{\mathbf{x}\colon \sum_{i=1}^n x_i \geq c(\theta_0)\},$$

where $c(\theta_0)$ is the unique solution of

$$\int_{c(\theta_0)/\theta_0}^{\infty} \frac{y^{n-1}}{(n-1)!}e^{-y}dy = 1-\alpha, \qquad 0 < \alpha < 1.$$

The UMA family of $(1-\alpha)$-level confidence sets is of the form

$$S(\mathbf{x}) = \{\theta\colon \mathbf{x} \in A(\theta)\}.$$

In the case $n = 1$, $c(\theta_0) = \theta_0 \log\frac{1}{1-\alpha}$ and $S(x) = \left[0, \frac{x}{-\log(1-\alpha)}\right]$.

Example 5. Let $X_1, X_2, \ldots, X_n$ be iid $U(0,\theta)$ RVs. In Problem 9.4.3 we asked the reader to show that the test

$$\phi(x) = \begin{cases} 1 & x_{(n)} > \theta_0 \text{ or } x_{(n)} < \theta_0 \alpha^{1/n}, \\ 0 & \text{otherwise} \end{cases}$$

is UMP size α test of $\theta = \theta_0$ against $\theta \neq \theta_0$. Then

$$A(\theta_0) = \{\mathbf{x}\colon \theta_0 \alpha^{1/n} \le x_{(n)} \le \theta_0\}$$

and it follows that $[x_{(n)}, x_{(n)}\alpha^{-1/n}]$ is a $(1-\alpha)$-level UMA confidence interval for θ.

The third method we consider is based on Bayesian analysis where we take into account any prior knowledge that the experimenter has about θ. This is reflected in the specification of the prior distribution $\pi(\theta)$ on Θ. Under this setup the claims of probability of coverage are based not on the distribution of $\mathbf{X}$ but on the conditional distribution of θ given $\mathbf{X} = \mathbf{x}$, the posterior distribution of θ.

Let Θ be the parameter set, and let the observable RV $\mathbf{X}$ have PDF (PMF) $f_\theta(\mathbf{x})$. Suppose that we consider θ as an RV with distribution $\pi(\theta)$ on Θ. Then $f_\theta(\mathbf{x})$ can be considered as the conditional PDF (PMF) of $\mathbf{X}$, given that the RV θ takes the value θ. Note that we are using the same symbol for the RV θ and the value that it assumes. We can determine the joint distribution of $\mathbf{X}$ and θ, the marginal distribution of $\mathbf{X}$, and also the conditional distribution of θ, given $\mathbf{X} = \mathbf{x}$ as usual. Thus the joint distribution is given by

$$f(\mathbf{x}, \theta) = \pi(\theta) f_\theta(\mathbf{x}), \tag{7}$$

and the marginal distribution of $\mathbf{X}$ by

$$g(\mathbf{x}) = \begin{cases} \sum \pi(\theta) f_\theta(\mathbf{x}) & \text{if } \pi \text{ is a PMF}, \\ \int \pi(\theta) f_\theta(\mathbf{x})\, d\theta & \text{if } \pi \text{ is a PDF}. \end{cases} \tag{8}$$

The conditional distribution of θ, given that $\mathbf{x}$ is observed, is given by

$$h(\theta \mid \mathbf{x}) = \frac{\pi(\theta) f_\theta(\mathbf{x})}{g(\mathbf{x})}, \qquad g(\mathbf{x}) > 0. \tag{9}$$

Given $h(\theta \mid \mathbf{x})$, it is easy to find functions $l(\mathbf{x}), u(\mathbf{x})$ such that

$$P\{l(\mathbf{X}) < \theta < u(\mathbf{X})\} \ge 1 - \alpha,$$

where

$$P\{l(\mathbf{X}) < \theta < u(\mathbf{X}) \mid \mathbf{X} = \mathbf{x}\} = \begin{cases} \int_l^u h(\theta \mid \mathbf{x})\, d\theta \\ \sum_l^u h(\theta \mid \mathbf{x}), \end{cases} \tag{10}$$

depending on whether h is a PDF or a PMF.

Definition 2. An interval $(l(\mathbf{x}), u(\mathbf{x}))$ that has probability at least $1-\alpha$ of including θ is called a $(1-\alpha)$-*level Bayes interval* for θ. Also $l(\mathbf{x})$ and $u(\mathbf{x})$ are called the *lower and upper limits* of the interval.

One can similarly define one-sided Bayes intervals or $(1-\alpha)$-level lower and upper Bayes limits.

Remark 5. We note that, under the Bayesian set-up, we can speak of the probability that θ lies in the interval $(l(\mathbf{x}), u(\mathbf{x}))$ with probability $1-\alpha$ because l and u are computed based on the posterior distribution of θ given $\mathbf{x}$. In order to emphasize this distinction between Bayesian and classical analysis, some authors prefer the term *credible sets* for Bayesian confidence sets.

Example 6. Let $X_1, X_2, \ldots, X_n$ be iid $\mathcal{N}(\mu, 1)$, $\mu \in \mathcal{R}$, and let the a priori distribution of μ be $\mathcal{N}(0,1)$. Then from Example 8.8.6 we know that $h(\mu \mid \mathbf{x})$ is

$$\mathcal{N}\left(\frac{\sum_1^n x_i}{n+1}, \frac{1}{n+1}\right).$$

Thus a $(1-\alpha)$-level Bayesian confidence interval is

$$\left(\frac{n\overline{x}}{n+1} - \frac{z_{\alpha/2}}{\sqrt{n+1}}, \frac{n\overline{x}}{n+1} + \frac{z_{\alpha/2}}{\sqrt{n+1}}\right)$$

A $(1-\alpha)$-level confidence interval for μ (treating μ as fixed) is a random interval with value

$$\left(\overline{x} - \frac{z_{\alpha/2}}{\sqrt{n}}, \overline{x} + \frac{z_{\alpha/2}}{\sqrt{n}}\right).$$

Thus the Bayesian interval is somewhat shorter in length. This is to be expected since we assumed more in the Bayesian case.

Example 7. Let $X_1, X_2, \ldots, X_n$ be iid $b(1,p)$ RVs, and let the prior distribution on $\Theta = (0,1)$ be $U(0,1)$. A simple computation shows that the posterior PDF of p, given x, is

$$h(p|x) = \begin{cases} \frac{p^{-\sum_1^n x_i}(1-p)^{n-\sum_1^n x_i}}{B\left(\sum_1^n x_i+1, n-\sum_1^n x_i+1\right)}, & 0 < p < 1 \\ 0, & \text{otherwise.} \end{cases}$$

Given a table of incomplete beta integrals and the observed value of $\sum_1^n x_i$, one can easily construct a Bayesian confidence interval for p.

Finally, we consider some large sample methods of constructing confidence intervals. Suppose $T(X) \sim \text{AN}(\theta, v(\theta)/n)$. Then

$$\sqrt{n}\frac{T(X)-\theta}{\sqrt{v(\theta)}} \xrightarrow{L} Z,$$

where $Z \sim N(0,1)$. Suppose further that there is a statistic $S(X)$ such that $S(X) \xrightarrow{P} v(\theta)$. Then, by Slutsky's theorem

$$\sqrt{n}\frac{T(\mathbf{X})-\theta}{\sqrt{S(\mathbf{X})}} \xrightarrow{L} Z,$$

and we can obtain an (approximate) $(1-\alpha)$-level confidence interval for θ by inverting the inequality

$$\left|\sqrt{n}\frac{T(\mathbf{X})-\theta}{\sqrt{S(\mathbf{X})}}\right| \leq z_{\alpha/2}.$$

Example 8. Let $X_1, X_2, \ldots, X_n$ be iid RVs with finite variance. Also, let $EX_i = \mu$ and $EX_i^2 = \sigma^2 + \mu^2$. From the CLT it follows that

$$\frac{\overline{X}-\mu}{\sigma/\sqrt{n}} \xrightarrow{L} Z,$$

where $Z \sim \mathcal{N}(0,1)$. Suppose that we want a $(1-\alpha)$-level confidence interval for μ when σ is not known. Since $S \xrightarrow{P} \sigma$, for large n the quantity $[\sqrt{n}(\overline{X}-\mu)/S]$ is approximately normally distributed with mean 0 and variance 1. Hence, for large n, we can find constants c_1, c_2 such that

$$P\left\{c_1 < \frac{\overline{X}-\mu}{S}\sqrt{n} < c_2\right\} = 1-\alpha.$$

In particular, we can choose $-c_1 = c_2 = z_{\alpha/2}$ to give

$$\left(\overline{x} - \frac{s}{\sqrt{n}} z_{\alpha/2}, \overline{x} + \frac{s}{\sqrt{n}} z_{\alpha/2}\right)$$

as an approximate $(1-\alpha)$-level confidence interval for μ.

Recall that if $\hat{\theta}$ is the MLE of θ and the conditions of Theorem 8.7.4 or 8.7.5 are satisfied (*caution:* See Remark 8.7.4), then

$$\frac{\sqrt{n}(\hat{\theta}-\theta)}{\sigma} \xrightarrow{L} \mathcal{N}(0,1) \qquad \text{as } n \to \infty,$$

where

$$\sigma^2 = \left[E_\theta\left\{\frac{\partial \log f_\theta(X)}{\partial \theta}\right\}^2\right]^{-1} = \frac{1}{I(\theta)}.$$

Then we can invert the statement

$$P_\theta\left\{-z_{\alpha/2} < \frac{\hat{\theta}-\theta}{\sigma}\sqrt{n} < z_{\alpha/2}\right\} \geq 1-\alpha$$

to give an approximate $(1-\alpha)$-level confidence interval for θ.

Yet another possible procedure has universal applicability and hence can be used for large or small samples. Unfortunately, however, this procedure usually yields confidence intervals that are much too large in length. The method employs the well-known Chebychev inequality (see Section 3.4):

$$P\left\{|X-EX|<\varepsilon\sqrt{\operatorname{var}(X)}\right\}>1-\frac{1}{\varepsilon^2}.$$

If $\hat{\theta}$ is an estimate of θ (not necessarily unbiased) with finite variance $\sigma^2(\theta)$, then by Chebychev's inequality

$$P\left\{|\hat{\theta}-\theta|<\varepsilon\sqrt{E(\hat{\theta}-\theta)^2}\right\}>1-\frac{1}{\varepsilon^2}.$$

It follows that

$$\left(\hat{\theta}-\varepsilon\sqrt{E(\hat{\theta}-\theta)^2},\hat{\theta}+\varepsilon\sqrt{E(\hat{\theta}-\theta)^2}\right)$$

is a $[1-(1/\varepsilon^2)]$-level confidence interval for θ. Under some mild consistency conditions one can replace the normalizing constant $\sqrt{[E(\hat{\theta}-\theta)^2]}$, which will be some function $\lambda(\theta)$ of θ, by $\lambda(\hat{\theta})$.

Note that the estimator $\hat{\theta}$ need not have a limiting normal law.

Example 9. Let $X_1,X_2,\ldots,X_n$ be iid $b(1,p)$ RVs, and it is required to find a confidence interval for p. We know that $E\overline{X}=p$ and

$$\operatorname{var}(\overline{X})=\frac{\operatorname{var}(X)}{n}=\frac{p(1-p)}{n}.$$

It follows that

$$P\left\{|\overline{X}-p|<\varepsilon\sqrt{\frac{p(1-p)}{n}}\right\}>1-\frac{1}{\varepsilon^2}.$$

Since $p(1-p)\leq\frac{1}{4}$, we have

$$P\left\{\overline{X}-\frac{1}{2\sqrt{n}}\varepsilon<p<\overline{X}+\frac{1}{2\sqrt{n}}\varepsilon\right\}>1-\frac{1}{\varepsilon^2}.$$

One can now choose ε and n or, if n is kept constant at a given number, ε to get the desired level.

Actually the confidence interval obtained above can be improved somewhat. We note that

$$P\left\{|\overline{X}-p|<\varepsilon\sqrt{\frac{p(1-p)}{n}}\right\}>1-\frac{1}{\varepsilon^2}.$$

so that

$$P\left\{|\overline{X}-p|^2 < \frac{\varepsilon^2 p(1-p)}{n}\right\} > 1-\frac{1}{\varepsilon^2}.$$

Now

$$|\overline{X}-p|^2 < \frac{\varepsilon^2}{n}p(1-p)$$

if and only if

$$\left(1+\frac{\varepsilon^2}{n}\right)p^2-\left(2\overline{X}+\frac{\varepsilon^2}{n}\right)p+\overline{X}^2 < 0.$$

This last inequality holds if and only if p lies between the two roots of the quadratic equation

$$\left(1+\frac{\varepsilon^2}{n}\right)p^2-\left(2\overline{X}+\frac{\varepsilon^2}{n}\right)p+\overline{X}^2 = 0.$$

The two roots are

$$\begin{aligned} p_1 &= \frac{2\overline{X}+(\varepsilon^2/n)-\sqrt{[2\overline{X}+(\varepsilon^2/n)]^2-4[1+(\varepsilon^2/n)]\overline{X}^2}}{2[1+(\varepsilon^2/n)]} \\ &= \frac{\overline{X}}{1+(\varepsilon^2/n)}+\frac{(\varepsilon^2/n)-\sqrt{4(\varepsilon^2/n)\overline{X}(1-\overline{X})+(\varepsilon^4/n^2)}}{2[1+(\varepsilon^2/n)]} \end{aligned}$$

and

$$\begin{aligned} p_2 &= \frac{2\overline{X}+(\varepsilon^2/n)+\sqrt{[2\overline{X}+(\varepsilon^2/n)]^2-4[1+(\varepsilon^2/n)]\overline{X}^2}}{2[1+(\varepsilon^2/n)]} \\ &= \frac{\overline{X}}{1+(\varepsilon^2/n)}+\frac{(\varepsilon^2/n)+\sqrt{4(\varepsilon^2/n)\overline{X}(1-\overline{X})+(\varepsilon^4/n^2)}}{2[1+(\varepsilon^2/n)]} \end{aligned}$$

It follows that

$$P\{p_1 < p < p_2\} > 1-\frac{1}{\varepsilon^2}.$$

Note that when n is large

$$p_1 \approx \overline{X}-\varepsilon\sqrt{\frac{\overline{X}(1-\overline{X})}{n}} \quad \text{and} \quad p_2 \approx \overline{X}+\varepsilon\sqrt{\frac{\overline{X}(1-\overline{X})}{n}},$$

as one should expect in view of the fact that $\overline{X} \to p$ with probability 1 and $\sqrt{[\overline{X}(1-\overline{X})/n]}$ estimates $\sqrt{[p(1-p)/n]}$. Alternatively, we could have used the CLT (or large-sample property of the MLE) to arrive at the same result but with ε replaced by $z_{\alpha/2}$.

Example 10. Let $X_1, X_2, \ldots, X_n$ be a sample from $U(0,\theta)$. We seek a confidence interval for the parameter θ. The estimator $\hat{\theta} = X_{(n)}$ is the MLE of θ, which is also sufficient for θ. From Example 5, $[X_{(n)}, \alpha^{-1/n}X_{(n)}]$ is a $(1-\alpha)$-level UMA confidence interval for θ.

Let us now apply the method of Chebychev's inequality to the same problem. We have

$$E_\theta X_{(n)} = \frac{n}{n+1}\theta$$

and

$$E_\theta (X_{(n)} - \theta)^2 = \theta^2 \frac{2}{(n+1)(n+2)}.$$

Thus

$$P\left\{\frac{|X_{(n)} - \theta|}{\theta}\sqrt{\frac{(n+1)(n+2)}{2}} < \varepsilon\right\} > 1 - \frac{1}{\varepsilon^2}.$$

Since $X_{(n)} \xrightarrow{P} \theta$, we replace θ by $X_{(n)}$ in the denominator, and, for moderately large n,

$$P\left\{\frac{|X_{(n)} - \theta|}{X_{(n)}}\sqrt{\frac{(n+1)(n+2)}{2}} < \varepsilon\right\} > 1 - \frac{1}{\varepsilon^2}.$$

It follows that

$$\left(X_{(n)} - \varepsilon X_{(n)} \frac{\sqrt{2}}{\sqrt{(n+1)(n+2)}}, X_{(n)} + \varepsilon X_{(n)} \frac{\sqrt{2}}{\sqrt{(n+1)(n+2)}}\right)$$

is a $[1-(1/\varepsilon^2)]$-confidence interval for θ. Choosing $1-(1/\varepsilon^2) = 1-\alpha$, or $\varepsilon = 1/\sqrt{\alpha}$, and noting that $1/\sqrt{[(n+1)(n+2)]} \approx 1/n$ for large n, and the fact that with probability 1, $X_{(n)} \le \theta$, we can use the approximate confidence interval

$$\left(X_{(n)}, X_{(n)}\left(1 + \frac{1}{n}\sqrt{\frac{2}{\alpha}}\right)\right)$$

for θ.

In the examples given above we see that, for a given confidence interval $1-\alpha$, a wide choice of confidence intervals is available. Clearly, the larger the interval, the better the chance of trapping a true parameter value. Thus the interval $(-\infty, +\infty)$, which ignores the data completely will include the real-valued parameter θ with confidence level 1. However, the larger the confidence interval, the less meaningful it is. Therefore, for a given confidence level $1-\alpha$, it is desirable to choose the shortest possible confidence interval. Since the length $\overline{\theta}-\underline{\theta}$, in general, is a random variable, one can show that a confidence interval of level $1-\alpha$ with uniformly minimum length among all such intervals does not exist in most cases. The alternative, to minimize $E_\theta(\overline{\theta}-\underline{\theta})$, is also quite unsatisfactory. In the next section we consider the problem of finding shortest-length confidence interval based on some suitable statistic.

PROBLEMS 11.3

1. A sample of size 25 from a normal population with variance 81 produced a mean of 81.2. Find a 0.95 level confidence interval for the mean μ.

2. Let $\overline{X}$ be the mean of a random sample of size n from $\mathcal{N}(\mu, 16)$. Find the smallest sample size n such that $(\overline{X}-1, \overline{X}+1)$ is a 0.90 level confidence interval for μ.

3. Let $X_1, X_2, \ldots, X_m$ and $Y_1, Y_2, \ldots, Y_n$ be independent random samples from $\mathcal{N}(\mu_1, \sigma^2)$ and $\mathcal{N}(\mu_2, \sigma^2)$, respectively. Find a confidence interval for $\mu_1-\mu_2$ at confidence level $1-\alpha$ when (a) σ is known and (b) σ is unknown.

4. Two independent samples, each of size 7, from normal populations with common unknown variance σ^2 produced sample means 4.8 and 5.4 and sample variances 8.38 and 7.62, respectively. Find a 0.95 level confidence interval for $\mu_1-\mu_2$, the difference between the means of samples 1 and 2.

5. In Problem 3 suppose that the first population has variance σ_1^2 and the second population has variance σ_2^2, where both σ_1^2, and σ_2^2 are known. Find a $(1-\alpha)$-level confidence interval for $\mu_1-\mu_2$. What happens if both σ_1^2 and σ_2^2 are unknown and unequal?

6. In Problem 5 find a confidence interval for the ratio σ_2^2/σ_1^2, both when μ_1, μ_2 are known and when μ_1, μ_2 are unknown. What happens if either μ_1 or μ_2 is unknown but the other is known?

7. Let $X_1, X_2, \ldots, X_n$ be a sample from a $G(1, \beta)$ distribution. Find a confidence interval for the parameter β with confidence level $1-\alpha$.

8. (a) Use the large-sample properties of the MLE to construct a $(1-\alpha)$-level confidence interval for the parameter θ in each of the following cases: (i) $X_1, X_2, \ldots, X_n$ is a sample from $G(1, 1/\theta)$ and (ii) $X_1, X_2, \ldots, X_n$ is a sample from $P(\theta)$.

(b) In part (a) use Chebychev's inequality to do the same.

9. For a sample of size 1 from the population

$$f_\theta(x) = \frac{2}{\theta^2}(\theta - x), \qquad 0 < x < \theta,$$

find a $(1-\alpha)$-level confidence interval for θ.

10. Let $X_1, X_2, \ldots, X_n$ be a sample from the uniform distribution on N points. Find an upper $(1-\alpha)$-level confidence bound for N, based on $\max(X_1, X_2, \ldots, X_n)$.

11. In Example 10 find the smallest n such that the length of the $(1-\alpha)$-level confidence interval $(X_{(n)}, \alpha^{-1/n}X_{(n)}) < d$, provided it is known that $\theta \le a$, where a is a known constant.

12. Let X and Y be independent RVs with PDFs $\lambda e^{-\lambda x}$ $(x > 0)$ and $\mu e^{-\mu y}$ $(y > 0)$, respectively. Find a $(1-\alpha)$-level confidence region for (λ, μ) of the form $\{(\lambda, \mu) \colon \lambda X + \mu Y \le k\}$.

13. Let $X_1, X_2, \ldots, X_n$ be a sample from $\mathcal{N}(\mu, \sigma^2)$, where σ^2 is known. Find a UMA $(1-\alpha)$-level upper confidence bound for μ.

14. Let $X_1, X_2, \ldots, X_n$ be a sample from a Poisson distribution with unknown parameter λ. Assuming that λ is a value assumed by a $G(\alpha, \beta)$ RV, find a Bayesian confidence interval for λ.

15. Let $X_1, X_2, \ldots, X_n$ be a sample from a geometric distribution with parameter θ. Assuming that θ has a priori PDF that is given by the density of a $B(\alpha, \beta)$ RV, find a Bayesian confidence interval for θ.

16. Let $X_1, X_2, \ldots, X_n$ be a sample from $\mathcal{N}(\mu, 1)$, and suppose that the a priori PDF for μ is $U(-1, 1)$. Find a Bayesian confidence interval for μ.

11.4 SHORTEST-LENGTH CONFIDENCE INTERVALS

We have already remarked that we can increase the confidence level by simply taking a larger-length confidence interval. Indeed, the worthless interval $-\infty < \theta < \infty$, which simply says that θ is a point on the real line, has confidence level 1. In practice, one would like to set the level at a given fixed number $1-\alpha$ $(0 < \alpha < 1)$ and, if possible, construct an interval as short in length as possible among all confidence intervals with the same level. Such an interval is desirable since it is more informative. We have already remarked that shortest-length confidence intervals do not always exist. In this section we will investigate the possibility of constructing shortest-length confidence intervals based on simple RVs. The discussion here is based on Guenther [37]. Theorem 11.3.1 is really the key to the following discussion.

Let $X_1, X_2, \ldots, X_n$ be a sample from a PDF $f_\theta(x)$, and $T(X_1, X_2, \ldots, X_n, \theta) = T_\theta$ be a pivot for θ. Also, let $\lambda_1 = \lambda_1(\alpha)$, $\lambda_2 = \lambda_2(\alpha)$ be chosen so that

$$P\{\lambda_1 < T_\theta < \lambda_2\} = 1-\alpha, \tag{1}$$

and suppose that (1) can be rewritten as

$$P\{\underline{\theta}(\mathbf{X}) < \theta < \overline{\theta}(\mathbf{X})\} = 1-\alpha. \tag{2}$$

For every T_θ, λ_1 and λ_2 can be chosen in many ways. We would like to choose λ_1 and λ_2 so that $\overline{\theta} - \underline{\theta}$ is minimum. Such an interval is a $(1-\alpha)$-level shortest-length confidence interval based on T_θ. It may be possible, however, to find another RV T_θ^* that may yield an even shorter interval. Therefore we are not asserting that the procedure, if it succeeds, will lead to a $(1-\alpha)$-level confidence interval that has shortest length among all intervals of this level. For T_θ we use the simplest RV that is a function of a sufficient statistic and θ.

Remark 1. An alternative to minimizing the length of the confidence interval is to minimize the expected length $E_\theta\{\overline{\theta}(\mathbf{X}) - \underline{\theta}(\mathbf{X})\}$. Unfortunately, this also is quite unsatisfactory since, in general, there does not exist a member of the class of all $(1-\alpha)$-level confidence intervals that minimizes $E_\theta\{\overline{\theta}(\mathbf{X}) - \underline{\theta}(\mathbf{X})\}$ for all θ. The procedures applied in finding the shortest-length confidence interval based on a pivot are also applicable in finding an interval that minimizes the expected length. We remark here that the restriction to unbiased confidence intervals is natural if we wish to minimize $E_\theta\{\overline{\theta}(\mathbf{X}) - \underline{\theta}(\mathbf{x})\}$. See Section 11.5 for definitions and further details.

Example 1. Let $X_1, X_2, \ldots, X_n$ be sample from $\mathcal{N}(\mu, \sigma^2)$, where σ^2 is known. Then $\overline{X}$ is sufficient for μ and take

$$T_\mu(\mathbf{X}) = \frac{\overline{X} - \mu}{\sigma/\sqrt{n}}.$$

Then

$$1-\alpha = P\left\{a < \frac{\overline{X}-\mu}{\sigma}\sqrt{n} < b\right\} = P\left\{\overline{X} - b\frac{\sigma}{\sqrt{n}} < \mu < \overline{X} - a\frac{\sigma}{\sqrt{n}}\right\}.$$

The length of this confidence interval is $(\sigma/\sqrt{n})(b-a)$. We wish to minimize $L = (\sigma/\sqrt{n})$ $(b-a)$ such that

$$\Phi(b) - \Phi(a) = \frac{1}{\sqrt{2\pi}}\int_a^b e^{-x^2/2}\,dx = \int_a^b \varphi(x)\,dx = 1-\alpha.$$

Here φ and Φ, respectively, are the PDF and DF of an $\mathcal{N}(0,1)$ RV. Thus

$$\frac{dL}{da} = \frac{\sigma}{\sqrt{n}}\left(\frac{db}{da} - 1\right)$$

and

$$\varphi(b)\frac{db}{da} - \varphi(a) = 0,$$

giving

$$\frac{dL}{da} = \frac{\sigma}{\sqrt{n}}\left[\frac{\varphi(a)}{\varphi(b)} - 1\right].$$

The minimum occurs when $\varphi(a) = \varphi(b)$, that is, when $a = b$ or $a = -b$. Since $a = b$ does not satisfy

$$\int_a^b \varphi(t)\,dt = 1-\alpha,$$

we choose $a = -b$. The shortest confidence interval based on T_μ is therefore the equals-tails interval,

$$\left(\overline{X} + z_{1-\alpha/2}\frac{\sigma}{\sqrt{n}}, \overline{X} + z_{\alpha/2}\frac{\sigma}{\sqrt{n}}\right) \text{ or } \left(\overline{X} - z_{\alpha/2}\frac{\sigma}{\sqrt{n}}, \overline{X} + z_{\alpha/2}\frac{\sigma}{\sqrt{n}}\right).$$

The length of this interval is $2z_{\alpha/2}(\sigma/\sqrt{n})$. In this case we can plan our experiment to give a prescribed confidence level and a prescribed length for the interval. To have level $1-\alpha$ and length $\leq 2d$, we choose the smallest n such that

$$d \geq z_{\alpha/2}\frac{\sigma}{\sqrt{n}} \qquad \text{or} \qquad n \geq z_{\alpha/2}^2\frac{\sigma^2}{d^2}.$$

This can also be interpreted as follows. If we estimate μ by $\overline{X}$, taking a sample of size $n \geq z_{\alpha/2}^2(\sigma^2/d^2)$, we are $100(1-\alpha)$ percent confident that the error in our estimate is at most d.

Example 2. In Example 1, suppose that σ is unknown. In that case we use

$$T_\mu(\mathbf{X}) = \frac{\overline{X}-\mu}{S}\sqrt{n}$$

as a pivot. T_μ has Student's t-distribution with $n-1$ d.f. Thus

$$1-\alpha = P\left\{a < \frac{\overline{X}-\mu}{S}\sqrt{n} < b\right\} = P\left\{\overline{X} - b\frac{S}{\sqrt{n}} < \mu < \overline{X} - a\frac{S}{\sqrt{n}}\right\}.$$

We wish to minimize

$$L = (b-a)\frac{S}{\sqrt{n}}$$

subject to

$$\int_a^b f_{n-1}(t)\,dt = 1-\alpha,$$

where $f_{n-1}(t)$ is the PDF of T_μ. We have

$$\frac{dL}{da} = \left(\frac{db}{da} - 1\right)\frac{S}{\sqrt{n}} \qquad \text{and} \qquad f_{n-1}(b)\frac{db}{da} - f_{n-1}(a) = 0,$$

giving

$$\frac{dL}{da} = \left[\frac{f_{n-1}(a)}{f_{n-1}(b)} - 1\right]\frac{S}{\sqrt{n}}.$$

It follows that the minimum occurs at $a = -b$ (the other solution, $a = b$, is not admissible). The shortest-length confidence interval based on T_μ is the equal-tails interval,

$$\left(\overline{X} - t_{n-1,\alpha/2}\frac{S}{\sqrt{n}}, \overline{X} + t_{n-1,\alpha/2}\frac{S}{\sqrt{n}}\right).$$

The length of this interval is $2t_{n-1,\alpha/2}(S/\sqrt{n})$, which, being random, may be arbitrarily large. Note that the same confidence interval minimizes the expected length of the interval, namely, $EL = (b-a)c_n(\sigma/\sqrt{n})$, where c_n is a constant determined from $ES = c_n\sigma$ and the minimum expected length is $2t_{n-1,\alpha/2}c_n(\sigma/\sqrt{n})$.

Example 3. Let $X_1, X_2, \ldots, X_n$ be iid $\mathcal{N}(\mu, \sigma^2)$ RVs. Suppose that μ is known and we want a confidence interval for σ^2. The obvious choice for a pivot T_{σ^2} is given by

$$T_{\sigma^2}(\mathbf{x}) = \frac{\sum_1^n (X_i-\mu)^2}{\sigma^2},$$

which has a chi-square distribution with n d.f. Now

$$P\left\{a < \frac{\sum_1^n (X_i-\mu)^2}{\sigma^2} < b\right\} = 1-\alpha,$$

so that

$$P\left\{\frac{\sum_1^n (X_i-\mu)^2}{b} < \sigma^2 < \frac{\sum_1^n (X_i-\mu^2)}{a}\right\} = 1-\alpha.$$

We wish to minimize

$$L = \left(\frac{1}{a} - \frac{1}{b}\right)\sum_1^n (X_i-\mu)^2$$

subject to

$$\int_a^b f_n(t)\,dt = 1-\alpha,$$

where f_n is the PDF of a chi-square RV with n d.f. We have

$$\frac{dL}{da} = -\left(\frac{1}{a^2} - \frac{1}{b^2}\frac{db}{da}\right)\sum_1^n (X_i-\mu)^2$$

and

$$\frac{db}{da} = \frac{f_n(a)}{f_n(b)},$$

so that

$$\frac{dL}{da} = -\left[\frac{1}{a^2} - \frac{1}{b^2}\frac{f_n(a)}{f_n(b)}\right]\sum_1^n (X_i-\mu)^2,$$

which vanishes if

$$\frac{1}{a^2} = \frac{1}{b^2}\frac{f_n(a)}{f_n(b)}.$$

Numerical results giving values of a and b to four significant places of decimals are available (see Tate and Klett [112]). In practice, the simpler equal-tails interval,

$$\left(\frac{\sum_{i=1}^{n}(X_i-\mu)^2}{\chi^2_{n,\alpha/2}}, \frac{\sum_{i=1}^{n}(X_i-\mu)^2}{\chi^2_{n,1-\alpha/2}}\right),$$

may be used.

If μ is unknown, we use

$$T_{\sigma^2}(\mathbf{X}) = \frac{\sum_1^n (X_i-\overline{X})^2}{\sigma^2} = (n-1)\frac{S^2}{\sigma^2}$$

as a pivot. T_{σ^2} has a $\chi^2(n-1)$ distribution. Proceeding as above, we can show that the shortest-length confidence interval based on T_{σ^2} is $((n-1)(S^2/b),(n-1)(S^2/a))$; here a and b are a solution of

$$P\{a < \chi^2(n-1) < b\} = 1-\alpha$$

and

$$a^2 f_{n-1}(a) = b^2 f_{n-1}(b),$$

where f_{n-1} is the PDF of a $\chi^2(n-1)$ RV. Numerical solutions due to Tate and Klett [112] may be used, but, in practice, the simpler equal-tails confidence interval,

$$\left(\frac{(n-1)S^2}{\chi^2_{n-1,\alpha/2}}, \frac{(n-1)S^2}{\chi^2_{n-1,1-\alpha/2}}\right)$$

is employed.

Example 4. Let $X_1,X_2,\ldots,X_n$ be a sample from $U(0,\theta)$. Then $X_{(n)}$ is sufficient for θ with density

$$f_n(y) = n\frac{y^{n-1}}{\theta^n}, \qquad 0<y<\theta.$$

The RV $T_\theta = X_{(n)}/\theta$ has PDF

$$h(t) = nt^{n-1}, \qquad 0<t<1.$$

Using T_θ as pivot, we see that the confidence interval is $(X_{(n)}/b, X_{(n)}/a)$ with length $L = X_{(n)}(1/a-1/b)$. We minimize L subject to

$$\int_a^b nt^{n-1}\,dt = b^n - a^n = 1-\alpha.$$

Now

$$(1-\alpha)^{1/n} < b \le 1$$

and

$$\frac{dL}{db} = X_{(n)}\left(-\frac{1}{a^2}\frac{da}{db} + \frac{1}{b^2}\right) = X_{(n)}\left(\frac{a^{n+1} - b^{n+1}}{b^2 a^{n+1}}\right) < 0,$$

so that the minimum occurs at $b = 1$. The shortest interval is therefore $(X_{(n)}, X_{(n)}/\alpha^{1/n})$. Note that

$$EL = \left(\frac{1}{a} - \frac{1}{b}\right) EX_{(n)} = \frac{n\theta}{n+1}\left(\frac{1}{a} - \frac{1}{b}\right),$$

which is minimized subject to

$$b^n - a^n = 1 - \alpha,$$

where $b = 1$ and $a = \alpha^{1/n}$. The expected length of the interval that minimizes EL is $[(1/\alpha^{1/n}) - 1][n\theta/(n+1)]$, which is also the expected length of the shortest confidence interval based on $X_{(n)}$.

Note that the length of the interval $(X_{(n)}, \alpha^{-1/n}X_{(n)})$ goes to 0 as $n \to \infty$.

For some results on asymptotically shortest-length confidence intervals, we refer the reader to Wilks [118, pp. 374–376].

PROBLEMS 11.4

1. Let $X_1, X_2, \ldots, X_n$ be a sample from

$$f_\theta(x) = \begin{cases} e^{-(x-\theta)} & \text{if } x > \theta, \\ 0 & \text{otherwise.} \end{cases}$$

Find the shortest-length confidence interval for θ at level $1-\alpha$ based on a sufficient statistic for θ.

2. Let $X_1, X_2, \ldots, X_n$ be a sample from $G(1, \theta)$. Find the shortest-length confidence interval for θ at level $1-\alpha$, based on a sufficient statistic for θ.

3. In Problem 11.3.9 how will you find the shortest-length confidence interval for θ at level $1-\alpha$, based on the statistic X/θ?

4. Let $T(\mathbf{X}, \theta)$ be a pivot of the form $T(\mathbf{X}, \theta) = T_1(\mathbf{X}) - \theta$. Show how one can construct a confidence interval for θ with fixed width d and maximum possible confidence coefficient. In particular, construct a confidence interval that has fixed width d and maximum possible confidence coefficient for the mean μ of a normal population with variance 1. Find the smallest size n for which this confidence interval has a confidence coefficient $\geq 1-\alpha$. Repeat the above in sampling from an exponential PDF

$$f_\mu(x) = e^{\mu-x} \text{ for } x > \mu \text{ and } f_\mu(x) = 0 \qquad \text{for } x \leq \mu.$$

(Desu [21])

5. Let $X_1, X_2, \ldots, X_n$ be a random sample from

$$f_\theta(x) = \frac{1}{2\theta}\exp\{-|x|/\theta\}, \quad x \in \mathcal{R}, \quad \theta > 0.$$

Find the shortest-length $(1-\alpha)$-level confidence interval for θ, based on the sufficient statistic $\sum_{i=1}^{n} |X_i|$.

6. In Example 4, let $R = X_{(n)} - X_{(1)}$. Find a $(1-\alpha)$-level confidence interval for θ of the form $(R, R/c)$. Compare the expected length of this interval to the one computed in Example 4.

7. Let $X_1, X_2, \ldots, X_n$ be a random sample from a Pareto PDF $f_\theta(x) = \theta/x^2$, $x > \theta$, and $= 0$ for $x \leq \theta$. Show that the shortest-length confidence interval for θ based on $X_{(1)}$ is $(X_{(1)}\alpha^{1/n}, X_{(1)})$. (Use $\theta/X_{(1)}$ as a pivot.)

8. Let $X_1, X_2, \ldots, X_n$ be a sample from PDF $f_\theta(x) = 1/(\theta_2 - \theta_1)$, $\theta_1 \leq x \leq \theta_2$, $\theta_1 < \theta_2$ and $= 0$ otherwise. Let $R = X_{(n)} - X_{(1)}$. Using $R/(\theta_2 - \theta_1)$ as a pivot for estimating $\theta_2 - \theta_1$, show that the shortest-length confidence interval is of the form $(R, R/c)$, where c is determined from the level as a solution of $c^{n-1}\{(n-1)c - n\} + \alpha = 0$ (Ferentinos [27]).

11.5 UNBIASED AND EQUIVARIANT CONFIDENCE INTERVALS

In Section 11.3 we studied test inversion as one of the methods of constructing confidence intervals. We showed that UMP tests lead to UMA confidence intervals. In Chapter 9 we saw that UMP tests generally do not exist. In such situations we either restrict consideration to smaller subclasses of tests by requiring that the test functions have some desirable properties, or we restrict the class of alternatives to those near the null parameter values.

In this section will follow a similar approach in constructing confidence intervals.

Definition 1. A family $\{S(\mathbf{x})\}$ of confidence sets for a parameter θ is said to be *unbiased* at confidence level $1-\alpha$ if

$$P_\theta\{S(\mathbf{X}) \text{ contains } \theta\} \geq 1-\alpha \tag{1}$$

and

$$P_\theta\{S(\mathbf{X}) \text{ contains } \theta'\} \leq 1-\alpha \qquad \text{for all } \theta, \theta' \in \Theta, \theta \neq \theta'. \tag{2}$$

If $S(\mathbf{X})$ is an interval satisfying (1) and (2), we call it a $(1-\alpha)$-level unbiased confidence interval. If a family of unbiased confidence sets at level $1-\alpha$ is UMA in the class of all

$(1-\alpha)$-level unbiased confidence sets, we call it a UMA unbiased (UMAU) family of confidence sets at level $1-\alpha$. In other words if $S^*(\mathbf{x})$ satisfies (1) and (2) and minimizes

$$P_\theta\{S(\mathbf{X}) \text{ contains } \theta'\} \qquad \text{for } \theta,\ \theta' \in \Theta,\ \ \theta \neq \theta'$$

among all unbiased families of confidence sets $S(\mathbf{X})$ at level $1-\alpha$, then $S^*(\mathbf{X})$ is a UMAU family of confidence sets at level $1-\alpha$.

Remark 1. Definition 1 says that a family $S(\mathbf{X})$ of confidence sets for a parameter θ is unbiased at level $1-\alpha$ if the probability of true coverage is at least $1-\alpha$ and that of false coverage is at most $1-\alpha$. In other words, $S(\mathbf{X})$ traps a true parameter value more often than it does a false one.

Theorem 1. Let $A(\theta_0)$ be the acceptance region of a UMP unbiased size α test of $H_0(\theta_0)\colon \theta=\theta_0$ against $H_1(\theta_0)\colon \theta \neq \theta_0$ for each θ_0. Then $S(\mathbf{x})=\{\theta\colon \mathbf{x} \in A(\theta)\}$ is a UMA unbiased family of confidence sets at level $1-\alpha$.

Proof. To see that $S(\mathbf{x})$ is unbiased we note that, since $A(\theta)$ is the acceptance region of an unbiased test,

$$P_\theta\{S(\mathbf{X}) \text{ contains } \theta'\} = P_\theta\{\mathbf{X} \in A(\theta')\} \leq 1-\alpha.$$

We next show that $S(\mathbf{X})$ is UMA. Let $S^*(\mathbf{x})$ be any other unbiased $(1-\alpha)$-level family of confidence sets, and write $A^*(\theta)=\{\mathbf{x}\colon S^*(\mathbf{x}) \text{ contains } \theta\}$. Then $P_\theta\{\mathbf{X} \in A^*(\theta')\} = P_\theta\{S^*(\mathbf{X}) \text{ contains } \theta'\} \leq 1-\alpha$, and it follows that $A^*(\theta)$ is the acceptance region of an unbiased size α test. Hence

$$\begin{aligned} P_\theta\{S^*(\mathbf{X}) \text{ contains } \theta'\} &= P_\theta\{\mathbf{X} \in A^*(\theta')\} \\ &\geq P_\theta\{\mathbf{X} \in A(\theta')\} \\ &= P_\theta\{S(\mathbf{X}) \text{ contains } \theta'\}. \end{aligned}$$

The inequality follows since $A(\theta)$ is the acceptance region of a UMP unbiased test. This completes the proof.

Example 1. Let $X_1, X_2, \ldots, X_n$ be a sample from $\mathcal{N}(\mu, \sigma^2)$ where both μ and σ^2 are unknown. For testing $H_0\colon \mu=\mu_0$ against $H_1\colon \mu \neq \mu_0$, it is known (Ferguson [28, p. 232]) that the t-test

$$\varphi(\mathbf{x}) = \begin{cases} 1, & \dfrac{|\sqrt{n}(\overline{x}-\mu_0)|}{s} > c, \\ 0, & \text{otherwise}, \end{cases}$$

where $\overline{x}=\sum x_i/n$ and $s^2=(n-1)^{-1}\sum(x_i-\overline{x})^2$ is UMP unbiased. We choose c from the size requirement

$$\alpha = P_{\mu=\mu_0}\left\{\left|\frac{\sqrt{n}(\overline{X}-\mu_0)}{S}\right| > c\right\},$$

so that $c = t_{n-1,\alpha/2}$. Thus

$$A(\mu_0) = \left\{ \mathbf{x}\colon \left| \frac{\sqrt{n}\,(\bar{x}-\mu_0)}{s} \right| \le t_{n-1,\alpha/2} \right\}$$

is the acceptance region of a UMP unbiased size α test of $H_0\colon \mu = \mu_0$ against $H_1 : \mu \neq \mu_0$. By Theorem 1, it follows that

$$\begin{aligned} S(\mathbf{x}) &= \{\mu\colon \mathbf{x} \in A(\mu)\} \\ &= \left\{ \bar{x} - \frac{s}{\sqrt{n}} t_{n-1,\alpha/2} \le \mu \le \bar{x} + \frac{s}{\sqrt{n}} t_{n-1,\alpha/2} \right\} \end{aligned}$$

is a UMA unbiased family of confidence sets at level $1-\alpha$.

If the measure of precision of a confidence interval is its expected length, one is naturally led to a consideration of unbiased confidence intervals. Pratt [81] has shown that the expected length of a confidence interval is the average of false coverage probabilities.

Theorem 2. Let Θ be an interval on the real line and f_θ be the PDF of $\mathbf{X}$. Let $S(\mathbf{X})$ be a family of $(1-\alpha)$-level confidence intervals of finite length, that is, let $S(\mathbf{X}) = (\underline{\theta}(\mathbf{X}), \overline{\theta}(\mathbf{X}))$, and suppose that $\overline{\theta}(\mathbf{X}) - \underline{\theta}(\mathbf{X})$ is (random) finite. Then

$$\int (\overline{\theta}(\mathbf{x}) - \underline{\theta}(\mathbf{x})) f_\theta(\mathbf{x})\, d\mathbf{x} = \int_{\theta' \neq \theta} P_\theta\{S(\mathbf{X}) \text{ contains } \theta'\}\, d\theta' \tag{3}$$

for all $\theta \in \Theta$.

Proof. We have

$$\overline{\theta} - \underline{\theta} = \int_{\underline{\theta}}^{\overline{\theta}} d\theta'.$$

Thus for all $\theta \in \Theta$

$$\begin{aligned} E_\theta\{\overline{\theta}(\mathbf{X}) - \underline{\theta}(\mathbf{X})\} &= E_\theta\left\{ \int_{\underline{\theta}}^{\overline{\theta}} d\theta' \right\} \\ &= \int f_\theta(\mathbf{x}) \left\{ \int_{\underline{\theta}}^{\overline{\theta}} d\theta' \right\} d\mathbf{x} \\ &= \int \left\{ \int_{\underline{\theta}}^{\overline{\theta}} f_\theta(\mathbf{x})\, d\mathbf{x} \right\} d\theta' \\ &= \int P_\theta\{S(\mathbf{X}) \text{ contains } \theta'\}\, d\theta' \\ &= \int_{\theta' \neq \theta} P_\theta\{S(\mathbf{X}) \text{ contains } \theta'\}\, d\theta'. \end{aligned}$$

Remark 2. If $S(\mathbf{X})$ is a family of UMAU $(1-\alpha)$-level confidence intervals, the expected length of $S(\mathbf{X})$ is minimal. This follows since the left-hand side of (3) is the expected length, if θ is the true value, of $S(\mathbf{X})$ and $P_\theta\{S(\mathbf{X}) \text{ contains } \theta'\}$ is minimal [because $S(\mathbf{X})$ is UMAU], by Theorem 1, with respect to all families of $1-\alpha$ unbiased confidence intervals uniformly in $\theta(\theta \neq \theta')$.

Since a reasonably complete discussion of UMP unbiased tests (see Section 9.5) is beyond the scope of this text, the following procedure for determining unbiased confidence intervals is sometimes quite useful (see Guenther [38]). Let $X_1, X_2, \ldots, X_n$ be a sample from an absolutely continuous DF with PDF $f_\theta(x)$ and suppose that we seek an unbiased confidence interval for θ. Following the discussion in Section 11.4, suppose that

$$T(X_1, X_2, \ldots, X_n, \theta) = T(\mathbf{X}, \theta) = T_\theta$$

is a pivot, and suppose that the statement

$$P\{\lambda_1(\alpha) < T_\theta < \lambda_2(\alpha)\} = 1-\alpha$$

can be converted to

$$P_\theta\{\underline{\theta}(\mathbf{X}) < \theta < \overline{\theta}(\mathbf{X})\} = 1-\alpha.$$

In order for $(\underline{\theta}, \overline{\theta})$ to be unbiased, we must have

$$P(\theta, \theta') = P_\theta\{\underline{\theta}(\mathbf{X}) < \theta' < \overline{\theta}(\mathbf{X})\} = 1-\alpha \qquad \text{if } \theta' = \theta \tag{4}$$

and

$$P(\theta, \theta') < 1-\alpha \qquad \text{if } \theta' \neq \theta. \tag{5}$$

If $P(\theta, \theta')$ depends only on a function γ of θ, θ', we may write

$$P(\gamma) \begin{cases} = 1-\alpha & \text{if } \theta' = \theta, \\ < 1-\alpha & \text{if } \theta' \neq \theta, \end{cases} \tag{6}$$

and it follows that $P(\gamma)$ has a maximum at $\theta' = \theta$.

Example 2. Let $X_1, X_2, \ldots, X_n$ be iid $\mathcal{N}(\mu, \sigma^2)$ RVs, and suppose that we desire an unbiased confidence interval for σ^2. Then

$$T(\mathbf{X}, \sigma^2) = \frac{(n-1)S^2}{\sigma^2} = T_\sigma$$

has a $\chi^2(n-1)$ distribution, and we have

$$P\left\{\lambda_1 < (n-1)\frac{S^2}{\sigma^2} < \lambda_2\right\} = 1-\alpha,$$

so that

$$P\left\{(n-1)\frac{S^2}{\lambda_2} < \sigma^2 < (n-1)\frac{S^2}{\lambda_1}\right\} = 1-\alpha.$$

Then

$$\begin{aligned} P(\sigma^2,\sigma'^2) &= P_{\sigma^2}\left\{(n-1)\frac{S^2}{\lambda_2} < \sigma'^2 < (n-1)\frac{S^2}{\lambda_1}\right\} \\ &= P\left\{\frac{T_\sigma}{\lambda_2} < \gamma < \frac{T_\sigma}{\lambda_1}\right\}, \end{aligned}$$

where $\gamma = \sigma'^2/\sigma^2$ and $T_\sigma \sim \chi^2(n-1)$. Thus

$$P(\gamma) = P\{\lambda_1\gamma < T_\sigma < \lambda_2\gamma\}.$$

Then

$$P(1) = 1-\alpha$$

and

$$P(\gamma) < 1-\alpha.$$

Thus we need λ_1, λ_2 such that

$$P(1) = 1-\alpha \tag{7}$$

and

$$\left.\frac{dP(\gamma)}{d\gamma}\right|_{\gamma=1} = \lambda_2 f_{n-1}(\lambda_2) - \lambda_1 f_{n-1}(\lambda_1) = 0, \tag{8}$$

where f_{n-1} is the PDF of T_σ. Equations (7) and (8) have been solved numerically for λ_1, λ_2 by several authors (see, for example, Tate and Klett [112]). Having obtained λ_1, λ_2 from (7) and (8), we have as the unbiased $(1-\alpha)$-level confidence interval

$$\left((n-1)\frac{S^2}{\lambda_2}, (n-1)\frac{S^2}{\lambda_1}\right). \tag{9}$$

Note that in this case the shortest-length confidence interval (based on T_σ) derived in Example 11.4.3, the usual equal-tails confidence interval, and (9) are all different. The length of the confidence interval (9), however, can be considerably greater than that of the shortest interval of Example 11.4.3. For large n all three sets of intervals are approximately the same.

Finally, let us briefly investigate how invariance considerations apply to confidence estimation. Let $\mathbf{X} = (X_1, X_2, \ldots, X_n) \sim f_\theta$, $\theta \in \Theta \subseteq \mathcal{R}$. Let $\mathcal{G}$ be a group of transformations on $\mathcal{X}$ which leaves $\mathcal{P} = \{f_\theta : \theta \in \Theta\}$ invariant. Let $S(\mathbf{X})$ be a $(1-\alpha)$-level confidence set for θ.

Definition 2. Let $\mathcal{P}$ be invariant under $\mathcal{G}$ and let $S(\mathbf{x})$ be a confidence set for θ. Then S is *equivariant* under $\mathcal{G}$, if for every $\mathbf{x} \in \mathcal{X}$, $\theta \in \Theta$ and $g \in \mathcal{G}$

$$S(\mathbf{x}) \in \theta \Leftrightarrow S(g(\mathbf{x})) \ni \bar{g}\theta. \tag{10}$$

Example 3. Let $X_1, X_2, \ldots, X_n$ be a sample from PDF

$$f_\theta(x) = \exp\{-(x-\theta)\}, \quad x > \theta$$

and $= 0$ if $x \le \theta$. Let $\mathcal{G} = \{\{a,1\} : a \in \mathcal{R}\}$, where $\{a,1\}\mathbf{x} = (x_1 + a, x_2 + a, \ldots, x_n + a)$ and $\mathcal{G}$ induces $\bar{\mathcal{G}} = \mathcal{G}$ on $\Theta = \mathcal{R}$. The family $\{f_\theta\}$ remains invariant under $\mathcal{G}$. Consider a confidence interval of the form

$$S(\mathbf{x}) = \{\theta : \bar{x} - c_1 \le \theta \le \bar{x} + c_2\}$$

where c_1, c_2 are constants. Then

$$S(\{a,1\}\mathbf{x}) = \{\theta : \bar{x} + a - c_1 \le \theta \le \bar{x} + a - c_2\}.$$

Clearly,

$$\begin{aligned} S(\mathbf{x}) \ni \theta &\Longleftrightarrow \bar{x} + a - c_1 \le \theta + a \le \bar{x} + a - c_2 \\ &\Longleftrightarrow S(\{a,1\}\mathbf{x}) \ni \bar{g}\theta \end{aligned}$$

and it follows that $S(\mathbf{x})$ is an equivariant confidence interval.

The most useful method of constructing invariant confidence intervals is test inversion. Inverting the acceptance region of invariant tests often leads to equivariant confidence intervals under certain conditions. Recall that a group $\mathcal{G}$ of transformations leaves a hypothesis testing problem invariant if $\mathcal{G}$ leaves both Θ_0 and Θ_1 invariant. For each $H_0 : \theta = \theta_0$, $\theta_0 \in \Theta$, we have a different group of transformations, $\mathcal{G}_{\theta_0}$, which leaves the problem of testing $\theta = \theta_0$ invariant. The equivariant confidence interval, on the other hand, must be equivariant with respect to $\mathcal{G}$, which is a much larger group since $\mathcal{G} \supset \mathcal{G}_{\theta_0}$ for all θ_0. The relationship between an equivariant confidence set and invariant tests is more complicated when the family $\mathcal{P}$ has a nuisance parameter τ.

Under certain conditions there is a relationship between equivariant confidence sets and associated invariant tests. Rather than pursue this relationship, we refer the reader to Ferguson [28, p. 262]; it is generally easy to check that (10) holds for a given confidence interval S to show that S is invariant. The following example illustrates this point.

Example 4. Let $X_1, X_2, \ldots, X_n$ be iid $\mathcal{N}(\mu, \sigma^2)$ RVs where both μ and σ^2 are unknown. In Example 9.5.3 we showed that the test

$$\phi(\mathbf{x}) = \begin{cases} 1 & \text{if } \sum_1^n (x_i - \overline{x})^2 \le \sigma_0^2 \chi^2_{n-1,1-\alpha} \\ 0 & \text{otherwise} \end{cases}$$

is UMP invariant, under translation group for testing $H_0 : \sigma^2 \ge \sigma_0^2$ against $H_1 : \sigma^2 < \sigma_0^2$. Then the acceptance region of ϕ is

$$A(\mathbf{x}) = \left\{ \mathbf{x} : \sum_1^n (x_i - \overline{x})^2 > \sigma_0^2 \chi^2_{n-1,1-\alpha} \right\}.$$

Clearly,

$$\mathbf{x} \in A(\mathbf{x}) \Longleftrightarrow \sigma_0^2 < (n-1)s^2 / \chi^2_{n-1,1-\alpha}$$

and it follows that

$$S(\mathbf{x}) = \{\sigma^2 : \sigma^2 < (n-1)s^2 / \chi^2_{n-1,1-\alpha}\}$$

is a $(1-\alpha)$-level confidence interval (upper confidence bound) for σ^2. We show that S is invariant with respect to the scale group. In fact

$$S(\{0,c\}\mathbf{x}) = \{\sigma^2 : \sigma^2 < c^2 (n-1)s^2 / \chi^2_{n-1,1-\alpha}\}$$

and

$$\sigma^2 < (n-1)s^2 / \chi^2_{n-1,1-\alpha} \Longleftrightarrow S(\{0,c\}\mathbf{x}) \ni \bar{g}\sigma^2 = \{0,c\}\sigma^2$$

and it follows that $S(\mathbf{x})$ is an equivariant confidence interval for σ^2.

PROBLEMS 11.5

1. Let $X_1, X_2, \ldots, X_n$ be a sample from $U(0, \theta)$. Show that the unbiased confidence intervals for θ based on the pivot $\max X_i / \theta$ coincides with the shortest-length confidence interval based on the same pivot.
2. Let $X_1, X_2, \ldots, X_n$ be a sample from $G(1, \theta)$. Find the unbiased confidence interval for θ based on the pivot $2\sum_{i=1}^n X_i / \theta$.
3. Let $X_1, X_2, \ldots, X_n$ be a sample from PDF

$$f_\theta(x) = \begin{cases} e^{-(x-\theta)} & \text{if } x > \theta \\ 0 & \text{otherwise.} \end{cases}$$

   Find the unbiased confidence interval based on the pivot $2n[\min X_i - \theta]$.
4. Let $X_1, X_2, \ldots, X_n$ be iid $\mathcal{N}(\mu, \sigma^2)$ RVs where both μ and σ^2 are unknown. Using the pivot $T_{\mu,\sigma} = \sqrt{n}(\overline{X} - \mu)/S$ show that the shortest-length unbiased $(1-\alpha)$-level confidence interval for μ is the equal-tails interval $(\overline{X} - t_{n-1,\alpha/2} S/\sqrt{n}, \overline{X} + t_{n-1,\alpha/2} S/\sqrt{n})$.

5. Let $X_1,X_2,\ldots,X_n$ be iid with PDF $f_\theta(x) = \theta/x^2$, $x \geq \theta$, and $= 0$ otherwise. Find the shortest-length $(1-\alpha)$-level unbiased confidence interval for θ based on the pivot $\theta/X_{(1)}$.
6. Let $X_1,X_2,\ldots,X_n$ be a random sample from a location family $\mathcal{P} = \{f_\theta(x) = f(x-\theta); \theta \in \mathcal{R}\}$. Show that a confidence interval of the form $S(\mathbf{x}) = \{\theta : T(\mathbf{x}) - c_1 \leq \theta \leq T(\mathbf{x}) + c_2\}$ where $T(\mathbf{x})$ is an equivariant estimate under location group is an equivariant confidence interval.
7. Let $X_1,X_2,\ldots,X_n$ be iid RVs with common scale PDF $f_\sigma(x) = \frac{1}{\sigma}f(x/\sigma)$, $\sigma > 0$. Consider the scale group $\mathcal{G} = \{\{0,b\} : b > 0\}$. If $T(\mathbf{x})$ is an equivariant estimate of σ, show that a confidence interval of the form

$$X(\mathbf{x}) = \left\{\sigma : c_1 \leq \frac{T(\mathbf{x})}{\sigma} \leq c_2\right\}$$

is equivariant.
8. Let $X_1,X_2,\ldots,X_n$ be iid RVs with PDF $f_\theta(x) = \exp\{-(x-\theta)\}$, $x > \theta$ and, $= 0$, otherwise. For testing $H_0 : \theta = \theta_0$ against $H_1 : \theta > \theta_0$, consider the (UMP) test

$$\phi(\mathbf{x}) = 1, \text{ if } X_{(1)} \geq \theta_0 - (\ell\text{n } \alpha)/n, \; = 0, \text{ otherwise.}$$

Is the acceptance region of this α-level test an equivariant $(1-\alpha)$-level confidence interval (lower bound) for θ with respect to the location group?

11.6 RESAMPLING: BOOTSTRAP METHOD

In many applications of statistical inference the investigator has a random sample from a population distribution DF F which may or may not be completely specified. Indeed the empirical data may not even fit any known distribution. The inference is typically based on some statistic such as $\overline{X}$, S^2, a percentile or some much more complicated statistic such as sample correlation coefficient or odds ratio. For this purpose we need to know the distribution of the statistic being used and/or its moments. Except for the simple situations such as those described in Chapter 6 this is not easy. And even if we are able to get a handle on it, it may be inconvenient to deal with it. Often, when the sample is large enough, one can resort to asymptotic approximations considered in Chapter 7. Alternatively, one can use computer-intensive techniques which have become quite popular in the last 25 years due to the availability of fast home or office laptops and desktops.

Suppose $x_1,x_2,\ldots,x_n$ is a random sample from a distribution F with unknown parameter $\theta(F)$, and let $\tilde{\theta}$ be an estimate of $\theta(F)$. What is the bias of $\tilde{\theta}$ and its SE? *Resampling* refers to sampling from $x_1,x_2,\ldots,x_n$ and using these samples to estimate the statistical properties of $\tilde{\theta}$. *Jackknife* is one such method where one uses subsets of the sample by excluding one or more observations at a time. For each of these subsamples an estimate $\tilde{\theta j}$ of θ is computed, and these estimates are then used to investigate the statistical properties of $\tilde{\theta}$.

The most commonly used resampling method is the *bootstrap*, introduced by Efron [22], where one draws random samples of size n, with replacement, from $x_1,x_2,\ldots,x_n$. This allows us to generate a large number of bootstrap samples and hence bootstrap estimates $\hat{\theta}_b$ of θ. This bootstrap distribution of $\hat{\theta}_b$ is then used to study the statistical properties of $\tilde{\theta}$.

Let $X^*_{b1}, X^*_{b2}, \ldots, X^*_{bn}$, $b = 1, 2, \ldots, B$, be iid RVs with common DF F^*_n, the empirical DF corresponding to the sample $x_1, x_2, \ldots, x_n$. Then $(X^*_{b1}, X^*_{b2}, \ldots, X^*_{bn})$ is called a *bootstrap sample*. Let θ be the parameter of interest associated with DF F and suppose we have chosen $\hat{\theta}$ to be an estimate of θ based on the sample $x_1, x_2, \ldots, x_n$. For each bootstrap sample let $\hat{\theta}_b$, $b = 1, 2, \ldots, B$, be the corresponding bootstrap estimate of θ. We can now study the statistical properties of $\hat{\theta}$ based on the distribution of the $\hat{\theta}_b$, $b = 1, 2, \ldots, B$, values. Let $\overline{\theta^*} = \Sigma_1^B \hat{\theta}_b / B$. Then the variance of $\hat{\theta}$ is estimated by the bootstrap variance.

$$\text{var}_{bs}(\hat{\theta}) = \text{var}(\hat{\theta}_b) = \frac{1}{B-1} \Sigma_{b=1}^{B} \left(\hat{\theta}_b - \overline{\theta}^*\right)^2. \qquad (1)$$

Similarly the bias of $\hat{\theta}$, $b(\theta) = \text{E}(\hat{\theta})$ - θ, is estimated by

$$\text{bias}_{bs}(\hat{\theta}) = \overline{\theta}^* - \hat{\theta}. \qquad (2)$$

Arranging the values of $\hat{\theta}_b$, $b = 1, 2, \ldots, B$, in increasing order of magnitude and then excluding $100\alpha/2$ percent smallest and largest values we get a $(1-\alpha)$-level confidence interval for θ. This is the so-called *percentile confidence interval*. One can also use this confidence interval to test hypotheses concerning θ.

Example 1. For this example we took a random sample of size 20 from a distribution on (.25, 1.25) with following results.

0.75	0.49	1.14	0.79	0.59	1.14	1.17	0.42	0.57	1.05
0.31	0.46	0.73	0.32	0.81	0.45	0.56	0.42	0.66	0.63

Suppose we wish to estimate the mean θ of the population distribution. For the sake of this illustration we use $\hat{\theta} = \bar{x}$ and use the bootstrap to estimate the SE of $\hat{\theta}$.

We took 1000 random samples, with replacement, of size 20 each from this sample with the following distribution of $\hat{\theta}_b$.

Interval	Frequency
0.49–0.56	6
0.53–0.57	29
0.57–0.61	109
0.61–0.65	200
0.65–0.69	234
0.69–0.73	229
0.73–0.77	123
0.77–0.81	59
0.81–0.85	10
0.85–0.89	2

The bootstrap estimate of θ is $\overline{\theta^*}$ =0.677 and that of the variance is 0.061. By excluding the smallest and the largest twenty-five vales of $\hat{\theta}_b$ a 95 percent confidence interval for θ is given by (0.564, 0.793). (We note that $\overline{x} = 0.673$ *and* $s^2 = 0.273$ so the $\text{SE}(\overline{x}) = .061$.)
Figure 1 show the frequency distribution of the bootstrap statistic $\hat{\theta}_b$.

It is natural to ask how well does the distribution of the bootstrap statistic $\hat{\theta}_b$ approximate the distribution of $\hat{\theta}$? The bootstrap approximation is often better when applied to the appropriately centered $\hat{\theta}$. Thus to estimate population mean θ bootstrap is applied to the centered sample mean $\overline{x} - \theta$. The corresponding bootstrapped version will then be $\overline{x}_b - \overline{x}$, where $\overline{x}_b$ is the sample mean of the bth bootstrap sample. Similarly if $\hat{\theta} = Z_{1/2} = \text{med}(X_1, X_2, \ldots, X_n)$ then the bootstrap approximation will be applied to the centered sample median $Z_{1/2} - F^{-1}(0.5)$. The bootstrap version will be then be $\text{med}(X^*_{b1}, X^*_{b2}, \ldots, X^*_{bn}) - Z_{1/2}$. Similarly , in estimation of the distribution of sample variance S^2, the bootstrap version will be applied to the ratio S^2/σ^2, where σ^2 is the variance of the DF F.

We have already considered the percentile method of constructing confidence intervals. Let us denote the αth percentile of the distribution of $\hat{\theta}_b$, b= $1, 2, \ldots, B$, by B_α. Suppose that the sampling distribution of $\hat{\theta} - \theta$ is approximated by the bootstrap distribution of $\hat{\theta}_b - \hat{\theta}$. Then the probability that $\hat{\theta} - \theta$ is covered by the interval $(B_{\alpha/2} - \hat{\theta}, B_{1-\alpha/2} + \hat{\theta})$ is approximately $(1 - \alpha)$. This is called a $(1 - \alpha)$-level *centered bootstrap percentile confidence interval* for θ.

Recall that in sampling from a normal distribution when both mean and the variance are unknown, a $(1 - \alpha)$-level confidence interval for the mean θ is based on t-statistic and is given by $(\overline{x} - t_{n-1,\alpha/2}, \overline{x} + t_{n-1,\alpha/2})$. For nonnormal distributions the bootstrap analog of the Student's t-statistic is the statistic $(\hat{\theta} - \theta)/(\hat{\sigma}/\sqrt{n})$. The bootstrap version is the statistic $T_b = (\hat{\theta}_b - \hat{\theta})/SE_b$, where SE_b is the SE computed from the bootstrap sample distribution. $A(1 - \alpha)$-level confidence interval is now easily constructed.

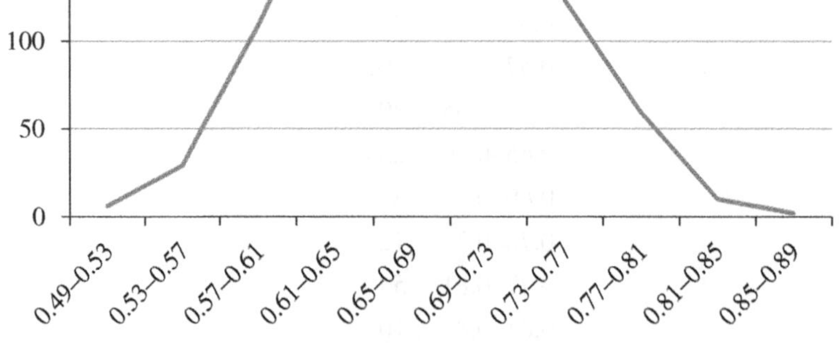

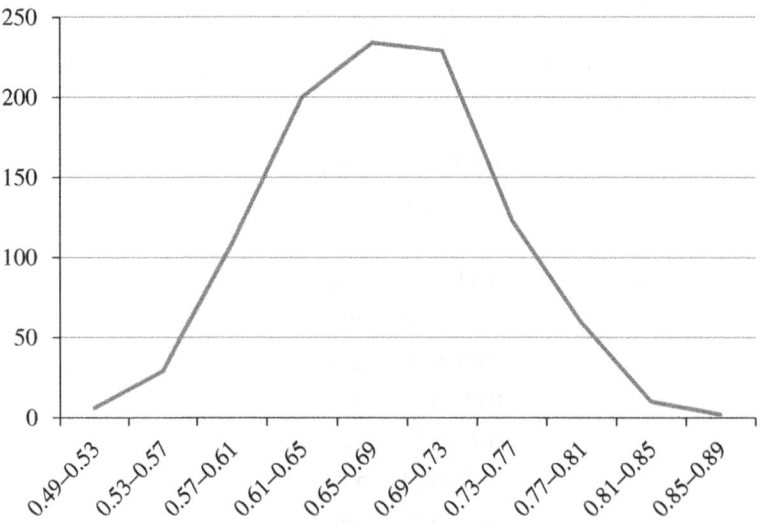

Fig. 1

In our discussion above we have assumed that $F(\theta)$ is completely unspecified. What if we know F except for the parameter θ? In that case we take bootstrap samples from the distribution $F(\hat{\theta})$.

We refer the reader to Efron and Tibshirani [23] for further details.

PROBLEMS 11.6

1. (a) Show that there are $\binom{2n-1}{n}$ distinct bootstrap samples of size n. [Hint: Problem 1.4.17.]

(b) What is the probability that a bootstrap sample is identical to the original samples?

(c) What is the most likely bootstrap sample to be drawn?

(d) What is the mean number of times that x_i appears in the bootstrap samples?

2. Let $x_1, x_2, \ldots, x_n$ be a random sample. Then $\hat{\mu} = \overline{x}$ is an estimate of the unknown mean μ. Consider the leave-one-out Jackknife sample. Let $\tilde{\mu}_i$ be the mean of the remaining $(n-1)$ observations when x_i is excluded:

(a) Show that $x_i = n\,\hat{\mu} - (n-1)\tilde{\mu}_i$.

(b) Now suppose we need to estimate a parameter θ and choose $\hat{\theta}$ to be an estimate from the sample. Imitating the Jackknife procedure for estimating μ we note that $\theta_i^* = n\hat{\theta} - (n-1)\tilde{\theta}_i$. What is the Jackknife estimate of θ? What is the Jackknife estimate of the bias of $\hat{\theta}$ and its variance?

3. Let $x_1, x_2, \ldots, x_n$ be a random sample from $\mathcal{N}(\theta,1)$ and suppose that $\overline{x}$ is an estimate of θ. Let $X_1^*, X_2^*, \ldots, X_n^*$ be a bootstrap sample from $\mathcal{N}(\overline{x},1)$. Show that both $\overline{X} - \theta$ and $\overline{X}^* - \overline{x}$ have the same $\mathcal{N}(0,1/n)$ distribution.

4. Consider the data set

2, 5, 3, 9.

Let $x_1^*, x_2^*, x_3^*, x_4^*$ be a bootstrap sample from this data set:

(a) Find the probability that the bootstrap mean equals 2.

(b) Find the probability that the maximum value of the bootstrap mean is 9.

(c) Find the probability that the bootstrap sample mean is 4.

12

GENERAL LINEAR HYPOTHESIS

12.1 INTRODUCTION

This chapter deals with the general linear hypothesis. In a wide variety of problems the experimenter is interested in making inferences about a vector parameter. For example, he may wish to estimate the mean of a multivariate normal or to test some hypotheses concerning the mean vector. The problem of estimation can be solved, for example, by resorting to the method of maximum likelihood estimation, discussed in Section 8.7. In this chapter we restrict ourselves to the so-called linear model problems and concern ourselves mainly with problems of hypotheses testing.

In Section 12.2 we formally describe the general model and derive a test in complete generality. In the next four sections we demonstrate the power of this test by solving four important testing problems. We will need a considerable amount of linear algebra in Section 12.2.

12.2 GENERAL LINEAR HYPOTHESIS

A wide variety of problems of hypotheses testing can be treated under a general setup. In this section we state the general problem and derive the test statistic and its distribution. Consider the following examples.

Example 1. Let $Y_1, Y_2, \ldots, Y_k$ be independent RVs with $EY_i = \mu_i, i = 1, 2, \ldots, k$, and common variance σ^2. Also, n_i observations are taken on $Y_i, i = 1, 2, \ldots, k$, and $\sum_{i=1}^{k} n_i = n$.

An Introduction to Probability and Statistics, Third Edition. Vijay K. Rohatgi and A.K. Md. Ehsanes Saleh.

It is required to test $H_0 : \mu_i = \mu_2 = \cdots = \mu_k$. The case $k = 2$ has already been treated in Section 10.4. Problems of this nature arise quite naturally, for example, in agricultural experiments where one is interested in comparing the average yield when k fertilizers are available.

Example 2. An experimenter observes the velocity of a particle moving along a line. He takes observations at given times $t_1, t_2, \ldots, t_n$. Let β_1 be the initial velocity of the particle and β_2 the acceleration; then the velocity at time t is given by $y = \beta_1 + \beta_2 t + \varepsilon$, where ε is an RV that is nonobservable (like an error in measurement). In practice, the experimenter does not know β_1 and β_2 and has to use the random observations $Y_1, Y_2, \ldots, Y_n$ made at times $t_1, t_2, \ldots, t_n$, respectively, to obtain some information about the unknown parameters β_1, β_2.

A similar example is the case when the relation between y and t is governed by

$$y = \beta_0 + \beta_1 t + \beta_2 t^2 + \varepsilon,$$

where t is a mathematical variable, $\beta_0, \beta_1, \beta_2$ are unknown parameters, and ε is a nonobservable RV. The experimenter takes observations $Y_1, Y_2, \ldots, Y_n$ at predetermined values $t_1, t_2, \ldots, t_n$, respectively, and is interested in testing the hypothesis that the relation is in fact linear, that is, $\beta_2 = 0$.

Examples of the type discussed above and their much more complicated variants can all be treated under a general setup. To fix ideas, let us first make the following definition.

Definition 1. Let $\mathbf{Y} = (Y_1, Y_2, \ldots, Y_n)'$ be a random column vector and $\mathbf{X}$ be an $n \times k$ matrix, $k < n$, of known constants $x_{ij}, i = 1, 2, \ldots, n;\ j = 1, 2, \ldots, k$. We say that the distribution of $\mathbf{Y}$ satisfies a *linear model* if

$$E\mathbf{Y} = \mathbf{X}\boldsymbol{\beta}, \tag{1}$$

where $\boldsymbol{\beta} = (\beta_1, \beta_2, \ldots, \beta_k)'$ is a vector of unknown (scalar) parameters $\beta_1, \beta_2, \ldots, \beta_k$. It is convenient to write

$$\mathbf{Y} = \mathbf{X}\boldsymbol{\beta} + \boldsymbol{\varepsilon}, \tag{2}$$

where $\boldsymbol{\varepsilon} = (\varepsilon_1, \varepsilon_2, \ldots, \varepsilon_n)'$ is a vector of nonobservable RVs with $E\varepsilon_j = 0, j = 1, 2, \ldots, n$. Relation (2) is known as a linear model. Then *general linear hypothesis* concerns $\boldsymbol{\beta}$, namely, that $\boldsymbol{\beta}$ satisfies H_0: $\mathbf{H}\boldsymbol{\beta} = \mathbf{0}$, where $\mathbf{H}$ is a known $r \times k$ matrix with $r \leq k$.

In what follows we will assume that $\varepsilon_1, \varepsilon_2, \ldots, \varepsilon_n$ are independent, normal RVs with common variance σ^2 and $E\varepsilon_j = 0, j = 1, 2, \ldots, n$. In view of (2), it follows that $Y_1, Y_2, \ldots, Y_n$ are independent normal RVs with

$$EY_i = \sum_{j=1}^{k} x_{ij}\beta_j \qquad \text{and} \qquad \text{var}(Y_i) = \sigma^2, \qquad i = 1, 2, \ldots, n. \tag{3}$$

We will assume that $\mathbf{H}$ is a matrix of full rank $r, r \leq k$, and $\mathbf{X}$ is a matrix of full rank $k < n$. Some remarks are in order.

Remark 1. Clearly, $\mathbf{Y}$ satisfies a linear model if the vector of means $E\mathbf{Y}=(EY_1, EY_2,\ldots,EY_n)'$ lies in a k-dimensional subspace generated by the linearly independent column vectors $\mathbf{x}_1,\mathbf{x}_2,\ldots,\mathbf{x}_k$ of the matrix $\mathbf{X}$. Indeed, (1) states that $E\mathbf{Y}$ is a linear combination of the known vectors $\mathbf{x}_1,\ldots,\mathbf{x}_k$. The general linear hypothesis H_0: $\mathbf{H}\boldsymbol{\beta}=\mathbf{0}$ states that the parameters $\beta_1,\beta_2,\ldots,\beta_k$ satisfy r independent homogeneous linear restrictions. It follows that, under H_0, $E\mathbf{Y}$ lies in a $(k-r)$-dimensional subspace of the k-space generated by $\mathbf{x}_1,\ldots,\mathbf{x}_k$.

Remark 2. The assumption of normality, which is conventional, is made to compute the likelihood ratio test statistic of H_0 and its distribution. If the problem is to estimate $\boldsymbol{\beta}$, no such assumption is needed. One can use the *principle of least squares* and estimate $\boldsymbol{\beta}$ by minimizing the sum of squares,

$$\sum_{i=1}^{n}\varepsilon_i^2=\varepsilon\varepsilon'=(\mathbf{Y}-\mathbf{X}\boldsymbol{\beta})'(\mathbf{Y}-\mathbf{X}\boldsymbol{\beta}). \tag{4}$$

The minimizing value $\hat{\boldsymbol{\beta}}(\mathbf{y})$ is known as a *least square estimate of* $\boldsymbol{\beta}$. This is not a difficult problem, and we will not discuss it here in any detail but will mention only that any solution of the so-called *normal equations*

$$\mathbf{X}'\mathbf{X}\boldsymbol{\beta}=\mathbf{X}'\mathbf{Y} \tag{5}$$

is a least square estimator. If the rank of $\mathbf{X}$ is $k(<n)$, then $\mathbf{X}'\mathbf{X}$, which has the same rank as $\mathbf{X}$, is a nonsingular matrix that can be inverted to give a unique least square estimator

$$\hat{\boldsymbol{\beta}}=(\mathbf{X}'\mathbf{X})^{-1}\mathbf{X}'\mathbf{Y}. \tag{6}$$

If the rank of $\mathbf{X}$ is $<k$, then $\mathbf{X}'\mathbf{X}$ is singular and the normal equations do not have a unique solution. One can show, for example, that $\hat{\boldsymbol{\beta}}$ is unbiased for $\boldsymbol{\beta}$, and if the Y_i's are uncorrelated with common variance σ^2, the variance–covariance matrix of the $\hat{\beta}_i$'s is given by

$$E\left\{\left(\hat{\boldsymbol{\beta}}-\boldsymbol{\beta}\right)\left(\hat{\boldsymbol{\beta}}-\boldsymbol{\beta}\right)'\right\}=\sigma^2(\mathbf{X}'\mathbf{X})^{-1}. \tag{7}$$

Remark 3. One can similarly compute the so-called *restricted least square estimator* of $\boldsymbol{\beta}$ by the usual method of Lagrange multipliers. For example, under H_0: $\mathbf{H}\boldsymbol{\beta}=\mathbf{0}$ one simply minimizes $(\mathbf{Y}-\mathbf{X}\boldsymbol{\beta})'\,(\mathbf{Y}-\mathbf{X}\boldsymbol{\beta})$ subject to $\mathbf{H}\boldsymbol{\beta}=\mathbf{0}$ to get the restricted least square estimator $\hat{\hat{\boldsymbol{\beta}}}$. The important point is that, if ε is assumed to be a multivariate normal RV with mean vector $\mathbf{0}$ and dispersion matrix $\sigma^2\mathbf{I}_n$, the MLE of $\boldsymbol{\beta}$ is the same as the least square estimator. In fact, one can show that $\hat{\beta}_i$ is the UMVUE of $\beta_i, i=1,2,\ldots,k$, by the usual methods.

Example 3. Suppose that a random variable Y is linearly related to a mathematical variable x that is not random (see Example 2). Let $Y_1,Y_2,\ldots,Y_n$ be observations made at different known values $x_1,x_2,\ldots,x_n$ of x. For example, $x_1,x_2,\ldots,x_n$ may represent different levels of fertilizer, and $Y_1,Y_2,\ldots,Y_n$, respectively, the corresponding yields of a crop. Also, $\varepsilon_1, \varepsilon_2, \ldots, \varepsilon_n$ represent unobservable RVs that may be errors of measurements. Then

$$Y_i = \beta_0 + \beta_1 x_i + \varepsilon_i, \qquad i = 1, 2, \ldots, n,$$

and we wish to test whether $\beta_1 = 0$, that the fertilizer levels do not affect the yield. Here

$$\mathbf{X} = \begin{pmatrix} 1 & x_1 \\ 1 & x_2 \\ \vdots & \vdots \\ 1 & x_n \end{pmatrix},$$

$$\boldsymbol{\beta} = (\beta_0, \beta_1)', \qquad \text{and} \qquad \boldsymbol{\varepsilon} = (\varepsilon_1, \varepsilon_2, \ldots, \varepsilon_n)'.$$

The hypothesis to be tested is $H_0: \beta_1 = 0$ so that, with $\mathbf{H} = (0, 1)$, the null hypothesis can be written as $H_0: \mathbf{H}\boldsymbol{\beta} = 0$. This is a problem of *linear regression*.

Similarly, we may assume that the regression of Y on x is *quadratic:*

$$Y = \beta_0 + \beta_1 x + \beta_2 x^2 + \varepsilon,$$

and we may wish to test that a linear function will be sufficient to describe the relationship, that is, $\beta_2 = 0$. Here $\mathbf{X}$ is the $n \times 3$ matrix

$$\mathbf{X} = \begin{pmatrix} 1 & x_1 & x_1^2 \\ 1 & x_2 & x_2^2 \\ \vdots & \vdots & \vdots \\ 1 & x_n & x_n^2 \end{pmatrix},$$

$$\boldsymbol{\beta} = (\beta_0, \beta_1, \beta_2)', \qquad \boldsymbol{\varepsilon} = (\varepsilon_1, \varepsilon_2, \ldots, \varepsilon_n)',$$

and $\mathbf{H}$ is the 1×3 matrix $(0, 0, 1)$.

In another example of regression, the Y's can be written as

$$Y = \beta_1 x_1 + \beta_2 x_2 + \beta_3 x_3 + \varepsilon,$$

and we wish to test the hypothesis that $\beta_1 = \beta_2 = \beta_3$. In this case, $\mathbf{X}$ is the matrix

$$\mathbf{X} = \begin{pmatrix} x_{11} & x_{12} & x_{13} \\ x_{21} & x_{22} & x_{23} \\ \vdots & \vdots & \vdots \\ x_{n1} & x_{n2} & x_{n3} \end{pmatrix},$$

and $\mathbf{H}$ may be chosen to be the 2×3 matrix

$$\mathbf{H} = \begin{pmatrix} 1 & 0 & -1 \\ 1 & -1 & 0 \end{pmatrix}.$$

Example 4. Another important example of the general linear hypothesis involves the *analysis of variance*. We have already derived tests of hypotheses regarding the equality of the means of two normal populations when the variances are equal. In practice, one is frequently interested in the equality of several means when the variances are the same, that is, one has k samples from $\mathcal{N}(\mu_1,\sigma^2),\ldots,\mathcal{N}(\mu_k,\sigma^2)$, where σ^2 is unknown and one wants to test $H_0\colon \mu_1 = \mu_2 = \cdots = \mu_k$ (see Example 1). Such a situation is of common occurrence in agricultural experiments. Suppose that k treatments are applied to experimental units (plots), the ith treatment is applied to n_i randomly chosen units, $i = 1,2,\ldots,k$, $\sum_{i=1}^{k} n_i = n$, and the observation y_{ij} represents some numerical characteristic (yield) of the jth experimental unit under the ith treatment. Suppose also that

$$Y_{ij} = \mu_i + \varepsilon_{ij}, \qquad j = 1,2,\ldots,n; \qquad i = 1,2,\ldots,k,$$

where ε_{ij} are iid $\mathcal{N}(0,\sigma^2)$ RVs. We are interested in testing $H_0\colon \mu_1 = \mu_2 = \cdots = \mu_k$. We write

$$\mathbf{Y} = (Y_{11}, Y_{12}, \ldots, Y_{1n_1}, Y_{21}, Y_{22}, \ldots, Y_{2n_2}, \ldots, Y_{k_1}, Y_{k_2}, \ldots, Y_{kn_k})',$$
$$\boldsymbol{\beta} = (\mu_1, \mu_2, \ldots, \mu_k)',$$
$$\mathbf{X} = \begin{pmatrix} \mathbf{1}_{n_1} & \mathbf{0} & \cdots & \mathbf{0} \\ \mathbf{0} & \mathbf{1}_{n_2} & \cdots & \mathbf{0} \\ \vdots & \vdots & \vdots\vdots\vdots & \vdots \\ \mathbf{0} & \mathbf{0} & \cdots & \mathbf{1}_{n_k} \end{pmatrix},$$

where $\mathbf{1}_{n_i} = (1,1,\ldots,1)'$ is the n_i-vector $(i = 1,2,\ldots,k)$, each of whose elements is unity. Thus $\mathbf{X}$ is $n \times k$. We can choose

$$\mathbf{H} = \begin{pmatrix} 1 & -1 & 0 & \cdots & 0 \\ 1 & 0 & -1 & \cdots & 0 \\ \vdots & \vdots & \vdots & \vdots\vdots\vdots & \vdots \\ 1 & 0 & 0 & \cdots & -1 \end{pmatrix}$$

so that $H_0\colon \mu_1 = \mu_2 = \cdots = \mu_k$ is of the form $\mathbf{H}\boldsymbol{\beta} = \mathbf{0}$. Here $\mathbf{H}$ is a $(k-1)\times k$ matrix.

The model described in this example is frequently referred to as a *one-way analysis of variance* model. This is a very simple example of an analysis of variance model. Note that the matrix $\mathbf{X}$ is of a very special type, namely, the elements of $\mathbf{X}$ are either 0 or 1. $\mathbf{X}$ is known as a *design matrix*.

Returning to our general model

$$\mathbf{Y} = \mathbf{X}\boldsymbol{\beta} + \boldsymbol{\varepsilon},$$

we wish to test the null hypothesis $H_0\colon \mathbf{H}\boldsymbol{\beta} = \mathbf{0}$. We will compute the likelihood ratio test and the distribution of the test statistic. In order to do so, we assume that $\boldsymbol{\varepsilon}$ has a multivariate normal distribution with mean vector $\mathbf{0}$ and variance–covariance matrix $\sigma^2\mathbf{I}_n$, where σ^2 is unknown and $\mathbf{I}_n$ is the $n \times n$ identity matrix. This means that $\mathbf{Y}$ has an n-variate normal distribution with mean $\mathbf{X}\boldsymbol{\beta}$ and dispersion matrix $\sigma^2\mathbf{I}_n$ for some $\boldsymbol{\beta}$ and some σ^2, both unknown. Here the parameter space Θ is the set of $(k+1)$-tuples $(\boldsymbol{\beta}', \sigma^2) = (\beta_1, \beta_2, \ldots, \beta_k, \sigma^2)$, and the joint PDF of the X's is given by

$$f_{\boldsymbol{\beta},\sigma^2}(y_1, y_2, \ldots, y_n) \tag{8}$$
$$= \frac{1}{(2\pi)^{n/2}\sigma^n} \exp\left\{-\frac{1}{2\sigma^2}\sum_{i=1}^{n}(y_i - \beta_1 x_{i1} - \cdots - \beta_k x_{ik})^2\right\}$$
$$= \frac{1}{(2\pi)^{n/2}\sigma^n} \exp\left\{-\frac{1}{2\sigma^2}(\mathbf{Y} - \mathbf{X}\boldsymbol{\beta})'(\mathbf{Y} - \mathbf{X}\boldsymbol{\beta})\right\}.$$

Theorem 1. Consider the linear model

$$\mathbf{Y} = \mathbf{X}\boldsymbol{\beta} + \boldsymbol{\varepsilon},$$

where $\mathbf{X}$ is an $n \times k$ matrix $((x_{ij}))$, $i = 1, 2, \ldots, n$, $j = 1, 2, \ldots, k$, of known constants and full rank $k < n, \boldsymbol{\beta}$ is a vector of unknown parameters $\beta_1, \beta_2, \ldots, \beta_k$ and $\boldsymbol{\varepsilon} = (\varepsilon_1, \varepsilon_2, \ldots, \varepsilon_n)$ is a vector of nonobservable independent normal RVs with common variance σ^2 and mean $E\boldsymbol{\varepsilon} = \mathbf{0}$. The likelihood ratio test for testing the linear hypothesis $H_0\colon \mathbf{H}\boldsymbol{\beta} = \mathbf{0}$, where $\mathbf{H}$ is an $r \times k$ matrix of full rank $r \leq k$, is to reject H_0 at level α if $F \geq F_\alpha$, where $P_{H_0}\{F \geq F_\alpha\} = \alpha$, and F is the RV given by

$$F = \frac{(\mathbf{Y} - \mathbf{X}\hat{\hat{\boldsymbol{\beta}}})'(\mathbf{Y} - \mathbf{X}\hat{\hat{\boldsymbol{\beta}}}) - (\mathbf{Y} - \mathbf{X}\hat{\boldsymbol{\beta}})'(\mathbf{Y} - \mathbf{X}\hat{\boldsymbol{\beta}})}{(\mathbf{Y} - \mathbf{X}\hat{\boldsymbol{\beta}})'(\mathbf{Y} - \mathbf{X}\hat{\boldsymbol{\beta}})}. \tag{9}$$

In (9), $\hat{\boldsymbol{\beta}}, \hat{\hat{\boldsymbol{\beta}}}$ are the MLE's of $\boldsymbol{\beta}$ under Θ and Θ_0, respectively. Moreover, the RV $[(n-k)/r]F$ has an F-distribution with $(r, n-k)$ d.f. under H_0.

Proof. The likelihood ratio test of $H_0\colon \mathbf{H}\boldsymbol{\beta} = \mathbf{0}$ is to reject H_0 if and only if $\lambda(\mathbf{y}) < c$, where

$$\lambda(\mathbf{y}) = \frac{\sup_{\boldsymbol{\theta}\in\Theta_0} f_{\boldsymbol{\beta},\sigma^2}(\mathbf{y})}{\sup_{\boldsymbol{\theta}\in\Theta} f_{\boldsymbol{\beta},\sigma^2}(\mathbf{y})}, \tag{10}$$

$\boldsymbol{\theta} = (\boldsymbol{\beta}', \sigma^2)'$, and $\Theta_0 = \{(\boldsymbol{\beta}', \sigma^2)'\colon \mathbf{H}\boldsymbol{\beta} = \mathbf{0}\}$. Let $\hat{\boldsymbol{\theta}} = (\hat{\boldsymbol{\beta}}', \hat{\sigma}^2)'$ be the MLE of $\boldsymbol{\theta}' \in \Theta$, and $\hat{\hat{\boldsymbol{\theta}}} = (\hat{\hat{\boldsymbol{\beta}}}', \hat{\hat{\sigma}}^2)'$ be the MLE of $\boldsymbol{\theta}$ under H_0, that is, when $\mathbf{H}\boldsymbol{\beta} = \mathbf{0}$. It is easily seen that $\hat{\boldsymbol{\beta}}$ is the value of $\boldsymbol{\beta}$ that minimizes $(\mathbf{y} - \mathbf{X}\boldsymbol{\beta})'(\mathbf{y} - \mathbf{X}\boldsymbol{\beta})$, and

$$\hat{\sigma}^2 = n^{-1}(\mathbf{y} - \mathbf{X}\hat{\boldsymbol{\beta}})'(\mathbf{y} - \mathbf{X}\hat{\boldsymbol{\beta}}). \tag{11}$$

Similarly, $\hat{\hat{\boldsymbol{\beta}}}$ is the value of $\boldsymbol{\beta}$ that minimizes $(\mathbf{y} - \mathbf{X}\boldsymbol{\beta})'(\mathbf{y} - \mathbf{X}\boldsymbol{\beta})$ subject to $\mathbf{H}\boldsymbol{\beta} = \mathbf{0}$, and

$$\hat{\hat{\sigma}}^2 = n^{-1}(\mathbf{y} - \mathbf{X}\hat{\hat{\boldsymbol{\beta}}})'(\mathbf{y} - \mathbf{X}\hat{\hat{\boldsymbol{\beta}}}). \tag{12}$$

It follows that

$$\lambda(\mathbf{y}) = \left(\frac{\hat{\sigma}^2}{\hat{\hat{\sigma}}^2}\right)^{n/2}. \tag{13}$$

The critical region $\lambda(\mathbf{y}) < c$ is equivalent to the region $\{\lambda(\mathbf{y})\}^{-2/n} < \{c\}^{-2/n}$, which is of the form

$$\frac{\hat{\hat{\sigma}}^2}{\hat{\sigma}^2} > c_1. \tag{14}$$

This may be written as

$$\frac{(\mathbf{y}-\mathbf{X}\hat{\hat{\beta}})'(\mathbf{y}-\mathbf{X}\hat{\hat{\beta}})}{(\mathbf{y}-\mathbf{X}\hat{\beta})'(\mathbf{y}-\mathbf{X}\hat{\beta})} > c_1 \tag{15}$$

or, equivalently, as

$$\frac{(\mathbf{y}-\mathbf{X}\hat{\hat{\beta}})'(\mathbf{y}-\mathbf{X}\hat{\hat{\beta}}) - (\mathbf{y}-\mathbf{X}\hat{\beta})'(\mathbf{y}-\mathbf{X}\hat{\beta})}{(\mathbf{y}-\mathbf{X}\hat{\beta})'(\mathbf{y}-\mathbf{X}\hat{\beta})} > c_1 - 1. \tag{16}$$

It remains to determine the distribution of the test statistic. For this purpose it is convenient to reduce the problem to the *canonical form*. Let V_n be the vector space of the observation vector $\mathbf{Y}, V_k$ be the subspace of V_n generated by the column vectors $\mathbf{x}_1, \mathbf{x}_2, \ldots, \mathbf{x}_k$ of $\mathbf{X}$, and V_{k-r} be the subspace of V_k in which $E\mathbf{Y}$ is postulated to lie under H_0. We change variables from $Y_1, Y_2, \ldots, Y_n$ to $Z_1, Z_2, \ldots, Z_n$, where $Z_1, Z_2, \ldots, Z_n$ are independent normal RVs with common variance σ^2 and means $EZ_i = \theta_i$, $i = 1, 2, \ldots, k$, $EZ_i = 0$, $i = k+1, \ldots, n$. This is done as follows. Let us choose an orthonormal basis of $k-r$ column vectors $\{\boldsymbol{\alpha}_i\}$ for V_{k-r}, say $\{\boldsymbol{\alpha}_{r+1}, \boldsymbol{\alpha}_{r+2}, \ldots, \boldsymbol{\alpha}_k\}$. We extend this to an orthonormal basis $\{\boldsymbol{\alpha}_1, \boldsymbol{\alpha}_2, \ldots, \boldsymbol{\alpha}_r, \boldsymbol{\alpha}_{r+1}, \ldots, \boldsymbol{\alpha}_k\}$ for V_k, and then extend once again to an orthonormal basis $\{\boldsymbol{\alpha}_1, \boldsymbol{\alpha}_2, \ldots, \boldsymbol{\alpha}_k, \boldsymbol{\alpha}_{k+1}, \ldots, \boldsymbol{\alpha}_n\}$ for V_n. This is always possible.

Let $z_1, z_2, \ldots, z_n$ be the coordinates of $\mathbf{y}$ relative to the basis $\{\boldsymbol{\alpha}_1, \boldsymbol{\alpha}_2, \ldots, \boldsymbol{\alpha}_n\}$. Then $z_i = \boldsymbol{\alpha}_i'\mathbf{y}$ and $\mathbf{z} = \mathbf{PY}$, where $\mathbf{P}$ is an orthogonal matrix with ith row $\boldsymbol{\alpha}_i'$. Thus $E\mathbf{Z}_i = E\boldsymbol{\alpha}_i'\mathbf{Y} = \boldsymbol{\alpha}_i'\mathbf{X}\boldsymbol{\beta}$, and $E\mathbf{Z} = \mathbf{PX}\boldsymbol{\beta}$. Since $\mathbf{X}\boldsymbol{\beta} \in V_k$ (Remark 1), it follows that $\boldsymbol{\alpha}_i'\mathbf{X}\boldsymbol{\beta} = 0$ for $i > k$. Similarly, under $H_0, \mathbf{X}\boldsymbol{\beta} \in V_{k-r} \subset V_k$, so that $\boldsymbol{\alpha}_i'\mathbf{X}\boldsymbol{\beta} = 0$ for $i \le r$. Let us write $\boldsymbol{\omega} = \mathbf{PX}\boldsymbol{\beta}$. Then $\omega_{k+1} = \omega_{k+2} = \cdots = \omega_n = 0$, and under $H_0, \omega_1 = \omega_2 = \cdots = \omega_r = 0$. Finally, from Corollary 2 of Theorem 6 it follows that $Z_1, Z_2, \ldots, Z_n$ are independent normal RVs with the same variance σ^2 and $EZ_i = \omega_i$, $i = 1, 2, \ldots, n$. We have thus transformed the problem to the following simpler canonical form:

$$\begin{cases} \Omega: & Z_i \text{ are independent } \mathcal{N}(\omega_i, \sigma^2), \quad i = 1, 2, \ldots, n, \\ & \omega_{k+1} = \omega_{k+2} = \cdots = \omega_n = 0, \\ H_0: & \omega_1 = \omega_2 = \cdots = \omega_r = 0. \end{cases} \tag{17}$$

Now

$$(\mathbf{y}-\mathbf{X}\boldsymbol{\beta})'(\mathbf{y}-\mathbf{X}\boldsymbol{\beta}) = (\mathbf{P}'\mathbf{z}-\mathbf{P}'\boldsymbol{\omega})'(\mathbf{P}'\mathbf{z}-\mathbf{P}'\boldsymbol{\omega}) \tag{18}$$

$$= (\mathbf{z}-\boldsymbol{\omega})'(\mathbf{z}-\boldsymbol{\omega})$$
$$= \sum_{i=1}^{k}(z_i-\omega_i)^2 + \sum_{i=k+1}^{n} z_i^2.$$

The quantity $(\mathbf{y}-\mathbf{X}\boldsymbol{\beta})'(\mathbf{y}-\mathbf{X}\boldsymbol{\beta})$ is minimized if we choose $\hat{\omega}_i = z_i$, $i = 1,2,\ldots,k$, so that

$$(\mathbf{y}-\mathbf{X}\hat{\boldsymbol{\beta}})'(\mathbf{y}-\mathbf{X}\hat{\boldsymbol{\beta}}) = \sum_{i=k+1}^{n} z_i^2. \tag{19}$$

Under $H_0, \omega_1 = \omega_2 = \cdots = \omega_r = 0$, so that $(\mathbf{y}-\mathbf{X}\boldsymbol{\beta})'(\mathbf{y}-\mathbf{X}\boldsymbol{\beta})$ will be minimized if we choose $\hat{\hat{\omega}}_i = z_i$, $i = r+1,\ldots,k$. Thus

$$(\mathbf{y}-\mathbf{X}\hat{\hat{\boldsymbol{\beta}}})'(\mathbf{y}-\mathbf{X}\hat{\hat{\boldsymbol{\beta}}}) = \sum_{i=1}^{r} z_i^2 + \sum_{i=k+1}^{n} z_i^2. \tag{20}$$

It follows that

$$F = \frac{\sum_{i=1}^{r} Z_i^2}{\sum_{i=k+1}^{n} Z_i^2}.$$

Now $\sum_{i=k+1}^{n} Z_i^2/\sigma^2$ has a $\chi^2(n-k)$ distribution, and, under H_0, $\sum_{i=1}^{r} Z_i^2/\sigma^2$ has a $\chi^2(r)$ distribution. Since $\sum_{i=1}^{r} Z_i^2$ and $\sum_{i=k+1}^{n} Z_i^2$ are independent, we see that $[(n-k)/r]F$ is distributed as $F(r,n-k)$ under H_0, as asserted. This completes the proof of the theorem.

Remark 4. In practice, one does not need to find a transformation that reduces the problem to the canonical form. As will be done in the following sections, one simply computes the estimators $\hat{\boldsymbol{\theta}}$ and $\hat{\hat{\boldsymbol{\theta}}}$ and then computes the test statistic in any of the equivalent forms (14), (15), or (16) to apply the F-test.

Remark 5. The computation of $\hat{\boldsymbol{\beta}}, \hat{\hat{\boldsymbol{\beta}}}$ is greatly facilitated, in view of Remark 3, by using the principle of least squares. Indeed, this was done in the proof of Theorem 1 when we reduced the problem of maximum likelihood estimation to that of minimization of sum of squares $(\mathbf{y}-\mathbf{X}\boldsymbol{\beta})'(\mathbf{y}-\mathbf{X}\boldsymbol{\beta})$.

Remark 6. The distribution of the test statistic under H_1 is easily determined. We note that $Z_i/\sigma \sim \mathcal{N}(\omega_i/\sigma, 1)$ for $i = 1,2,\ldots,r$, so that $\sum_{i=1}^{r} Z_i^2/\sigma^2$ has a noncentral chi-square distribution with r d.f. and noncentrality parameter $\delta = \sum_{i=1}^{r} \omega_i^2/\sigma^2$. It follows that $[(n-k)/r]F$ has a noncentral F-distribution with d.f. $(r, n-k)$ and noncentrality parameter δ. Under $H_0, \delta = 0$, so that $[(n-k)/r]F$ has a central $F(r,n-k)$ distribution. Since $\sum_{i=1}^{r} \omega_i^2 = \sum_{i=1}^{r}(EZ_i)^2$, it follows from (19) and (20) that if we replace each observation Y_i by its expected value in the numerator of (16), we get $\sigma^2\delta$.

Remark 7. The general linear hypothesis makes use of the assumption of common variance. For instance, in Example 4, $Y_{ij} \sim \mathcal{N}(\mu_i, \sigma^2)$, $j = 1,2,\ldots,k$. Let us suppose that $Y_{ij} \sim \mathcal{N}(\mu_i, \sigma_i^2)$, $i = 1,2,\ldots,k$. Then we need to test that $\sigma_1 = \sigma_2 = \cdots = \sigma_k$ before we can apply Theorem 1. The case $k = 2$ has already been considered in Section 10.3. For the case where $k > 2$ one can show that a UMP unbiased test does not exist. A large-sample approximation is described in Lehmann [64, pp. 376–377]. It is beyond the scope of this book to consider the effects of departures from the underlying assumptions. We refer the reader to Scheffé [101, Chapter 10], for a discussion of this topic.

Remark 8. The general linear model (GLM) is widely used in social sciences where Y is often referred to as the response (or dependent) variable and X as the explanatory (or independent) variable. In this language the GLM "predicts" a response variable from a linear combination of one or more explanatory variables. It should be noted that dependent and independent in this context do not have the same meaning as in Chapter 4. Moreover, dependence does not imply causality.

PROBLEMS 12.2

1. Show that any solution of the normal equations (5) minimizes the sum of squares $(\mathbf{Y}-\mathbf{X}\boldsymbol{\beta})'(\mathbf{Y}-\mathbf{X}\boldsymbol{\beta})$.
2. Show that the least square estimator given in (6) is an unbiased estimator of $\boldsymbol{\beta}$. If the RVs Y_i are uncorrelated with common variance σ^2, show that the covariance matrix of the $\hat{\beta}_i$'s is given by (7).
3. Under the assumption that ε [in model (2)] has a multivariate normal distribution with mean 0 and dispersion matrix $\sigma^2 I_n$ show that the least square estimators and the MLE's of $\boldsymbol{\beta}$ coincide.
4. Prove statements (11) and (12).
5. Determine the expression for the least squares estimator of $\boldsymbol{\beta}$ subject to $\mathbf{H}\boldsymbol{\beta} = \mathbf{0}$

12.3 REGRESSION ANALYSIS

In this section we study regression analysis, which is a tool to investigate the interrelationship between two or more variables. Typically, in its simplest form a response random variable Y is hypothesized to be related to one or more explanatory nonrandom variables x_i's. Regression analysis with a single explanatory RV is known as *simple regression* and if, in addition, the relationship is thought to be linear, it is called *simple linear regression* (Example 12.2.3). In the case where several explanatory variables x_i's are involved the regression is referred to as *multiple linear regression.* Regression analysis is widely used in forecasting and prediction. Again this is a special case of GLM.

This section is divided into three subsections. The first subsection deals with multiple linear regression where the RV Y is of the continuous type. In the next two subsections we study the case when Y is either Bernoulli or a count variable.

12.3.1 Multiple Linear Regression

It is convenient to write GLM in the form

$$\mathbf{Y} = \beta_0 \mathbf{1}_n + \mathbf{X}\boldsymbol{\beta} + \varepsilon, \tag{1}$$

where $\mathbf{Y}, \mathbf{X}, \varepsilon$, and $\boldsymbol{\beta}$ are as in Equation (12.2.1), and $\mathbf{1}_n$ is the column $n \times 1$ unit vector $(1, 1, \ldots, 1)$. The parameter β_0 is usually referred to as the intercept whereas $\boldsymbol{\beta}$ is known as the slope vector with k parameters. The least estimator (LSE) of β_0 and $\boldsymbol{\beta}$ are easily obtained by minimizing.

$$\sum_{i=1}^{n} (y_i - \beta_0 - \mathbf{x}_i'\boldsymbol{\beta})^2, \; \mathbf{x}_i = (x_{i1}, x_{i2}, \ldots, x_{ik})', \; i = 1, 2, \ldots, n, \tag{2}$$

resulting in $k+1$ normal equations

$$\begin{aligned} &\bar{y} = \beta_0 + \beta_1\bar{x}_1 + \beta_2\bar{x}_2 + \cdots + \beta_k\bar{x}_k = \beta_0 + \boldsymbol{\beta}'\bar{\mathbf{x}}, \; \bar{\mathbf{x}} = \sum_{i=1}^{n} \frac{\mathbf{x}_i}{n} \\ &\mathbf{S}_{xx} = \sum_{i=1}^{n} (\mathbf{x}_i - \bar{\mathbf{x}})(\mathbf{x}_i - \bar{\mathbf{x}})', \mathbf{S}_{xy} = \sum_{i=1}^{n} (\mathbf{x}_i - \bar{\mathbf{x}})y_i \\ &\mathbf{S}_{xx}\hat{\boldsymbol{\beta}} = \mathbf{S}_{xy} \end{aligned} \tag{3}$$

or

$$\begin{aligned} &\hat{\boldsymbol{\beta}} = \mathbf{S}_{xx}^{-1}\mathbf{S}_{xy} \text{ and } \hat{\beta}_0 = \bar{y} - \hat{\boldsymbol{\beta}}'\bar{\mathbf{x}} \\ &E(\hat{\beta}_0) = \beta_0, \; E(\hat{\boldsymbol{\beta}}) = \boldsymbol{\beta} \end{aligned} \tag{4}$$

and

$$\operatorname{Cov}(\hat{\beta}_0, \hat{\boldsymbol{\beta}}') = \frac{\sigma^2}{n} \begin{pmatrix} 1 + n\,\bar{\mathbf{x}}'\,\mathbf{S}_{xx}^{-1}\,\bar{\mathbf{x}} & n\,\bar{\mathbf{x}}\,\mathbf{S}_{xx}^{-1} \\ n\,\bar{\mathbf{x}}\,\mathbf{S}_{xx}^{-1} & n\,\mathbf{S}_{xx}^{-1} \end{pmatrix} \tag{5}$$

An unbiased estimate of σ^2 is given by

$$\hat{\sigma}^2 = \frac{1}{n-k-1} \left(\mathbf{Y} - \hat{\beta}_0\mathbf{1}_n - \mathbf{X}\hat{\boldsymbol{\beta}}\right)' \left(Y - \hat{\beta}_0\mathbf{1_n} - \mathbf{X}\hat{\boldsymbol{\beta}}\right). \tag{6}$$

Let us now consider the simple linear regression model

$$\mathbf{y} = \beta_0\mathbf{1_n} + \mathbf{X}\beta + \epsilon. \tag{7}$$

The LSEs of $(\beta_0, \beta)'$ is given by

$$\begin{pmatrix} \hat{\beta}_0 \\ \hat{\beta} \end{pmatrix} = \begin{pmatrix} \bar{y} - \hat{\beta}\bar{x} \\ \frac{\sum (x_i - \bar{x})y_i}{\sum (x_i - \bar{x})^2} \end{pmatrix} \tag{8}$$

and

$$\hat{\sigma}^2 = \frac{1}{n-2} \sum_{n=1}^{n} \left(y_i - \bar{y} - \hat{\beta}(x_i - \bar{x})\right)^2. \tag{9}$$

The covariance matrix is given by

$$\operatorname{Cov}\begin{pmatrix}\hat{\beta}_0\\ \hat{\beta}\end{pmatrix}=\frac{\sigma^2}{n}\begin{pmatrix}1+n\,\frac{n\bar{x}^2}{s_n} & -\frac{n\bar{x}}{s_n}\\ -\frac{n\bar{x}}{s_n} & \frac{n}{s_n}\end{pmatrix}, \tag{10}$$

where $s_n^2=\sum_{i=1}^{n}(x_i-\bar{x})^2$

Let us now verify these results using the maximum likelihood method.

Clearly, $Y_1,Y_2,\ldots,Y_n$ are independent normal RVs with $EY_i=\beta_0+\beta_1x_i$ and $\operatorname{var}(Y_i)=\sigma^2$, $i=1,2,\ldots,n$, and $\mathbf{Y}$ is an n-variate normal random vector with mean $\mathbf{X}\boldsymbol{\beta}$ and variance $\sigma^2\mathbf{I}_n$. The joint PDF of $\mathbf{Y}$ is given by

$$f(\mathbf{y};\beta_0,\beta_1,\sigma^2)=\frac{1}{(2\pi)^{n/2}}\frac{1}{\sigma^n}\exp\left\{-\frac{1}{2\sigma^2}\sum_{i=1}^{n}(y_i-\beta_0-\beta_1x_i)^2\right\}. \tag{11}$$

It easily follows that the MLE's for β_0,β_1, and σ^2 are given by

$$\hat{\beta}_0=\frac{\sum_{i=1}^{n}Y_i}{n}-\hat{\beta}_1\bar{x}, \tag{12}$$

$$\hat{\beta}_1=\frac{\sum_{i=1}^{n}(x_i-\bar{x})(Y_i-\overline{Y})}{\sum_{i=1}^{n}(x_i-\bar{x})^2} \tag{13}$$

and

$$\hat{\sigma}^2=\frac{1}{n}\sum_{i=1}^{n}(Y_i-\hat{\beta}_0-\hat{\beta}_1x_i)^2, \tag{14}$$

where $\bar{x}=n^{-1}\sum_{i=1}^{n}x_i$.

If we wish to test $H_0\colon\beta_1=0$, we take $\mathbf{H}=(0,1)$, so that the model is a special case of the general linear hypothesis with $k=2$, $r=1$. Under H_0 the MLE's are

$$\hat{\hat{\beta}}_0=\overline{Y}=\frac{\sum_{i=1}^{n}Y_i}{n} \tag{15}$$

and

$$\hat{\hat{\sigma}}^2=\frac{1}{n}\sum_{i=1}^{n}(Y_i-\overline{Y})^2. \tag{16}$$

Thus

$$\begin{aligned}F&=\frac{\sum_{i=1}^{n}(Y_i-\overline{Y})^2-\sum_{i=1}^{n}(Y_i-\overline{Y}+\hat{\beta}_1\bar{x}-\hat{\beta}_1x_i)^2}{\sum_{i=1}^{n}(Y_i-\overline{Y}+\hat{\beta}_1\bar{x}-\hat{\beta}_1x_i)^2}\\ &=\frac{\hat{\beta}_1^2\sum_{i=1}^{n}(x_i-\bar{x})^2}{\sum_{i=1}^{n}(Y_i-\overline{Y}+\hat{\beta}_1\bar{x}-\hat{\beta}_1x_i)^2}.\end{aligned} \tag{17}$$

From Theorem 12.2.1, the statistic $[(n-2)/1]F$ has a central $F(1,n-2)$ distribution under H_0. Since $F(1,n-2)$ is the square of a $t(n-2)$, the likelihood ratio test rejects H_0 if

$$|\hat{\beta}_1| \left\{ \frac{(n-2)\sum_{i=1}^{n}(x_i-\overline{x})^2}{\sum_{i=1}^{n}(Y_i-\overline{Y}+\hat{\beta}_1\overline{x}-\hat{\beta}_1 x_i)^2} \right\}^{1/2} > c_0, \tag{18}$$

where c_0 is computed from t-tables for $n-2$ d.f.

For testing $H_0: \beta_0 = 0$, we choose $\mathbf{H} = (1,0)$ so that the model is again a special case of the general linear hypothesis. In this case

$$\hat{\hat{\beta}}_1 = \frac{\sum_{i=1}^{n} x_i Y_i}{\sum_{i=1}^{n} x_i^2}$$

and

$$\hat{\hat{\sigma}}^2 = \frac{1}{n}\sum_{i=1}^{n}(Y_i - \hat{\hat{\beta}}_1 x_i)^2. \tag{19}$$

It follows that

$$F = \frac{\sum_{i=1}^{n}(Y_i-\hat{\hat{\beta}}_1 x_i)^2 - \sum_{i=1}^{n}(Y_i-\overline{Y}+\hat{\beta}_1\overline{x}-\hat{\beta}_1 x_i)^2}{\sum_{i=1}^{n}(Y_i-\overline{Y}+\hat{\beta}_1\overline{x}-\hat{\beta}_1 x_i)^2}, \tag{20}$$

and since

$$\begin{aligned}\hat{\hat{\beta}}_1 &= \frac{\sum_{i=1}^{n} x_i Y_i}{\sum_{i=1}^{n} x_i^2} = \frac{\sum_{i=1}^{n}(x_i-\overline{x})(Y_i-\overline{Y}) + n\overline{x}\overline{Y}}{\sum_{i=1}^{n} x_i^2} \\ &= \frac{\hat{\beta}_1\sum_{i=1}^{n}(x_i-\overline{x})^2 + n\overline{x}(\hat{\beta}_0+\hat{\beta}_1\overline{x})}{\sum_{i=1}^{n} x_i^2} \\ &= \hat{\beta}_1 + \frac{n\hat{\beta}_0\overline{x}}{\sum_{i=1}^{n} x_i^2},\end{aligned} \tag{21}$$

we can write the numerator of F as

$$\begin{aligned}&\sum_{i=1}^{n}(Y_i-\hat{\hat{\beta}}_1 x_i)^2 - \sum_{i=1}^{n}(Y_i-\overline{Y}+\hat{\beta}_1\overline{x}-\hat{\beta}_1 x_i)^2 \\ &\quad = \sum_{i=1}^{n}\left(Y_i-\hat{\beta}_1 x_i+\hat{\beta}_1\overline{x}-\overline{Y}+\overline{Y}-\hat{\beta}_1\overline{x}-\frac{n\hat{\beta}_0\overline{x}x_i}{\sum_{i=1}^{n} x^2}\right)^2 \\ &\quad\quad - \sum_{i=1}^{n}(Y_i-\overline{Y}+\hat{\beta}_1\overline{x}-\hat{\beta}_1 x_i)^2 \\ &\quad = \sum_{i=1}^{n}\left(\overline{Y}-\hat{\beta}_1\overline{x}-\frac{n\hat{\beta}_0\overline{x}x_i}{\sum_{i=1}^{n} x_i^2}\right)^2 + 2\sum_{i=1}^{n}(Y_i-\hat{\beta}_1 x_i+\hat{\beta}_1\overline{x}-\overline{Y})\end{aligned} \tag{22}$$

$$\cdot\left(\overline{Y}-\hat{\beta}_1\overline{x}-\frac{n\hat{\beta}_0\overline{x}x_i}{\sum_{i=1}^n x_i^2}\right)$$
$$=\frac{\hat{\beta}_0^2 n\sum_{i=1}^n (x_i-\overline{x})^2}{\sum_{i=1}^n x_i^2}.$$

It follows from Theorem 12.2.1 that the statistic

$$\frac{\hat{\beta}_0\sqrt{n\sum_{i=1}^n (x_i-\overline{x})^2/\sum_{i=1}^n x_i^2}}{\sqrt{\sum_{i=1}^n (Y_i-\overline{Y}+\hat{\beta}_1\overline{x}-\hat{\beta}_1 x_i)^2/(n-2)}} \tag{23}$$

has a central t-distribution with $n-2$ d.f. under $H_0\colon \beta_0=0$. The rejection region is therefore given by

$$\frac{|\hat{\beta}_0|\sqrt{n\sum_{i=1}^n (x_i-\overline{x})^2/\sum_{i=1}^n x_i^2}}{\sqrt{\sum_{i=1}^n (Y_i-\hat{\beta}_0-\hat{\beta}_1 x_i)^2/(n-2)}} > c_0, \tag{24}$$

where c_0 is determined from the tables of $t(n-2)$ distribution for a given level of significance α.

For testing $H_0\colon \beta_0=\beta_1=0$, we choose $\mathbf{H}=\begin{pmatrix}1 & 0\\ 0 & 1\end{pmatrix}$, so that the model is again a special case of the general linear hypothesis with $r=2$. In this case

$$\hat{\hat{\sigma}}^2=\frac{1}{n}\sum_{i=1}^n Y_i^2 \tag{25}$$

and

$$F=\frac{\sum_{i=1}^n Y_i^2-\sum_{i=1}^n (Y_i-\overline{Y}+\hat{\beta}_1\overline{x}-\hat{\beta}_1 x_i)^2}{\sum_{i=1}^n (Y_i-\overline{Y}+\hat{\beta}_1\overline{x}-\hat{\beta}x_i)^2} \tag{26}$$
$$=\frac{n\overline{Y}^2+\hat{\beta}_1^2\sum_{i=1}^n (x_i-\overline{x})^2}{\sum_{i=1}^n (Y_i-\hat{\beta}_0-\hat{\beta}_1 x_i)^2}$$
$$=\frac{n(\hat{\beta}_0+\hat{\beta}_1\overline{x})^2+\hat{\beta}_1^2\sum_{i=1}^n (x_i-\overline{x})^2}{\sum_{i=1}^n (Y_i-\hat{\beta}_0-\hat{\beta}_1 x_i)^2}.$$

From Theorem 12.2.1, the statistic $[(n-2)/2]F$ has a central $F(2,n-2)$ distribution under $H_0\colon \beta_0=\beta_1=0$. It follows that the α-level rejection region for H_0 is given by

$$\frac{n-2}{2}F > c_0, \tag{27}$$

where F is given by (26), and c_0 is the upper α percent point under the $F(2,n-2)$ distribution.

Remark 1. It is quite easy to modify the analysis above to obtain tests of null hypotheses $\beta_0 = \beta_0'$, $\beta_1 = \beta_1'$, and $(\beta_0, \beta_1)' = (\beta_0', \beta_1')'$, where β_0', β_1' are given real numbers (Problem 4).

Remark 2. The confidence intervals for β_0, β_1 are also easily obtained. One can show that a $(1-\alpha)$-level confidence interval for β_0 is given by

$$\left(\hat{\beta}_0 - t_{n-2,\alpha/2}\sqrt{\frac{\sum_{i=1}^n x_i^2 \sum_{i=1}^n (Y_i - \hat{\beta}_0 - \hat{\beta}_1 x_i)^2}{n(n-2)\sum_{i=1}^n (x_i - \bar{x})^2}}, \right. \tag{28}$$

$$\left. \hat{\beta}_0 + t_{n-2,\alpha/2}\sqrt{\frac{\sum_{i=1}^n x_i^2 \sum_{i=1}^n (Y_i - \hat{\beta}_0 - \hat{\beta}_1 x_i)^2}{n(n-2)\sum_{i=1}^n (x_i - \bar{x})^2}} \right)$$

and that for β_1 is given by

$$\left(\hat{\beta}_1 - t_{n-2,\alpha/2}\sqrt{\frac{\sum_{i=1}^n (Y_i - \hat{\beta}_0 - \hat{\beta}_1 x_i)^2}{(n-2)\sum_{i=1}^n (x_i - \bar{x})^2}}, \right. \tag{29}$$

$$\left. \hat{\beta}_1 + t_{n-2,\alpha/2}\sqrt{\frac{\sum_{i=1}^n (Y_i - \hat{\beta}_0 - \hat{\beta}_1 x_i)^2}{(n-2)\sum_{i=1}^n (x_i - \bar{x})^2}} \right).$$

Similarly, one can obtain confidence sets for $(\beta_0, \beta_1)'$ from the likelihood ratio test of $(\beta_0, \beta_1)' = (\beta_0', \beta_1')'$. It can be shown that the collection of sets of points $(\beta_0, \beta_1)'$ satisfying

$$\frac{(n-2)[n(\hat{\beta}_0 - \beta_0)^2 + 2n\bar{x}(\hat{\beta}_0 - \beta_0)(\hat{\beta}_1 - \beta_1) + \sum_{i=1}^n x_i^2(\hat{\beta}_1 - \beta_1)^2]}{2\sum_{i=1}^n (Y_i - \hat{\beta}_0 - \hat{\beta}_1 x_i)^2} \tag{30}$$
$$\leq F_{2,n-2,\alpha}$$

is a $(1-\alpha)$-level collection of confidence sets (ellipsoids) for $(\beta_0, \beta_1)'$ centered at $(\hat{\beta}_0, \hat{\beta}_1)'$.

Remark 3. Sometimes interest lies in constructing a confidence interval on the unknown linear regression function $E\{Y \mid x_0\} = \beta_0 + \beta_1 x_0$ for a given value of x, or on a value of Y, given $x = x_0$. We assume that x_0 is a value of x distinct from $x_1, x_2, \ldots, x_n$. Clearly, $\hat{\beta}_0 + \hat{\beta}_1 x_0$ is the maximum likelihood estimator of $\beta_0 + \beta_1 x_0$. This is also the best linear unbiased estimator. Let us write $\hat{E}\{Y \mid x_0\} = \hat{\beta}_0 + \hat{\beta}_1 x_0$. Then

$$\begin{aligned}\hat{E}\{Y \mid x_0\} &= \overline{Y} - \hat{\beta}_1 \bar{x} + \hat{\beta}_1 x_0 \\ &= \overline{Y} + (x_0 - \bar{x})\frac{\sum_{i=1}^n (x_i - \bar{x})(Y_i - \overline{Y})}{\sum_{i=1}^n (x_i - \bar{x})^2},\end{aligned}$$

which is clearly a linear function of normal RVs Y_i. It follows that $\hat{E}\{Y \mid x_0\}$ is also normally distributed with mean $E(\hat{\beta}_0 + \hat{\beta}_1 x_0) = \beta_0 + \beta_1 x_0$ and variance

$$\mathrm{var}\{\hat{E}\{Y \mid x_0\}\} = E\{\hat{\beta}_0 - \beta_0 + \hat{\beta}_1 x_0 - \beta_1 x_0\}^2 \tag{31}$$

$$= \operatorname{var}(\hat{\beta}_0) + x_0^2 \operatorname{var}(\hat{\beta}_1) + 2x_0 \operatorname{cov}(\hat{\beta}_0, \hat{\beta}_1)$$
$$= \sigma^2 \left[\frac{1}{n} + \frac{(\overline{x} - x_0)^2}{\sum_{i=1}^n (x_i - \overline{x})^2} \right].$$

(See Problem 6.) It follows that

$$\frac{\hat{\beta}_0 + \hat{\beta}_1 x_0 - \beta_0 - \beta_1 x_0}{\sigma\{(1/n) + [(\overline{x} - x_0)^2 / \sum_{i=1}^n (x_i - \overline{x})^2]\}^{1/2}} \tag{32}$$

is $\mathcal{N}(0, 1)$. But σ is not known, so that we cannot use (32) to construct a confidence interval for $E\{Y \mid x_0\}$. Since $n\hat{\sigma}^2/\sigma^2$ is a $\chi^2(n-2)$ RV and $n\hat{\sigma}^2/\sigma^2$ is independent of $\hat{\beta}_0 + \hat{\beta}_1 x_0$ (why?), it follows that

$$\sqrt{n-2}\frac{\hat{\beta}_0 + \hat{\beta}_1 t_0 - \beta_0 - \beta_1 x_0}{\hat{\sigma}\{1 + n[(\overline{x} - x_0)^2 / \sum_{i=1}^n (x_i - \overline{x})^2]\}^{1/2}} \tag{33}$$

has a $t(n-2)$ distribution. Thus, a $(1-\alpha)$-level confidence interval for $\beta_0 + \beta_1 x_0$ is given by

$$\left(\hat{\beta}_0 + \hat{\beta}_1 x_0 - t_{n-2,\alpha/2}\hat{\sigma}\sqrt{\frac{n}{n-2}\left[\frac{1}{n} + \frac{(\overline{x} - x_0)^2}{\sum_{i=1}^n (x_i - \overline{x})^2}\right]}, \right. \tag{34}$$
$$\left. \hat{\beta}_0 + \hat{\beta}_1 x_0 + t_{n-2,\alpha/2}\hat{\sigma}\sqrt{\frac{n}{n-2}\left[\frac{1}{n} + \frac{(\overline{x} - x_0)^2}{\sum_{i=1}^n (x_i - \overline{x})^2}\right]} \right).$$

In a similar manner one can show (Problem 7) that

$$\left(\hat{\beta}_0 + \hat{\beta}_1 x_0 - t_{n-2,\alpha/2}\hat{\sigma}\sqrt{\frac{n}{n-2}\left[\frac{n+1}{n} + \frac{(\overline{x} - x_0)^2}{\sum_{i=1}^n (x_i - \overline{x})^2}\right]}, \right. \tag{35}$$
$$\left. \hat{\beta}_0 + \hat{\beta}_1 x_0 + t_{n-2,\alpha/2}\hat{\sigma}\sqrt{\frac{n}{n-2}\left[\frac{n+1}{n} + \frac{(\overline{x} - x_0)^2}{\sum_{i=1}^n (x_i - \overline{x})^2}\right]} \right)$$

is a $(1-\alpha)$-level confidence interval for $Y_0 = \beta_0 + \beta_1 x_0 + \varepsilon$, that is, for the estimated value Y_0 of Y at x_0.

Remark 4. The simple regression model (2) considered above can be generalized in many directions. Thus we may consider EY as a polynomial in x of a degree higher than 1, or we may regard EY as a function of several variables. Some of these generalizations will be taken up in the problems.

Remark 5. Let $(X_1, Y_1), (X_2, Y_2), \ldots, (X_n, Y_n)$ be a sample from a bivariate normal population with parameters $EX = \mu_1$, $EY = \mu_2$, $\operatorname{var}(X) = \sigma_1^2$, $\operatorname{var}(Y) = \sigma_2^2$, and $\operatorname{cov}(X, Y) = \rho$.

In Section 6.6 we computed the PDF of the sample correlation coefficient R and showed (Remark 6.6.4) that the statistic

$$T = R\sqrt{\frac{n-2}{1-R^2}} \tag{36}$$

has a $t(n-2)$ distribution, provided that $\rho = 0$. If we wish to test $\rho = 0$, that is, the independence of two jointly normal RVs, we can base a test on the statistic T. Essentially, we are testing that the population covariance is 0, which implies that the population regression coefficients are 0. Thus we are testing, in particular, that $\beta_1 = 0$. It is therefore not surprising that (36) is identical with (18). We emphasize that we derived (36) for a bivariate normal population, but (18) was derived by taking the X's as fixed and the distribution of Y's as normal. Note that for a bivariate normal population $E\{Y \mid x\} = \mu_2 + \rho(\sigma_2/\sigma_1)(x-\mu_1)$ is linear, in consistency with our model (1) or (2).

Example 1. Let us assume that the following data satisfy a linear regression model:

$$Y_i = \beta_0 + \beta_1 x_i + \varepsilon_i.$$

x	0	1	2	3	4	5
y	0.475	1.007	0.838	−0.618	1.0378	0.943.

Let us test the null hypothesis that $\beta_1 = 0$. We have

$$\overline{x} = 2.5, \quad \sum_{i=0}^{5}(x_i - \overline{x})^2 = 17.5, \qquad \overline{y} = 0.671,$$

$$\sum_{i=0}^{5}(x_i - \overline{x})(y_i - \overline{y}) = 0.9985,$$

$$\hat{\beta}_1 = 0.0571, \qquad \hat{\beta}_0 = \overline{y} - \hat{\beta}_1\overline{x} = 0.5279,$$

$$\sum_{i=0}^{5}(y_i - \hat{\beta}_0 - \hat{\beta}_1 x_i)^2 = 2.3571,$$

and

$$|\hat{\beta}_1|\sqrt{\frac{(n-2)\sum(x_i-\overline{x})^2}{\sum(y_i - \hat{\beta}_0 - \hat{\beta}_1 x_i)^2}} = 0.3106.$$

Since $t_{n-2,\alpha/2} = t_{4,0.025} = 2.776 > 0.3106$, we accept H_0 at level $\alpha = 0.05$.

Let us next find a 95 percent confidence interval for $E\{Y \mid x = 7\}$. This is given by (34). We have

$$t_{n-2,\alpha/2}\hat{\sigma}\sqrt{\frac{n}{n-2}\left[\frac{1}{n} + \frac{(\overline{x}-x_0)^2}{\sum(x_i-\overline{x})^2}\right]} = 2.776\sqrt{\frac{2.3571}{6}}\sqrt{\frac{6}{4}\left(\frac{1}{6} + \frac{20.25}{17.5}\right)}$$
$$= 2.3707,$$

$$\begin{aligned}\hat{\beta}_0 + \hat{\beta}_1 x_0 &= 0.5279 + 0.0571 \times 7\\ &= 0.9276,\end{aligned}$$

so that the 95 percent confidence interval is $(-1.4431, 3.2983)$.

(The data were produced from Table ST6, of random numbers with $\mu = 0$, $\sigma = 1$, by letting $\beta_0 = 1$ and $\beta_1 = 0$ so that $E\{Y \mid x\} = \beta_0 + \beta_1 x = 1$, which surely lies in the interval.)

12.3.2 Logistic and Poisson Regression

In the regression model considered above Y is a continuous type RV. However, in a wide variety of problems Y is either binary or a count variable. Thus in a medical study Y may be the presence or absence of a disease such as diabetes. How do we modify linear regression model to apply in this case? The idea here is to choose a function of $E(Y)$ so that in Section 12.3.1

$$f(E(Y)) = \mathbf{X}\boldsymbol{\beta}.$$

This can be accomplished by choosing the function f to be the logarithm of the odds ratio

$$f(p) = \log\left(\frac{p}{1-p}\right), \tag{37}$$

where $p = P(Y = 1)$ so that $E(Y) = p$. It follows that

$$p = E(Y) = P(Y = 1) = \frac{\exp(\mathbf{X}\boldsymbol{\beta})}{1 + \exp(\mathbf{X}\boldsymbol{\beta})} \tag{38}$$

so that logistic regression models the logarithm of odds ratio as a linear function of RVs X_i. The term *logistic regression* derives from the fact that the function $e^{\mathrm{x}}/(1 + e^{\mathrm{x}})$ is known as the *logistic function.*

For simplicity we will only consider the simple linear regression model case so that

$$\mathbf{E}(Y_i) = \pi_i(\beta_0 + \beta \mathrm{x}_i),\ i = 1, 2, \ldots, n,\ 0 < \pi_i(\beta_0 + \beta \mathrm{x}_i) < 1. \tag{39}$$

Choosing the logistic distribution as

$$\pi_i = \pi_i(\beta_0 + \beta \mathrm{x}_i) = \frac{\exp(\beta_0 + \beta x_i)}{1 + \exp(\beta_0 + \beta x_i)}, \tag{40}$$

let $Y_1, Y_2, \ldots, Y_n$ be iid binary RVs taking values 0 or 1. Then the joint PMF of $Y_1, Y_2, \ldots, Y_n$ is given by

$$\begin{aligned}L(\beta_0, \beta \mid \mathbf{x}) &= \prod_{i=1}^{n} \left\{\pi_i^{y_i}(1 - \pi_i)^{1-y_i}\right\}\\ &= \prod_{i=1}^{n}(1 - \pi_i)\left\{\exp \sum_{i=1}^{n} y_i \log\left(\frac{\pi_i}{1 - \pi_i}\right)\right\} \end{aligned}\tag{41}$$

and the log likelihood function by

$$\log L(\beta_0,\beta|\mathbf{x}) = n\bar{y}\beta_0 + \beta\sum_{i=1}^{x} x_i y_i - \sum_{i=1}^{n}\log\{1+\exp(\beta_0+\beta x_i)\}. \tag{42}$$

It is easy to see that

$$\begin{aligned}\frac{\partial \log L}{\partial \beta_0} &= n\bar{y} - \sum_{i=1}^{n}\pi_i = 0,\\ \frac{\partial \log L}{\partial \beta} &= \sum_{i=1}^{n} x_i y_i - \sum_{i=1}^{n} x_i\pi_i = 0.\end{aligned} \tag{43}$$

Since the likelihood equations are nonlinear in the parameters, the MLEs of β_0 and β are obtained numerically by using Newton–Raphson method.

Let $\hat{\beta}_0$ and $\hat{\beta}$ be the MLE of β_0 and β, respectively. From section 8.7 we note that the variance of $\hat{\beta}$ is given by

$$\text{var}(\hat{\beta}) = \sum_{i=1}^{n} x_i^2\pi_i(1-\pi_i), \tag{44}$$

so that the standard error (SE) of $\hat{\beta}$ is its square root. For large n, the so-called Wald statistic $Z = \hat{\beta}/\text{SE}(\hat{\beta})$ has an approximate $N(0,1)$ distribution under $H_0 : \beta = 0$. Thus we reject H_0 at level α if $|z| > z_{\alpha/2}$. One can use $\hat{\beta} \pm z_{\alpha/2}\,\text{SE}(\hat{\beta})$ as a $(1-\alpha)$-level confidence interval for β.

Yet another choice for testing H_0 is to use the LRT statistics $-2\log\lambda$ (see Theorem 10.2.3). Under $H_0, -2\log\lambda$ has a chi-square distribution with 1 d.f. Here

$$\lambda = \frac{L(\hat{\beta}_0, 0|\mathbf{x})}{L(\hat{\beta}_0,\hat{\beta}|\mathbf{x})}. \tag{45}$$

In (40) we chose the DF of a logistic RV. We could have chosen some other DF such as $\phi(x)$, the DF of a $N(0,1)$ RV. In that case we have $\beta_0 + \beta x = \phi(x)$. The resulting model is called *probit regression*.

We finally consider the case when the RV Y is a count of rare events and has Poisson distribution with parameter λ. Clearly, the GLM is not directly applicable. Again we only consider the linear regression model case. Let $\text{Y}_i, i = 1,2,\ldots,k$, be independent $P(\lambda_i)$ RVs where $\lambda_i = \exp(\beta_0 + \text{x}_i\beta_1)$, so that

$$\theta_i = \log\lambda_i = \beta_0 + \text{x}_i\beta_1.$$

The log likelihood function is given by

$$\log L(\beta_0,\beta_1;y_1,\ldots,y_n) = \sum_{i=1}^{n}\{y_i\theta_i - e^{\theta_i} - \log(y_i!)\}. \tag{46}$$

In order to find the MLEs of β_0 and β_1 we need to solve the likelihood equations

$$\frac{\partial \log L}{\partial \beta_0} = \sum_{i=1}^{n} \{y_i - \theta_i\} = 0$$

$$\frac{\partial \log L}{\partial \beta_1} = \sum_{i=1}^{n} \{x_i y_i - x_i \theta_i\} = 0, \tag{47}$$

which are nonlinear in β_0 and β_1. The most common method of obtaining the MLEs is to apply the iteratively weighted least squares algorithm.

Once the MLEs of β_0 and β_1 are computed, one can compute the SEs of the estimates by using methods of Section 8.7. Using the SE($\hat{\beta}_1$), for example, one can test hypothesis concerning β_1 or construct $(1-\alpha)$-level confidence interval for β_1.

For a detailed discussion of Geometric and Poisson regression we refer Agresti [1]. A wide variety of software is available, which can be used to carry out the computations required.

PROBLEMS 12.3

1. Prove statements (12), (13), and (14).
2. Prove statements (15) and (16).
3. Prove statement (19).
4. Obtain tests of null hypotheses $\beta_0 = \beta_0', \beta_1 = \beta_1'$, and $(\beta_0, \beta_1)' = (\beta_0', \beta_1')'$, where β_0', β_1' are given real numbers.
5. Obtain the confidence intervals for β_0 and β_1 as given in (28) and (29), respectively.
6. Derive the expression for $\text{var}\{\hat{E}\{Y \mid x_0\}\}$ as given in (31).
7. Show that the interval given in (35) is a $(1-\alpha)$-level confidence interval for $Y_0 = \beta_0 + \beta_1 x_0 + \varepsilon$, the estimated value of Y at x_0.
8. Suppose that the regression of Y on the (mathematical) variable x is a quadratic

$$Y_i = \beta_0 + \beta_1 x_i + \beta_2 x_i^2 + \varepsilon_i,$$

where $\beta_0, \beta_1, \beta_2$ are unknown parameters, $x_1, x_2, \ldots, x_n$ are known values of x, and $\varepsilon_1, \varepsilon_2, \ldots, \varepsilon_n$ are unobservable RVs that are assumed to be independently normally distributed with common mean 0 and common variance σ^2 (see Example 12.2.3). Assume that the coefficient vectors $(x_1^k, x_2^k, \ldots, x_n^k)$, $k = 0, 1, 2$, are linearly independent. Write the normal equations for estimating the β's and derive the generalized likelihood ratio test of $\beta_2 = 0$.
9. Suppose that the Y's can be written as

$$Y_i = \beta_1 x_{i1} + \beta_2 x_{i2} + \beta_3 x_{i3} + \varepsilon_i,$$

where x_{i1}, x_{i2}, x_{i3} are three mathematical variables, and ε_i are iid $\mathcal{N}(0,1)$ RVs. Assuming that the matrix $\mathbf{X}$ (see Example 3) is of full rank, write the normal equations and derive the likelihood ratio test of the null hypothesis $H_0\colon \beta_1 = \beta_2 = \beta_3$.

10. The following table gives the weight Y (grams) of a crystal suspended in a saturated solution against the time suspended T (days):

Time T	0	1	2	3	4	5	6
Weight Y	0.4	0.7	1.1	1.6	1.9	2.3	2.6

(a) Find the linear regression line of Y on T.

(b) Test the hypothesis that $\beta_0 = 0$ in the linear regression model $Y_i = \beta_0 + \beta_1 T_i + \varepsilon_i$.

(c) Obtain a 0.95 level confidence interval for β_0.

11. Let $o_i = \pi_i/(1-\pi_i)$ be the odds ratio corresponding to x_i, $i = 1, 2, \ldots, n$. By considering the ratio o_{i+1}/o_i, how will you interpret the value of the slope parameter β_1?

12. Do the same for parameter β_1 in the Poisson regression model by considering the ratio λ_{i+1}/λ_1.

12.4 ONE-WAY ANALYSIS OF VARIANCE

In this section we return to the problem of one-way analysis of variance considered in Examples 12.2.1 and 12.2.4. Consider the model

$$Y_{ij} = \mu_i + \varepsilon_{ij}, \qquad j = 1, 2, \ldots, n_i; \qquad i = 1, 2, \ldots, k, \tag{1}$$

as described in Example 12.2.4. In matrix notation we write

$$\mathbf{Y} = \mathbf{X}\boldsymbol{\beta} + \boldsymbol{\varepsilon}, \tag{2}$$

where

$$\begin{aligned}
\mathbf{Y} &= (Y_{11}, Y_{12}, \ldots, Y_{1n_1}, Y_{21}, Y_{22}, \ldots, Y_{2n_2}, \ldots, Y_{k1}, Y_{k2}, \ldots, Y_{kn_k})', \\
\boldsymbol{\beta} &= (\mu_1, \mu_2, \ldots, \mu_k)', \\
\mathbf{X} &= \begin{pmatrix} \mathbf{1}_{n_1} & \mathbf{0} & \cdots & \mathbf{0} \\ \vdots & \vdots & \vdots\vdots\vdots & \vdots \\ \mathbf{0} & \mathbf{0} & \cdots & \mathbf{1}_{n_k} \end{pmatrix}, \\
\boldsymbol{\varepsilon} &= (\varepsilon_{11}, \varepsilon_{12}, \ldots, \varepsilon_{1n_1}, \varepsilon_{21}, \varepsilon_{22}, \ldots, \varepsilon_{2n_2}, \ldots, \varepsilon_{k1}, \varepsilon_{k2}, \ldots, \varepsilon_{kn_k})'.
\end{aligned}$$

As in Example 12.2.4, $\mathbf{Y}$ is a vector of n-observations ($n = \sum_{i=1}^{k} n_i$), whose components Y_{ij} are subject to random error $\varepsilon_{ij} \sim \mathcal{N}(0, \sigma^2)$, $\boldsymbol{\beta}$ is a vector of k unknown parameters, and $\mathbf{X}$ is a design matrix. We wish to find a test of $H_0\colon \mu_1 = \mu_2 = \cdots = \mu_k$ against all alternatives. We may write H_0 in the form $\mathbf{H}\boldsymbol{\beta} = \mathbf{0}$, where $\mathbf{H}$ is a $(k-1) \times k$ matrix of rank $(k-1)$, which can be chosen to be

$$\mathbf{H} = \begin{pmatrix} 1 & -1 & 0 & \cdots & 0 \\ 1 & 0 & -1 & \cdots & 0 \\ \vdots & \vdots & \vdots & \vdots & \vdots \\ 1 & 0 & 0 & \cdots & -1 \end{pmatrix}.$$

Let us write $\mu_1 = \mu_2 = \cdots \mu_k = \mu$ under H_0. The joint PDF of $\mathbf{Y}$ is given by

$$f(\mathbf{y}; \mu_1, \mu_2, \ldots, \mu_k, \sigma^2) = \left(\frac{1}{2\pi\sigma^2}\right)^{n/2} \exp\left\{-\frac{1}{2\sigma^2}\sum_{i=1}^{k}\sum_{j=1}^{n_i}(y_{ij} - \mu_i)^2\right\}, \tag{3}$$

and, under H_0, by

$$f(\mathbf{x}; \mu, \sigma^2) = \left(\frac{1}{2\pi\sigma^2}\right)^{n/2} \exp\left\{-\frac{1}{2\sigma^2}\sum_{i=1}^{k}\sum_{j=1}^{n_i}(y_{ij} - \mu)^2\right\}. \tag{4}$$

It is easy to check that the MLEs are

$$\hat{\mu}_i = \frac{\sum_{j=1}^{n_i} y_{ij}}{n_i} = \overline{y}_{i\cdot}, \qquad i = 1, 2, \ldots, k, \tag{5}$$

$$\hat{\sigma}^2 = \frac{\sum_{i=1}^{k}\sum_{j=1}^{n_i}(y_{ij} - \overline{y}_{i\cdot})^2}{n}, \tag{6}$$

$$\hat{\hat{\mu}} = \frac{\sum_{i=1}^{k}\sum_{j=1}^{n_i} y_{ij}}{n} = \overline{y}, \tag{7}$$

and

$$\hat{\hat{\sigma}}^2 = \frac{\sum_{i=1}^{k}\sum_{j=1}^{n_i}(y_{ij} - \overline{y})^2}{n}. \tag{8}$$

By Theorem 12.2.1, the likelihood ratio test is to reject H_0 if

$$\frac{\sum_{i=1}^{k}\sum_{j=1}^{n_i}(Y_{ij} - \overline{Y})^2 - \sum_{i=1}^{k}\sum_{j=1}^{n_i}(Y_{ij} - \overline{Y}_{i\cdot})^2}{\sum_{i=1}^{k}\sum_{j=1}^{n_i}(Y_{ij} - \overline{Y}_{i\cdot})^2}\left(\frac{n-k}{k-1}\right) \geq F_0, \tag{9}$$

where F_0 is the upper α percent point in the $F(k-1, n-k)$ distribution. Since

$$\begin{aligned} \sum_{i=1}^{k}\sum_{j=1}^{n_i}(Y_{ij} - \overline{Y})^2 &= \sum_{i=1}^{k}\sum_{j=1}^{n_i}(Y_{ij} - \overline{Y}_{i\cdot} + \overline{Y}_{i\cdot} - \overline{Y})^2 \qquad (10) \\ &= \sum_{i=1}^{k}\sum_{j=1}^{n_i}(Y_{ij} - \overline{Y}_{i\cdot})^2 + \sum_{i=1}^{k} n_i(\overline{Y}_{i\cdot} - \overline{Y})^2, \end{aligned}$$

we may rewrite (9) as

$$\frac{\sum_{i=1}^{k} n_i(\overline{Y}_{i\cdot} - \overline{Y})^2/(k-1)}{\sum_{i=1}^{k}\sum_{j=1}^{n_i}(Y_{ij} - \overline{Y}_{i\cdot})^2/(n-k)} \geq F_0. \tag{11}$$

It is usual to call the sum of squares in the numerator of (11) the *between sum of squares* (BSS) and the sum of squares in the denominator of (11) *the within sum of squares* (WSS). The results are conveniently displayed in a so-called *analysis of variance table* in the following form.

One-Way Analysis of Variance

Source Variation	Sum of Squares	Degrees of Freedom	Mean Sum of Squares	F-Ratio
Between	$\text{BSS} = \sum_{i=1}^{k} n_i(\overline{Y}_{i\cdot} - \overline{Y})^2$	$k-1$	$\text{BSS}/(k-1)$	$\dfrac{\text{BSS}/(k-1)}{\text{WSS}/(n-k)}$
Within	$\text{WSS} = \sum_{i=1}^{k}\sum_{j=1}^{n_i}(Y_{ij} - \overline{Y}_{i\cdot})^2$	$n-k$	$\text{WSS}/(n-k)$	
Mean	$n\overline{Y}^2$	1		
Total	$\text{TSS} = \sum_{i=1}^{k}\sum_{j=1}^{n_i} Y_{ij}^2$	n		

The third row, designated "Mean," has been included to make the total of the second column add up to the total sum of squares (TSS), $\sum_{i=1}^{k}\sum_{j=1}^{n_i} Y_{ij}^2$.

Example 1. The lifetimes (in hours) of samples from three different brands of batteries were recorded with the following results:

Brand		
Y_1	Y_2	Y_3
40	60	60
30	40	50
50	55	70
50	65	65
30		75
		40

We wish to test whether the three brands have different average lifetimes. We will assume that the three samples come from normal populations with common (unknown) standard deviation σ.

From the data $n_1 = 5$, $n_2 = 4$, $n_3 = 6$, $n = 15$, and

$$\bar{y}_1 = \frac{200}{5} = 40, \qquad \bar{y}_2 = \frac{220}{4} = 55, \qquad \bar{y}_3 = \frac{360}{6} = 60,$$

$$\sum_{i=1}^{5}(y_{1i} - \bar{y}_1)^2 = 400, \qquad \sum_{i=1}^{4}(y_{2i} - \bar{y}_2)^2 = 350, \qquad \sum_{i=1}^{6}(y_{3i} - \bar{y}_3)^2 = 850.$$

Also, the grand mean is

$$\bar{y} = \frac{200 + 220 + 360}{15} = \frac{780}{15} = 52.$$

Thus

$$\begin{aligned} \text{BSS} &= 5(40 - 52)^2 + 4(55 - 52)^2 + 6(60 - 52)^2 \\ &= 1140, \\ \text{WSS} &= 400 + 350 + 850 = 1600. \end{aligned}$$

Analysis of Variance

Source	SS	d.f.	MSS	F-Ratio
Between	1140	2	570	$570/133.33 = 4.28$
Within	1600	12	133.33	

Choosing $\alpha = 0.05$, we see that $F_0 = F_{2,12,0.05} = 3.89$. Thus we reject $H_0: \mu_1 = \mu_2 = \mu_3$ at level $\alpha = 0.05$.

Example 2. Three sections of the same elementary statistics course were taught by three instructors. The final grades of students were recorded as follows:

Instructor		
I	II	III
95	88	68
33	78	79
48	91	91
76	51	71
89	85	87
82	77	68

(continued)

Instructor		
I	II	III
60	31	79
77	62	16
	96	35
	81	

Let us test the hypothesis that the average grades given by the three instructors are the same at level $\alpha = 0.05$.

From the data $n_1 = 8$, $n_2 = 10$, $n_3 = 9$, $n = 27$, $\bar{y}_1 = 70$, $\bar{y}_2 = 74$, $\bar{y}_3 = 66$, $\sum_{i=1}^{8}(y_{1i}-\bar{y}_1)^2 = 3168$, $\sum_{i=1}^{10}(y_{2i}-\bar{y}_2)^2 = 3686$, $\sum_{i=1}^{9}(y_{3i}-\bar{y}_3)^2 = 4898$. Also, the grand mean is

$$\bar{y} = \frac{560+740+594}{27} = \frac{1894}{27} = 70.15.$$

Thus

$$\begin{aligned} \text{BSS} &= 8(0.15)^2 + 10(3.85)^2 + 9(4.15)^2 \\ &= 303.4075 \\ \text{WSS} &= 3168 + 3686 + 4898 \\ &= 11,752. \end{aligned}$$

Analysis of Variance

Source	SS	d.f.	MSS	F-Ratio
Between	303.41	2	151.70	151.70/489.67
Within	11,752.00	24	489.67	

We therefore cannot reject the null hypothesis that the average grades given by the three instructors are the same.

PROBLEMS 12.4

1. Prove statements (5), (6), (7), and (8).
2. The following are the coded values of the amounts of corn (in bushels per acre) obtained from four varieties, using unequal number of plots for the different varieties:

$$A: \quad 2,1,3,2$$
$$B: \quad 3,4,2,3,4,2$$

C: 6, 4, 8
D: 7, 6, 7, 4

Test whether there is a significant difference between the yields of the varieties.

3. A consumer interested in buying a new car has reduced his search to six different brands: D, F, G, P, V, T. He would like to buy the brand that gives the highest mileage per gallon of regular gasoline. One of his friends advises him that he should use some other method of selection, since the average mileages of the six brands are the same, and offers the following data in support of her assertion.

Distance Traveled (Miles) per Gallon of Gasoline

	Brand					
Car	D	F	G	P	V	T
1	42	38	28	32	30	25
2	35	33	32	36	35	32
3	37	28	35	27	25	24
4		37	37	26	30	
5				28	30	
6				19		

Should the consumer accept his friend's advice?

4. The following data give the ages of entering freshmen in independent random samples from three different universities.

University		
A	B	C
17	16	21
19	16	23
20	19	22
21		20
18		19

Test the hypothesis that the average ages of entering freshman at these universities are the same.

5. Five cigarette manufacturers claim that their product has low tar content. Independent random samples of cigarettes are taken from each manufacturer and the following tar levels (in milligrams) are recorded.

Brand	Tar Level (mg)
A	4.2, 4.8, 4.6, 4.0, 4.4
B	4.9, 4.8, 4.7, 5.0, 4.9, 5.2
C	5.4, 5.3, 5.4, 5.2, 5.5
D	5.8, 5.6, 5.5, 5.4, 5.6, 5.8
E	5.9, 6.2, 6.2, 6.8, 6.4, 6.3

Can the differences among the sample means be attributed to chance?

6. The quantity of oxygen dissolved in water is used as a measure of water pollution. Samples are taken at four locations in a lake and the quantity of dissolved oxygen is recorded as follows (lower reading corresponds to greater pollution):

Location	Quantity of Dissolved Oxygen (%)
A	7.8, 6.4, 8.2, 6.9
B	6.7, 6.8, 7.1, 6.9, 7.3
C	7.2, 7.4, 6.9, 6.4, 6.5
D	6.0, 7.4, 6.5, 6.9, 7.2, 6.8

Do the data indicate a significant difference in the average amount of dissolved oxygen for the four locations?

12.5 TWO-WAY ANALYSIS OF VARIANCE WITH ONE OBSERVATION PER CELL

In many practical problems one is interested in investigating the effects of two factors that influence an outcome. For example, the variety of grain and the type of fertilizer used both affect the yield of a plot or the score on a standard examination is influenced by the size of the class and the instructor.

Let us suppose that two factors affect the outcome of an experiment. Suppose also that one observation is available at each of a number of levels of these two factors. Let $Y_{ij}(i=1,2,\ldots,a; j=1,2,\ldots,b)$ be the observation when the first factor is at the ith level, and the second factor at the jth level. Assume that

$$Y_{ij}=\mu+\alpha_i+\beta_j+\varepsilon_{ij}, \qquad i=1,2,\ldots,a; \qquad j=1,2,\ldots,b, \tag{1}$$

where α_i is the effect of the ith level of the first factor, β_j is the effect of the jth level of the second factor, and ε_{ij} is the random error, which is assumed to be normally distributed with mean 0 and variance σ^2. We will assume that the ε_{ij}'s are independent. It follows that Y_{ij} are independent normal RVs with means $\mu+\alpha_i+\beta_j$ and variance σ^2. There is no loss of generality in assuming that $\sum_{i=1}^{a}\alpha_i=\sum_{j=1}^{b}\beta_j=0$, for, if $\mu_{ij}=\mu'+\alpha_i'+\beta_j'$, we can write

$$\begin{aligned}\mu_{ij}&=(\mu'+\overline{\alpha}'+\overline{\beta}')+(\alpha_i'-\overline{\alpha}')+(\beta_j'-\overline{\beta}')\\&=\mu+\alpha_i+\beta_j\end{aligned}$$

and $\sum_{i=1}^{a}\alpha_i = 0$, $\sum_{j=1}^{b}\beta_j = 0$. Here we have written $\overline{\alpha}'$ and $\overline{\beta}'$ for the means of α_i''s and β_j''s, respectively. Thus Y_{ij} may denote the yield from the use of the ith variety of some grain and the jth type of some fertilizer. The two hypotheses of interest are

$$\alpha_1 = \alpha_2 = \cdots = \alpha_a = 0 \qquad \text{and} \qquad \beta_1 = \beta_2 = \cdots = \beta_b = 0.$$

The first of these, for example, says that the first factor has no effect on the outcome of the experiment.

In view of the fact that $\sum_{i=1}^{a}\alpha_i = 0$ and $\sum_{j=1}^{b}\beta_j = 0$, $\alpha_a = -\sum_{i=1}^{a-1}\alpha_i$, $\beta_b = -\sum_{j=1}^{b-1}\beta_j$, and we can write our model in matrix notation as

$$\mathbf{Y} = \mathbf{X}\boldsymbol{\beta} + \boldsymbol{\varepsilon}, \tag{2}$$

where

$$\begin{aligned}
\mathbf{Y} &= (Y_{11}, Y_{12}, \ldots, Y_{1b}, Y_{21}, Y_{22}, \ldots, Y_{2b}, \ldots, Y_{a1}, Y_{a2}, \ldots, Y_{ab})',\\
\boldsymbol{\beta} &= (\mu, \alpha_1, \alpha_2, \ldots, \alpha_{a-1}, \beta_1, \beta_2, \ldots, \beta_{b-1})',\\
\boldsymbol{\varepsilon} &= (\varepsilon_{11}, \varepsilon_{12}, \ldots, \varepsilon_{1b}, \varepsilon_{21}, \varepsilon_{22}, \ldots, \varepsilon_{2b}, \ldots, \varepsilon_{a1}, \varepsilon_{a2}, \ldots, \varepsilon_{ab})',
\end{aligned}$$

and

$$\mathbf{X} = \left(\begin{array}{c|cccc|cccc}
\mu & \alpha_1 & \alpha_2 & \cdots & \alpha_{a-1} & \beta_1 & \beta_2 & \cdots & \beta_{b-1}\\
\hline
1 & 1 & 0 & \cdots & 0 & 1 & 0 & \cdots & 0\\
1 & 1 & 0 & \cdots & 0 & 0 & 1 & \cdots & 0\\
\cdot & \cdot & \cdot & \cdots & \cdot & \cdot & \cdot & \cdots & \cdot\\
\cdot & \cdot & \cdot & \cdots & \cdot & \cdot & \cdot & \cdots & \cdot\\
\cdot & \cdot & \cdot & \cdots & \cdot & \cdot & \cdot & \cdots & \cdot\\
1 & 1 & 0 & \cdots & 0 & 0 & 0 & \cdots & 1\\
1 & 1 & 0 & \cdots & 0 & -1 & -1 & \cdots & -1\\
1 & 0 & 1 & \cdots & 0 & 1 & 0 & \cdots & 0\\
1 & 0 & 1 & \cdots & 0 & 0 & 1 & \cdots & 0\\
\cdot & \cdot & \cdot & \cdots & \cdot & \cdot & \cdot & \cdots & \cdot\\
\cdot & \cdot & \cdot & \cdots & \cdot & \cdot & \cdot & \cdots & \cdot\\
\cdot & \cdot & \cdot & \cdots & \cdot & \cdot & \cdot & \cdots & \cdot\\
1 & 0 & 1 & \cdots & 0 & 0 & 0 & \cdots & 1\\
1 & 0 & 1 & \cdots & 0 & -1 & -1 & \cdots & -1\\
\cdot & \cdot & \cdot & \cdots & \cdot & \cdot & \cdot & \cdots & \cdot\\
\cdot & \cdot & \cdot & \cdots & \cdot & \cdot & \cdot & \cdots & \cdot\\
\cdot & \cdot & \cdot & \cdots & \cdot & \cdot & \cdot & \cdots & \cdot\\
1 & -1 & -1 & \cdots & -1 & 1 & 0 & \cdots & 0\\
1 & -1 & -1 & \cdots & -1 & 0 & 1 & \cdots & 0\\
\cdot & \cdot & \cdot & \cdots & \cdot & \cdot & \cdot & \cdots & \cdot\\
\cdot & \cdot & \cdot & \cdots & \cdot & \cdot & \cdot & \cdots & \cdot\\
\cdot & \cdot & \cdot & \cdots & \cdot & \cdot & \cdot & \cdots & \cdot\\
1 & -1 & -1 & \cdots & -1 & 0 & 0 & \cdots & 1\\
1 & -1 & -1 & \cdots & -1 & -1 & -1 & \cdots & -1
\end{array}\right)$$

The vector of unknown parameters $\boldsymbol{\beta}$ is $(a+b-1)\times 1$ and the matrix $\mathbf{X}$ is $ab\times(a+b-1)$ (b blocks of a rows each). We leave the reader to check that $\mathbf{X}$ is of full rank, $a+b-1$. The hypothesis $H_\alpha: \alpha_1=\alpha_2=\cdots=\alpha_a=0$ or $H_\beta: \beta_1=\beta_2=\cdots=\beta_b=0$ can easily be put into the form $\mathbf{H}\boldsymbol{\beta}=\mathbf{0}$. For example, for H_β we can choose $\mathbf{H}$ to be the $(b-1)\times(a+b-1)$ matrix of full rank $b-1$, given by

$$\mathbf{H}=\left(\begin{array}{c|cccc|cccc}
\mu & \alpha_1 & \alpha_2 & \cdots & \alpha_{a-1} & \beta_1 & \beta_2 & \cdots & \beta_{b-1}\\
\hline
0 & 0 & 0 & \cdots & 0 & 1 & 0 & \cdots & 0\\
0 & 0 & 0 & \cdots & 0 & 0 & 1 & \cdots & 0\\
\cdot & \cdot & \cdot & \cdots & \cdot & \cdot & \cdot & \cdots & \cdot\\
\cdot & \cdot & \cdot & \cdots & \cdot & \cdot & \cdot & \cdots & \cdot\\
\cdot & \cdot & \cdot & \cdots & \cdot & \cdot & \cdot & \cdots & \cdot\\
0 & 0 & 0 & \cdots & 0 & 0 & 0 & \cdots & 1
\end{array}\right).$$

Clearly, the model described above is a special case of the general linear hypothesis, and we can use Theorem 12.2.1 to test H_β.

To apply Theorem 12.2.1 we need the estimators $\hat{\mu}_{ij}$ and $\hat{\hat{\mu}}_{ij}$. It is easily checked that

$$\hat{\mu}=\frac{\sum_{i=1}^{a}\sum_{j=1}^{b}y_{ij}}{ab}=\overline{y} \tag{3}$$

and

$$\hat{\alpha}_i=\overline{y}_{i\cdot}-\overline{y}, \qquad \hat{\beta}_j=\overline{y}_{\cdot j}-\overline{y}, \tag{4}$$

where $\overline{y}_{i\cdot}=\sum_{j=1}^{b}y_{ij}/b$, $\overline{y}_{\cdot j}=\sum_{i=1}^{a}y_{ij}/a$. Also, under H_β, for example,

$$\hat{\hat{\mu}}=\overline{y} \qquad \text{and} \qquad \hat{\hat{\alpha}}_i=\overline{y}_{i\cdot}-\overline{y}. \tag{5}$$

In the notation of Theorem 12.2.1, $n=ab$, $k=a+b-1$, $r=b-1$, so that $n-k=ab-a-b+1=(a-1)(b-1)$, and

$$F=\frac{\sum_{i=1}^{a}\sum_{j=1}^{b}(Y_{ij}-\overline{Y}_{i\cdot})^2-\sum_{i=1}^{a}\sum_{j=1}^{b}(Y_{ij}-\overline{Y}_{i\cdot}-\overline{Y}_{\cdot}j+\overline{Y})^2}{\sum_{i=1}^{a}\sum_{j=1}^{b}(Y_{ij}-\overline{Y}_{i\cdot}-\overline{Y}_{\cdot j}+\overline{Y})^2}. \tag{6}$$

Since

$$\begin{aligned}
\sum_{i=1}^{a}\sum_{j=1}^{b}(Y_{ij}-\overline{Y}_{i\cdot})^2 &= \sum_{i=1}^{a}\sum_{j=1}^{b}\{(Y_{ij}-\overline{Y}_{i\cdot}-\overline{Y}_{\cdot j}+\overline{Y})+(\overline{Y}_{\cdot j}-\overline{Y})\}^2 \\
&= \sum_{i=1}^{a}\sum_{j=1}^{b}(Y_{ij}-\overline{Y}_{i\cdot}-\overline{Y}_{\cdot j}+\overline{Y})^2+a\sum_{j=1}^{b}(\overline{Y}_{\cdot j}-\overline{Y})^2,
\end{aligned} \tag{7}$$

we may write

$$F = \frac{a\sum_{j=1}^{b}(\overline{Y}_{\cdot j} - \overline{Y})^2}{\sum_{i=1}^{a}\sum_{j=1}^{b}(Y_{ij} - \overline{Y}_{i\cdot} - \overline{Y}_{\cdot j} + \overline{Y})^2}. \tag{8}$$

It follows that, under H_β, $(a-1)F$ has a central $F(b-1,(a-1)(b-1))$ distribution.

The numerator of F in (8) measures the variability between the means $\overline{Y}_{\cdot j}$, and the denominator measures the variability that exists once the effects due to the two factors have been subtracted.

If H_α is the null hypothesis to be tested, one can show that under H_α the MLEs are

$$\hat{\hat{\mu}} = \overline{y} \qquad \text{and} \qquad \hat{\hat{\beta}}_j = \overline{y}_{\cdot j} - \overline{y}. \tag{9}$$

As before, $n = ab$, $k = a + b - 1$, but $r = a - 1$. Also,

$$F = \frac{\sum_{i=1}^{a}\sum_{j=1}^{b}(Y_{ij} - \overline{Y}_{\cdot j})^2 - \sum_{i=1}^{a}\sum_{j=1}^{b}(Y_{ij} - \overline{Y}_{i\cdot} - \overline{Y}_{\cdot j} + \overline{Y})^2}{\sum_{i=1}^{a}\sum_{j=1}^{b}(Y_{ij} - \overline{Y}_{i\cdot} - \overline{Y}_{\cdot j} + \overline{Y})^2}, \tag{10}$$

which may be rewritten as

$$F = \frac{b\sum_{i=1}^{a}(\overline{Y}_{i\cdot} - \overline{Y})^2}{\sum_{i=1}^{a}\sum_{j=1}^{b}(Y_{ij} - \overline{Y}_{i\cdot} - \overline{Y}_{\cdot j} + \overline{Y})^2}. \tag{11}$$

It follows that, under H_α, $(b-1)F$ has a central $F(a-1,(a-1)(b-1))$ distribution. The numerator of F in (11) measures the variability between the means $\overline{Y}_{i\cdot}$.

If the data are put into the following form:

β / α		Levels of Factor 2				
		1	2		b	Row Means
	1	Y_{11},	Y_{12},	$\cdots$,	Y_{1b}	$\overline{Y}_{1\cdot}$
Levels	2	Y_{21},	Y_{22},	$\cdots$,	Y_{2b}	$\overline{Y}_{2\cdot}$
of	·	·	·	···	·	·
Factor 1	·	·	·	···	·	·
	·	·	·	···	·	·
	a	Y_{a1},	Y_{a2},	$\cdots$,	Y_{ab}	$\overline{Y}_{a\cdot}$
Column Means		$\overline{Y}_{\cdot 1}$,	$\overline{Y}_{\cdot 2}$,	$\cdots$,	$\overline{Y}_{\cdot b}$	$\overline{Y}$

so that the rows represent various levels of factor 1, and the columns, the levels of factor 2, one can write

$$\begin{aligned}\text{between sum of squares for rows} &= b\sum_{i=1}^{a}(\overline{Y}_{i\cdot} - \overline{Y})^2\\ &= \text{sum of squares for factor 1}\\ &= \text{SS}_1.\end{aligned}$$

Similarly,

$$\begin{aligned}\text{between sum of squares for columns} &= a\sum_{j=1}^{b}(\overline{Y}_{\cdot j}-\overline{Y})^2\\ &= \text{sum of squares for factor 2}\\ &= \text{SS}_2.\end{aligned}$$

It is usual to write error or residual sum of squares (SSE) for the denominator of (8) or (11). These results are conveniently presented in an analysis of variance table as follows.

Two-Way Analysis of Variance Table with One Observation per Cell

Source of Variation	Sum of Squares	Degrees of Freedom	Mean Square	F-Ratio
Rows	SS_1	$a-1$	$\text{MS}_1=\text{SS}_1/(a-1)$	MS_1/MSE
Columns	SS_2	$b-1$	$\text{MS}_2=\text{SS}_2/(b-1)$	MS_2/MSE
Error	SSE	$(a-1)(b-1)$	$\text{MSE}=\text{SSE}/(a-1)(b-1)$	
Mean	$ab\overline{Y}^2$	1	$ab\overline{Y}^2$	
Total	$\sum_{i=1}^{a}\sum_{j=1}^{b}Y_{ij}^2$	ab	$\sum_{i=1}^{a}\sum_{j=1}^{b}Y_{ij}^2/ab$	

Example 1. The following table gives the yield (pounds per plot) of three varieties of wheat, obtained with four different kinds of fertilizers.

	Variety of Wheat		
Fertilizer	A	B	C
α	8	3	7
β	10	4	8
γ	6	5	6
δ	8	4	7

Let us test the hypothesis of equality in the average yields of the three varieties of wheat and the null hypothesis that the four fertilizers are equally effective.

In our notation, $b=3, a=4, \overline{y}_1.=6, \overline{y}_2.=7.33, \overline{y}_3.=5.67, \overline{y}_4.=6.33, \overline{y}_{\cdot 1}=8, \overline{y}_{\cdot 2}=4, \overline{y}_{\cdot 3}=7, \overline{y}=6.33$.

Also,

$$\begin{aligned}\text{SS}_1 &= \text{sum of squares due to fertilizer}\\ &= 3[(0.33)^2+1^2+(0.66)^2+0^2]\\ &= 4.67;\end{aligned}$$

$$\begin{aligned} SS_2 &= \text{sum of squares due to variety of wheat} \\ &= 4[(1.67)^2 + (2.33)^2 + (0.67)^2] \\ &= 34.67 \end{aligned}$$

$$\begin{aligned} SSE &= \sum_{i=1}^{4}\sum_{j=1}^{3}(y_{ij} - \bar{y}_{i\cdot} - \bar{y}_{\cdot j} + \bar{y})^2 \\ &= 7.33 \end{aligned}$$

The results are shown in the following table:

Analysis of Variance

Source	SS	d.f.	MS	F-Ratio
Variety of wheat	34.67	2	17.33	14.2
Fertilizer	4.67	3	1.56	1.28
Error	7.33	6	1.22	
Mean	481.33	1	481.33	
Total	528.00	12	44.00	

Now $F_{2,6,0.05} = 5.14$ and $F_{3,6,0.05} = 4.76$. Since $14.2 > 5.14$, we reject H_β, that there is equality in the average yield of the three varieties; but, since $1.28 \not> 4.76$, we accept H_α, that the four fertilizers are equally effective.

PROBLEMS 12.5

1. Show that the matrix $\mathbf{X}$ for the model defined in (2) is of full rank, $a + b - 1$.
2. Prove statements (3), (4), (5), and (9).
3. The following data represent the units of production per day turned out by four different brands of machines used by four machinists:

	Machinist			
Machine	A_1	A_2	A_3	A_4
B_1	15	14	19	18
B_2	17	12	20	16
B_3	16	18	16	17
B_4	16	16	15	15

Test whether the differences in the performances of the machinists are significant and also whether the differences in the performances of the four brands of machines are significant. Use $\alpha = 0.05$.

4. Students were classified into four ability groups, and three different teaching methods were employed. The following table gives the mean for four groups:

	Teaching Method		
Ability Group	*A*	*B*	*C*
1	15	19	14
2	18	17	12
3	22	25	17
4	17	21	19

Test the hypothesis that the teaching methods yield the same results. That is, that the teaching methods are equally effective.

5. The following table shows the yield (pounds per plot) of four varieties of wheat, obtained with three different kinds of fertilizers.

	Variety of Wheat			
Fertilizer	*A*	*B*	*C*	*D*
α	8	3	6	7
β	10	4	5	8
γ	8	4	6	7

Test the hypotheses that the four varieties of wheat yield the same average yield and that the three fertilizers are equally effective.

12.6 TWO-WAY ANALYSIS OF VARIANCE WITH INTERACTION

The model described in Section 12.5 assumes that the two factors act independently, that is, are *additive*. In practice this is an assumption that needs testing. In this section we allow for the possibility that the two factors might jointly affect the outcome, that is, there might be so-called *interactions*. More precisely, if Y_{ij} is the observation in the (i,j)th cell, we will consider the model

$$Y_{ij} = \mu + \alpha_i + \beta_j + \gamma_{ij} + \varepsilon_{ij}, \tag{1}$$

where $\alpha_i (i = 1, 2, \ldots, a)$ represent row effects (or effects due to factor 1), $\beta_j (j = 1, 2, \ldots, b)$ represent column effects (or effects due to factor 2), and γ_{ij} represent interactions or joint effects. We will assume that ε_{ij} are independently $\mathcal{N}(0, \sigma^2)$. We will further assume that

$$\sum_{i=1}^{a} \alpha_i = 0 = \sum_{j=1}^{b} \beta_j \quad \text{and} \quad \sum_{j=1}^{b} \gamma_{ij} = 0 \quad \text{for all } i, \sum_{i=1}^{a} \gamma_{ij} = 0 \quad \text{for all } j. \tag{2}$$

The hypothesis of interest is

$$H_0: \gamma_{ij} = 0 \qquad \text{for all } i,j. \tag{3}$$

One may also be interested in testing that all α's are 0 or that all β's are 0 in the presence of interactions γ_{ij}.

We first note that (2) is not restrictive since we can write

$$Y_{ij} = \mu' + \alpha'_i + \beta'_j + \gamma'_{ij} + \varepsilon_{ij},$$

where α'_i, β'_j, and γ'_{ij} do not satisfy (2), as

$$\begin{aligned} Y_{ij} &= \mu' + \overline{\alpha}' + \overline{\beta}' + \overline{\gamma}' + (\alpha'_i - \overline{\alpha}' + \overline{\gamma}'_{i\cdot} - \overline{\gamma}') + (\beta'_j - \overline{\beta}' + \overline{\gamma}'_{\cdot j} - \overline{\gamma}') \\ &\quad + (\gamma'_{ij} - \overline{\gamma}'_{i\cdot} - \overline{\gamma}'_{\cdot j} + \overline{\gamma}') + \varepsilon_{ij}, \end{aligned}$$

and then (2) is satisfied by choosing

$$\begin{aligned} \mu &= \mu' + \overline{\alpha}' + \overline{\beta}' + \overline{\gamma}', \\ \alpha_i &= \alpha'_i - \overline{\alpha}' + \overline{\gamma}'_{i\cdot} - \overline{\gamma}', \\ \beta_j &= \beta'_j - \overline{\beta}' + \overline{\gamma}'_{\cdot j} - \overline{\gamma}', \\ \gamma_{ij} &= \gamma'_{ij} - \overline{\gamma}'_{i\cdot} - \overline{\gamma}'_{\cdot j} + \overline{\gamma}'. \end{aligned}$$

Here

$$\overline{\alpha}' = a^{-1}\sum_{i=1}^{a} \alpha'_i, \qquad \overline{\beta}' = b^{-1}\sum_{j=1}^{b} \beta'_j, \qquad \overline{\gamma}'_{i\cdot} = b^{-1}\sum_{j=1}^{b} \gamma'_{ij},$$

$$\overline{\gamma}'_{\cdot j} = a^{-1}\sum_{i=1}^{a} \gamma'_{ij}, \qquad \text{and} \qquad \overline{\gamma}' = (ab)^{-1}\sum_{i=1}^{a}\sum_{j=1}^{b} \gamma'_{ij}.$$

Next note that, unless we replicate, that is, take more than one observation per cell, there are no degrees of freedom left to estimate the error SS (see Remark 1).

Let Y_{ijs} be the sth observation when the first factor is at the ith level, and the second factor at the jth level, $i = 1,2,\ldots,a$, $j = 1,2,\ldots,b$, $s = 1,2,\ldots,m(>1)$. Then the model becomes as follows:

	Levels of Factor 2			
Levels of Factor 1	1	2	$\cdots$	b
1	y_{111}	y_{121}	$\cdots$	y_{1b1}
	.	.	$\cdots$	.
	.	.	$\cdots$	.
	.	.	$\cdots$	.
	y_{11m}	y_{12m}	$\cdots$	y_{1bm}

(continued)

	Levels of Factor 2			
Levels of Factor 1	1	2	⋯	b
2	y_{211}	y_{221}	⋯	y_{2b1}
	.	.	⋯	.
	.	.	⋯	.
	.	.	⋯	.
	y_{21m}	y_{22m}	⋯	y_{2bm}
	.	.	⋯	.
	.	.	⋯	.
	.	.	⋯	.
a	y_{a11}	y_{a21}	⋯	y_{ab1}
	.	.	⋯	.
	.	.	⋯	.
	.	.	⋯	.
	y_{a1m}	y_{a2m}	⋯	y_{abm}

$$Y_{ijs} = \mu + \alpha_i + \beta_j + \gamma_{ij} + \varepsilon_{ijs}, \tag{4}$$

$i = 1,2,\ldots,a$, $j = 1,2,\ldots,b$, and $s = 1,2,\ldots,m$, where ε_{ijs}'s are independent $\mathcal{N}(0,\sigma^2)$. We assume that $\sum_{i=1}^{a}\alpha_i = \sum_{j=1}^{b}\beta_j = \sum_{i=1}^{a}\gamma_{ij} = \sum_{j=1}^{b}\gamma_{ij} = 0$. Suppose that we wish to test $H_\alpha\colon \alpha_1 = \alpha_2 = \cdots = \alpha_a = 0$. We leave the reader to check that model (4) is then a special case of the general linear hypothesis with $n = abm$, $k = ab$, $r = a-1$, and $n-k = ab(m-1)$.

Let us write

$$\begin{aligned}\overline{Y} &= \frac{\sum_{i=1}^{a}\sum_{j=1}^{b}\sum_{s=1}^{m} Y_{ijs}}{n}, \overline{Y}_{ij\cdot} = \frac{\sum_{s=1}^{m} Y_{ijs}}{m},\\ \overline{Y}_{i\cdot\cdot} &= \frac{\sum_{j=1}^{b}\sum_{s=1}^{m} Y_{ijs}}{mb}, \overline{Y}_{\cdot j\cdot} = \frac{\sum_{i=1}^{a}\sum_{s=1}^{m} Y_{ijs}}{am}.\end{aligned} \tag{5}$$

Then it can be easily checked that

$$\begin{cases}\hat{\mu} = \hat{\hat{\mu}} = \overline{Y}, \qquad \hat{\alpha}_i = \overline{Y}_{i\cdot\cdot} - \overline{Y}, \qquad \hat{\beta}_j = \hat{\hat{\beta}}_j = \overline{Y}_{\cdot j\cdot} - \overline{Y},\\ \hat{\gamma}_{ij} = \hat{\hat{\gamma}}_{ij} = \overline{Y}_{ij\cdot} - \overline{Y}_{i\cdot\cdot} - \overline{Y}_{\cdot j\cdot} + \overline{Y}.\end{cases} \tag{6}$$

It follows from Theorem 12.2.1 that

$$F = \frac{\sum_i\sum_j\sum_s(Y_{ijs} - \overline{Y}_{ij\cdot} + \overline{Y}_{i\cdot\cdot} - \overline{Y})^2 - \sum_i\sum_j\sum_s(Y_{ijs} - \overline{Y}_{ij\cdot})^2}{\sum_i\sum_j\sum_s(Y_{ijs} - \overline{Y}_{ij\cdot})^2}. \tag{7}$$

Since

$$\begin{aligned}\sum_i\sum_j\sum_s(Y_{ijs}-\overline{Y}_{ij\cdot}+\overline{Y}_{i\cdot\cdot}-\overline{Y})^2\\ =\sum_i\sum_j\sum_s(Y_{ijs}-\overline{Y}_{ij\cdot})^2+\sum_i\sum_j\sum_s(\overline{Y}_{i\cdot\cdot}-\overline{Y})^2,\end{aligned}$$

we can write (7) as

$$F=\frac{bm\sum_i(\overline{Y}_{i\cdot\cdot}-\overline{Y})^2}{\sum_i\sum_j\sum_s(Y_{ijs}-\overline{Y}_{ij\cdot})^2}. \tag{8}$$

Under H_α the statistic $[ab(m-1)/(a-1)]F$ has the central $F(a-1,ab(m-1))$ distribution, so that the likelihood ratio test rejects H_α if

$$\frac{ab(m-1)}{a-1}\frac{mb\sum_i(\overline{Y}_{i\cdot\cdot}-\overline{Y})^2}{\sum_i\sum_j\sum_s(Y_{ijs}-\overline{Y}_{ij\cdot})^2}>c. \tag{9}$$

A similar analysis holds for testing $H_\beta: \beta_1=\beta_2=\cdots=\beta_b$.

Next consider the test of hypothesis $H_\gamma: \gamma_{ij}=0$ for all i,j, that is, that the two factors are independent and the effects are additive. In this case $n=abm$, $k=ab$, $r=(a-1)(b-1)$, and $n-k=ab(m-1)$. It can be shown that

$$\hat{\hat{\mu}}=\overline{Y},\qquad \hat{\hat{\alpha}}_i=\overline{Y}_{i\cdot\cdot}-\overline{Y},\qquad \text{and } \hat{\hat{\beta}}_j=\overline{Y}_{\cdot j\cdot}-\overline{Y}. \tag{10}$$

Thus

$$F=\frac{\sum_i\sum_j\sum_s(Y_{ijs}-\overline{Y}_{i\cdot\cdot}-\overline{Y}_{\cdot j\cdot}+\overline{Y})^2-\sum_i\sum_j\sum_s(Y_{ijs}-\overline{Y}_{ij\cdot})^2}{\sum_i\sum_j\sum_s(Y_{ijs}-\overline{Y}_{ij\cdot})^2}. \tag{11}$$

Now

$$\begin{aligned}\sum_i\sum_j\sum_s(Y_{ijs}-\overline{Y}_{i\cdot\cdot}-\overline{Y}_{\cdot j\cdot}+\overline{Y})^2\\ =\sum_i\sum_j\sum_s(Y_{ijs}-\overline{Y}_{ij\cdot}+\overline{Y}_{ij\cdot}-\overline{Y}_{i\cdot\cdot}-\overline{Y}_{\cdot j\cdot}+\overline{Y})^2\\ =\sum_i\sum_j\sum_s(Y_{ijs}-\overline{Y}_{ij\cdot})^2+\sum_i\sum_j\sum_s(\overline{Y}_{ij\cdot}-\overline{Y}_{i\cdot\cdot}-\overline{Y}_{\cdot j\cdot}+\overline{Y})^2,\end{aligned}$$

so that we may write

$$F=\frac{\sum_i\sum_j\sum_s(\overline{Y}_{ij\cdot}-\overline{Y}_{i\cdot\cdot}-\overline{Y}_{\cdot j\cdot}+\overline{Y})^2}{\sum_i\sum_j\sum_s(Y_{ijs}-\overline{Y}_{ij\cdot})^2}. \tag{12}$$

Under H_γ, the statistic $\{(m-1)ab/[(a-1)(b-1)]\}F$ has the $F((a-1)(b-1), ab(m-1))$ distribution. The likelihood ratio test rejects H_γ if

$$\frac{(m-1)ab}{(a-1)(b-1)} \frac{m\sum_i\sum_j(\overline{Y}_{ij\cdot} - \overline{Y}_{i\cdot\cdot} - \overline{Y}_{\cdot j\cdot} + \overline{Y})^2}{\sum_i\sum_j\sum_s(Y_{ijs} - \overline{Y}_{ij\cdot})^2} > c. \tag{13}$$

Let us write

$$\begin{aligned}
\text{SS}_1 &= \text{sum of squares due to factor 1 (row sum of squares)} \\
&= bm\sum_{i=1}^{a}(\overline{Y}_{i\cdot\cdot} - \overline{Y})^2, \\
\text{SS}_2 &= \text{sum of squares due to factor 2 (column sum of squares)} \\
&= am\sum_{j=1}^{b}(\overline{Y}_{\cdot j\cdot} - \overline{Y})^2, \\
\text{SSI} &= \text{sum of squares due to interaction} \\
&= m\sum_{i=1}^{a}\sum_{j=1}^{b}(\overline{Y}_{ij\cdot} - \overline{Y}_{i\cdot\cdot} - \overline{Y}_{\cdot j\cdot} + \overline{Y})^2,
\end{aligned}$$

and

$$\begin{aligned}
\text{SSE} &= \text{sum of squares due to error (residual sum of squares)} \\
&= \sum_{i=1}^{a}\sum_{j=1}^{b}\sum_{s=1}^{m}(Y_{ijs} - \overline{Y}_{ij\cdot})^2.
\end{aligned}$$

Then we may summarize the above results in the following table.

Two-Way Analysis of Variance Table with Interaction

Source of Variation	Sum of Squares	Degrees of Freedom	Mean Square	F-Ratio
Rows	SS_1	$a-1$	$\text{MS}_1 = \text{SS}_1/(a-1)$	MS_1/MSE
Columns	SS_2	$b-1$	$\text{MS}_2 = \text{SS}_2/(b-1)$	MS_2/MSE
Interaction	SSI	$(a-1)(b-1)$	$\text{MSI} = \text{SSI}/(a-1)(b-1)$	MSI/MSE
Error	SSE	$ab(m-1)$	$\text{MSE} = \text{SSE}/ab(m-1)$	
Mean	$abm\overline{X}^2$	1	$abm\overline{X}^2$	
Total	$\sum_{i=1}^{a}\sum_{j=1}^{b}\sum_{s=1}^{m} Y_{ijs}^2$	abm	$\sum_{i=1}^{a}\sum_{j=1}^{b}\sum_{s=1}^{m} Y_{ijs}^2/abm$	

Remark 1. Note that, if $m = 1$, there are no d.f. associated with the SSE. Indeed, SSE $= 0$ if $m = 1$. Hence, we cannot make tests of hypotheses when $m = 1$, and for this reason we assume $m > 1$.

Example 1. To test the effectiveness of three different teaching methods, three instructors were randomly assigned 12 students each. The students were then randomly assigned to the different teaching methods and were taught exactly the same material. At the conclusion of the experiment, identical examinations were given to the students with the following results in regard to grades.

	Instructor		
Teaching Method	I	II	III
1	95	60	86
	85	90	77
	74	80	75
	74	70	70
2	90	89	83
	80	90	70
	92	91	75
	82	86	72
3	70	68	74
	80	73	86
	85	78	91
	85	93	89

From the data the table of means is as follows:

	$\overline{y}_{ij\cdot}$			$\overline{y}_{i\cdot\cdot}$
	82	75	77	78.0
	86	89	75	83.3
	80	78	85	81.0
$\overline{y}_{\cdot j\cdot}$	82.7	80.7	79.0	$\overline{y} = 80.8$

Then

$$\begin{aligned}
\text{SS}_1 &= \text{sum of squares due to methods} \\
&= bm\sum_{i=1}^{a}(\bar{y}_{i\cdot\cdot}-\bar{y})^2 \\
&= 3\times 4\times 14.13 = 169.56, \\
\text{SS}_2 &= \text{sum of squares due to instructors} \\
&= am\sum_{j=1}^{b}(\bar{y}_{\cdot j\cdot}-\bar{y})^2 \\
&= 3\times 4\times 6.86 = 82.32, \\
\text{SSI} &= \text{sum of squares due to interaction} \\
&= m\sum_{i=1}^{3}\sum_{j=1}^{3}(\bar{y}_{ij\cdot}-\bar{y}_{i\cdot\cdot}-\bar{y}_{\cdot j\cdot}+\bar{y})^2 \\
&= 4\times 140.45 = 561.80, \\
\text{SSE} &= \text{residual sum of squares} \\
&= \sum_{i=1}^{3}\sum_{j=1}^{3}\sum_{s=1}^{4}(y_{ijs}-\bar{y}_{ij\cdot})^2 = 1830.00.
\end{aligned}$$

Analysis of Variance

Source	SS	d.f.	MSS	F-Ratio
Methods	169.56	2	84.78	1.25
Instructors	82.32	2	41.16	0.61
Interactions	561.80	4	140.45	2.07
Error	1830.00	27	67.78	

With $\alpha = 0.05$, we see from the tables that $F_{2,27,0.05} = 3.35$ and $F_{4,27,0.05} = 2.73$, so that we cannot reject any of the three hypotheses that the three methods are equally effective, that the three instructors are equally effective, and that the interactions are all 0.

PROBLEMS 12.6

1. Prove statement (6).
2. Obtain the likelihood ratio test of the null hypothesis $H_\beta: \beta_1 = \beta_2 = \cdots = \beta_b = 0$.
3. Prove statement (10).

4. Suppose that the following data represent the units of production turned out each day by three different machinists, each working on the same machine for three different days:

	Machinist		
Machine	A	B	C
B_1	15, 15, 17	19, 19, 16	16, 18, 21
B_2	17, 17, 17	15, 15, 15	19, 22, 22
B_3	15, 17, 16	18, 17, 16	18, 18, 18
B_4	18, 20, 22	15, 16, 17	17, 17, 17

Using a 0.05 level of significance, test whether (a) the differences among the machinists are significant, (b) the differences among the machines are significant, and (c) the interactions are significant.

5. In an experiment to determine whether four different makes of automobiles average the same gasoline mileage, a random sample of two cars of each make was taken from each of four cities. Each car was then test run on 5 gallons of gasoline of the same brand. The following table gives the number of miles traveled.

	Automobile Make			
Cities	A	B	C	D
Cleveland	92.3, 104.1	90.4, 103.8	110.2, 115.0	120.0, 125.4
Detroit	96.2, 98.6	91.8, 100.4	112.3, 111.7	124.1, 121.1
San Francisco	90.8, 96.2	90.3, 89.1	107.2, 103.8	118.4, 115.6
Denver	98.5, 97.3	96.8, 98.8	115.2, 110.2	126.2, 120.4

Construct the analysis of variance table. Test the hypothesis of no automobile effect, no city effect, and no interactions. Use $\alpha = 0.05$.

13

NONPARAMETRIC STATISTICAL INFERENCE

13.1 INTRODUCTION

In all the problems of statistical inference considered so far, we assumed that the distribution of the random variable being sampled is known except, perhaps, for some parameters. In practice, however, the functional form of the distribution is seldom, if ever, known. It is therefore desirable to devise methods that are free of this assumption concerning distribution. In this chapter we study some procedures that are commonly referred to as *distribution-free* or *nonparametric methods*. The term "distribution-free" refers to the fact that no assumptions are made about the underlying distribution except that the distribution function being sampled is absolutely continuous. The term "nonparametric" refers to the fact that there are no parameters involved in the traditional sense of the term "parameter" used thus far. To be sure, there is a parameter which indexes the family of absolutely continuous DFs, but it is not numerical and hence the parameter set cannot be represented as a subset of $\mathcal{R}_n$, for any $n \geq 1$. The restriction to absolutely continuous distribution functions is a simplifying assumption that allows us to use the probability integral transformation (Theorem 5.3.1) and the fact that ties occur with probability 0.

Section 13.2 is devoted to the problem of unbiased (nonparametric) estimation. We develop the theory of U-statistics since many estimators and test statistics may be viewed as U-statistics. Sections 13.3 through 13.5 deal with some common hypotheses testing problems. In Section 13.6 we investigate applications of order statistics in nonparametric methods. Section 13.7 considers underlying assumptions in some common parametric problems and the effect of relaxing these assumptions.

An Introduction to Probability and Statistics, Third Edition. Vijay K. Rohatgi and A.K. Md. Ehsanes Saleh.

13.2 *U*-STATISTICS

In Chapter 6 we encountered several nonparametric estimators. For example, the empirical DF defined in Section 6.3 as an estimator of the population DF is distribution-free, and so also are the sample moments as estimators of the population moments. These are examples of what are known as *U–statistics* which lead to unbiased estimators of population characteristics. In this section we study the general theory of U-statistics. Although the thrust of this investigation is unbiased estimation, many of the U-statistics defined in this section may be used as test statistics.

Let $X_1, X_2, \ldots, X_n$ be iid RVs with common law $\mathcal{L}(X)$, and let $\mathcal{P}$ be the class of all possible distributions of X that consists of the absolutely continuous or discrete distributions, or subclasses of these.

Definition 1. A statistic $T(\mathbf{X})$ is *sufficient for the family of distributions* $\mathcal{P}$ if the conditional distribution of $\mathbf{X}$, given $T = t$, is the same whatever the true $F \in \mathcal{P}$.

Example 1. Let $X_1, X_2, \ldots, X_n$ be a random sample from an absolutely continuous DF, and let $\mathbf{T} = (X_{(1)}, \ldots, X_{(n)})$ be the order statistic. Then

$$f(\mathbf{x} \mid \mathbf{T} = \mathbf{t}) = (n!)^{-1},$$

and we see that $\mathbf{T}$ is sufficient for the family of absolutely continuous distributions on $\mathcal{R}$.

Definition 2. A family of distributions $\mathcal{P}$ is *complete* if the only unbiased estimator of 0 is the zero function itself, that is,

$$E_F h(\mathbf{X}) = 0 \qquad \text{for all } F \in \mathcal{P} \Rightarrow h(\mathbf{x}) = 0$$

for all $\mathbf{x}$ (except for a null set with respect to each $F \in \mathcal{P}$).

Definition 3. A statistic $T(\mathbf{X})$ is said to be *complete in relation to a class of distributions* $\mathcal{P}$ if the class of induced distributions of T is complete.

We have already encountered many examples of complete statistics or complete families of distributions in Chapter 8.

The following result is stated without proof. For the proof we refer to Fraser [32, pp. 27–30, 139–142].

Theorem 1. The order statistic $(X_{(1)}, X_{(2)}, \ldots, X_{(n)})$ is a complete sufficient statistic provided that the iid RVs $X_1, X_2, \ldots, X_n$ are of either the discrete or the continuous type.

Definition 4. A real-valued parameter $g(F)$ is said to be *estimable* if it has an unbiased estimator, that is, if there exists a statistic $T(\mathbf{X})$ such that

$$E_F T(\mathbf{X}) = g(F) \qquad \text{for all } F \in \mathcal{P}. \tag{1}$$

Example 2. If $\mathcal{P}$ is the class of all distributions for which the second moment exists, $\overline{X}$ is an unbiased estimator of $\mu(F)$, the population mean. Similarly, $\mu_2(F) = \text{var}_F(X)$ is also estimable, and an unbiased estimator is $S^2 = \sum_1^n (X_i - \overline{X})^2/(n-1)$. We would like to know whether $\overline{X}$ and S^2 are UMVUEs. Similarly, $F(x)$ and $P_F(X_1 + X_2 > 0)$ are estimable for $F \in \mathcal{P}$.

Definition 5. The *degree m* $(m \geq 1)$ of an estimable parameter $g(F)$ is the smallest sample size for which the parameter is estimable, that is, it is the smallest m such that there exists an unbiased estimator $T(X_1, X_2, \ldots, X_m)$ with

$$E_F T = g(F) \qquad \text{for all } F \in \mathcal{P}.$$

Example 3. The parameter $g(F) = P_F\{X > c\}$, where c is a known constant, has degree 1. Also, $\mu(F)$ is estimable with degree 1 (we assume that there is at least one $F \in \mathcal{P}$ such that $\mu(F) \neq 0$), and $\mu_2(F)$ is estimable with degree $m = 2$, since $\mu_2(F)$ cannot be estimated (unbiasedly) by one observation only. At least two observations are needed. Similarly, $\mu^2(F)$ has degree 2, and $P(X_1 + X_2 > 0)$ also is of degree 2.

Definition 6. An unbiased estimator of a parameter based on the smallest sample size (equal to degree m) is called a *kernel*.

Example 4. Clearly X_i $1 \leq i \leq n$ is a kernel of $\mu(F)$; $T(X_i) = 1$, if $X_i > c$, and $= 0$ if $X_i \leq c$ is a kernel of $P(X > c)$. Similarly, $T(X_i, X_j) = 1$ if $X_i + X_j > 0$, and $=0$ otherwise is a kernel of $P(X_i + X_j > 0)$, $X_i X_j$ is a kernel of $\mu^2(F)$ and $X_i^2 - X_i X_j$ is a kernel of $\mu_2(F)$.

Lemma 1. There exists a symmetric kernel for every estimable parameter.

Proof. If $T(X_1, X_2, \ldots, X_m)$ is a kernel of $g(F)$, so also is

$$T_s(X_1, X_2, \ldots, X_m) = \frac{1}{m!} \sum_P T(X_{i_1}, X_{i_2}, \ldots, X_{i_m}), \tag{2}$$

where the summation P is over all $m!$ permutations of $\{1, 2, \ldots, m\}$.

Example 5. A symmetric kernel for $\mu_2(F)$ is

$$\begin{aligned} T_s(X_i, X_j) &= \tfrac{1}{2}\{T(X_i, X_j) + T(X_j, X_i)\} \\ &= \tfrac{1}{2}(X_i - X_j)^2, \qquad i, j = 1, 2, \ldots, n \ (i \neq j). \end{aligned}$$

Definition 7. Let $g(F)$ be an estimable parameter of degree m, and let $X_1, X_2, \ldots, X_n$ be a sample of size n, $n \geq m$. Corresponding to any kernel $T(X_{i_1}, \ldots, X_{i_m})$ of $g(F)$, we define a *U-statistic* for the sample by

$$U(X_1, X_2, \ldots, X_n) = \binom{n}{m}^{-1} \sum_C T_s(X_{i_1}, \ldots, X_{i_m}), \tag{3}$$

where the summation C is over all $\binom{n}{m}$ combinations of m integers $(i_1, i_2, \ldots, i_m)$ chosen from $\{1, 2, \ldots, n\}$, and T_s is the symmetric kernel defined in (2).

Clearly, the U-statistic defined in (3) is symmetric in the X_i's, and

$$E_F U(\mathbf{X}) = g(F) \qquad \text{for all } F. \tag{4}$$

Moreover, $U(\mathbf{X})$ is a function of the complete sufficient statistic $X_{(1)}, X_{(2)}, \ldots, X_{(n)}$. It follows from Theorem 8.4.6 that it is UMVUE of its expected value.

Example 6. For estimating $\mu(F)$, the U-statistic is $n^{-1}\sum_1^n X_i$. For estimating $\mu_2(F)$, a symmetric kernel is

$$T_s(X_{i_1}, X_{i_2}) = \tfrac{1}{2}(X_{i_1} - X_{i_2})^2, \qquad i_1 = 1, 2, \ldots, n \ (i_1 \neq i_2),$$

so that the corresponding U-statistic is

$$\begin{aligned} U(\mathbf{X}) &= \binom{n}{2}^{-1} \sum_{i_1 < i_2} \frac{1}{2}(X_{i_1} - X_{i_2})^2 \\ &= \frac{1}{n-1}\sum_1^n (X_i - \overline{X})^2 \\ &= S^2. \end{aligned}$$

Similarly, for estimating $\mu^2(F)$, a symmetric kernel is $T_s(X_{i_1}, X_{i_2}) = X_{i_1} X_{i_2}$, and the corresponding U-statistic is

$$U(\mathbf{X}) = \frac{1}{\binom{n}{2}} \sum_{i<j} X_i X_j = \frac{1}{n(n-1)} \sum_{i \neq j} X_i X_j.$$

For estimating $\mu^3(F)$, a symmetric kernel is $T_s(X_{i_1}, X_{i_2}, X_{i_3}) = X_{i_1} X_{i_2} X_{i_3}$ so that the corresponding U-statistic is

$$\begin{aligned} U(\mathbf{X}) &= \binom{n}{3}^{-1} \sum\sum\sum_{i<j<k} X_i X_j X_k \\ &= \frac{1}{n(n-1)(n-2)} \sum_{i \neq j \neq k} X_i X_j X_k. \end{aligned}$$

For estimating $F(x)$ a symmetric kernel is $I_{[X_i \leq x]}$ so the corresponding U-statistic is

$$U(\mathbf{X}) = \frac{1}{n}\sum_{i=1}^n I_{[X_i \leq x]} = F_n^*(x),$$

and for estimating $P(X > 0)$ the U-statistic is

$$U(\mathbf{X}) = \frac{1}{n}\sum_{i=1}^n I_{[X_i > 0]} = 1 - F_n^*(0).$$

Finally, for estimating $P(X_1 + X_2 > 0)$ the U-statistic is

$$U(\mathbf{X}) = \frac{1}{\binom{n}{2}} \sum_{i<j} I_{[X_i+X_j>0]}.$$

Theorem 2. The variance of the U-statistic defined in (3) is given by

$$\operatorname{var} U(\mathbf{X}) = \frac{1}{\binom{n}{m}} \sum_{c=1}^{m} \binom{m}{c} \binom{n-m}{m-c} \zeta_c, \tag{5}$$

where

$$\zeta_c = \operatorname{cov}_F \{T_s(X_{i_1}, \ldots, X_{i_m}), T_s(X_{j_i}, \ldots, X_{j_m})\}$$

with m, the degree of $g(F)$, and c is the common number of integers in the sets $\{i_1, \ldots, i_m\}$ and $\{j_1, \ldots, j_m\}$. (For $c = 0$, the two statistics $T(X_{i_1}, \ldots, X_{i_m})$ and $T(X_{j_1}, \ldots, X_{j_m})$ are independent and have zero covariance.)

Proof. Clearly

$$\operatorname{var} U(\mathbf{X}) \\ = \frac{1}{\left[\binom{n}{m}\right]^2} \sum\sum E_F [\{T_s(X_{i_1}, \ldots, X_{i_m}) - g(F)\}\{T_s(X_{j_1}, \ldots, X_{j_m}) - g(F)\}].$$

Let c be the number of common integers in $\{i_1, i_2, \ldots, i_m\}$ and $\{j_2, j_2, \ldots, j_m\}$. Then c takes values $0, 1, \ldots, m$ and for $c = 0$, $T_s(X_{i_1}, \ldots, X_{i_m})$ and $T_s(X_{j_1}, \ldots, X_{j_m})$ are independent. It follows that

$$\operatorname{var} U(\mathbf{X}) = \frac{1}{\left[\binom{n}{m}\right]^2} \sum_{c=1}^{m} \binom{n}{m} \binom{m}{c} \binom{n-m}{m-c} \zeta_c, \tag{6}$$

which is (5). The counting argument from (6) to (7) is as follows: First we select integers $\{i_1, \ldots, i_m\}$ from $\{1, 2, \ldots, n\}$ in $\binom{n}{m}$ ways. Next we select the integers in $\{j_1, \ldots, j_m\}$. This is done by selecting first the c integers that will be in $\{i_1, \ldots, i_m\}$ (hence common to both sets) and then the $m-c$ integers from $n-m$ integers which will not be $\{j_1, \ldots, j_m\}$. Note that $\zeta_0 = 0$ from independence.

Example 7. Consider the U-statistic estimator $\overline{X}$ of $g(F) = \mu(F)$ in Example 6. Here $m = 1$, $T(x) = x$, and $\zeta_1 = \operatorname{var}(X_1) = \sigma^2$ so that $\operatorname{var}(\overline{X}) = \sigma^2/n$.

For the parameter $g(F) = \mu_2(F)$, $U(\mathbf{X}) = S^2$. In this case, $m = 2$, $T_s(X_{i_1}, X_{i_2}) = (X_{i_1} - X_{i_2})^2/2$ so

$$\operatorname{var} U(\mathbf{X}) = \frac{1}{\binom{n}{2}} \{2(n-2)\zeta_1 + \zeta_2\},$$

where

$$\zeta_2 = E_F \left\{ \frac{1}{4} (X_{i_1} - X_{i_2})^4 - \sigma^4 \right\} = \frac{\mu_4 + \sigma^4}{2}$$

and

$$\zeta_1 = \text{cov}\left\{\frac{1}{2}(X_{i_1} - X_{i_2})^2, \frac{1}{2}(X_{i_1} - X_{j_2})^2\right\}$$

where $i_2 \neq j_2$. Then

$$\zeta_1 = \frac{(\mu_4 - \sigma^4)}{4}$$

and

$$\begin{aligned}\text{var}\, U(\mathbf{X}) = \text{var}(S^2) &= \frac{2}{n(n-1)}\left[\frac{(n-2)(\mu_4 - \sigma^4)}{2} + \frac{\mu_4 + \sigma^4}{2}\right] \\ &= \frac{1}{n}\left\{\mu_4 - \frac{n-3}{n-1}\sigma^4\right\},\end{aligned}$$

which agrees with Corollary 2 to Theorem 6.3.4.

For the parameter $g(F) = F(x)$, $\text{var}\, U(\mathbf{X}) = F(x)(1 - F(x))/n$, and for $g(F) = P_F(X_1 + X_2 > 0)$

$$\text{var}\, U(\mathbf{X}) = \frac{1}{n(n-1)}\{4(n-2)\zeta_1 + 2\zeta_2\},$$

where

$$\zeta_1 = P_F(X_1 + X_2 > 0,\ X_1 + X_3 > 0) - P_F^2(X_1 + X_2 > 0)$$

and

$$\begin{aligned}\zeta_2 &= P_F(X_1 + X_2 > 0) - P_F^2(X_1 + X_2 > 0) \\ &= P_F(X_1 + X_2 > 0)P_F(X_1 + X_2 \leq 0).\end{aligned}$$

Corollary to Theorem 2. Let U be the U-statistic for a symmetric kernel $T_s(X_1, X_2, \ldots, X_m)$. Suppose $E_F[T_s(X_1, \ldots, X_m)]^2 < \infty$. Then

$$\lim_{n \to \infty}\{n\,\text{var}\, U(\mathbf{X})\} = m^2\zeta_1. \tag{7}$$

Proof. It is easily shown that $0 \leq \zeta_c \leq \zeta_m$ for $1 \leq c \leq m$. It follows from the hypothesis $\zeta_m = \text{var}[T_s(X_1, \ldots, X_m)]^2 < \infty$ and (5) that $\text{var}\, U(\mathbf{X}) < \infty$. Now

$$\begin{aligned}n\frac{\binom{m}{c}\binom{n-m}{m-c}}{\binom{n}{m}}\zeta_c &= \frac{(m!)^2 n}{c![(m-c)!]^2}\frac{[(n-m)!]^2}{n!(n-2m+c)!}\zeta_c \\ &= \frac{(m!)^2}{c![(m-c)!]^2}n\frac{(n-m)(n-m-1)\cdots(n-2m+c+1)}{n(n-1)\cdots(n-m+1)}\zeta_c.\end{aligned}$$

Now note that the numerator has $m-c+1$ factors involving n, while the denominator has m such factors so that for $c>1$, the ratio involving n goes to 0 as $n\to\infty$. For $c=1$, this ratio $\to 1$ and

$$\operatorname{var} U(\mathbf{X}) \longrightarrow \frac{(m!)^2}{((m-1)!)^2}\zeta_1 = m^2\zeta_1$$

as $n\to\infty$.

Example 8. In Example 7, $n\operatorname{var}(\overline{X}) \equiv \sigma^2$ and

$$n\operatorname{var}(S^2) \longrightarrow 2^2\zeta_1 = \mu_4 - \sigma^4$$

as $n\to\infty$.

Finally we state, without proof, the following result due to Hoeffding [45], which establishes the asymptotic normality of a suitably centered and normed U-statistic. For proof we refer to Lehmann [61, pp. 364–365] or Randles and Wolfe [85, p. 82].

Theorem 3. Let $X_1, X_2, \ldots, X_n$ be a random sample from a DF F and let $g(F)$ be an estimable parameter of degree m with symmetric kernel $T_s(X_1, X_2, \ldots, X_m)$.

If $E_F\{T_s(X_1,X_2,\ldots,X_m)\}^2 < \infty$ and U is the U-statistic for g (as defined in (3)), then $\sqrt{n}(U(\mathbf{X}) - g(F)) \xrightarrow{L} \mathcal{N}(0, m^2\zeta_1)$, provided

$$\zeta_1 = \operatorname{cov}_F\{T_s(X_{i_1},\ldots,X_{i_m}), T_s(X_{j_1},\ldots,X_{j_m})\} > 0.$$

In view of the corollary to Theorem 2, it follows that $(U-g(F))/\sqrt{\operatorname{var}(U)} \xrightarrow{L} \mathcal{N}(0,1)$, provided $\zeta_1 > 0$.

Example 9 (*Example 7 continued*). Clearly, $\sqrt{n}(\overline{X}-\mu)/\sigma \xrightarrow{L} \mathcal{N}(0,1)$ as $n\to\infty$ since $\zeta_1 = \sigma^2 > 0$.

For the parameter $g(F) = \mu_2(F)$, $\operatorname{var} U(\mathbf{X}) = \operatorname{var}(S^2) = \frac{1}{n}\left\{\mu_4 - \frac{n-3}{n-1}\sigma^4\right\}$, $\zeta_1 = (\mu_4 - \sigma^4)/4 > 0$ so it follows from Theorem 3 that

$$\sqrt{n}(S^2 - \sigma^2) \xrightarrow{L} \mathcal{N}(0, \mu^4 - \sigma^4).$$

The concept of U-statistics can be extended to multiple random samples. We will restrict ourselves to the case of two samples. Let $X_1, X_2, \ldots, X_{n_1}$ and $Y_1, Y_2, \ldots, Y_{n_2}$ be two independent random samples from DFs F and G, respectively.

Definition 8. A parameter $g(F,G)$ is estimable of degrees (m_1, m_2) if m_1 and m_2 are the smallest sample sizes for which there exists a statistic $T(X_1,\ldots,X_{m_1}; Y_1,\ldots,Y_{m_2})$ such that

$$E_{F,G} T(X_1,\ldots,X_{m_1}; Y_1,\ldots,Y_{m_2}) = g(F,G) \tag{8}$$

for all $F, G \in \mathcal{P}$.

The statistic T in Definition 8 is called a *kernel* of g and a symmetrized version of T, T_s is called a *symmetric kernel* of g. Without loss of generality therefore we assume that the two-sample kernel T in (9) is a symmetric kernel.

Definition 9. Let $g(F,G)$, $F,G \in \mathcal{P}$ be an estimable parameter of degree (m_1,m_2). Then, a (two-sample) U-statistic estimate of g is defined by

$$U(\mathbf{X};\mathbf{Y}) = \binom{n_1}{m_1}^{-1}\binom{n_2}{m_2}^{-1}\sum_{\mathbf{i}\in A}\sum_{\mathbf{j}\in \mathbf{B}} T\left(X_{i_1},\ldots,X_{i_{m_1}};Y_{j_1},\ldots,Y_{j_{m_2}}\right), \tag{9}$$

where A and B are collections of all subsets of m_1 and m_2 integers chosen without replacement from the sets $\{1,2,\ldots,n_1\}$ and $\{1,2,\ldots,n_2\}$ respectively.

Example 10. Let $X_1,X_2,\ldots,X_{n_1}$ and $Y_1,Y_2,\ldots,Y_{n_1}$ be two independent samples from DFs F and G, respectively. Let $g(F,G) = P(X<Y) = \int_{-\infty}^{\infty} F(x)g(x)dx = \int_{-\infty}^{\infty} P(Y>y)f(y)dy$, where f and g are the respective PDFs of F and G. Then

$$T(X_i;Y_j) = \begin{cases} 1, & \text{if } X_i < Y_j \\ 0, & \text{if } X_i \geq Y_j \end{cases}$$

is an unbiased estimator of g. Clearly, g has degree $(1,1)$ and the two-sample U-statistic is given by

$$U(\mathbf{X};\mathbf{Y}) = \frac{1}{n_1 n_2}\sum_{i=1}^{n_1}\sum_{j=1}^{n_2} T(X_i;Y_j).$$

Theorem 4. The variance of the two-sample U-statistic defined in (10) is given by

$$\operatorname{var} U(\mathbf{X};\mathbf{Y}) = \frac{1}{\binom{n_1}{m_1}\binom{n_2}{m_2}}\sum_{c=0}^{m_1}\sum_{d=0}^{m_2}\binom{m_1}{c}\binom{n_1-m_1}{m_1-c}\binom{m_2}{d}\binom{n_2-m_2}{m_2-d}\zeta_{c,d}, \tag{10}$$

where $\zeta_{c,d}$ is the covariance between $T\left(X_{i_1},\ldots,X_{i_{m_1}};Y_{j_1},\ldots,Y_{j_{m_1}}\right)$ and $T(X_{k_1},\ldots,X_{k_{m_1}}; Y_{\ell_1},\ldots,Y_{\ell_{m_2}})$ with exactly c X's and d Y's in common.

Corollary. Suppose $E_{F,G}T^2(X_1,\ldots,X_{m_1};Y_1,\ldots,Y_{m_2}) < \infty$ for all $F,G \in \mathcal{P}$. Let $N = n_1 + n_2$ and suppose $n_1,n_2,N \to \infty$ such that $n_1/N \to \lambda$, $n_2/N \to 1-\lambda$. Then

$$\lim_{N\to\infty} N \operatorname{var} U(\mathbf{X};\mathbf{Y}) = \frac{m_1^2}{\lambda}\zeta_{1,0} + \frac{m_2^2}{1-\lambda}\zeta_{0,1}. \tag{11}$$

The proofs of Theorem 4 and its corollary parallel those of Theorem 2 and its corollary and are left to the reader.

Example 11. For the U-statistic in Example 10

$$E_{F,G}U^2(\mathbf{X};\mathbf{Y}) = \frac{1}{n_1^2 n_2^2}\sum\sum\sum\sum E_{F,G}\left\{T(X_i;Y_j)T(X_k;Y_\ell)\right\}.$$

Now

$$E_{F,G}\{T(X_i;Y_j)T(X_k;Y_\ell)\} = P(X_i < Y_j, X_k < Y_\ell)$$
$$= \begin{cases} \int_{-\infty}^{\infty} F(x)g(x)dx & \text{for } i=k,\ j=\ell, \\ \int_{-\infty}^{\infty} [1-G(x)]^2 f(x)dx & \text{for } i=k,\ j\neq\ell, \\ \int_{-\infty}^{\infty} F^2(x)g(x)dx & \text{for } i\neq k,\ j=\ell, \\ \left[\int_{-\infty}^{\infty} F(x)g(x)dx\right]^2 & \text{for } i\neq k,\ j\neq\ell, \end{cases}$$

where f and g are PDFs of F and G, respectively. Moreover,

$$\zeta_{1,0} = \int_{-\infty}^{\infty} [1-G(x)]^2 f(x)dx - [g(F,G)]^2$$

and

$$\zeta_{0,1} = \int_{-\infty}^{\infty} F^2(X)g(x)dx - [g(F,G)]^2.$$

It follows that

$$\operatorname{var} U(\mathbf{X};\mathbf{Y}) = \frac{1}{n_1 n_2}\{g(F,G)[1-g(F,G)] + (n_1-1)\zeta_{1,0} + (n_2-1)\zeta_{0,1}\}.$$

In the special case when $F = G$, $g(F,G) = 1/2$, $\zeta_{1,0} = \zeta_{0,1} = 1/3 - 1/4 = 1/12$, and $\operatorname{var} U = (n_1+n_2+1)/[12n_1n_2]$.

Finally we state, without proof, the two-sample analog of Theorem 3 which establishes the asymptotic normality of the two-sample U-statistic defined in (10).

Theorem 5. Let $X_1, X_2, \ldots, X_{n_1}$ and $Y_1, Y_2, \ldots, Y_{n_2}$ be independent random samples from DFs F and G, respectively, and let $g(F,G)$ be an estimable parameter of degree (m_1, m_2). Let $T(X_1,\ldots,X_{m_1}; Y_1,\ldots,Y_{m_2})$ be a symmetric kernel for g such that $ET^2 < \infty$. Then

$$\sqrt{n_1+n_2}\{U(\mathbf{X};\mathbf{Y}) - g(F,G)\} \xrightarrow{L} \mathcal{N}(0,\sigma^2),$$

where $\sigma^2 = \dfrac{m_1^2\zeta_{1,0}}{\lambda} + \dfrac{m_2^2\zeta_{0,1}}{1-\lambda}$, provided $\sigma^2 > 0$ and $0 < \lambda = \lim_{N\to\infty}(m_1/N) = \lambda < 1$, $N = n_1 + n_2$.

We see that $(U-g)/\sqrt{\operatorname{var} U} \xrightarrow{L} \mathcal{N}(0,1)$, provided $\sigma^2 > 0$.

For the proof of Theorem 5 we refer to Lehmann [61, p. 364], or Randles and Wolfe [85, p. 92].

Example 11 (Continued). In Example 11 we saw that in the special case when $F = G$, $\zeta_{1,0} = \zeta_{0,1} = 1/12$, and $\operatorname{var} U = (n_1+n_2+1)/[12n_1n_2]$. It follows that

$$\frac{U(\mathbf{X};\mathbf{Y}) - (1/2)}{\sqrt{(n_1+n_2+1)/[12n_1n_2]}} \xrightarrow{L} \mathcal{N}(0,1).$$

PROBLEMS 13.2

1. Let $(\mathcal{R}, \mathfrak{B}, P_\theta)$ be a probability space, and let $\mathcal{P} = \{P_\theta : \theta \in \Theta\}$. Let A be a Borel subset of $\mathcal{R}$, and consider the parameter $d(\theta) = P_\theta(A)$. Is d estimable? If so, what is the degree? Find the UMVUE for d, based on a sample of size n, assuming that $\mathcal{P}$ is the class of all continuous distributions.
2. Let $X_1, X_2, \ldots, X_m$ and $Y_1, Y_2, \ldots, Y_n$ be independent random samples from two absolutely continuous DFs. Find the UMVUEs of (a) $E\{XY\}$ and (b) $\text{var}(X+Y)$.
3. Let $(X_1, Y_1), (X_2, Y_2), \ldots, (X_n, Y_n)$ be a random sample from an absolutely continuous distribution. Find the UMVUEs of (a) $E(XY)$ and (b) $\text{var}(X+Y)$.
4. Let $T(X_1, X_2, \ldots, X_n)$ be a statistic that is symmetric in the observations. Show that T can be written as a function of the order statistic. Conversely, if $T(X_1, X_2, \ldots, X_n)$ can be written as a function of the order statistic, T is symmetric in the observations.
5. Let $X_1, X_2, \ldots, X_n$ be a random sample from an absolutely continuous DF F, $F \in \mathcal{P}$. Find U-statistics for $g_1(F) = \mu^3(F)$ and $g_2(F) = \mu_3(F)$. Find the corresponding expressions for the variance of the U-statistic in each case.
6. In Example 3, show that $\mu_2(F)$ is not estimable with one observation. That is, show that the degree of $\mu_2(F)$ where $F \in \mathcal{P}$, the class of all distributions with finite second moment, is 2.
7. Show that for $c = 1, 2, \ldots, m$, $0 \le \zeta_c \le \zeta_m$.
8. Let $X_1, X_2, \ldots, X_n$ be a random sample from an absolutely continuous DF F, $F \in \mathcal{P}$. Let

$$g(F) = E_F|X_1 - X_2|.$$

Find the U-statistic estimator of $g(F)$ and its variance.

13.3 SOME SINGLE-SAMPLE PROBLEMS

Let $X_1, X_2, \ldots, X_n$ be a random sample from a DF F. In Section 13.2 we studied properties of U-statistics as nonparametric estimators of parameters $g(F)$. In this section we consider some nonparametric tests of hypotheses. Often the test statistic may be viewed as a function of a U-statistic.

13.3.1 Goodness-of-Fit Problem

The problem of fit is to test the hypothesis that the sample comes from a specified DF F_0 against the alternative that it is from some other DF F, where $F(x) \neq F_0(x)$ for some $x \in \mathcal{R}$. In Section 10.3 we studied the chi-square test of goodness of fit for testing $H_0 : X_i \sim F_0$. Here we consider the Kolmogorov–Smirnov test of H_0. Since H_0 concerns the underlying DF of the X's, it is natural to compare the U-statistic estimator of $g(F) = F(x)$ with the specified DF F_0 under H_0. The U-statistic for $g(F) = F(x)$ is the empirical DF $F_n^*(x)$.

Definition 1. Let $X_1, X_2, \ldots, X_n$ be a sample from a DF F, and let F_n^* be a corresponding empirical DF. The statistic

$$D_n = \sup_x |F_n^*(x) - F(x)| \tag{1}$$

is called the *(two-sided) Kolmogorov–Smirnov statistic.* We write

$$D_n^+ = \sup_n [F_n^*(x) - F(x)] \tag{2}$$

and

$$D_n^- = \sup_x [F(x) - F_n^*(x)] \tag{3}$$

and call D_n^+, D_n^- the *one-sided Kolmogorov–Smirnov statistic.*

Theorem 1. The statistics D_n, D_n^-, D_n^+ are distribution-free for any continuous DF F.

Proof. Clearly, $D_n = \max(D_n^+, D_n^-)$. Let $X_{(1)} \le X_{(2)} \le \cdots \le X_{(n)}$ be the order statistics of $X_1, X_2, \ldots, X_n$, and define $X_{(0)} = -\infty$, $X_{(n+1)} = +\infty$. Then

$$F_n^*(x) = \frac{i}{n} \qquad \text{for } X_{(i)} \le x < X_{(i+1)}, \quad i = 0, 1, 2, \ldots, n,$$

and we have

$$\begin{aligned} D_n^+ &= \max_{0 \le i \le n} \sup_{X_{(i)} \le x < X_{(i+1)}} \left\{ \frac{i}{n} - F(x) \right\} \\ &= \max_{0 \le i \le n} \left\{ \frac{i}{n} - \inf_{X_{(i)} \le x < X_{(i+1)}} F(x) \right\} \\ &= \max_{0 \le i \le n} \left\{ \frac{i}{n} - F(X_{(i)}) \right\} \\ &= \max \left\{ \max_{1 \le i \le n} \left[\frac{i}{n} - F(X_{(i)}) \right], 0 \right\}. \end{aligned}$$

Since $F(X_{(i)})$ is the ith-order statistic of a sample from $U(0,1)$ irrespective of what F is, as long as it is continuous, we see that the distribution of D_n^+ is independent of F. Similarly,

$$D_n^- = \max \left\{ \max_{1 \le i \le n} \left[F(X_{(i)}) - \frac{i-1}{n} \right], 0 \right\},$$

and the result follows.

Without loss of generality, therefore, we assume that F is the DF of a $U(0,1)$ RV.

Theorem 2. If F is continuous, then

$$P\left\{D_n \le v + \frac{1}{2n}\right\} = \begin{cases} 0 & \text{if } v \le 0, \\ \displaystyle\int_{(1/2n)-v}^{v+(1/2n)} \int_{(3/2n)-v}^{v+(3/2n)} \cdots & \\ \displaystyle\int_{[(2n-1)/2n]-v}^{v+[(2n-1)/2n]} f(u_1, u_2, \ldots, u_n) \cdot \prod_n^1 du_i & \text{if } 0 < v < \dfrac{2n-1}{2n}, \\ 1 & \text{if } v \ge \dfrac{2n-1}{2n}, \end{cases} \tag{4}$$

where

$$f(u_1, u_2, \ldots, u_n) = \begin{cases} n!, & 0 < u_1 < \cdots < u_n < 1, \\ 0, & \text{otherwise}, \end{cases} \tag{5}$$

is the joint PDF of the set of order statistics for a sample of size n from $U(0,1)$.

We will not prove this result here. Let $D_{n,\alpha}$ be the upper α-percent point of the distribution of D_n, that is, $P\{D_n > D_{n,\alpha}\} \le \alpha$. The exact distribution of D_n for selected values of n and α has been tabulated by Miller [74], Owen [79], and Birnbaum [9]. The large-sample distribution of D_n was derived by Kolmogorov [53], and we state it without proof.

Theorem 3. Let F be any continuous DF. Then for every $z \ge 0$

$$\lim_{n\to\infty} P\{D_n \le zn^{-1/2}\} = L(z), \tag{6}$$

where

$$L(z) = 1 - 2\sum_{i=1}^{\infty} (-1)^{i-1} e^{-2i^2z^2}. \tag{7}$$

Theorem 3 can be used to find d_α such that $\lim_{n\to\infty} P\{\sqrt{n}D_n \le d_\alpha\} = 1-\alpha$. Tables of d_α for various values of α are also available in Owen [79].

The statistics D_n^+ and D_n^- have the same distribution because of symmetry, and their common distribution is given by the following theorem.

Theorem 4. Let F be a continuous DF. Then

$$P\{D_n^+ \le z\} = \begin{cases} 0 & \text{if } z \le 0, \\ \displaystyle\int_{1-z}^{1} \int_{[(n-1)/n]-z}^{u_n} \cdots \int_{(2/n)-z}^{u_3} & \\ \displaystyle\times \int_{(1/n)-z}^{u_2} f(u_1, u_2, \ldots, u_n) \prod_{i=1}^{n} du_i & \text{if } 0 < z < 1, \\ 1 & \text{if } z \ge 1, \end{cases} \tag{8}$$

where f is given by (5).

Proof. We leave the reader to prove Theorem 4.

Tables for the critical values $D^+_{n,\alpha}$, where $P\{D^+_n > D^+_{n,\alpha}\} \le \alpha$, are also available for selected values of n and α; see Birnbaum and Tingey [8]. Table ST7 at the end of this book gives $D^+_{n,\alpha}$ and $D_{n,\alpha}$ for some selected values of n and α. For large samples Smirnov [108] showed that

$$\lim_{n\to\infty} P\{\sqrt{n}D^+_n \le z\} = 1 - e^{-2z^2}, \qquad z \ge 0. \tag{9}$$

In fact, in view of (9), the statistic $V_n = 4nD^{+2}_n$ has a limiting $\chi^2(2)$ distribution, for $4nD^{+2}_n \le 4z^2$ if and only if $\sqrt{n}D^+_n \le z$, $z \ge 0$, and the result follows since

$$\lim_{n\to\infty} P\{V_n \le z^2\} = 1 - e^{-2z^2}, \qquad z \ge 0,$$

so that

$$\lim_{n\to\infty} P\{V_n \le x\} = 1 - e^{-x/2}, \qquad x \ge 0,$$

which is the DF of a $\chi^2(2)$ RV.

Example 1. Let $\alpha = 0.01$, and let us approximate $D^+_{n,\alpha}$. We have $\chi^2_{2,0.01} = 9.21$. Thus $X_n = 9.21$, yielding

$$D^+_{n,0.01} = \sqrt{\frac{9.21}{4n}} = \frac{3.03}{2\sqrt{n}}.$$

If, for example, $n = 9$, then $D^+_{n,0.01} = 3.03/6 = 0.50$. Of course, the approximation is better for large n.

The statistic D_n and its one-sided analogs can be used in testing $H_0\colon X \sim F_0$ against $H_1 : X \sim F$, where $F_0(x) \ne F(x)$ for some x.

Definition 2. To test $H_0\colon F(x) = F_0(x)$ for all x at level α, the *Kolmogorov–Smirnov test* rejects H_0 if $D_n > D_{n,\alpha}$. Similarly, it rejects $F(x) \ge F_0(x)$ for all x if $D^-_n > D^+_{n,\alpha}$ and rejects $F(x) \le F_0(x)$ for all x at level α if $D^+_n > D^+_{n,\alpha}$.

For large samples we can approximate by using Theorem 3 or (9) to obtain an approximate α-level test.

Example 2. Let us consider the data in Example 10.3.3, and apply the Kolmogorov–Smirnov test to determine the goodness of the fit. Rearranging the data in increasing order of magnitude, we have the following result:

x	$F_0(x)$	$F^*_{20}(x)$	$i/20 - F_0(x_{(i)})$	$F_0(x_{(i)}) - (i-1)/20$
−1.787	0.0367	$\frac{1}{20}$	0.0133	0.0367
−1.229	0.1093	$\frac{2}{20}$	−0.0093	0.0593
−0.525	0.2998	$\frac{3}{20}$	−0.1498	0.1998
−0.513	0.3050	$\frac{4}{20}$	−0.1050	0.1550
−0.508	0.3050	$\frac{5}{20}$	−0.0550	0.1050
−0.486	0.3121	$\frac{6}{20}$	−0.0121	0.0621
−0.482	0.3156	$\frac{7}{20}$	0.0344	0.0156
−0.323	0.3745	$\frac{8}{20}$	0.0255	0.0245
−0.261	0.3974	$\frac{9}{20}$	0.0526	−0.0026
−0.068	0.4721	$\frac{10}{20}$	0.0279	0.0221
−0.057	0.4761	$\frac{11}{20}$	0.0739	−0.0239
0.137	0.5557	$\frac{12}{20}$	0.0443	0.0057
0.464	0.6772	$\frac{13}{20}$	−0.0272	0.0772
0.595	0.7257	$\frac{14}{20}$	−0.0257	0.0757
0.881	0.8106	$\frac{15}{20}$	−0.0606	0.1106
0.906	0.8186	$\frac{16}{20}$	−0.0186	0.0686
1.046	0.8531	$\frac{17}{20}$	−0.0031	0.0531
1.237	0.8925	$\frac{18}{20}$	0.0075	0.0425
1.678	0.9535	$\frac{19}{20}$	−0.0035	0.0535
2.455	0.9931	1	0.0069	0.0431

From Theorem 1,

$$D^-_{20} = 0.1998,\ D^+_{20} = 0.0739,\ \text{and } D_{20} = \max(D^+_{20}, D^-_{20}) = 0.1998.$$

Let us take $\alpha = 0.05$. Then $D_{20,0.05} = 0.294$. Since $0.1998 < 0.294$, we accept H_0 at the 0.05 level of significance.

It is worthwhile to compare the chi-square test of goodness of fit and the Kolmogorov–Smirnov test. The latter treats individual observations directly, whereas the former discretizes the data and sometimes loses information through grouping. Moreover, the Kolmogorov–Smirnov test is applicable even in the case of very small samples, but the chi-square test is essentially for large samples.

The chi-square test can be applied when the data are discrete or continuous, but the Kolmogorov–Smirnov test assumes continuity of the DF. This means that the latter test provides a more refined analysis of the data. If the distribution is actually discontinuous, the Kolmogorov–Smirnov test is conservative in that it favors H_0.

We next turn our attention to some other uses of the Kolmogorov–Smirnov statistic. Let $X_1, X_2, \ldots, X_n$ be a sample from a DF F, and let F_n^* be the sample DF. The estimate F_n^* of F for large n should be close to F. Indeed,

$$P\left\{|F_n^*(x) - F(x)| \le \frac{\lambda\sqrt{F(x)[1-F(x)]}}{\sqrt{n}}\right\} \ge 1 - \frac{1}{\lambda^2}, \tag{10}$$

and, since $F(x)[1-F(x)] \le \frac{1}{4}$, we have

$$P\left\{|F_n^*(x) - F(x)| \le \frac{\lambda}{2\sqrt{n}}\right\} \ge 1 - \frac{1}{\lambda^2}. \tag{11}$$

Thus F_n^* can be made close to F with high probability by choosing λ and large enough n. The Kolmogorov–Smirnov statistic enables us to determine the smallest n such that the error in estimation never exceeds a fixed value ε with a large probability $1-\alpha$. Since

$$P\{D_n \le \varepsilon\} \ge 1 - \alpha, \tag{12}$$

$\varepsilon = D_{n,\alpha}$; and, given ε and α, we can read n from the tables. For large n, we can use the asymptotic distribution of D_n and solve $d_\alpha = \varepsilon\sqrt{n}$ for n.

We can also form confidence bounds for F. Given α and n, we first find $D_{n,\alpha}$ such that

$$P\{D_n > D_{n,\alpha}\} \le \alpha, \tag{13}$$

which is the same as

$$P\left\{\sup_x |F_n^*(x) - F(x)| \le D_{n,\alpha}\right\} \ge 1 - \alpha.$$

Thus

$$P\{|F_n^*(x) - F(x)| \le D_{n,\alpha} \quad \text{for all } x\} \ge 1 - \alpha. \tag{14}$$

Define

$$L_n(x) = \max\{F_n^*(x) - D_{n,\alpha}, 0\} \tag{15}$$

and

$$U_n(x) = \min\{F_n^*(x) + D_{n,\alpha}, 1\}. \tag{16}$$

Then the region between $L_n(x)$ and $U_n(x)$ can be used as a confidence band for $F(x)$ with associated confidence coefficient $1-\alpha$.

Example 3. For the data on the standard normal distribution of Example 2, let us form a 0.90 confidence band for the DF. We have $D_{20,0.10} = 0.265$. The confidence band is, therefore, $F_{20}^*(x) \pm 0.265$ as long as the band is between 0 and 1.

13.3.2 Problem of Location

Let $X_1, X_2, \ldots, X_n$ be a sample of size n from some unknown DF F. Let p be a positive real number, $0 < p < 1$, and let $\mathfrak{z}_p(F)$ denote the quantile of order p for the DF F. In the following analysis we assume that F is absolutely continuous. The problem of location is to test $H_0\colon \mathfrak{z}_p(F) = \mathfrak{z}_0$, $\mathfrak{z}_0$ a given number, against one of the alternatives $\mathfrak{z}_p(F) > \mathfrak{z}_0$, $\mathfrak{z}_p < \mathfrak{z}_0$, and $\mathfrak{z}_p \neq \mathfrak{z}_0$. The problem of location and symmetry is to test $H_0'\colon \mathfrak{z}_{0.5}(F) = \mathfrak{z}_0$, and F is symmetric against $H_1'\colon \mathfrak{z}_{0.5}(F) \neq \zeta_0$ or F is not symmetric.

We consider two tests of location. First, we describe the sign test.

13.3.2.1 The Sign Test Let $X_1, X_2, \ldots, X_n$ be iid RVs with common PDF f. Consider the hypothesis testing problem

$$H_0\colon \mathfrak{z}_p(f) = \mathfrak{z}_0 \qquad \text{against } H_1\colon \mathfrak{z}_p(f) > \mathfrak{z}_0, \tag{17}$$

where $\mathfrak{z}_p(f)$ is the quantile of order p of PDF f, $0 < p < 1$. Let $g(F) = P(X_i > \mathfrak{z}_0) = P(X_i - \mathfrak{z}_0 > 0)$. Then the corresponding U-statistic is given by

$$nU(\mathbf{X}) = R^+(\mathbf{X}),$$

the number of positive elements in $X_1 - \mathfrak{z}_0, X_2 - \mathfrak{z}_0, \ldots, X_n - \mathfrak{z}_0$. Clearly, $P(X_i = \mathfrak{z}_0) = 0$. Fraser [32, pp. 167–170] has shown that a UMP test of H_0 against H_1 is given by

$$\varphi(\mathbf{x}) = \begin{cases} 1, & R^+(\mathbf{x}) > c, \\ \gamma, & R^+(\mathbf{x}) = c, \\ 0, & R^+(\mathbf{x}) < c, \end{cases} \tag{18}$$

where c and γ are chosen from the size restriction

$$\alpha = \sum_{i=c+1}^{n} \binom{n}{R^+(\mathbf{x})} (1-p)^{R^+(\mathbf{x})} p^{n-R^+(\mathbf{x})} + \gamma \binom{n}{c} (1-p)^c p^{n-c}. \tag{19}$$

Note that, under H_0, $\mathfrak{z}_p(f) = \mathfrak{z}_0$, so that $P_{H_0}(X \leq \mathfrak{z}_0) = p$ and $R^+(\mathbf{X}) \sim b(n, 1-p)$. The same test is UMP for $H_0\colon \mathfrak{z}_p(f) \leq \mathfrak{z}_0$ against $H_1\colon \mathfrak{z}_p(f) > \mathfrak{z}_0$. For the two-sided case, Fraser [32, p. 171] shows that the two-sided sign test is UMP unbiased.

If, in particular, $\mathfrak{z}_0$ is the median of f, then $p = 1/2$ under H_0. In this case one can also use the sign test to test $H_0\colon \text{med}(X) = \mathfrak{z}_0$, F is symmetric.

For large n one can use the normal approximation to binomial to find c and γ in (19).

Example 4. Entering college freshmen have taken a particular high school achievement test for many years, and the upper quartile ($p = 0.75$) is well established at a score of 195. A particular high school sent 12 of its graduates to college, where they took the examination and obtained scores of 203, 168, 187, 235, 197, 163, 214, 233, 179, 185, 197, 216. Let us test the null hypothesis H_0 that $\mathfrak{z}_{0.75} \leq 195$ against $H_1\colon \mathfrak{z}_{0.75} > 195$ at the $\alpha = 0.05$ level.

We have to find c and γ such that

$$\sum_{i=c+1}^{12} \binom{12}{i} \left(\frac{1}{4}\right)^i \left(\frac{3}{4}\right)^{12-i} + \gamma \binom{12}{c} \left(\frac{1}{4}\right)^c \left(\frac{3}{4}\right)^{12-c} = 0.05.$$

From the table of cumulative binomial distribution (Table ST1) for $n = 12$, $p = \frac{1}{4}$, we see that $c = 6$. Then γ is given by

$$0.0142 + \gamma \binom{12}{6} \left(\frac{1}{4}\right)^6 \left(\frac{3}{4}\right)^6 = 0.05.$$

Thus

$$\gamma = \frac{0.0358}{0.0402} = 0.89.$$

In our case the number of positive signs, $x_i - 195$, $i = 1, 2, \ldots, 12$, is 7, so we reject H_0 that the upper quartile is ≤ 195.

Example 5. A random sample of size 8 is taken from a normal population with mean 0 and variance 1. The sample values are -0.465, 0.120, -0.238, -0.869, -1.016, 0.417, 0.056, 0.561. Let us test hypothesis $H_0\colon \mu = -1.0$ against $H_1: \mu > -1.0$. We should expect to reject H_0 since we know that it is false. The number of observations, $x_i - \mu_0 = x_i + 1.0$, that are ≥ 0 is 7. We have to find c and γ such that

$$\sum_{i=c+1}^{8} \binom{8}{i} \left(\frac{1}{2}\right)^8 + \gamma \binom{8}{c} \left(\frac{1}{2}\right)^8 = 0.05, \text{ say,}$$

that is,

$$\sum_{i=c+1}^{8} \binom{8}{i} + \gamma \binom{8}{c} = 12.8.$$

We see that $c = 6$ and $\gamma = 0.13$. Since the number of positive $x_i - \mu_0$ is > 6, we reject H_0.

Let us now apply the parametric test here. We have

$$\overline{x} = -\frac{1.434}{8} = -0.179.$$

Since $\sigma = 1$, we reject H_0 if

$$\begin{aligned} \overline{x} > \mu_0 + \frac{1}{\sqrt{n}} z_\alpha &= -1.0 + \frac{1}{\sqrt{8}} 1.64 \\ &= -0.42. \end{aligned}$$

Since $-0.179 > -0.42$, we reject H_0.

The single-sample sign test described above can easily be modified to apply to sampling from a bivariate population. Let $(X_1, Y_1), (X_2, Y_2), \ldots, (X_n, Y_n)$ be a random sample from a bivariate population. Let $Z_i = X_i - Y_i$, $i = 1, 2, \ldots, n$, and assume that Z_i has an absolutely continuous DF. Then one can test hypotheses concerning the order parameters of Z by using the sign test. A hypothesis of interest here is that Z has a given median $\mathfrak{z}_0$. Without loss of generality let $\mathfrak{z}_0 = 0$. Then $H_0\colon \text{med}(Z) = 0$, that is, $P\{Z > 0\} = P\{Z < 0\} = \frac{1}{2}$. Note that $\text{med}(Z)$ is not necessarily equal to $\text{med}(X) - \text{med}(Y)$, so that H_0 is not that $\text{med}(X) = \text{med}(Y)$ but that $\text{med}(Z) = 0$. The sign test is UMP against one-sided alternatives and UMP unbiased against two-sided alternatives.

Example 6. We consider an example due to Hahn and Nelson [40], in which two measuring devices take readings on each of 10 test units. Let X and Y, respectively, be the readings on a test unit by the first and second measuring devices. Let $X = A + \varepsilon_1$, $Y = A + \varepsilon_2$, where A, ε_1, ε_2, respectively, are the contributions to the readings due to the test unit and to the first and the second measuring devices. Let A, ε_1, ε_2 be independent with $EA = \mu$, $\text{var}(A) = \sigma_a^2$, $E\varepsilon_1 = E\varepsilon_2 = 0$, $\text{var}(\varepsilon_1) = \sigma_1^2$, $\text{var}(\varepsilon_2) = \sigma_2^2$, so that X and Y have common mean μ and variances $\sigma_1^2 + \sigma_a^2$ and $\sigma_2^2 + \sigma_a^2$, respectively. Also, the covariance between X and Y is σ_a^2. The data are as follows:

	Test unit									
	1	2	3	4	5	6	7	8	9	10
First device, X	71	108	72	140	61	97	90	127	101	114
Second device, Y	77	105	71	152	88	117	93	130	112	105
$Z = X - Y$	−6	3	1	−8	−17	−20	−3	−3	−11	9

Let us test the hypothesis $H_0\colon \text{med}(Z) = 0$. The number of Z_i's > 0 is 3. We have

$$P\{\text{number of } Z_i\text{'s} > 0 \text{ is } \le 3 \mid H_0\} = \sum_{k=0}^{3} \binom{10}{k} \left(\frac{1}{2}\right)^{10} = 0.172.$$

Using the two-sided sign test, we cannot reject H_0 at level $\alpha = 0.05$, since $0.172 > 0.025$. The RVs Z_i can be considered to be distributed normally, so that under H_0 the common mean of Z_i's is 0. Using a paired comparison t-test on the data, we can show that $t = -0.88$ for 9 d.f., so we cannot reject the hypothesis of equality of means of X and Y at level $\alpha = 0.05$.

Finally, we consider the Wilcoxon signed-ranks test.

13.3.2.2 The Wilcoxon Signed-Ranks Test The sign test for median and symmetry loses information since it ignores the magnitude of the difference between the observations and the hypothesized median. The Wilcoxon signed-ranks test provides an alternative test of location (and symmetry) that also takes into account the magnitudes of these differences.

Let $X_1, X_2, \ldots, X_n$ be iid RVs with common absolutely continuous DF F, which is symmetric about the median $\mathfrak{z}_{1/2}$. The problem is to test $H_0\colon \mathfrak{z}_{1/2} = \mathfrak{z}_0$ against the usual one- or two-sided alternatives. Without loss of generality, we assume that $\mathfrak{z}_0 = 0$. Then $F(-x) = 1 - F(x)$ for all $x \in \mathcal{R}$. To test $H_0\colon F(0) = \frac{1}{2}$ or $\mathfrak{z}_{1/2} = 0$, we first arrange $|X_1|, |X_2|, \ldots, |X_n|$ in increasing order of magnitude, and assign ranks $1, 2, \ldots, n$, keeping track of the original signs of X_i. For example, if $n = 4$ and $|X_2| < |X_4| < |X_1| < |X_3|$, the rank of $|X_1|$ is 3, of $|X_2|$ is 1, of $|X_3|$ is 4, and of $|X_4|$ is 2.

Let

$$\begin{cases} T^+ = \text{the sum of the ranks of positive } X_i\text{'s}, \\ T^- = \text{the sum of the ranks of negative } X_i\text{'s}. \end{cases} \tag{20}$$

Then, under H_0, we expect T^+ and T^- to be the same. Note that

$$T^+ + T^- = \sum_1^n i = \frac{n(n+1)}{2}, \tag{21}$$

so that T^+ and T^- are linearly related and offer equivalent criteria. Let us define

$$Z_i = \begin{cases} 1 & \text{if } X_i > 0 \\ 0 & \text{if } X_i < 0 \end{cases}, \quad i = 1, 2, \ldots, n, \tag{22}$$

and write $R(|X_i|) = R_i^+$ for the rank of $|X_i|$. Then $T^+ = \sum_{i=1}^n R_i^+ Z_i$ and $T^- = \sum_{i=1}^n (1 - Z_i) R_i^+$. Also,

$$\begin{aligned} T^+ - T^- &= -\sum_{i=1}^n R_i^+ + 2\sum_{i=1}^n Z_i R_i^+ \\ &= 2\sum_{i=1}^n R_i^+ Z_i - \frac{n(n+1)}{2}. \end{aligned} \tag{23}$$

The statistic T^+ (or T^-) is known as the *Wilcoxon statistic*. A large value of T^+ (or, equivalently, a small value of T^-) means that most of the large deviations from 0 are positive, and therefore we reject H_0 in favor of the alternative, $H_1: \mathfrak{z}_{1/2} > 0$.

A similar analysis applies to the other two alternatives. We record the results as follows:

		Test
H_0	H_1	Reject H_0 if
$\mathfrak{z}_{1/2} = 0$	$\mathfrak{z}_{1/2} > 0$	$T^+ > c_1$
$\mathfrak{z}_{1/2} = 0$	$\mathfrak{z}_{1/2} < 0$	$T^+ < c_2$
$\mathfrak{z}_{1/2} = 0$	$\mathfrak{z}_{1/2} \neq 0$	$T^+ < c_3$ or $T^+ > c_4$

We now show how the Wilcoxon signed-ranks test statistic is related to the U-statistic estimate of $g_2(F) = P_F(X_1 + X_2 > 0)$. Recall from Example 13.2.6 that the corresponding U-statistic is

$$U_2(\mathbf{X}) = \binom{n}{2}^{-1} \sum_{1 \le i < j \le n} I_{[X_i + X_j > 0]}. \tag{24}$$

First note that

$$\sum_{1 \le i \le j \le n} I_{[X_i + X_j > 0]} = \sum_{j=1}^{n} I_{[X_j > 0]} + \sum_{1 \le i < j \le n} I_{[X_i + X_j > 0]}. \tag{25}$$

Next note that for $i < j$, $X_{(i)} + X_{(j)} > 0$ if and only if $X_{(j)} > 0$ and $|X_{(i)}| < |X_{(j)}|$. It follows that $\sum_{i=1}^{j} I_{[X_{(i)} + X_{(j)} > 0]}$ is the signed-rank of $X_{(j)}$. Consequently,

$$\begin{aligned} T^+ &= \sum_{j=1}^{n} \sum_{i=1}^{j} I_{[X_{(i)} + X_{(j)} > 0]} = \sum_{1 \le i \le j \le n} I_{[X_i + X_j > 0]} \\ &= \sum_{j=1}^{n} I_{[X_j > 0]} + \sum_{1 \le i < j \le n} I_{[X_i + X_j > 0]} \\ &= nU_1(\mathbf{X}) + \binom{n}{2} U_2(\mathbf{X}), \end{aligned} \tag{26}$$

where U_1 is the U-statistic for $g_1(F) = P_F(X_1 > 0)$.

We next compute the distribution of T^+ for small samples. The distribution of T^+ is tabulated by Kraft and Van Eeden [55, pp. 221–223].

Let

$$Z_{(i)} = \begin{cases} 1 & \text{if the } |X_j| \text{ that has rank } i \text{ is } > 0 \\ 0 & \text{otherwise.} \end{cases}$$

Note that $T^+ = 0$ if all differences have negative signs, and $T^+ = n(n+1)/2$ if all differences have positive signs. Here a difference means a difference between the observations and the postulated value of the median. T^+ is completely determined by the indicators $Z_{(i)}$, so that the sample space can be considered as a set of 2^n n-tuples $(z_1, z_2, \ldots, z_n)$, where each z_i is 0 or 1. Under H_0, $\mathfrak{z}_{1/2} = \mathfrak{z}_0$ and each arrangement is equally likely. Thus

$$\begin{aligned} P_{H_0}\{T^+ = t\} &= \frac{\{\text{number of ways to assign } + \text{ or } - \text{ signs to integers } 1, 2, \ldots, n \text{ so that the sum is } t\}}{2^n} \\ &= \frac{n(t)}{2^n}, \text{ say.} \end{aligned} \tag{27}$$

Note that every assignment has a conjugate assignment with plus and minus signs interchanged so that for this conjugate, T^+ is given by

$$\sum_1^n i(1-Z_{(i)}) = \frac{n(n+1)}{2} - \sum_1^n iZ_{(i)}. \tag{28}$$

Thus under H_0 the distribution of T^+ is symmetric about the mean $n(n+1)/4$.

Example 7. Let us compute the null distribution for $n=3$. $E_{H_0}T^+ = n(n+1)/4 = 3$, and T^+ takes values from 0 to $n(n+1)/2 = 6$:

Value of T^+	Ranks Associated with Positive Differences	$n(t)$
6	1, 2, 3	1
5	2, 3	1
4	1, 3	1
3	1, 2; 3	2

so that

$$P_{H_0}\{T^+ = t\} = \begin{cases} \frac{1}{8}, & t = 4,5,6,0,1,2, \\ \frac{2}{8}, & t = 3, \\ 0, & \text{otherwise.} \end{cases} \tag{29}$$

Similarly, for $n = 4$, one can show that

$$P_{H_0}\{T^+ = t\} = \begin{cases} \frac{1}{16}, & t = 0,1,2,8,9,10, \\ \frac{2}{16}, & t = 3,4,5,6,7, \\ 0, & \text{otherwise.} \end{cases} \tag{30}$$

An alternative procedure would be to use the MGF technique. Under H_0, the RVs $iZ_{(i)}$ are independent and have the PMF

$$P\{iZ_{(i)} = i\} = P\{iZ_{(i)} = 0\} = \tfrac{1}{2}.$$

Thus

$$\begin{aligned} M(t) &= Ee^{tT^+} \\ &= \prod_{i=1}^n \left(\frac{e^{it}+1}{2}\right). \end{aligned} \tag{31}$$

We express $M(t)$ as a sum of terms of the form $\alpha_j e^{jt}/2^n$. The PMF of T^+ can then be determined by inspection. For example, in the case $n = 4$, we have

$$M(t) = \prod_{i=1}^{4} \left(\frac{e^{it}+1}{2}\right) = \left(\frac{e^{t}+1}{2}\right)\left(\frac{e^{2t}+1}{2}\right)\left(\frac{e^{3t}+1}{2}\right)\left(\frac{e^{4t}+1}{2}\right)$$

$$= \tfrac{1}{4}(e^{3t}+e^{2t}+e^{t}+1)\left(\frac{e^{3t}+1}{2}\right)\left(\frac{e^{4t}+1}{2}\right) \tag{32}$$

$$= \tfrac{1}{8}(e^{6t}+e^{5t}+e^{4t}+2e^{3t}+e^{2t}+e^{t}+1)\left(\frac{e^{4t}+1}{2}\right) \tag{33}$$

$$= \tfrac{1}{16}(e^{10t}+e^{9t}+e^{8t}+2e^{7t}+2e^{6t}+2e^{5t}+2e^{4t}+2e^{3t}+e^{2t}+e^{t}+1). \tag{34}$$

This method gives us the PMF of T^+ for $n = 2$, $n = 3$, and $n = 4$ immediately. Quite simply,

$$P_{H_0}\{T^+ = j\} = \text{coefficient of } e^{jt} \text{ in the expansion of } M(t), j = 0, 1, \ldots, n(n+1)/2. \tag{35}$$

See Problem 3.3.12 for the PGF of T^+.

Example 8. Let us return to the data of Example 5 and test $H_0\colon \mathfrak{z}_{1/2} = \mu = -1.0$ against $H_1\colon \mathfrak{z}_{1/2} > -1.0$. Ranking $|x_i - \mathfrak{z}_{1/2}|$ in increasing order of magnitude, we have

0.016 <	0.131 <	0.535 <	0.762 <	1.056 <	1.120 <	1.417 <	1.561
5	4	1	3	7	2	6	8

Thus

$$r_1 = 3, \quad r_2 = 6, \quad r_3 = 4, \quad r_4 = 2,$$
$$r_5 = 1, \quad r_6 = 7, \quad r_7 = 5, \quad r_8 = 8$$

and

$$T^+ = 3+6+4+2+7+5+8 = 35.$$

From Table ST10, H_0 is rejected at level $\alpha = 0.05$ if $T^+ \geq 31$. Since $35 > 31$, we reject H_0.

Remark 1. The Wilcoxon test statistic can also be used to test for symmetry. Let $X_1, X_2, \ldots, X_n$ be iid observations on an RV with absolutely continuous DF F. We set the null hypothesis as

$$H_0\colon \mathfrak{z}_{1/2} = \mathfrak{z}_0, \text{ and DF } F \text{ is symmetric about } \mathfrak{z}_0.$$

The alternative is

$$H_1\colon \mathfrak{z}_{1/2} \neq \mathfrak{z}_0 \text{ and } F \text{ symmetric, or } F \text{ asymmetric.}$$

The test is the same since the null distribution of T^+ is the same.

Remark 2. If we have n independent pairs of observations $(X_1,Y_1),(X_2,Y_2),,\ldots,(X_n,Y_n)$ from a bivariate DF, we form the differences $Z_i=X_i-Y_i$, $i=1,2,\ldots,n$. Assuming that $Z_1,Z_2,\ldots,Z_n$ are (independent) observations from a population of differences with absolutely continuous DF F that is symmetric with median $\mathfrak{z}_{1/2}$, we can use the Wilcoxon statistic to test $H_0\colon \mathfrak{z}_{1/2}=\mathfrak{z}_0$.

We present some examples.

Example 9. For the data of Example 10.3.3 let us apply the Wilcoxon statistic to test $H_0\colon \mathfrak{z}_{1/2}=0$ and F is symmetric against $H_1\colon \mathfrak{z}_{1/2}\neq 0$ and F symmetric or F not symmetric.

The absolute values, when arranged in increasing order of magnitude, are as follows:

$$\begin{array}{cccccccccc} 0.057< & 0.068< & 0.137< & 0.261< & 0.323< & 0.464< & 0.482< & 0.486< & 0.508< & 0.513 \\ 13 & 5 & 2 & 17 & 4 & 1 & 11 & 15 & 20 & 7 \end{array}$$

$$\begin{array}{cccccccccc} <0.525 & <0.595 & <0.881 & <0.906 & <1.046 & <1.229 & <1.237 & <1.678 & <1.787 & <2.455 \\ 8 & 9 & 10 & 6 & 19 & 14 & 18 & 12 & 16 & 3 \end{array}$$

Thus

$$\begin{array}{llllll} r_1=6, & r_2=3, & r_3=20, & r_4=5, & r_5=2, & r_6=14, \\ r_7=10, & r_8=11, & r_9=12, & r_{10}=13, & r_{11}=7, & r_{12}=18, \\ r_{13}=1, & r_{14}=16, & r_{15}=8, & r_{16}=19, & r_{17}=4, & r_{18}=17, \\ r_{19}=15, & r_{20}=9, & & & & \end{array}$$

and

$$T^+=6+3+20+14+12+13+18+17+15=118.$$

From Table ST10 we see that H_0 cannot be rejected even at level $\alpha=0.20$.

Example 10. Returning to the data of Example 6, we apply the Wilcoxon test to the differences $Z_i=X_i-Y_i$. The differences are $-6, 3, 1, -8, -17, -20, -3, -3, -11, 9$. To test $H_0\colon \mathfrak{z}_{1/2}=0$ against $H_1\colon \mathfrak{z}_{1/2}\neq 0$, we rank the absolute values of z_i in increasing order to get

$$1<3=3=3<6<8<9<11<17<20$$

and

$$T^+=1+2+7=10.$$

Here we have assigned ranks 2, 3, 4 to observations $+3, -3, -3$. (If we assign rank 4 to observation 3, then $T^+=12$ without appreciably changing the result.)

From Table ST10, we reject H_0 at $\alpha=0.05$ if either $T^+>46$ or $T^+<9$. Since $T^+>9$ and <46, we accept H_0. Note that hypothesis H_0 was also accepted by the sign test.

For large samples we use the normal approximation. In fact, from (26) we see that

$$\frac{\sqrt{n}(T^+ - ET^+)}{\binom{n}{2}} = \frac{n^{3/2}}{\binom{n}{2}}(U_1 - EU_1) + \sqrt{n}(U_2 - EU_2).$$

Clearly, $U_1 - EU_1 \xrightarrow{P} 0$ and since $n^{3/2}\Big/\binom{n}{2} \to 0$, the first term $\to 0$ in probability as $n \to \infty$. By Slutsky's theorem (Theorem 7.2.15) it follows that

$$\frac{\sqrt{n}}{\binom{n}{2}}(T^+ - ET^+) \quad \text{and} \quad \sqrt{n}(U_2 - EU_2)$$

have the same limiting distribution. From Theorem 13.2.3 and Example 13.2.7 it follows that $\sqrt{n}(U_2 - EU_2)$, and hence $(T^+ - ET^+)\sqrt{n}/\binom{n}{2}$, has a limiting normal distribution with mean 0 and variance

$$4\zeta_1 = 4P_F(X_1 + X_2 > 0, X_1 + X_3 > 0) - 4P_F^2(X_1 + X_2 > 0).$$

Under H_0, the RVs $iZ_{(i)}$ are independent $b(1, 1/2)$ so

$$E_{H_0}T^+ = \frac{n(n+1)}{4} \quad \text{and} \quad \text{var}_{H_0}T^+ = \left(\frac{1}{2}\right)\left(\frac{1}{2}\right)\sum_{i=1}^{n} i^2 = \frac{n(n+1)(2n+1)}{24}.$$

Also, under H_0, F is continuous and symmetric so

$$P_F(X_1 + X_2 > 0) = \int_{-\infty}^{\infty} P_F(X_1 > -x)f(x)dx = \frac{1}{2}$$

and

$$P_F(X_1 + X_2 > 0, X_1 + X_3 > 0) = \int_{-\infty}^{\infty} [P_F(X_1 > -x)]^2 f(x)dx = \frac{1}{3}$$

Thus $4\zeta_1 = 4/3 - 4/4 = 1/3$ so that

$$\frac{T^+ - E_{H_0}T^+}{\binom{n}{2}\sqrt{\frac{1}{3n}}} \xrightarrow{L} \mathcal{N}(0,1).$$

However,

$$\frac{(\text{var}_{H_0}T^+)^{1/2}}{\binom{n}{2}\sqrt{\frac{1}{3n}}} = \frac{[n(n+1)(2n+4)/24]^{1/2}}{\frac{n(n-1)}{2}\sqrt{\frac{1}{3n}}} \to 1$$

as $n \to \infty$. Consequently, under H_0

$$T^+ \sim \text{AN}\left(\frac{n(n+1)}{4}, \frac{n(n+1)(2n+1)}{24}\right).$$

Thus, for large enough n we can determine the critical values for a test based on T^+ by using normal approximation.

As an example, take $n = 20$. From Table ST10 the P-value associated with $t^+ = 140$ is 0.10. Using normal approximation

$$P_{H_0}(T^+ > 140) \approx P\left(Z > \frac{140 - 105}{27.45}\right) = P(Z > 1.28) = 0.10003$$

PROBLEMS 13.3

1. Prove Theorem 4.
2. A random sample of size 16 from a continuous DF on $[0, 1]$ yields the following data: 0.59, 0.72, 0.47, 0.43, 0.31, 0.56, 0.22, 0.90, 0.96, 0.78, 0.66, 0.18, 0.73, 0.43, 0.58, 0.11. Test the hypothesis that the sample comes from $U[0, 1]$.
3. Test the goodness of fit of normality for the data of Problem 10.3.6, using the Kolmogorov–Smirnov test.
4. For the data of Problem 10.3.6 find a 0.95 level confidence band for the distribution function.
5. The following data represent a sample of size 20 from $U[0, 1]$: 0.277, 0.435, 0.130, 0.143, 0.853, 0.889, 0.294, 0.697, 0.940, 0.648, 0.324, 0.482, 0.540, 0.152, 0.477, 0.667, 0.741, 0.882, 0.885, 0.740. Construct a .90 level confidence band for $F(x)$.
6. In Problem 5 test the hypothesis that the distribution is $U[0, 1]$. Take $\alpha = 0.05$.
7. For the data of Example 2 test, by means of the sign test, the null hypothesis $H_0: \mu = 1.5$ against $H_1: \mu \neq 1.5$.
8. For the data of Problem 5 test the hypothesis that the quantile of order $p = 0.20$ is 0.20.
9. For the data of Problem 10.4.8 use the sign test to test the hypothesis of no difference between the two averages.
10. Use the sign test for the data of Problem 10.4.9 to test the hypothesis of no difference in grade-point averages.
11. For the data of Problem 5 apply the signed-rank test to test $H_0: \mathfrak{z}_{1/2} = 0.5$ against $H_1: \mathfrak{z}_{1/2} \neq 0.5$.
12. For the data of Problems 10.4.8 and 10.4.9 apply the signed-rank test to the differences to test $H_0: \mathfrak{z}_{1/2} = 0$ against $H_1: \mathfrak{z}_{1/2} \neq 0$.

13.4 SOME TWO-SAMPLE PROBLEMS

In this section we consider some two-sample tests. Let $X_1, X_2, \ldots, X_m$ and $Y_1, Y_2, \ldots, Y_n$ be independent samples from two absolutely continuous distribution functions F_X and F_Y, respectively. The problem is to test the null hypothesis H_0: $F_X(x) = F_Y(x)$ for all $x \in \mathcal{R}$ against the usual one- and two-sided alternatives.

Tests of H_0 depend on the type of alternative specified. We state some of the alternatives of interest even though we will not consider all of these in this text.

I Location alternative: $F_Y(x) = F_X(x-\theta)$, $\theta \neq 0$.
II Scale alternative: $F_Y(x) = F_X(x/\sigma)$, $\sigma > 0$.
III Lehmann alternative: $F_Y(x) = 1-[1-F_X(x)]^{\theta+1}$, $\theta+1 > 0$.
IV Stochastic alternative: $F_Y(x) \geq F_X(x)$ for all x, and $F_Y(x) > F_X(x)$ for at least one x.
V General alternative: $F_Y(x) \neq F_X(x)$ for some x.

Some comments are in order. Clearly I through IV are special cases of V. Alternatives I and II show differences in F_X and F_Y in location and scale, respectively. Alternative III states that $P(Y > x) = [P(X > x)]^{\theta+1}$. In the special case when θ is an integer it states that Y has the same distribution as the smallest of the $\theta+1$ of X-variables. A similar alternative to test that is sometimes used is $F_Y(x) = [F_X(x)]^{\alpha}$ for some $\alpha > 0$ and all x. When α is an integer, this states that Y is distributed as the largest of the α X-variables. Alternative IV refers to the relative magnitudes of X's and Y's. It states that

$$P(Y \leq x) \geq P(X \leq x) \quad \text{for all } x,$$

so that

$$P(Y > x) \leq P(X > x), \tag{1}$$

for all x. In other words, X's tend to be larger than the Y's.

Definition 1. We say that a continuous RV X is *stochastically larger* than a continuous RV Y if inequality (1) is satisfied for all x with strict inequality for some x.

A similar interpretation may be given to the one-sided alternative $F_X > F_Y$. In the special case where both X and Y are normal RVs with means μ_1, μ_2 and common variance σ^2, $F_X = F_Y$ corresponds to $\mu_1 = \mu_2$ and $F_X > F_Y$ corresponds to $\mu_1 < \mu_2$

In this section we consider some common two-sample tests for location (Case I) and stochastic ordering (Case IV) alternatives. First, note that a test of stochastic ordering may also be used as a test of less restrictive location alternatives since, for example, $F_X > F_Y$ corresponds to larger Y's and hence larger location for Y. Second, we note that the chi-square test of homogeneity described in Section 10.3 can be used to test general alternatives (Case V) $H_1 : F(x) \neq G(x)$ for some x. Briefly, one partitions the real line into Borel sets $A_1, A_2, \ldots, A_k$. Let

$$p_{i1} = P(X_j \in A_i) \quad \text{and} \quad p_{i2} = P(Y_j \in A_i),$$

$i = 1, 2, \ldots, k$. Under $H_0 : F = G, p_{i1} = p_{i2}, i = 1, 2, \ldots, k$, which is the problem of testing equality of two independent multinomial distributions discussed in Section 10.3.

We first consider a simple test of location. This test, based on the sample median of the combined sample, is a test of the equality of medians of the two DFs. It will tend to accept $H_0 : F = G$ even if the shapes of F and G are different as long as their medians are equal.

13.4.1 Median Test

The combined sample $X_1,X_2,\ldots,X_m$, $Y_1,Y_2,\ldots,Y_n$ is ordered and a sample median is found. If $m+n$ is odd, the median is the $[(m+n+1)/2]$th value in the ordered arrangement. *If* $m+n$ is even, the median is any number between the two middle values. Let V be the number of observed values of X that are less than or equal to the sample median for the combined sample. If V is large, it is reasonable to conclude that the actual median of X is smaller than the median of Y. One therefore rejects $H_0: F=G$ in favor of $H_1: F(x)\geq G(x)$ for all x and $F(x)>G(x)$ for some x if V is too large, that is, if $V\geq c$. If, however, the alternative is $F(x)\leq G(x)$ for all x and $F(x)<G(x)$ for some x, the median test rejects H_0 if $V\leq c$.

For the two-sided alternative that $F(x)\neq G(x)$ for some x, we use the two-sided test.

We next compute the null distribution of the RV V. If $m+n=2p$, p a positive integer, then

$$P_{H_0}\{V=v\}=P_{H_0}\{\text{exactly } v \text{ of the } X_i\text{'s are} \leq \text{combined median}\}$$
$$=\begin{cases}\dfrac{\binom{m}{v}\binom{n}{p-v}}{\binom{m+n}{p}}, & v=0,1,2,\ldots,m,\\ 0, & \text{otherwise.}\end{cases} \tag{2}$$

Here $0\leq V\leq \min(m,p)$. If $m+n=2p+1$, $p>0$, is an integer, the $[(m+n+1)/2]$th value is the median in the combined sample, and

$$P_{H_0}\{V=v\}=P\{\text{exactly } v \text{ of the } X_i\text{'s are below the } (p+1)\text{th value in the ordered arrangement}\}$$
$$=\begin{cases}\dfrac{\binom{m}{v}\binom{n}{p-v}}{\binom{m+n}{p}}, & v=0,1,\ldots,\min(m,p),\\ 0, & \text{otherwise.}\end{cases} \tag{3}$$

Remark 1. Under H_0 we expect $(m+n)/2$ observations above the median and $(m+n)/2$ below the median. One can therefore apply the chi-square test with 1 d.f. to test H_0 against the two-sided alternative.

Example 1. The following data represent lifetimes (hours) of batteries for two different brands:

Brand A:	40	30	40	45	55	30
Brand B:	50	50	45	55	60	40

The combined ordered sample is 30, 30, 40, 40, 40, 45, 45, 50, 50, 55, 55, 60. Since $m+n=3$ is even, the median is 45. Thus

$$\begin{aligned} v &= \text{number of observed values of } X \text{ that are less than or equal to } 45 \\ &= 5. \end{aligned}$$

Now

$$P_{H_0}\{V \geq 5\} = \frac{\binom{6}{5}\binom{6}{1}}{\binom{12}{6}} + \frac{\binom{6}{6}\binom{6}{0}}{\binom{12}{6}} \approx 0.04.$$

Since $P_{H_0}\{V \geq 5\} > 0.025$, we cannot reject H_0 that the two samples come from the same population.

We now consider two tests of the stochastic alternatives. As mentioned earlier they may also be used as tests of location.

13.4.2 Kolmogorov–Smirnov Test

Let $X_1, X_2, \ldots, X_m$ and $Y_1, Y_2, \ldots, Y_n$ be independent random samples from continuous DFs F and G, respectively. Let F_m^* and G_n^*, respectively, be the empirical DFs of the X's and the Y's. Recall the F_m^* is the U-statistic for F and G_n^*, that for G. Under H_0: $F(x) = G(x)$ for all x, we expect a reasonable agreement between the two sample DFs. We define

$$D_{m,n} = \sup_x |F_m^*(x) - G_n^*(x)|. \tag{4}$$

Then $D_{m,n}$ may be used to test H_0 against the two-sided alternative H_1: $F(x) \neq G(x)$ for some x. The test rejects H_0 at level α if

$$D_{m,n} \geq D_{m,n,\alpha}, \tag{5}$$

where $P_{H_0}\{D_{m,n} \geq D_{m,n,\alpha}\} \leq \alpha$.

Similarly, one can define the one-sided statistics

$$D_{m,n}^+ = \sup_x [F_m^*(x) - G_n^*(x)] \tag{6}$$

and

$$D_{m,n}^- = \sup_x [G_n^*(x) - F_m^*(x)], \tag{7}$$

to be used against the one-sided alternatives

$$\begin{aligned} &G(x) \leq F(x) \quad \text{for all } x \qquad \text{and} \qquad G(x) < F(x) \quad \text{for some } x \\ &\text{with rejection region } D_{m,n}^+ \geq D_{m,n,\alpha}^+ \end{aligned} \tag{8}$$

and

$$\begin{array}{l} F(x) \le G(x) \quad \text{for all } x \qquad \text{and} \qquad F(x) < G(x) \quad \text{for some } x \\ \text{with rejection region } D_{m,n}^{-} \ge D_{m,n,\alpha}^{-}, \end{array} \tag{9}$$

respectively.

For small samples tables due to Massey [72] are available. In Table ST9, we give the values of $D_{m,n,\alpha}$ and $D_{m,n,\alpha}^{+}$ for some selected values of m, n, and α. Table ST8 gives the corresponding values for the $m = n$ case.

For large samples we use the limiting result due to Smirnov [107]. Let $N = mn/(m+n)$. Then

$$\lim_{m,n\to\infty} P\{\sqrt{N}D_{m,n}^{+} \le \lambda\} = \begin{cases} 1 - e^{-2\lambda^2}, & \lambda > 0, \\ 0, & \lambda \le 0, \end{cases} \tag{10}$$

$$\lim_{m,n\to\infty} P\{\sqrt{N}D_{m,n} \le \lambda\} = \begin{cases} \sum_{j=-\infty}^{\infty} (-1)^j e^{-2j^2\lambda^2}, & \lambda > 0, \\ 0, & \lambda \le 0. \end{cases} \tag{11}$$

Relations (10) and (11) give the distribution of $D_{m,n}^{+}$ and $D_{m,n}$, respectively, under H_0: $F(x) = G(x)$ for all $x \in \mathcal{R}$.

Example 2. Let us apply the test to data from Example 1. Do the two brands differ with respect to average life?

Let us first apply the Kolmogorov–Smirnov test to test H_0 that the population distribution of length of life for the two brands is the same.

x	$F_6^*(x)$	$G_6^*(x)$	$\lvert F_6^*(x) - G_6^*(x)\rvert$
30	$\frac{2}{6}$	0	$\frac{2}{6}$
40	$\frac{4}{6}$	$\frac{1}{6}$	$\frac{3}{6}$
45	$\frac{5}{6}$	$\frac{2}{6}$	$\frac{3}{6}$
50	$\frac{5}{6}$	$\frac{4}{6}$	$\frac{1}{6}$
55	1	$\frac{5}{6}$	$\frac{1}{6}$
60	1	1	0

$$D_{6,6} = \sup_x |F_6^*(x) - G_6^*(x)| = \frac{3}{6}.$$

From Table ST8, the critical value for $m = n = 6$ at level $\alpha = 0.05$ is $D_{6,6,0.05} = \frac{4}{6}$. Since $D_{6,6} \not> D_{6,6,0.05}$, we accept H_0 that the population distribution for the length of life for the two brands is the same.

Let us next apply the two-sample t-test. We have $\bar{x} = 40$, $\bar{y} = 50$, $s_1^2 = 90$, $s_2^2 = 50$, $s_p^2 = 70$. Thus

$$t = \frac{40 - 50}{\sqrt{70}\sqrt{\frac{1}{6} + \frac{1}{6}}} = -2.08.$$

Since $t_{10,0.025} = 2.2281$, we accept the hypothesis that the two samples come from the same (normal) population.

The second test of stochastic ordering alternatives we consider is the Mann–Whitney–Wilcoxon test which can be viewed as a test based on a U-statistic.

13.4.3 The Mann–Whitney–Wilcoxon Test

Let $X_1, X_2, \ldots, X_m$ and $Y_1, Y_2, \ldots, Y_n$ be independent samples from two continuous DFs, F and G, respectively. As in Example 13.2.10, let

$$T(X_i; Y_j) = \begin{cases} 1, & \text{if } X_i < Y_j \\ 0, & \text{if } X_i \geq Y_j, \end{cases}$$

for $i = 1, 2, \ldots, m$, $j = 1, 2, \ldots, n$. Recall that $T(X_i; Y_j)$ is an unbiased estimator of $g(F, G) = P_{F,G}(X < Y)$ and the two sample U-statistic for g is given by $U_1(\mathbf{X}; \mathbf{Y}) = (m, n)^{-1} \sum_{i=1}^{m} \sum_{j=1}^{n} T(X_i; Y_j)$. For notational convenience, let us write

$$U = mnU_1(\mathbf{X}; \mathbf{Y}) = \sum_{i=1}^{m} \sum_{j=1}^{n} T(X_i; Y_j). \tag{12}$$

Then U is the number of values of $X_1, X_2, \ldots, X_m$ that are smaller than each of $Y_1, Y_2, \ldots, Y_n$. The statistic U is called the *Mann–Whitney statistic*. An alternative equivalent form using Wilcoxon scores is the linear rank statistic given by

$$W = \sum_{j=1}^{n} Q_j, \tag{13}$$

where Q_j = rank of Y_j among the combined $m + n$ observations. Indeed,

$$Q_j = \text{rank of } Y_j = (\#\text{ of } X_i\text{'s} < Y_j) + \text{ rank of } Y_j \text{ in } Y\text{'s}.$$

Thus

$$W = \sum_{j=1}^{n} Q_j = U + \sum_{j=1}^{n} j = U + \frac{n(n+1)}{2} \tag{14}$$

so that U and W are equivalent test statistics. Hence the name *Mann–Whitney–Wilcoxon Test*. We will restrict attention to U as the test statistic.

Example 3. Let $m = 4$, $n = 3$, and suppose that the combined sample when ordered is as follows:

$$x_2 < x_1 < y_3 < y_2 < x_4 < y_1 < x_3.$$

Then $U = 7$, since there are three values of $x < y_1$, two values of $x < y_2$, and two values of $x < y_3$. Also, $W = 13$ so $U = 13 - 3(4)/2 = 7$.

Note that $U = 0$ if all the X_i's are larger than all the Y_j's and $U = mn$ if all the X_i's are smaller than all the Y_j's, because then there are m X's $< Y_1$, m X's $< Y_2$, and so on. Thus $0 \le U \le mn$. If U is large, the values of Y tend to be larger than the values of X (Y is stochastically larger than X), and this supports the alternative $F(x) \ge G(x)$ for all x and $F(x) > G(x)$ for some x. Similarly, if U is small, the Y values tend to be smaller than the X values, and this supports the alternative $F(x) \le G(x)$ for all x and $F(x) < G(x)$ for some x. We summarize these results as follows:

H_0	H_1	Reject H_0 if
$F = G$	$F \ge G$	$U \ge c_1$
$F = G$	$F \le G$	$U \le c_2$
$F = G$	$F \ne G$	$U \ge c_3$ or $U \le c_4$

To compute the critical values we need the null distribution of U. Let

$$p_{m,n}(u) = P_{H_0}\{U = u\}. \tag{15}$$

We will set up a difference equation relating $p_{m,n}$ to $p_{m-1,n}$ and $p_{m,n-1}$. If the observations are arranged in increasing order of magnitude, the largest value can be either an x value or a y value. Under H_0, all $m+n$ values are equally likely, so the probability that the largest value will be an x value is $m/(m+n)$ and that it will be a y value is $n/(m+n)$.

Now, if the largest value is an x, it does not contribute to U, and the remaining $m-1$ values of x and n values of y can be arranged to give the observed value $U = u$ with probability $p_{m-1,n}(u)$. If the largest value is a Y, this value is larger than all the m x's. Thus, to get $U = u$, the remaining $n-1$ values of Y and m values of x contribute $U = u - m$. It follows that

$$p_{m,n}(u) = \frac{m}{m+n} p_{m-1,n}(u) + \frac{n}{m+n} p_{m,n-1}(u-m). \tag{16}$$

If $m = 0$, then for $n \ge 1$

$$p_{0,n}(u) = \begin{cases} 1 & \text{if } u = 0, \\ 0 & \text{if } u > 0. \end{cases} \tag{17}$$

If $n = 0$, $m \geq 1$, then

$$p_{m,0}(u) = \begin{cases} 1 & \text{if } u = 0, \\ 0 & \text{if } u > 0, \end{cases} \tag{18}$$

and

$$p_{m,n}(u) = 0 \qquad \text{if } u < 0, \quad m \geq 0, \quad n \geq 0. \tag{19}$$

For small values of m and n one can easily compute the null PMF of U. Thus, if $m = n = 1$, then

$$p_{1,1}(0) = \tfrac{1}{2}, \qquad p_{1,1}(1) = \tfrac{1}{2}.$$

If $m = 1$, $n = 2$, then

$$p_{1,2}(0) = p_{1,2}(1) = p_{1,2}(2) = \tfrac{1}{3}.$$

Tables for critical values are available for small values of m and n, $m \leq n$. See, for example, Auble [3] or Mann and Whitney [71]. Table ST11 gives the values of u_α for which $P_{H_0}\{U > u_\alpha\} \leq \alpha$ for some selected values of m, n, and α.

If m, n are large we can use the asymptotic normality of U. In Example 13.2.11 we showed that, under H_0,

$$\frac{U/(mn) - \frac{1}{2}}{\sqrt{(m+n+1)/(12mn)}} \xrightarrow{L} \mathcal{N}(0,1)$$

as $m, n \to \infty$ such that $m/(m+n) \to$ constant. The approximation is fairly good for $m, n \geq 8$.

Example 4. Two samples are as follows:

Values of X_i: $1, 2, 3, 5, 7, 9, 11, 18$

Values of Y_i: $4, 6, 8, 10, 12, 13, 14, 15, 19$

Thus $m = 8$, $n = 9$, and $U = 3+4+5+6+7+7+7+7+8 = 54$. The (exact) p-value $P_{H_0}(U \geq 54) = 0.046$, so we reject H_0 at (two-tailed) level $\alpha = 0.1$. Let us apply the normal approximation. We have

$$E_{H_0}U = \frac{8 \cdot 9}{2} = 36, \qquad \text{var}_{H_0}(U) = \frac{8 \cdot 9}{12}(8+9+1) = 108,$$

and

$$Z = \frac{54 - 36}{\sqrt{108}} = \frac{18}{6\sqrt{3}} = \sqrt{3} = 1.732.$$

We note that $P(Z > 1.73) = 0.042$.

PROBLEMS 13.4

1. For the data of Example 4 apply the median test.
2. Twelve 4-year-old boys and twelve 4-year-old girls were observed during two 15-minute play sessions, and each child's play during these two periods was scored as follows for incidence and degree of aggression:

$$\text{Boys: } 86, 69, 72, 65, 113, 65, 118, 45, 141, 104, 41, 50$$
$$\text{Girls: } 55, 40, 22, 58, 16, 7, 9, 16, 26, 36, 20, 15$$

Test the hypothesis that there were sex differences in the amount of aggression shown, using (a) the median test and (b) the Mann-Whitney-Wilcoxon test (Siegel [105]).

3. To compare the variability of two brands of tires, the following mileages (1000 miles) were obtained for eight tires of each kind:

$$\text{Brand } A\text{: } 32.1, 20.6, 17.8, 28.4, 19.6, 21.4, 19.9, 30.1$$
$$\text{Brand } B\text{: } 19.8, 27.6, 30.8, 27.6, 34.1, 18.7, 16.9, 17.9$$

Test the null hypothesis that the two samples come from the same population, using the Mann–Whitney–Wilcoxon test.

4. Use the data of Problem 2 to apply the Kolmogorov–Smirnov test.
5. Apply the Kolmogorov–Smirnov test to the data of Problem 3.
6. Yet another test for testing $H_0 : F = G$ against general alternatives is the so-called *runs test*. A *run* is a succession of one or more identical symbols which are preceeded and followed by a different symbol (or no symbol). The *length* of a run is the number of like symbols in a run. The total number of runs, R, in the combined sample of X's and Y's when arranged in increasing order can be used as a test of H_0. Under H_0 the X and Y symbols are expected to be well-mixed. A small value of R supports $H_1 : F \neq G$. A test based on R is appropriate only for two-sided (general) alternatives. Tables of critical values are available. For large samples, one uses normal approximation: $R \sim \text{AN}\left(1 + \frac{2mn}{m+n}, \frac{2mn(2mn-m-n)}{(m+n-1)(m+n)^2}\right)$.

   (a) Let $R_1 = \#$ of X-runs, $R_2 = \#Y$-runs, and $R = R_1 + R_2$. Under H_0, show that

$$P(R_1 = r_1, R_2 = r_2) = k\frac{\binom{m-1}{r_1-1}\binom{n-1}{r_2-1}}{\binom{m+n}{m}},$$

   where $k = 2$ if $r_1 = r_2$, $= 1$ if $|r_1 - r_2| = 1$, $r_1 = 1, 2, \ldots, m$ and $r_2 = 1, 2, \ldots, n$.

   (b) Show that

$$P_{H_0}(R_1 = r_1) = \frac{\binom{m-1}{r_1-1}\binom{n+1}{r_1}}{\binom{m+n}{m}}, \quad 0 \leq r_1 \leq m.$$

7. Fifteen 3-year-old boys and 15 3-year-old girls were observed during two sessions of recess in a nursery school. Each child's play was scored for incidence and degree of aggression as follows:

Boys:	96	65	74	78	82	121	68	79	111	48	53	92	81	31	40
Girls:	12	47	32	59	83	14	32	15	17	82	21	34	9	15	51

Is there evidence to suggest that there are sex differences in the incidence and amount of aggression? Use both Mann–Whitney–Wilcoxon and runs tests.

13.5 TESTS OF INDEPENDENCE

Let X and Y be two RVs with joint DF $F(x,y)$, and let F_1 and F_2, respectively, be the marginal DFs of X and Y. In this section we study some tests of the hypothesis of independence, namely,

$$H_0: F(x,y) = F_1(x)F_2(y) \qquad \text{for all } (x,y) \in \mathcal{R}_2$$

against the alternative

$$H_1: F(x,y) \neq F_1(x)F_2(y) \qquad \text{for some } (x,y).$$

If the joint distribution function F is bivariate normal, we know that X and Y are independent if and only if the correlation coefficient $\rho = 0$. In this case, the test of independence is to test $H_0: \rho = 0$.

In the nonparametric situation the most commonly used test of independence is the chi-square test, which we now study.

13.5.1 Chi-square Test of Independence—Contingency Tables

Let X and Y be two RVs, and suppose that we have n observations on (X,Y). Let us divide the space of values assumed by X (the real line) into r mutually exclusive intervals $A_1, A_2, \ldots, A_r$. Similarly, the space of values of Y is divided into c disjoint intervals $B_1, B_2, \ldots, B_c$. As a rule of thumb, we choose the length of each interval in such a way that the probability that $X(Y)$ lies in an interval is approximately $(1/r)(1/c)$. Moreover, it is desirable to have n/r and n/c at least equal to 5. Let X_{ij} denote the number of pairs (X_k, Y_k), $k = 1,2,\ldots,n$, that lie in $A_i \times B_j$, and let

$$p_{ij} = P\{(X,Y) \in A_i \times B_j\} = P\{X \in A_i \text{ and } Y \in B_j\}, \tag{1}$$

where $i = 1,2,\ldots,r$, $j = 1,2,\ldots,c$. If each p_{ij} is known, the quantity

$$\sum_{i=1}^{r}\sum_{j=1}^{c}\left[\frac{(X_{ij} - np_{ij})^2}{np_{ij}}\right] \tag{2}$$

has approximately a chi-square distribution with $rc-1$ d.f., provided that n is large (see Theorem 10.3.2.). If X and Y are independent, $P\{(X,Y)\in A_i\times B_j\}=P\{X\in A_i\}P\{Y\in B_j\}$. Let us write $p_{i\cdot}=P\{X\in A_i\}$ and $p_{\cdot j}=P\{Y\in B_j\}$. Then under H_0: $p_{ij}=p_{i\cdot}p_{\cdot j}$, $i=1,2,\ldots,r$, $j=1,2,\ldots,c$. In practice, p_{ij} will not be known. We replace p_{ij} by their estimates. Under H_0, we estimate $p_{i\cdot}$ by

$$\hat{p}_{i\cdot}=\frac{\sum_{j=1}^{c}X_{ij}}{n},\qquad i=1,2,\ldots,r, \tag{3}$$

and $p_{\cdot j}$ by

$$\hat{p}_{\cdot j}=\sum_{i=1}^{r}\frac{X_{ij}}{n},\qquad j=1,2,\ldots,c. \tag{4}$$

Since $\sum_{j=1}^{c}\hat{p}_{\cdot j}=1=\sum_{1}^{r}\hat{p}_{i\cdot}$, we have estimated only $r-1+c-1=r+c-2$ parameters. It follows (see Theorem 10.3.4) that the RV

$$U=\sum_{i=1}^{r}\sum_{j=1}^{c}\left[\frac{(X_{ij}-n\hat{p}_{i\cdot}\hat{p}_{\cdot j})^2}{n\hat{p}_{i\cdot}\hat{p}_{\cdot j}}\right] \tag{5}$$

is asymptotically distributed as χ^2 with $rc-1-(r+c-2)=(r-1)(c-1)$ d.f., under H_0. The null hypothesis is rejected if the computed value of U exceeds $\chi^2_{(r-1)(c-1),\alpha}$.

It is frequently convenient to list the observed and expected frequencies of the rc events $A_i\times B_j$ in an $r\times c$ table, called a *contingency table*, as follows:

Observed Frequencies, O_{ij}				Expected Frequencies, E_{ij}			
	B_1	$B_2\cdots B_c$			B_1	$B_2\ \cdots\ B_c$	
A_1	X_{11}	$X_{12}\cdots X_{1c}$	$\sum X_{1j}$	A_1	$np_{1\cdot}p_{\cdot 1}$	$np_{1\cdot}p_{\cdot 2}\cdots np_{1\cdot}p_{\cdot c}$	$np_{1\cdot}$
A_2	X_{21}	$X_{22}\cdots X_{2c}$	$\sum X_{2j}$	A_2	$np_{2\cdot}p_{\cdot 1}$	$np_{2\cdot}p_{\cdot 2}\cdots np_{2\cdot}p_{\cdot c}$	$np_{2\cdot}$
.	.	.....	.	.	.	. ...	
.	.	.....	.	.	.	. ...	
.	.	.....	.	.	.	. ...	
A_r	X_{r1}	$X_{r2}\cdots X_{rc}$	$\sum X_{rj}$	A_r	$np_{r\cdot}p_{\cdot 1}$	$np_{r\cdot}p_{\cdot 2}\cdots np_{r\cdot}p_{\cdot c}$	$np_{r\cdot}$
	$\sum X_{i1}$	$\sum X_{i2}\ \ \sum X_{ic}$	n		$np_{\cdot 1}$	$np_{\cdot 2}\ \ \ np_{\cdot c}$	n

Note that the X_{ij}'s in the table are frequencies. Once the category $A_i\times B_j$ is determined for an observation (X,Y), numerical values of X and Y are irrelevant. Next, we need to compute the expected frequency table. This is done quite simply by multiplying the row and column totals for each pair (i,j) and dividing the product by n. Then we compute the quantity

$$\sum_i \sum_j \frac{(E_{ij} - O_{ij})^2}{E_{ij}}$$

and compare it with the tabulated χ^2 value. In this form the test can be applied even to qualitative data. $A_1, A_2, \ldots, A_r$ and $B_1, B_2, \ldots, B_c$ represent the two attributes, and the null hypothesis to be tested is that the attributes A and B are independent.

Example 1. The following are the results for a random sample of 400 employed individuals:

Length of time (years) with the Same Company	Annual Income (dollars)			
	Less than 40,000	40,000–75,000	More than 75,000	Total
<5	50	75	25	150
5–10	25	50	25	100
10 or more	25	75	50	150
	100	200	100	400

If X denotes the length of service with the same company, and Y, the annual income we wish to test the hypothesis that X and Y are independent. The expected frequencies are as follows:

Time (years) with the Same Company	Expected Frequencies			
	$<$40,000	40–75,000	$\geq$75,000	Total
<5	37.5	75	37.5	150
5–10	25	50	25	100
≥ 10	37.5	75	37.5	150
	100	200	100	400

Thus

$$\begin{aligned} U &= \frac{(12.5)^2}{37.5} + \frac{0}{25} + \frac{(12.5)^2}{37.5} + 0 + 0 + 0 + \frac{(12.5)^2}{37.5} + 0 + \frac{(12.5)^2}{37.5} \\ &= 16.66. \end{aligned}$$

The number of degrees of freedom is $(3-1)(3-1) = 4$, and $\chi^2_{4,0.05} = 9.488$. Since $16.66 > 9.488$, we reject H_0 at level 0.05 and conclude that length of service with a company is not independent of annual income.

13.5.2 Kendall's Tau

Let $(X_1,Y_1),(X_2,Y_2),\ldots,(X_n,Y_n)$ be a sample from a bivariate population.

Definition 1. For any two pairs (X_i,Y_i) and (X_j,Y_j) we say that the relation is *perfect concordance (or agreement)* if

$$X_i < X_j \text{ whenever } Y_i < Y_j \qquad \text{or} \qquad X_i > X_j \text{ whenever } Y_i > Y_j \tag{6}$$

and that the relation is *perfect discordance (or disagreement)* if

$$X_i > X_j \text{ whenever } Y_i < Y_j \qquad \text{or} \qquad X_i < X_j \text{ whenever } Y_i > Y_j. \tag{7}$$

Writing π_c and π_d for the probability of perfect concordance and of perfect discordance, respectively, we have

$$\pi_c = P\{(X_j - X_i)(Y_j - Y_i) > 0\} \tag{8}$$

and

$$\pi_d = P\{(X_j - X_i)(Y_j - Y_i) < 0\}, \tag{9}$$

and, if the marginal distributions of X and Y are continuous,

$$\begin{aligned}\pi_c &= [P\{Y_i < Y_j\} - P\{X_i > X_j \text{ and } Y_i < Y_j\}] \\ &\quad + [P\{Y_i > Y_j\} - P\{X_i < X_j \text{ and } Y_i > Y_j\}] \\ &= 1 - \pi_d.\end{aligned} \tag{10}$$

Definition 2. The measure of association between the RVs X and Y defined by

$$\tau = \pi_c - \pi_d \tag{11}$$

is known as *Kendall's tau*.

If the marginal distributions of X and Y are continuous, we may rewrite (11), in view of (10), as follows:

$$\tau = 1 - 2\pi_d = 2\pi_c - 1. \tag{12}$$

In particular, if X and Y are independent and continuous RVs, then

$$P\{X_i < X_j\} = P\{X_i > X_j\} = \tfrac{1}{2},$$

since then $X_i - X_j$ is a symmetric RV. Then

$$\begin{aligned}\pi_c &= P\{X_i < X_j\}P\{Y_i < Y_j\} + P\{X_i > X_j\}P\{Y_i > Y_j\} \\ &= P\{X_i > X_j\}P\{Y_i < Y_j\} + P\{X_i < X_j\}P\{Y_i > Y_j\} \\ &= \pi_d,\end{aligned}$$

and it follows that $\tau = 0$ for independent continuous RVs.

Note that, in general, $\tau = 0$ does not imply independence. However, for the bivariate normal distribution $\tau = 0$ if and only if the correlation coefficient ρ, between X and Y, is 0, so that $\tau = 0$ if and only if X and Y are independent (Problem 6).

Let

$$\psi((x_1, y_1), (x_2, y_2)) = \begin{cases} 1, & (y_2 - y_1)(x_2 - x_1) > 0, \\ 0, & \text{otherwise}. \end{cases} \tag{13}$$

Then $E\psi((X_1, Y_1), (X_2, Y_2)) = \tau_c = (1+\tau)/2$, and we see that τ_c is estimable of degree 2, with symmetric kernel ψ defined in (13). The corresponding one-sample U-statistic is given by

$$U((X_1, Y_1), \ldots, (X_n, Y_n)) = \binom{n}{2}^{-1} \sum_{1 \le i < j \le n} \psi((X_i, Y_i), (X_j, Y_j)). \tag{14}$$

Then the corresponding estimator of Kendall's tau is

$$T = 2U - 1 \tag{15}$$

and is called *Kendall's sample correlation coefficient.*

Note that $-1 \le T \le 1$. To test H_0 that X and Y are independent against H_1 : X and Y are dependent, we reject H_0 if $|T|$ is large. Under H_0, $\tau = 0$, so that the null distribution of T is symmetric about 0. Thus we reject H_0 at level α if the observed value of T, t, satisfies $|t| > t_{\alpha/2}$, where $P\{|T| \ge t_{\alpha/2} \mid H_0\} = \alpha$.

For small values of n the null distribution can be directly evaluated. Values for $4 \le n \le 10$ are tabulated by Kendall [51]. Table ST12 gives the values of S_α for which $P\{S > S_\alpha\} \le \alpha$, where $S = \binom{n}{2} T$ for selected values of n and α.

For a direct evaluation of the null distribution we note that the numerical value of T is clearly invariant under all order-preserving transformations. It is therefore convenient to order X and Y values and assign them ranks. If we write the pairs from the smallest to the largest according to, say, X values, then the number of pairs of values of $1 \le i < j \le n$ for which $Y_j - Y_i > 0$ is the number of concordant pairs, P.

Example 2. Let $n = 4$, and let us find the null distribution of T. There are 4! different permutations of ranks of Y:

Ranks of X values:	1	2	3	4
Ranks of Y values:	a_1	a_2	a_3	a_4

where (a_1, a_2, a_3, a_4) is one of the 24 permutations of 1,2,3,4. Since the distribution is symmetric about 0, we need only compute one half of the distribution.

P	T	Number of Permutations	$P_{H_0}\{T=t\}$
0	−1.00	1	$\frac{1}{24}$
1	−0.67	3	$\frac{3}{24}$
2	−0.33	5	$\frac{5}{24}$
3	0.00	6	$\frac{6}{24}$

Similarly, for $n=3$ the distribution of T under H_0 is as follows:

P	T	Number of Permutations	$P_{H_0}\{T=t\}$
0	−1.00	1: $(3,2,1)$	$\frac{1}{6}$
1	−0.33	2: $(2,3,1),(3,1,2)$	$\frac{2}{6}$

Example 3. Two judges rank four essays as follows:

	Essay			
Judge	1	2	3	4
$1, X$	3	4	2	1
$2, Y$	3	1	4	2

To test H_0 : rankings of the two judges are independent, let us arrange the rankings of the first judge from 1 to 4. Then we have:

Judge $1, X$:	1	2	3	4
Judge $2, Y$:	2	4	3	1

P = number of pairs of rankings for Judge 2 such that for $j > i$, $Y_j - Y_i > 0 = 2$ [the pairs $(2,4)$ and $(2,3)$], and

$$t = \frac{2 \cdot 2}{\binom{4}{2}} - 1 = -0.33.$$

Since

$$P_{H_0}\{|T| \geq 0.33\} = \frac{18}{24} = 0.75,$$

we cannot reject H_0.

For large n we can use an extension of Theorem 13.3.3 to bivariate case to conclude that $\sqrt{n}(U - \tau_c) \xrightarrow{L} \mathcal{N}(0, 4\zeta_1)$, where

$$\zeta_1 = \operatorname{cov}\{\psi((X_1,Y_1),(X_2,Y_2)), \psi((X_1,Y_1),(X_3,Y_3))\}.$$

Under H_0, it can be shown that

$$\frac{3\sqrt{n(n-1)}}{\sqrt{2(2n+5)}}T \xrightarrow{L} \mathcal{N}(0,1).$$

See, for example, Kendall [51], Randles and Wolfe [85], or Gibbons [35]. Approximation is good for $n \geq 8$.

13.5.3 Spearman's Rank Correlation Coefficient

Let $(X_1, Y_1), (X_2, Y_2), \ldots, (X_n, Y_n)$ be a sample from a bivariate population. In Section 6.3 we defined the sample correlation coefficient by

$$R = \frac{\sum_{i=1}^{n}(X_i - \overline{X})(Y_i - \overline{Y})}{\left\{\sum_{i=1}^{n}(X_i - \overline{X})^2 \sum_{i=1}^{n}(Y_i - \overline{Y})^2\right\}^{1/2}}, \tag{16}$$

where

$$\overline{X} = n^{-1}\sum_{i=1}^{n} X_i \qquad \text{and} \qquad \overline{Y} = n^{-1}\sum_{i=1}^{n} Y_i.$$

If the sample values $X_1, X_2, \ldots, X_n$ and $Y_1, Y_2, \ldots, Y_n$ are each ranked from 1 to n in increasing order of magnitude separately, and if the X's and Y's have continuous DFs, we get a unique set of rankings. The data will then reduce to n pairs of rankings. Let us write

$$R_i = \text{rank}(X_i) \qquad \text{and} \qquad S_i = \text{rank}(Y_i)$$

then R_i and $S_i \in \{1, 2, \ldots, n\}$. Also,

$$\sum_{1}^{n} R_i = \sum_{1}^{n} S_i = \frac{n(n+1)}{2}, \tag{17}$$

$$\overline{R} = n^{-1}\sum_{1}^{n} R_i = \frac{n+1}{2}, \qquad \overline{S} = n^{-1}\sum_{1}^{n} S_i = \frac{n+1}{2}, \tag{18}$$

and

$$\sum_{1}^{n}(R_i - \overline{R})^2 = \sum_{1}^{n}(S_i - \overline{S})^2 = \frac{n(n^2-1)}{12}. \tag{19}$$

Substituting in (16), we obtain

$$\begin{aligned} R &= \frac{12\sum_{i=1}^{n}(R_i - \overline{R})(S_i - \overline{S})}{n^3 - n} \\ &= \frac{12\sum_{1}^{n} R_i S_i}{n(n^2-1)} - \frac{3(n+1)}{n-1}. \end{aligned} \tag{20}$$

Writing $D_i = R_i - S_i = (R_i - \overline{R}) - (S_i - \overline{S})$, we have

$$\begin{aligned}\sum_{i=1}^{n} D_i^2 &= \sum_{i=1}^{n}(R_i - \overline{R})^2 + \sum_{i=1}^{n}(S_i - \overline{S})^2 - 2\sum_{i=1}^{n}(R_i - \overline{R})(S_i - \overline{S}) \\ &= \frac{1}{6}n(n^2-1) - 2\sum_{i=1}^{n}(R_i - \overline{R})(S_i - \overline{S}),\end{aligned}$$

and it follows that

$$R = 1 - \frac{6\sum_{i=1}^{n} D_i^2}{n(n^2-1)}. \tag{21}$$

The statistic R defined in (20) and (21) is called *Spearman's rank correlation coefficient* (see also Example 4.5.2).

From (20) we see that

$$\begin{aligned}ER &= \frac{12}{n(n^2-1)} E\left(\sum_{i=1}^{n} R_i S_i\right) - \frac{3(n+1)}{n-1} \\ &= \frac{12}{n^2-1} E(R_i S_i) - \frac{3(n+1)}{n-1}.\end{aligned} \tag{22}$$

Under H_0, the RVs X and Y are independent, so that the ranks R_i and S_i are also independent. It follows that

$$E_{H_0}(R_i S_i) = ER_i ES_i = \left(\frac{n+1}{2}\right)^2$$

and

$$E_{H_0} R = \frac{12}{n^2-1}\left(\frac{n+1}{2}\right)^2 - \frac{3(n+1)}{n-1} = 0. \tag{23}$$

Thus we should reject H_0 if the absolute value of R is large, that is, reject H_0 if

$$|R| > R_\alpha, \tag{24}$$

where $P_{H_0}\{|R| > R_\alpha\} \le \alpha$. To compute R_α we need the null distribution of R. For this purpose it is convenient to assume, without loss of generality, that $R_i = i$, $i = 1, 2, \ldots, n$. Then $D_i = i - S_i$, $i = 1, 2, \ldots, n$. Under H_0, X and Y being independent, the $n!$ pairs (i, S_i) of ranks are equally likely. It follows that

$$\begin{aligned}P_{H_0}\{R = r\} &= (n!)^{-1} \times (\text{number of pairs for which } R = r) \\ &= \frac{n_r}{n!}, \text{ say.}\end{aligned} \tag{25}$$

Note that $-1 \le R \le 1$, and the extreme values can occur only when either the rankings match, that is, $R_i = S_i$, in which case $R = 1$, or $R_i = n + 1 - S_i$, in which case $R = -1$.

Moreover, one need compute only one half of the distribution, since it is symmetric about 0 (Problem 7).

In the following example we will compute the distribution of R for $n = 3$ and 4. The exact complete distribution of $\sum_{i=1}^{n} D_i^2$, and hence R, for $n \le 10$ has been tabulated by Kendall [51]. Table ST13 gives the values of R_α for some selected values of n and α.

Example 4. Let us first enumerate the null distribution of R for $n = 3$. This is done in the following table:

(s_1, s_2, s_3)	$\sum_{i=1}^{n} is_i$	$r = \frac{12\sum_1^n is_i}{n(n^2-1)} - \frac{3(n+1)}{n-1}$
$(1,2,3)$	14	1.0
$(1,3,2)$	13	0.5
$(2,1,3)$	13	0.5

Thus

$$P_{H_0}\{R = r\} = \begin{cases} \frac{1}{6}, & r = 1.0, \\ \frac{2}{6}, & r = \ \ 0.5, \\ \frac{2}{6}, & r = -0.5, \\ \frac{1}{6}, & r = -1.0. \end{cases}$$

Similarly, for $n = 4$ we have the following:

(s_1, s_2, s_3, s_4)	$\sum_1^n is_i$	r	n_r	$P_{H_0}\{R = r\}$
$(1,2,3,4)$	30	1	1	$\frac{1}{24}$
$\begin{cases}(1,3,2,4),(2,1,3,4)\\(1,2,4,3)\end{cases}$	29	0.8	3	$\frac{3}{24}$
$(2,1,4,3)$	28	0.6	1	$\frac{1}{24}$
$\begin{cases}(1,3,4,2),(1,4,2,3),(2,3,1,4)\\(3,1,2,4)\end{cases}$	27	0.4	4	$\frac{4}{24}$
$(1,4,3,2),(3,2,1,4)$	26	0.2	2	$\frac{2}{24}$
	25	0.0	2	$\frac{2}{24}$

The last value is obtained from symmetry.

Example 5. In Example 3, we see that

$$r = \frac{12 \times 23}{4 \times 15} - \frac{3 \times 5}{3} = -0.4.$$

Since $P_{H_0}\{|R| \geq 0.4\} = 18/24 = 0.75$, we cannot reject H_0 at $\alpha = 0.05$ or $\alpha = 0.10$.

For large samples it is possible to use a normal approximation. It can be shown (see, e.g., Fraser [32, pp. 247–248]) that under H_0 the RV

$$Z = \left(12\sum_{i=1}^{n} R_i S_i - 3n^3\right) n^{-5/2}$$

or, equivalently,

$$Z = R\sqrt{n-1}$$

has approximately a standard normal distribution. The approximation is good for $n \geq 10$.

PROBLEMS 13.5

1. A sample of 240 men was classified according to characteristics A and B. Characteristic A was subdivided into four classes A_1, A_2, A_3, and A_4, while B was subdivided into three classes B_1, B_2, and B_3, with the following result:

	A_1	A_2	A_3	A_4	
B_1	12	25	32	11	80
B_2	17	18	22	23	80
B_3	21	17	16	26	80
	50	60	70	60	240

Is there evidence to support the theory that A and B are independent?

2. The following data represent the blood types and ethnic groups of a sample of Iraqi citizens:

	Blood Type			
Ethnic Group	O	A	B	AB
Kurd	531	450	293	226
Arab	174	150	133	36
Jew	42	26	26	8
Turkoman	47	49	22	10
Ossetian	50	59	26	15

Is there evidence to conclude that blood type is independent of ethnic group?

3. In a public opinion poll, a random sample of 500 American adults across the country was asked the following question: "Do you believe that there was a concerted

effort to cover up the Watergate scandal? Answer yes, no, or no opinion." The responses according to political beliefs were as follows:

Political	Response			
Affiliation	Yes	No	No Opinion	
Republican	45	75	30	150
Independent	85	45	20	150
Democrat	140	30	30	200
	270	150	80	500

Test the hypothesis that attitude toward the Watergate cover-up is independent of political party affiliation.

4. A random sample of 100 families in Bowling Green, Ohio, showed the following distribution of home ownership by family income:

	Annual Income (dollars)		
Residential Status	Less than 30,000	30,000–50,000	50,000 or Above
Home Owner	10	15	30
Renter	8	17	20

Is home ownership in Bowling Green independent of family income?

5. In a flower show the judges agreed that five exhibits were outstanding, and these were numbered arbitrarily from 1 to 5. Three judges each arranged these five exhibits in order of merit, giving the following rankings:

Judge A:	5	3	1	2	4
Judge B:	3	1	5	4	2
Judge C:	5	2	3	1	4

Compute the average values of Spearman's rank correlation coefficient R and Kendall's sample tau coefficient T from the three possible pairs of rankings.

6. For the bivariate normally distributed RV (X, Y) show that $\tau = 0$ if and only if X and Y are independent. [*Hint:* Show that $\tau = (2/\pi)\sin^{-1}\rho$, where ρ is the correlation coefficient between X and Y.]

7. Show that the distribution of Spearman's rank correlation coefficient R is symmetric about 0 under H_0.

8. In Problem 5 test the null hypothesis that rankings of judge A and judge C are independent. Use both Kendall's tau and Spearman's rank correlation tests.

9. A random sample of 12 couples showed the following distribution of heights:

	Height (in.)			Height (in.)	
Couple	Husband	Wife	Couple	Husband	Wife
1	80	72	7	74	68
2	70	60	8	71	71
3	73	76	9	63	61
4	72	62	10	64	65
5	62	63	11	68	66
6	65	46	12	67	67

(a) Compute T.

(b) Compute R.

(c) Test the hypothesis that the heights of husband and wife are independent, using T as well as R. In each case use the normal approximation.

13.6 SOME APPLICATIONS OF ORDER STATISTICS

In this section we consider some applications of order statistics. We are mainly interested in three applications, namely, tolerance intervals for distributions, coverages, and confidence interval estimates for quantiles and location parameters.

Definition 1. Let F be a continuous DF. A *tolerance interval for* F with *tolerance coefficient* γ is a random interval such that the probability is γ that this random interval covers at least a specific percentage $(100p)$ of the distribution.

Let $X_1, X_2, \ldots, X_n$ be a sample of size n from F, and let $X_{(1)}, X_{(2)}, \ldots, X_{(n)}$ be the corresponding set of order statistics. If the end points of the tolerance interval are two-order statistics $X_{(r)}, X_{(s)}$, $r < s$, we have

$$P\{P\{X_{(r)} < X < X_{(s)}\} \geq p\} = \gamma. \tag{1}$$

Since F is continuous, $F(X)$ is $U(0,1)$, and we have

$$\begin{aligned} P\{X_{(r)} < X < X_{(s)}\} &= P\{X < X_{(s)}\} - P\{X \leq X_{(r)}\} \\ &= F(X_{(s)}) - F(X_{(r)}) \\ &= U_{(s)} - U_{(r)}, \end{aligned} \tag{2}$$

where $U_{(r)}, U_{(s)}$ are the order statistics from $U(0,1)$. Thus (1) reduces to

$$P\{U_{(s)} - U_{(r)} \geq p\} = \gamma. \tag{3}$$

The statistic $V = U_{(s)} - U_{(r)}$, $1 \leq r < s \leq n$, is called the *coverage of the interval* $(X_{(r)}, X_{(s)})$. More precisely, the differences $V_k = F(X_{(k)}) - F(X_{(k-1)}) = U_{(k)} - U_{(k-1)}$, for $k = 1, 2, \ldots, n+1$, where $U_{(0)} = -\infty$ and $U_{(n+1)} = 1$, are called *elementary coverages.*

Since the joint PDF of $U_{(1)}, U_{(2)}, \ldots, U_{(n)}$ is given by

$$f(u_1, u_2, \ldots, u_n) = \begin{cases} n!, & 0 < u_1 < u_2 < \cdots < u_n, \\ 0, & \text{otherwise,} \end{cases}$$

the joint PDF of $V_1, V_2, \ldots, V_n$ is easily seen to be

$$h(v_1, v_2, \ldots, v_n) = \begin{cases} n!, & v_i \geq 0,\ i = 1, 2, \ldots, n,\ \sum_1^n v_i \leq 1 \\ 0, & \text{otherwise.} \end{cases} \tag{4}$$

Note that h is symmetric in its arguments. Consequently, V_i's are exchangeable RVs and the distribution of every sum of r, $r < n$, of these coverages is the same and, in particular, it is the distribution of $U_{(r)} = \sum_{j=1}^{r} V_j$, namely,

$$g_r(u) = \begin{cases} n\binom{n-1}{r-1} u^{r-1}(1-u)^{n-r}, & 0 < u < 1 \\ 0, & \text{otherwise.} \end{cases} \tag{5}$$

The common distribution of elementary coverages is

$$g_1(u) = n(1-u)^{n-1},\ 0 < u < 1,\ = 0,\ \text{otherwise.}$$

Thus $EV_i = 1/(n+1)$ and $\sum_{i=1}^{r} EV_i = r/(n+1)$. This may be interpreted as follows: The order statistics $X_{(1)}, X_{(2)}, \ldots, X_{(n)}$ partition the area under the PDF in $n+1$ parts such that each part has the same average (expected) area.

The sum of any r successive elementary coverages $V_{i+1}, V_{i+1}, \ldots, V_{i+r}$ is called an *r-coverage*. Clearly

$$\sum_{j=1}^{r} V_{i+j} = U_{(i+r)} - U_{(i)}, \quad i + r \leq n, \tag{6}$$

and, in particular, $U_{(s)} - U_{(r)} = \sum_{j=r+1}^{s} V_j$. Since V's are exchangeable it follows that

$$U_{(s)} - U_{(r)} \stackrel{d}{=} U_{(s-r)} \tag{7}$$

with PDF

$$g_{s-r}(u) = n\binom{n-1}{s-r-1} u^{s-r-1}(1-u)^{n-s+r}, \quad 0 < u < 1.$$

From (3), therefore,

$$\gamma = \int_p^1 g_{s-r}(u)du = \sum_{i=0}^{s-r-1} \binom{n}{i} p^i (1-p)^{n-i}, \tag{8}$$

where the last equality follows from (5.3.48). Given n, p, γ it may not always be possible to find $s-r$ to satisfy (8).

Example 1. Let $s=n$ and $r=1$. Then

$$\gamma=\sum_{i=0}^{n-2}\binom{n}{i}p^i(1-p)^{n-i}=1-p^n-np^{n-1}(1-p).$$

If $p=0.8$, $n=5$, $r=1$, then

$$\gamma=1-(0.8)^5-5(0.8)^4(0.2)=0.263.$$

Thus the interval $(X_{(1)},X_{(5)})$ in this case defines a 26 percent tolerance interval for 0.80 probability under the distribution (of X).

Example 2. Let X_1,X_2,X_3,X_4,X_5 be a sample from a continuous DF F. Let us find r and s, $r<s$, such that $(X_{(r)},X_{(s)})$ is a 90 percent tolerance interval for 0.50 probability under F. We have

$$0.90=P\left\{U\geq\frac{1}{2}\right\}=\sum_{i=0}^{s-r-1}\binom{5}{i}\left(\frac{1}{2}\right)^5.$$

It follows that, if we choose $s-r=4$, then $\gamma=0.81$; and if we choose $s-r=5$, then $\gamma=0.969$. In this case, we must settle for an interval with tolerance coefficient 0.969, exceeding the desired value 0.90.

In general, given p, $0<p<1$, it is possible to choose a sufficiently large sample of size n and a corresponding value of $s-r$ such that with probability $\geq\gamma$ an interval of the form $(X_{(r)},X_{(s)})$ covers at least $100p$ percent of the distribution. If $s-r$ is specified as a function of n, one chooses the smallest sample size n.

Example 3. Let $p=\frac{3}{4}$ and $\gamma=0.75$. Suppose that we want to choose the smallest sample size required such that $(X_{(2)},X_{(n)})$ covers at least 75 percent of the distribution. Thus we want the smallest n to satisfy

$$0.75\leq\sum_{i=0}^{n-3}\binom{n}{i}\left(\frac{3}{4}\right)^i\left(\frac{1}{4}\right)^{n-i}.$$

From Table ST1 of binomial distributions we see that $n=14$.

We next consider the use of order statistics in constructing confidence intervals for population quantiles. Let X be an RV with a continuous DF F, $0<p<1$. Then the quantile of order p satisfies

$$F(\mathfrak{z}_p)=p. \tag{9}$$

Let $X_1, X_2, \ldots, X_n$ be n independent observations on X. Then the number of X_i's $< \mathfrak{z}_p$ is an RV that has a binomial distribution with parameters n and p. Similarly, the number of X_i's that are at least $\mathfrak{z}_p$ has a binomial distribution with parameters n and $1-p$.

Let $X_{(1)}, X_{(2)}, \ldots, X_{(n)}$ be the set of order statistics for the sample. Then

$$P\{X_{(r)} \le \mathfrak{z}_p\} = P\{\text{At least } r \text{ of the } X_i\text{'s} \le \mathfrak{z}_p\}$$
$$= \sum_{i=r}^{n} \binom{n}{i} p^i (1-p)^{n-i}. \tag{10}$$

Similarly

$$P\{X_{(s)} \ge \mathfrak{z}_p\} = P\{\text{At least } n-s+1 \text{ of the } X_i\text{'s} \ge \mathfrak{z}_p\}$$
$$= P\{\text{At most } s-1 \text{ of the } X_i\text{'s} < \mathfrak{z}_p\}$$
$$= \sum_{i=0}^{s-1} \binom{n}{i} p^i (1-p)^{n-i}. \tag{11}$$

It follows from (10) and (11) that

$$P\{X_{(r)} \le \mathfrak{z}_p \le X_{(s)}\} = P\{X_{(s)} \ge \mathfrak{z}_p\} - P\{X_{(r)} > \mathfrak{z}_p\}$$
$$= P\{X_{(r)} \le \mathfrak{z}_p\} - 1 + P\{X_{(s)} \ge \mathfrak{z}_p\}$$
$$= \sum_{i=r}^{n} \binom{n}{i} p^i (1-p)^{n-i} + \sum_{i=0}^{s-1} \binom{n}{i} p^i (1-p)^{n-i} - 1$$
$$= \sum_{i=r}^{s-1} \binom{n}{i} p^i (1-p)^{n-i}. \tag{12}$$

It is easy to determine a confidence interval for $\mathfrak{z}_p$ from (12), once the confidence level is given. In practice, one determines r and s such that $s-r$ is as small as possible, subject to the condition that the level is $1-\alpha$.

Example 4. Suppose that we want a confidence interval for the median ($p = \frac{1}{2}$), based on a sample of size 7 with confidence level 0.90. It suffices to find r and s, $r < s$, such that

$$\sum_{i=r}^{s-1} \binom{7}{i} \left(\frac{1}{2}\right)^7 \ge 0.90.$$

By trial and error, using the probability distribution $b(7, \frac{1}{2})$ we see that we can choose $s = 7, r = 2$ or $r = 1, s = 6$; in either case $s - r$ is minimum $(= 5)$, and the confidence level is at least 0.92.

Example 5. Let us compute the number of observations required for $(X_{(1)}, X_{(n)})$ to be a 0.95 level confidence interval for the median, that is, we want to find n such that

$$P\{X_{(1)} \le \mathfrak{z}_{1/2} \le X_{(n)}\} \ge 0.95.$$

It suffices to find n such that

$$\sum_{i=1}^{n-1} \binom{n}{i} \left(\frac{1}{2}\right)^n \geq 0.95.$$

It follows from Table ST1 that $n = 6$.

Finally we consider applications of order statistics to constructing confidence intervals for a location parameter. For this purpose we will use the method of test inversion discussed in Chapter 11. We first consider confidence estimation based on the sign test of location.

Let $X_1, X_2, \ldots, X_n$ be a random sample from a symmetric, continuous DF $F(x-\theta)$ and suppose we wish to find a confidence interval for θ. Let $R^+(\mathbf{X}-\theta_0) = \#$ of X_i's $> \theta_0$, be the sign-test statistic for testing $H_0 : \theta = \theta_0$ against $H_1 : \theta \neq \theta_0$. Clearly, $R^+(\mathbf{X}-\theta_0) \sim b(n, 1/2)$ under H_0. The sign-test rejects H_0 if

$$\min\{R^+(\mathbf{X}-\theta_0),\ R^+(\theta_0-\mathbf{X})\} \leq c \tag{13}$$

for some integer c to be determined from the level of the test. Let $r = c+1$. Then any value of θ is acceptable provided it is greater than the rth smallest observation and smaller than the rth largest observation, giving as confidence interval

$$X_{(r)} < \theta < X_{(n+1-r)}. \tag{14}$$

If we want level $1-\alpha$ to be associated with (14), we choose c so that the level of test (13) is α.

Example 6. The following 12 observations come from a symmetric, continuous DF $F(x-\theta)$:

$$-223, -380, -94, -179, 194, 25, -177, -274, -496, -507, -20, 122.$$

We wish to obtain a 95% confidence interval for θ. Sign test rejects H_0 if $R^+(\mathbf{X}) \geq 9$ or ≤ 2 at level 0.05. Thus

$$P\{3 < R^+(\mathbf{X}-\theta) < 10\} = 1 - 2(0.0193) = 0.9614 \geq 0.95.$$

It follows that a 95% confidence interval for θ is given by $(X_{(3)}, X_{(10)})$ or $(-380, 25)$.

We next consider the Wilcoxon signed-ranks test of $H_0 : \theta = \theta_0$ to construct a confidence interval for θ. The test statistic in this case is $T^+ =$ sum of ranks of positive $(X_i - \theta_0)$'s in the ordered $|X_i - \theta_0|$'s. From (13.3.4)

$$\begin{aligned} T^+ &= \sum_{1 \leq i \leq j \leq n} I_{[X_i + X_j > 2\theta_0]} \\ &= \text{number of } \frac{X_i + X_j}{2} > \theta_0. \end{aligned}$$

Let $T_{ij} = (X_i + X_j)/2$, $1 \le i \le j \le n$ and order the $N = \binom{n+1}{2}$ T_{ij}'s in increasing order of magnitude

$$T_{(1)} < T_{(2)} < \cdots < T_{(N)}.$$

Then using the argument that converts (13) to (14) we see that a confidence interval for θ is given by

$$T_{(r)} < \theta < T_{(N+1-r)}. \tag{15}$$

Critical values c are taken from Table ST10.

Example 7. For the data in Example 6, the Wilcoxon signed-rank test rejects $H_0 : \theta = \theta_0$ at level 0.05 if $T^+ > 64$ or $T^+ < 14$. Thus

$$P\{14 \le T^+(\mathbf{X} - \theta_0) \le 64\} \ge 0.95.$$

It follows that a 95% confidence interval for θ is given by $[T_{(14)}, T_{(64)}] = [-336.5, -20]$.

PROBLEMS 13.6

1. Find the smallest values of n such that the intervals (a) $(X_{(1)}, X_{(n)})$ and (b) $(X_{(2)}, X_{(n-1)})$ contain the median with probability ≥ 0.90.
2. Find the smallest sample size required such that $(X_{(1)}, X_{(n)})$ covers at least 90 percent of the distribution with probability ≥ 0.98.
3. Find the relation between n and p such that $(X_{(1)}, X_{(n)})$ covers at least $100\,p$ percent of the distribution with probability $\ge 1 - p$.
4. Given γ, δ, p_0, p_1 with $p_1 > p_0$, find the smallest n such that

$$P\{F(X_{(s)}) - F(X_{(r)}) \ge p_0\} \ge \gamma$$

and

$$P\{F(X_{(s)}) - F(X_{(r)}) \ge p_1\} \le \delta.$$

Find also $s - r$.
[*Hint:* Use the normal approximation to the binomial distribution.]

5. In Problem 4 find the smallest n and the associated value of $s - r$ if $\gamma = 0.95$, $\delta = 0.10$, $p_1 = 0.75$, $p_0 = 0.50$.
6. Let $X_1, X_2, \ldots, X_7$ be a random sample from a continuous DF F. Compute:
   (a) $P(X_{(1)} < \mathfrak{z}_{.5} < X_{(7)})$.
   (b) $P(X_{(2)} < \mathfrak{z}_{.3} < X_{(5)})$.
   (c) $P(X_{(3)} < \mathfrak{z}_{.8} < X_{(6)})$.
7. Let $X_1, X_2, \ldots, X_n$ be iid with common continuous DF F.

(a) What is the distribution of

$$F((X_{n-1}) - F(X_{(j)}) + F(X_{(i)}) - F(X_{(2)})$$

for $2 \leq i < j \leq n-1$?

(b) What is the distribution of $[F(X_{(n)}) - F(X_{(2)})]/[F(X_{(n)}) - F(X_{(1)})]$.

13.7 ROBUSTNESS

Most of the statistical inference problems treated in this book are parametric in nature. We have assumed that the functional form of the distribution being sampled is known except for a finite number of parameters. It is to be expected that any estimator or test of hypothesis concerning the unknown parameter constructed on this assumption will perform better than the corresponding nonparametric procedure, provided that the underlying assumptions are satisfied. It is therefore of interest to know how well the parametric optimal tests or estimators constructed for one population perform when the basic assumptions are modified. If we can construct tests or estimators that perform well for a variety of distributions, for example, there would be little point in using the corresponding nonparametric method unless the assumptions are seriously violated.

In practice, one makes many assumptions in parametric inference, and any one or all of these may be violated. Thus one seldom has accurate knowledge about the true underlying distribution. Similarly, the assumption of mutual independence or even identical distribution may not hold. Any test or estimator that performs well under modifications of underlying assumptions is usually referred to as *robust*.

In this section we will first consider the effect that slight variation in model assumptions have on some common parametric estimators and tests of hypotheses. Next we will consider some corresponding nonparametric competitors and show that they are quite robust.

13.7.1 Effect of Deviations from Model Assumptions on Some Parametric Procedures

Let us first consider the effect of contamination on sample mean as an estimator of the population mean.

The most commonly used estimator of the population mean μ is the sample mean $\overline{X}$. It has the property of unbiasedness for all populations with finite mean. For many parent populations (normal, Poisson, Bernoulli, gamma, etc.) it is a complete sufficient statistic and hence a UMVUE. Moreover, it is consistent and has asymptotic normal distribution whenever the conditions of the central limit theorem are satisfied. Nevertheless, the sample mean is affected by extreme observations, and a single observation that is either too large or too small may make $\overline{X}$ worthless as an estimator of μ. Suppose, for example, that $X_1, X_2, \ldots, X_n$ is a sample from some normal population. Occasionally something happens to the system, and a wild observation is obtained that is, suppose one is sampling from $\mathcal{N}(\mu, \sigma^2)$, say, 100α percent of the time and from $\mathcal{N}(\mu, k\sigma^2)$, where $k > 1$, $(1-\alpha)100$

percent of the time. Here both μ and σ^2 are unknown, and one wishes to estimate μ. In this case one is really sampling from the density function

$$f(x) = \alpha f_0(x) + (1-\alpha) f_1(x), \tag{1}$$

where f_0 is the PDF of $\mathcal{N}(\mu, \sigma^2)$, and f_1, the PDF of $\mathcal{N}(\mu, k\sigma^2)$. Clearly,

$$\overline{X} = \frac{\sum_1^n X_i}{n} \tag{2}$$

is still unbiased for μ. If α is nearly 1, there is no problem since the underlying distribution is nearly $\mathcal{N}(\mu, \sigma^2)$, and $\overline{X}$ is nearly the UMVUE of μ with variance σ^2/n. If $1-\alpha$ is large (that is, not nearly 0), then, since one is sampling from f, the variance of X_1 is σ^2 with probability α and is $k\sigma^2$ with probability $1-\alpha$, and we have

$$\text{var}_\sigma(\overline{X}) = \frac{1}{n}\text{var}(X_1) = \frac{\sigma^2}{n}[\alpha + (1-\alpha)k]. \tag{3}$$

If $k(1-\alpha)$ is large, $\text{var}_\sigma(\overline{X})$ is large and we see that even an occasional wild observation makes $\overline{X}$ subject to a sizable error. The presence of an occasional observation from $\mathcal{N}(\mu, k\sigma^2)$ is frequently referred to as *contamination*. The problem is that we do not know, in practice, the distribution of the wild observations and hence we do not know the PDF f. It is known that the sample median is a much better estimator than the mean in the presence of extreme values. In the contamination model discussed above, if we use $Z_{1/2}$, the sample median of the X_i's, as an estimator of μ (which is the population median), then for large n

$$E(Z_{1/2} - \mu)^2 = \text{var}(Z_{1/2}) \approx \frac{1}{4n}\frac{1}{[f(\mu)]^2}. \tag{4}$$

(See Theorem 7.5.2 and Remark 7.5.7.) Since

$$\begin{aligned} f(\mu) &= \alpha f_0(\mu) + (1-\alpha) f_1(\mu) \\ &= \frac{\alpha}{\sigma\sqrt{2\pi}} + (1-\alpha)\frac{1}{\sigma\sqrt{2\pi k}} = \left(\alpha + \frac{1-\alpha}{\sqrt{k}}\right)\frac{1}{\sigma\sqrt{2\pi}}, \end{aligned}$$

we have

$$\text{var}(Z_{1/2}) \approx \frac{\pi\sigma^2}{2n}\frac{1}{\{\alpha + [(1-\alpha)/\sqrt{k}]\}^2}. \tag{5}$$

As $k \to \infty$, $\text{var}(Z_{1/2}) \approx \pi\sigma^2/(2n\alpha^2)$. If there is no contamination, $\alpha = 1$ and $\text{var}(Z_{1/2}) \approx \pi\sigma^2/2n$. Also,

$$\frac{\pi\sigma^2/2n\alpha^2}{\pi\sigma^2/2n} = \frac{1}{\alpha^2},$$

which will be close to 1 if α is close to 1. Thus the estimator $Z_{1/2}$ will not be greatly affected by how large k is, that is, how wild the observations are. We have

$$\frac{\mathrm{var}(\overline{X})}{\mathrm{var}(Z_{1/2})} = \frac{2}{\pi}[\alpha+(1-\alpha)k]\left[\alpha+\frac{(1-\alpha)}{\sqrt{k}}\right]^2 \to \infty \qquad \text{as } k\to\infty.$$

Indeed, $\mathrm{var}(\overline{X}) \to \infty$ as $k \to \infty$, whereas $\mathrm{var}(Z_{1/2}) \to \pi\sigma^2/(2n\alpha^2)$ as $k \to \infty$. One can check that, when $k = 9$ and $\alpha \approx 0.915$, the two variances are (approximately) equal. As k becomes larger than 9 or α smaller than 0.915, $Z_{1/2}$ becomes a better estimator of μ than $\overline{X}$.

There are other flaws as well. Suppose, for example, that $X_1, X_2, \ldots, X_n$ is a sample from $U(0,\theta)$, $\theta > 0$. Then both $\overline{X}$ and $T(\mathbf{X}) = (X_{(1)} + X_{(n)})/2$, where $X_{(1)} = \min(X_1,\ldots,X_n)$, $X_{(n)} = \max(X_1,\ldots,X_n)$, are unbiased for $EX = \theta/2$. Also, $\mathrm{var}_\theta(\overline{X}) = \mathrm{var}(X)/n = \theta^2/[12n]$, and one can show that $\mathrm{var}(T) = \theta^2/[2(n+1)(n+2)]$. It follows that the efficiency of $\overline{X}$ relative to that of T is

$$\mathrm{eff}_\theta(\overline{X} \mid T) = \frac{\mathrm{var}_\theta(T)}{\mathrm{var}_\theta(\overline{X})} = \frac{6n}{(n+1)(n+2)} < 1 \qquad \text{if } n > 2.$$

In fact, $\mathrm{eff}_\theta(\overline{X} \mid T) \to 0$ as $n \to \infty$, so that in sampling from a uniform parent $\overline{X}$ is much worse than T, even for moderately large values of n.

Let us next turn our attention to the estimation of standard deviation. Let $X_1, X_2, \ldots, X_n$ be a sample from $\mathcal{N}(\mu,\sigma^2)$. Then the MLE of σ is

$$\hat{\sigma} = \left\{\sum_{i=1}^{n}\frac{(X_i-\overline{X})^2}{n}\right\}^{1/2} = \left(\frac{n-1}{n}\right)^{1/2} S. \tag{6}$$

Note that the lower bound for the variance of any unbiased estimator for σ is $\sigma^2/2n$. Although $\hat{\sigma}$ is not unbiased, the estimator

$$S_1 = \sqrt{\frac{n}{2}}\frac{\Gamma[(n-1)/2]}{\Gamma(n/2)}\hat{\sigma} = \sqrt{\frac{n-1}{2}}\frac{\Gamma[(n-1)/2]}{\Gamma(n/2)}S \tag{7}$$

is unbiased for σ. Also,

$$\begin{aligned}\mathrm{var}(S_1) &= \sigma^2\left\{\frac{n-1}{2}\left(\frac{\Gamma[(n-1)/2]}{\Gamma(n/2)}\right)^2 - 1\right\}\\ &= \frac{\sigma^2}{2n} + O\left(\frac{1}{n^2}\right).\end{aligned} \tag{8}$$

Thus the efficiency of S_1 (relative to the estimator with least variance $= \sigma^2/2n$) is

$$\frac{\sigma^2/2n}{\mathrm{var}(S_1)} = \frac{1}{1+\sigma^2 O\left(\frac{2}{n}\right)} < 1$$

and $\to 1$ as $n \to \infty$. For small n, the efficiency of S_1 is considerably smaller than 1. Thus, for $n = 2$, $\text{eff}(S_1) = 1/[2(\pi - 2)] = 0.438$ and, for $n = 3$, $\text{eff}(S_1) = \pi/[6(4-\pi)] = 0.61$.

Yet another estimator of σ is the sample mean deviation

$$S_2 = \frac{1}{n}\sum_{i=1}^{n} |X_i - \overline{X}|. \tag{9}$$

Note that

$$E\left\{\sqrt{\frac{\pi}{2}}\frac{1}{n}\sum_{i=1}^{n}|X_i - \mu|\right\} = \sqrt{\frac{\pi}{2}}E|X_i - \mu| = \sigma,$$

and

$$\text{var}\left\{\sqrt{\frac{\pi}{2}}\frac{1}{n}\sum_{i=1}^{n}|X_i - \mu|\right\} = \frac{\pi - 2}{2n}\sigma^2. \tag{10}$$

If n is large enough so that $\overline{X} \approx \mu$, we see that $S_3 = \sqrt{(\pi/2)}\,S_2$ is nearly unbiased for σ with variance $[(\pi-2)/2n]\sigma^2$. The efficiency of S_3 is

$$\frac{\sigma^2(2n)}{\sigma^2[(\pi-2)/(2n)]} = \frac{1}{\pi - 2} < 1.$$

For large n, the efficiency of S_1 relative to S_3 is

$$\frac{\text{var}(S_3)}{\text{var}(S_1)} = \frac{[(\pi-2)/(2n)]\sigma^2}{\sigma^2/(2n) + O(1/n^2)} = \pi - 2 + \frac{\pi - 2}{O(2/n)} > 1.$$

Now suppose that there is some contamination. As before, let us suppose that for a proportion α of the time we sample from $\mathcal{N}(\mu, \sigma^2)$ and for a proportion $1-\alpha$ of the time we get a wild observation from $\mathcal{N}(\mu, k\sigma^2)$, $k > 1$. Assuming that both μ and σ^2 are unknown, suppose that we wish to estimate σ. In the notation used above, let

$$f(x) = \alpha f_0(x) + (1-\alpha) f_1(x),$$

where f_0 is the PDF of $\mathcal{N}(\mu, \sigma^2)$, and f_1, the PDF of $\mathcal{N}(\mu, k\sigma^2)$. Let us see how even small contamination can make the maximum likelihood estimate $\hat{\sigma}$ of σ quite useless.

If $\hat{\theta}$ is the MLE of θ, and φ is a function of θ, then $\varphi(\hat{\theta})$ is the MLE of $\varphi(\theta)$. In view of (7.5.7) we get

$$E(\hat{\sigma} - \sigma)^2 \approx \frac{1}{4\sigma^2}E(\hat{\sigma}^2 - \sigma^2)^2. \tag{11}$$

Using Theorem 7.3.5, we see that

$$E(\hat{\sigma}^2 - \sigma^2)^2 \approx \frac{\mu_4 - \mu_2^2}{n} \tag{12}$$

(dropping the other two terms with n^2 and n^3 in the denominator), so that

$$E(\hat{\sigma}-\sigma)^2 \approx \frac{1}{4\sigma^2 n}(\mu_4-\mu_2^2). \tag{13}$$

For the density f, we see that

$$\mu_4 = 3\sigma^4[\alpha+k^2(1-\alpha)] \tag{14}$$

and

$$\mu_2 = \sigma^2[\alpha+k(1-\alpha)]. \tag{15}$$

It follows that

$$E\{\hat{\sigma}-\sigma\}^2 \approx \frac{\sigma^2}{4n}\left\{3[\alpha+k^2(1-\alpha)]-[\alpha+k(1-\alpha)]^2\right\}. \tag{16}$$

If we are interested in the effect of very small contamination, $\alpha \approx 1$ and $1-\alpha \approx 0$. Assuming that $k(1-\alpha)\approx 0$, we see that

$$\begin{aligned} E\{\hat{\sigma}-\sigma\}^2 &\approx \frac{\sigma^2}{4n}\{3[1+k^2(1-\alpha)]-1\} \\ &= \frac{\sigma^2}{2n}\left[1+\tfrac{3}{2}k^2(1-\alpha)\right]. \end{aligned} \tag{17}$$

In the normal case, $\mu_4 = 3\sigma^4$ and $\mu_2^2=\sigma^4$, so that from (11)

$$E\{\hat{\sigma}-\sigma\}^2 \approx \frac{\sigma^2}{2n}.$$

Thus we see that the mean square error due to a small contamination is now multiplied by a factor $[1+\frac{3}{2}k^2(1-\alpha)]$. If, for example, $k=10$, $\alpha=0.99$, then $1+\frac{3}{2}k^2(1-\alpha)=\frac{5}{2}$. If $k=10$, $\alpha=0.98$, then $1+\frac{3}{2}k^2(1-\alpha)=4$, and so on.

A quick comparison with S_3 shows that, although S_1 (or even $\hat{\sigma}$) is a better estimator of σ than S_3 if there is no contamination, S_3 becomes a much better estimator in the presence of contamination as k becomes large.

Next we consider the effect of deviation from model assumptions on tests of hypotheses. One of the most commonly used tests in statistics is Student's t-test for testing the mean of a normal population when the variance is unknown. Let $X_1,X_2,\ldots,X_n$ be a sample from some population with mean μ and finite variance σ^2. As usual, let $\overline{X}$ denote the sample mean, and S^2, the sample variance. If the population being sampled is normal, the t-test rejects $H_0\colon \mu=\mu_0$ against $H_1:\mu\neq\mu_0$ at level α if $|\overline{x}-\mu_0| > t_{n-1,\alpha/2}(s/\sqrt{n})$. If n is large, we replace $t_{n-1,\alpha/2}$ by the corresponding critical value, $z_{\alpha/2}$, under the standard normal law. If the sample does not come from a normal population, the statistic $T=[(\overline{X}-\mu_0)/S]\sqrt{n}$ is no longer distributed as a $t(n-1)$ statistic. If, however, n is sufficiently large, we know that T has an asymptotic normal distribution irrespective of the population being sampled, as long as it has a finite variance. Thus, for large n, the distribution of T is independent of the form of the population, and the t-test is stable. The same considerations apply to testing the difference between two means when the two variances are equal. Although we assumed that n is sufficiently large for Slutsky's result (Theorem 7.2.15) to hold, empirical investigations have shown that the test based on Student's statistic is robust. Thus a significant value of t may not be interpreted to mean a departure from normality of the observations. Let us next consider the effect of departure from independence on the t-distribution. Suppose that the observations $X_1, X_2, \ldots, X_n$ have a multivariate normal distribution with $EX_i = \mu$, $\text{var}(X_i) = \sigma^2$, and ρ as the common correlation coefficient between any X_i and X_j, $i \neq j$. Then

$$E\overline{X} = \mu \quad \text{and} \quad \text{var}(\overline{X}) = \frac{\sigma^2}{n}[1 + (n-1)\rho], \tag{18}$$

and since X_i's are exchangeable it follows from Remark 6.3.1 that

$$ES^2 = \sigma^2(1-\rho). \tag{19}$$

For large n, the statistic $\sqrt{n}(\overline{X} - \mu_0)/S$ will be asymptotically distributed as $\mathcal{N}(0, 1 + n\rho/(1-\rho))$, instead of $\mathcal{N}(0,1)$. Under H_0, $\rho = 0$ and $T^2 = n(\overline{X} - \mu_0)^2/S^2$ is distributed as $F(1, n-1)$. Consider the ratio

$$\frac{nE(\overline{X} - \mu_0)^2}{ES^2} = \frac{\sigma^2[1 + (n-1)\rho]}{\sigma^2(1-\rho)} = 1 + \frac{n\rho}{1-\rho}. \tag{20}$$

The ratio equals 1 if $\rho = 0$ but is > 0 for $\rho > 0$ and $\to \infty$ as $\rho \to 1$. It follows that a large value of T is likely to occur when $\rho > 0$ and is large, even though μ_0 is the true value of the mean. Thus a significant value of t may be due to departure from independence, and the effect can be serious.

Next, consider a test of the null hypothesis $H_0 \colon \sigma = \sigma_0$ against $H_1 : \sigma \neq \sigma_0$. Under the usual normality assumptions on the observations $X_1, X_2, \ldots, X_n$, the test statistic used is

$$V = \frac{(n-1)S^2}{\sigma^2} = \frac{\sum_{i=1}^{n}(X_i - \overline{X})^2}{\sigma^2}, \tag{21}$$

which has a $\chi^2(n-1)$ distribution under H_0. The usual test is to reject H_0 if

$$V_0 = \frac{(n-1)S^2}{\sigma_0^2} > \chi^2_{n-1,\alpha/2} \quad \text{or} \quad V_0 < \chi^2_{n-1,1-\alpha/2}. \tag{22}$$

Let us suppose that $X_1, X_2, \ldots, X_n$ are not normal. It follows from Corollary 2 of Theorem 7.3.4 that

$$\text{var}(S^2) = \frac{\mu_4}{n} + \frac{3-n}{n(n-1)}\mu_2^2, \tag{23}$$

so that

$$\text{var}\left(\frac{S^2}{\sigma^2}\right) = \frac{1}{n}\frac{\mu_4}{\sigma^4} + \frac{3-n}{n(n-1)}. \tag{24}$$

Writing $\gamma_2 = (\mu_4/\sigma^4) - 3$, we have

$$\text{var}\left(\frac{S^2}{\sigma^2}\right) = \frac{\gamma_2}{n} + \frac{2}{n-1} \tag{25}$$

when the X_i's are not normal, and

$$\text{var}\left(\frac{S^2}{\sigma^2}\right) = \frac{2}{n-1} \tag{26}$$

when the X_i's are normal ($\gamma_2 = 0$). Now $(n-1)S^2 = \sum_{i=1}^{n}(X_i - \overline{X})^2$ is the sum of n identically distributed but dependent RVs $(X_j - \overline{X})^2, j = 1, 2, \ldots, n$. Using a version of the central limit theorem for dependent RVs (see, e.g., Cramér [17, p. 365]), it follows that

$$\left(\frac{n-1}{2}\right)^{-1/2}\left(\frac{S^2}{\sigma^2} - 1\right),$$

under H_0, is asymptotically $\mathcal{N}(0, 1 + (\gamma_2/2))$, and not $\mathcal{N}(0,1)$ as under the normal theory. As a result the size of the test based on the statistic V_0 will be different from the stated level of significance if γ_2 differs greatly from 0. It is clear that the effect of violation of the normality assumption can be quite serious on inferences about variances, and the chi-square test is not robust.

In the above discussion we have used somewhat crude calculations to investigate the behavior of the most commonly used estimators and test statistics when one or more of the underlying assumptions are violated. Our purpose here was to indicate that some tests or estimators are robust whereas others are not. The moral is clear: One should check carefully to see that the underlying assumptions are satisfied before using parametric procedures.

13.7.2 Some Robust Procedures

Let $X_1, X_2, \ldots, X_n$ be a random sample from a continuous PDF $f(x - \theta)$, $\theta \in \mathcal{R}$ and assume that f is symmetric about θ. We shall be interested in estimation or tests of hypotheses concerning θ. Our objective is to find procedures that perform well for several different types of distributions but do not have to be optimal for any particular distribution. We will call such procedures *robust*. We first consider estimation of θ.

The estimators fall under one of the following three types:

1. Estimators that are functions of $\mathbf{R} = (R_1, R_2, \ldots, R_n)$, where R_j is the rank of X_j, are known as *R-estimators*. Hodges and Lehmann [44] devised a method of deriving such estimators from rank tests. These include the sample median $\tilde{X}$ (based on the sign test) and $W = \text{med}\{(X_i + X_j)/2,\ 1 \le i \le j \le n\}$ based on the Wilcoxon signed-rank test.
2. Estimators of the form $\sum_{i=1}^{n} a_i X_{(i)}$ are called *L-estimators*, being linear combinations of order statistics. This class includes the median, the mean, and the trimmed mean obtained by dropping a prespecified proportion of extreme observations.

3. Maximum likelihood type estimators obtained as solutions to certain equations $\sum_{j=1}^{n} \psi(X_j - \theta) = 0$ are called *M-estimators*. The function $\psi(t) = -f'(t)/f(t)$ gives MLEs.

Definition 1. Let $k = [n\alpha]$ be the largest integer $\leq n\alpha$ where $0 < \alpha < 1/2$. Then the estimator

$$\overline{X}_\alpha = \sum_{j=k+1}^{n-k} \frac{X_{(j)}}{n-2k} \tag{27}$$

is called a *trimmed-mean*.

Two extreme examples of trimmed means are the sample mean $\overline{X}(\alpha = 0)$ and the median $\tilde{X}$ when all except the central (n odd) or the two central (n even) observations are excluded.

Example 1. Consider the following sample of size 15 taken from a symmetric distribution.

0.97	0.66	0.73	0.78	1.30	0.58	0.79	0.94
0.52	0.52	0.83	1.25	1.47	0.96	0.71	

Suppose $\alpha = 0.10$. Then $k = [n\alpha] = 1$ and

$$\overline{x}_{0.10} = \frac{\sum_{j=2}^{14} x_{(j)}}{15-2} = 0.85.$$

Here $\overline{x} = 0.867$, $\underset{1 \leq j \leq 15}{\text{med}}\, x_j = x_{(8)} = 0.79$.

We will limit this discussion to four estimators of location, namely, the sample median, trimmed mean, sample mean, and Hodges–Lehmann type estimator based on Wilcoxon signed-rank test. In order to compare the performance of two procedures A and B we will use a (large sample) measure of relative efficiency due to Pitman. Pitman's *asymptotic relative efficiency* (ARE) of procedure B relative to procedure A is the limit of the ratio of sample sizes n_A/n_B, where n_A, n_B are sample sizes needed for procedures A and B to perform equivalently with respect to a specified criterion. For example, suppose $\{T_{n(A)}\}$ and $\{T_{n(B)}\}$ are two sequences of estimators for $\psi(\theta)$ such that

$$T_{n(A)} \sim \text{AN}\left(\psi(\theta), \frac{\sigma_A^2(\theta)}{n(A)}\right),$$

and

$$T_{n(B)} \sim \text{AN}\left(\psi(\theta), \frac{\sigma_B^2(\theta)}{n(B)}\right).$$

Suppose further that A and B perform equivalently if their asymptotic variances are the same, that is,

$$\frac{\sigma_A^2(\theta)}{n(A)} \approx \frac{\sigma_B^2(\theta)}{n(B)}.$$

Then

$$\frac{n(A)}{n(B)} \longrightarrow \frac{\sigma_A^2(\theta)}{\sigma_B^2(\theta)}.$$

Clearly, different performance measures may lead to different measures of ARE.

Similarly if procedures A and B lead to two sequences of tests, then ARE is the limiting ratio of the sample sizes needed by the tests to reach a certain power β_0 against the same alternative and at the same limiting level α.

Accordingly, let $e(B,A)$ denote the ARE of B relative to A. If $e(B,A) = 1/2$ say, then procedure A requires (approximately) half as many observations as procedure B. We will write $e_F(B,A)$, whenever necessary to indicate the dependence of ARE on the underlying DF F.

For detailed discussion of Pitman efficiency we refer to Lehmann [61, pp. 371–380], Lehmann [63, section 5.2], Serfling [102, chapter 10], Randles and Wolfe [85, chapter 5], and Zacks [121]. The expressions for AREs of median and the Hodges-Lehmann estimators of location parameter θ with respect to the sample mean $\overline{X}$ are

$$e_F(\tilde{X},\overline{X}) = 4\sigma_F^2 f(0), \tag{28}$$

$$e_F(W,\overline{X}) = 12\sigma_F^2 \left[\int_{-\infty}^{\infty} f^2(x)dx\right]^2, \tag{29}$$

where f is the PDF corresponding to F. In order to get $e_F(\tilde{X}, W)$ we use the fact that

$$\begin{aligned} e_F(\tilde{X}, W) &= \frac{e_F(\tilde{X},\overline{X})}{e_F(W,\overline{X})} \\ &= \frac{f(0)}{3\left[\int_{-\infty}^{\infty} f^2(x)dx\right]^2}. \end{aligned} \tag{30}$$

Bickel [5] showed that

$$e_F(\overline{X}_\alpha,\overline{X}) = \frac{\sigma_F^2}{\sigma_\alpha^2}, \tag{31}$$

where

$$\sigma_\alpha^2 = \frac{2}{(1-2\alpha)^2}\left[\int_0^{\mathfrak{z}_{1-\alpha}} t^2 f(t)dt + \alpha \mathfrak{z}_{1-\alpha}\right] \tag{32}$$

and $\mathfrak{z}_\alpha$ is the unique αth percentile of F. It is clear from (32) that no closed form expression for $e_F(\overline{X}_\alpha,\overline{X})$ is possible for most DFs F.

In the following table we give the AREs for some selected F.

ARE Computations for Selected F

F	$e(\tilde{X},\overline{X})$	$e(W,\overline{X})$	$e(\tilde{X},W)$
$U(-1/2,1/2)$	1/3	1	1/3
$\mathcal{N}(0,1)$	$2/\pi = 0.637$	$3/\pi = 0.955$	2/3
Logistic, $f(x) = e^{-x}(1+e^{-x})^{-1}$	$\pi^2/12 = 0.822$	1.10	0.748
Double Exponential, $f(x) = (1/2)\exp(-\|x\|)$	2	1.5	4/3
$\mathcal{C}(0,1)$	∞	∞	4/3

It can be shown that $e_F(\tilde{X},\overline{X}) \geq 1/3$ for all symmetric F, so $\tilde{X}$ is quite inefficient compared to $\overline{X}$ for $U(-1/2,1/2)$. Even for normal f, $\tilde{X}$ would require 157 observations to achieve the same accuracy that $\overline{X}$ achieves with 100 observations. For heavier tailed distributions, however, $\tilde{X}$ provides more protection that $\overline{X}$.

The values of $e(W,\overline{X})$, on the other hand, are quite high for most F and, in fact, $e_F(W,\overline{X}) \geq 0.864$ for all symmetric F. Even for normal F one loses little (4.5%) in using W instead of $\overline{X}$. Thus W is more robust as an estimator of θ.

A look at the values of $e(\tilde{X},W)$ shows that $\tilde{X}$ is worse than W for distributions with light-tails but does slightly better than W for heavier-tailed F.

Let us now compare the AREs of $\overline{X}_\alpha$, $\overline{X}$, and W. The following AREs for selected α are due to Bickel [5].

ARE Comparisons

	$\alpha = 0.01$		$\alpha = 0.05$	
F	$e(\overline{X}_\alpha,\overline{X})$	$e(W,\overline{X}_\alpha)$	$e(\overline{X}_\alpha,\overline{X})$	$e(W,\overline{X}_\alpha)$
Uniform	0.96	1.04	0.83	1.20
Normal	0.995	0.96	0.97	0.985
Double Exponential	1.06	1.41	1.21	1.24
Cauchy	∞	6.72	∞	2.67

We note that $\overline{X}_\alpha$ performs quite well compared to $\overline{X}$. In fact, for normal distribution the efficiency is quiet close to 1 so there is little loss in using $\overline{X}_\alpha$. For heavier-tailed distributions $\overline{X}_\alpha$ is preferable. For small values of α, it should be noted that $\overline{X}_\alpha$ does not differ much from $\overline{X}$. Nevertheless, $\overline{X}_\alpha$ is more robust; it cannot do much worse than $\overline{X}$ but can do much better. Compared to Hodges–Lehmann estimator, $\overline{X}_\alpha$ does not perform as well. It (W) provides better protection against outliers (heavy tails) and gives up little in the normal case.

Finally we consider testing $H_0 : \theta = \theta_0$ against $H_1 : \theta > \theta_0$. Recall that $X_1, X_2, \ldots, X_n$ are iid with common continuous symmetric DF $F(x-\theta)$, $\theta \in \mathcal{R}$ and PDF $f(x-\theta)$. Suppose $\sigma_F^2 = \text{Var}(X_1) < \infty$. Let S denotes the sign test based on the statistic $R^+(\mathbf{X}) = \sum_{i=1}^n I_{[X_i > \theta_0]}$, W denotes the Wilcoxon signed-rank test based on the statistic $T^+(\mathbf{X}) = \sum_{1\le i\le j\le n} I_{[X_i+X_j>2\theta_0]}$, M denotes the test based on the Z-statistic $Z = \sqrt{n}((\overline{X}-\theta_0)/\sigma_F$, and t denotes the student's t-test based on the statistic $\sqrt{n}(\overline{X}-\theta_0)/S$, where S^2 is the sample variance.

First note that $e(T,M)=1$. Next we note that $e_F(S,t)=e_F(\tilde{X},\overline{X})$, $e_F(W,t)=e_F(W,\overline{X})$ so that AREs are the same as given in (28), (29), and (30) and values of ARE given in the table for various F remain the same for corresponding tests.

Similar remarks apply as in the case of estimation of θ. Sign test is not as efficient as the Wilcoxon signed-rank test. But for heavier-tailed distributions such as Cauchy and double exponential sign test does better than the Wilcoxon signed-rank test.

PROBLEMS 13.7

1. Let $(X_1,X_2,\ldots,X_n)$ be jointly normal with $EX_i=\mu$, $\text{var}(X_i)=\sigma^2$, and $\text{cov}(X_i,X_j)=\rho\sigma^2$ if $|i-j|=1$, $i\neq j$, and $=0$ otherwise.

(a) Show that

$$\text{var}(\overline{X}) = \frac{\sigma^2}{n}\left[1+2\rho\left(1-\frac{1}{n}\right)\right]$$

and

$$E(S^2)=\sigma^2\left(1-\frac{2\rho}{n}\right).$$

(b) Show that the t-statistic $\sqrt{n}(\overline{X}-\mu)/S$ is asymptotically normally distributed with mean 0 and variance $1+2\rho$. Conclude that the significance of t is overestimated for positive values of ρ and underestimated for $\rho<0$ in large samples.

(c) For finite n, consider the statistic

$$T^2=\frac{n(\overline{X}-\mu)^2}{S^2}.$$

Compare the expected values of the numerator and the denominator of T^2 and study the effect of $\rho\neq 0$ to interpret significant t values (Scheffé [101, p. 338].)

2. Let $X_1,X_2,\ldots,X_n$ be a random sample from $G(\alpha,\beta)$, $\alpha>0$, $\beta>0$:

(a) Show that

$$\mu_4 = 3\alpha(\alpha+2)/\beta^4.$$

(b) Show that

$$\text{var}\left\{(n-1)\frac{S^2}{\sigma^2}\right\}\approx(n-1)\left(2+\frac{6}{\alpha}\right).$$

(c) Show that the large sample distribution of $(n-1)S^2/\sigma^2$ is normal.

(d) Compare the large-sample test of $H_0: \sigma=\sigma_0$ based on the asymptotic normality of $(n-1)S^2/\sigma^2$ with the large-sample test based on the same statistic when the observations are taken from a normal population. In particular, take $\alpha=2$.

3. Let $X_1, X_2, \ldots, X_m$ and $Y_1, Y_2, \ldots, Y_n$ be two independent random samples from populations with means μ_1 and μ_2, and variances σ_1^2 and σ_2^2, respectively. Let $\overline{X}, \overline{Y}$ be the two sample means, and S_1^2, S_2^2 be the two sample variances. Write $N = m + n$, $R = m/n$, and $\theta = \sigma_1^2/\sigma_2^2$. The usual normal theory test of H_0: $\mu_1 - \mu_2 = \delta_0$ is the t-test based on the statistic

$$T = \frac{\overline{X} - \overline{Y} - \delta_0}{S_p(1/m + 1/n)^{1/2}},$$

where

$$S_p^2 = \frac{(m-1)S_1^2 + (n-1)S_2^2}{m+n-2}.$$

Under H_0, the statistic T has a t-distribution with $N-2$ d.f., provided that $\sigma_1^2 = \sigma_2^2$.

Show that the asymptotic distribution of T in the nonnormal case is $\mathcal{N}(0, (\theta + R)(1 + R\theta)^{-1})$ for large m and n. Thus, if $R = 1$, T is asymptotically $\mathcal{N}(0, 1)$ as in the normal theory case assuming equal variances, even though the two samples come from nonnormal populations with unequal variances. Conclude that the test is robust in the case of large, equal sample sizes (Scheffé [101, p. 339]).

4. Verify the computations in the table above using the expressions of ARE in (28), (29), and (30).

5. Suppose F is a $G(\alpha, \beta)$ r.v. Show that

$$e(W, \overline{X}) = \frac{3\alpha\Gamma^2(2\alpha)}{2^{4(\alpha-1)}(2\alpha-1)^2\{\Gamma(\alpha)\}^4}.$$

(Note that F is not symmetric.)

6. Suppose F has PDF

$$f(x) = \frac{\Gamma(m)}{\Gamma(1/2)\Gamma((m-1)/2)(1+x^2)^m}, \quad -\infty < x < \infty,$$

for $m \geq 1$. compute $e(\tilde{X}, \overline{X})$, $e(W, \overline{X})$, and $e(\tilde{X}, W)$. (From Problem 3.2.3, $E|X|^k < \infty$ if $k < m - 1/2$.)

FREQUENTLY USED SYMBOLS AND ABBREVIATIONS

$\Rightarrow$	implies
$\Leftrightarrow$	implies and is implied by
$\rightarrow$	converges to
$\uparrow, \downarrow$	increasing, decreasing
$\not\uparrow, \not\downarrow$	nonincreasing, nondecreasing
$\Gamma(x)$	gamma function
$\overline{\lim}, \underline{\lim}, \lim$	limit superior, limit inferior, limit
$\mathcal{R}, \mathcal{R}_n$	real line, n-dimensional Euclidean space
$\mathfrak{B}, \mathfrak{B}_n$	Borel σ-field on $\mathcal{R}$, Borel σ-field on $\mathcal{R}_n$
I_A	indicator function of set A
$\varepsilon(x)$	$= 1$ if $x \geq 0$, and $= 0$ if $x < 0$
μ	EX, expected value
m_n	EX^n, $n \geq 0$ integral
β_α	$E\|X\|^\alpha$, $\alpha > 0$
μ_k	$E(X - EX)^k$, $k \geq 0$ integral
σ^2	$= \mu_2$, variance
f', f'', f'''	first, second, third derivative of f
$\sim$	distributed as
$\approx$	asymptotically (or approximately) equal to

An Introduction to Probability and Statistics, Third Edition. Vijay K. Rohatgi and A.K. Md. Ehsanes Saleh.

$\xrightarrow{L}$	convergence in law
$\xrightarrow{P}$	convergence in probability
$\xrightarrow{\text{a.s.}}$	convergence almost surely
$\xrightarrow{r}$	convergence in rth mean
RV	random variable
DF	distribution function
PDF	probability density function
PMF	probability mass function
PGF	probability generating function
MGF	moment generating function
d.f.	degrees of freedom
BLUE	best linear unbiased estimate
MLE	maximum likelihood estimate
MVUE	minimum variance unbiased estimate
UMA	uniformly most accurate
UMVUE	uniformly minimum variance unbiased estimate
UMAU	uniformly most accurate unbiased
MP	most powerful
UMP	uniformly most powerful
GLM	general linear model
i.o.	infinitely often
iid	independent, identically distributed
SD	standard deviation
SE	standard error
MLR	monotone likelihood ratio
MSE	mean square error
WLLN	weak law of large numbers
SLLN	strong law of large numbers
CLT	central limit theorem
SPRT	sequential probability ratio test
$b(1,p)$	Bernoulli with parameter p
$b(n,p)$	binomial with parameters n,p
$NB(r;p)$	negative binomial with parameters r,p
$P(\lambda)$	Poisson with parameter λ
$U[a,b]$	uniform on $[a,b]$
$G(\alpha,\beta)$	gamma with parameters α,β
$B(\alpha,\beta)$	beta with parameters α,β
$\chi^2(n)$	chi-square with d.f. n
$\mathcal{C}(\mu,\theta)$	Cauchy with parameters μ,θ

$\mathcal{N}(\mu, \sigma^2)$	normal with mean μ, variance σ^2
$t(n)$	Student's t with n d.f.
$F(m,n)$	F-distribution with (m,n) d.f.
z_α	$100(1-\alpha)$th percentile of $\mathcal{N}(0,1)$
$\chi^2_{n,\alpha}$	$100(1-\alpha)$th percentile of $\chi^2(n)$
$t_{n,\alpha}$	$100(1-\alpha)$th percentile of $t(n)$
$F_{m,n,\alpha}$	$100(1-\alpha)$th percentile of $F(m,n)$
$\text{AN}(\mu_n, \sigma_n^2)$	asymptotically normal
GLR	generalized likelihood ratio
MRE	minimum risk equivariant
$\ell\text{n}\, x$	logarithm (to base e) of x
$\exp(X)$	exponential
LMP	locally most powerful
$\mathcal{L}(x)$	law or distribution of RV X
$\text{b}(\delta,.)$	bias in estimator δ
iid	independent, identically distributed

REFERENCES

1. A. Agresti, *Categorical Data Analysis*, 3rd ed., Wiley, New York, 2012.
2. T. W. Anderson, Maximum likelihood estimates for a multivariate normal distribution when some observations are missing, *J. Am. Stat. Assoc.* 52 (1957), 200–203.
3. J. D. Auble, Extended tables for the Mann–Whitney statistic, *Bull. Inst. Educ. Res.* 1 (1953), No. i–iii, 1–39.
4. D. Bernstein, Sur une propriété charactéristique de la loi de Gauss, *Trans. Leningrad Polytech. Inst.* 3 (1941), 21–22.
5. P. J. Bickel, On some robust estimators of location, *Ann. Math. Stat.* 36 (1965), 847–858.
6. P. Billingsley, *Probability and Measure*, 2nd ed., Wiley, New York, 1986.
7. D. Birkes, Generalized likelihood ratio tests and uniformly most powerful tests, *Am. Stat.* 44 (1990), 163–166.
8. Z. W. Birnbaum and F. H. Tingey, One-sided confidence contours for probability distribution functions, *Ann. Math. Stat.* 22 (1951), 592–596.
9. Z. W. Birnbaum, Numerical tabulation of the distribution of Kolmogorov's statistic for finite sample size, *J. Am. Stat. Assoc.*, 17 (1952), 425–441.
10. D. Blackwell, Conditional expectation and unbiased sequential estimation, *Ann. Math. Stat.* 18 (1947), 105–110.
11. J. Boas, A note on the estimation of the covariance between two random variables using extra information on the separate variables, *Stat. Neerl.* 21 (1967), 291–292.
12. D. G. Chapman and H. Robbins, Minimum variance estimation without regularity assumptions, *Ann. Math. Stat.* 22 (1951), 581–586.
13. S. D. Chatterji, Some elementary characterizations of the Poisson distribution, *Am. Math. Mon.* 70 (1963), 958–964.

An Introduction to Probability and Statistics, Third Edition. Vijay K. Rohatgi and A.K. Md. Ehsanes Saleh.

14. K. L. Chung and P. Erdös, On the application of the Borel–Cantelli lemma, *Trans. Am. Math. Soc.* 72 (1952), 179–186.
15. K. L. Chung, *A Course in Probability Theory*, Harcourt, Brace & World, New York, 1968.
16. H. Cramér, Über eine Eigenschaft der normalen Verteilungsfunktion, *Math. Z.* 41 (1936), 405–414.
17. H. Cramér, *Mathematical Methods of Statistics*, Princeton University Press, Princeton, N.J., 1946.
18. H. Cramér, A contribution to the theory of statistical estimation, *Skand. Aktuarietidskr.* 29 (1946), 85–94.
19. J. H. Curtiss, A note on the theory of moment generating functions, *Ann. Math. Stat.* 13 (1942), 430–433.
20. D. A. Darmois, Sur diverses propriétés charactéristique de la loi de probabilité de Laplace-Gauss, *Bull. Int. Stat. Inst.* 23 (1951), part II, 79–82.
21. M. M. Desu, Optimal confidence intervals of fixed width, *Am. Stat.* 25 (1971), No. 2, 27–29.
22. B. Efron, Bootstrap methods: Another look at the jackknife. *Ann. Stat.* 7 (1979), 1–26.
23. B. Efron and R. J. Tibshirani, *An Introduction to Bootstrap*, Chapman Hall, New York, 1993.
24. W. Feller, Über den Zentralen Granzwertsatz der Wahrscheinlichkeitsrechnung, *Math. Z.* 40 (1935), 521–559; 42 (1937), 301–312.
25. W. Feller, *An Introduction to Probability Theory and Its Applications*, Vol. 1, 3rd ed., Wiley, New York, 1968.
26. W. Feller, *An Introduction to Probability Theory and Its Applications*, Vol. 2, 2nd ed., Wiley, New York, 1971.
27. K. K. Ferentinos, Shortest confidence intervals for families of distributions involving truncation parameters, *Am. Stat.* 44 (1990), 40–41.
28. T. S. Ferguson, *Mathematical Statistics*, Academic Press, New York, 1967.
29. T. S. Ferguson, *A Course in Large Sample Theory*, Chapman & Hall, London, 1996.
30. R. A. Fisher, On the mathematical foundations of theoretical statistics, *Phil. Trans. R. Soc.* A222 (1922), 309–386.
31. M. Fisz, *Probability Theory and Mathematical Statistics*, 3rd ed., Wiley, 1963.
32. D. A. S. Fraser, *Nonparametric Methods in Statistics*, Wiley, New York, 1965.
33. D. A. S. Fraser, *The Structure of Inference*, Wiley, New York, 1968.
34. M. Fréchet, Sur l'extension de certaines evaluations statistiques au cas de petits echantillons, *Rev. Inst. Int. Stat.* 11 (1943), 182–205.
35. J. D. Gibbons, *Nonparametric Statistical Inference*, Dekker, New York, 1985.
36. B. V. Gnedenko, Sur la distribution limite du terme maximum d'une série aléatoire, *Ann. Math.* 44 (1943), 423–453.
37. W. C. Guenther, Shortest confidence intervals, *Am. Stat.* 23 (1969), No. 1, 22–25.
38. W. C. Guenther, Unbiased confidence intervals, *Am. Stat.* 25 (1971), No. 1, 51–53.
39. E. J. Gumbel, Distributions à plusieurs variables dont les marges sont données, *C. R. Acad. Sci. Paris* 246 (1958), 2717–2720.
40. J. H. Hahn and W. Nelson, A problem in the statistical comparison of measuring devices, *Technometrics* 12 (1970), 95–102.
41. P. R. Halmos and L. J. Savage, Application of the Radon–Nikodym theorem to the theory of sufficient statistics, *Ann. Math. Stat.* 20 (1949), 225–241.
42. P. R. Halmos, *Measure Theory*, Van Nostrand, New York, 1950.

43. J. L. Hodges and E. L. Lehmann, Some problems in minimax point estimation, *Ann. Math. Stat.* 21 (1950), 182–197.

44. J. L. Hodges, Jr. and E. L. Lehmann, Estimates of location based on rank tests, *Ann. Math. Stat.* 34 (1963), 598–611.

45. W. Hoeffding, A class of statistics with asymptotically normal distribution, *Ann. Math. Stat.* 19 (1948), 293–325.

46. D. Hogben, The distribution of the sample variance from a two-point binomial population, *Am. Stat.* 22 (1968), No. 5, 30.

47. V. S. Huzurbazar, The likelihood equation consistency, and maxima of the likelihood function, *Ann. Eugen.* (London) 14 (1948), 185–200.

48. M. Kac, *Lectures in Probability*, The Mathematical Association of America, Washinton, D.C., 1964–1965.

49. W. C. M. Kallenberg et al., *Testing Statistical Hypotheses: Worked Solutions*, CWI, Amsterdam, 1980.

50. J. F. Kemp, A maximal distribution with prescribed marginals, *Am. Math. Mon.* 80 (1973), 83.

51. M. G. Kendall, *Rank Correlation Methods*, 3rd ed., Charles Griffin, London, 1962.

52. J. Kiefer, On minimum variance estimators, *Ann. Math. Stat.* 23 (1952), 627–629.

53. A. N. Kolmogorov, Sulla determinazione empirica di una legge di distribuzione, *G. Inst. Ital. Attuari* 4 (1933), 83–91.

54. A. N. Kolmogorov and S. V. Fomin, *Elements of the Theory of Functions and Functional Analysis*, Vol. 2, Graylock Press, Albany, N.Y., 1961.

55. C. H. Kraft and C. Van Eeden, *A Nonparametric Introduction to Statistics*, Macmillan, New York, 1968.

56. W. Kruskal, Note on a note by C. S. Pillai, *Am. Stat.* 22 (1968), No. 5, 24–25.

57. G. Kulldorf, On the condition for consistency and asymptotic efficiency of maximum likelihood estimates, *Skand. Aktuarietidskr.* 40 (1957), 129–144.

58. R. G. Laha and V. K. Rohatgi, *Probability Theory*, Wiley, New York, 1979.

59. J. Lamperti, Density of random variable, *Am. Math. Mon.* 66 (1959), 317.

60. J. Lamperti and W. Kruskal, Solution by W. Kruskal to the problem "Poisson Distribution" posed by J. Lamperti, *Am. Math. Mon.* 67 (1960), 297–298.

61. E. L. Lehmann, *Nonparametrics: Statistical Methods Based on Ranks*, Holden-Day, San Francisco, C.A., 1975.

62. E. L. Lehmann, An interpretation of completeness and Basu's theorem, *J. Am. Stat. Assoc.* 76 (1981), 335–340.

63. E. L. Lehmann, *Theory of Point Estimation*, Wiley, New York, 1983.

64. E. L. Lehmann, *Testing Statistical Hypotheses*, 2nd ed., Wiley, New York, 1986.

65. E. L. Lehmann and H. Scheffé, Completeness, similar regions, and unbiased estimation, *Sankhyā*, Ser. A, 10 (1950), 305–340.

66. M. Loève, *Probability Theory*, 4th ed., Springer-Verlag, New York, 1977.

67. E. Lukacs, A characterization of the normal distribution, *Ann. Math. Stat.* 13 (1942), 91–93.

68. E. Lukacs, Characterization of populations by properties of suitable statistics, *Proc. Third Berkeley Symp.* 2 (1956), 195–214.

69. E. Lukacs, *Characteristic Functions*, 2nd ed., Hafner, New York, 1970.

70. E. Lukacs and R. G. Laha, *Applications of Characteristic Functions*, Hafner, New York, 1964.

71. H. B. Mann and D. R. Whitney, On a test whether one of two random variables is stochastically larger than the other, *Ann. Math. Stat.* 18 (1947), 50–60.
72. F. J. Massey, Distribution table for the deviation between two sample cumulatives, *Ann. Math. Stat.* 23 (1952), 435–441.
73. M. V. Menon, A characterization of the Cauchy distribution, *Ann. Math. Stat.* 33 (1962), 1267–1271.
74. L. H. Miller, Table of percentage points of Kolmogorov statistics, *J. Am. Stat. Assoc.* 51 (1956), 111–121.
75. M. G. Natrella, *Experimental Statistics*, Natl. Bur. Stand. Handb. 91, Washington, D.C., 1963.
76. J. Neyman and E. S. Pearson, On the problem of the most efficient tests of statistical hypotheses, *Phil. Trans. R. Soc.* A231 (1933), 289–337.
77. J. Neyman and E. L. Scott, Consistent estimates based on partially consistent observations, *Econometrica* 16 (1948), 1–32.
78. E. H. Oliver, A maximum likelihood oddity, *Am. Stat.* 26 (1972), No. 3, 43–44.
79. D. B. Owen, *Handbook of Statistical Tables*, Addison-Wesley, Reading, M.A., 1962.
80. E. J. G. Pitman and E. J. Williams, Cauchy-distributed functions of Cauchy variates, *Ann. Math. Stat.* 38 (1967), 916–918.
81. J. W. Pratt, Length of confidence intervals, *J. Am. Stat. Assoc.* 56 (1961), 260–272.
82. B. J. Prochaska, A note on the relationship between the geometric and exponential distributions, *Am. Stat.* 27 (1973), 27.
83. P. S. Puri, On a property of exponential and geometric distributions and its relevance to multivariate failure rate, *Sankhyā*, Ser. A, 35 (1973), 61–68.
84. D. A. Raikov, On the decomposition of Gauss and Poisson laws (in Russian), *Izv. Akad. Nauk. SSSR, Ser. Mat.* 2 (1938), 91–124.
85. R. R. Randles and D. A. Wolfe, *Introduction to the Theory of Nonparametric Statistics*, Krieger, Melbourne, F.L., 1991.
86. C. R. Rao, Information and the accuracy attainable in the estimation of statistical parameters, *Bull. Calcutta Math. Soc.* 37 (1945), 81–91.
87. C. R. Rao, Sufficient statistics and minimum variance unbiased estimates, *Proc. Cambridge Phil. Soc.* 45 (1949) 213–218.
88. C. R. Rao, *Linear Statistical Inference and Its Applications*, 2nd ed., Wiley, New York, 1973.
89. S. C. Rastogi, Note on the distribution of a test statistic, *Am. Stat.* 23 (1969), 40–41.
90. V. K. Rohatgi, *Statistical Inference*, Wiley, New York, 1984.
91. V. K. Rohatgi, On the moments of $F(X)$ when F is discrete, *J. Stat. Comp. Simul.* 29 (1988), 340–343.
92. V. I. Romanovsky, On the moments of the standard deviations and of the correlation coefficient in samples from a normal population, *Metron* 5 (1925), No. 4, 3–46.
93. L. Rosenberg, Nonnormality of linear combinations of normally distributed random variables, *Am. Math. Mon.* 72 (1965), 888–890.
94. J. Roy and S. Mitra, Unbiased minimum variance estimation in a class of discrete distributions, *Sankhyā* 18 (1957), 371–378.
95. R. Roy, Y. LePage, and M. Moore, On the power series expansion of the moment generating function, *Am. Stat.* 28 (1974), 58–59.
96. H. L. Royden, *Real Analysis*, 2nd ed., Macmillan, New York, 1968.
97. Y. D. Sabharwal, A sequence of symmetric Bernoulli trials, *SIAM Rev.* 11 (1969), 406–409.

98. A. Sampson and B. Spencer, Sufficiency, minimal sufficiency, and lack thereof, *Am. Stat.* 30 (1976) 34–35. Correction, 31 (1977), 54.
99. P. A. Samuelson, How deviant can you be? *J. Am. Stat. Assoc.* 63 (1968), 1522–1525.
100. H. Scheffé, A useful convergence theorem for probability distributions, *Ann. Math. Stat.* 18 (1947), 434–438.
101. H. Scheffé, *The Analysis of Variance*, Wiley, New York, 1961.
102. R. J. Serfling, *Approximation Theorems of Mathematical Statistics*, Wiley, New York, 1979.
103. D. N. Shanbhag and I. V. Basawa, On a characterization property of the multinomial distribution, *Ann. Math. Stat.* 42 (1971), 2200.
104. L. Shepp, Normal functions of normal random variables, *SIAM Rev.* 4 (1962), 255–256.
105. A. E. Siegel, Film-mediated fantasy aggression and strength of aggression drive, *Child Dev.* 27 (1956), 365–378.
106. V. P. Skitovitch, Linear forms of independent random variables and the normal distribution law, *Izv. Akad. Nauk. SSSR. Ser. Mat.* 18 (1954), 185–200.
107. N. V. Smirnov, On the estimation of the discrepancy between empirical curves of distributions for two independent samples (in Russian), *Bull. Moscow Univ.* 2 (1939), 3–16.
108. N. V. Smirnov, Approximate laws of distribution of random variables from empirical data (in Russian), *Usp. Mat. Nauk.* 10 (1944), 179–206.
109. R. C. Srivastava, Two characterizations of the geometric distribution, *J. Am. Stat. Assoc.* 69 (1974), 267–269.
110. S. M. Stigler, Completeness and unbiased estimation, *Am. Stat.* 26 (1972), 28–29.
111. P. T. Strait, A note on the independence and conditional probabilities, *Am. Stat.* 25 (1971), No. 2, 17–18.
112. R. F. Tate and G. W. Klett, Optimum confidence intervals for the variance of a normal distribution, *J. Am. Stat. Assoc.* 54 (1959), 674–682.
113. W. A. Thompson, Jr., *Applied Probability*, Holt, Rinehart and Winston, New York, 1969.
114. H. G. Tucker, *A Graduate Course in Probability*, Academic Press, New York, 1967.
115. A. Wald, Note on the consistency of the maximum likelihood estimate, *Ann. Math. Stat.* 20 (1949), 595–601.
116. G. N. Watson, *A Treatise on the Theory of Bessel Functions*, 2nd ed., Cambridge University Press, Cambridge, 1966.
117. D. V. Widder, *Advanced Calculus*, 2nd ed., Prentice-Hall, Englewood Cliffs, N.J., 1961.
118. S. S. Wilks, *Mathematical Statistics*, Wiley, New York, 1962.
119. J. Wishart, The generalized product-moment distribution in samples from a normal multivariate population, *Biometrika* 20A (1928), 32–52.
120. C. K. Wong, A note on mutually independent events. *Am. Stat.* 26 (1972), 27.
121. S. Zacks, *The Theory of Statistical Inference*, Wiley, New York, 1971.
122. P. W. Zehna, Invariance of maximum likelihood estimation, *Ann. Math. Stat.* 37 (1966), 755.

STATISTICAL TABLES

ST1	Cumulative Binomial Probabilities
ST2	Tail Probability Under Standard Normal Distribution
ST3	Critical Values Under Chi-Square Distribution
ST4	Student's t-Distribution
ST5	F-Distribution: 5% and 1% Points for the Distribution of F
ST6	Random Normal Numbers, $\mu = 0$ and $\sigma = 1$
ST7	Critical Values of the Kolmogorov–Smirnov One-Sample Test Statistic
ST8	Critical Values of the Kolmogorov–Smirnov Test Statistics for Two Samples of Equal Size
ST9	Critical Values of the Kolmogorov–Smirnov Test Statistics for Two Samples of Unequal Size
ST10	Critical Values of the Wilcoxon Signed–Rank Test Statistic
ST11	Critical Values of the Mann–Whitney–Wilcoxon Test Statistic
ST12	Critical Points of Kendall's Tau Statistics
ST13	Critical Values of Spearman's Rank Correlation Statistic

An Introduction to Probability and Statistics, Third Edition. Vijay K. Rohatgi and A.K. Md. Ehsanes Saleh.

Table ST1. Cumulative Binomial Probabilities, $\sum_{x=0}^{r} \binom{n}{x} p^x (1-p)^{n-x}$, $r = 0, 1, 2, \ldots, n-1$

		p								
n	r	0.01	0.05	0.10	0.20	0.25	0.30	0.333	0.40	0.50
2	0	0.9801	0.9025	0.8100	0.6400	0.5625	0.4900	0.4444	0.3600	0.2500
	1	0.9999	0.9975	0.9900	0.9600	0.9375	0.9100	0.8888	0.8400	0.7500
3	0	0.9703	0.8574	0.7290	0.5120	0.4219	0.3430	0.2963	0.2160	0.1250
	1	0.9997	0.9928	0.9720	0.8960	0.8438	0.7840	0.7407	0.6480	0.5000
	2	1.0000	0.9999	0.9990	0.9920	0.9844	0.9730	0.9629	0.9360	0.8750
4	0	0.9606	0.8145	0.6561	0.4096	0.3164	0.2401	0.1975	0.1296	0.0625
	1	0.9994	0.9860	0.9477	0.8192	0.7383	0.6517	0.5926	0.4742	0.3125
	2	1.0000	0.9995	0.9963	0.9728	0.9492	0.9163	0.8889	0.8198	0.6875
	3		1.0000	0.9999	0.9984	0.9961	0.9919	0.9877	0.9734	0.9375
5	0	0.9510	0.7738	0.5905	0.3277	0.2373	0.1681	0.1317	0.0778	0.0312
	1	0.9990	0.9774	0.9185	0.7373	0.6328	0.5283	0.4609	0.3370	0.1874
	2	1.0000	0.9988	0.9914	0.9421	0.8965	0.8370	0.7901	0.6826	0.4999
	3		0.9999	0.9995	0.9933	0.9844	0.9693	0.9547	0.9130	0.8124
	4		1.0000	1.0000	0.9997	0.9990	0.9977	0.9959	0.9898	0.9686
6	0	0.9415	0.7351	0.5314	0.2621	0.1780	0.1176	0.0878	0.0467	0.0156
	1	0.9986	0.9672	0.8857	0.6553	0.5340	0.4201	0.3512	0.2333	0.1094
	2	1.0000	0.9977	0.9841	0.9011	0.8306	0.7442	0.6804	0.5443	0.3438
	3		0.9998	0.9987	0.9830	0.9624	0.9294	0.8999	0.8208	0.6563
	4		0.9999	0.9999	0.9984	0.9954	0.9889	0.9822	0.9590	0.8907
	5		1.0000	1.0000	0.9999	0.9998	0.9991	0.9987	0.9959	0.9845
7	0	0.9321	0.6983	0.4783	0.2097	0.1335	0.0824	0.0585	0.0280	0.0078
	1	0.9980	0.9556	0.6554	0.5767	0.4450	0.3294	0.2633	0.1586	0.0625
	2	1.0000	0.9962	0.8503	0.8520	0.7565	0.6471	0.5706	0.4199	0.2266
	3		0.9998	0.9743	0.9667	0.9295	0.8740	0.8267	0.7102	0.5000
	4		1.0000	0.9973	0.9953	0.9872	0.9712	0.9547	0.9037	0.7734
	5			0.9998	0.9996	0.9987	0.9962	0.9931	0.9812	0.9375
	6			1.0000	1.0000	0.9999	0.9998	0.9995	0.9984	0.9922
8	0	0.9227	0.6634	0.4305	0.1678	0.1001	0.0576	0.0390	0.0168	0.0039
	1	0.9973	0.9427	0.8131	0.5033	0.3671	0.2553	0.1951	0.1064	0.0352
	2	0.9999	0.9942	0.9619	0.7969	0.6786	0.5518	0.4682	0.3154	0.1445
	3	1.0000	0.9996	0.9950	0.9437	0.8862	0.8059	0.7413	0.5941	0.3633
	4		1.0000	0.9996	0.9896	0.9727	0.9420	0.9120	0.8263	0.6367
	5			1.0000	0.9988	0.9958	0.9887	0.9803	0.9502	0.8555
	6				1.0000	0.9996	0.9987	0.9974	0.9915	0.9648
	7					1.0000	0.9999	0.9998	0.9993	0.9961
9	0	0.9135	0.6302	0.3874	0.1342	0.0751	0.0404	0.0260	0.0101	0.0020
	1	0.9965	0.9287	0.7748	0.4362	0.3004	0.1960	0.1431	0.0706	0.0196
	2	0.9999	0.9916	0.9470	0.7382	0.6007	0.4628	0.3772	0.2318	0.0899
	3	1.0000	0.9993	0.9916	0.9144	0.8343	0.7296	0.6503	0.4826	0.2540
	4		0.9999	0.9990	0.9805	0.9511	0.9011	0.8551	0.7334	0.5001

(*Continued*)

n	r	p								
		0.01	0.05	0.10	0.20	0.25	0.30	0.333	0.40	0.50
	5		1.0000	0.9998	0.9970	0.9900	0.9746	0.9575	0.9006	0.7462
	6			0.9999	0.9998	0.9987	0.9956	0.9916	0.9749	0.9103
	7			1.0000	1.0000	0.9999	0.9995	0.9989	0.9961	0.9806
	8					1.0000	0.9999	0.9998	0.9996	0.9982
10	0	0.9044	0.5987	0.3487	0.1074	0.0563	0.0282	0.0173	0.0060	0.0010
	1	0.9958	0.9138	0.7361	0.3758	0.2440	0.1493	0.1040	0.0463	0.0108
	2	1.0000	0.9884	0.9298	0.6778	0.5256	0.3828	0.2991	0.1672	0.0547
	3		0.9989	0.9872	0.8791	0.7759	0.6496	0.5592	0.3812	0.1719
	4		0.9999	0.9984	0.9672	0.9219	0.8497	0.7868	0.6320	0.3770
	5		1.0000	0.9999	0.9936	0.9803	0.9526	0.9234	0.8327	0.6231
	6			1.0000	0.9991	0.9965	0.9894	0.9803	0.9442	0.8282
	7				0.9999	0.9996	0.9984	0.9966	0.9867	0.9454
	8				1.0000	1.0000	0.9998	0.9996	0.9973	0.9893
	9						1.0000	0.9999	0.9999	0.9991
11	0	0.8954	0.5688	0.3138	0.0859	0.0422	0.0198	0.0116	0.0036	0.0005
	1	0.9948	0.8981	0.6974	0.3221	0.1971	0.1130	0.0752	0.0320	0.0059
	2	0.9998	0.9848	0.9104	0.6174	0.4552	0.3128	0.2341	0.1189	0.0327
	3	1.0000	0.9984	0.9815	0.8389	0.7133	0.5696	0.4726	0.2963	0.1133
	4		0.9999	0.9972	0.9496	0.8854	0.7897	0.7110	0.5328	0.2744
	5		1.0000	0.9997	0.9884	0.9657	0.9218	0.8779	0.7535	0.5000
	6			1.0000	0.9981	0.9924	0.9784	0.9614	0.9007	0.7256
	7				0.9998	0.9988	0.9947	0.9912	0.9707	0.8867
	8				1.0000	0.9999	0.9994	0.9986	0.9941	0.9673
	9					1.0000	0.9999	0.9999	0.9993	0.9941
	10						1.0000	1.0000	1.0000	0.9995
12	0	0.8864	0.5404	0.2824	0.0687	0.0317	0.0139	0.0077	0.0022	0.0002
	1	0.9938	0.8816	0.6590	0.2749	0.1584	0.0850	0.0540	0.0196	0.0032
	2	0.9998	0.9804	0.8892	0.5584	0.3907	0.2528	0.1811	0.0835	0.0193
	3	1.0000	0.9978	0.9744	0.7946	0.6488	0.4925	0.3931	0.2254	0.0730
	4	1.0000	0.9998	0.9957	0.9806	0.8424	0.7237	0.6315	0.4382	0.1939
	5	1.0000	1.0000	0.9995	0.9961	0.9456	0.8822	0.8223	0.6652	0.3872
	6			1.0000	0.9994	0.9858	0.9614	0.9336	0.8418	0.6128
	7				0.9999	0.9972	0.9905	0.9812	0.9427	0.8062
	8				1.0000	0.9996	0.9983	0.9962	0.9848	0.9270
	9					10000	0.9998	0.9995	0.9972	0.9807
	10						1.0000	0.9999	0.9997	0.9968
	11							1.0000	1.0000	0.9998
13	0	0.8775	0.5134	0.2542	0.0550	0.0238	0.0097	0.0052	0.0013	0.0000
	1	0.9928	0.8746	0.6214	0.2337	0.1267	0.0637	0.0386	0.0126	0.0017
	2	0.9997	0.9755	0.8661	0.5017	0.3326	0.2025	0.1388	0.0579	0.0112
	3	1.0000	0.9969	0.9659	0.7473	0.5843	0.4206	0.3224	0.1686	0.0462
	4		0.9997	0.9936	0.9009	0.7940	0.6543	0.5521	0.3531	0.1334
	5		1.0000	0.9991	0.9700	0.9198	0.8346	0.7587	0.5744	0.2905

Table ST1. (*Continued*)

		p								
n	r	0.01	0.05	0.10	0.20	0.25	0.30	0.333	0.40	0.50
	6			0.9999	0.9930	0.9757	0.9376	0.8965	0.7712	0.5000
	7			1.0000	0.9988	0.9944	0.9818	0.9654	0.9024	0.7095
	8				0.9998	0.9990	0.9960	0.9912	0.9679	0.8666
	9				1.0000	0.9999	0.9994	0.9984	0.9922	0.9539
	10					1.0000	0.9999	0.9998	0.9987	0.9888
	11						1.0000	1.0000	0.9999	0.9983
	12								1.0000	0.9999
14	0	0.8687	0.4877	0.2288	0.0440	0.0178	0.0068	0.0034	0.0008	0.0000
	1	0.9916	0.8470	0.5847	0.1979	0.1010	0.0475	0.0274	0.0081	0.0009
	2	0.9997	0.9700	0.8416	0.4480	0.2812	0.1608	0.1054	0.0398	0.0065
	3	1.0000	0.9958	0.9559	0.6982	0.5214	0.3552	0.2612	0.1243	0.0287
	4		0.9996	0.9908	0.8702	0.7416	0.5842	0.4755	0.2793	0.0898
	5		1.0000	0.9986	0.9562	0.8884	0.7805	0.6898	0.4859	0.2120
	6			0.9998	0.9884	0.9618	0.9067	0.8506	0.6925	0.3953
	7			1.0000	0.9976	0.9897	0.9686	0.9424	0.8499	0.6048
	8				0.9996	0.9979	0.9917	0.9826	0.9417	0.7880
	9				1.0000	0.9997	0.9984	0.9960	0.9825	0.9102
	10					1.0000	0.9998	0.9993	0.9961	0.9713
	11						1.0000	0.9999	0.9994	0.9936
	12							1.0000	0.9999	0.9991
	13									0.9999
15	0	0.8601	0.4633	0.2059	0.0352	0.0134	0.0048	0.0023	0.0005	0.0000
	1	0.9904	0.8291	0.5491	0.1672	0.0802	0.0353	0.0194	0.0052	0.0005
	2	0.9996	0.9638	0.8160	0.3980	0.2361	0.1268	0.0794	0.0271	0.0037
	3	1.0000	0.9946	0.9444	0.6482	0.4613	0.2969	0.2092	0.0905	0.0176
	4		0.9994	0.9873	0.8358	0.6865	0.5255	0.4041	0.2173	0.0592
	5		1.0000	0.9978	0.9390	0.8516	0.7216	0.6184	0.4032	0.1509
	6			0.9997	0.9820	0.9434	0.8689	0.7970	0.6098	0.3036
	7			1.0000	0.9958	0.9827	0.9500	0.9118	0.7869	0.5000
	8				0.9992	0.9958	0.9848	0.9692	0.9050	0.6964
	9				0.9999	0.9992	0.9964	0.9915	0.9662	0.8491
	10				1.0000	0.9999	0.9993	0.9982	0.9907	0.9408
	11					1.0000	0.9999	0.9997	0.9981	0.9824
	12						1.0000	1.0000	0.9997	0.9963
	13								1.0000	0.9995
	14									1.0000

Source: For $n = 2$ through 10, adapted with permission from E. Parzen, *Modern Probability Theory and Its Applications*, John Wiley, New York, 1962. For $n = 11$ through 15, adapted with permission from *Tables of Cumulative Binomial Probability Distribution*, Harvard University Press, Cambridge, M.A., 1955.

Table ST2. Tail Probability Under Standard Normal Distribution[a]

z	0.00	0.01	0.02	0.03	0.04	0.05	0.06	0.07	0.08	0.09
0.0	0.5000	0.4960	0.4920	0.4880	0.4840	0.4801	0.4761	0.4721	0.4681	0.4641
0.1	0.4602	0.4562	0.4522	0.4483	0.4443	0.4404	0.4364	0.4325	0.4286	0.4247
0.2	0.4207	0.4168	0.4129	0.4090	0.4052	0.4013	0.3974	0.3936	0.3897	0.3859
0.3	0.3821	0.3783	0.3745	0.3707	0.3669	0.3632	0.3594	0.3557	0.3520	0.3483
0.4	0.3446	0.3409	0.3372	0.3336	0.3300	0.3264	0.3228	0.3192	0.3156	0.3121
0.5	0.3085	0.3050	0.3015	0.2981	0.2946	0.2912	0.2877	0.2843	0.2810	0.2776
0.6	0.2743	0.2709	0.2676	0.2643	0.2611	0.2578	0.2546	0.2514	0.2483	0.2451
0.7	0.2420	0.2389	0.2358	0.2327	0.2297	0.2266	0.2231	0.2206	0.2177	0.2148
0.8	0.2119	0.2090	0.2061	0.2033	0.2005	0.1977	0.1949	0.1922	0.1984	0.1867
0.9	0.1841	0.1814	0.1788	0.1762	0.1736	0.1711	0.1685	0.1660	0.1635	0.1611
1.0	0.1587	0.1562	0.1539	0.1515	0.1492	0.1469	0.1446	0.1423	0.1401	0.1379
1.1	0.1357	0.1335	0.1314	0.1292	0.1271	0.1251	0.1230	0.1210	0.1190	0.1170
1.2	0.1151	0.1131	0.1112	0.1093	0.1075	0.1056	0.1038	0.1020	0.1003	0.0985
1.3	0.0968	0.0951	0.0934	0.0918	0.0901	0.0885	0.0869	0.0853	0.0838	0.0823
1.4	0.0808	0.0793	0.0778	0.0764	0.0749	0.0735	0.0721	0.0708	0.0694	0.0681
1.5	0.0668	0.0655	0.0643	0.0630	0.0618	0.0606	0.0594	0.0582	0.0571	0.0559
1.6	0.0548	0.0537	0.0526	0.0516	0.0505	0.0495	0.0485	0.0475	0.0465	0.0455
1.7	0.0446	0.0436	0.0427	0.0418	0.0409	0.0401	0.0392	0.0384	0.0375	0.0367
1.8	0.0359	0.0351	0.0344	0.0336	0.0329	0.0322	0.0314	0.0307	0.0301	0.0294
1.9	0.0287	0.0281	0.0274	0.0268	0.0262	0.0256	0.0250	0.0244	0.0239	0.0233
2.0	0.0228	0.0222	0.0217	0.0212	0.0207	0.0202	0.0197	0.0192	0.0188	0.0183
2.1	0.0179	0.0174	0.0170	0.0166	0.0162	0.0158	0.0154	0.0150	0.0146	0.0143
2.2	0.0139	0.0136	0.0132	0.0129	0.0125	0.0122	0.0119	0.0116	0.0113	0.0110
2.3	0.0107	0.0104	0.0102	0.0099	0.0096	0.0094	0.0091	0.0089	0.0087	0.0084
2.4	0.0082	0.0080	0.0078	0.0075	0.0073	0.0017	0.0069	0.0068	0.0066	0.0064
2.5	0.0062	0.0060	0.0059	0.0057	0.0055	0.0054	0.0052	0.0051	0.0049	0.0048
2.6	0.0047	0.0045	0.0044	0.0043	0.0041	0.0040	0.0039	0.0038	0.0037	0.0036
2.7	0.0035	0.0034	0.0033	0.0032	0.0031	0.0030	0.0029	0.0028	0.0027	0.0026
2.8	0.0026	0.0025	0.0024	0.0023	0.0023	0.0022	0.0021	0.0021	0.0020	0.0019
2.9	0.0019	0.0018	0.0018	0.0017	0.0016	0.0016	0.0015	0.0015	0.0014	0.0014
3.0	0.0013	0.0013	0.0013	0.0012	0.0012	0.0011	0.0011	0.0011	0.0010	0.0010

Source: Adapted with permission from P. G. Hoel, *Introduction to Mathematical Statistics*, 4th ed., Wiley, New York, 1971, p. 391.

[a]This table gives the probability that the standard normal variable Z will exceed a given positive value z, that is, $P\{Z > z_\alpha\} = \alpha$. The probabilities for negative values of z are obtained by symmetry.

Table ST3. Critical Values Under Chi-Square Distribution[a]

Degrees of Freedom	α												
	0.99	0.98	0.95	0.90	0.80	0.70	0.50	0.30	0.20	0.10	0.05	0.02	0.01
1	0.000157	0.000628	0.00393	0.0158	0.0642	0.148	0.455	1.074	1.642	2.706	3.841	5.412	6.635
2	0.0201	0.0404	0.103	0.211	0.446	0.713	1.386	2.408	3.219	4.605	5.991	7.824	9.210
3	0.115	0.185	0.352	0.584	1.005	1.424	2.366	3.665	4.642	6.251	7.815	9.837	11.341
4	0.297	0.429	0.711	1.064	1.649	2.195	3.357	4.878	5.989	7.779	9.488	11.668	13.277
5	0.554	0.752	1.145	1.610	2.343	3.000	4.351	6.064	7.289	9.236	11.070	13.388	15.086
6	0.872	1.134	1.635	2.204	3.070	3.828	5.348	7.231	8.558	10.645	12.592	15.033	16.812
7	1.239	1.564	2.167	2.833	3.822	4.671	6.346	8.383	9.803	12.017	14.067	16.622	18.475
8	1.646	2.032	2.733	3.490	4.594	5.527	7.344	9.524	11.030	13.362	15.507	18.168	20.090
9	2.088	2.532	3.325	4.168	5.380	6.393	8.343	10.656	12.242	14.684	16.919	19.679	21.666
10	2.558	3.059	3.940	4.865	6.179	7.267	9.342	11.781	13.442	15.987	18.307	21.161	23.209
11	3.053	3.609	4.575	5.578	6.989	8.148	10.341	12.899	14.631	17.275	19.675	22.618	24.725
12	3.571	4.178	5.226	6.304	7.807	9.034	11.340	14.011	15.812	18.549	21.026	24.054	26.217
13	4.107	4.765	5.892	7.042	8.634	9.926	12.340	15.119	16.985	19.812	22.362	25.472	27.688
14	4.660	5.368	6.571	7.790	9.467	10.821	13.339	16.222	18.151	12.064	23.685	26.873	29.141
15	5.229	5.985	7.261	8.547	10.307	11.721	14.339	17.322	19.311	22.307	24.996	28.259	30.578
16	5.812	6.614	7.962	9.312	11.152	12.624	15.338	18.418	20.465	23.542	26.296	29.633	32.000
17	6.408	7.255	8.672	10.085	12.002	13.531	16.338	19.511	21.615	24.669	27.587	30.995	33.409
18	7.015	7.906	9.390	10.865	12.857	14.440	17.338	20.601	22.760	25.989	28.869	32.346	34.805
19	7.633	8.567	10.117	11.651	13.716	15.352	18.338	21.689	23.900	27.204	30.144	33.687	36.191
20	8.260	9.237	10.851	12.443	14.578	16.266	19.337	22.775	25.038	28.412	31.410	35.020	37.566
21	8.897	9.915	11.591	13.240	15.445	17.182	20.337	23.858	26.171	29.615	32.671	36.343	38.932
22	9.542	10.600	12.338	14.041	16.314	18.101	21.337	24.939	27.301	30.813	33.924	37.659	40.289
23	10.196	11.293	13.091	14.848	17.187	19.021	22.337	26.018	28.429	32.007	35.172	38.968	41.638
24	10.856	11.992	13.848	15.659	18.062	19.943	23.337	27.096	29.553	33.196	36.415	40.270	42.980
25	11.524	12.697	14.611	16.473	18.940	20.867	24.337	28.172	30.675	34.382	37.652	41.566	44.314
26	12.198	13.409	15.379	17.292	19.820	21.792	25.336	29.246	31.795	35.563	38.885	42.856	45.642
27	12.879	14.125	16.151	18.114	20.703	22.719	26.336	30.319	32.912	36.741	40.113	44.140	46.963
28	13.565	14.847	16.928	18.939	21.588	23.647	27.336	31.391	34.027	37.916	41.337	45.419	48.278
29	14.256	15.574	17.708	19.768	22.475	24.577	28.336	32.461	35.139	39.087	42.557	46.693	49.588
30	14.953	16.306	18.493	20.599	23.364	25.508	29.336	33.530	36.250	40.256	43.773	47.962	50.892

Source: Reproduced from *Statistical Methods for Research Workers*, 14th ed., 1972, with the permission of the Estate of R. A. Fisher, and Hafner Press.

[a]For degrees of freedom greater than 30 ...

Table ST4. Student's t-Distribution[a]

	α				
n	0.10	0.05	0.025	0.01	0.005
1	3.078	6.314	12.706	31.821	63.657
2	1.886	2.920	4.303	6.965	9.925
3	1.638	2.353	3.182	4.541	5.841
4	1.533	2.132	2.776	3.747	4.604
5	1.476	2.015	2.571	3.365	4.032
6	1.440	1.943	2.447	3.143	3.707
7	1.415	1.895	2.365	2.998	3.499
8	1.397	1.860	2.306	2.896	3.355
9	1.383	1.833	2.262	2.821	3.250
10	1.372	1.812	2.228	2.764	3.169
11	1.363	1.796	2.201	2.718	3.106
12	1.356	1.782	2.179	2.681	3.055
13	1.350	1.771	2.160	2.650	3.012
14	1.345	1.761	2.145	2.624	2.977
15	1.341	1.753	2.131	2.602	2.947
16	1.337	1.746	2.120	2.583	2.921
17	1.333	1.740	2.110	2.567	2.898
18	1.330	1.734	2.101	2.552	2.878
19	1.328	1.729	2.093	2.539	2.861
20	1.325	1.725	2.086	2.528	2.845
21	1.323	1.721	2.080	2.518	2.831
22	1.321	1.717	2.074	2.508	2.819
23	1.319	1.714	2.069	2.500	2.807
24	1.318	1.711	2.064	2.492	2.797
25	1.316	1.708	2.060	2.485	2.787
26	1.315	1.706	2.056	2.479	2.779
27	1.314	1.703	2.052	2.473	2.771
28	1.313	1.701	2.048	2.467	2.763
29	1.311	1.699	2.045	2.462	2.756
30	1.310	1.697	2.042	2.457	2.750
40	1.303	1.684	2.021	2.423	2.704
60	1.296	1.671	2.000	2.390	2.660
120	1.289	1.658	1.980	2.358	2.617
∞	1.282	1.645	1.960	2.326	2.576

Source: P. G. Hoel, *Introduction to Mathematical Statistics*, 4th ed., Wiley, New York, 1971, p. 393. Reprinted by permission of John Wiley & Sons, Inc.

[a]The first column lists the number of degrees of freedom (n). The headings of the other columns give probabilities (α) for t to exceed the entry value. Use symmetry for negative t values.

Table ST5. ***F*-Distribution: 5% (Lightface Type) and 1% (Boldface Type) Points for the Distribution of *F***

Degrees of Freedom for Denominator (n)	Degrees of Freedom for Numerator (m) 1	2	3	4	5	6	7	8	9	10	11	12	14	16	20	24	30	40	50	75	100	200	500	∞
1	161	200	216	225	230	234	237	239	241	242	243	244	245	246	248	249	250	251	252	253	253	254	254	254
	4052	**4999**	**5403**	**5625**	**5764**	**5859**	**5928**	**5981**	**6022**	**6056**	**6082**	**6106**	**6142**	**6169**	**6208**	**6234**	**6258**	**6286**	**6302**	**6323**	**6334**	**6352**	**6361**	**6366**
2	18.51	19.00	19.16	19.25	19.30	19.33	19.36	19.37	19.38	19.39	19.40	19.41	19.42	19.43	19.44	19.45	19.46	19.47	19.47	19.48	19.49	19.49	19.50	19.50
	98.49	**99.01**	**99.17**	**99.25**	**99.30**	**99.33**	**99.34**	**99.36**	**99.38**	**99.40**	**99.41**	**99.42**	**99.43**	**99.44**	**99.45**	**99.46**	**99.47**	**99.48**	**99.48**	**99.49**	**99.49**	**99.49**	**99.50**	**99.50**
3	10.13	9.55	9.28	9.12	9.01	8.94	8.88	8.84	8.81	8.78	8.76	8.74	8.71	8.69	8.66	8.64	8.62	8.60	8.58	8.57	8.56	8.54	8.54	8.53
	34.12	**30.81**	**29.46**	**28.71**	**28.24**	**27.91**	**27.67**	**27.49**	**27.34**	**27.23**	**27.13**	**27.05**	**26.92**	**26.83**	**26.69**	**26.60**	**26.50**	**26.41**	**26.30**	**26.27**	**26.23**	**26.18**	**26.14**	**26.12**
4	7.71	6.94	6.59	6.39	6.26	6.16	6.09	6.04	6.00	5.96	5.93	5.91	5.87	5.84	5.80	5.77	5.74	5.71	5.70	5.68	5.66	5.65	5.64	5.63
	21.20	**18.00**	**16.69**	**15.98**	**15.52**	**15.21**	**14.98**	**14.80**	**14.66**	**14.54**	**14.45**	**14.37**	**14.24**	**14.15**	**14.02**	**13.93**	**13.83**	**13.74**	**13.69**	**13.61**	**13.57**	**13.52**	**13.48**	**13.46**
5	6.61	5.79	5.41	5.19	5.05	4.95	4.88	4.82	4.78	4.74	4.70	4.68	4.64	4.60	4.56	4.53	4.50	4.46	4.44	4.42	4.40	4.38	4.37	4.36
	16.26	**13.27**	**12.06**	**11.39**	**10.97**	**10.67**	**10.45**	**10.27**	**10.15**	**10.05**	**9.96**	**9.89**	**9.77**	**9.68**	**9.55**	**9.47**	**9.38**	**9.29**	**9.24**	**9.17**	**9.13**	**9.07**	**9.04**	**9.02**
6	5.99	5.14	4.76	4.53	4.39	4.28	4.21	4.15	4.10	4.06	4.03	4.00	3.96	3.92	3.87	3.84	3.81	3.77	3.75	3.72	3.71	3.69	3.68	3.67
	13.74	**10.92**	**9.78**	**9.15**	**8.75**	**8.47**	**8.26**	**8.10**	**7.98**	**7.87**	**7.79**	**7.72**	**7.60**	**7.52**	**7.39**	**7.31**	**7.23**	**7.14**	**7.09**	**7.02**	**6.99**	**6.94**	**6.90**	**6.88**
7	5.59	4.74	4.35	4.12	3.97	3.87	3.79	3.73	3.68	3.63	3.60	3.57	3.52	3.49	3.44	3.41	3.38	3.34	3.32	3.29	3.28	3.25	3.24	3.23
	12.25	**9.55**	**8.45**	**7.85**	**7.46**	**7.19**	**7.00**	**6.84**	**6.71**	**6.62**	**6.54**	**6.47**	**6.35**	**6.27**	**6.15**	**6.07**	**5.98**	**5.90**	**5.85**	**5.78**	**5.75**	**5.70**	**5.67**	**5.65**
8	5.32	4.46	4.07	3.84	3.69	3.58	3.50	3.44	3.39	3.34	3.31	3.28	3.23	3.20	3.15	3.12	3.08	3.05	3.03	3.00	2.98	2.96	2.94	2.93
	11.26	**8.65**	**7.59**	**7.01**	**6.63**	**6.37**	**6.19**	**6.03**	**5.91**	**5.82**	**5.74**	**5.67**	**5.56**	**5.48**	**5.36**	**5.28**	**5.20**	**5.11**	**5.06**	**5.00**	**4.96**	**4.91**	**4.88**	**4.86**
9	5.12	4.26	3.86	3.63	3.48	3.37	3.29	3.23	3.18	3.13	3.10	3.07	3.02	2.98	2.93	2.90	2.86	2.82	2.80	2.77	2.76	2.73	2.72	2.71
	10.56	**8.02**	**6.99**	**6.42**	**6.06**	**5.80**	**5.62**	**5.47**	**5.35**	**5.26**	**5.18**	**5.11**	**5.00**	**4.92**	**4.80**	**4.73**	**4.64**	**4.56**	**4.51**	**4.45**	**4.41**	**4.36**	**4.33**	**4.31**
10	4.96	4.10	3.71	3.48	3.33	3.22	3.14	3.07	3.02	2.97	2.94	2.91	2.86	2.82	2.77	2.74	2.70	2.67	2.64	2.61	2.59	2.56	2.55	2.54
	10.04	**7.56**	**6.55**	**5.99**	**5.64**	**5.39**	**5.21**	**5.06**	**4.95**	**4.85**	**4.78**	**4.71**	**4.60**	**4.52**	**4.41**	**4.33**	**4.25**	**4.17**	**4.12**	**4.05**	**4.01**	**3.96**	**3.93**	**3.91**
11	4.84	3.98	3.59	3.36	3.20	3.09	3.01	2.95	2.90	2.96	2.82	2.79	2.74	2.70	2.65	2.61	2.57	2.53	2.50	2.47	2.45	2.42	2.41	2.40
	9.65	**7.20**	**6.22**	**5.67**	**5.32**	**5.07**	**4.88**	**4.74**	**4.63**	**4.54**	**4.46**	**4.40**	**4.29**	**4.21**	**4.10**	**4.02**	**3.94**	**3.86**	**3.80**	**3.74**	**3.70**	**3.66**	**3.62**	**3.60**
12	4.75	3.88	3.49	3.26	3.11	3.00	2.92	2.85	2.80	2.76	2.72	2.69	2.64	2.60	2.54	2.50	2.46	2.42	2.40	2.36	2.35	2.32	2.31	2.30
	9.33	**6.93**	**5.95**	**5.41**	**5.06**	**4.82**	**4.65**	**4.50**	**4.39**	**4.30**	**4.22**	**4.16**	**4.05**	**3.98**	**3.86**	**3.78**	**3.70**	**3.61**	**3.56**	**3.49**	**3.46**	**3.41**	**3.38**	**3.36**
13	4.67	3.80	3.41	3.18	3.02	2.92	2.84	2.77	2.72	2.67	2.63	2.60	2.55	2.51	2.46	2.42	2.38	2.34	2.32	2.28	2.26	2.24	2.22	2.21
	9.07	**6.70**	**5.74**	**5.20**	**4.86**	**4.62**	**4.44**	**4.30**	**4.19**	**4.10**	**4.02**	**3.96**	**3.85**	**3.78**	**3.67**	**3.59**	**3.51**	**3.42**	**3.37**	**3.30**	**3.27**	**3.21**	**3.18**	**3.16**

14	4.60	3.74	3.34	3.11	2.96	2.85	2.77	2.70	2.65	2.60	2.56	2.53	2.48	2.44	2.39	2.35	2.31	2.27	2.24	2.21	2.19	2.16	2.14	2.13
	8.86	**6.51**	**5.56**	**5.03**	**4.69**	**4.46**	**4.28**	**4.14**	**4.03**	**3.94**	**3.86**	**3.80**	**3.70**	**3.62**	**3.51**	**3.43**	**3.34**	**3.26**	**3.21**	**3.14**	**3.11**	**3.06**	**3.02**	**3.00**
15	4.54	3.68	3.29	3.06	2.90	2.79	2.70	2.64	2.59	2.55	2.51	2.48	2.43	2.39	2.33	2.29	2.25	2.21	2.18	2.15	2.12	2.10	2.08	2.07
	8.68	**6.36**	**5.42**	**4.89**	**4.56**	**4.32**	**4.14**	**4.00**	**3.89**	**3.80**	**3.73**	**3.67**	**3.56**	**3.48**	**3.36**	**3.29**	**3.20**	**3.12**	**3.07**	**3.00**	**2.97**	**2.92**	**2.89**	**2.87**
16	4.49	3.63	3.24	3.01	2.85	2.74	2.66	2.59	2.54	2.49	2.45	2.42	2.37	2.33	2.28	2.24	2.20	2.16	2.13	2.09	2.07	2.04	2.02	2.01
	8.53	**6.23**	**5.29**	**4.77**	**4.44**	**4.20**	**4.03**	**3.89**	**3.78**	**3.69**	**3.61**	**3.55**	**3.45**	**3.37**	**3.25**	**3.18**	**3.10**	**3.01**	**2.96**	**2.89**	**2.86**	**2.80**	**2.77**	**2.75**
17	4.45	3.59	3.20	2.96	2.81	2.70	2.62	2.55	2.50	2.45	2.41	2.38	2.33	2.29	2.23	2.19	2.15	2.11	2.08	2.04	2.02	1.99	1.97	1.96
	8.40	**6.11**	**5.18**	**4.67**	**4.34**	**4.10**	**3.93**	**3.79**	**3.68**	**3.59**	**3.52**	**3.45**	**3.35**	**3.27**	**3.16**	**3.08**	**3.00**	**2.92**	**2.86**	**2.79**	**2.76**	**2.70**	**2.67**	**2.65**
18	4.41	3.55	3.16	2.93	2.77	2.66	2.58	2.51	2.46	2.41	2.37	2.34	2.29	2.25	2.19	2.15	2.11	2.07	2.04	2.00	1.98	1.95	1.93	1.92
	8.28	**6.01**	**5.09**	**4.58**	**4.25**	**4.01**	**3.85**	**3.71**	**3.60**	**3.51**	**3.44**	**3.37**	**3.27**	**3.19**	**3.07**	**3.00**	**2.91**	**2.83**	**2.78**	**2.71**	**2.68**	**2.62**	**2.59**	**2.57**
19	4.38	3.52	3.13	2.90	2.74	2.63	2.55	2.48	2.43	2.38	2.34	2.31	2.26	2.21	2.15	2.11	2.07	2.02	2.00	1.96	1.94	1.91	1.90	1.88
	8.18	**5.93**	**5.01**	**4.50**	**4.17**	**3.94**	**3.77**	**3.63**	**3.52**	**3.43**	**3.36**	**3.30**	**3.19**	**3.12**	**3.00**	**2.92**	**2.84**	**2.76**	**2.70**	**2.63**	**2.60**	**2.54**	**2.51**	**2.49**
20	4.35	3.49	3.10	2.87	2.71	2.60	2.52	2.45	2.40	2.35	2.31	2.28	2.23	2.18	2.12	2.08	2.04	1.99	1.96	1.92	1.90	1.87	1.85	1.84
	8.10	**5.85**	**4.94**	**4.43**	**4.10**	**3.87**	**3.71**	**3.56**	**3.45**	**3.37**	**3.30**	**3.23**	**3.13**	**3.05**	**2.94**	**2.86**	**2.77**	**2.69**	**2.63**	**2.56**	**2.53**	**2.47**	**2.44**	**2.42**
21	4.32	3.47	3.07	2.84	2.68	2.57	2.49	2.42	2.37	2.32	2.28	2.25	2.20	2.15	2.09	2.05	2.00	1.96	1.93	1.89	1.87	1.84	1.82	1.81
	8.02	**5.78**	**4.87**	**4.37**	**4.04**	**3.81**	**3.65**	**3.51**	**3.40**	**3.31**	**3.24**	**3.17**	**3.07**	**2.99**	**2.88**	**2.80**	**2.72**	**2.63**	**2.58**	**2.51**	**2.47**	**2.42**	**2.38**	**2.36**
22	4.30	3.44	3.05	2.82	2.66	2.55	2.47	2.40	2.35	2.30	2.26	2.23	2.18	2.13	2.07	2.03	1.98	1.93	1.91	1.87	1.84	1.81	1.80	1.78
	7.94	**5.72**	**4.82**	**4.31**	**3.99**	**3.76**	**3.59**	**3.45**	**3.35**	**3.26**	**3.18**	**3.12**	**3.02**	**2.94**	**2.83**	**2.75**	**2.67**	**2.58**	**2.53**	**2.46**	**2.42**	**2.37**	**2.33**	**2.31**
23	4.28	3.42	3.03	2.80	2.64	2.53	2.45	2.38	2.32	2.28	2.24	2.20	2.14	2.10	2.04	2.00	1.96	1.91	1.88	1.84	1.82	1.79	1.77	1.76
	7.88	**5.66**	**4.76**	**4.26**	**3.94**	3.71	**3.54**	**3.41**	**3.30**	**3.21**	**3.14**	**3.07**	**2.97**	**2.89**	**2.78**	**2.70**	**2.62**	**2.53**	**2.48**	**2.41**	**2.37**	**2.32**	**2.28**	**2.26**
24	4.26	3.40	3.01	2.78	2.62	2.51	2.43	2.36	2.30	2.26	2.22	2.18	2.13	2.09	2.02	1.98	1.94	1.89	1.86	1.82	1.80	1.76	1.74	1.73
	7.82	**5.61**	**4.72**	**4.22**	**3.90**	**3.67**	**3.50**	**3.36**	**3.25**	**3.17**	**3.09**	**3.03**	**2.93**	**2.85**	**2.74**	**2.66**	**2.58**	**2.49**	**2.44**	**2.36**	**2.33**	**2.27**	**2.23**	**2.21**
25	4.24	3.38	2.99	2.76	2.60	2.49	2.41	2.34	2.28	2.24	2.20	2.16	2.11	2.06	2.00	1.96	1.92	1.87	1.84	1.80	1.77	1.74	1.72	1.71
	7.77	**5.57**	**4.68**	**4.18**	**3.86**	**3.63**	**3.46**	**3.32**	**3.21**	**3.13**	**3.05**	**2.99**	**2.89**	**2.81**	**2.70**	**2.62**	**2.54**	**2.45**	**2.40**	**2.32**	**2.29**	**2.23**	**2.19**	**2.17**
26	4.22	3.37	2.89	2.74	2.59	2.47	2.39	2.32	2.27	2.22	2.18	2.15	2.10	2.05	1.99	1.95	1.90	1.85	1.82	1.78	1.76	1.72	1.70	1.69
	7.72	**5.53**	**4.64**	**4.14**	**3.82**	**3.59**	**3.42**	**3.29**	**3.17**	**3.09**	**3.02**	**2.96**	**2.86**	**2.77**	**2.66**	**2.58**	**2.50**	**2.41**	**2.36**	**2.28**	**2.25**	**2.19**	**2.15**	**2.13**

Table ST5. *(Continued)*

Degrees of Freedom for Denominator (n)	Degrees of Freedom for Numerator, m																						
	1	2	3	4	5	6	7	8	9	10	11	12	14	16	20	24	30	40	50	75	100	200	500	∞
27	4.21	3.35	2.96	2.73	2.57	2.46	2.37	2.30	2.25	2.20	2.16	2.13	2.08	2.03	1.97	1.93	1.88	1.84	1.80	1.76	1.74	1.71	1.68	1.67
	7.68	**5.49**	**4.60**	**4.11**	**3.79**	**3.56**	**3.39**	**3.26**	**3.14**	**3.06**	**2.98**	**2.93**	**2.83**	**2.74**	**2.63**	**2.55**	**2.47**	**2.38**	**2.33**	**2.25**	**2.21**	**2.16**	**2.12**	**2.10**
28	4.20	3.34	2.95	2.71	2.56	2.44	2.36	2.29	2.24	2.19	2.15	2.12	2.06	2.02	1.96	1.91	1.87	1.81	1.78	1.75	1.72	1.69	1.67	1.65
	7.64	**5.45**	**4.57**	**4.07**	**3.76**	**3.53**	**3.36**	**3.23**	**3.11**	**3.03**	**2.95**	**2.90**	**2.80**	**2.71**	**2.60**	**2.52**	**2.44**	**2.35**	**2.30**	**2.22**	**2.18**	**2.13**	**2.09**	**2.06**
29	4.18	3.33	2.93	2.70	2.54	2.43	2.35	2.28	2.22	2.18	2.14	2.10	2.05	2.00	1.94	1.90	1.85	1.80	1.77	1.73	1.71	1.68	1.65	1.64
	7.60	**5.52**	**4.54**	**4.04**	**3.73**	**3.50**	**3.33**	**3.20**	**3.08**	**3.00**	**2.92**	**2.87**	**2.77**	**2.68**	**2.57**	**2.49**	**2.41**	**2.32**	**2.27**	**2.19**	**2.15**	**2.10**	**2.06**	**2.03**
30	4.17	3.32	2.92	2.69	2.53	2.42	2.34	2.27	2.21	2.16	2.12	2.09	2.04	1.99	1.93	1.89	1.84	1.79	1.76	1.72	1.69	1.66	1.64	1.62
	7.56	**5.39**	**4.51**	**4.02**	**3.70**	**3.47**	**3.30**	**3.17**	**3.06**	**2.98**	**2.90**	**2.84**	**2.74**	**2.66**	**2.55**	**2.47**	**2.38**	**2.29**	**2.24**	**2.16**	**2.13**	**2.07**	**2.03**	**2.01**
32	4.15	3.30	2.90	2.67	2.51	2.40	2.32	2.25	2.19	2.14	2.10	2.07	2.02	1.97	1.91	1.86	1.82	1.76	1.74	1.69	1.67	1.64	1.61	1.59
	7.50	**5.34**	**4.46**	**3.97**	**3.66**	**3.42**	**3.25**	**3.12**	**3.01**	**2.94**	**2.86**	**2.80**	**2.70**	**2.62**	**2.51**	**2.42**	**2.34**	**2.25**	**2.20**	**2.12**	**2.08**	**2.02**	**1.98**	**1.96**
34	4.13	3.28	2.88	2.65	2.49	2.38	2.30	2.23	2.17	2.12	2.08	2.05	2.00	1.95	1.89	1.84	1.80	1.74	1.71	1.67	1.64	1.61	1.59	1.57
	7.44	**5.29**	**4.42**	**3.93**	**3.61**	**3.38**	**3.21**	**3.08**	**2.97**	**2.89**	**2.82**	**2.76**	**2.66**	**2.58**	**2.47**	**2.38**	**2.30**	**2.21**	**2.15**	**2.08**	**2.04**	**1.98**	**1.94**	**1.91**
36	4.11	3.26	2.86	2.63	2.48	2.36	2.28	2.21	2.15	2.10	2.06	2.03	1.89	1.93	1.87	1.82	1.78	1.72	1.69	1.65	1.62	1.59	1.56	1.55
	7.39	**5.25**	**4.38**	**3.89**	**3.58**	**3.35**	**3.18**	**3.04**	**2.94**	**2.86**	**2.78**	**2.72**	**2.62**	**2.54**	**2.43**	**2.35**	**2.26**	**2.17**	**2.12**	**2.04**	**2.00**	**1.94**	**1.90**	**1.87**
38	4.10	3.25	2.85	2.62	2.46	2.35	2.26	2.19	2.14	2.09	2.05	2.02	1.96	1.92	1.85	1.80	1.76	1.71	1.67	1.63	1.60	1.57	1.54	1.53
	7.35	**5.21**	**4.34**	**3.86**	**3.54**	**3.32**	**3.15**	**3.02**	**2.91**	**2.82**	**2.75**	**2.69**	**2.59**	**2.51**	**2.30**	**2.32**	**2.22**	**2.14**	**2.08**	**2.00**	**1.97**	**1.90**	**1.86**	**1.84**
40	4.08	3.23	2.84	2.61	2.45	2.34	2.25	2.18	2.12	2.07	2.04	2.00	1.95	1.90	1.84	1.79	1.74	1.69	1.66	1.61	1.59	1.55	1.53	1.51
	7.31	**5.18**	**4.31**	**3.83**	**3.51**	**3.29**	**3.12**	**2.99**	**2.88**	**2.80**	**2.73**	**2.66**	**2.56**	**2.49**	**2.37**	**2.29**	**2.20**	**2.11**	**2.05**	**1.97**	**1.94**	**1.88**	**1.84**	**1.81**
42	4.07	3.22	2.83	2.59	2.44	2.32	2.24	2.17	2.11	2.06	2.02	1.99	1.94	1.89	1.82	1.78	1.73	1.68	1.64	1.60	1.57	1.54	1.51	1.49
	7.27	**5.15**	**4.29**	**3.80**	**3.49**	**3.26**	**3.10**	**2.96**	**2.86**	**2.77**	**2.70**	**2.64**	**2.54**	**2.46**	**2.35**	**2.26**	**2.17**	**2.08**	**2.02**	**1.94**	**1.91**	**1.85**	**1.80**	**1.78**
44	4.06	3.21	2.82	2.58	2.43	2.31	2.23	2.16	2.10	2.05	2.01	1.98	1.92	1.88	1.81	1.76	1.72	1.66	1.63	1.58	1.56	1.52	1.50	1.48
	7.24	**5.12**	**4.26**	**3.78**	**3.46**	**3.24**	**3.07**	**2.94**	**2.84**	**2.75**	**2.68**	**2.62**	**2.52**	**2.44**	**2.32**	**2.24**	**2.15**	**2.06**	**2.00**	**1.92**	**1.88**	**1.82**	**1.78**	**1.75**
46	4.05	3.20	2.81	2.57	2.42	2.30	2.22	2.14	2.09	2.04	2.00	1.97	1.91	1.87	1.80	1.75	1.71	1.65	1.62	1.57	1.54	1.51	1.48	1.46
	7.21	**5.10**	**4.24**	**3.76**	**3.44**	**3.22**	**3.05**	**2.92**	**2.82**	**2.73**	**2.66**	**2.60**	**2.50**	**2.42**	**2.40**	**2.22**	**2.13**	**2.04**	**1.98**	**1.90**	**1.86**	**1.80**	**1.76**	**1.72**
48	4.04	3.19	2.80	2.56	2.41	2.30	2.21	2.14	2.08	2.03	1.99	1.96	1.90	1.86	1.79	1.74	1.70	1.64	1.61	1.56	1.53	1.50	1.47	1.45
	7.19	**5.08**	**4.22**	**3.74**	**3.42**	**3.20**	**3.04**	**2.90**	**2.80**	**2.71**	**2.64**	**2.58**	**2.48**	**2.40**	**2.28**	**2.20**	**2.11**	**2.02**	**1.96**	**1.88**	**1.84**	**1.78**	**1.73**	**1.70**

50	4.03	3.18	2.79	2.56	2.40	2.29	2.20	2.13	2.07	2.02	1.98	1.95	1.90	1.85	1.78	1.74	1.69	1.63	1.60	1.55	1.52	1.48	1.46	1.44
	7.17	**5.06**	**4.20**	**3.72**	**3.41**	**3.18**	**3.02**	**2.88**	**2.78**	**2.70**	**2.62**	**2.56**	**2.46**	**2.39**	**2.26**	**2.18**	**2.10**	**2.00**	**1.94**	**1.86**	**1.82**	**1.76**	**1.71**	**1.68**
55	4.02	3.17	2.78	2.54	2.38	2.27	2.18	2.11	2.05	2.00	1.97	1.93	1.88	1.83	1.76	1.72	1.67	1.61	1.58	1.52	1.50	1.46	1.43	1.41
	7.12	**5.01**	**4.16**	**3.68**	**3.37**	**3.15**	**2.98**	**2.85**	**2.75**	**2.66**	**2.59**	**2.53**	**2.43**	**2.35**	**2.23**	**2.15**	**2.06**	**1.96**	**1.90**	**1.82**	**1.78**	**1.71**	**1.66**	**1.64**
60	4.00	3.15	2.76	2.52	2.37	2.25	2.17	2.10	2.04	1.99	1.95	1.92	1.86	1.81	1.75	1.70	1.65	1.59	1.56	1.50	1.48	1.44	1.41	1.39
	7.08	**4.98**	**4.13**	**3.65**	**3.34**	**3.12**	**2.95**	**2.82**	**2.72**	**2.63**	**2.56**	**2.50**	**2.40**	**2.32**	**2.20**	**2.12**	**2.03**	**1.93**	**1.87**	**1.79**	**1.74**	**1.68**	**1.63**	**1.60**
65	3.99	3.14	2.75	2.51	2.36	2.24	2.15	2.08	2.02	1.98	1.94	1.90	1.85	1.80	1.73	1.68	1.63	1.57	1.54	1.49	1.46	1.42	1.39	1.37
	7.04	**4.95**	**4.10**	**3.62**	**3.31**	**3.09**	**2.93**	**2.79**	**2.70**	**2.61**	**2.54**	**2.47**	**2.30**	**2.37**	**2.18**	**2.09**	**2.00**	**1.90**	**1.84**	**1.76**	**1.71**	**1.64**	**1.60**	**1.56**
70	3.98	3.13	2.74	2.50	2.35	2.32	2.14	2.07	2.01	1.97	1.93	1.89	1.84	1.79	1.72	1.67	1.62	1.56	1.53	1.47	1.45	1.40	1.37	1.35
	7.01	**4.92**	**4.08**	**3.60**	**3.29**	**3.07**	**2.91**	**2.77**	**2.67**	**2.59**	**2.51**	**2.45**	**2.35**	**2.28**	**2.15**	**2.07**	**1.98**	**1.88**	**1.82**	**1.74**	**169**	**1.63**	**1.56**	**1.53**
80	3.96	3.11	2.72	2.48	2.33	2.21	2.12	2.05	1.99	1.95	1.91	1.88	1.82	1.77	1.70	1.65	1.60	1.54	1.51	1.45	1.42	1.38	1.35	1.32
	6.96	**4.88**	**4.04**	**3.56**	**3.25**	**3.04**	**2.87**	**2.74**	**2.64**	**2.55**	**2.48**	**2.41**	**2.32**	**2.24**	**2.11**	**2.03**	**1.94**	**1.84**	**1.78**	**1.70**	**1.65**	**1.57**	**1.52**	**1.49**
100	3.94	3.09	2.70	2.46	2.30	2.19	2.10	2.03	1.97	1.92	1.88	1.85	1.79	1.75	1.68	1.63	1.57	1.51	1.48	1.42	1.39	1.34	1.30	1.28
	6.90	**4.82**	**3.98**	**3.51**	**3.20**	**2.99**	**2.82**	**2.69**	**2.59**	**2.51**	**2.43**	**2.36**	**2.26**	**2.19**	**2.06**	**1.98**	**1.89**	**1.79**	**1.73**	**1.64**	**1.59**	**1.51**	**1.46**	**1.43**
125	3.92	3.07	2.68	2.44	2.29	2.17	2.08	2.01	1.95	1.90	1.86	1.83	1.77	1.72	1.65	1.60	1.55	1.49	1.45	1.39	1.36	1.31	1.27	1.25
	6.84	**4.78**	**3.94**	**3.47**	**3.17**	**2.95**	**2.79**	**2.65**	**2.56**	**2.47**	**2.40**	**2.33**	**2.23**	**2.15**	**2.03**	**1.94**	**1.85**	**1.75**	**1.68**	**1.59**	**1.54**	**1.46**	**1.40**	**1.37**
150	3.91	3.06	2.67	2.43	2.27	2.16	2.07	2.00	1.94	1.89	1.85	1.82	1.76	1.71	1.64	1.59	1.54	1.47	1.44	1.37	1.34	1.29	1.25	1.22
	6.81	**4.75**	**3.91**	**3.44**	**3.13**	**2.92**	**2.76**	**2.62**	**2.53**	**2.44**	**2.37**	**2.30**	**2.20**	**2.12**	**2.00**	**1.91**	**1.83**	**1.72**	**1.66**	**1.56**	**1.51**	**1.43**	**1.37**	**1.33**
200	3.89	3.04	2.65	2.41	2.26	2.14	2.05	1.98	1.92	1.87	1.83	1.80	1.74	1.69	1.62	0.157	1.52	1.45	1.42	1.35	1.32	1.26	1.22	1.19
	6.76	**4.71**	**3.88**	**3.41**	**3.11**	**2.90**	**2.73**	**2.60**	**2.50**	**2.41**	**2.34**	**2.28**	**1.17**	**2.09**	**1.97**	**1.88**	**1.79**	**1.69**	**1.62**	**1.53**	**1.48**	**1.39**	**1.33**	**1.28**
400	3.86	3.02	2.62	2.39	2.23	2.12	2.03	1.96	190	1.85	1.81	1.78	1.72	1.67	1.60	1.54	1.49	1.42	1.38	1.32	1.28	1.22	1.16	1.13
	6.70	**4.66**	**3.83**	**3.36**	**3.06**	**2.85**	**2.69**	**2.55**	**2.46**	**2.37**	**2.29**	**2.23**	**2.12**	**2.04**	**1.92**	**1.84**	**1.74**	**1.64**	**1.57**	**1.47**	**1.42**	**1.32**	**1.24**	**1.19**
1000	3.85	3.00	2.61	2.38	2.22	2.10	2.02	1.95	1.89	1.84	1.80	1.76	1.70	1.65	1.58	1.53	1.47	1.41	1.36	1.30	1.26	1.19	1.13	1.08
	6.66	**4.62**	**3.80**	**3.34**	**3.04**	**2.82**	**2.66**	**2.53**	**2.43**	**2.34**	**2.26**	**2.20**	**2.09**	**2.01**	**1.89**	**1.81**	**1.71**	**1.61**	**1.54**	**1.44**	**1.38**	**1.28**	**1.19**	**1.11**
∞	3.84	2.99	2.60	2.37	2.21	2.09	2.01	1.94	1.88	1.83	1.79	1.75	1.69	1.64	1.57	1.52	1.46	1.40	1.35	1.28	1.24	1.17	1.11	1.00
	6.64	**4.60**	**3.78**	**3.32**	**3.02**	**2.80**	**2.64**	**2.51**	**2.41**	**2.32**	**2.24**	**2.18**	**2.07**	**1.99**	**1.87**	**1.79**	**1.69**	**1.59**	**1.52**	**1.41**	**1.36**	**1.25**	**1.15**	**1.00**

Source: Reprinted by permission from George W. Snedecor and William G. Cochran, *Statistical Methods*, 6th ed., © 1967 by Iowa State University Press, Ames, I.A.

Table ST6. Random Normal Numbers, $\mu = 0$ and $\sigma = 1$

1	2	3	4	5	6	7	8	9	10
0.464	0.137	2.455	−0.323	−0.068	0.290	−0.288	1.298	0.241	−0.957
0.060	−2.526	−0.531	−0.194	0.543	−1.558	0.187	−1.190	0.022	0.525
1.486	−0.354	−0.634	0.697	0.926	1.375	0.785	−0.963	−0.853	−1.865
1.022	−0.472	1.279	3.521	0.571	−1.851	0.194	1.192	−0.501	−0.273
1.394	−0.555	0.046	0.321	2.945	1.974	−0.258	0.412	0.439	−0.035
0.906	−0.513	−0.525	0.595	0.881	−0.934	1.579	0.161	−1.885	0.371
1.179	−1.055	0.007	0.769	0.971	0.712	1.090	−0.631	−0.255	−0.702
−1.501	−0.488	−0.162	−0.136	1.033	0.203	0.448	0.748	−0.423	−0.432
−0.690	0.756	−1.618	−0.345	−0.511	−2.051	−0.457	−0.218	0.857	−0.465
1.372	0.225	0.378	0.761	0.181	−0.736	0.960	−1.530	−0.260	0.120
−0.482	1.678	−0.057	−1.229	−0.486	0.856	−0.491	−1.983	−2.830	−0.238
−1.376	−0.150	1.356	−0.561	−0.256	−0.212	0.219	0.779	0.953	−0.869
−1.010	0.598	−0.918	1.598	0.065	0.415	−0.169	0.313	−0.973	−1.016
−0.005	−0.899	0.012	−0.725	1.147	−0.121	1.096	0.481	−1.691	0.417
1.393	1.163	−0.911	1.231	−0.199	−0.246	1.239	−2.574	−0.558	0.056
−1.787	−0.261	1.237	1.046	−0.508	−1.630	−0.146	−0.392	−0.627	0.561
−0.105	−0.357	−1.384	0.360	−0.992	−0.116	−1.698	−2.832	−1.108	−2.357
−1.339	1.827	−0.959	0.424	0.969	−1.141	−1.041	0.362	−1.726	1.956
1.041	0.535	0.731	1.377	0.983	−1.330	1.620	−1.040	0.524	−0.281
0.279	−2.056	0.717	−0.873	−1.096	−1.396	1.047	0.089	−0.573	0.932
−1.805	−2.008	−1.633	0.542	0.250	−0.166	0.032	0.079	0.471	−1.029
−1.186	1.180	1.114	0.882	1.265	−0.202	0.151	−0.376	−0.310	0.479
0.658	−1.141	1.151	−1.210	0.927	0.425	0.290	−0.902	0.610	2.709
−0.439	0.358	−1.939	0.891	−0.227	0.602	0.873	−0.437	−0.220	−0.057
−1.399	−0.230	0.385	−0.649	−0.577	0.237	−0.289	0.513	0.738	−0.300
0.199	0.208	−1.083	−0.219	−0.291	1.221	1.119	0.004	−2.015	−0.594
0.159	0.272	−0.313	0.084	−2.828	−0.430	−0.792	−1.275	−0.623	−1.047
2.273	0.606	0.606	−0.747	0.247	1.291	0.063	−1.793	−0.699	−1.347
0.041	−0.307	0.121	0.790	−0.584	0.541	0.484	−0.986	0.481	0.996
−1.132	−2.098	0.921	0.145	0.446	−1.661	1.045	−1.363	−0.586	−1.023
0.768	0.079	−1.473	0.034	−2.127	0.665	0.084	−0.880	−0.579	0.551
0.375	−1.658	−0.851	0.234	−0.656	0.340	−0.086	−0.158	−0.120	0.418
−0.513	−0.344	0.210	−0.736	1.041	0.008	0.427	−0.831	0.191	0.074
0.292	−0.521	1.266	−1.206	−0.899	0.110	−0.528	−0.813	0.071	0.524
1.026	2.990	−0.574	−0.491	−1.114	1.297	−1.433	−1.345	−3.001	0.479
−1.334	1.278	−0.568	−0.109	−0.515	−0.566	2.923	0.500	0.359	0.326
−0.287	−0.144	−0.254	0.574	−0.451	−1.181	−1.190	−0.318	−0.094	1.114
0.161	−0.886	−0.921	−0.509	1.410	−0.518	0.192	−0.432	1.501	1.068
−1.346	0.193	−1.202	0.394	−1.045	0.843	0.942	1.045	0.031	0.772
1.250	−0.199	−0.288	1.810	1.378	0.584	1.216	0.733	0.402	0.226
0.630	−0.537	0.782	0.060	0.499	−0.431	1.705	1.164	0.884	−0.298
0.375	−1.941	0.247	−0.491	0.665	−0.135	−0.145	−0.498	0.457	1.064

(*Continued*)

1	2	3	4	5	6	7	8	9	10
−1.420	0.489	−1.711	−1.186	0.754	−0.732	−0.066	1.006	−0.798	0.162
−0.151	−0.243	−0.430	−0.762	0.298	1.049	1.810	2.885	−0.768	−0.129
−0.309	0.531	0.416	−1.541	1.456	2.040	−0.124	0.196	0.023	−1.204
0.424	−0.444	0.593	0.993	−0.106	0.116	0.484	−1.272	1.066	1.097
0.593	0.658	−1.127	−1.407	−1.579	−1.616	1.458	1.262	0.736	−0.916
0.862	−0.885	−0.142	−0.504	0.532	1.381	0.022	−0.281	−0.342	1.222
0.235	−0.628	−0.023	−0.463	−0.899	−0.394	−0.538	1.707	−0.188	−1.153
−0.853	0.402	0.777	0.833	0.410	−0.349	−1.094	0.580	1.395	1.298

Source: From tables of the RAND Corporation, by permission.

Table ST7. Critical Values of the Kolmogorov–Smirnov One-Sample Test Statistic[a]

One-Sided Test:											
$\alpha=$	0.10	0.05	0.025	0.01	0.005	$\alpha=$	0.10	0.05	0.025	0.01	0.005
Two-Sided Test:											
$\alpha=$	0.20	0.10	0.05	0.02	0.01	$\alpha=$	0.20	0.10	0.05	0.02	0.01
$n=1$	0.900	0.950	0.975	0.990	0.995	$n=21$	0.226	0.259	0.287	0.321	0.344
2	0.684	0.776	0.842	0.900	0.929	22	0.221	0.253	0.281	0.314	0.337
3	0.565	0.636	0.708	0.785	0.829	23	0.216	0.247	0.275	0.307	0.330
4	0.493	0.565	0.624	0.689	0.734	24	0.212	0.242	0.269	0.301	0.323
5	0.447	0.509	0.563	0.627	0.669	25	0.208	0.238	0.264	0.295	0.317
6	0.410	0.468	0.519	0.577	0.617	26	0.204	0.233	0.259	0.290	0.311
7	0.381	0.436	0.483	0.538	0.576	27	0.200	0.229	0.254	0.284	0.305
8	0.358	0.410	0.454	0.507	0.542	28	0.197	0.225	0.250	0.279	0.300
9	0.339	0.387	0.430	0.480	0.513	29	0.193	0.221	0.246	0.275	0.295
10	0.323	0.369	0.409	0.457	0.489	30	0.190	0.218	0.242	0.270	0.290
11	0.308	0.352	0.391	0.437	0.468	31	0.187	0.214	0.238	0.266	0.285
12	0.296	0.338	0.375	0.419	0.449	32	0.184	0.211	0.234	0.262	0.281
13	0.285	0.325	0.361	0.404	0.432	33	0.182	0.208	0.231	0.258	0.277
14	0.275	0.314	0.349	0.390	0.418	34	0.179	0.205	0.227	0.254	0.273
15	0.266	0.304	0.338	0.377	0.404	35	0.177	0.202	0.224	0.251	0.269
16	0.258	0.295	0.327	0.366	0.392	36	0.174	0.199	0.221	0.247	0.265
17	0.250	0.286	0.318	0.355	0.381	37	0.172	0.196	0.218	0.244	0.262
18	0.244	0.279	0.309	0.346	0.371	38	0.170	0.194	0.215	0.241	0.258
19	0.237	0.271	0.301	0.337	0.361	39	0.168	0.191	0.213	0.238	0.255
20	0.232	0.265	0.294	0.329	0.352	40	0.165	0.189	0.210	0.235	0.252
						Approximation for $n > 40$	$\frac{1.07}{\sqrt{n}}$	$\frac{1.22}{\sqrt{n}}$	$\frac{1.36}{\sqrt{n}}$	$\frac{1.52}{\sqrt{n}}$	$\frac{1.63}{\sqrt{n}}$

Source: Adapted by permission from Table 1 of Leslie H. Miller, Table of Percentage points of Kolmogrov statistics, *J. Am. Stat. Assoc.* 51 (1956), 111–121.

[a]This table gives the values of $D^+_{n,\alpha}$ and $D_{n,\alpha}$ for which $\alpha \geq P\{D^+_n > D^+_{n,\alpha}\}$ and $\alpha \geq P\{D_n > D_{n,\alpha}\}$ for some selected values of n and α.

Table ST8. Critical Values of the Kolmogorov–Smirnov Test Statistic for Two Samples of Equal Size[a]

One-Sided Test: $\alpha =$	0.10	0.05	0.025	0.01	0.005	$\alpha =$	0.10	0.05	0.025	0.01	0.005
Two-Sided Test: $\alpha =$	**0.20**	**0.10**	**0.05**	**0.02**	**0.01**	**$\alpha =$**	**0.20**	**0.10**	**0.05**	**0.02**	**0.01**
$n = 3$	2/3	2/3				$n = 20$	6/20	7/20	8/20	9/20	10/20
4	3/4	3/4	3/4			21	6/21	7/21	8/21	9/21	10/21
5	3/5	3/5	4/5	4/5	4/5	22	7/22	8/22	8/22	10/22	10/22
6	3/6	4/6	4/6	5/6	5/6	23	7/23	8/23	9/23	10/23	10/23
7	4/7	4/7	5/7	5/7	5/7	24	7/24	8/24	9/24	10/24	11/24
8	4/8	4/8	5/8	5/8	6/8	25	7/25	8/25	9/25	10/25	11/25
9	4/9	5/9	5/9	6/9	6/9	26	7/26	8/26	9/26	10/26	11/26
10	4/10	5/10	6/10	6/10	7/10	27	7/27	8/27	9/27	11/27	11/27
11	5/11	5/11	6/11	7/11	7/11	28	8/28	9/28	10/28	11/28	12/28
12	5/12	5/12	6/12	7/12	7/12	29	8/29	9/29	10/29	11/29	12/29
13	5/13	6/13	6/13	7/13	8/13	30	8/30	9/30	10/30	11/30	12/30
14	5/14	6/14	7/14	7/14	8/14	31	8/31	9/31	10/31	11/31	12/31
15	5/15	6/15	7/15	8/15	8/15	32	8/32	9/32	10/32	12/32	12/32
16	6/16	6/16	7/16	8/16	9/16	34	8/34	10/34	11/34	12/34	13/34
17	6/17	7/17	7/17	8/17	9/17	36	9/36	10/36	11/36	12/36	13/36
18	6/18	7/18	8/18	9/18	9/18	38	9/38	10/38	11/38	13/38	14/38
19	6/19	7/19	8/19	9/19	9/19	40	9/40	10/40	12/40	13/40	14/40
						Approximation for $n > 40$:	$\frac{1.52}{\sqrt{n}}$	$\frac{1.73}{\sqrt{n}}$	$\frac{1.92}{\sqrt{n}}$	$\frac{2.15}{\sqrt{n}}$	$\frac{2.30}{\sqrt{n}}$

Source: Adapted by permission from Tables 2 and 3 of Z. W. Birnbaum and R. A. Hall, Small sample distributions for multisample statistics of the Smirnov type, *Ann. Math. Stat.* 31 (1960), 710–720.

[a]This table gives the values of $D^+_{n,n,\alpha}$ and $D_{n,n,\alpha}$ for which $\alpha \geq P\{D^+_{n,n} > D^+_{n,n,\alpha}\}$ and $\alpha \geq P\{D_{n,n} > D_{n,n,\alpha}\}$ for some selected values of n and α.

Table ST9. Critical Values of the Kolmogorov–Smirnov Test Statistic for Two Samples of Unequal Size[a]

One-Sided Test:	$\alpha =$	0.10	0.05	0.025	0.01	0.005
Two-Sided Test:	$\alpha =$	0.20	0.10	0.05	0.02	0.01
$N_1 = 1$	$N_2 = 9$	17/18				
	10	9/10				
$N_1 = 2$	$N_2 = 3$	5/6				
	4	3/4				
	5	4/5	4/5			
	6	5/6	5/6			
	7	5/7	6/7			
	8	3/4	7/8	7/8		
	9	7/9	8/9	8/9		
	10	7/10	4/5	9/10		
$N_1 = 3$	$N_2 = 4$	3/4	3/4			
	5	2/3	4/5	4/5		
	6	2/3	2/3	5/6		
	7	2/3	5/7	6/7	6/7	
	8	5/8	3/4	3/4	7/8	
	9	2/3	2/3	7/9	8/9	8/9
	10	3/5	7/10	4/5	9/10	9/10
	12	7/12	2/3	3/4	5/6	11/12
$N_1 = 4$	$N_2 = 5$	3/5	3/4	4/5	4/5	
	6	7/12	2/3	3/4	5/6	5/6
	7	17/28	5/7	3/4	6/7	6/7
	8	5/8	5/8	3/4	7/8	7/8
	9	5/9	2/3	3/4	7/9	8/9
	10	11/20	13/20	7/10	4/5	4/5
	12	7/12	2/3	2/3	3/4	5/6
	16	9/16	5/8	11/16	3/4	13/16
$N_1 = 5$	$N_2 = 6$	3/5	2/3	2/3	5/6	5/6
	7	4/7	23/35	5/7	29/35	6/7
	8	11/20	5/8	27/40	4/5	4/5
	9	5/9	3/5	31/45	7/9	4/5
	10	1/2	3/5	7/10	7/10	4/5
	15	8/15	3/5	2/3	11/15	11/15
	20	1/2	11/20	3/5	7/10	3/4

Table ST9. (*Continued*)

One-Sided Test:	$\alpha =$	0.10	0.05	0.025	0.01	0.005
Two-Sided Test:	$\alpha =$	0.20	0.10	0.05	0.02	0.01
$N_1 = 6$	$N_2 = 7$	23/42	4/7	29/42	5/7	5/6
	8	1/2	7/12	2/3	3/4	3/4
	9	1/2	5/9	2/3	13/18	7/9
	10	1/2	17/30	19/30	7/10	11/15
	12	1/2	7/12	7/12	2/3	3/4
	18	4/9	5/9	11/18	2/3	13/18
	24	11/24	1/2	7/12	5/8	2/3
$N_1 = 7$	$N_2 = 8$	27/56	33/56	5/8	41/56	3/4
	9	31/63	5/9	40/63	5/7	47/63
	10	33/70	39/70	43/70	7/10	5/7
	14	3/7	1/2	4/7	9/14	5/7
	28	3/7	13/28	15/28	17/28	9/14
$N_1 = 8$	$N_2 = 9$	4/9	13/24	5/8	2/3	3/4
	10	19/40	21/40	23/40	27/40	7/10
	12	11/24	1/2	7/12	5/8	2/3
	16	7/16	1/2	9/16	5/8	5/8
	32	13/32	7/16	1/2	9/16	19/32
$N_1 = 9$	$N_2 = 10$	7/15	1/2	26/45	2/3	31/45
	12	4/9	1/2	5/9	11/18	2/3
	15	19/45	22/45	8/15	3/5	29/45
	18	7/18	4/9	1/2	5/9	11/18
	36	13/36	5/12	17/36	19/36	5/9
$N_1 = 10$	$N_2 = 15$	2/5	7/15	1/2	17/30	19/30
	20	2/5	9/20	1/2	11/20	3/5
	40	7/20	2/5	9/20	1/2	
$N_1 = 12$	$N_2 = 15$	23/60	9/20	1/2	11/20	7/12
	16	3/8	7/16	23/48	13/24	7/12
	18	13/36	5/12	17/36	19/36	5/9
	20	11/30	5/12	7/15	31/60	17/30
$N_1 = 15$	$N_2 = 20$	7/20	2/5	13/30	29/60	31/60
$N_1 = 16$	$N_2 = 20$	27/80	31/80	17/40	19/40	41/80
Large-sample approximation		$1.07\sqrt{\frac{m+n}{mn}}$	$1.22\sqrt{\frac{m+n}{mn}}$	$1.36\sqrt{\frac{m+n}{mn}}$	$1.52\sqrt{\frac{m+n}{mn}}$	$1.63\sqrt{\frac{m+n}{mn}}$

Source: Adapted by permission from F. J. Massey, Distribution table for the deviation between two sample cumulatives, *Ann. Math. Stat.* 23 (1952), 435–441.

[a]This table gives the values of $D^+_{m,n,\alpha}$ and $D_{m,n,\alpha}$ for which $\alpha \geq P\{D^+_{m,n} > D^+_{m,n,\alpha}\}$ and $\alpha \geq P\{D_{m,n} > D_{m,n,\alpha}\}$ for some selected values of N_1 = smaller sample size, N_2 = larger sample size, and α.

Table ST10. Critical Values of the Wilcoxon Signed-Ranks Test Statistic[a]

	α			
n	0.01	0.025	0.05	0.10
3	6	6	6	6
4	10	10	10	9
5	15	15	14	12
6	21	20	18	17
7	27	25	24	22
8	34	32	30	27
9	41	39	36	34
10	49	46	44	40
11	58	55	52	48
12	67	64	60	56
13	78	73	69	64
14	89	84	79	73
15	100	94	89	83
16	112	106	100	93
17	125	118	111	104
18	138	130	123	115
19	152	143	136	127
20	166	157	149	140

Source: Adapted by permission from Table 1 of R. L. McCornack, Extended tables of the Wilcoxon matched pairs signed-rank statistics, *J. Am. Stat. Assoc.* 60 (1965), 864–871.

[a]This table gives values of t_α for which $P\{T^+ > t_\alpha\} \leq \alpha$ for selected values of n and α. Critical values in the lower tail may be obtained by symmetry from the equation $t_{1-\alpha} = n(n+1)/2 - t_\alpha$.

Table ST11. Critical Values of the Mann–Whitney–Wilcoxon Test Statistic[a]

		n								
m	α	2	3	4	5	6	7	8	9	10
2	0.01	4	6	8	10	12	14	16	18	20
	0.025	4	6	8	10	12	14	15	17	19
	0.05	4	6	8	9	11	13	14	16	18
	0.10	4	5	7	8	10	12	13	15	16
3	0.01		9	12	15	18	20	20	25	28
	0.025		9	12	14	16	19	21	24	26
	0.05		8	11	13	15	18	20	22	25
	0.10		7	10	12	14	16	18	21	23
4	0.01			16	19	22	26	29	32	36
	0.025			15	18	21	24	27	31	34
	0.05			14	17	20	23	26	29	32
	0.10			12	15	18	21	24	26	29
5	0.01				23	27	31	35	39	43
	0.025				22	26	29	33	37	41
	0.05				20	24	28	31	35	38
	0.10				19	22	26	29	32	36
6	0.01					32	37	41	46	51
	0.025					30	35	39	43	48
	0.05					28	33	37	41	45
	0.10					26	30	34	38	42
7	0.01						42	48	53	58
	0.025						40	45	50	55
	0.05						37	42	47	52
	0.10						35	39	44	48
8	0.01							54	60	66
	0.025							50	56	62
	0.05							48	53	59
	0.10							44	49	55
9	0.01								66	73
	0.025								63	69
	0.05								59	65
	0.10								55	61
10	0.01									80
	0.025									76
	0.05									72
	0.10									67

Source: Adapted by permission from Table 1 of L. R. Verdooren, Extended tables of critical values for Wilcoxon's test statistic, *Biometrika* 50 (1963), 177–186, with the kind permission of Professor E. S. Pearson, the author, and the *Biometrika* Trustees.

[a]This table gives values of u_α for which $P\{U > u_\alpha\} \leq \alpha$ for some selected values of *m*, *n*, and α. Critical values in the lower tail may be obtained by symmetry from the equation $u_{1-\alpha} = mn - u_\alpha$.

Table ST12. Critical Points of Kendall's Tau Test Statistic[a]

	α			
n	0.100	0.050	0.025	0.01
3	3	3	3	3
4	4	4	6	6
5	6	6	8	8
6	7	9	11	11
7	9	11	13	15
8	10	14	16	18
9	12	16	18	22
10	15	19	21	25

Source: Adapted by permission from Table 1, p. 173, of M. G. Kendall, *Rank Correlation Methods*, 3rd ed., Griffin, London, 1962. For values of $n \geq 11$, see W. J. Conover, *Practical Nonparametric Statistics*, John Wiley, New York, 1971, p. 390.

[a]This table gives the values of S_α for which $P\{S > S_\alpha\} \leq \alpha$, where $S = \binom{n}{2}T$, for some selected values of α and n. Values in the lower tail may be obtained by symmetry, $S_{1-\alpha} = -S_\alpha$.

Table ST13. Critical Values of Spearman's Rank Correlation Statistic[a]

	α			
n	0.01	0.025	0.05	0.10
3	1.000	1.000	1.000	1.000
4	1.000	1.000	0.800	0.800
5	0.900	0.900	0.800	0.700
6	0.886	0.829	0.771	0.600
7	0.857	0.750	0.679	0.536
8	0.810	0.714	0.619	0.500
9	0.767	0.667	0.583	0.467
10	0.721	0.636	0.552	0.442

Source: Adapted by permission from Table 2, pp. 174–175, of M. G. Kendall, *Rank Correlation Methods*, 3rd ed., Griffin, London, 1962. For values of $n \geq 11$, see W. J. Conover, *Practical Nonparametric Statistics*, John Wiley, New York, 1971, p. 391.

[a]This table gives the values of R_α for which $P\{R > R_\alpha\} \leq \alpha$ for some selected values of n and α. Critical values in the lower tail may be obtained by symmetry, $R_{1-\alpha} = -R_\alpha$.

ANSWERS TO SELECTED PROBLEMS

Problems 1.3

1. (a) Yes; (b) yes; (c) no. 2. (a) Yes; (b) no; (c) no.
6. (a) 0.9; (b) 0.05; (c) 0.95. 7. 1/16. 8. $\frac{1}{3}+\frac{2}{9}\ell n2 = 0.487$.

Problems 1.4

3. $\binom{R}{n}\binom{W}{n-r}\Big/\binom{N}{n}$ 4. 352146 5. $(n-k+1)!/n!$

6. $1 - {}_7P_5/_75$ 8. $\binom{n+k-r}{n-r}\Big/\binom{n+k}{k}$ 9. $1-\sum_{i=1}^{n-k}\binom{2i}{i}\Big/\binom{2n}{n-k}$

12. (a) $4\Big/\binom{52}{5}$ (b) $9(4)\Big/\binom{52}{5}$ (c) $13\binom{48}{1}\Big/\binom{52}{5}$

(d) $13\binom{4}{3}12\binom{4}{2}\Big/\binom{52}{5}$ (e) $\left[4\binom{13}{5}-9(4)-4\right]\Big/\binom{52}{5}$

(f) $[10(4)^5-4-9(4)]\Big/\binom{52}{5}$ (g) $13\binom{12}{2}\binom{4}{3}4^2\Big/\binom{52}{5}$

(h) $\binom{13}{2}\binom{4}{2}\binom{4}{2}\binom{44}{1}\Big/\binom{52}{5}$ (i) $\binom{13}{1}\binom{4}{3}\binom{12}{3}4^3\Big/\binom{52}{5}$

An Introduction to Probability and Statistics, Third Edition. Vijay K. Rohatgi and A.K. Md. Ehsanes Saleh.

Problems 1.5

3. $\alpha(pb)^r \sum_{\ell=0}^{\infty} \binom{r+\ell}{\ell} [p(1-b)]^{\ell}$ 4. $p/(2-p)$

5. $\sum_{j=0}^{N} (j/N)^{n+1} \Big/ \sum_{j=0}^{N} (j/N)^n \simeq \frac{n+1}{n+2}$ for large N 6. $n = 4$

10. $r/(r+g)$ 11. (a) 1/4; (b) 1/3   12. 0.08

13. (a) 173/480 (b) 108/173; 15/173   14. 0.0872

Problems 1.6

1. $1/(2-p)$; $(1-p)/(2-p)$ 4. $p^2(1-p)^2[3-7p(1-p)]$

12. For any two disjoint intervals $I_1, I_2 \subseteq (a,b)$, $\ell(I_1)\ell(I_2) = (b-a)\ell(I_1 \cap I_2)$, where $\ell(I)$ = length of interval I.

13. (a) $p_n = \begin{cases} 8/36 & \text{if } n = 1 \\ 2\left(\frac{27}{36}\right)^{n-2}\left(\frac{3}{6}\right)^2 + 2\left(\frac{26}{36}\right)^{n-2}\left(\frac{4}{36}\right)^2 + 2\left(\frac{25}{36}\right)^{n-2}\left(\frac{5}{36}\right)^2, n \geq 2 \end{cases}$

(b) $22/45$

(c) $12/36$; $2\left(\frac{27}{36}\right)^{n-2}\left(\frac{9}{36}\right)\left(\frac{3}{16}\right) + 2\left(\frac{26}{36}\right)^{n-2}\left(\frac{10}{36}\right)\left(\frac{4}{36}\right) + 2\left(\frac{25}{36}\right)^{n-2}\left(\frac{11}{36}\right)\left(\frac{5}{36}\right)$ for $n = 2, 3, \ldots$.

Problems 2.2

3. Yes; yes

4. ϕ; $\{(1,1,1,1,2),(1,1,1,2,1),(1,1,2,1,1),(1,2,1,1,1),(2,1,1,1,1)\}$; $\{(6,6,6,6,6)\}$; $\{(6,6,6,6,6),(6,6,6,6,5),(6,6,6,5,6),(6,6,5,6,6),(6,5,6,6,6),(5,6,6,6,6)\}$

5. Yes; $(1/4,1/2) \cup (3/4,1)$

Problems 2.3

1.

x	0	1	2	3
$P(X = x)$	1/8	3/8	3/8	1/8

$F(x) = 0$, $x < 0$, $= 1/8$, $0 \leq x < 1$; $= 1/2$, $1 \leq x < 2$; $= 5/8$, $2 \leq x < 3$; $= 1$, $x \geq 3$

3. (a) Yes; (b) yes; (c) yes; yes

Problems 2.4

1. $(1-p)^{n+1} - (1-p)^{N+1}$, $N \geq n$

2. (b) $\frac{1}{\pi(1+x^2)}$; (c) $1/x^2$; (d) e^{-x}

3. Yes; $F_\theta(x) = 0$ $x \leq 0$, $= 1 - e^{-\theta x} - \theta x e^{-\theta x}$ for $x > 0$; $P(X \geq 1) = 1 - F_\theta(1)$

4. Yes; $F(x) = 0, x \leq 0$; $= 1 - \left(1 + \frac{x}{\theta+1}\right)e^{-x/\theta}$ for $x > 0$

6. $F(x) = e^x/2$ for $x \leq 0$, $= 1 - e^{-x}/2$ for $x > 0$

8. (c), (d), and (f)

9. Yes; (a) $1/2, 0 < x < 1$, $1/4$ for $2 < x < 4$; (b) $1/(2\theta)$, $|x| \leq \theta$; (c) $xe^{-x}, x > 0$; (d) $(x-1)/4$ for $1 \leq x < 3$, and $P(X = 3) = 1/2$; (e) $2xe^{-x^2}, x > 0$

10. If $S(x) = 1 - F(x) = P(X > x)$, then $S'(x) = -f(x)$

Problems 2.5

2. $X \stackrel{d}{=} 1/X$
4. $\theta[1-\exp(-2\pi\theta)]\sqrt{1-y^2}\left[e^{-\theta \text{ arc cos } y}+e^{-2\pi\theta+\theta \text{ arc cos } y}\right], |y| \le 1;$
$\begin{cases} \theta\exp\{-\theta \text{ arctan } z\}[(1+z^2)(1-e^{-\theta\pi}]^{-1}, & z>0 \\ \theta\exp\{-\pi\theta-\text{arctan } z\}[(1+z^2)(1-e^{-\theta\pi})]^{-1}, & z<0 \end{cases}$
10. $f_{|X|}(y) = 2/3$ for $0<y<1$, $= 1/3$ for $1<y<2$
12. (a) $0, y<0; F(0)$ for $-1 \le y < 1$, and 1 for $y \ge 1$;
(b) $= 0$ if $y<-b$, $= F(-b)$ if $y=-b$, $= F(y)$ if $-b \le y < b$, $= 1$ if $y \ge b$;
(c) $= F(y)$ if $y<-b$, $= F(-b)$ if $-b \le y < 0$, $= F(b)$ if $0 \le y < b$, $= F(y)$.
if $y \ge b$.

Problems 3.2

3. $EX^{2r} = 0 \quad$ if $2r < 2m-1$ is an odd integer,
$= \dfrac{\Gamma\left(m-r+\frac{1}{2}\right)\Gamma\left(r+\frac{1}{2}\right)}{\Gamma\left(\frac{1}{2}\right)\Gamma\left(\frac{m-1}{2}\right)} \quad$ if $2r < 2m-1$ is an even integer
9. $\mathfrak{z}_p = a(1-v)/v$, where $v = (1-p)^{1/k}$
10. Binomial: $\alpha_3 = (q-p)/\sqrt{npq}$, $\alpha_4 = 3+(1-6pq)/3npq$
Poisson: $\alpha_3 = \lambda^{-1/2}$, $\alpha_4 = 3+1/\lambda$.

Problems 3.3

1. (b) $e^{-\lambda}(e^{\lambda s}-1)/(1-e^{-\lambda})$; (c) $p[1-(qs)^{N+1}]/[(1-qs)(1-q^{N+1})]$, $s<1/q$.
6. $f(\theta s)/f(\theta)$; $f(\theta e^t)/f(\theta)$.

Problems 3.4

3. For any $\sigma^2 > 0$ take $P(X=x) = \frac{\sigma^2}{\sigma^2+x^2}$, $P\left(X=-\frac{\sigma^2}{x}\right) = \frac{x^2}{\sigma^2+x^2}$, $x \ne 0$.
5. $P\left(X^2 = \frac{\sigma^4 K^2-\mu_4}{K^2\sigma^2-\sigma^2}\right) = \frac{\sigma^4[K^2-1]^2}{\mu_4+K^4\sigma^4-2K^2\sigma^4} \quad 1<K<\sqrt{2}$
$P(X^2 = K^2\sigma^2) = \frac{\mu_4-\sigma^4}{\mu_4+K^4\sigma^4-2K^2\sigma^4}$.

Problems 4.2

1. No 4. 1/6; 0. 7. Marginals negative binomial, so also conditionals.
8. $h(y|x) = \frac{1}{2}(c^2+x^2)/(c^2+x^2+y^2)^{3/2}$.
9. $X \sim B(p_1, p_2+p_3)$; $Y/(1-x) \sim B(p_2, p_3)$.
10. $X \sim G(\alpha, 1/\beta)$, $Y \sim G(\alpha+\gamma, 1/\beta)$, $X/y \sim B(\alpha,\gamma)$, $Y-x \sim G(\gamma, 1/\beta)$.
14. $P(X \le 7) = 1-e^{-7}$ 15. 1/24; 15/16. 17. 1/6.

Problems 4.3

3. No; Yes; No. 10. $= 1-a/(2b)$ if $a<b$, $= b/(2a)$ if $a>b$.
11. $\lambda/(\lambda+\mu)$; 1/2.

Problems 4.4

2. (b) $f_{V|U}(v|u) = 1/(2u)$, $|v|<u$, $u>0$.
6. $P(X=x, M=m) = \pi(1-\pi)^m[1-(1-\pi)^{m+1}]$ if $x=m$, $= \pi^2(1-\pi)^{m+x}$ if $x<m$. $P(M=m) = 2\pi(1-\pi)^m - \pi(2-\pi)(1-\pi)^{2m}$, $m \ge 0$.

7. $f_X(x) = \lambda^k e^{-\lambda}/k!,\ k \le x < k+1,\ k = 0,1,2,\ldots$
11. $f_U(u) = 3u^2/(1+u)^4,\ u > 0.$

13. (a) $F_{U,V}(u,v) = \left[1 - \exp\left(-\frac{u^2}{2\sigma^2}\right)\right]\left(\frac{\pi+2v}{2\pi}\right)$ if $u > 0,\ |v| \le \pi/2,$
$= 1 - \exp[1 - u^2/(2\sigma^2)]$ if $u > 0,\ v > \pi/2,\ = 0$ elsewhere.
(b) $f(u,v) = \frac{1}{\sqrt{\pi}} e^{-u^2} \frac{v^{1/2-1} e^{-v/2}}{\Gamma(1/2)\sqrt{2}}.$

Problems 4.5

2. $EX^kY^\ell = \frac{2^{\ell+1}}{(k+3)(\ell+1)} + \frac{2^{\ell+2}}{3(k+2)(\ell+2)}.$ 3. $\text{cov}(X,Y) = 0$; X, Y dependent.
15. $M_{U,V}(u,v) = (1-2v)^{-1}\exp\{u^2/(1-2v)\}$ for $v < 1/2$; $\rho(U,V) = 0$; no.
18. $\rho_{Z,W} = (\sigma_2^2 - \sigma_1^2)\sin\theta\cos\theta/\sqrt{\text{var}(Z)\,\text{var}(W)}.$
21. If U has PDF f, then $EX^m = EU^m/(m+1)$ for $m \ge 0$; $\rho = \frac{1}{2} - \frac{EU^2}{\frac{8}{3}\text{var}(U) + \frac{2}{3}(EU)^2}.$

Problems 4.6

1. $\mu + \sigma\left[f\left(\frac{a-\mu}{\sigma}\right) - f\left(\frac{b-\mu}{\sigma}\right)\right]/\Phi\left(\frac{b-\mu}{\sigma}\right) - \Phi\left(\frac{a-\mu}{\sigma}\right)\right]$ where Φ is the standard normal DF.
2. (a) $2(1+X)$. 3. $E\{X|y\} = \mu_1 + \rho\frac{\sigma_1}{\sigma_2}(y - \mu_2)$. 4. $E(\text{var}\{Y|X\})$.
6. 4/9. 7. (a) 1; (b) 1/4. 8. $x^k/(k+1)$; $1/(1+k)^2$.

Problems 4.7

5. (a) $\left(\sum_{j=1}^{n} 1/j\right)/\beta$; (b) $\frac{n}{n+1}$.

Problems 5.2

5. $F_Y(y) = \binom{y}{M}/\binom{N}{M}$, $P(Y = y) = \binom{y-1}{M-1}/\binom{N}{M}$, $y \ge M+1$, and
$P(Y = M) = 1/\binom{N}{M}$. $P(x_1,\ldots,x_m|Y = y) = \frac{(y-m)!}{(y-1)!M}$, $0 < x_i \le y$,
$i = 1,\ldots,j$, $x_i \ne x_j$ for $i \ne j$.
9. $P(Y_1 = x) = qp^x + pq^x,\ x \ge 1$. $P(Y_2 = x) = p^2q^{x-1} + q^2p^{x-1},\ x \ge 1$
$P(Y_n = x) = P(Y_1 = x)$ for n odd; $= P(Y_2 = x)$ for n even.

Problems 5.3

2. (a) $P\left\{F(X) = \sum_{k=0}^{x}\binom{n}{k}p^k(1-p)^{n-k}\right\} = \binom{n}{x}p^x(1-p)^{n-x},\ x = 0,1,\ldots,n.$

13. $\mathcal{C}\left(\sum_{i=1}^{n}\frac{|a_i|}{a_i^2+b_i^2}, \sum_{i=1}^{n}\frac{b_i}{a_i^2+b_i^2}\right)$

22. $X/|Y| \sim \mathcal{C}(1,0)$; $(2/\pi)(1+z^2)^{-1},\ 0 < z < \infty$.
27. (a) t/α^2; (c) $= 0$ if $t \le \theta$, $= \alpha/t$ if $t > \theta$; (d) $(\alpha/\beta)t^{\alpha-1}$.
29. (b) $1/(2\sqrt{\pi})$; $1/2$.

Problems 5.4

1. (a) $\mu_1 = 4$; $\mu_2 = 15/4$, $\rho = -3/4$; (b) $\mathcal{N}(6 - \frac{9}{16}x, \frac{63}{16})$; (c) 0.3191.
4. $B\mathcal{N}(a\mu_1 + b, c\mu_2 + d, a^2\sigma_1^2, c^2\sigma_2^2, \rho)$. 6. $\tan^2\theta = EX^2/EY^2$. 7. $\sigma_1^2 = \sigma_2^2$.

Problems 6.2

1. $P(\overline{X} = 0) = P(\overline{X} = 1) = 1/8$, $P(\overline{X} = 1/3) = P(\overline{X} = 2/3) = 3/8$
$P(S^2 = 0) = 1/4$, $P(S^2 = 1/3) = 3/4$.
2.

$\overline{x}$	1	1.5	2	2.5	3	3.5	4	4.5	5	5.5	6
$p(\overline{x})$	1/36	2/36	3/36	4/36	5/36	6/36	5/36	4/36	3/36	2/36	1/36

Problems 6.3

1. $\{F(\min(x, y)) - F(x)F(y)\}/n$.
6. $E(S^2)^k = \frac{\sigma^2}{(n-1)^k}(n-1)(n+2)\cdots(n+2k-3)$, $k \ge 1$.
9. (a) $P(\overline{X} = t) = e^{-n\lambda}(n\lambda)^{tn}/(tn)!$, $t = 0, 1/n, 2/n, \ldots$; (b) $\mathcal{C}(1, 0)$;
(c) $\Gamma(nm/2, 2/n)$. 10. (b) $2/\sqrt{\alpha n}$; $3 + 6/(\alpha n)$.
11. $0, 1, 0, E(\overline{X}_n - 0.5)^4/(144n^2)$. 12. $\text{var}(S^2) = \frac{1}{n}\left(\lambda + \frac{2n\lambda^2}{n-1}\right) > \text{var}(\overline{X})$.

Problems 6.4

2. $n(m+\delta)/[m(n-2)]$; $2n^2\{(m+\delta)^2 + (n-2)(m+2\delta)\}/[m^2(n-2)^2(n-4)]$.
3. $\delta\sqrt{\frac{n}{2}}\frac{\Gamma(\frac{n-1}{2})}{\Gamma(\frac{n}{2})}$, $n > 1$; $\frac{n}{n-2}(1+\delta^2) - \left(\delta\sqrt{\frac{n}{2}}\frac{\Gamma(\frac{n-1}{2})}{\Gamma(\frac{n}{2})}\right)^2$, $n > 2$.
11. $2m^{m/2}n^{n/2}(n + me^{2z})^{-(m+n)/2}e^{zm}/B\left(\frac{m}{2}, \frac{n}{2}\right)$, $-\infty < z < \infty$.

Problems 6.5

1. $t(n-1)$ 2. $t(m+n-2)$ 3. $\left(\frac{2\sigma^2}{n-1}\right)^k\Gamma\left(\frac{n-1}{2} + k\right)/\Gamma\left(\frac{n-1}{2}\right)$.

Problems 6.6

3. $[2\pi(1-\rho^2)]^{-1/2}\left[1 + \frac{y_1^2 + y_2^2 - 2\rho y_1 y_2}{n(1-\rho^2)}\right]^{-(\frac{n}{2}+1)}$; both $\sim t(n)$.
4. $\sqrt{n-1}T \sim t(n-1)$.

Problems 7.2

1. No. 2. Yes
3. $Y_n \to Y \sim F(y) = 0$ if $y < 0$, $= 1 - e^{-y/\theta}$ if $y \ge 0$.
4. $F(y) = 0$ if $y \le 0$, $= 1 - e^{-y}$ if $y > 0$.
9. $\mathcal{C}(1, 0)$ 12. No
13. (a) $\exp(-x^{-\alpha})$, $x > 0$; $EX^k = \Gamma(1 - k/\alpha)$, $k < \alpha$.
(b) $\exp(-e^{-x})$, $-\infty < x < \infty$; $M(t) = \Gamma(1-t)$, $t < 1$.
(c) $\exp\{-(-x)^\alpha\}$, $x < 0$; $EX^k = (-1)^k\Gamma(1 + k/\alpha)$, $k > -\alpha$.
20. (a) Yes; No (b) Yes; No.

Problems 7.3

3. Yes; $A_n = n(n+1)\mu/2$, $B_n = \sigma\sqrt{n(n+1)(2n+1)/6}$
5. (a) $M_n(t) \to 0$ as $n \to \infty$; no. (b) $M_n(t)$ diverges as $n \to \infty$
(c) Yes (d) Yes (e) $M_n \to e^{t^2/4}$; no.

Problems 7.4

1. (a) No; (b) No. 2. No. 3. For $\alpha < 1/2$. 7. (a) Yes; (b) No.

Problems 7.5

4. Degenerate at β. 5. Degenerate at 0.
6. For $\rho \geq 0$, $\mathcal{N}(0, \sqrt{\rho})$, and for $\rho < 0$, $S_n/n \xrightarrow{L}$ degenerate.

Problems 7.6

1. (b) No; (c) Yes; (d) No.
2. $\mathcal{N}(0,1)$. 3. $\mathcal{N}(0, \sigma^2/\beta^2)$. 4. 163. 8. 0.0926; 1.92

Problems 7.7

1. (a) $\mathcal{AN}(\mu^2, 4\mu^2\sigma_n^2)$ for $\mu \neq 0$, $\overline{X}^2/\sigma_n^2 \xrightarrow{L} \chi^2(1)$ for $\mu = 0$, $\sigma_n^2 = \sigma^2/n$.
   (b) For $\mu \neq 0$, $1/\overline{X} \sim \mathcal{AN}(1/\mu, \sigma_n^2/\mu^4)$; for $\mu = 0$, $\sigma_n/\overline{X}_n \xrightarrow{L} 1/\mathcal{N}(0,1)$.
   (c) For $\mu \neq 0$, $\ell n|\overline{X}| \sim \mathcal{AN}(\ell n|\mu|, \sigma_n^2/\mu^2)$; for $\mu = 0$, $\ell n(|\overline{X}|/\sigma_n) \xrightarrow{L} \ell n|\mathcal{N}(0,1)|$.
   (d) $\mathcal{AN}(e^{\mu}, e^{2\mu}\sigma_n^2)$.
2. $c = 1/2$ and $\sqrt{X} \sim \mathcal{AN}(\sqrt{\lambda}, 1/4)$.

Problems 8.3

2. No. 7. $f_{\theta_2}(x)/f_{\theta_1}(x)$. 9. No. 10. No.
11. (b) $X_{(n)}$; (e) $(\overline{X}, S^2)$; (g) $\left(\prod_1^n X_i, \prod_1^n (1 - X_i)\right)$ (h) $X((_{1}), X_{(2)}, \ldots, X_{(n)})$.

Problems 8.4

2. $\left(\frac{n-1}{2}\right)^p \frac{\Gamma\left(\frac{n-1}{2}\right)}{\Gamma\left(\frac{n+p-1}{2}\right)} S^p$; $\left(\frac{n-1}{2}\right)^{p/2} \frac{\Gamma\left(\frac{n+p-1}{2}\right)}{\Gamma\left(\frac{n+2p-1}{2}\right)} S$.
3. $S_1^2 = \frac{n-1}{n+1} S^2$; $\text{var}(S_1^2) = \left(\frac{n-1}{n+1}\right)^2 \frac{2\sigma^4}{n-1} < \text{var}(S^2) = \frac{2\sigma^4}{n-1}$; 4. No; 5. No.
6. (a) $\binom{n-s}{t-s} / \binom{n}{t}$, $0 \leq s \leq t \leq n$, $t = \sum_1^n x_i$; (b) $= \binom{s}{t} / \binom{n}{t}$ if $0 \leq t < s$,
$= 2/\binom{n}{t}$ if $t = s$, and $\binom{n-s}{t-s} / \binom{n}{t}$ if $s + 1 \leq t \leq n$.
9. $\binom{t+n-2}{t} / \binom{t+n-1}{t}$, $t = \Sigma x_i$. 11. (a) NX/n; (b) No.
12. $t = \Sigma_1^n x_i$, $1 - \left(1 - \frac{t_0}{t}\right)^{n-1}$ if $t > t_0$, and 1 if $t \leq t_0$.
13. (a) With $t = \sum_1^n x_j$, $\sum_{j=0}^t \frac{t!}{j!} n^{j-t}$; (b) $\frac{t!}{(t-s)!} n^{-s}$, $t \geq s$ (c) $(1 - 1/n)^t$;
   (d) $(1 - 1/n)^{t-1}[1 + \frac{t-1}{n}]$.
14. With $t = x_{(n)}$, $[t^n \psi(t) - (t-1)^n \psi(t-1)]/[t^n - (t-1)^n]$, $t \geq 1$.
15. With $t = \sum_1^n x_j$, $\binom{t}{k} \left(\frac{1}{n}\right)^k \left(1 - \frac{1}{n}\right)^{t-k}$.

Problems 8.5

1. (a), (c), (d) Yes; (b) No. 2. $0.64761/n^2$.
3. $n^{-1}\sup_{x\neq 0}\{x^2/[e^{x^2}-1]\}$. 5. $2\theta(1-\theta)/n$

Problems 8.6

2. $\hat{\beta}=(n-1)S^2/(n\overline{X})$, $\hat{\alpha}=\overline{X}/\hat{\beta}$ 3. $\hat{\mu}=\overline{X}$, $\hat{\sigma}^2=(n-1)S^2/n$.
4. $\hat{\alpha}=\overline{X}(\overline{X}-\overline{X^2})[\overline{X^2}-\overline{X}^2]^{-1}$, $\overline{X^2}=\sum_1^n X_i^2/n$ $\hat{\beta}=(1-\overline{X})(\overline{X}-\overline{X^2})[\overline{X^2}-\overline{X}^2]^{-1}$.
5. $\hat{\mu}=\ell\text{n}\{\overline{X}^2/[\overline{X^2}]^{1/2}\}$, $\hat{\sigma}^2=\ell\text{n}\{\overline{X^2}/\overline{X}^2\}$, $\overline{X^2}=\sum_1^n X_i^2/n$.

Problems 8.7

1. (a) $\text{med}(X_j)$; (b) $X_{(1)}$; (c) $n/\sum_1^n X_j^\alpha$; (d) $-n/\sum_1^n \ell\text{n}(1-X_j)$.
2. (a) X/n; (b) $\hat{\theta}_n=1/2$ if $\overline{X}\leq 1/2$, $=\overline{X}$ if $1/2\leq\overline{X}\leq 3/4$, $=3/4$ if $\overline{X}\geq 3/4$;
(c) $\hat{\theta}=\begin{cases}\hat{\theta}_0, & \text{if } \overline{X}\geq 0\\ \hat{\theta}_1, & \text{if } \overline{X}\leq 0\end{cases}$ where $\hat{\theta}_0=-\frac{\overline{X}}{2}+\sqrt{\overline{X^2}+(\frac{\overline{X}}{2})^2}$,
$\hat{\theta}_1=-\frac{\overline{X}}{2}-\sqrt{\overline{X^2}+(\frac{\overline{X}}{2})^2}$, $\overline{X^2}=\sum X_1^2/n$;
(d) $\hat{\theta}=\frac{n_3}{n_1+n_3}$ if $n_1,n_3>0$; = any value in (0,1) if $n_1=n_3=0$;
no mle if $n_1=0, n_3\neq 0$; no mle if $n_1\neq 0, n_3=0$;
(e) $\hat{\theta}=-\frac{1}{2}+\frac{1}{2}\sqrt{1+4\overline{X^2}}$; (f) $\hat{\theta}=X$.
3. $\hat{\mu}=-\Phi^{-1}(m/n)$.
4. (a) $\hat{\alpha}=X_{(1)}$, $\hat{\beta}=\sum_1^n(X_i-\hat{\alpha})/n$; (b) $\Delta=P_{\alpha,\beta}(X_1\geq 1)=e^{(\alpha-1)\beta}$ $\alpha\leq 1$, $=1$, $\alpha\geq 1$. $\hat{\Delta}=1$ if $\hat{\alpha}\geq 1$, $=\exp\{(\hat{\alpha}-1)/\hat{\beta}\}$ if $\hat{\alpha}<1$.
5. $\hat{\theta}=1/\overline{X}$. 6. $\hat{\mu}=\Sigma\ell\text{n}X_i/n$, $\hat{\sigma}^2=\sum_1^n(\ell\text{n}X_i-\hat{\mu})^2/n$.
8. (a) $\hat{N}=\frac{M+1}{M}X_{(M)}-1$; (b) $X_{(M)}$.
9. $\hat{\mu}_i=\sum_{j=1}^n X_{ij}/n=\overline{X}_i$, $i=1,2,\ldots,s$ $\hat{\sigma}^2=\Sigma\Sigma(X_{ij}-\overline{X}_i)^2/(ns)$.
11. $\hat{\mu}=\overline{X}$, 13. $d(\hat{\theta})=(X/n)^2$. 15. $\hat{\mu}=\max(\overline{X},0)$.
16. $\hat{p}_j=X_j/n$, $j=1,2,\ldots,k-1$.

Problems 8.8

2. (a) $(\Sigma x_i+1)/(n+1)$; (b) $\left(\frac{n+1}{n+2}\right)^{\Sigma x_i+1}$. 3. $\overline{X}$. 5. X/n.
6. $(X+1)(X+n)/[(n+2)(n+3)]$. 8. $(\alpha+n)\max(a,X_{(n)})/(\alpha+n-1)$.

Problems 8.9

5. (c) $(n+2)\left[(X_{(n)}/2)^{-(n+1)}-(X_{(1)})^{-(n+1)}\right]/\{(n+1)[(X_{(n)}/2)^{-(n+2)}-(X_{(1)})^{-(n+2)}]\}$
10. $(\Sigma X_i)^k\Gamma(n+k)/\Gamma(n+2k)$

Problems 9.2

1. 0.019, 0.857. 2. $k=\mu_0+\sigma z_\alpha/\sqrt{n}$; $1-\Phi\left(z_\alpha-\frac{\mu_1-\mu_0}{\sigma}\sqrt{n}\right)$.
5. $\exp(-2)$; $\exp(-2/\theta)$, $\theta\geq 1$.

Problems 9.3

1. $\phi(x) = 1$ if $x < \theta_0(1-\sqrt{1-\alpha}) = 0$ otherwise.
4. $\phi(x) = 1$ if $||x|-1| > k$. 5. $\phi(\mathbf{x}) = 1$ if $x_{(1)} > c = \theta_0 - \ell n(\alpha^{1/n})$.
11. If $\theta_0 < \theta_1$, $\phi(\mathbf{x}) = 1$ if $x_{(1)} > \theta_0\alpha^{-1/n}$, and if $\theta_1 < \theta_0$, then $\phi(\mathbf{x}) = 1$ if $x_{(1)} < \theta_0(1-\alpha^{1/n})^{-1}$.
12. $\phi(x) = 1$ if $x < \sqrt{\alpha}/2$ or $> 1-\sqrt{\alpha}/2$.

Problems 9.4

1. (a), (b), (c), (d) have MLR in ΣX_j; (e) and (f) in $\prod_1^n X_j$
4. Yes. 5. Yes; yes.

Problems 9.5

1. $\phi(x_1,x_2) = 1$ if $|x_1 - x_2| > c, = 0$ otherwise, $c = \sqrt{2}z_{\alpha/2}$.
2. $\phi(\mathbf{x}) = 1$ if $\Sigma x_i > k$. Choose k from $\alpha = P_{\lambda_0}\left(\sum_1^n X_i > k\right)$.

Problems 9.6

3. $\phi(\mathbf{x}) = 1$ if (no. of x_i's $> 0-$ no. of x_i's $< 0) > k$.

Problems 10.2

2. $Y = \#$ of x_1, x_2 in sample, $Y < c_1$ or $Y > c_2$. 3. $X < c_1$ or $> c_2$.
4. $S^2 > c_1$ or $< c_2$. 5. (a) $X_{(n)} > N_0$; (b) $X_{(n)} > N_0$ or $< c$.
6. $|X - \theta_0/2| > c$. 7. (a) $\overline{X} < c_1$ or $> c_2$; (b) $\overline{X} > c$.
11. $X_{(1)} > \theta_0 - \ell n(\alpha)^{1/n}$. 12. $X_{(1)} > \theta_0\alpha^{-1/n}$.

Problems 10.3

1. Reject at $\alpha = 0.05$. 3. Do not reject $H_0 : p_1 = p_2 = p_3 = p_4$ at 0.05 level.
4. Reject H_0 at $\alpha = 0.05$. 5. Reject at 0.10 but not at 0.05 level.
7. Do not reject H_0 at $\alpha = 0.05$. 8. Do not reject H_0 at $\alpha = 0.05$.
10. $U = 15.41$. 12. P-value = 0.5447.

Problems 10.4

1. $t = -4.3$, reject H_0 at $\alpha = 0.02$. 2. $t = 1.64$, do not reject H_0.
5. $t = 5.05$. 6. Reject H_0 at $\alpha = 0.05$. 7. Reject H_0. 8. Reject H_0.

Problems 10.5

1. Do not reject $H_0 : \sigma_1 = \sigma_2$ at $\alpha = 0.10$.
3. Do not reject H_0 at $\alpha = 0.05$. 4. Do not reject H_0.

Problems 10.6

2. (a) $\phi(\mathbf{x}) = 1$ if $\Sigma x_i = 5, = 0.12$ if $\Sigma x_i = 4, = 0$ otherwise;
   (b) Minimax rule rejects H_0 if $\Sigma x_i = 4$ or 5, and with probability 1/16 if $\Sigma x_i = 3$;
   (c) Bayes rule rejects H_0 if $\Sigma x_i \geq 2$.

3. Reject H_0 if $\bar{x} \leq (1-1/n)\ell n2$
$\beta(1) = P(Y \leq (n-1)\ell n2)$, $\beta(2) = P(Z \leq (n-1)\ell n2)$, where $Y \sim G(n,1)$ and $Z \sim G(n,1/2)$

Problems 11.3

1. (77.7, 84.7). 2. $n = 42$. 7. $\left(\frac{2\Sigma X_i}{\chi^2_{2n,\alpha/2}}, 2\Sigma X_i/\chi^2_{2n,1-\alpha/2}\right)$.
9. $(2X/(2-\lambda_1), 2X/(2-\lambda_2))$, $\lambda_2^2 - \lambda_1^2 = 4(1-\alpha)$. 10. $[\alpha^{1/n}N]$.
11. $n \geq \frac{\ell n(1/\alpha)}{[\ell n(1+d/X_{(n)})]}$.
12. Choose k from $\alpha = (k+1)e^{-k}$. 13. $\overline{X} + z_\alpha \sigma/\sqrt{n}$
14. $(\Sigma X_i^2/c_2, \Sigma X_i^2/c_1)$, where $\int_{c_1}^{c_2} \chi_n^2(y)dy = 1-\alpha$ and $\int_{c_1}^{c_2} y\chi_n^2(y)dy = n(1-\alpha)$.
15. Posterior $B(n+\alpha, \Sigma x_i + \beta - n)$.
16. $h(\mu|\mathbf{x}) = \sqrt{\frac{n}{2\pi}} \exp\{-\frac{n}{2}(\mu-\bar{x})^2\}[\Phi(\sqrt{n}(1-\bar{x})) - \Phi(-\sqrt{n}(1+\bar{x}))]$, where Φ is standard normal DF.

Problems 11.4

1. $(X_{(1)} - \chi^2_{2,\alpha}/(2n), X_{(1)})$.
2. $(2n\overline{X}/b, 2n\overline{X}/a)$, choose a, b from $\int_a^b \chi^2_{2n}(u)du = 1-\alpha$, and $a^2\chi^2_{2n}(a) = b^2\chi^2_{2n}(b)$, where $\chi^2_v(x)$ is the PDF of $\chi^2(v)$ RV.
3. $(X/(1-b), X/(1-a))$, choose a, b from $1-\alpha = b^2 - a^2$ and $a(1-a)^2 = b(1-b)^2$.
4. $n = [4z^2_{1-\alpha/2}/d^2] + 1$; $n > (1/\alpha)\ell n(1/\alpha)$.

Problems 11.5

1. $(X_{(n)}, \alpha^{-1/n}X_{(n)})$.
2. $(2\Sigma X_i/\lambda_2, 2\Sigma X_i/\lambda_1)$, where λ_1, λ_2 are solutions of $\lambda_1 f_{2n\alpha}(\lambda_1) = \lambda_2 f_{2n\alpha}(\lambda_2)$ and $P(1) = 1-\alpha$, f_v is $\chi^2(v)$ PDF.
3. $(X_{(1)} - \frac{\chi^2_{2,\alpha}}{2n}, X_{(1)})$. 5. $(\alpha^{1/n}X_{(1)}, X_{(1)})$. 8. Yes.

Problems 12.3

4. Reject $H_0 : \alpha_0 = \alpha_0'$ if $\frac{|\hat{\alpha}_0 - \alpha_0'|\sqrt{n\Sigma(t_i-\bar{t})^2/\Sigma t_i^2}}{\sqrt{\Sigma(Y_i - \hat{\alpha}_0 - \hat{\alpha}_1 t_i)^2/(n-2)}} > c_0$.
8. Normal equations $\hat{\beta}_0\Sigma x_i^k + \hat{\beta}_1\Sigma x_i^{k+1} + \hat{\beta}_2\Sigma x_i^{k+2} = \Sigma Y_i x_i^k$, $k = 0,1,2$.
Reject $H_0 : \beta_2 = 0$ if $\{|\hat{\beta}_2|/\sqrt{c_1^2}\}/\sqrt{\Sigma(Y_i - \hat{\beta}_0 - \hat{\beta}_1 x_i - \hat{\beta}_2 x_i^2)\}} > c_0$, where $\hat{\beta}_2 = \Sigma c_i Y_i$ and $\hat{\beta}_0 = \overline{Y} - \hat{\beta}_1\bar{x}$, $\hat{\beta}_1 = \Sigma(x_i - \bar{x})(Y_i - \overline{Y})/\Sigma(x_i - \bar{x})^2$.
10. (a) $\hat{\beta}_0 = 0.28$, $\hat{\beta}_1 = 0.411$; (b) $t = 4.41$, reject H_0.

Problems 12.4

2. $F = 10.8$. 3. Reject at $\alpha = 0.05$ but not at $\alpha = 0.01$.
4. BSS = 28.57, WSS = 26, reject at $\alpha = 0.05$ but not at 0.01.
5. $F = 56.45$. 6. $F = 0.87$.

Problems 12.5

4. SS Methods = 50, SS Ability = 64.56, ESS = 25.44; reject H_0 at $\alpha = 0.05$, not at 0.01.
5. $F_{\text{variety}} = 24.00$.

Problems 12.6

2. Reject H_0 if $\frac{am\sum_1^b(\bar{y}_{.j.}-\bar{y})}{\Sigma\Sigma\Sigma(y_{ijs}-\bar{y}_{ij.})^2} > c$.
4. SS_1 (machines) = 2.786, d.f. = 3; SSI = 73.476, d.f. = 6;
   SS_2 (machines) = 27.054, d.f. = 2; SSE = 41.333, d.f. = 24.
5.

Cities	3	227.27	4.22
Auto	3	3695.94	68.66
Interactions	9	9.28	0.06
Error	16	287.08	

Problems 13.2

1. d is estimable of degree 1; (number of x_i's in $A)/n$.
2. (a) $(mn)^{-1}\Sigma X_i \Sigma Y_j$; (b) $S_1^2 + S_2^2$.
3. (a) $\Sigma X_i Y_i/n$; (b) $\Sigma(X_i + Y_i - \overline{X} - \overline{Y})^2/(n-1)$.

Problems 13.3

3. Do not reject H_0. 7. Reject H_0. 10. Do not reject H_0 at 0.05 level.
11. $T^+ = 133$, do not reject H_0.
12. (Second part) $T^+ = 9$, do not reject H_0 at $\alpha = 0.05$.

Problems 13.4

1. Do not reject H_0. 2. (a) Reject; (b) Reject.
3. $U = 29$, reject H_0. 5. $d = 1/4$, do not reject H_0.
7. $t = 313.5$, $z = 3.73$, reject; $r = 10$ or 12, do not reject at $\alpha = 0.05$.

Problems 13.5

1. Reject H_0 at $\alpha = 0.05$. 4. Do not reject H_0 at $\alpha = 0.05$.
9. (a) $t = 1.21$; (b) $r = 0.62$; (c) Reject H_0 in each case.

Problems 13.6

1. (a) 5; (b) 8. 3. $p^{n-2}(n+p-np) \leq 1$.
4. $n \geq (z_{1-\gamma}\sqrt{p_0(1-p_0)} - z_{1-\delta}\sqrt{p_1(1-p_1)})^2/(p_1-p_0)^2$.

Problems 13.7

1. (c) $E\{n(\overline{X}-\mu)^2\}/ES^2 = 1 + 2\rho(1-2\rho/n)^{-1}$; ratio = 1 if $\rho = 0$, > 1 for $\rho > 0$.
2. Chi-square test based on (c) is not robust for departures from normality.

AUTHOR INDEX

A. Agresti, 553
T.W. Anderson, 400
J.D. Auble, 606

I.V. Basawa, 178, 192
D. Bernstein, 220
P.J. Bickel, 633, 634
P. Billingsley, 315
D. Birkes, 465
Z.W. Birnbaum, 586, 587
D. Blackwell, 364, 368
J. Boas, 371

D.G. Chapman, 377, 381
S.D. Chatterji, 181, 188, 195
K.L. Chung, 45, 53, 155, 315
W. G. Cochran, 657
W. J. Conover, 665
H. Cramér, 220, 277, 281, 317, 372, 386, 397, 398, 474, 475, 631
J.H. Curtiss, 86, 317

D.A. Darmois, 221
M.M. Desu, 523

B. Efron, 530, 533
P. Erdös, 155

W. Feller, 76, 157, 219, 315, 325, 327
K.K. Ferentinos, 523
T.S. Ferguson, 334, 455, 475, 524, 528
R.A. Fisher, 227, 356, 374, 383
M. Fisz, 322
S.V. Fomin, 4
D.A.S. Fraser, 345, 409, 576, 590, 617
M. Fréchet, 372

J.D. Gibbons, 614
B.V. Gnedenko, 302
W.C. Guenther, 517, 526
E.J. Gumbel, 115

J.H. Hahn, 592
R. A. Hall, 660
P.R. Halmos, 4, 345
J.L. Hodges, 412, 414, 631
W. Hoeffding, 581
P. G. Hoel, 651, 653
D. Hogben, 258
V.S. Huzurbazar, 398

An Introduction to Probability and Statistics, Third Edition. Vijay K. Rohatgi and A.K. Md. Ehsanes Saleh.

M. Kac, 34
W.C.M. Kallenberg, 460
J.F. Kemp, 227
M.G. Kendall, 612, 614, 616
J. Kiefer, 377, 381
G.W. Klett, 521, 527
A.N. Kolmogorov, 2, 4, 586
C.H. Kraft, 594
W. Kruskal, 154, 188
G. Kulldorf, 398

R.G. Laha, 239, 315
J. Lamperti, 188, 226
E.L. Lehmann, 242, 345, 350, 353, 354, 356, 412, 414, 449, 455, 460, 543, 581, 583, 631, 633
Y. LePage, 92
M. Loève, 321
E. Lukacs, 88, 207, 221, 239, 317

H.B. Mann, 606
F.J. Massey, 603
M.V. Menon, 216
L.H. Miller, 586
S. Mitra, 369
M. Moore, 92

M.G. Natrella, 481, 482, 489
W. Nelson, 592
J. Neyman, 400, 438

E.H. Oliver, 392
D.B. Owen, 586

E. Parzen, 650
E.S. Pearson, 438
E.J.G. Pitman, 216
J.W. Pratt, 525
B.J. Prochaska, 227, 301
P.S. Puri, 195

D.A. Raikov, 187
R.R. Randles, 461, 581, 583, 614, 633
C.R. Rao, 290, 364, 372
S.C. Rastogi, 282
H. Robbins, 377, 381
V.K. Rohatgi, 226, 315, 328
V.I. Romanovsky, 277
L. Rosenberg, 233
J. Roy, 369
R. Roy, 92
H.L. Royden, 4

Y.D. Sabharwal, 175
A. Sampson, 354
P.A. Samuelson, 248
L. J. Savage, 345
H. Scheffé, 288, 353, 354, 543, 635, 636
E.L. Scott, 400
R.J. Serfling, 334, 633
D.N. Shanbhag, 178, 192
L. Shepp, 226
A.E. Siegel, 607
V.P. Skitovitch, 221
N.V. Smirnov, 587
G. W. Snedecor, 657
B. Spencer, 354
R.C. Srivastava, 195
S.M. Stigler, 366
P.T. Strait, 36

R.F. Tate, 521, 527
R.J. Tibshirani, 533
F.H. Tingey, 587
W.A. Thompson, Jr. 216
H.G. Tucker, 101

C. Van Eeden, 594
L. R. Verdooren, 664

A. Wald, 399
G.N. Watson, 356
D.R. Whitney, 606
D.V. Widder, 86, 454
S.S. Wilks, 470, 522
E.J. Williams, 216
J. Wishart, 277
D.A. Wolfe, 461, 581, 583, 614, 633
C.K. Wong, 32

S. Zacks, 633
P.W. Zehna, 396

SUBJECT INDEX

Absolutely continuous df, 47, 49, 53, 135, 335, 336, 576, 590
Actions, 401, 397
Admissible decision rule, 416
Analysis of variance, 539
  one-way, 539, 554
  table, 555
  two-way, 560
  two-way with interaction, 566, 570
Ancillary statistic, 355
Assignment of probability, 7, 13
  equally likely, 1, 7, 20
  on finite sample spaces, 20
  random, 13
  uniform, 7, 20
Asymptotic distribution,
  of *r*th order-statistic, 335
  of sample moments, 328
  of sample quantile, 336
Asymptotic relative efficiency(Pitman's), 632
Asymptotically efficient estimator, 382
Asymptotically normal, 332
Asymptotically normal estimator, 332
  best, 341
  consistent, 341
Asymptotically unbiased estimator, 341
At random, 1, 16

Banach's matchbox problem, 180
Bayes,
  risk, 403
  rule, 28, 403
  solution, 404
Behrens-Fisher problem, 486
  Welch approximation, 486
Bernoulli random variable, 174
Bernoulli trials, 174
Bertrand's paradox, 17
Best asymptotically normal estimator, 341
Beta distribution, 210
  bivariate, 113
  MGF, 212
  moments, 211
Beta function, 210
Bias of an estimator, 339, 360
Biased estimator, 359
Binomial coefficient, 79
Binomial distribution, 78, 176
  bounds for tail probability, 193
  central term, 193
  characterization, 178

Binomial distribution (*cont'd*)
coefficient of skewness, 82
generalized to multinomial, 190
Kurtosis, 82
mean, 78
MGF, 177
moments, 78, 82
PGF, 95, 177
relation to negative binomial, 180
tail probability as incomplete beta function, 213
variance, 78
Blackwell-Rao theorem, 364
Bonferroni's inequality, 10
Boole's inequality, 11
Bootstrap,
method, 530
sample, 530
Borel-Cantelli lemma, 309
Borel-measurable functions, of an rv, 55, 69, 117
Buffon's needle problem, 14

Canonical form, 541
Cauchy distribution, 68, 80, 213
bivariate, 113
characterization, 216
characteristic function, 215, 320
mean does not exist, 215
MGF does not exist, 215
moments, 214
as ratio of two normal, 221
as stable distribution, 216
Cauchy-Schwarz inequality, 153
Central limit theorem, 321
applications of, 327
Chapman, Robbins and Kiefer inequality, 377
for discrete uniform, 378
for normal, 379
for uniform, 378
Characteristic function, 87
of multiple RVs, 136
properties, 136
Chebychev-Bienayme inequality, 94
Chebychev's inequality, 94
improvement of, 95
Chi-square distribution, central, 206, 261
MGF, 207, 262
moments, 207, 262
as square of normal, 221
noncentral, 264
MGF, 264
moments, 264
Chi-square test(s), 472
as a goodness of fit, 476
for homogeneity, 479
for independence, 608
one-tailed, 472
robustness, 631
for testing equality of proportions, 473
for testing parameters of multinomial, 475
for testing variance, 472
two-tailed, 472
Combinatorics, 20
Complete, family of distributions, 347
Complete families, binomial, 348
chi-square, 348
discrete uniform, 358
hypergeometric, 358
uniform, 348
Complete sufficient statistic, 347, 576
for Bernoulli, 348
for exponential family, 350
for normal, 351
for uniform, 349
Concordance, 611
Conditional, DF, 108
distribution, 107
PDF, 109
PMF, 108
probability, 26
Conditional expectation, 158
properties of, 158
Confidence, bounds, 500
coefficient, 500
estimation problem, 500
Confidence interval, 499
Bayesian, 511
equivariant, 527
expected length of, 517
general method(s) of construction, 504
level of, 500
length of, 500
percentile, 531
for location parameter, 623
for the parameter of, Bernoulli, 513
discrete uniform, 516
exponential, 509

normal, 502–503
uniform, 509, 515
for quantile of order p, 621
shortest-length, 516
from tests of hypotheses, 507
UMA family, 502
UMAU family, 524
for normal mean, 524
for normal variance, 526
unbiased, 523
using Chebychev's inequality, 513
using CLT, 512
using properties of MLE's, 513
Conjugate prior distribution, 408
natural, 408
Confidence set, 501
for mean and variance of normal, 502
UMA family of, 502
UMAU family of, 524
unbiased, 523
Consistent estimator, 340
asymptotically normal, 341
in rth mean, 340
strong and weak, 340
Contaminated normal, 625
Contingency table, 608
Continuity correction, 328
Continuity theorem, 317
Continuous type distributions, 49
Convergence, a.s., 294
in distribution = weak, 286
in law, 286
of MGFs, 316–317
modes of, 285
of moments, 287
of PDFs, 287
of PMFs, 287–288
in probability, 288
in rth mean, 292
Convolution of DFs, 135
Correlation, 144
Correlation coefficient, 144, 277
properties, 145
Countable additivity, 7
Covariance, 144
sample, 277
Coverage, elementary, 619
r-coverage, 620
probability, 619
Credible sets, 511
Critical region, 431

Decision function, 401
Degenerate RV, 173
Degrees of freedom when pooling classes, 479
Delta method, 332
Density function, probability, 49, 104
Design matrix, 539
Diachotomous trials, 174
Discordance, 611
Discrete distributions, 173
Discrete uniform distribution, 175
Dispersion matrix = variance – covariance matrix, 328
Distribution, conditional, 107
conjugate prior, 408
of a function of an RV, 55
induced, 59
a posteriori, 404
a priori, 403
of sample mean, 257
of sample median, 259
of sample quantile, 167, 336
of sample range, 162, 326
Distribution function, 43
continuity points of a, 43, 50
of a continuous type RV, 49
convolution, 135
decomposition of a, 53
discontinuity points of a, 43
of a discrete type RV, 47
of a function of an RV, 56
of an RV, 43
of multiple RVs, 100, 102
Domain of attraction, 321

Efficiency of an estimate, 382
relative, 382
Empirical DF = sample DF, 249
Equal likelihood, 1
Equivalent RVs, 119
Estimable function, 360
Estimable parameter, 576, 581
degree, 577, 581
kernel, 577, 582
Estimator, 338
equivariant, 340, 420

Estimator (*cont'd*)
Hodges-Lehmann, 631
least squares, 537
minimum risk equivariant, 422
Pitman, 424, 426
point, 338
Event, 3
certain, 8
elementary = simple, 3
disjoint = mutually exclusive, 7, 33
independent, 31
null, 3
Exchangeable random variables, 120, 149, 255
Expectation, conditional, 158
properties, 158
Expected value = mean = mathematical expectation, 68
of a function of RV, 67, 136
of product of RVs, 148
of sum of RVs, 147
Exponential distribution, 206
characterizations, 208
memoryless property of, 207
MGF, 206
moments, 206
Exponential family, 242
k-parameter, 242
natural parameters of, 243
one-parameter, 240
Extreme value distribution, 224

Factorial moments, 79
Factorization criterion, 344
Finite mixture density function, 225
Finite population correction, 256
Fisher Information, 375
Fisher's Z-statistic, 270
Fitting of distribution, binomial, 482
geometric, 482
normal, 477
Poisson, 478
Fréchet, Cramér, and Rao inequality, 374
Fréchet, Cramér, and Rao lower bound, 375
binomial, 376
exponential, 385
normal, 385
one-parameter exponential family, 377
Poisson, 375

F-distribution, central, 267
moments of, 267
noncentral, 269
moments of, 269
F-test(s), 489
of general linear hypothesis, 540
as generalized likelihood ratio test, 540
for testing equality of variances, 440

Gamma distribution, 203
bivariate, 113
characterizations, 207
MGF, 205
moments, 206
relation with Poisson, 208
Gamma function, 202
General linear hypothesis, 536
canonical form, 541
estimation in, 536
GLR test of, 540
General linear model, 536
Generalized Likelihood ratio test, 464
asymptotic distribution, 470
F-test as, 468
for general linear hypothesis, 540
for parameter of, binomial, 465
for simple vs. simple hypothesis, 464
bivariate normal, 471
discrete uniform, 471
exponential, 472
normal, 466
Generating functions, 83
moment, 85
probability, 83
Geometric distribution, 84, 180
characterizations, 182
memoryless property of, 182
MGF, 180
moments, 180
order statistic, 164
PGF, 84
Glivenko-Cantelli theorem, 322
Goodness-of-fit problem, 584

Hazard(=failure rate) function, 227
Helmert orthogonal matrix, 274
Hodges-Lehmann estimators, 631
Hölder's inequality, 153

Hypergeometric distribution, 184
  bivariate, 113
  mean and variance, 184
Hypothesis, tests of, 429
  alternative, 430
  composite, 430
  null, 430
  parametric, 430
  simple, 430
  tests of, 430

Identically distributed RVs, 119
Implication rule, 11
Inadmissible decision rule, 416
Independence and correlation, 145
Independence of events, 115
  complete = mutual, 118
  pairwise, 118
Independence of RVs, 114–121
  complete = mutual, 118
  pairwise, 118
Independent, identically distributed rv's, 119
  sequence of, 119
Indicator function, 41
Induced distribution, 59
Infinitely often, 309
Interections, 566
Invariance, of hypothesis testing problem, 455
  principle, 455
Invariant,
  decision problem, 419
  family of distributions, 418
  function, 420, 455
  location, 421
  location-scale, 421
  loss function, 420
  maximal, 505
  scale, 420
  statistic, 420
Invariant, class of distributions, 419
  estimators, 420
  maximal, 422, 455
  tests, 455
Inverse Gaussian PDF, 228

Jackknife, 533
Joint, DF, 100–102
  PDF, 104
  PMF, 103
Jump, 47, 103
Jump point, of a DF, 47, 103

Kendall's sample tau, 612
  distribution of, 612
  generating function, 92
Kendall's tau coefficient, 611
Kendall's tau test, 612
Kernel, symmetric, 577, 582
Kolmogorov's, inequality, 312
  strong law of large numbers, 315
Kolmogorov-Smirnov one sample statistic, 584
  for confidence bounds of DF, 587
  distribution, 585–587
Kolmogorov-Smirnov test, 602
  comparison with chi-square test, 588
  one-sample, 587
  two-sample, 603
Kolmogorov-Smirnov two sample statistic, 601
  distribution, 603
Kronecker lemma, 313
Kurtosis, coefficient of, 83

Laplace = double exponential distribution, 91, 224
  MGF, 87
Least square estimation, 537
  principle, 537
  restricted, 537
L'Hospital rule, 323
Likelihood,
  equal, 1
  equation, 389
  equivalent, 353
  function, 389
Limit, inferior, 11
  set, 11
  superior, 11
Lindeberg central limit theorem, 325
Lindeberg-Levy CLT, 323
Lindeberg condition, 324
Linear combinations of RVs, 147
  mean and variance, 147, 149
Linear dependence, 145
Linear model, 536

Linear regression model, 538, 543
confidence intervals, 545
estimation, 543
testing of hypotheses, 545–546
Locally most powerful test, 459
Location family, 196
Location-scale family, 196
Logistic distribution, 223
Logistic function, 551
Logistic regression, 550
Lognormal distribution, 88, 222
Loss function, 339, 401
Lower bound for variance, Chapman, Robbins and Kiefer inequality, 377
Fréchet, Cramér and Rao inequality, 372
Lyapunov condition, 326
Lyapunov inequality, 96

Maclaurin expansion of an mgf, 86
Mann-Whitney statistic, 604
moments, 582
null distribution, 605
Mann-Whitney-Wilcoxon test, 605
Marginal,
DF, 107
PDF, 106
PMF, 105
Markov's inequality, 94
Maximal invariant statistic, 422, 455
function of, 457
Maximum likelihood estimation, principle of, 389
Maximum likelihood estimator, 389
asymptotic normality, 397–398
consistency, 397
as a function of sufficient statistic, 394
invariance property, 396
Maximum likelihood estimation method applied to, Bernoulli, 392
binomial, 399
bivariate normal, 395
Cauchy, 399
discrete uniform, 390
exponential, 396
gamma, 393
hypergeometric, 391
normal, 390
Poisson, 399
uniform, 391, 394
Mean square error, 339, 362
Median, 80, 82
Median test, 600
Memoryless property,
of exponential, 207
of geometric, 182
Method of finding distribution,
CF or MGF, 90, 137
DF, 56, 124
transformations 128
Methods of finding confidence interval
Bayes, 511
for large samples, 511
pivot, 504
test inversion, 507
Method of moments, 386
applied to, beta, 388
binomial, 387
gamma, 388
lognormal, 388
normal, 388
Poisson, 386
uniform, 387
Minimal sufficient statistic, 354
for beta, 358
for gamma, 358
for geometric, 358
for normal, 355
for Poisson, 358
for uniform, 354, 358
Minimax, estimator, 402
principle, 402
solution, 492
Minimax estimation for parameter of,
Bernoulli, 402
binomial, 412
hypergeometric, 414
Minimum mean square error estimator, 339
for variance of normal, 368
Minimum risk equivariant estinator, 421
for location parameter, 424
for scale parameter, 425
Mixing proportions, 225
Minkowski inequality, 153
Mixture density function, 224–225
Moment, about origin, 70
absolute, 70
central, 77
condition, 73
Factorial, 79
of conditional distribution, 158

of DF, 70
of functions of multiple RVs, 136
inequalities, 93
lemma, 74–75
non-existence of order, 75
of sample covariance, 257
of sample mean, 253
of sample variance, 253–254
Moment generating function, 85
continuity theorem for, 317
differentiation, 86
existence, 87
expansion, 86
limiting, 316
of linear combinations, 139
and moments, 86
of multiple RVs, 136
of sample mean, 256
series expansion, 86
of sum of independent RVs, 139
uniqueness, 86
Monotone likelihood ratio, 446
for hypergeometric, 448
for one-parameter exponential family, 447
UMP test for families with, 447
for uniform, 446
Most efficient estimator, 382
asymptotically, 382
as MLE, 395
Most powerful test, 432
for families with MLR, 446
as a function of sufficient statistic, 440
invariant, 456
Neyman-Pearson, 438
similar, 433
unbiased, 432
uniformly, 432
Multidimentional RV = multiple RV, 99
Multinomial coefficient, 23
Multinomial distribution, 190
MGF, 190
moments, 191
Multiple RV, 99
continuous type, 104
discrete type, 103
functions of, 123
Multiple regression, 543
Multiplication rule, 27
Multivariate hypergeometric distribution, 192
Multivariate negative binomial distribution, 193
Multivariate normal, 234
dispersion matrix, 236

Natural parameters, 243
Negative binomial (=Pascal or waiting time) distribution, 178–179
bivariate, 113
central term, 194
mean and variance, 179
MGF, 179
Negative hypergeometric distribution, 186
mean and variance, 186
Neyman-Pearson lemma, 438
Neyman-Pearson lemma applied to,
Bernoulli, 442
normal, 444
Noncentral, chi-square distribution, 263
F-distribution, 269
t-distribution, 266
Noncentrality parameter, of chi-square, 263
F-distribution, 269
t-distribution, 266
Noninformative prior, 409
Nonparametric = distribution-free estimation, 576–577
methods, 576
Nonparametric unbiased estimation, 576
of population mean, 578
of population variance, 578
Normal approximation, to binomial, 328
to Poisson, 330
Normal distribution = Gaussian law, 87, 216
bivariate, 228
characteristic function, 87
characterizations, 219, 221, 238
contaminated, 625, 628
folded, 426
as limit of binomial, 321, 328
as limit of chi-square, 322
as limit of Poisson, 330
MGF, 217
moments, 217–218
multivariate, 234
singular, 232
as stable distribution, 321
standard, 216

Normal distribution = Gaussian law (*cont'd*)
tail probability, 219
truncated, 111
Normal equations, 537

Odds, 8
Order statistic, 164
is complete and sufficient, 576
joint PDF, 165
joint marginal PDF, 168
*k*th, 164
marginal PDF, 167
uses, 619
moments, 169
Ordered samples, 21
Orders of magnitude, o and O notation, 318

Parameter(s), of a distribution, 67, 196, 576
estimable, 576
location, 196
location-scale, 196
order, 79
scale, 196
shape, 196
space, 338
Parametric statistical hypothesis, 430
alternative, 430
composite, 430
null, 430
problem of testing, 430
simple, 430
Parametric statistical inference, 245
Pareto distribution, 82, 222
Partition, 351
coarser, 352
finer, 352
minimal sufficient, 353
reduction of a, 352
sets, 351
sub-, 352
sufficient, 351
Percentile confidence interval, 531
centered percentile confidence interval, 532
Permutation, 21
Pitman estimator, 24
location, 426
scale, 426
Pitman's asymptotic relative efficiency, 632
Pivot, 504
Point estimator, 338, 340
Point estimation, problem of, 338
Poisson DF, as incomplete gamma, 209
Poisson distribution, 57, 83, 186
central term, 194
characterizations, 187
coefficient of skewness, 82
kurtosis, 82
as limit of binomial, 194
as limit of negative binomial, 194
mean and variance, 187
MGF, 187
moments, 82
PGF, 187
truncated, 111
Poisson regression, 553
Polya distribution, 185
Pooled sample variance, 485
Population, 245
Population distribution, 246
Posterior probability, 29
Principle of,
equivariance, 420
inclusion-exclusion, 9
invariance, 456
least squares, 537
Probability, 7
addition rule, 9
axioms, 7
conditional, 26
continuity of, 13
countable additivity of, 7
density function, 49
distribution, 42
equally likely assignment, 7, 21
on finite sample spaces, 20
generating function, 83
geometric, 13
integral transformation, 200
mass function, 47
measure, 7
monotone, 8
multiplication rule, 27
posterior and prior, 29
principle of inclusion-exclusion, 9
space, 8
subadditivity, 9
tail, 72

total, 28
uniform assignment of, 7, 21
Probability integral transformation, 200
Probit regression, 552
Problem,
of location, 590
of location and symmetry, 590
of moments, 88
P-value, 437, 481, 599

Quadratic form, 228
Quantile of order $p = (100p)$th percentile, 79

Random, 13
Random experiment = statistical experiment, 3
Random interval, 500
coverage of, 619
Random sample, 13, 246
from a finite population, 13
from a probability distribution, 13, 246
Random sampling, 246
Random set, family of, 500
Random variable(s), 40
bivariate, 103
continuous type, 49, 104
discrete type, 47
degenerate, 48
equivalent, 119
exchangeable 120, 149, 255
functions of a, 55
multiple = multivariate, 99
standardized, 78
symmetric, 69
symmetrized, 121
truncated, 110
uncorrelated, 145
Range, 168
Rank correlation coefficient, 614
Rayleigh distribution, 224
Realization of a sample, 246
Rectangular distribution, 199
Regression, 543
coefficient, 277
linear, 544
logistic, 551
model, 543
multiple, 543
Poisson, 552
probit, 552
Regularity conditions of FCR inequality, 372
Resampling, 530
Risk function, 339, 402
Robust estimator(s), 631
Robust test(s), 634
Robustness, of chi-square test, 631
of sample mean as an estimator, 628
of sample standard deviation as an estimator, 628
of Student's t-test, 629
Robust procedure, defined, 625, 631
Rules of counting, 21–24
Run, 607
Run test, 607

Sample, 245–246
correlation coefficient, 251
covariance, 251
DF, 250
mean, 247
median, 251
distribution of, 260
MGF, 251
moments, 250–251
ordered, 21
point, 3
quantile of order p, 251, 342
random, 246
regression coefficient, 282
space, 3
statistic(s), 246, 249
standard deviation, 248
standard error, 256
variance, 247
Sampling with and without replacement, 21, 247
Sampling from bivariate normal, 276
distribution of sample correlation coefficient, 277
distribution of sample regression coefficient, 277
independence of sample mean vector and dispersion matrix, 277
Sampling from univariate normal, 271
distribution of sample variance, 273
independence of $\bar{X}$ and S^2, 273
Scale family, 196

Sequence of events, 11
  limit inferior, 11
  limit set, 11
  limit superior, 11
  nondecreasing, 12
  nonincreasing, 12
Set function, 7
Shortest-length confidence interval(s), 517
  for the mean of normal, 518–519
  for the parameter of exponential, 523
  for the parameter of uniform, 521
  for the variance of normal, 519
σ-field, 3
  choice of, 3
  generated by a class = smallest, 40
Sign test, 590
Similar tests, 454
Single-sample problem(s), 584
  of fit, 584
  of location, 590
    and symmetry, 590
Skewness, coefficient of, 82
Slow variation, function of, 76
Slutsky's theorem, 298
Spearman's rank correlation coefficient, 614
  distribution, 615
Stable distribution, 216, 321
Standard deviation, 77
Standard error, 256
Standardized RV, 78
Statistic of order k, 164
  marginal PDF, 167
Stirling's approximation, 194
Stochastically larger, 600
Strong law of large numbers, 308
  Borel's, 315
  Kolmogorov's, 315
Student's t-distribution, central, 265
  bivariate, 282
  moments, 267
  noncentral, 267
    moments, 267
Student's t- statistic, 265
Student's t- test, 484–485
  as generalized likelihood ratio test, 467
  for paired observations, 486
  robustness of, 630
Substitution principle, 386
  estimator, 386
Sufficient statistic, 343
  factorization criterion, 344
  joint, 345
Sufficient statistic for, Bernoulli, 345
  beta, 356
  discrete uniform, 346
  gamma, 356
  lognormal, 357
  normal, 346
  Poisson, 343
  uniform, 346
Support, of a DF, 50, 103
Survival function = reliability function, 227
Symmetric DF or RV, 50, 103
Symmetrization, 121
Symmetrized rv, 121
Symmetry, center of, 73

Tail probabilities, 72
Test(s),
  α-similar, 453
  chi-square, 470
  critical = rejection region, 431
  critical function, 431
  of hypothesis, 431
  F-, 489
  invariant, 453
  level of significance, 431
  locally most powerful, 459
  most powerful, 432
  nonrandomized, 432
  one-tailed, 484
  power function, 432
  randomized, 432
  similar, 453
  size, 432
  statistic, 433
  Student's t, 506
  two tailed, 484
  unbiased, 484
  uniformly most powerful, 432
Testing the hypothesis of, equality of several normal means, 539
  goodness-of- fit, 482, 584
  homogeneity, 479
  independence, 608
Tests of hypothesis, Bayes, 507
  GLR, 463
  minimax, 491
  Neyman-Pearson, 438

Tests of location, 590
  sign test, 590
  Wilcoxon signed-rank, 592
Tolerance coefficient and interval, 619
Total probability rule, 28
Transformation, 55
  of continuous type, 58, 124, 128
  of discrete type, 58, 135
  Helmert, 274
  Jacobian of, 128
  not one-to-one, 165
  one-to-one, 56, 129
Triangular distribution, 52
Trimmed mean, 632
Trinomial distribution, 191
Truncated distribution, 110
Truncated RVs, 110
Truncation, 110
Two-point distribution, 174
Two-sample problems, 599
Types of error in testing hypotheses, 431

Unbiased confidence interval(s), 523
  general method of construction, 524
  for mean of normal, 524
  for parameter of exponential, 529
  for parameter of uniform, 529
  for variance of normal, 526
Unbiased estimator, 339
  best linear, 361
  and complete sufficient statistic, 365
  LMV, 361
  and sufficient statistic, 364
  UMV, 361
Unbiased estimation for parameter of,
  Bernoulli, 365, 364
  bivariate normal, 368
  discrete uniform, 369
  exponential, 369
  hypergeometric, 369
  negative binomial, 368
  normal, 365
  Poisson, 363
Unbiased test, 453
  for mean of normal, 454
  and similar test, 453
  UMP, 453
Uncorrelated RVs, 145
Uniform distribution, 56, 197
  characterization, 201
  discrete, 72, 175
  generating samples, 201
  MGF, 199
  moments, 199
  statistic of order k, 168, 213
  truncated, 111
UMP test(s)
  α-similar, 453
  invariant, 457
  unbiased, 453
U-statistic, 576
  for estimating mean and variance, 578
  one-sample, 576
  two-sample, 581

Variance, 77
  properties of, 77
  of sum of RVs, 148
Variance stablizing transformations, 333

Weak law of large numbers, 303, 306
  centering and norming constants, 303
Weibull distribution, 223
Welch approximate t-test, 486
Wilcoxon signed-rank test, 592
Wilcoxon statistic, 593
  distribution, 594, 597
  generating function, 93
  moments, 597
Winsorization, 112

WILEY SERIES IN PROBABILITY AND STATISTICS
ESTABLISHED BY WALTER A. SHEWHART AND SAMUEL S. WILKS

Editors: *David J. Balding, Noel A. C. Cressie, Garrett M. Fitzmaurice, Geof H. Givens, Harvey Goldstein, Geert Molenberghs, David W. Scott, Adrian F. M. Smith, Ruey S. Tsay, Sanford Weisberg*
Editors Emeriti: *J. Stuart Hunter, Iain M. Johnstone, Joseph B. Kadane, Jozef L. Teugels*

The *Wiley Series in Probability and Statistics* is well established and authoritative. It covers many topics of current research interest in both pure and applied statistics and probability theory. Written by leading statisticians and institutions, the titles span both state-of-the-art developments in the field and classical methods.

Reflecting the wide range of current research in statistics, the series encompasses applied, methodological and theoretical statistics, ranging from applications and new techniques made possible by advances in computerized practice to rigorous treatment of theoretical approaches.

This series provides essential and invaluable reading for all statisticians, whether in academia, industry, government, or research.

† ABRAHAM and LEDOLTER · Statistical Methods for Forecasting
AGRESTI · Analysis of Ordinal Categorical Data, *Second Edition*
AGRESTI · An Introduction to Categorical Data Analysis, *Second Edition*
AGRESTI · Categorical Data Analysis, *Third Edition*
AGRESTI · *Foundations of Linear and Generalized Linear Models*
ALSTON, MENGERSEN and PETTITT (editors) · Case Studies in Bayesian Statistical Modelling and Analysis
ALTMAN, GILL, and McDONALD · Numerical Issues in Statistical Computing for the Social Scientist
AMARATUNGA and CABRERA · Exploration and Analysis of DNA Microarray and Protein Array Data
AMARATUNGA, CABRERA, and SHKEDY · Exploration and Analysis of DNA Microarray and Other High-Dimensional Data, *Second Edition*
ANDĚL · Mathematics of Chance
ANDERSON · An Introduction to Multivariate Statistical Analysis, *Third Edition*
* ANDERSON · The Statistical Analysis of Time Series
ANDERSON, AUQUIER, HAUCK, OAKES, VANDAELE, and WEISBERG · Statistical Methods for Comparative Studies
ANDERSON and LOYNES · The Teaching of Practical Statistics
ARMITAGE and DAVID (editors) · Advances in Biometry
ARNOLD, BALAKRISHNAN, and NAGARAJA · Records
* ARTHANARI and DODGE · Mathematical Programming in Statistics
AUGUSTIN, COOLEN, DE COOMAN and TROFFAES (editors) · Introduction to Imprecise Probabilities
* BAILEY · The Elements of Stochastic Processes with Applications to the Natural Sciences
BAJORSKI · Statistics for Imaging, Optics, and Photonics
BALAKRISHNAN and KOUTRAS · Runs and Scans with Applications
BALAKRISHNAN and NG · Precedence-Type Tests and Applications
BARNETT · Comparative Statistical Inference, *Third Edition*
BARNETT · Environmental Statistics
BARNETT and LEWIS · Outliers in Statistical Data, *Third Edition*

BARTHOLOMEW, KNOTT, and MOUSTAKI · Latent Variable Models and Factor Analysis: A Unified Approach, *Third Edition*
BARTOSZYNSKI and NIEWIADOMSKA-BUGAJ · Probability and Statistical Inference, *Second Edition*
BASILEVSKY · Statistical Factor Analysis and Related Methods: Theory and Applications
BATES and WATTS · Nonlinear Regression Analysis and Its Applications
BECHHOFER, SANTNER, and GOLDSMAN · Design and Analysis of Experiments for Statistical Selection, Screening, and Multiple Comparisons
BEH and LOMBARDO · Correspondence Analysis: Theory, Practice and New Strategies
BEIRLANT, GOEGEBEUR, SEGERS, TEUGELS, and DE WAAL · Statistics of Extremes: Theory and Applications
BELSLEY · Conditioning Diagnostics: Collinearity and Weak Data in Regression
† BELSLEY, KUH, and WELSCH · Regression Diagnostics: Identifying Influential Data and Sources of Collinearity
BENDAT and PIERSOL · Random Data: Analysis and Measurement Procedures, *Fourth Edition*
BERNARDO and SMITH · Bayesian Theory
BHAT and MILLER · Elements of Applied Stochastic Processes, *Third Edition*
BHATTACHARYA and WAYMIRE · Stochastic Processes with Applications
BIEMER, GROVES, LYBERG, MATHIOWETZ, and SUDMAN · Measurement Errors in Surveys
BILLINGSLEY · Convergence of Probability Measures, *Second Edition*
BILLINGSLEY · Probability and Measure, *Anniversary Edition*
BIRKES and DODGE · Alternative Methods of Regression
BISGAARD and KULAHCI · Time Series Analysis and Forecasting by Example
BISWAS, DATTA, FINE, and SEGAL · Statistical Advances in the Biomedical Sciences: Clinical Trials, Epidemiology, Survival Analysis, and Bioinformatics
BLISCHKE and MURTHY (editors) · Case Studies in Reliability and Maintenance
BLISCHKE and MURTHY · Reliability: Modeling, Prediction, and Optimization
BLOOMFIELD · Fourier Analysis of Time Series: An Introduction, *Second Edition*
BOLLEN · Structural Equations with Latent Variables
BOLLEN and CURRAN · Latent Curve Models: A Structural Equation Perspective
BONNINI, CORAIN, MAROZZI and SALMASO · Nonparametric Hypothesis Testing: Rank and Permutation Methods with Applications in R
BOROVKOV · Ergodicity and Stability of Stochastic Processes
BOSQ and BLANKE · Inference and Prediction in Large Dimensions
BOULEAU · Numerical Methods for Stochastic Processes
* BOX and TIAO · Bayesian Inference in Statistical Analysis
BOX · Improving Almost Anything, *Revised Edition*
* BOX and DRAPER · Evolutionary Operation: A Statistical Method for Process Improvement
BOX and DRAPER · Response Surfaces, Mixtures, and Ridge Analyses, *Second Edition*
BOX, HUNTER, and HUNTER · Statistics for Experimenters: Design, Innovation, and Discovery, *Second Editon*
BOX, JENKINS, and REINSEL · Time Series Analysis: Forecasting and Control, *Fourth Edition*
BOX, LUCEÑO, and PANIAGUA-QUIÑONES · Statistical Control by Monitoring and Adjustment, *Second Edition*
* BROWN and HOLLANDER · Statistics: A Biomedical Introduction
CAIROLI and DALANG · Sequential Stochastic Optimization

CASTILLO, HADI, BALAKRISHNAN, and SARABIA · Extreme Value and Related Models with Applications in Engineering and Science
CHAN · Time Series: Applications to Finance with R and S-Plus®, *Second Edition*
CHARALAMBIDES · Combinatorial Methods in Discrete Distributions
CHATTERJEE and HADI · Regression Analysis by Example, *Fourth Edition*
CHATTERJEE and HADI · Sensitivity Analysis in Linear Regression
CHEN · The Fitness of Information: Quantitative Assessments of Critical Evidence
CHERNICK · Bootstrap Methods: A Guide for Practitioners and Researchers, *Second Edition*
CHERNICK and FRIIS · Introductory Biostatistics for the Health Sciences
CHILÈS and DELFINER · Geostatistics: Modeling Spatial Uncertainty, *Second Edition*
CHIU, STOYAN, KENDALL and MECKE · Stochastic Geometry and Its Applications, *Third Edition*
CHOW and LIU · Design and Analysis of Clinical Trials: Concepts and Methodologies, *Third Edition*
CLARKE · Linear Models: The Theory and Application of Analysis of Variance
CLARKE and DISNEY · Probability and Random Processes: A First Course with Applications, *Second Edition*
* COCHRAN and COX · Experimental Designs, *Second Edition*
COLLINS and LANZA · Latent Class and Latent Transition Analysis: With Applications in the Social, Behavioral, and Health Sciences
CONGDON · Applied Bayesian Modelling, *Second Edition*
CONGDON · Bayesian Models for Categorical Data
CONGDON · Bayesian Statistical Modelling, *Second Edition*
CONOVER · Practical Nonparametric Statistics, *Third Edition*
COOK · Regression Graphics
COOK and WEISBERG · An Introduction to Regression Graphics
COOK and WEISBERG · Applied Regression Including Computing and Graphics
CORNELL · A Primer on Experiments with Mixtures
CORNELL · Experiments with Mixtures, Designs, Models, and the Analysis of Mixture Data, *Third Edition*
COX · A Handbook of Introductory Statistical Methods
CRESSIE · Statistics for Spatial Data, *Revised Edition*
CRESSIE and WIKLE · Statistics for Spatio-Temporal Data
CSÖRGŐ and HORVÁTH · Limit Theorems in Change Point Analysis
DAGPUNAR · Simulation and Monte Carlo: With Applications in Finance and MCMC
DANIEL · Applications of Statistics to Industrial Experimentation
DANIEL · Biostatistics: A Foundation for Analysis in the Health Sciences, *Eighth Edition*
* DANIEL · Fitting Equations to Data: Computer Analysis of Multifactor Data, *Second Edition*
DASU and JOHNSON · Exploratory Data Mining and Data Cleaning
DAVID and NAGARAJA · Order Statistics, *Third Edition*
DAVINO, FURNO and VISTOCCO · Quantile Regression: Theory and Applications
* DEGROOT, FIENBERG, and KADANE · Statistics and the Law
DEL CASTILLO · Statistical Process Adjustment for Quality Control
DEMARIS · Regression with Social Data: Modeling Continuous and Limited Response Variables
DEMIDENKO · Mixed Models: Theory and Applications with R, *Second Edition*

DENISON, HOLMES, MALLICK, and SMITH · Bayesian Methods for Nonlinear Classification and Regression
DETTE and STUDDEN · The Theory of Canonical Moments with Applications in Statistics, Probability, and Analysis
DEY and MUKERJEE · Fractional Factorial Plans
DILLON and GOLDSTEIN · Multivariate Analysis: Methods and Applications
* DODGE and ROMIG · Sampling Inspection Tables, *Second Edition*
* DOOB · Stochastic Processes
DOWDY, WEARDEN, and CHILKO · Statistics for Research, *Third Edition*
DRAPER and SMITH · Applied Regression Analysis, *Third Edition*
DRYDEN and MARDIA · Statistical Shape Analysis
DUDEWICZ and MISHRA · Modern Mathematical Statistics
DUNN and CLARK · Basic Statistics: A Primer for the Biomedical Sciences, *Fourth Edition*
DUPUIS and ELLIS · A Weak Convergence Approach to the Theory of Large Deviations
EDLER and KITSOS · Recent Advances in Quantitative Methods in Cancer and Human Health Risk Assessment
* ELANDT-JOHNSON and JOHNSON · Survival Models and Data Analysis
ENDERS · Applied Econometric Time Series, *Third Edition*
† ETHIER and KURTZ · Markov Processes: Characterization and Convergence
EVANS, HASTINGS, and PEACOCK · Statistical Distributions, *Third Edition*
EVERITT, LANDAU, LEESE, and STAHL · Cluster Analysis, *Fifth Edition*
FEDERER and KING · Variations on Split Plot and Split Block Experiment Designs
FELLER · An Introduction to Probability Theory and Its Applications, Volume I, *Third Edition,* Revised; Volume II, *Second Edition*
FITZMAURICE, LAIRD, and WARE · Applied Longitudinal Analysis, *Second Edition*
* FLEISS · The Design and Analysis of Clinical Experiments
FLEISS · Statistical Methods for Rates and Proportions, *Third Edition*
† FLEMING and HARRINGTON · Counting Processes and Survival Analysis
FUJIKOSHI, ULYANOV, and SHIMIZU · Multivariate Statistics: High-Dimensional and Large-Sample Approximations
FULLER · Introduction to Statistical Time Series, *Second Edition*
† FULLER · Measurement Error Models
GALLANT · Nonlinear Statistical Models
GEISSER · Modes of Parametric Statistical Inference
GELMAN and MENG · Applied Bayesian Modeling and Causal Inference from ncomplete-Data Perspectives
GEWEKE · Contemporary Bayesian Econometrics and Statistics
GHOSH, MUKHOPADHYAY, and SEN · Sequential Estimation
GIESBRECHT and GUMPERTZ · Planning, Construction, and Statistical Analysis of Comparative Experiments
GIFI · Nonlinear Multivariate Analysis
GIVENS and HOETING · Computational Statistics
GLASSERMAN and YAO · Monotone Structure in Discrete-Event Systems
GNANADESIKAN · Methods for Statistical Data Analysis of Multivariate Observations, *Second Edition*
GOLDSTEIN · Multilevel Statistical Models, *Fourth Edition*
GOLDSTEIN and LEWIS · Assessment: Problems, Development, and Statistical Issues
GOLDSTEIN and WOOFF · Bayes Linear Statistics

GRAHAM · Markov Chains: Analytic and Monte Carlo Computations
GREENWOOD and NIKULIN · A Guide to Chi-Squared Testing
GROSS, SHORTLE, THOMPSON, and HARRIS · Fundamentals of Queueing Theory, *Fourth Edition*
GROSS, SHORTLE, THOMPSON, and HARRIS · Solutions Manual to Accompany Fundamentals of Queueing Theory, *Fourth Edition*
* HAHN and SHAPIRO · Statistical Models in Engineering
HAHN and MEEKER · Statistical Intervals: A Guide for Practitioners
HALD · A History of Probability and Statistics and their Applications Before 1750
† HAMPEL · Robust Statistics: The Approach Based on Influence Functions
HARTUNG, KNAPP, and SINHA · Statistical Meta-Analysis with Applications
HEIBERGER · Computation for the Analysis of Designed Experiments
HEDAYAT and SINHA · Design and Inference in Finite Population Sampling
HEDEKER and GIBBONS · Longitudinal Data Analysis
HELLER · MACSYMA for Statisticians
HERITIER, CANTONI, COPT, and VICTORIA-FESER · Robust Methods in Biostatistics
HINKELMANN and KEMPTHORNE · Design and Analysis of Experiments, Volume 1: Introduction to Experimental Design, *Second Edition*
HINKELMANN and KEMPTHORNE · Design and Analysis of Experiments, Volume 2: Advanced Experimental Design
HINKELMANN (editor) · Design and Analysis of Experiments, Volume 3: Special Designs and Applications
HOAGLIN, MOSTELLER, and TUKEY · Fundamentals of Exploratory Analysis of Variance
* HOAGLIN, MOSTELLER, and TUKEY · Exploring Data Tables, Trends and Shapes
* HOAGLIN, MOSTELLER, and TUKEY · Understanding Robust and Exploratory Data Analysis
HOCHBERG and TAMHANE · Multiple Comparison Procedures
HOCKING · Methods and Applications of Linear Models: Regression and the Analysis of Variance, *Third Edition*
HOEL · Introduction to Mathematical Statistics, *Fifth Edition*
HOGG and KLUGMAN · Loss Distributions
HOLLANDER, WOLFE, and CHICKEN · Nonparametric Statistical Methods, *Third Edition*
HOSMER and LEMESHOW · Applied Logistic Regression, *Second Edition*
HOSMER, LEMESHOW, and MAY · Applied Survival Analysis: Regression Modeling of Time-to-Event Data, *Second Edition*
HUBER · Data Analysis: What Can Be Learned From the Past 50 Years
HUBER · Robust Statistics
† HUBER and RONCHETTI · Robust Statistics, *Second Edition*
HUBERTY · Applied Discriminant Analysis, *Second Edition*
HUBERTY and OLEJNIK · Applied MANOVA and Discriminant Analysis, *Second Edition*
HUITEMA · The Analysis of Covariance and Alternatives: Statistical Methods for Experiments, Quasi-Experiments, and Single-Case Studies, *Second Edition*
HUNT and KENNEDY · Financial Derivatives in Theory and Practice, *Revised Edition*
HURD and MIAMEE · Periodically Correlated Random Sequences: Spectral Theory and Practice
HUSKOVA, BERAN, and DUPAC · Collected Works of Jaroslav Hajek— with Commentary
HUZURBAZAR · Flowgraph Models for Multistate Time-to-Event Data

JACKMAN · Bayesian Analysis for the Social Sciences
† JACKSON · A User's Guide to Principle Components
JOHN · Statistical Methods in Engineering and Quality Assurance
JOHNSON · Multivariate Statistical Simulation
JOHNSON and BALAKRISHNAN · Advances in the Theory and Practice of Statistics: A Volume in Honor of Samuel Kotz
JOHNSON, KEMP, and KOTZ · Univariate Discrete Distributions, *Third Edition*
JOHNSON and KOTZ (editors) · Leading Personalities in Statistical Sciences: From the Seventeenth Century to the Present
JOHNSON, KOTZ, and BALAKRISHNAN · Continuous Univariate Distributions, Volume 1, *Second Edition*
JOHNSON, KOTZ, and BALAKRISHNAN · Continuous Univariate Distributions, Volume 2, *Second Edition*
JOHNSON, KOTZ, and BALAKRISHNAN · Discrete Multivariate Distributions
JUDGE, GRIFFITHS, HILL, LÜTKEPOHL, and LEE · The Theory and Practice of Econometrics, *Second Edition*
JUREK and MASON · Operator-Limit Distributions in Probability Theory
KADANE · Bayesian Methods and Ethics in a Clinical Trial Design
KADANE AND SCHUM · A Probabilistic Analysis of the Sacco and Vanzetti Evidence
KALBFLEISCH and PRENTICE · The Statistical Analysis of Failure Time Data, *Second Edition*
KARIYA and KURATA · Generalized Least Squares
KASS and VOS · Geometrical Foundations of Asymptotic Inference
† KAUFMAN and ROUSSEEUW · Finding Groups in Data: An Introduction to Cluster Analysis
KEDEM and FOKIANOS · Regression Models for Time Series Analysis
KENDALL, BARDEN, CARNE, and LE · Shape and Shape Theory
KHURI · Advanced Calculus with Applications in Statistics, *Second Edition*
KHURI, MATHEW, and SINHA · Statistical Tests for Mixed Linear Models
* KISH · Statistical Design for Research
KLEIBER and KOTZ · Statistical Size Distributions in Economics and Actuarial Sciences
KLEMELÄ · Smoothing of Multivariate Data: Density Estimation and Visualization
KLUGMAN, PANJER, and WILLMOT · Loss Models: From Data to Decisions, *Third Edition*
KLUGMAN, PANJER, and WILLMOT · Loss Models: Further Topics
KLUGMAN, PANJER, and WILLMOT · Solutions Manual to Accompany Loss Models: From Data to Decisions, *Third Edition*
KOSKI and NOBLE · Bayesian Networks: An Introduction
KOTZ, BALAKRISHNAN, and JOHNSON · Continuous Multivariate Distributions, Volume 1, *Second Edition*
KOTZ and JOHNSON (editors) · Encyclopedia of Statistical Sciences: Volumes 1 to 9 with Index
KOTZ and JOHNSON (editors) · Encyclopedia of Statistical Sciences: Supplement Volume
KOTZ, READ, and BANKS (editors) · Encyclopedia of Statistical Sciences: Update Volume 1
KOTZ, READ, and BANKS (editors) · Encyclopedia of Statistical Sciences: Update Volume 2
KOWALSKI and TU · Modern Applied U-Statistics
KRISHNAMOORTHY and MATHEW · Statistical Tolerance Regions: Theory, Applications, and Computation

KROESE, TAIMRE, and BOTEV · Handbook of Monte Carlo Methods
KROONENBERG · Applied Multiway Data Analysis
KULINSKAYA, MORGENTHALER, and STAUDTE · Meta Analysis: A Guide to Calibrating and Combining Statistical Evidence
KULKARNI and HARMAN · An Elementary Introduction to Statistical Learning Theory
KUROWICKA and COOKE · Uncertainty Analysis with High Dimensional Dependence Modelling
KVAM and VIDAKOVIC · Nonparametric Statistics with Applications to Science and Engineering
LACHIN · Biostatistical Methods: The Assessment of Relative Risks, *Second Edition*
LAD · Operational Subjective Statistical Methods: A Mathematical, Philosophical, and Historical Introduction
LAMPERTI · Probability: A Survey of the Mathematical Theory, *Second Edition*
LAWLESS · Statistical Models and Methods for Lifetime Data, *Second Edition*
LAWSON · Statistical Methods in Spatial Epidemiology, *Second Edition*
LE · Applied Categorical Data Analysis, *Second Edition*
LE · Applied Survival Analysis
LEE · Structural Equation Modeling: A Bayesian Approach
LEE and WANG · Statistical Methods for Survival Data Analysis, *Fourth Edition*
LePAGE and BILLARD · Exploring the Limits of Bootstrap
LESSLER and KALSBEEK · Nonsampling Errors in Surveys
LEYLAND and GOLDSTEIN (editors) · Multilevel Modelling of Health Statistics
LIAO · Statistical Group Comparison
LIN · Introductory Stochastic Analysis for Finance and Insurance
LINDLEY · Understanding Uncertainty, *Revised Edition*
LITTLE and RUBIN · Statistical Analysis with Missing Data, *Second Edition*
LLOYD · The Statistical Analysis of Categorical Data
LOWEN and TEICH · Fractal-Based Point Processes
MAGNUS and NEUDECKER · Matrix Differential Calculus with Applications in Statistics and Econometrics, *Revised Edition*
MALLER and ZHOU · Survival Analysis with Long Term Survivors
MARCHETTE · Random Graphs for Statistical Pattern Recognition
MARDIA and JUPP · Directional Statistics
MARKOVICH · Nonparametric Analysis of Univariate Heavy-Tailed Data: Research and Practice
MARONNA, MARTIN and YOHAI · Robust Statistics: Theory and Methods
MASON, GUNST, and HESS · Statistical Design and Analysis of Experiments with Applications to Engineering and Science, *Second Edition*
McCULLOCH, SEARLE, and NEUHAUS · Generalized, Linear, and Mixed Models, *Second Edition*
McFADDEN · Management of Data in Clinical Trials, *Second Edition*
* McLACHLAN · Discriminant Analysis and Statistical Pattern Recognition
McLACHLAN, DO, and AMBROISE · Analyzing Microarray Gene Expression Data
McLACHLAN and KRISHNAN · The EM Algorithm and Extensions, *Second Edition*
McLACHLAN and PEEL · Finite Mixture Models
McNEIL · Epidemiological Research Methods
MEEKER and ESCOBAR · Statistical Methods for Reliability Data
MEERSCHAERT and SCHEFFLER · Limit Distributions for Sums of Independent Random Vectors: Heavy Tails in Theory and Practice
MENGERSEN, ROBERT, and TITTERINGTON · Mixtures: Estimation and Applications

MICKEY, DUNN, and CLARK · Applied Statistics: Analysis of Variance and Regression, *Third Edition*
* MILLER · Survival Analysis, *Second Edition*
MONTGOMERY, JENNINGS, and KULAHCI · Introduction to Time Series Analysis and Forecasting, *Second Edition*
MONTGOMERY, PECK, and VINING · Introduction to Linear Regression Analysis, *Fifth Edition*
MORGENTHALER and TUKEY · Configural Polysampling: A Route to Practical Robustness
MUIRHEAD · Aspects of Multivariate Statistical Theory
MULLER and STOYAN · Comparison Methods for Stochastic Models and Risks
MURTHY, XIE, and JIANG · Weibull Models
MYERS, MONTGOMERY, and ANDERSON-COOK · Response Surface Methodology: Process and Product Optimization Using Designed Experiments, *Third Edition*
MYERS, MONTGOMERY, VINING, and ROBINSON · Generalized Linear Models. With Applications in Engineering and the Sciences, *Second Edition*
NATVIG · Multistate Systems Reliability Theory With Applications
† NELSON · Accelerated Testing, Statistical Models, Test Plans, and Data Analyses
† NELSON · Applied Life Data Analysis
NEWMAN · Biostatistical Methods in Epidemiology
NG, TAIN, and TANG · Dirichlet Theory: Theory, Methods and Applications
OKABE, BOOTS, SUGIHARA, and CHIU · Spatial Tesselations: Concepts and Applications of Voronoi Diagrams, *Second Edition*
OLIVER and SMITH · Influence Diagrams, Belief Nets and Decision Analysis
PALTA · Quantitative Methods in Population Health: Extensions of Ordinary Regressions
PANJER · Operational Risk: Modeling and Analytics
PANKRATZ · Forecasting with Dynamic Regression Models
PANKRATZ · Forecasting with Univariate Box-Jenkins Models: Concepts and Cases
PARDOUX · Markov Processes and Applications: Algorithms, Networks, Genome and Finance
PARMIGIANI and INOUE · Decision Theory: Principles and Approaches
* PARZEN · Modern Probability Theory and Its Applications
PEÑA, TIAO, and TSAY · A Course in Time Series Analysis
PESARIN and SALMASO · Permutation Tests for Complex Data: Applications and Software
PIANTADOSI · Clinical Trials: A Methodologic Perspective, *Second Edition*
POURAHMADI · Foundations of Time Series Analysis and Prediction Theory
POURAHMADI · High-Dimensional Covariance Estimation
POWELL · Approximate Dynamic Programming: Solving the Curses of Dimensionality, *Second Edition*
POWELL and RYZHOV · Optimal Learning
PRESS · Subjective and Objective Bayesian Statistics, *Second Edition*
PRESS and TANUR · The Subjectivity of Scientists and the Bayesian Approach
PURI, VILAPLANA, and WERTZ · New Perspectives in Theoretical and Applied Statistics
† PUTERMAN · Markov Decision Processes: Discrete Stochastic Dynamic Programming
QIU · Image Processing and Jump Regression Analysis
* RAO · Linear Statistical Inference and Its Applications, *Second Edition*
RAO · Statistical Inference for Fractional Diffusion Processes

RAUSAND and HØYLAND · System Reliability Theory: Models, Statistical Methods, and Applications, *Second Edition*
RAYNER, THAS, and BEST · Smooth Tests of Goodnes of Fit: Using R, *Second Edition*
RENCHER and SCHAALJE · Linear Models in Statistics, *Second Edition*
RENCHER and CHRISTENSEN · Methods of Multivariate Analysis, *Third Edition*
RENCHER · Multivariate Statistical Inference with Applications
RIGDON and BASU · Statistical Methods for the Reliability of Repairable Systems
* RIPLEY · Spatial Statistics
* RIPLEY · Stochastic Simulation
ROHATGI and SALEH · An Introduction to Probability and Statistics, *Third Edition*
ROLSKI, SCHMIDLI, SCHMIDT, and TEUGELS · Stochastic Processes for Insurance and Finance
ROSENBERGER and LACHIN · Randomization in Clinical Trials: Theory and Practice
ROSSI, ALLENBY, and McCULLOCH · Bayesian Statistics and Marketing
† ROUSSEEUW and LEROY · Robust Regression and Outlier Detection
ROYSTON and SAUERBREI · Multivariate Model Building: A Pragmatic Approach to Regression Analysis Based on Fractional Polynomials for Modeling Continuous Variables
* RUBIN · Multiple Imputation for Nonresponse in Surveys
RUBINSTEIN and KROESE · Simulation and the Monte Carlo Method, *Second Edition*
RUBINSTEIN and MELAMED · Modern Simulation and Modeling
RUBINSTEIN, RIDDER, and VAISMAN · Fast Sequential Monte Carlo Methods for Counting and Optimization
RYAN · Modern Engineering Statistics
RYAN · Modern Experimental Design
RYAN · Modern Regression Methods, *Second Edition*
RYAN · Sample Size Determination and Power
RYAN · Statistical Methods for Quality Improvement, *Third Edition*
SALEH · Theory of Preliminary Test and Stein-Type Estimation with Applications
SALTELLI, CHAN, and SCOTT (editors) · Sensitivity Analysis
SCHERER · Batch Effects and Noise in Microarray Experiments: Sources and Solutions
* SCHEFFE · The Analysis of Variance
SCHIMEK · Smoothing and Regression: Approaches, Computation, and Application
SCHOTT · Matrix Analysis for Statistics, *Second Edition*
SCHOUTENS · Levy Processes in Finance: Pricing Financial Derivatives
SCOTT · Multivariate Density Estimation
SCOTT · Multivariate Density Estimation: Theory, Practice, and Visualization
* SEARLE · Linear Models
† SEARLE · Linear Models for Unbalanced Data
† SEARLE · Matrix Algebra Useful for Statistics
† SEARLE, CASELLA, and McCULLOCH · Variance Components
SEARLE and WILLETT · Matrix Algebra for Applied Economics
SEBER · A Matrix Handbook For Statisticians
† SEBER · Multivariate Observations
SEBER and LEE · Linear Regression Analysis, *Second Edition*
† SEBER and WILD · Nonlinear Regression
SENNOTT · Stochastic Dynamic Programming and the Control of Queueing

Systems
* SERFLING · Approximation Theorems of Mathematical Statistics
SHAFER and VOVK · Probability and Finance: It's Only a Game!
SHERMAN · Spatial Statistics and Spatio-Temporal Data: Covariance Functions and Directional Properties
SILVAPULLE and SEN · Constrained Statistical Inference: Inequality, Order, and Shape Restrictions
SINGPURWALLA · Reliability and Risk: A Bayesian Perspective
SMALL and McLEISH · Hilbert Space Methods in Probability and Statistical Inference
SRIVASTAVA · Methods of Multivariate Statistics
STAPLETON · Linear Statistical Models, *Second Edition*
STAPLETON · Models for Probability and Statistical Inference: Theory and Applications
STAUDTE and SHEATHER · Robust Estimation and Testing
STOYAN · Counterexamples in Probability, *Second Edition*
STOYAN and STOYAN · Fractals, Random Shapes and Point Fields: Methods of Geometrical Statistics
STREET and BURGESS · The Construction of Optimal Stated Choice Experiments: Theory and Methods
STYAN · The Collected Papers of T. W. Anderson: 1943–1985
SUTTON, ABRAMS, JONES, SHELDON, and SONG · Methods for Meta-Analysis in Medical Research
TAKEZAWA · Introduction to Nonparametric Regression
TAMHANE · Statistical Analysis of Designed Experiments: Theory and Applications
TANAKA · Time Series Analysis: Nonstationary and Noninvertible Distribution Theory
THOMPSON · Empirical Model Building: Data, Models, and Reality, *Second Edition*
THOMPSON · Sampling, *Third Edition*
THOMPSON · Simulation: A Modeler's Approach
THOMPSON and SEBER · Adaptive Sampling
THOMPSON, WILLIAMS, and FINDLAY · Models for Investors in Real World Markets
TIERNEY · LISP-STAT: An Object-Oriented Environment for Statistical Computing
and Dynamic Graphics
TROFFAES and DE COOMAN · Lower Previsions
TSAY · Analysis of Financial Time Series, *Third Edition*
TSAY · An Introduction to Analysis of Financial Data with R
TSAY · Multivariate Time Series Analysis: With R and Financial Applications
UPTON and FINGLETON · Spatial Data Analysis by Example, Volume II: Categorical and Directional Data
† VAN BELLE · Statistical Rules of Thumb, *Second Edition*
VAN BELLE, FISHER, HEAGERTY, and LUMLEY · Biostatistics: A Methodology for the Health Sciences, *Second Edition*
VESTRUP · The Theory of Measures and Integration
VIDAKOVIC · Statistical Modeling by Wavelets
VIERTL · Statistical Methods for Fuzzy Data
VINOD and REAGLE · Preparing for the Worst: Incorporating Downside Risk in Stock Market Investments
WALLER and GOTWAY · Applied Spatial Statistics for Public Health Data
WEISBERG · Applied Linear Regression, *Fourth Edition*

WEISBERG · Bias and Causation: Models and Judgment for Valid Comparisons
WELSH · Aspects of Statistical Inference
WESTFALL and YOUNG · Resampling-Based Multiple Testing: Examples and Methods for *p*-Value Adjustment
* WHITTAKER · Graphical Models in Applied Multivariate Statistics
WINKER · Optimization Heuristics in Economics: Applications of Threshold Accepting
WOODWORTH · Biostatistics: A Bayesian Introduction
WOOLSON and CLARKE · Statistical Methods for the Analysis of Biomedical Data, *Second Edition*
WU and HAMADA · Experiments: Planning, Analysis, and Parameter Design Optimization, *Second Edition*
WU and ZHANG · Nonparametric Regression Methods for Longitudinal Data Analysis
YAKIR · Extremes in Random Fields
YIN · Clinical Trial Design: Bayesian and Frequentist Adaptive Methods
YOUNG, VALERO-MORA, and FRIENDLY · Visual Statistics: Seeing Data with Dynamic Interactive Graphics
ZACKS · Examples and Problems in Mathematical Statistics
ZACKS · Stage-Wise Adaptive Designs
* ZELLNER · An Introduction to Bayesian Inference in Econometrics
ZELTERMAN · Discrete Distributions—Applications in the Health Sciences
ZHOU, OBUCHOWSKI, and McCLISH · Statistical Methods in Diagnostic Medicine, *Second Edition*

*Now available in a lower priced paperback edition in the Wiley Classics Library.
†Now available in a lower priced paperback edition in the Wiley–Interscience Paperback Series.

Printed and bound by CPI Group (UK) Ltd, Croydon, CR0 4YY

16/04/2025

14658367-0005